AF890529

Ancient Fishes and their Living Relatives

A Tribute to John G. Maisey

Alan Pradel, John S. S. Denton,
and Philippe Janvier
(editors)

Ancient Fishes and their Living Relatives

A Tribute to John G. Maisey

Alan Pradel, John S. S. Denton, and Philippe Janvier
(editors)

Verlag Dr. Friedrich Pfeil · München 2021
ISBN 978-3-89937-269-4

Bibliographic information published by the Deutsche Nationalbibliothek

The Deutsche Nationalbibliothek lists this publication in the Deutsche Nationalbibliografie;
detailed bibliographic data are available in the Internet at
http://dnb.dnb.de.

*Support for this volume was generously provided
by the Herbert and Evelyn Axelrod Research Fund in Vertebrate Paleontology
at the American Museum of Natural History.*

Cover photograph

Details of 2nd dorsal fin, anal fin, and caudal fin of *Mesiteia emiliae* Kramberger, 1884.
(Pfeil, this volume, pp. 101–182, pl. 12, p. 144)

Copyright © 2021 by Verlag Dr. Friedrich Pfeil, München
Alle Rechte vorbehalten – All rights reserved.
No part of this publication may be reproduced, stored in a retrieval system, or transmitted in any form or by any means,
electronic, mechanical, photocopying or otherwise, without the prior permission of the copyright owner.
Applications for such permission, with a statement of the purpose and extent of the reproduction,
should be addressed to the Publisher, Verlag Dr. Friedrich Pfeil, Wolfratshauser Str. 27, 81379 München, Germany.

Druckvorstufe: Verlag Dr. Friedrich Pfeil, München
Druck: PBtisk a.s., Příbram I – Balonka

ISBN 978-3-89937-269-4

Printed in the European Union

Verlag Dr. Friedrich Pfeil, Wolfratshauser Straße 27, 81379 München, Germany
Tel.: +49 (0)89 5528600-0 • Fax: +49 (0)89 5528600-4 • E-Mail: info@pfeil-verlag.de
www.pfeil-verlag.de

Contents

Preface by the editors

When we announced the preparation of this volume, we quickly received inquiries from around the world. Colleagues, collaborators, friends, and former students, working on topics ranging from the analytical estimation of evolutionary trees, to Brazilian paleobiogeography, and from the inference of anatomical modules to the descriptive anatomy and development of chondrichthyans, holocephalans, and bony fishes offered to contribute. This worldwide response from so diverse a range of colleagues is a testament not only to John's esteem in the paleoichthyology community but also to the breadth of his knowledge base and his own history of helping those sharing these broad interests. Although the major research focus of John's career has been the discovery and description of Paleozoic chondrichthyans, John believed in the necessity of studying Recent taxa to better inform the description of fossils, and many of his studies of Recent taxa, especially the broadnose sevengill shark *Notorynchus*, are now considered classic works in chondrichthyan paleontology. His holistic thinking has populated the academic literature, often in ways quietly predating the primary works on the subjects. John's academic training and influences, as well as his interests, are presented more extensively in a chapter of this volume.

The diversity of the topics represented here reflects the spectacular evolution of the field of paleontology during the last quarter-century. These advances are largely technical, resting on new means of observation and data analysis. Paleontology's shift toward digitization of fossils using X-ray tomographic methods has made many previously inaccessible specimens amenable to description. John was among the earliest adopters of tomographic methods in paleontology, and several chapters of this volume present descriptions of specimens derived from two varieties of tomographic information: axial computed tomography (μCT), and synchrotron microtomography (SRμCT). This influx of new data often led John to express concern about the utility of specimens (especially fossils), the consequences of their often incomplete data, and the effective units of analysis. Several chapters of this volume highlight subjects related to John's interest in the inference of evolutionary trees and the placement and identification of unusual taxa, as well as questions of data and methods usage and development.

John's career has been bookended by studies of histology, from his doctoral thesis on dorsal fin spine morphology to his recent collaborative analysis on the histology of tessellated calcified cartilage across the 'acanthodian'-chondrichthyan divide, two studies separated by nearly fifty years. One chapter of this volume summarizes the tantalizing acanthodian tissue diversity, and several others utilize histological fine structures in their anatomical descriptions.

John is notable for his knowledge of anatomy and its developmental underpinnings and also of geological and geochemical processes, and one chapter in this volume presents such a holistic treatment of fishes from a faunal location in the United States. John's contributions to this field are notably characterized by transformative research on the Cretaceous fish faunas of Brazil, and the significance of these contributions is the subject of a chapter of this volume. Some of John's fieldwork led to the discovery of preserved soft tissues in fossil fishes within concretions or in 'Lagerstatte'. The phylogenetic importance of such soft tissue data had long been advocated by one of John's mentors, the Late Colin Patterson, and the fine-scale resolution of these fossils played a role in John's later interest in imaging technology.

Over a career spanning nearly half a century, John found delight and good humor in what he did. One of our number recalls struggling with John over the puzzle presented by the peculiar Devonian chondrichthyan *Pucapampella* from Bolivia, doggedly assembling rubber casts of its isolated cranial cartilage elements preserved in nodules, and sharing excitement to seemingly rival that of early nineteenth century paleontologists progressively unravelling the anatomy of an entirely new form. Another recalls John's bemusement at having been identified as a coelacanth expert, following his description of mawsoniid specimens from Brazil, and continually receiving manuscript review requests on the subject for the next twenty years. And, there was his editorial reaction to an imprecise word choice in a manuscript given to John for comment, who returned it with a small cartoon shark scribbled in the margin, and the word bubble, "only two teef [sic]? Am I cute?"

We have been fortunate to have been the beneficiaries of John's knowledge and spirit, and this volume honors him with two new taxa named for him. The contributions in this volume hope to represent some of the breadth of John's erudition and influence, which reflect the value of classical training, patience, and humility to which we all aspire.

The editors

Ancient Fishes and their Living Relatives: a Tribute to John G. Maisey. Alan Pradel, John S. S. Denton & Philippe Janvier (eds.): pp. 9–18, 8 figs., 1 tab.
© 2021 by Verlag Dr. Friedrich Pfeil, München, Germany – ISBN 978-3-89937-269-4

John G. Maisey – a Biographical Sketch

Maria da Gloria P. de Carvalho*

Introduction

John Maisey was born in London on June 28th, 1949, the only child of Albert John and Norah Maisey. His parents married during the Second World War, when his father worked at Fairey Aviation, designing electrical systems for naval military aircraft during the week, and serving in the Home Guard and playing trumpet in his brother's swing band on weekends. His father taught John the basics of draughtsmanship (technical drawing) and photography. John's mother was a professional typist, and introduced him to the QWERTY keyboard at an early age. John attended Norwood Green Primary School in Southall, Middlesex until 1960, and then Isleworth Grammar School (1960–1967).

John found his first fossil in an unlikely manner at the age of 11, whilst walking home from school. Lorries laden with gravel from nearby pits bounced daily up the street where he lived, dropping stones as they passed. One day he picked up a nicely shaped but chipped pebble, which he eventually identified as a flint endocast of a heart-urchin (*Micraster*). From then on, he knew he wanted to be a paleontologist. Over the next few years, he persuaded his parents (who knew nothing about geology or paleontology) to take him on holidays to various places in Britain where he could prospect for fossils. Much of his early collecting was in the Jurassic-Eocene of Southern England and the Paleozoic of Wales.

Besides collecting fossils, John enjoyed painting and drawing. He won his first diploma (at the age of ten) in an art competition sponsored by the Brooke Bond Tea Company, with a painting of the *Diplodocus* skeleton at the (then) British Museum of Natural History, and spent his prize money (a 7s/6d book token) on "The Story of Evolution" by Julian Huxley.
The rigid academic system at his Grammar School forbade John from combining subjects deemed to be 'arts' and 'sciences', but his teachers recognized his determination to study paleontology and he was permitted to study Biology (a 'Science') and Geography (the closest subject to Geology offered at school, but considered an 'Art') as well as Artwork for his 'A-Levels' (University entrance qualifications). Having passed the Art examination in just one year (1966), he enrolled in Geology classes at a local technical College, taking both years of the course concurrently. John thus obtained 'A-Level' certificates in Art, Biology, Geography and Geology and was offered places to study Geology at Oxford, Reading, and Exeter universities in 1967. He chose Exeter, where he found the academic environment well-suited to his interests. His principal course was Geology, with Zoology for two years and Botany for one, and he was graduated in 1970 with Honours.

In the summer of 1969, in his final Undergraduate year, he worked as an intern for Dr. Roger Miles at the British Museum (Natural History), assisting Ian Macadie with acid preparation, thin sectioning, and conservation of Devonian fossils from the Gogo Formation of Western Australia, an experience which would pave the way for his future career in Paleoichthyology.

In 1970, he was offered places at the University of Bristol and at University College London to study for his Ph.D., as well as a place at the University of Leicester School of Museum

Fig. 1. John Maisey with his parents in 1960, Brighton Beach (Southern England).

Fig. 2. John Maisey with his mother in North Wales, c. 1963.

* 1. American Museum of Natural History, 79th. Street and Central Park West, New York, NY 10024-5192, U.S.A.
2. Instituto de Geociencias, Departamento de Geologia, Universidade Federal do Rio de Janeiro, Brazil, CEP 21.910.000.

Fig. 3. John Maisey's Graduation portrait, Exeter University, England, June 1970. ▷

▷▷ **Fig. 4.** John Maisey in his office at the AMNH, c. 1986.

Studies. He chose University College London, as it provided him with an opportunity to study his favorite group, the shark-like fishes (chondrichthyans). His advisors were Kenneth Kermack (mostly known for his work on early mammals) plus two paleoichthyologists: Colin Patterson (NHM), and Brian Gardiner (Queen Elizabeth College).

John was awarded his Ph.D in 1974, although his funding at UCL terminated in 1973; between 1973 and 1975, he worked at the University of Exeter as Demonstrator in Geology, teaching paleontology and stratigraphic principles. From 1975–1978, he taught Geology and Biology at St. Alban's College of Further Education in Hertfordshire, often using his own fossil collection for student labs. John received his teacher certification in 1978 at Hatfield Polytechnic, and also taught for the Open University as a part-time Tutor in Geology (1976–1978).

Following the retirement of Dr. Bobb Schaeffer from the American Museum of Natural History, John was interviewed and offered the position of Assistant Curator in Vertebrate Paleontology at the Museum in 1978. Although he accepted the position, his appointment to the museum was delayed 11 months until he received his 'Green Card'. During the interim, his Geology course in St Albans College was canceled and he took the somewhat risky step of resigning. Fortunately, thanks to his father's training in draughtsmanship, John obtained temporary employment at a local drawing office of the Lancer-Boss Fork-Lift Truck Company in Leighton Buzzard (Bedfordshire), until he was able to emigrate to the U.S.A. in May of 1979.

John Maisey was promoted to Associate Curator in 1984 and full Curator in 1989, a rank he held until his retirement from the AMNH in June 2018.

Besides his scientific interests, John is also an avid model shipbuilder, with a special interest in 17th and 18th century naval architecture and maritime history.

John Maisey's Scientific Contributions

At the time of his retirement in June 2018, John Maisey had published 136 scientific and technical papers (including 16 systematic descriptions and 13 technical chapters in "Santana Fossils"), 31 non-technical articles, six encyclopedia entries, 10 reviews, and numerous published abstracts. He has also been involved with five book projects, two as sole author, two as a co-author, and one as an editor (see citations below and the bibliography at the end of this article). More than half (73) of his publications to date involve the chondrichthyan fishes. He has presented over 120 oral contributions (including more than 40 invited lectures) at research meetings internationally. Over the course of his career, he named and described 12 new species, 17 new genera and two new families (Table 1). Whilst at the AMNH, he obtained National Science Foundation funding to reorganize the Fossil Fish Collection (1987-1990), developed a new permanent exhibition of fossil fishes in the museum's 4th floor Halls of Vertebrate Evolution during the mid-1990's, and participated in field expeditions to Alaska, New York, Kansas, Oklahoma, Texas, Tennessee, Kentucky, New Brunswick (Canada), Brazil, Venezuela, the Falkland Islands, and Western Australia. He reconstructed the jaws and dentition of the giant 'megalodon' shark on permanent exhibit at the AMNH, and was also a consultant on the USNM 'megalodon' jaws reconstruction. In 2020, John was a consultant on extinct sharks for a temporary exhibition at the AMNH, and supervised the development of several life-sized models including *Doliodus*, *Heliocoprion*, and 'megalodon'.

Research on the chondrichthyan dermal skeleton. John Maisey has published 13 papers on chondrichthyan fin spines (12 as sole author and one collaboratively), including one on modern elasmobranch fin spines, three on Mesozoic hybodont and 'neoselachian' fin spines, and nine on Paleozoic examples (including a series of papers in which he revised the 'catch-all' taxon *Ctenacanthus*).

John Maisey's doctoral research on Recent and fossil chondrichthyan fin spines brought him into contact with Wolf-Ernst Reif (University of Tübingen), who was then completing his doctoral thesis on Recent and fossil shark teeth and denticles. Reif sparked John's early interest in developmental patterning of the dermal skeleton in shark-like fishes and also encouraged him to publish some of his preliminary findings about fin spines in Germany. Reif's influence was still evident decades later, in John's collaboration with John Denton, where Reif's 'inhibitory field theory' was revisited in the light of reaction-diffusion (R-D) systems (Maisey & Denton, 2016). John described the dentition in the Lower Devonian stem chondrichthyan *Doliodus* and discovered evidence of monognathic and dignathic dental

Fig. 5. John Maisey with one of his ship models, c. 1996.

patterning like that of modern elasmobranchs (Maisey et al. 2013).

Research on chondrichthyan endoskeletal morphology. John Maisey has authored or co-authored 22 papers on skeletal morphology in Paleozoic chondrichthyans, 19 on Mesozoic ones, and six on Cenozoic and modern ones. In addition, he has published 12 broader works on elasmobranch, holocephalan, and chondrichthyan anatomy and phylogeny. In 2010-2015, he managed the morphological component of the NSF Chondrichthyan Tree of Life project headed by Dr. Gavin Naylor (then based in Charleston, S.C.), generating high-resolution scan data of modern sharks, rays and chimaeroids at the AMNH for subsequent image processing. With Postdoctoral Fellow John Denton, he participated in a collaborative international synchrotron-based investigation of cranial anatomy in the pocket shark, *Mollisquama* (Denton et al. 2018).

John Maisey's interest in endoskeletal morphology began with a study of cranial morphology of the Early Cretaceous hybodont *'Hybodus' basanus* (= *Egertonodus* Maisey, 1987b), using well-preserved three-dimensional specimens in the British Museum (Natural History). He began this project while he was teaching at St Alban's College and completed it in New York, where he benefitted greatly from discussions with Bobb Schaeffer (who was then still in residence at the AMNH, completing his study of cranial anatomy in Permian xenacanth sharks). He also collaborated with John Dick (then working on Paleozoic sharks from Scotland at the University of Newcastle upon Tyne). In a series of papers, John Maisey clarified the phylogenetic status of hybodonts as (in modern terms) a sister group to the elasmobranch crown (Maisey 1978, 1982, 1983, 1985, 1986, 1987a, 1989, 2011, Maisey & Carvalho 1997, Lane & Maisey 2012).

John Maisey's chondrichthyan research in the late 1990's coincided with the advent of computerized tomography in paleontology and related improvements in imaging software. His first papers to include tomographic data were published in 2001 (Maisey 2001b,c, Maisey & Anderson 2001). His first completely scan-based paper (Maisey 2004a) described in detail the cranial morphology in the modern broadnose sevengill shark, using specimens originally studied by J. Frank Daniel in the early 20th Century. Then, again using tomographic data, John re-investigated the famous three-dimensional Devonian braincase of *'Cladodus' wildungensis* originally described decades earlier by Walter Gross in Germany. John went on to describe an equally spectacular braincase of a Late Mississippian symmoriiform (Maisey 2005, 2007). Working on synchrotron scan data with AMNH Postdoctoral Fellow Alan Pradel and others, he co-authored the first detailed description of the visceral skeleton in a Paleozoic chondrichthyan (*Ozarcus*; Pradel et al. 2014).

For almost a decade, John Maisey has collaborated with Philippe Janvier (MNHN, Paris) and Alan Pradel in an analysis of the bizarre Lower Devonian 'pucapampellid' stem chondrichthyans, based on studies involving literally hundreds of fragmentary specimens from Bolivia (some of which were scanned in New York and Paris; Janvier & Maisey 2010, Maisey et al. 2019). In other collaborations, he investigated the cranium, jaws, dentition, pectoral fins and dermal spines in *Doliodus*, a Lower Devonian 'transitional fossil' linking acanthodians with 'conventional' chondrichthyans (Maisey et al. 2009, Maisey et al. 2014, Maisey et al. 2017, Burrow et al. 2017). With Paul Tafforeau (ESRF, Grenoble), Alan Pradel and Philippe Janvier, he co-authored a paper on cranial morphology of the enigmatic stem gnathostome *Ramirosuarezia* (Pradel et al. 2009), collected from the same Lower Devonian beds that yielded *Pucapampella*. He also participated in several collaborative works on other Paleozoic chondrichthyans (Maisey & Melo 2005, Ginter & Maisey 2007, Pradel et al. 2009, Pradel et al. 2011) and has contributed to several papers in which chondrichthyan and/or vertebrate anatomy, phylogeny, and distributions were investigated (Janvier & Maisey 2010, Khonsari et al. 2013, Maisey 1975, 1980, 1984a,c, 1987, 1988, 2001, 2004a,b, 2008, 2012; Maisey & Lane 2010). His current paleontological research on chondrichthyans focuses on the evolutionary relationships of extinct relatives of the elasmobranch and holocephalan crown groups and on the evolution of tessellated calcified cartilage and fin spines within the chondrichthyan stem.

Fig. 6. John Maisey with new AMNH reconstruction of 'megalodon' jaws for the Hall of Vertebrate Origins, c. 1996.

Research on Cretaceous Fishes. John Maisey also worked extensively on Early Cretaceous fishes from South America, Africa and North America. His contributions include 12 papers on actinopterygians, four on coelacanths, and eight on other topics, as well as over 30 chapters of the "Santana Fossils" volume published in 1991 and sections in "Discovering Fossil Fishes" (Maisey 1996). Besides his systematic contributions on Cretaceous fishes, he also addressed the paleogeographic distribution of Early Cretaceous fishes in Western Gondwana, drawing upon tectonic and other geological data in a series of papers, in which he argued that several non-marine Cretaceous taxa with a modern 'transatlantic' distribution were entirely 'African' during the Lower Cretaceous, separated from the rest of South America by an interior seaway (Maisey 1993, 2000, 2011). He also investigated predator-prey relationships and trophic hierarchies among Santana Formation fishes using stomach-contents data retrieved from acid-prepared specimens (Maisey 1994), and contributed to descriptions of marine arthropods recovered from fish stomach contents, including a new genus and species of fossil shrimp and a crustacean zoea larva (the first known fossil example; Maisey & de Carvalho 1995).

In 1993, John Maisey became a Corresponding Member of the Brazilian Academy of Sciences, and was appointed in 2011 as Adjunct Professor and Pesquisador (Researcher) in the Department of Geosciences at the Federal University of Rio de Janeiro, where he has ongoing involvement in two projects on the paleontology of North-Eastern Brazil.

Besides his work in South America, John collaborated with Louis Taverne (Belgium) in describing a Cretaceous notopterid cranium from Morocco (Taverne & Maisey 1999) and a revision of the pycnodontiform *Stenoprotome* from Lebanon (Taverne et al. 2015). He has also worked with Mee-Mann Chang (IVPP, Beijing) on a comparison of Cretaceous and Eocene clupeomorphs from China and the Americas (Chang & Maisey 2003).

Table 1. Taxa described by John Maisey (in date order).

Elasmobranchii inc. sed.	*Breviacanthus* Maisey, 1976a	gen.
Elasmobranchii inc. sed.	*Squalogaleus woodwardi* Maisey, 1976b	gen. sp.
Chondrichthyes inc. sed.	*Amelacanthus* Maisey, 1982a	gen.
Chondrichthyes inc. sed.	*Wodnika borealis* Maisey, 1982a	sp.
Heterodontidae	*Heterodontus tuberculatus* Maisey, 1982c	sp.
Hexanchiformes	*Notidanoides* Maisey, 1986a	gen.
Hybodontiformes	*Egertonodus* Maisey, 1987b	sp.
Actinistia	*Axelrodichthys araripensis* Maisey, 1986c	gen. sp.
Hybodontiformes	*Hamiltonichthys mapesi* Maisey, 1989a	gen. sp.
Clupeomorpha	*Santanaclupea silvasantosi* Maisey, 1993a	gen. sp.
Decapoda	*Paleomattea deliciosa* Maisey & Carvalho, 1995	gen. sp.
Chondrichthyes inc. sed.	*Cladodoides* Maisey, 2001c.	gen.
Teleostei inc. sed.	*Araripichthys axelrodi* Maisey & Moody, 2001	sp.
Myliobatiformes	*Asterotrygon maloneyi* Carvalho et al., 2004	gen. sp.
Gnathostomata inc. sed.	*Ramirosuarezia boliviana* Pradel et al., 2009	gen. sp.
Heslerodidae	Maisey, 2010	fam.
Heslerodidae?	*Avonacanthus* Maisey, 2010	gen.
Symmoriiformes	*Kawichthys moodiei* Pradel et al., 2011	gen. sp.
Symmoriiformes	*Ozarcus mapesae* Pradel et al., 2014	gen. sp.
Chondrichthyes inc. sed.	*Gogoselachus lynnbeazleyae* Long et al., 2015	gen. sp.
Pucapampellidae	*Gydoselache oosthuizeni* Maisey et al., 2019	fam. gen. sp.
Squatiniformes	*Cretasquatina americana* Maisey et al., 2020	gen. sp.

Preparation Techniques. John Maisey enjoyed 'hands-on' preparation of fossils throughout his career. During his summer internship at the British Museum (Natural History) in 1969, he learned the protocols of transfer acid preparation, with Ian Macadie. They also attempted to extract and identify organic proteins from fossil bone using traditional 'wet' chromatographic techniques. Their results were inconclusive and never published, but they nevertheless identified trace amounts of amino acids and also found that toluidine blue stained some Gogo arthrodire bone fragments, but not the surrounding matrix, suggesting the presence of protein molecules.

As part of John's Graduate research in University College London, he became proficient in the use of a special (and very expensive!) 'Capco' annular saw. Acquired by his advisor, Kenneth Kermack, the 'Capco' was originally developed for industrial cutting of quartz and germanium crystals, but John developed a novel embedding technique which allowed him to use the machine to investigate the structure, histology and growth of modern chondrichthyan fin spines, without the need for decalcification of hard tissues prior to serial sectioning (Maisey 1975). As a result, he was able to prepare serial thin-sections of 'bony' skeletal material with pre-stained soft tissues still attached. Despite the fact that each finished slice was somewhat thick (approx. 80 microns), he could nevertheless study directly the relationships of hard and scleroblastic 'soft' tissues throughout the length of fin spines in modern sharks and chimaeroids. He was also able to prepare serial sections of fossil fin spines at intervals of 0.3–0.5 mm, which provided data essential for his comparison of Recent and fossil material. Although the 'Capco' cutter could operate automatically, it required frequent attention and kicked up a fine spray of coolant (vegetable oil; Kermack disallowed water, which he considered a corrosion threat to his investment!). Consequently, John invariably emerged from his cutting sessions with an oily sheen that would make a duck envious!

Thanks to his undergraduate internship at the BM(NH), John was well versed in the use of organic (acetic, formic) acids in fossil preparation and the 'transfer' embedding technique developed in that Museum by Harry Toombs and Arthur Rixon. That experience would prove crucial in his eventual work on Cretaceous fishes from Brazil. However, he also experimented with other novel chemical preparation techniques, including thioglycollic acid, which he used to prepare a specimen of *'Hybodus' basanus* in the AMNH collection (thereby revealing its almost complete dentition), as well as a concoction of sodium salts previously used by mineralogists to remove ferric iron from mineral specimens. This technique worked well on fossil fishes embedded in laminated limestones from the Crato Formation of Brazil (Maisey et al. 1989).

John Maisey was among the first scientists to recognize the potential of computerized tomography in paleontology, and began using it extensively for his research in the late 20th century and early 21st century (first based on scans obtained from Tim Rowe's lab in Austin, TX, then at the AMNH). He also worked closely with Philippe Janvier and Alan Pradel at the MNHN in Paris on specimens scanned there and at the European Synchrotron Radiation Facility in Grenoble.

Students and Postdoctoral Fellows. Dr. Gavin Naylor worked as a Postdoctoral Fellow with John Maisey and Les Marcus on variability in *Carcharhinus* teeth. Research Fellow Stanley Blum assisted John in writing parts of the 'Santana Fossils' volume (published in 1991) on Cretaceous fishes and, with Ivy Rutzky (John's Technician and subsequently his Scientific Assistant), developed the first computerized data-base for the Fossil Fish Collection (c. 1987, one of the first in the museum). Postdoctoral Fellow Marcelo de Carvalho joined the Maisey lab in the late 1990's, and together they described two previously misidentified specimens of *Protospinax* that John had discovered decades earlier in the old Alexander Agassiz collection at the MCZ (de Carvalho & Maisey, 1996). They also published a brief account of cranial anatomy in *Tribodus* (Maisey & Carvalho 1997), based on a spectacular three-dimensional specimen prepared the 'old-fashioned' way (i.e., mechanically) by Technician Bob Evander. John Maisey and Marcelo de Carvalho also collaborated with Lance Grande (Field Museum) on a revision of stingray phylogeny (Carvalho et al. 2004). Postdoctoral Fellow Arnaud Filleul joined John Maisey in 1991 to study some of the smaller teleosts from the Santana Formation, including *Santanichthys* (possibly the earliest stem characiform; Filleul & Maisey 2004). CUNY Graduate student Jennifer Lane described the cranial, visceral and pectoral anatomy of *Tribodus*, revised John's earlier work on *'Hybodus' basanus* using computerized tomography, and compared skeletal labyrinth morphology in modern and fossil chondrichthyans. Postdoctoral Fellow Amy Balanoff scanned and imaged modern elasmobranchs at the AMNH for the 'Tree of Life' project. Postdoctoral Fellow Alan Pradel tackled image processing head-on, generating spectacular 3-D reconstitutions of Paleozoic chondrichthyan heads, modern chimaeroids and elasmobranchs, using scans made at the ESRF in Grenoble. AMNH/RGGS Graduate student Allison Bronson investigated several well-preserved Late Mississippian chondrichthyans from Arkansas, analyzed jaw mobility in the Permian xenacanth *Orthacanthus*, collaborated with John in describing gigantic Pennsylvanian 'supersharks' from Texas, and redescribed the enigmatic Upper Cretaceous elasmobranch *Platylithophycus* (previously described as a calcareous green alga and a coleoid!) Postdoctoral Fellow John Denton conducted a synchrotron-based investigation of the modern pocket shark *Mollisquama*, and was also instrumental in applying mathematical reaction-diffusion models to dermal denticle patterning in *Tribodus* (the first time this had been attempted in paleontology). John Maisey has

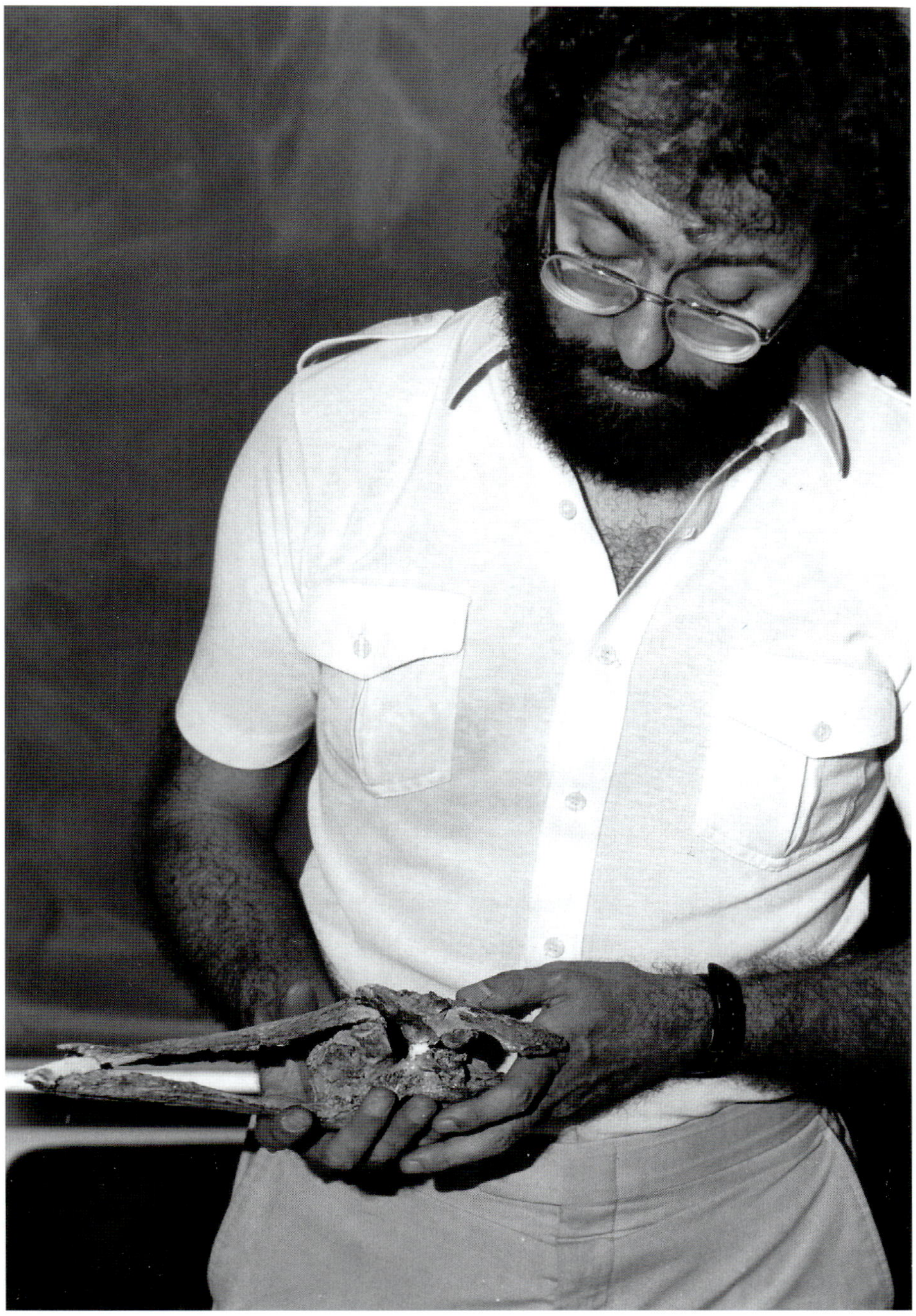

Fig. 7. John Maisey with an acid-prepared cranium of *Axelrodichthys*, AMNH June 1986.

Fig. 8. John Maisey with Hans-Peter Schultze, Gloria Arratia and Lance Grande at the Western Interior Paleontological Society Lagerstätten Symposium, Denver, Colorado, March 2019.

also supervised several undergraduate interns at the AMNH, and has served on graduate student committees in the United States, Canada, France and Brazil.

Acknowledgements

The author thanks John Denton for accepting my suggestion to write this summary about John Maisey's professional career, and for his patient revision of the ms. Much of the information about John Maisey's early life was recounted to me by his mother, who was always immensely proud of him; thanks largely to her drive and persistence, he was able to follow his chosen path in life. I am immensely grateful to all of John's colleagues, support staff, volunteers, and students at the AMNH for the help, support, and intellectual stimulation they provided to him over almost 40 years at the museum. It is not possible to name them all here, but I am sure he deeply appreciates all their endeavors. I am also grateful to all the researchers who have worked with him over the years, not only for their collegiality and friendship, but also for sometimes encouraging him to work outside his 'comfort zone'! I would also like to thank the late Dr. Herbert Axelrod and his wife Evelyn, who helped support much of John's research through the endowed Herbert and Evelyn Axelrod Research Chair in Vertebrate Paleontology at the AMNH, which he held from its inception in 1998 until his retirement in 2018. Their generosity is greatly appreciated. Finally, I am sure that John will join me in thanking the AMNH for the almost four decades of institutional support it provided him and (most importantly) for the academic freedom it continues to provide for all its research staff.

John Maisey's publications

Peer-reviewed articles

Maisey, J. G. 1970. A rediscovery of some mammalian remains from the Hamstead beds of the Isle of Wight, and a warning to those who hoard fossils. Tertiary Times, I: 71.

Maisey, J. G. 1975a. The interrelationships of phalacanthous selachians. Neues Jahrbuch für Mineralogie, Monatshefte, 9: 553–567.

Maisey, J. G. 1975b. A serial sectioning technique for fossils and hard tissues. Curator, 18: 140–147.

Maisey, J. G. 1976a. The Middle Jurassic selachian fish *Breviacanthus*, n.g. Neues Jahrbuch für Mineralogie, Monatshefte, 7: 432–438.

Maisey, J. G. 1976b. The Jurassic selachian fish *Protospinax* Woodward. Palaeontology, 19: 733–747.

Maisey, J. G. 1977a. The fossil selachian fishes *Palaeospinax* Egerton and *Nemacanthus* Agassiz. Biological Journal of the Linnean Society of London, 60: 259–273.

Maisey, J. G. 1977b. Structural notes on a cladoselachian dorsal spine. Neues Jahrbuch für Mineralogie, Monatshefte, I: 47-55.

Maisey, J. G. 1978a. Growth and form of finspines in hybodont sharks. Palaeontology, 23: 657–666.

Maisey, J. G. 1978b. Preservation and prefossilization of fossil finspines. Neues Jahrbuch für Mineralogie, Monatshefte, 10: 595–599.

Maisey, J. G. 1979. Finspine morphogenesis in squalid and heterodontid sharks. Zoological Journal of the Linnean Society of London, 66: 161–183.

Maisey, J. G. 1980. An evaluation of jaw suspension in sharks. American Museum Novitates, 2706: 1–17.

Dick, J. R. F. & J. G. Maisey. 1980. The Scottish lower Carboniferous shark *Onychoselache traquairi*. Palaeontology, 23: 363–374.

Maisey, J. G. 1981. Studies on the Paleozoic selachian genus *Ctenacanthus* Ag. No. 1. Historical review and revised diagnosis of *Ctenacanthus* with a list of referred taxa. American Museum Novitates, 2718: 1–22.

Maisey, J. G. 1982a. Studies on the Paleozoic selachian genus *Ctenacanthus* Ag. No. 2. *Bythiacanthus* St. John and Worthen, *Amelacanthus*, g. nov., *Eunemacanthus* St. John and Worthen, *Sphenacanthus* Ag. and *Wodnika*. American Museum Novitates, 2721: 1–24.

Maisey, J. G. 1982b. The anatomy and interrelationships of Mesozoic hybodont sharks. American Museum Novitates, 2724: 1–48.

Maisey, J. G. 1982c. Fossil hornshark finspines (Elasmobranchii; Heterodontidae) with notes on a new species (*Heterodontus tuberculatus*). Neues Jahrbuch für Mineralogie, Abhandlungen, 164: 393–413.

Maisey, J. G. 1983a. Cranial anatomy of *Hybodus basanus* Egerton from the Lower Cretaceous of England. American Museum Novitates, 2758: 1–64.

Maisey, J. G. 1983b. Some Pennsylvanian chondrichthyan spines from Nebraska. Transactions of the Nebraska Academy of Sciences, 11: 81-84.

Maisey, J. G. 1984a. Studies on the Paleozoic selachian genus *Ctenacanthus* Ag. No. 3. Nominal species referred to *Ctenacanthus*. American Museum Novitates, 2774: 1–20.

Maisey, J. G. 1984b. Chondrichthyan phylogeny: a look at the evidence. Journal of Vertebrate Paleontology, 4: 359–371.

Maisey, J. G. 1984c. Higher elasmobranch phylogeny and biostratigraphy. Zoological Journal of the Linnean Society of London, 82: 33–54.

Maisey, J. G. & K. Wolfram. 1984. "*Notidanus*." – In: N. Eldredge and S. Stanley (eds.), Living Fossils, pp. 170–180. New York, Springer.

Maisey, J. G. 1985a. Cranial morphology of the fossil elasmobranch *Synechodus dubrisiensis*. American Museum Novitates, 2804: 1–28.

Maisey, J. G. 1985b. Relationships of the megamouth shark, *Megachasma*. Copeia, 1985: 228–231.

Maisey, J. G. 1986a. The Upper Jurassic hexanchoid elasmobranch *Notidanoides*, n. g. Neues Jahrbuch für Mineralogie, Abhandlungen, 172: 83–106.

Maisey, J. G. 1986b. Anatomical revision of the fossil shark *Hybodus fraasi* (Chondrichthyes: Elasmobranchii). American Museum Novitates, 2357: 1–16.

Maisey, J. G. 1986c. Coelacanths from the Lower Cretaceous of Brazil. American Museum Novitates, 2866: 1–30.

Maisey, J. G. 1986d. Heads and tails: a chordate phylogeny. Cladistics, 2: 201–256.

Maisey, J. G. 1987a. Notes on the structure and phylogeny of vertebrate otoliths. Copeia, 1987: 495–499.

Maisey, J. G. 1987b. Cranial anatomy of the Lower Jurassic shark *Hybodus reticulatus* (Chondrichthyes: Elasmobranchii), with comments on hybodontid systematics. American Museum Novitates, 2878: 1–39.

Maisey, J. G. 1987c. Chondrichthyes II: Mesozoic and Cenozoic Elasmobranchii, by H. Cappetta (review). Cladistics, 3: 386–389.

Maisey, J. G. 1987d. The Ancestry of the Vertebrates, by R. P. S. Jefferies (review). Quarterly Review of Biology, 62: 450–451.

Maisey, J. G. 1988a. Response to Schultze. Copeia, 1988: 259–260.

Maisey, J. G. 1988b. Phylogeny of early vertebrate skeletal induction and ossification patterns. – In: Hecht, M., B. Wallace & G. Prance (eds.), Evolutionary Biology, 22: 1–36.

Maisey, J. G. 1989a. *Hamiltonichthys mapesi*, g. and sp. nov. (Chondrichthyes; Elasmobranchii), from the Upper Pennsylvanian of Kansas. American Museum Novitates, 2931: 1–42.

Maisey, J. G. 1989b. The hybodont shark from Hamilton Quarry. – In: G. Mapes and R. H. Mapes (eds.), Regional geology and paleontology of Upper Paleozoic Hamilton Quarry area in south-eastern Kansas. 22nd Annual Meeting, South-Central Section. Geological Society of America, pp. 177–180.

Maisey, J. G. 1989c. Visceral skeleton and musculature in a late Devonian shark. Journal of Vertebrate Paleontology, 9: 174–190.

Maisey, J. G. 1989d. Sharks of the order Carcharhiniformes, by L. J. V. Compagno (review). American Scientist, 77: 493.

Maisey, J. G., S. Blum & I.S. Rutzky. 1989. A method for chemical reduction and removal of ferric iron applied to vertebrate fossils. Journal of Vertebrate Paleontology, 9: 119–121.

McNamara, K. J., G. W. Kendrick, A. Baynes, E. Frankel, A. Gale, R. Holst, J. Long & J. G. Maisey. 1989. Brolgas on the Lyndon River. Western Australian Naturalist, 18: 24.

Antunes, M. T, J. G. Maisey, M. M. Marques, B. Schaeffer & K. S. Thomson. 1990. Triassic fishes from the Cassange Depression (R. P. de Angola). Ciências da Terra, Numero Especial: 1–64.

Grimaldi, D. and J. G. Maisey. 1990. Introduction. – In: D. Grimaldi (ed.), Fossil insects from the Santana Formation, Lower Cretaceous, of Brazil. Bulletin of the American Museum of Natural History, 195: 5–14.

Maisey, J. G. 1990. Stratigraphy and depositional environment of the Crato Member (Santana Formation, Lower Cretaceous of Brazil). – In: D. Grimaldi (ed.), Fossil insects from the Santana Formation, Lower Cretaceous, of Brazil. Bulletin of the American Museum of Natural History, 195: 15–19.

Maisey, J. G. & S. Blum. 1990. Elopomorph fishes of the Santana Formation. – In: Campos, D. A., et al. (eds.), Atlas do 1 Simposio a bacia do Araripe e bacias interiores do Nordeste, pp. 351–373.

Crane, P. & J. G. Maisey. 1991. Fossil plants – In: Maisey, J. G. (ed.), Santana Fossils: An Illustrated Atlas. Neptune NJ, TFH Publications, pp. 414–421.

Kellner, A. W. A., R. G. Martins-Neto & , J. G. Maisey. 1991. Undetermined feather. – In Maisey, J. G. (ed.), Santana Fossils: An Illustrated Atlas. Neptune NJ, TFH Publications, pp. 376–377.

Maisey, J. G. 1991a. Introduction. – In: Maisey, J. G. (ed.), Santana Fossils: An Illustrated Atlas. Neptune NJ, TFH Publications, pp. 9–10.

Maisey, J. G. 1991b. The occurrence. – In: Maisey, J. G. (ed.), Santana Fossils: An Illustrated Atlas. Neptune NJ, TFH Publications, pp. 11–21.

Maisey, J. G. 1991c. The changing patterns of life. – In: Maisey, J. G. (ed.), Santana Fossils: An Illustrated Atlas. Neptune NJ, TFH Publications, pp. 22–25.

Maisey, J. G. 1991d. Continents adrift. – In: Maisey, J. G. (ed.), Santana Fossils: An Illustrated Atlas. Neptune NJ, TFH Publications, pp. 26–30.

Maisey, J. G. 1991e. Cretaceous sedimentary basins in Brazil. – In: Maisey, J. G. (ed.), Santana Fossils: An Illustrated Atlas. Neptune NJ, TFH Publications, pp. 31–33.

Maisey, J. G. 1991f. Succession within the Araripe Basin. – In: Maisey, J. G. (ed.), Santana Fossils: An Illustrated Atlas. Neptune NJ, TFH Publications, pp. 34–35.

Maisey, J. G. 1991g. Zonation of the Brazilian Lower Cretaceous. – In: Maisey, J. G. (ed.), Santana Fossils: An Illustrated Atlas. Neptune NJ, TFH Publications, pp. 36–39.

Maisey, J. G. 1991h. An ocean is formed. – In: Maisey, J. G. (ed.), Santana Fossils: An Illustrated Atlas. Neptune NJ, TFH Publications, pp. 44–56.

Maisey, J. G. 1991i. Fossil forensics. – In: Maisey, J. G. (ed.), Santana Fossils: An Illustrated Atlas. Neptune NJ, TFH Publications, pp. 57–89.

Maisey, J. G. 1991j. Local collecting and preparation methods. – In: Maisey, J. G. (ed.), Santana Fossils: An Illustrated Atlas. Neptune NJ, TFH Publications, pp. 90–97.

Maisey, J. G. 1991k. *Rhinobatos*. – In: Maisey, J. G. (ed.), Santana Fossils: An Illustrated Atlas. Neptune NJ, TFH Publications, pp. 110–115.

Maisey, J. G. 1991l. *Araripelepidotes*. – In: Maisey, J. G. (ed.), Santana Fossils: An Illustrated Atlas. Neptune NJ, TFH Publications, pp. 118–123.

Maisey, J. G. 1991m. *Calamopleurus*. – In: Maisey, J. G. (ed.), Santana Fossils: An Illustrated Atlas. Neptune NJ, TFH Publications, pp. 139–155.

Maisey, J. G. 1991n. *Oshunia*. – In: Maisey, J. G. (ed.), Santana Fossils: An Illustrated Atlas. Neptune NJ, TFH Publications, pp. 157–168.

Maisey, J. G. 1991o. *Ophiopsis*. – In: Maisey, J. G. (ed.), Santana Fossils: An Illustrated Atlas. Neptune NJ, TFH Publications, p. 169.

Maisey, J. G. 1991p. *Vinctifer*. – In: Maisey, J. G. (ed.), Santana Fossils: An Illustrated Atlas. Neptune NJ, TFH Publications, pp. 170–189.

Maisey, J. G. 1991q. *Cladocyclus*. – In: Maisey, J. G. (ed.), Santana Fossils: An Illustrated Atlas. Neptune NJ, TFH Publications, pp. 190–207.

Maisey, J. G. 1991r. *Araripichthys*. – In: Maisey, J. G. (ed.), Santana Fossils: An Illustrated Atlas. Neptune NJ, TFH Publications, pp. 208–217.

Maisey, J. G. 1991s. *Rhacolepis*. – In: Maisey, J. G. (ed.), Santana Fossils: An Illustrated Atlas. Neptune NJ, TFH Publications, pp. 248–257.

Maisey, J. G. 1991t. *Notelops*. – In: Maisey, J. G. (ed.), Santana Fossils: An Illustrated Atlas. Neptune NJ, TFH Publications, pp. 258–271.

Maisey, J. G. 1991u. "*Leptolepis*". – In: Maisey, J. G. (ed.), Santana Fossils: An Illustrated Atlas. Neptune NJ, TFH Publications, pp. 272–273.

Maisey, J. G. 1991v. Undetermined Santana clupeiform. – In: Maisey, J. G. (ed.), Santana Fossils: An Illustrated Atlas. Neptune NJ, TFH Publications, pp. 297–302.

Maisey, J. G. 1991w. *Axelrodichthys*. – In: Maisey, J. G. (ed.), Santana Fossils: An Illustrated Atlas. Neptune NJ, TFH Publications, pp. 303–315.

Maisey, J. G. 1991x. *Mawsonia*. – In: Maisey, J. G. (ed.), Santana Fossils: An Illustrated Atlas. Neptune NJ, TFH Publications, pp. 316–323.

Maisey, J. G. 1991y. Undetermined Santana frog. – In: Maisey, J. G. (ed.), Santana Fossils: An Illustrated Atlas. Neptune NJ, TFH Publications, pp. 324–325.

Maisey, J. G. 1991z. Echinodermata. – In: Maisey, J. G. (ed.), Santana Fossils: An Illustrated Atlas. Neptune NJ, TFH Publications, p. 378.

Maisey, J. G. 1991aa. Arachnids. – In: Maisey, J. G. (ed.), Santana Fossils: An Illustrated Atlas. Neptune NJ, TFH Publications, p. 406.

Maisey, J. G. 1991bb. Molluscs. – In: Maisey, J. G. (ed.), Santana Fossils: An Illustrated Atlas. Neptune NJ, TFH Publications, p. 413.

Maisey, J. G. & Martins-Neto, R. G. 1991. Crustaceans. – In: Maisey, J. G. (ed.), Santana Fossils: An Illustrated Atlas. Neptune NJ, TFH Publications, pp. 407–412.

Maisey, J. G. & S. Blum. 1991. *Paraelops*. – In: Maisey, J. G. (ed.), Santana Fossils: An Illustrated Atlas. Neptune NJ, TFH Publications, pp. 238–247.

Maisey, J. G., I. Rutzky, S. Blum & W. Elvers. 1991. Laboratory preparation techniques. – In: Maisey, J. G. (ed.), Santana Fossils: An Illustrated Atlas. Neptune NJ, TFH Publications, pp. 98–105.

Nursall, J. R. & J. G. Maisey. 1991. *Neoproscinetes*. – In: Maisey, J. G. (ed.), Santana Fossils: An Illustrated Atlas. Neptune NJ, TFH Publications, pp. 125–136.

Maisey, J. G. 1992. The biology of *Latimeria chalumnae* and evolution of coelacanths, by J. A. Musick, M. K. Bruton & E. K. Balon (eds.), Kluwer Academic Publishers, Norwell (review). Copeia, 1992: 926–927.

Maisey, J. G. 1993a. A new clupeomorph fish from the Santana Formation (Albian) of NE Brazil. American Museum Novitates, 3076: 1–15.

Maisey, J. G. 1993b. Tectonics, the Santana lagerstatten, and the implications for Late Gondwanan biogeography. – In: P. Goldblatt (ed.). Biological relationships between Africa and South America. New Haven, Yale University Press, pp. 435–454.

Maisey, J. G. 1993c. Sharks in question: the Smithsonian answer book (review). American Scientist, 79: 464.

Kellner, A. W. A., J. G. Maisey & R. Evander. 1994. Ein Besonderes Fossil: Die Henkersmahlzeit eines Fisches aus der Kreide. Palaeontologische Zeitschrift, 68: 1–3.

Kellner, A. W. A, J. G. Maisey & D. A. Campos. 1994. Fossil down feather from the Lower Cretaceous of Brazil. Palaeontology, 37: 489–492.

Maisey J. G. 1994a. Gnathostomes. – In: D. R. Prothero & R. M. Schoch (eds.). Major Features of Vertebrate Evolution. Paleontological Society Short Course 7: 38–56.

Maisey, J. G. & J. Moody. 1994b. New Cretaceous marine vertebrate assemblages from north-western Venezuela and their significance. Journal of Vertebrate Paleontology, 14: 1–8.

Maisey, J. G. 1994c. Predator-prey relationships and trophic level reconstruction in a fossil fish community. Environmental Biology of Fishes, 40: 1–22.

Maisey, J. G. & M. G. P. de Carvalho. 1995. First records of fossil sergestid decapods and fossil brachyuran crab larvae (Arthropoda, Crustacea), with remarks on some supposed palaemonid fossils, from the Santana Formation (Aptian-Albian, NE Brazil). American Museum Novitates, 3132: 1–17.

Carvalho M. R. & J. G. Maisey. 1996. Phylogenetic relationships of the Late Jurassic shark *Protospinax* Woodward 1919 (Chondrichthyes, Elasmobranchii). – In: Arratia, G. & G. Viohl (eds.), Mesozoic Fishes – Systematics and Paleontology. Munich, Verlag Pfeil, pp. 9–46.

Maisey, J. G., B. G. Gardiner and D. T. J. Littlewood. 1996. Interrelationships of basal neopterygians. – In: M. L. Stiassny, L. R. Parenti & G. D. Johnson (eds.), Interrelationships of fishes. San Diego, Academic Press, pp. 117–146 .

Maisey, J. G. 1997. Early Vertebrates, by P. Janvier, Oxford University Press (review). Nature, 386: 568.

Maisey, J. G. & M. R. de Carvalho. 1997. A new look at old sharks. Nature, 385: 779–780.

Maisey, J. G., M. Arai & J. H. G. de Melo. 1997. First low latitude deep water fossil record of hexanchoid sharks from the Vraconian (uppermost Albian, Lower Cretaceous) of the Campos Basin, offshore eastern Brazil. Anais da Academia Brasileira de Ciências, 69: 141–142.

Maisey, J. G. 1999a. The supraotic bone in neopterygian fishes (Osteichthyes, Actinopterygii). American Museum Novitates, 3267: 1–52.

Maisey, J. G. 1999b. The origin of animal body plans: a study in evolutionary developmental biology, by W. Arthur (review). Cladistics, 15: 437–438.

Maisey, J. G., D. A. Campos & A. W. A. Kellner. 1999. A new specimen of *Obaichthys* (Lepisosteidae) from the Romualdo Member (Albian), Santana Formation, Northeast Brazil. Boletim de Resumos XVI Congresso de Paleontologia, 1999: 62–63.

Maisey, J. G. & L. Taverne. 1999. A notopterid skull (Teleostei, Osteoglossomorpha) from the continental Early Cretaceous of Southern Morocco. American Museum Novitates, 3260: 1–12.

Maisey, J. G. 2000. Continental break-up and the distribution of fishes in Western Gondwana during the early Cretaceous. Cretaceous Research, 21: 281–314.

Maisey, J. G. 2001a. A primitive chondrichthyan braincase from the Middle Devonian of Bolivia. – In: Ahlberg, P. (ed.), Major events in Early Vertebrate Evolution: Paleontology, Phylogeny and Development. Routledge, London, pp. 263–288.

Maisey, J. G. 2001b. Remarks on the inner ear of elasmobranchs and its interpretation from skeletal labyrinth morphology. Journal of Morphology, 250: 236–264.

Maisey, J. G. 2001c. CT-scan reveals new cranial features in Devonian chondrichthyan *"Cladodus" wildungensis*. Journal of Vertebrate Paleontology, 21: 807–810.

Maisey, J. G. & E. Anderson. 2001. A primitive chondrichthyan braincase from the Early Devonian of South Africa. Journal of Vertebrate Paleontology, 21: 702–713.

Maisey, J. G. & J. Moody. 2001. A review of the problematic extinct teleost fish *Araripichthys*, with a description of a new species from the Lower Cretaceous of Venezuela. American Museum Novitates, 3324: 1–27.

Maisey, J. G. 2002. Handbook of Paleoichthyology Vol. IV. Chondrichthyes III. Holocephali, by B. J. Stahl (review). Journal of Vertebrate Paleontology, 22: 188–189.

Maisey, J. G., M. P. G. de Carvalho & L. Borghi. 2002. Lower Devonian fish remains from the Falkland Islands. Journal of Vertebrate Paleontology, 22: 708–711.

Chang, M.-M. & J. G. Maisey. 2003. Redescription of *Elimma branneri* and *Diplomystus shengliensis* and relationships between some basal clupeomorphs. American Museum Novitates, 3404: 1–35.

Carvalho, M. R. de, J. G. Maisey & L. Grande. 2004. Freshwater stingrays of the Green River Formation of Wyoming (early Eocene), with a description of a new genus and species and an analysis of its phylogenetic relationships (Chondrichthyes: Myliobatiformes). Bulletin of the American Museum of Natural History, 284: 1–136.

Filleul, A. & J. G. Maisey. 2004. Redescription of *Santanichthys diasii* (Otophysi, Characiformes) from the Albian of the Santana Formation and comments on its implications for otophysan relationships. American Museum Novitates, 3455: 1–21.

Maisey, J. G. 2004a. Morphology of the braincase in the broadnose sevengill shark *Notorynchus* (Elasmobranchii, Hexanchiformes), based on CT scanning. American Museum Novitates, 3429: 1–52.

Maisey, J. G. 2004b. Endocranial morphology in fossil and Recent chondrichthyans. – In: Arratia, G., R. Cloutier & M. Wilson (eds.), Recent advances in the Origin and Radiation of Early Vertebrates. Munich, Verlag Pfeil, pp. 139–170.

Maisey, J. G., G. Naylor & D. Ward. 2004. Mesozoic elasmobranchs, neoselachian phylogeny, and the rise of modern neoselachian diversity. – In: Arratia, G. and Tintori, A. (eds.), Mesozoic Fishes III. Systematics, Paleoenvironments and Biodiversity. Munich, Verlag Pfeil, pp. 17–36.

Maisey, J. G. 2005. Braincase of the Upper Devonian shark *Cladodoides wildungensis* (Chondrichthyes, Elasmobranchii), with observations on the braincase in early chondrichthyans. Bulletin of the American Museum of Natural History, 288: 1–103.

Maisey, J. G. & J. H. G. Melo. 2005. Some Middle Devonian (Eifelian- Givetian) fossil fish remains from the Pimenteira Formation of the Parnaiba Basin, Northeast Brazil. Archivos do Museu Nacional, 63: 495–505.

Maisey, J. G. 2007. The braincase in Paleozoic symmoriiform and cladoselachian sharks. Bulletin of the American Museum of Natural History, 307: 1–122.

Ginter, M. & J. G. Maisey. 2007. The braincase and jaws of *Cladodus* from the Lower Carboniferous of Scotland. Palaeontology, 50: 305–322.

Carvalho, M. S. S. de & J. G. Maisey. 2008. New occurrence of *Mawsonia* (Sarcopterygii: Actinistia) from the Early Cretaceous of the Sanfranciscana Basin, Minas Gerais, southeastern Brazil. – In: Cavin, L., Longbottom, A. & Richter, M. (eds.), Fishes and the break-up of Pangaea. Geological Society of London, Special Publications 295: 109–144.

Maisey, J. G. 2008a. The postorbital palatoquadrate articulation in elasmobranchs. Journal of Morphology, 269: 1022–1040.

Maisey, J. G. 2008b. Some observations on *Denaea fournieri* (Chondrichthyes, Symmoriiformes) from the Lower Carboniferous of Belgium. Acta Geologica Polonica, 58: 185–190.

Maisey, J. G. 2009. The spine-brush complex in symmoriiform sharks (Chondrichthyes; Symmoriiformes), with comments on dorsal fin modularity. Journal of Vertebrate Paleontology, 29: 14–24.

Lane, J. A. & J. G. Maisey. 2009. Pectoral anatomy of *Tribodus limae* (Elasmobranchii: Hybodontiformes), from the Lower Cretaceous of Northeastern Brazil. Journal of Vertebrate Paleontology, 29: 25–38.

Maisey, J. G., R. Miller & S. Turner. 2009. The braincase of the chondrichthyan *Doliodus* from the Lower Devonian Campbellton Formation of New Brunswick, Canada. Acta Zoologica, 90 Suppl. 1: 109–122.

Pradel, A, J. G. Maisey, P. Tafforeau & P. Janvier. 2009. An enigmatic gnathostome vertebrate skull from the Middle Devonian of Bolivia. Acta Zoologica, 90: 123–133.

Pradel, A., M. Langer, J. G. Maisey, D. Geffard-Kuriyama, P. Cloetens & P. Janvier. 2009. Skull and brain of a 300 million-year-old fish revealed by synchrotron holotomography. Proceedings of the National Academy of Sciences of the United States of America, 106: 5224–5228. DOI: 10.1073/pnas.0807047106).

Forey, P. L. & J. G. Maisey. 2010. Structure and relationships of *Brannerion* (Albuloidei), a Lower Cretaceous teleost from Brazil. – In: G. Arratia and A. Tintori (eds.), Mesozoic Fishes III. Systematics, paleoenvironments and biodiversity. pp. 17–36. Munich, Verlag Pfeil.

Janvier, P. & J. G. Maisey. 2010. The Devonian vertebrates of South America and their biogeographical relationships. – In: D. K. Elliott, J. G. Maisey, X. Yu & D. Miao (eds.), Morphology, Phylogeny and Biogeography of Fossil Fishes: A Symposium in Honor of Meemann Chang. pp. 431–459. Munich, Verlag Pfeil.

Maisey, J. G. 2010. Heslerodidae (Chondrichthyes, Elasmobranchii), a new family of Paleozoic phalacanthous sharks. Kirtlandia, 57: 13–21.

Maisey J. G. & J.A. Lane. 2010. Labyrinth morphology and the evolution of low-frequency phonoreception in elasmobranchs. Comptes rendus Palevol, 9: 289–309.

Maisey, J. G. 2011a. Northeastern Brazil: out of Africa? – In: I. S. Carvalho, N. K. Srivastava, O. Strohschoen Jr. & C. C. Lana (eds.), Paleontologia: Cenários de Vida, 4: 515–529. Rio de Janeiro, Editoria Interciência.

Maisey, J. G. 2011b. The braincase of the Middle Triassic shark *Acronemus tuberculatus* (Bassani 1886). Palaeontology, 54: 417–428.

Pradel, A., P. Tafforeau, J. G. Maisey & P. Janvier. 2011. A New Paleozoic Symmoriiformes (Chondrichthyes) from the Late Carboniferous of Kansas (USA) and cladistic analysis of early chondrichthyans. PLoS ONE, 6: e24938. DOI: 10.1371/journal. pone. 0024938.

Dutel, H., J. G. Maisey, D. R. Schwimmer, P. Janvier, M. Herbin & G. Clement. 2012. The giant Crecaeous coelacanth (Actinistia, Sarcopterygii) *Megalocoelacanthus dobiei* Schwimmer, Stewart & Williams, 1994, and its bearing on latimerioidei interrelationships. PLoS ONE, 7: e49911. DOI: 10.1371/journal. pone. 0049911.

Lane, J. A. & J. G. Maisey. 2012. The visceral skeleton and jaw suspension in the durophagous hybodontid shark *Tribodus limae* from the Lower Cretaceous of Brazil. Journal of Paleontology, 86: 886–905.

Maisey, J. G. 2012. What is an 'elasmobranch'? The impact of palaeontology in understanding elasmobranch phylogeny and evolution. Journal of Fish Biology, 80: 918–951.

Soto, M., M. S. de Carvalho, J. G. Maisey, D. Perea & J. Da Silva. 2012. Coelacanth remains from the Late Jurassic–?earliest Cretaceous of Uruguay: The southernmost occurrence of the Mawsoniidae. Journal of Vertebrate Paleontology, 32: 530–537.

Khonsari, R. H., M.Seppala, A. Pradel, H. Dutel, G. Clement, O. Lebedev, S. Ghafoor, M. Rothova, A. Tucker, J. G. Maisey, C-M. Fan, A. Ohazama, P. Tafforeau, P. Janvier, M. T. Cobourne & P.T. Sharpe. 2013. The buccohypophyseal canal is an ancestral vertebrate trait maintained by modulation in sonic hedgehog signaling. BioMedCentral Biology, 2013, 11:27 http://www.biomedcentral.com/1741-7007/11/27.

Maisey, J. G. & V. G. Springer. 2013. Chondrocranial morphology of the salmon shark, *Lamna ditropis*, and the porbeagle, *L. nasus*. Copeia, 2013: 378–389.

Maisey, J. G., S. Turner, G. J. P. Naylor & R. Miller. 2013. Dental patterning in the earliest sharks: implications for tooth evolution. Journal of Morphology, 275: 586–596. DOI: 10.1002/jmor.20242.

Pradel, A, D. Didier, D. Casane, P. Tafforeau & J. G. Maisey. 2013. Holocephalan embryo provides new information on the evolution of the glossopharyngeal nerve, metotic fissure and parachordal plate in gnathostomes. PLoS ONE, 8: e66988. DOI: 10.1371/journal. pone. 0066988.

Maisey, J. G. 2014. The diversity of tessellated calcification in modern and extinct chondrichthyans. Revue de Paléobiologie, 32: 355–371.

Pradel, A., J. G. Maisey, P. Tafforeau, R. H. Mapes & J. Mallatt. 2014. A Palaeozoic shark with osteichthyan-like branchial arches. Nature. DOI: 10.1038/nature13195 (4 pp. + supplementary online data).

Lindoso, M. L., J. G. Maisey & I. S. S. Carvalho. 2015. Ichthyofauna from the Codo Formation, Lower Cretaceous, Aptian, Parnaiba Basin, Northeastern Brazil and their paleobiogeographical and paleoecological significance. Paleogeography, Paleoclimatology, Paleoecology, 447: 53–64.

Long J. A., C. J. Burrow, M. Ginter, J. G. Maisey, K. M. Trinajstic, M. I. Coates, G. C. Young & T. J. Senden. 2015. First shark from the Late Devonian (Frasnian) Gogo Formation, Western Australia sheds new light on the development of tessellated calcified cartilage. PLoS ONE, 10: e0126066. DOI: 10.1371/journal. pone.0126066.

Maisey, J. G. 2015. Mr. Mawson's fossils. – In: Z. Johanson, P. M. Barrett, M. Richter & M. Smith (eds.) Arthur Smith Woodward: His Life and Influence on Modern Vertebrate Palaeontology. Geological Society London, Special Publications, 430: 219–233.

Taverne, L., J. G. Maisey & L. Capasso. 2015. A third longirostrine gladiopycnodontid fish genus (Pycnodontiformes) from the marine Late Cretaceous of Lebanon. Paleontos, 28: 99–104.

Maisey J. G. & J. S. S. Denton. 2016. Dermal denticle patterning in the Cretaceous hybodont shark *Tribodus limae* (Euselachii, Hybodontiformes), and its implications for the evolution of patterning in the chondrichthyan dermal skeleton. Journal of Vertebrate Paleontology, 36:5, e1179200, DOI: 10.1080/02724634.2016.1179200

Pradel A., J. G. Maisey, R. H. Mapes & I. Kruta. 2016. First evidence of an intercalar bone in the braincase of "palaeonisciform" actinopterygians, with a virtual reconstruction of a new braincase of *Lawrenciella* Poplin, 1984, from the Carboniferous of Oklahoma. Geodiversitas, 38: 489–504. DOI: 10.5252/g2016n4a2

Soler-Gijón R., F. J. Poyato-Ariza, J. G. Maisey & J. A. Lane. 2016. Chondrichthyes – In: Poyato-Ariza, F. J. & A. D. Busclioni (eds.), Las Hoyas: A Cretaceous Wetland. pp. 103–113. Munich, Verlag Pfeil.

Burrow C. J., S. Turner, J. G. Maisey, S. Desbiens & R. F. Miller. 2017. Spines of *Doliodus latispinosus* (Whiteaves) comb. nov. from the Early Devonian of eastern Canada. Canadian Journal of Earth Sciences, 54, 1248–1262, 10.1139/cjes-2017-0059

Maisey J. G., R. F. Miller, A. Pradel, J. S. S. Denton, A. Bronson & P. Janvier. 2017. The pectoral skeleton of *Doliodus*: bridging the 'acanthodian'-chondrichthyan divide. American Museum Novitates, 3875: 1–15.

Maisey J. G., A. W. Bronson, R. R. Williams & M. McKinzie. 2017. A Pennsylvanian 'Supershark' from Texas. Journal of Vertebrate Paleontology, 2017; e1325369

Bronson A. W. & J. G. Maisey. 2018. The taxonomic journey of *Platylithophycus cretaceum*. Journal of Paleontology, 92: 743–750. DOI: 10.1017/jpa.2018.14

Bronson, A. W., R. Mapes, J. G. Maisey. 2018. Chondrocranial morphology of *Carcharopsis wortheni* (Chondrichthyes; Euselachii incertae sedis), based on new material from the Fayetteville Shale (Upper Mississippian, Middle Chesterian). Papers in Palaeontology. DOI: 10.1002/spp2.1110

Denton, J. S. S., J. G. Maisey, M. Grace, A. Pradel, M. H. Doosey, H. L. Bart, Jr. & G. J. P. Naylor. 2018. Cranial morphology in *Mollisquama* sp. (Squaliformes; Dalatiidae) and patterns of cranial evolution in dalatiid sharks. Journal of Anatomy, 233: 15–32. DOI: 10.1111/joa.12823..

Maisey J. G., P. Janvier, A. Pradel, J. S. S. Denton, A. W. Bronson, R. Miller & C. Burrow. 2019. *Doliodus* and pucapampellids: contrasting perspectives on stem chondrichthyan morphology. – In: Johanson, Z., C. Underwood & M. Richter (eds.), Evolution and Development of Fishes (Cambridge University Press, 87–109.

Maisey, J. G., D. Ehret & J. S. S. Denton. 2020. A new genus of Late Cretaceous angel shark (Elasmobranchii; Squatinidae), with comments on squatinid phylogeny. American Museum Novitates, 3954: 1–29.

Witten, P. E., A. Huysseune, J. G. Maisey, C. Winkler & Z. Gong. 2021. A boost for fish skeletal research (Editorial). Journal of Fish Biology, 98: 903–905

Maisey, J. G., J. S. S. Denton, C. Burrow & A. Pradel. 2021. Architectural and ultrastructural features of tessellated calcified cartilage in modern and extinct chondrichthyan fishes. Journal of Fish Biology, 98: 919–941

Books

Maisey, J. G. (ed.) 1991. Santana Fossils: An Illustrated Atlas. 459 pp. TFH Publications, Neptune, NJ.

Maisey, J. G., E. S. Gaffney, M. A. Norell, M. Posen & L. Dingus. 1995. The Hall of Vertebrate Origins: a guide to fishes, amphibians, turtles, lizards, crocodiles, and pterosaurs. 92 pp. New York, American Museum of Natural History.

Maisey, J. G. 1996. Discovering Fossil Fishes. 223 pp. Henry Holt, New York (hardback, reprinted in 2000 as soft cover edition).

Elliott, D. K., J. G. Maisey, X. Yu & D. Miao (eds.), 2010. Morphology, phylogeny and paleobiogeography of fossil fishes, honoring Meemann Chang. 422 pp. Verlag Pfeil, Munich.

Maisey, J. G. 2017. Sharks. 24 pp. New York, AMNH/Sterling Children's Books.

Encyclopedia entries

Maisey, J. G. 1982. Osteichthyes (Fossils). McGraw Hill Encyclopedia of Science and Technology, 5th ed., pp. 692–694. McGraw-Hill, New York.

Maisey, J. G. 1982. Selachii. McGraw Hill Encyclopedia of Science and Technology, 5th ed., p. 246. McGraw-Hill, New York.

Maisey, J. G. 2000. Shark. McGraw-Hill Yearbook of Science and Technology 2001, pp. 343–344. McGraw-Hill, New York.

Maisey, J. G. 2006. Chimaeriformes. McGraw-Hill Encycopedia of Science and Technology 2006, pp. 89–91. McGraw-Hill, New York.

Maisey, J. G. 2011. The fossil collector. In H. R. Axelrod & B. Degen (eds.), Discus. Ruhmannsfelden, Verlag Degen Mediahouse, pp. 16–19.

Maisey, J. G. 2015. Shark history. McGraw-Hill Encycopedia of Science and Technology 2015, pp. 294–296. McGraw-Hill, New York.

Maisey, J. G. 2019. Chondrichthyes. – In: S. Elias & D. Alderton (eds.), Encyclopedia of Geology 2nd Ed. Elsevier, Oxford.

Popular articles

Maisey, J. G. 1986. The 110-million-year-old fossil fishes from Brazil. Tropical Fish Hobbyist May 1986, pp. 26–36.
Maisey, J. G. 1987. New fossil coelacanth named after Dr. Herbert R. Axelrod. Tropical Fish Hobbyist, November 1987, pp. 76–84.
Maisey, J. G. 1987. Evolution of the shark. In Stevens, J. D. (ed.), Sharks. pp. 14–17. Weldon, Sydney.
Maisey, J. G. & M.G.P. de Carvalho. 1992. Saving the dinosaurs and other fossils. Tropical Fish Hobbyist Nov. 1992: 116–130.
Maisey, J. G. 1993. Fish Fossils and evolution. Tropical Fish Hobbyist Jan. 1993: 92–96
Maisey, J. G. 1993. The enigmatic *Cichlasoma woodringi*. Tropical Fish Hobbyist Feb. 1993: 160–161.
Maisey, J. G. 1993. What is *Tharrhias?* Tropical Fish Hobbyist Mar. 1993: 132–134.
Maisey, J. G. 1993. *Dastilbe* by the thousands. Tropical Fish Hobbyist Apr. 1993: 131–133.
Maisey, J. G. 1993. *Vinctifer*: not an armored catfish. Tropical Fish Hobbyist June 1993: 164–165.
Maisey, J. G. 1993. *Araripichthys*, a scale-finned oddity. Tropical Fish Hobbyist Aug. 1993: 164–166.
Maisey, J. G. 1993. The world's oldest anchovy? Tropical Fish Hobbyist Dec. 1993: 164–165.
Maisey, J. G. 1994. *Semionotus* species flocks. Tropical Fish Hobbyist May 1994: 172–175.
Maisey, J. G. 1994. *Calamopleurus*: an ancient bowfin. Tropical Fish Hobbyist June 1994: 204–206.
Maisey, J. G. 1994. The oldest bonefish? Tropical Fish Hobbyist July 1994: 204–206.
Maisey, J. G. 1994. Ancient tarpon-like fishes. Tropical Fish Hobbyist Aug. 1994: 112–116.
Maisey, J. G. 1994. Branner's fish. Tropical Fish Hobbyist Sept. 1994: 195–198.
Maisey, J. G. 1995. All fins and no feet. Tropical Fish Hobbyist Aug. 1995: 224–226.
Maisey, J. G. 1995. Ancient coelacanth giants. Tropical Fish Hobbyist Sept. 1995: 227–229.
Maisey, J. G. 1996. Fossil surgeonfishes. Tropical Fish Hobbyist April 1996: 124–129.
Maisey, J. G. 1996. Clam-crushing fishes from the Age of Dinosaurs. Tropical Fish Hobbyist May 1996: 182–184.
Maisey, J. G. 1996. Fossil Fishes: so much diversity, so little change. Natural History June 1996: 39–40.
Maisey, J. G. 1996. Lungfishes: Living fossils. Tropical Fish Hobbyist Oct. 1996: 136–138.
Maisey, J. G. 2001. Uncovering the secrets of ancient sharks. The Fossil Times (News Publication of AMNH), Spring issue: pp. 1,3.
Maisey, J. G. 2011. The fossil collector. In Axelrod, H. R. & B. Degen (eds.), Discus. Ruhmannsfelden, Verlag Degen Mediahouse, pp. 16–19.

Other publications

1980–1989. Journal of Vertebrate Paleontology Newsletter – editor/contributor for AMNH.
1973. Illustrations in: Ward, D. J., The English Palaeogene chimaeroid fishes. Proceedings of the Geologists Association 84: 315–330.

Ancient Fishes and their Living Relatives: a Tribute to John G. Maisey. Alan Pradel, John S. S. Denton & Philippe Janvier (eds.): pp. 19–36, 9 figs., 2 apps.
© 2021 by Verlag Dr. Friedrich Pfeil, München, Germany – ISBN 978-3-89937-269-4

The intriguing †*Atacamichthys* fish from the Middle Jurassic of Chile – an amiiform or a teleosteomorph?

Gloria Arratia*, Hans-Peter Schultze*, Soledad Gouiric-Cavalli** and Claudio Quezada-Romegialli***

Abstract

The Oxfordian El Profeta Formation in northern Chile was an important place of teleosteomorph or stem teleostean diversification in the southern Paleo-Pacific Ocean. Among the diverse fish fauna recovered in this formation is †*Atacamichthys greeni*, which exhibits an interesting combination of amiiform and teleostean morphological features. As a consequence, the fish has been interpreted as a "caturid-like fish" or Halecostomi incertae sedis or a teleost incertae sedis in the past, but a phylogenetic analysis was missing. This paper re-examines the morphological characters of †*Atacamichthys*, and the phylogenetic position of the fish is investigated using a pre-existing data matrix whose characters were re-polarized using more distant phylogenetic outgroup taxa, e.g., †*Birgeria* and *Polypterus*. The new results suggest †*Atacamichthys greeni* is a teleosteomorph or stem teleost placed above the phylogenetic level of †*Prohalecites* and below that of †Pholidophoriformes. Changes in the outgroup taxa did not affect the position of †*Atacamichthys*. †*Atacamichthys greeni* is suggested as a member of †Atacamichthyidae new family, which is supported by a combination of morphological characters differentiating it from other Jurassic families worldwide and increasing the endemism of the fish faunas of the Paleo-Pacific Ocean during the Jurassic.

Introduction

Among the endemic fish fauna characterizing the Jurassic of Chile, there are a few fishes that still remain a mystery. One is a notable, well-preserved large caudal fin with some remarkable anatomical characters, which is simply identified as Teleostei sp. 1 until more specimens become available (Arratia 1991, Arratia & Schultze 1999, Arratia 2015a). Another unusual fish is †*Atacamichthys greeni* Arratia & Schultze, 1987, the only fish known from the Quebrada del Profeta, Cordillera de Domeyko, Atacama Desert (Fig. 2B) that has amioid-type scales (sensu Schultze 1966, fig. 3a in p. 145, Arratia 2015b, fig. 15B; see description below), diplospondylous vertebral chordacentra (Arratia & Schultze 1987, figs. 9–12), and a suite of characters that are also found in neopterygian taxa such as amiiforms, pachycormiforms, and the pholidophoriform †*Eurycormus*. Consequently, †*Atacamichthys* was interpreted as a "caturid-like fish" by Arratia (1987), as a halecostome incertae sedis by Arratia & Schultze (1987, p. 1: taxonomic assignment) and as a teleost incertae sedis based on a non-computer generated cladogram (Arratia & Schultze 1987, fig. 20 and Fig. 1 herein) where †*Atacamichthys* was interpreted as part of Teleostei sensu lato (that would correspond to Teleosteomorpha of Arratia 2001), and later as a stem teleost by Pinna (1996), an interpretation based on the description by Arratia & Schultze (1987).

As with other fishes from the Oxfordian of El Profeta Formation, †*Atacamichthys greeni* (Fig. 2C) is preserved in calcareous concretions (in bituminous shales) that commonly are missing the anterior tip of the head and the posterior body, including the caudal fin. The fishes show outstanding preservation that includes soft tissues, such as cartilage, muscles, blood vessels, and notochord, as has been described for the teleosts †*Protoclupea chilensis*, †*Domeykos profetaensis*, †*Chongichthys dentatus*, †*Varasichthys ariasi*, and other recovered fossils (Arratia 1981, Arratia & Schultze 1985, Schultze 1989, Arratia & Schultze 1999).

The classification of the most advanced actinopterygians has undergone dramatic changes in the last 50 years. In 1973, the old term Holostei was erased from the actinopterygian classification when C. Patterson invalidated it and replaced it with a new classification including the Divisions Ginglymodi and Halecostomi, the latter with two main taxa, the Subdivision Halecomorphi (e.g., Amiidae, †Caturidae) and the Subdivision Teleostei (e.g., †Pachycormidae, †Aspidorhynchidae, †Pholidophoridae). Thus, the first study of †*Atacamichthys greeni* was done in this period, when the teleosts were interpreted as part of the Division Halecostomi. Patterson's (1973) interpretation was changed in 2010, when L. Grande resurrected the Holostei. During the last 32 years extensive new knowledge in what we currently understand as holosteans (e.g., Grande & Bemis 1998, Grande 2010, Thies & Waschkewitz 2015, Xu &

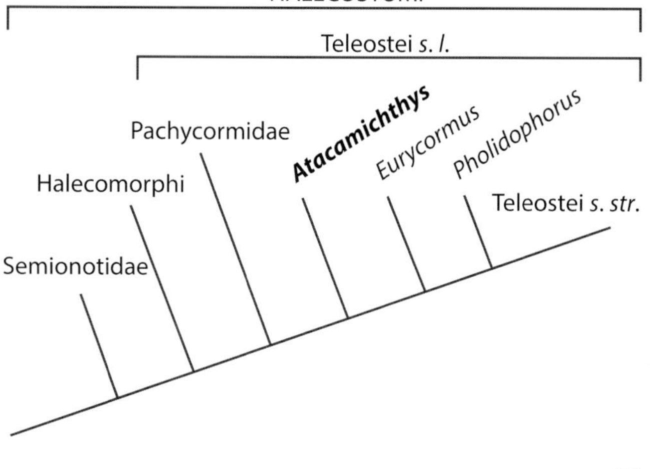

Fig. 1. Phylogenetic hypothesis concerning the possible phylogenetic position of †*Atacamichthys* (after Arratia & Schultze 1987).

* Biodiversity Institute, University of Kansas, Dyche Hall, 1345 Jayhawk Blvd, Lawrence, Kansas 66045, U.S.A.; garratia@ku.edu; hp1937@ku.edu (corresponding authors)

** División Paleontología Vertebrados, Museo de La Plata, Facultad de Ciencias Naturales y Museo, Universidad Nacional de La Plata, Paseo del Bosque s/n, B1900FWA, La Plata, Buenos Aires, Argentina; and Consejo de Investigaciones Científicas y Técnicas, Godoy Cruz 2290 C1425FQB, Ciudad Autónoma de Buenos Aires, Argentina

*** Departamento de Biología, Facultad de Ciencias Naturales y Exactas, Universidad de Playa Ancha, Valparaíso, Chile; and Núcleo Milenio INVASAL, Concepción, Chile

Fig. 2. Landscape of a section of Quebrada del Profeta where specimens of †*Atacamichthys greeni* were found. **A.** South America with location of Cordillera de Domeyko in a section where specimens of †*Atacamichthys* were recovered. **B.** G. Arratia and A. Quincio collecting fossils in Quebrada del Profeta in 1978. **C.** †*Atacamichthys greeni*, specimen LBUCH 248.

Ma 2018) and teleosteomorphs (e.g., Tintori 1990, Arratia & Lambert 1996, Brito 1997, Arratia 1997, 1999, 2000, 2013, 2017, Friedman et al. 2010, Arratia & Schultze 2013, Nelson et al. 2016) has accumulated, especially for teleosteomorphs. Currently, amiiforms are interpreted in the Division Halecomorphi (Infraclass Holostei) and the teleosteomorphs in the Infraclass Teleosteomorpha (e.g., Nelson et al. 2016, Betancur-R. et al. 2017). There are differences in the use of high-levels ranks among authors, such as infraclass or division, but such differences are meaningless for this study. Based on a re-examination of the material and original description, the main goal of this contribution is to reevaluate the taxonomy of †*Atacamichthys* and to study its phylogenetic relationships under the present understanding of the amiiforms and teleosteomorphs.

Material and method

Institutional Abbreviations

Studied specimens are deposited in the following institutions: ANSP, Academy of Natural Sciences, Philadelphia, Pennsylvania, USA; BGHan, Bundesanstalt für Geowissenschaften und Rohstoffe, Niedersächsisches Landesamt für Bodenforschung, Hannover, Germany; SNSB-BSPG, Bayerische Staatssammlung für Paläontologie und historische Geologie, München, Germany; CAS(SU), California Academy of Sciences, Department of Ichthyology, San Francisco, California, USA; CM, Carnegie Museum, Pittsburgh, Pennsylvania, USA; FMNH, Field Museum of Natural History, Chicago, Illinois, USA; GPIT, Institut und Museum für Geologie und Paläontologie, Tübingen, Germany; GZG, Geoscience Center, Georg-August Universität, Göttingen, Germany; Innsb, Department of Geology, University of Innsbruck, Innsbruck, Austria; IVVP, Institute of Vertebrate Palaeontology and Palaeoanthropology, Beijing, China; JFBM, James Ford Bell Museum, Saint Paul, Minnesota, USA; JME, Jura Museum Eichstätt (SOS indicates that the fish was recovered in the Solnhofen Limestone), Eichstätt, Germany; KU:KUI, Division of Ichthyology, Natural History Museum, University of Kansas, Lawrence, Kansas, USA; KUVP, Division of Vertebrate Paleontology, Natural History Museum, University of Kansas, Lawrence, Kansas, USA; LBUCH, Laboratorio de Biología, Universidad de Chile, Sede Santiago-Sur, Chile; MB, Museum für Naturkunde, Humboldt Universität, Berlin, Germany; MCSNB, Museo Civico di Scienze Naturali, Bergamo, Italy; MCSNIO, Civico Museo Insubrico di Storia Naturele, Unduno-Olona, Italy; MCZ, University Museum of Zoology, Harvard University, Cambridge, Massachusetts, USA; NHM, Natural History Museum, London, UK; SMNH, Department of Paleozoology, Naturhistoriska Riksmuseet, Stockholm, Sweden; SMNS, Staatliches Museum für Naturkunde, Stuttgart, Germany; TCWC, Texas Cooperative Wildlife Collection, Department of Wildlife and Fisheries Science, Texas A&M University, College Station, Texas, USA.

Material Studied

The specimens are listed following a taxonomic arrangement and ordered alphabetically within each taxon. Extinct taxa are identified with the sign "†". Abbreviations: cl&st, cleared and stained specimen; sk, skeleton.

Outgroup

†**Birgeriiformes** – †*Birgeria groenlandica* from Stensiö (1932), and Nielsen (1949); †*Birgeria nielseni* from Lehman (1952).

†**Pholidopleuriformes** – †*Australosomus kochi* from Stensiö (1932), and Nielsen (1949) (= †*Australosomus simplex*); †*Australosomus merlei* from Lehman (1952).

Scanilepiformes – *Polypterus bichir*; MB uncatalogued: 1 cl&st; MCZ 48572, 1 dry sk; MCZ 85766, 1 dry sk; SMNH uncatalogued, histological slides. *Polypterus ornatipinnis*, KU:KUI 23123, 1 cl&st; *Calamoichthys* sp. KU:KUI 21392, 1 cl&st. (For systematic position see Giles et al. 2017 and Schultze 2018).

Ingroup

Neopterygii
Holostei

Amiiformes – *Amia calva*: KU:KUI 6883, 5 cl&st; KU:KUI 21261, 1 sk; KU:KUI 21290, 4 cl&st; and KU 21607, 1 cl&st. †*Caturus elongatus*: MB f.3851. †*Caturus furcatus*: MB f.2931. †*Caturus brevicostatus*: MB f.3849; and MB f. 3850. †*Caturus* sp.: JME SOS 3344.

Lepisosteiformes – *Atractosteus spatula*: KU:KUI 18537; KU:KUI 18540; KU:KU, 18545; and KU:KU 18548, dry crania. *Lepisosteus osseus*: KU:KUI 3651, 3 cl&st; KU:KUI 3677, 1 cl&st; KU:KUI 8530, 1 cl&st; KU:KUI 12645, 1 cl&st; KU:KUI 16246, 12 cl&st; and KU:KUI 17935, 1 cl&st. *Lepisosteus platostomus*: KU:KUI 3138, 1 cl&st; and KU:KUI 16142 1 cl&st.

Teleosteomorpha

†**Aspidorhynchiformes** – †*Aspidorhynchus acutirostris*: MB. f.358; MB. f.3529; MB. f.3554; and MB f.3566. †*Aspidorhynchus* sp.: JME ETT 2006-2. †*Belonostomus muensteri*: MB. f.3544. †*Belonostomus tenuirostris*: JME SOS 2339; and JME SOS 2844. †*Belonostomus* sp.: SNSB-BSPG 1956 I 422.

†**Pachycormiformes** – †*Asthenocormus titanius*: JME SOS 542; and JME SOS 3556. †*Euthynotus* sp.: SMNH P 1037. †*Hypsocormus insignis*: SNSB-BSPG AS VI 4a,b (holotype); SMS P 5698. †*Orthocormus cornutus*: SenkM 1863 (holotype); JME SOS 3460; and JME SOS 3571a-b. †*Orthocormus roeperi*: SNSB-BSPG 1993 XVIII-VFKO B16. †*Pachycormus macropterus*: GPIT/OS/00775. †*Pachycormus macropterus*: GPIT/OS/00774. †*Pachycormus* sp.: SMS P 6151; SMNH P 6152; and MB f.2349. †*Protosphyraena tenuis*: KUVP 49419. †*Protosphyraena* cf. *P. perniciosa*: KUVP 67877. †*Protosphyraena* sp.: KUVP 49418; and KUVP 55000. †*Pseudoasthenocormus retrodorsalis*: SNSB-BSPG 1956 I 361. †*Sauropsis depressus*: CM 4766A/B (holotype). †*Sauropsis longimanus*: SNSB-BSPG AS VII 1089 (holotype). †*Sauropsis* sp.: JME SOS 3336a–b; JME 2181a–b; and MB f.2348.

†**Atacamichthyidae** new fam. – †*Atacamichthys greeni*: Holotype LBUCH 012478b (part) and KUVP 86777 (counterpart); LBUCH 245; LBUCH 246–248; LBUCH 370678b (part) and KUVP 86778 (counterpart); KUVP 86779; KUVP 86780; KUVP 86781, displaced head bones and vertebrae; KUVP 86778; KUVP 86782; KUVP 86783; and KUVP 86784 histological sections of abdominal vertebrae of specimen KUVP 86778 plus other cross sections of vertebrae; MB 3952.

†**Prohaleciteiformes** – †*Prohalecites porroi*:MCSNIO P 15L; MCSNIO P 331; MCSNIO P 342; MCSNIO P 335; MCSNIO P 343; MCSNIO P 364; MCSNIO P 378-1; MCSNIO P 384; MCSNIO P 407; MCSNIO P 413; MCSNIO P 658; and many others. †*Prohalecites* indet.: MCSNIO P 378-1; MCSNIO P 657; and MCSNIO P 658.

Teleostei incertae sedis

†**"Pholidophoriformes"** – †*Jianligichthys serratus*: IVVP 6353; IVVP V6353-1.-2; IVVP V6354. †*"Pholidophorus" armatus*: MB f. 7010. †*Pholidophorus dorsalis*: Innsb. 124. †*"Pholidophorus" furcatus*: MCZ 9863. †*"Pholidophorus" germanicus*: MCZ 9762. †*"Pholidophorus" granulatus*: NHM P 3605; NHM P 6378e; and NHM P6379. †*"Pholidophorus" hartmani*: NHM P 605; NHM P 3591a; and NHM 22527. †*"Pholidophorus" latus*: GPIT/OS/836. †*"Pholidophorus" latus*: GPIT/OS/837. †*"Pholidophorus" leptocephalus*: MCZ 9761. †*"Pholidophorus" limbatus*: MCZ 9763. †*"Pholidophorus" ornatus*: NHM P6378 †*Pholidophorus pusillus*: Innsb. 115. †*"Pholidophorus"?*: MCZ 9764; MCZ 13661. †*"Pholidophorus"* sp.: MCZ 10317. And many other specimens identified as †*"Pholidophorus"* from the Jurassic of Europe.

Teleostei (apomorphy-based group)

†**Ankylophoridae** (or †*Siemensichthys* Group) – †*Ankylophorus similis*: NHM 1083.

†***Siemensichthys macrocephalus***: SNSB-BSPG AS I 1134; JME SOS 2812; MB f.7017; MB f.7018a–b; MB f. 7019; GPIT/OS/83a; and GPIT/OS/835. †*Siemensichthys siemensi*: JME SCHA 82.

†**Dorsetichthyiformes** – †Dorsetichthyidae: †*Dorsetichthys* (= †*Pholidophorus*) *bechei*: FMNH 2137; MB f.3504; MB f.19219; SMNS P 944; MCZ 9873; MCZ 6301; and MCZ 12056.

†**Leptolepidiformes** – †Leptolepididae: †*Leptolepis coryphaenoides*: SMNS 87381; SMNS 87382; SMNS 87383; BGHan 1931-4; BGHan 1956-8; BGHan 1957-2; BGHan 1957-5; BGHan 1960 (acid-prepared specimens); and many uncatalogued articulated and disarticulated specimens. Many disarticulated bones prepared for SEM.

†**Ascalaboidiformes** – †Ascalaboidae: †*Ascalabos voithii*: JME SOS 537 (neotype); CM 9491; CMMH 9491; JME 537; JME SOS 1961/122; JME SOS 2362; JME SOS 2363; JME SOS 2364; JME SOS 2365; JME SOS 2458; JME SOS 2483; JME SOS 2496; JME SOS 2497; JME SOS 2886 (peels of scales); and many other specimens deposited at the JME; NHM 3672, NHM 3673a; NHM 37062; NHM 37080; SHL, collection of the late H. Leich (Bochum, Germany) currently deposited in Tierpark und Fossilium Bochum (Bochum); and SMNH P5683. †*Tharsis dubius*: SNSB-BSPG 1964 XXIII 280; CM 4845; FMNH 25076; FMNH 25124; and JME, many specimens from different localities.

†**Crossognathiformes** – †Varasichthyidae: †*Domeykos profetaensis*: LBUCH 12-260972a–b; and LBUCH 01277-13a–3b. †*Protoclupea atacamensis*: LBUCH 1-250277a. †*Protoclupea chilensis*: R-396a, R396b; and LBUCH 190179a-b. †*Varasichthys ariasi*: LBUCH 16-260972a–b; LBUCH 012378a; LBUCH 020778a; LBUCH 020778b; and many other specimens of †*Varasichthys*.

†**Pholidophoriformes** – †Pholidophoridae: †*Lombardichthys gervasutti* (= †*Pholidophorus gervasutti*): MCSNB 4723a–c (holotype); MCSNB 4306a–b (no 3406a–b in Arratia, 2013); MCSNB 3462d; MCSNB 4300a–c; MCSNB 4301a–d; MCSNB 4302a–b; MCSNB 4303a–d; MCSNB 4304a–d; MCSNB 4305; MCSNB 4306a–b; MCSNB 4308; MCSNB 4309; MCSNB 4310a–g; MCSNB 4311; MCSNB 4313; MCSNB 4317a–c; MCSNB 4318a; MCSNB 4329a–b; MCSNB 4330; MCSNB 4332; MCSNB 4334; MCSNB 4340; MCSNB 4345a–b; MCSNB 4346a–d; MCSNB 4416; MCSNB 4418; MCSNB 4425; MCSNB 4438; MCSNB 4453; MCSNB 4455; MCSNB 4469a–b; MCSNB 4470a–b; MCSNB 4473; MCSNB 4708c; MCSNB 4710; and MCSNB 4726a–b. †*Pholidoctenus serianus*: MCSNB 3067 (holotype); MCSNB 2875 to 2877; MCSNB 3012; MCSNB 3034 to 3063; MCSNB 3064a-m; MCSNB 3065, 3066, 3068 to 3070; MCSNB 3095; MCSNB 3097 to 3090; MCSNB 3312 to 3315; MCSNB 3357; and MCSNB 3373 to 3378. †*Pholidorhynchodon malzanii*: MCSNB 3385 (holotype); MCSNB 3161; MCSNB 3164; MCSNB 3243; MCSNB 3244; MCSNB 3274; MCSNB 3276; MCSNB 3281; MCSNB 3284; MCSNB 3285; MCSNB 3286; MCSNB 3287; MCSNB 3353; MCSNB 3381; MCSNB 3386; MCSNB 3285; MCSNB 3286; MCSNB 3392; MCSNB 3848; MCSNB 3859; MCSNB 3862; MCSNB 3887; and MCSNB 3914. †Eurycormidae: †*Eurycormus speciosus*: SNSB-BSPG AS V510; SNSB-BSPG 1960 XVIII 106; JME SOS 2339; JME SOS 2341; JME 3004; JME 4615; SMNS 80144/37; SMNS 87316/3; SMNS 87848/21; SMNS 88987/3; SMNS 95445/11; and SMNS 95445/12.

Elopocephala (or crown-group Teleostei)
Elopomorpha

Elopiformes – *Elops hawaiensis*: CAS(SU) 35105, partially disarticulated sk; OS 5105, 2 cl&st leptocephalous larvae. *Elops saurus*: ANSP 147401, 2 cl&st; CAS(SU) 10847, sk; TCWC 0503.1; TCWC 0782; TCWC 2452.2, 5 cl&st; UNC 82/8, 2 cl&st *Megalops cyprinoides*: CAS 145216, 2 cl&st.

Osteoglossomorpha

Hiodontiformes – *Hiodon alosoides*: JFBM 43312, 1 sk; JFBM 43306, 1 sk; KU:KUI 9618, 7 cl&st; KU:KUI 9661, 2 cl&st; KU:KUI 13993, 2 cl&st. *Hiodon tergisus*: KU:KUI 9662, 3 cl&st.

Phylogenetic methodology. The phylogenetic analysis was conducted using PAUP (Phylogenetic Analysis Using Parsimony) software (version 4.0a build 165) for both 32-bit Microsoft Windows and Macintosh (Swofford 2000) and tested with the on-line version of PAUP* 4 (version 3.99.165.0). The phylogenetic analysis used part of the list of characters and coding of Arratia et al. (2019), which is an expanded version of Arratia (2013), with a revision of all characters that are coded based on the new outgroups (Appendix 1), and the ingroup is restricted to six holostean taxa and 25 teleosteomorphs. The character matrix was constructed using MacClade (Appendix 2). All characters are unweighted, unordered, and considered to be independent of one another. The outgroups used in the analysis are the actinopterygians †*Birgeria* and *Polypterus*. An additional analysis was executed including only two fossil taxa in the outgroup, †*Australosomus* and †*Birgeria*.

Anatomical terminology. The terminology of the skull roof bones follows Westoll (1943), Jollie (1962), and Schultze (2008). To avoid confusion, all figures show in square brackets the names of bones in the traditional terminology, e. g., parietal bone [= frontal]: pa [fr]. The terminology of the vertebral column follows Arratia et al. (2001) and Arratia (2015b), whereas that of the caudal endoskeletal elements and caudal skeletal types (e. g., polyural or diural) follows Nybelin (1963), Schultze & Arratia (1988, 1989, 2013), and Arratia & Schultze (1992). "True" uroneurals are modified ural neural arches; "uroneural-like" elements are modified preural neural arches; these terms are included in the phylogenetic analysis, and their usage follows Patterson (1973), Arratia & Lambers (1996), and Arratia & Schultze (2013). The terms fin rays, scutes, fulcra, procurrent rays, epaxial rudimentary rays, and principal rays follow definitions provided by Arratia (2008, 2009). The terminology of scales, ganoid, amioid and cycloid types follows Schultze (1966, 1996, 2015).

Analysis of some morphological characters

For a complete description of †*Atacamichthys greeni* see Arratia & Schultze (1987). Some anatomical features of the fish, which may have another interpretation than the one offered in 1987, are presented below. For further explanations, see list of characters in Appendix 1.

Cranium

Skull roof shape and fusion patterns. The skull roof of †*Atacamichthys* is narrow in its orbital region, expanding broadly posteriad, with an elegant curvature at its lateral margin, making the transition between orbital and postorbital regions smooth (Figs. 3, 4B). The parietal [= frontal of traditional terminology] bones have their narrowest point anteriorly, at the region that was possibly articulating with a rostral bone, which is not preserved. The parietal broadens slightly posteriorly, covering about 89 % of the skull roof, making this the largest bone of the skull roof in comparison to amiids, caturids, parasemionotids, and teleosteomorphs (Fig. 4A, C–F). In †*Atacamichthys*, the posterior expansion of the skull roof is enhanced by the enlargement of the posterior part of the parietal bone and the lateral expansion of the dermopterotic, which forms the lateral margin and most of the posterior margin of the skull roof. In amiiforms as well as in parasemionotids and caturids, the parietals are broad in the orbital region, keeping an almost constant broadening or become narrower posteriad, whereas in †*Solnhofenamia* the parietals are narrower anteriorly as in †*Atacamichthys*, but the arrangement and sizes of bones differ posteriorly. In †*Dorsetichthys bechei* and †*Leptolepis coryphaenoides*, the posterior expansion of the skull roof is mainly due to the posterior expansion of the parietal bones that is accompanied by the lateral expansion of the pterotics (Fig. 4E–F; Arratia 2013, fig. 97).

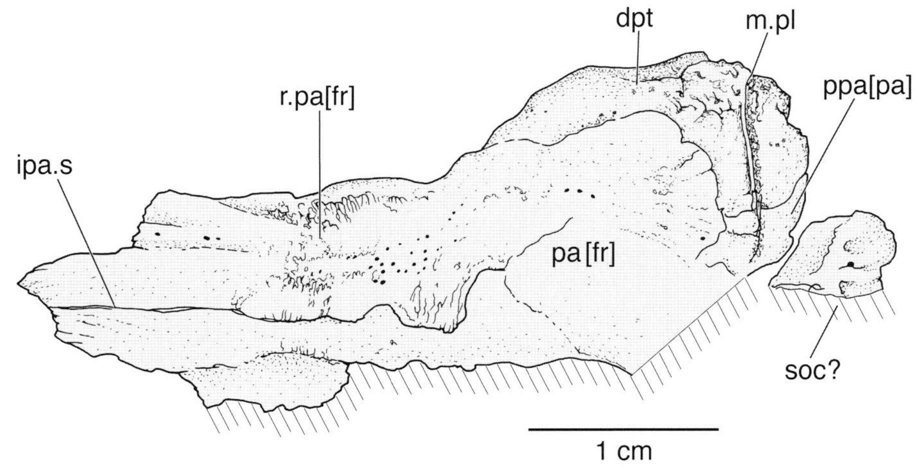

Fig. 3. †*Atacamichthys greeni*, dorsal view of skull roof; broken areas are represented with oblique lines (KUVP 86777). Abbreviations: dpt, dermopterotic; ipa.s, interparietal suture; m.pl, middle pit-line; pa [fr], parietal bone [= frontal of traditional terminology]; note that right and left parietal are fused posteriorly; ppa [pa], postparietal bone [= parietal of traditional terminology]; r.pa [fr], right parietal bone; soc? supraoccipital?

In †*Atacamichthys*, both parietal bones (Figs. 3, 4B) are sutured anteriorly by an almost straight suture that is deeply undulated at the level of the posterior corner of the orbits, and both bones become fused posteriorly. Fusion among skull roof bones is a common feature in pholidophoriforms (Arratia 2013, figs. 10–14, 32, 48, 59, 61, 62, 78–81) where fusion may involve all cranial bones into a single plate. Fusion may include some bones in aspidorhynchiforms, especially those of the preorbital and orbital regions (Brito 1997). Fusion of postparietals [= parietal bones of traditional terminology] also have been described for some adult specimens of *Amia calva* (Grande & Bemis 1998, figs. 12F, 13F).

Dermopterotic. This bone (Figs. 3, 4B) is well-developed in †*Atacamichthys*, with a significant section partially covered by the posterior region of the parietal bone. The main postorbital expansion of the skull roof is due to the enlargement of the dermopterotic. Each dermopterotic sutures medially with a small postparietal [= parietal bone of the traditional terminology].

Supraoccipital. An incompletely preserved bone (Fig. 3) placed posteriorly to the postparietals and slightly displaced with respect to the parietal and epiotic was identified as a supraoccipital in †*Atacamichthys* by Arratia & Schultze (1987). The bone looks too massive for a supraoccipital, and though uncertain, we do not have any other possible identification for it. The presence of a supraoccipital bone has been traditionally interpreted as a synapomorphy of teleosts (Patterson 1975, 1977) being known in †*Dorsetichthys bechei* and more advanced teleosts (e. g., Patterson 1975, Arratia 1997, 1999, 2013, 2017). However, a supraoccipital is unknown in stem teleosts, such as aspidorhynchiforms, †*Prohalecites*, and pholidophoriforms, including †*Eurycormus* (Arratia 1999, 2013, 2017, Arratia & Schultze 2007).

Supraorbital canal and pit-lines. The trajectory of the supraorbital canal (Figs. 3, 6) is limited to the parietal bone. Thus, the canal does not continue into the postparietal and does not have a lateral branch extending into the dermosphenotic. The supraorbital canal (as well as the otic canal in the dermopterotic) is a ramified canal exiting to the bony surface through many small, round pores, which is the condition found in other fishes, such as *Amia calva* among amiiforms and †*Eurycormus speciosus* among stem teleosts.

Anteriorly, the supraorbital canal continues, probably, in the incomplete nasal bone that is broken along its mid-region. The nasal bone (Fig. 6) is a rectangular bone that is broader than in the previous restoration of †*Atacamichthys* (Arratia & Schultze 1987, fig. 2), fitting in the anterior margin of the parietal. According to the position of the nasal, supraorbital, and antorbital bones, the nasal is excluded from the orbit. Due to the incomplete preservation of this region, the position of the nostrils is unknown.

The middle pit-line (Figs. 3, 4A) is positioned in a deep, continuous groove, extending from the postparietal to the lateral margin of the dermopterotic. A short anterior pit-line extends from the middle pit-line to the anterior margin of the postparietal.

Orbit and circumorbital bones. The orbit is small, about 19 % of head length (Figs. 5, 6). The circumorbital series of bones is re-evaluated below. As already stated in the original description, an elongate, slightly bent supraorbital bone is present (Arratia & Schultze 1987, fig. 7), and an almost triangular antorbital is well preserved, with the characteristic branching of the antorbital branch of the infraorbital canal that is one of the diagnostic features of †*Atacamichthys* (Arratia & Schultze 1987 and below). This condition differentiates the antorbital of †*Atacamichthys* from that of holosteans, which have a tube-like anterior extension. Infraorbitals 1 and 2 still remain unknown, whereas the so-called infraorbitals 3 and 4 and suborbitals are poorly preserved. Two narrow pieces of bones (Fig. 6) at the postero-ventral margin of the orbit that carry the infraorbital canal were interpreted as infraorbitals 3 and 4 by Arratia & Schultze (1987). Infraorbital 3 is slightly expanded posteriorly but its posterior region is not preserved so that its whole size and shape remains unknown. Infraorbital 4 is a narrow bone with its posterior margin crenulated. It is assumed that the dermosphenotic was small because of the small space left between the dorsal margin of infraorbital 4 and the lateral margin of the skull roof. Posterior to those bones there are two bony plates that were interpreted as two suborbitals by Arratia & Schultze (1987). The dorsal suborbital is smaller than the ventral one, and its incompletely preserved posterior margin extends onto the anterior margin of the opercle in specimen KUVP 86777. The ventral suborbital is squarish, with its posterior margin incomplete and not covering the quadrate region (if this would be preserved); however, and by comparison with other fishes, it could be possible that the size of the lower suborbital was larger than the interpretation offered in the previous reconstruction of the fish or that a third, small suborbital was present, but it is not preserved.

Sclerotic bones. Sclerotic bones are represented by two small pieces in specimen KUVP 86777; with such incomplete information, we are unable to make further interpretations concerning the number of elements. Sclerotic bones are absent in a few fishes, such as in the Amiinae (*Amia calva* and †*A. pattersoni*; Grande & Bemis 1998, p. 577: character 14) and lepisosteids.

Maxilla. The maxilla is short in comparison to the lower jaw (Figs. 5, 6). Its posterior margin just reaches the anterior margin of infraorbital 3. The bone is narrow and expands slightly posteriorly, which is similar to the pattern shown by many amiiforms. The posterior margin of the maxilla seems to be notched, closer to the condition in †*Solnhofenamia* (see Grande & Bemis 1998, fig. 243); however, a closer examination shows that the margin seems to be partially broken. The oral margin of the maxilla bears conic teeth that are slightly smaller than those of the premaxilla.

Supramaxilla. Only one supramaxilla (Fig. 6) is present. The bone is elongate and covers the last two thirds of the dorsal margin of the maxilla. Having one supramaxilla is interpreted as a synapomorphy of halecomorphs (Grande & Bemis 1998); however, one large supramaxilla covering almost the whole maxillary blade is also found in some ionoscopiforms and teleosteomorphs, such as in †*Siemensichthys siemensi* (Arratia 2000). A single supramaxilla is also found in pachycormiforms, but the position of the bone is postero-dorsal or posterior in relation to the posterior tip of the maxilla.

Lower jaw. The lower jaw (Figs. 5, 6) is long and massive, and the oral margin is covered with a row of conical teeth with no space between them. The articulation between the lower jaw and the quadrate is posterior to the posterior margin of the orbit, so that the lower jaw is long, about 67 % of the head length. The jaw protrudes anteriad to the level of the premaxilla, but the anterior margin of the dentary is restored in Figure 6, because this margin is not preserved in any concretion so that the jaw could still be slightly longer. The dorso-posterior margin of the jaw is damaged so that the extension of the surangular is unknown. A large prearticular bone is preserved in the medial aspect of the jaw; it reaches close to the mandibular symphysis. No coronoid bones are present in the small space left at the anterior medial side of the dentary.

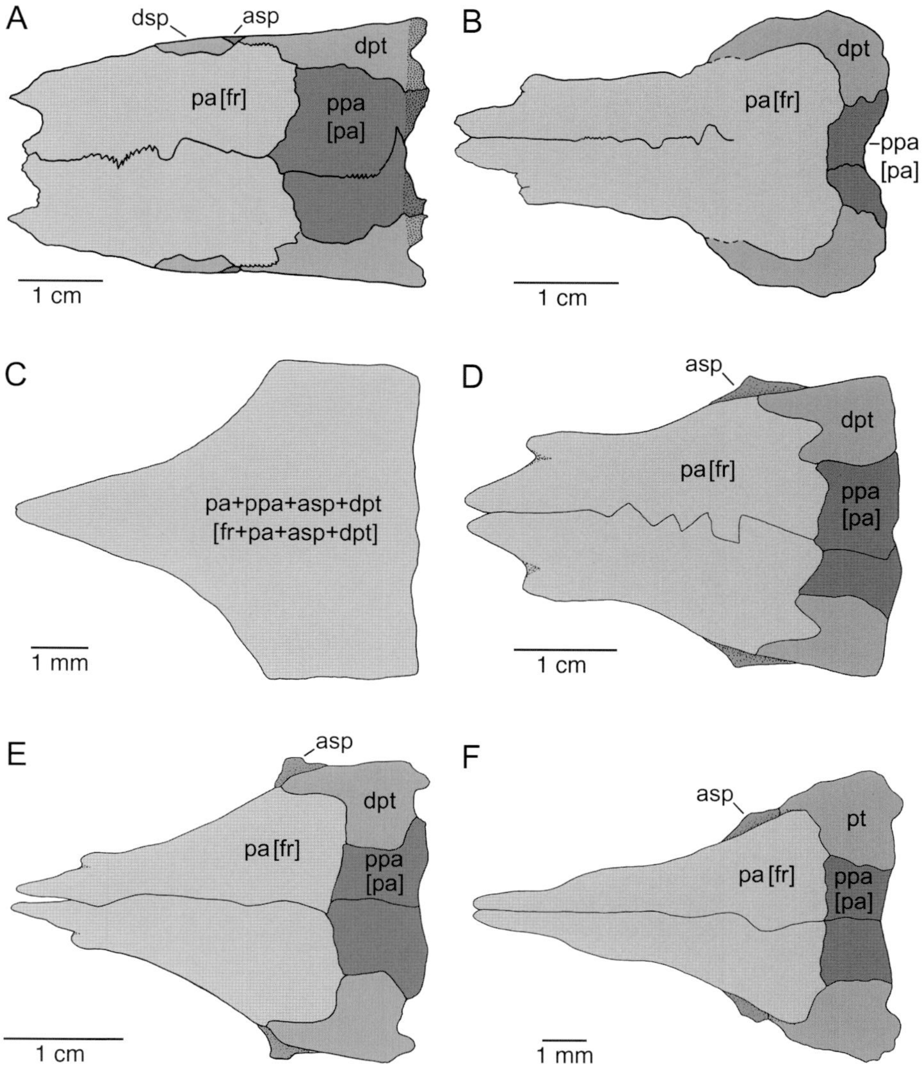

Fig. 4. Diagrammatic skull roofs of an amiid and some teleosteomorphs in dorsal view illustrating changes in size and relation of bones. **A.** *Amia calva*. **B.** †*Atacamichthys greeni*. **C.** †*Pholidoctenus serianus*. **D.** †*Siemensichthys macrocephalus*. **E.** †*Dorsetichthys bechei*. **F.** †*Leptolepis coryphaenoides*. A supraoccipital bone is not illustrated in B and D–F. Abbreviations: asp, autosphenotic; dpt, dermopterotic; dsp, dermosphenotic; pa [fr], parietal bone [= frontal of traditional terminology]; ppa [pa], postparietal bone [= parietal of traditional terminology]; pa+ppa+asp+dpt [= fr+pa+asp+dpt], fused parietal, postparietal, autosphenotic, and dermopterotic [= fused frontal, parietal, autosphenotic, and dermopterotic of traditional terminology]; pt, pterotic. C to F from Arratia (2013).

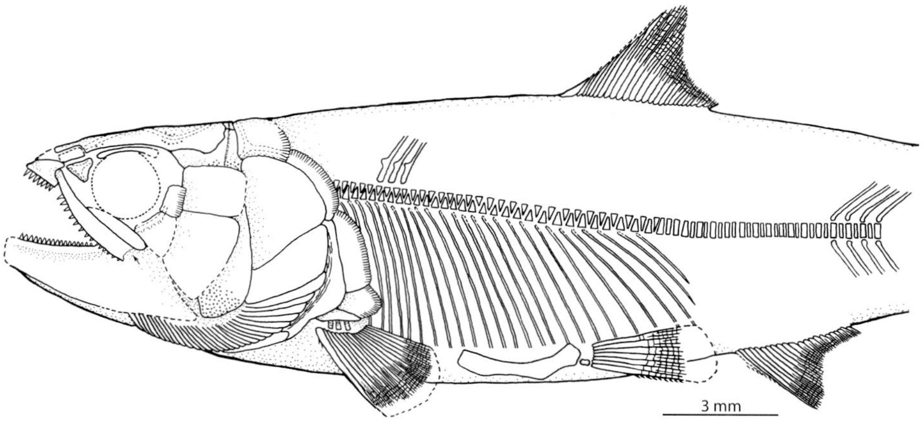

Fig. 5. †*Atacamichthys greeni*, restoration in lateral view. A small arrow points to the broken posterior margin of the maxilla.

Suspensorium. The hyomandibula is nicely exposed in specimen KUVP 86777. The articular region with the braincase is strongly inclined anteriorly, with apparently two articular regions, a larger, anterior one that articulates with the autosphenotic, and a smaller one with the dermopterotic. The main hyomandibular shaft is oriented straight and is broad, with even broader anterior and posterior flanges. The ventral part of the shaft of the hyomandibula articulates with a broad, but apparently short, symplectic. It is unclear whether the symplectic articulates with the lower jaw or not because of incomplete preservation.

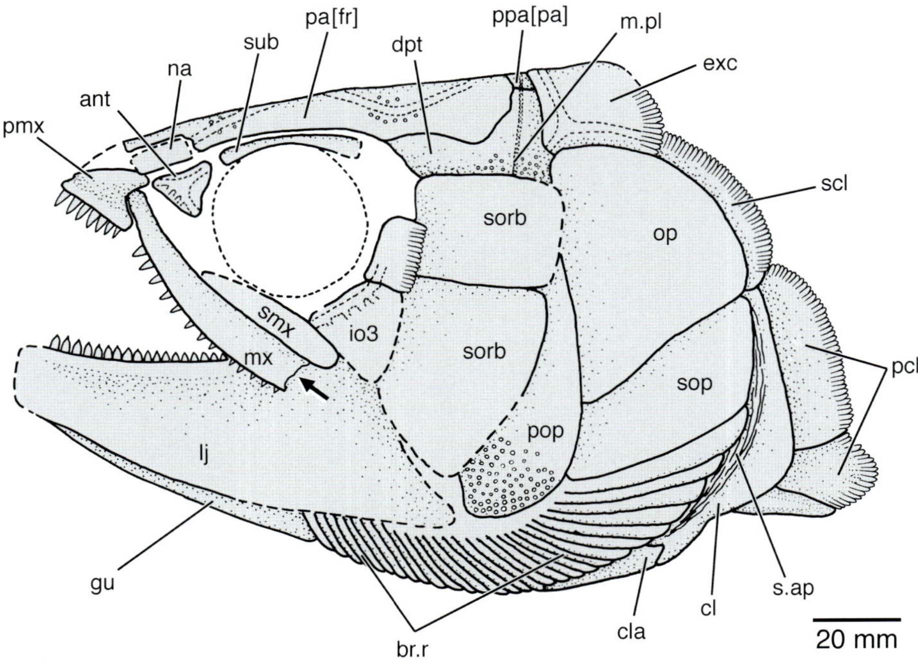

◁ **Fig. 6.** †*Atacamichthys greeni*, restoration of head in lateral view. A small arrow points to the broken posterior margin of the maxilla. Abbreviations: ant, antorbital; br.r, branchiostegal rays; cl, cleithrum; cla, clavicle; dpt, dermopterotic; exc, extrascapular; gu, gular plate; io3, infraorbital 3; lj, lower jaw; m.pl, middle pit-line; mx, maxilla; na, nasal; pa[fr], parietal bone [= frontal of traditional terminology]; op, opercle; pcl, postcleithrum; pmx, premaxilla; ppa[pa], postparietal bone [= parietal bone of traditional terminology]; s.ap, serrated appendage; scl, supracleithrum; smx, supramaxilla; sop, subopercle; sorb, suborbitals; sub, supraorbital bone.

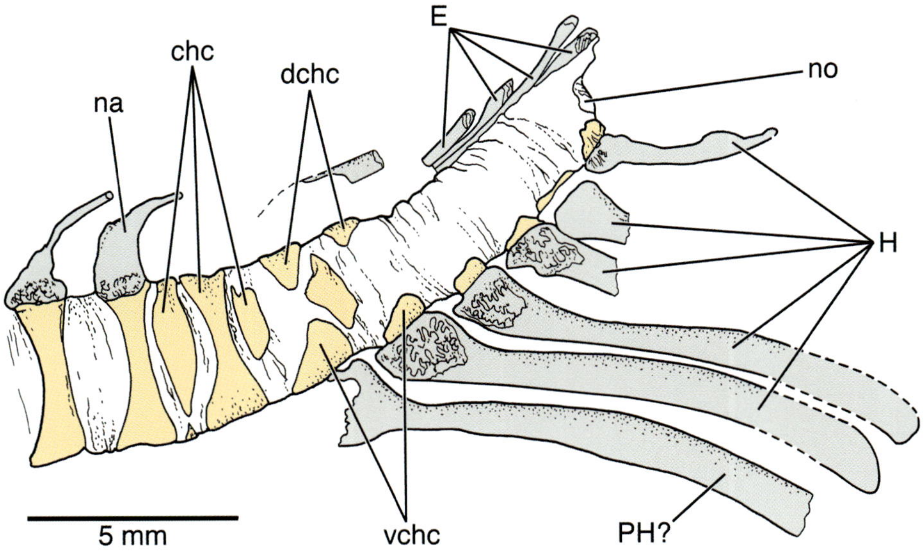

Fig. 7. †*Atacamichthys greeni*, part of the caudal endoskeleton in lateral view. Abbreviations: chc, chordacentra; dchc, dorsal hemichordacentra; E, epurals; H, hypurals; na, neural arches; no, notochord; PH?, parhypural?; vchc, ventral hemichordacentra. (Slightly modified from Arratia & Schultze 1999).

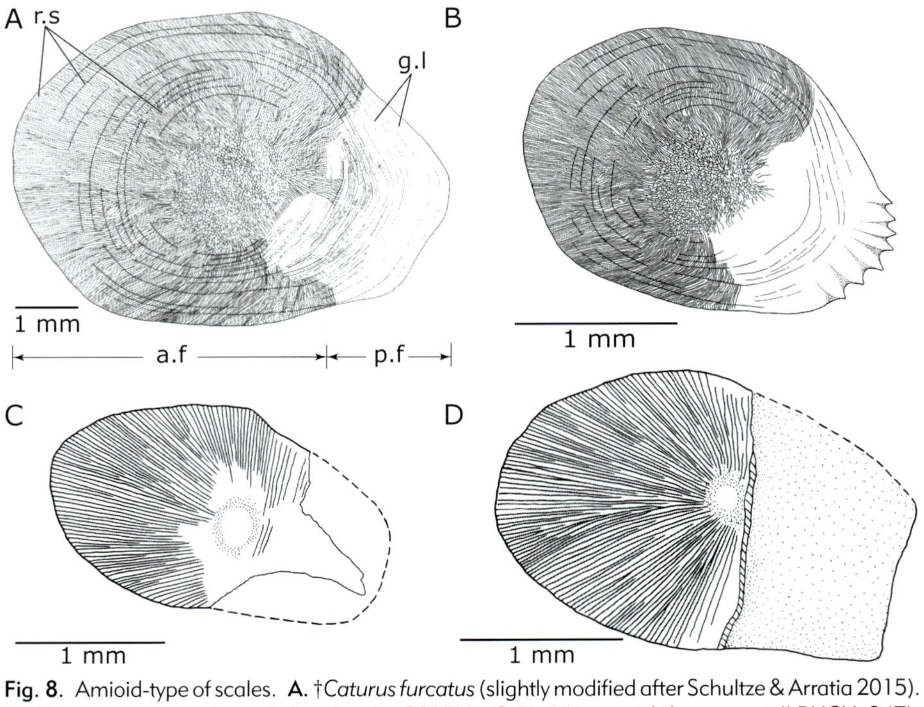

Fig. 8. Amioid-type of scales. **A.** †*Caturus furcatus* (slightly modified after Schultze & Arratia 2015). **B.** †*Eurycormus speciosus* (after Arratia 2015b). **C, D.** †*Atacamichthys greeni* (LBUCH 247). Abbreviations: a.f, anterior field; g.l, growth lines; p.f, posterior field; r.s, radial striae.

Lower section of the hyoid arch. The anterior ceratohyal (Arratia & Schultze 1987, fig. 7) is a large, compact bone narrowed anteriorly, with its middle and posterior sections deeper and without notches (or foramina) at the dorsal margin. The bone is as long or even slightly longer than the lower jaw. In contrast, the posterior ceratohyal is very small. The shape and size of the ceratohyals are similar to those in *Amia calva* (Arratia & Schultze 1990, fig. 2A, Grande & Bemis 1998, fig. 54), †*Caturus*, †*Solnhofenamia*, and †*Watsonulus* among halecomorphs. Similar bones, but even deeper, are present in †*Eurycormus*. The ceratohyals are almost unknown in Triassic pholidophorids, except for †*Lombardichthys gervasutti*, which has a tiny, narrow, bar-shaped anterior ceratohyal that retains a large portion of cartilage anteriorly, but does not have an independent hypohyal region.

The anterior articulatory region for the hypohyal comprises the whole anterior margin of the ceratohyal in †*Atacamichthys*. In one specimen, the anterior ceratohyal articulates with only one hypohyal that is as deep as the anterior margin of the ceratohyal, whereas in another specimen, the anterior margin is deeper and articulates ventrally with a large hypohyal and dorsally with a small hypohyal retaining a cartilaginous area. Thus, more specimens are needed to clarify the hypohyal region in †*Atacamichthys*. One hypohyal articulates with the anterior ceratohyal in amiids, †*Caturus* and †*Watsonulus* among halecomorphs and in aspidorhynchiforms, Triassic pholidophorids, and †*Eurycormus* among teleosteomorphs. Two hypohyals articulate with the anterior ceratohyal in †*Leptolepis coryphaenoides* and more advanced teleosts.

Branchiostegal rays. †*Atacamichthys* has approximately 30 branchiostegal rays (Figs. 5, 6), with the anterior ones very short, and the posterior ones longer and covering part of the cleithrum. A similar number of branchiostegal rays is found in pachycormiforms.

Preopercle. The bone is elongate and narrow in †*Atacamichthys*, becoming slightly expanded ventrally, without distinct dorsal and ventral arms (Fig. 6). It is placed vertically in front of the opercle and subopercle. An interopercle has not been observed below or slightly postero-ventral to the preopercle. Slightly similar preopercular crescentic shapes can be found in halecomorphs and in Triassic pholidophorids. One important difference with the latter is that the preopercular canal in †*Atacamichthys* ramifies into numerous branches that open to the bony surface through small pores that give the bone a characteristic pitted surface.

Vertebral column

As stated in Arratia & Schultze (1987), the vertebral column (Fig. 5) of †*Atacamichthys* is formed by monospondylous chordacentra in the abdominal region and diplospondylous chordacentra (see also Arratia & Schultze 1987, fig. 10) in most of the caudal region, except for the last preural and ural vertebrae, which have monospondylous chordacentra (Fig. 7). There are great differences among the fishes studied, some of them with a persistent functional notochord or aspondylous vertebrae (e.g., caturids, †*Eurypoma*, †*Watsonulus*, and pachycormiforms) or chordacentral type of vertebrae (e.g., aspidorhynchiforms, pholidophoriforms, and †*Atacamichthys*) or ossified type of vertebrae (e.g., amiids).

Pectoral girdle

In general, the dermal pectoral girdle (Figs. 5, 6) of †*Atacamichthys* is characterized by the presence of a large supracleithrum almost as large as the cleithrum and with the lateral line positioned in its upper third. The external surface closer to the anterior border separating the main shaft of the cleithrum and its anterior lamina is densely covered with long denticulated bony ridges – the serrated appendage. The denticulated ridges in †*Atacamichthys* seems to be fused or immovably attached to the lateral surface of the cleithrum and they do not extend onto the lateral surface of the supracleithrum. The well-developed denticulated ridges are set closer together forming a compact appendage (see Arratia & Schultze 1987, fig. 13). In the available specimens, the antero-ventral part of the cleithrum ends sharply, resembling the condition present in Triassic pholidophorids, with a clavicle articulating with the antero-ventral margin of the cleithrum (Arratia 2013, figs. 20, 69; 2017). The region anterior to the antero-ventral part of the cleithrum is damaged in most specimens of †*Atacamichthys*, except for LBUCH 248, which shows an elongate clavicle (Fig. 5).

Proximal radials. There are three elongate proximal radials preserved in †*Atacamichthys*, but their total count is unknown. This count is closer to the four proximal radials characterizing *Vinctifer* (Brito 1997) and †*Leptolepis coryphaenoides* plus more advanced teleosts (Jessen 1972, Patterson 1977, Arratia 1997, 1999, 2017).

Pelvic girdle

The pelvic girdle or basipterygium in †*Atacamichthys* is a long, squarish, flat, plate-like bone that is as long as the fin (Fig. 5). The plate is gently curved, with its anterior margin straight. The pelvic plate resembles that of *Amia*, †*Solnhofenamia* and †*Caturus* (Grande & Bemis 1998, fig. 91). However, some pholidophorids (where the pelvic plate is known) also present a broad and elongate plate (Arratia 2013, fig. 55). The basipterygium of †*Leptolepis coryphaenoides* and more advanced teleosts is an elongate, triangular, plate-like bone.

Caudal endoskeleton

Part of a caudal endoskeleton (Fig. 7) interpreted as possibly belonging to †*Atacamichthys greeni* was illustrated by Arratia & Schultze (1999, fig. 13). The specimen is interpreted as †*Atacamichthys* because of its amioid type of scales and structure of the vertebral column and dorsal and anal fins. The caudal endoskeleton preserved the preural vertebrae, showing very short neural spines. In contrast, no arches or any other neural elements are present in the ural region. The hypural region is incomplete. After re-studying the specimen, the four elements previously interpreted as uroneurals by Arratia & Schultze (1999) are re-interpreted as epurals, because they are unpaired narrow bones keeping remnants of cartilage at their proximal and distal tips.

Scales

The scales are strongly imbricated, with their two anterior thirds covered by the preceding ones. The scales (Fig. 8C, D) are of amioid-type, with their focus placed closer to the posterior border than the anterior one. The anterior field is covered with radiating striae (radial striae sensu Schultze 1966) that converge close to the focus. Convergent lines are also lateral to the focus. The posterior field may be densely covered by growth lines, placed almost parallel to the margin of the posterior field of the scale or the lines may be few and discontinuous. The scales of †*Atacamichthys* are slightly more elongated than those of amiiforms (e.g., †*Caturus* and *Amia*; Fig. 8A) and the pholidophoriform †*Eurycormus* (Fig. 8B). In addition, the density of the striae in the anterior field seem to be lesser in †*Atacamichthys* than in amiiforms and †*Eurycormus*.

†*Atacamichthys* and its relationships: Phylogenetic Analysis

To assess the phylogenetic relationships of †*Atacamichthys greeni*, a cladistic analysis of 35 taxa and 130 characters was performed; †*Birgeria* and *Polypterus* were used as outgroup taxa. For the list of characters, see Appendix 1, and for the coding of characters, see Appendix 2.

The parsimony analysis was performed using PAUP 4.0a (Swofford 2000) and replicated with the latest on-line version of PAUP (April, 2019), using ACCTRAN character-state optimization, a heuristic search using a random addition sequence with 1000 replicates, and the tree bisection and reconnection (TBR) branch-swapping algorithm. The parsimony analysis recovered one tree of 349 steps (retention index = 0.765; consistency index = 0.481; homoplasy index = 0.519).

Figure 9 represents the topology of the only tree found. The topology differs from that of Arratia et al. (2019, fig. 12) where pachycormiforms and aspidorhynchiforms form a monophyletic group at Node A, and with the addition of nodes F (†*Atacamichthys* plus more advanced teleosteomorphs) and Node K (†*Longileptolepis* plus more advanced teleosts). For some characters with a few question marks, the parsimony analysis predicted potential synapomorphies at certain nodes. These assumptions are identified below. Characters supporting Nodes C to G are presented below and are listed in the Figure 9 caption. For other nodes see the caption of Figure 9. An asterisk (*) identifies a uniquely derived character.

Node A (Neopterygii) represents the branching of holosteans and teleosteomorphs. This node is supported by seven synapomorphies, four of which are interpreted as uniquely derived: foramen for glossopharyngeal nerve positioned in prootic or prootic-exoccipital suture (Ch. 34[1]*); symplectic present (Ch. 126[1]*); most dorsal fin rays equal in number with their supporting pterygiophores (Ch. 128[1]*); and most anal fin rays equal in number with their supporting pterygiophores (Ch. 129[1]*). Homoplastic features supporting this node are: lateral margin of nasal bone not sutured with antorbital (Ch. 25[1]); one supramaxillary bone present (Ch. 57[1]); interopercle present (Ch. 83[1]); and dorsal or epaxial leading margin of caudal fin with basal fulcra (Ch. 114[1]).

Node B (lepisosteids + amiiforms = Holostei) is supported by nine synapomorphies, two of which are uniquely derived: autosphenotic with a small dermal component present (Ch. 13[1]*); and a tube-like canal in the anterior arm of antorbital present (Ch. 39[1]*). Other synapomorphies are: orbitosphenoid region/bone absent or reduced in size (Ch. 15[1]); paired vomer (in adult individuals) (Ch. 28[0]); parasphenoid with small teeth (Ch. 29[1]); supraorbital and infraorbital canals join within the dermosphenotic (Ch. 36[2]); supramaxillary bone or most posterior supramaxilla placed on the dorsal margin of maxilla (Ch. 58[0]); leading margin of dorsal fin bearing fringing fulcra (Ch. 102[1]); and arrangement of hypurals and caudal-fin rays with each hypural normally articulated with one caudal ray (Ch. 113[1]).

Node C (Teleosteomorpha or stem Teleostei) stands at the branching of †Pachycormiformes plus more advanced teleosteomorphs. This node is supported by ten synapomorphies, five of which are uniquely derived: nasal bones separated from each other by parietal [= frontal] bones (Ch. 23[2]); a complete ring of two sclerotic bones (oriented anterior and posterior to eye) (Ch. 48[2]); posterior or postero-ventral margin of quadrate markedly thickened and with a diminutive process distally (Ch. 73[1]); four pectoral proximal radials (Ch. 93[1]*); all or a few preural neural spines enlarged and expanded laterally (uroneural-like bones; Ch. 110[1]); olfactory organ with accessory nasal sacs (Ch. 122[1]*); craniotemporal muscle present (Ch. 123[1]*); heart with two arterial valves present (Ch. 124[1]*); and muscles at the basal arteria absent (Ch. 125[1]*); and quadratojugal absent (Ch. 127[1]). The parsimony analysis predicts that characters 122[1]*, 123[1]*, 124[1]*, and 125[1]* that are present in extant teleosts were also present at this phylogenetic level.

Node D stands at the branching of †Aspidorhynchiformes plus more advanced teleosteomorphs and is supported by the following synapomorphies: orbitosphenoid region/bone absent or reduced in size (Ch. 15[1]), which is a homoplastic feature that also occurs at node B; maxilla moderately long, extending below orbit (Ch. 53[1]); lower jaw without coronoid bones (Ch. 63[1]); serrated appendages are long, toothed elements covering the whole medial surface of cleithrum (Ch. 91[2]); leading margin of dorsal fin with fringing fulcra (Ch. 102[1]); ural neural arches modified as uroneurals (Ch. 111[1]); and ventral hypaxial leading margin of caudal fin with accessory fulcra irregularly intercalated between hypaxial fringing fulcra (Ch. 116[1]).

Node E stands at the trichotomy formed by †*Prohalecites* and more advanced teleosts. This node is supported by 14 synapomorphies, two of which are uniquely derived: skull roof with orbital region slightly narrower than postorbital region (Ch. 3[1]); premaxilla with ascending process (Ch. 50[1]*); mobile premaxilla (Ch. 52[1]*); supramaxillary bone or most posterior supramaxilla dorsal to maxilla (Ch. 58[0]); position of symplectic medial to posterior margin of quadrate (Ch. 75[1]); crescent-shaped preopercle present (Ch. 82[1]); diplospondylous mid-caudal centra (in adults) (Ch. 87[0]); abdominal neural arches with epineural processes (Ch. 88[1]); leading margin of pectoral fin with fringing fulcra (Ch. 97[1]); leading margin of pelvic fin with fringing fulcra (Ch. 99[1]); leading margin of anal fin with fringing fulcra (Ch. 104[1]); neural spine of preural centrum 2 shorter than neural spine of preural centrum 3 (Ch. 106[1]); all or almost all preural neural spines are unmodified (Ch. 110[0]); and epaxial lobe of the caudal fin is with rudimentary ray (Ch. 115[1]).

Node F represents the branching of †*Atacamichthys* plus more advanced teleosts and is weakly supported by four synapomorphies: interparietal [= interfrontal] suture undulated or wavy in its middle section (Ch. 22[2]); supraorbital canal with branched tubules (Ch. 35[1]); articular bone fused with angular and retroarticular (Ch. 60[1]); and three or four epurals present (Ch. 109[1]).

Node G and upper nodes (Fig. 9) have the same topology as in Arratia (2017, fig. 9), Arratia et al. (2019, fig. 12), and Bean & Arratia (2019, fig. 29), with the addition of †*Longileptolepis* at Node K. The clade formed by †Pholidophoriformes [†*Eurycormus* + †Pholidophoridae] plus more advanced teleosts stands at Node G. This node is supported by the following characters: two supramaxillary bones present (Ch. 57[2]); quadrate-mandibular articulation placed below the posterior half of orbit (Ch. 59[1]); oral margin of dentalosplenial or dentary with a notch (= "leptolepid" notch; Ch. 62[1]); dentary with a well-developed, protruding lateral bony ridge separating dental and splenial regions (Ch. 66[1]); a different-shaped preopercle (not crescent-shaped; Ch. 82[0]); and the first pectoral fin-ray is a compound ray fused with basal fulcra (Ch. 95[1]).

Node H stands at the branching of the †*Siemensichthys* Group plus more advanced teleosts and is supported by eight synapomorphies, three of which are uniquely derived: supraoccipital bone present (Ch. 16[1]*); maxilla elongate, extending behind orbit (Ch. 53[0]); elongated postero-dorsal or postero-ventral process of quadrate (depending on the angle of the quadrate) present (Ch. 72[1]*); posterior or postero-ventral margin of quadrate markedly ossified into a postero-dorsal (or ventral depending on angle of quadrate) process partially separated by a deep notch from main body of the bone (Ch. 73[2]); hyoid arch with dorsal and ventral hypohyals articulated with anterior ceratohyal (Ch. 77[1]*); pectoral girdle without clavicle (Ch. 92[1]); epaxial lobe of the caudal fin without rudimentary ray (Ch. 115[0]); and ventral hypaxial leading margin of caudal fin without accessory fulcra (Ch. 116[0]).

To test the phylogenetic position of †*Atacamichthys*, a second analysis was performed, but this time two fossil taxa, †*Australosomus* and †*Birgeria*, were considered as outgroups. One tree was obtained (tree length: 341; consistency index: 0.490; retention index: 0.774; homoplasy index: 0.510). This tree differs with respect to the first analysis (Fig. 9) in that pachycormiforms stand below the clade formed by holosteans + teleosteomorphs, and this position is supported by one homoplastic character: supramaxillary bone or most posterior supramaxilla posterior or dorso-posterior to maxilla (Ch. 58[1]). In both trees †*Atacamichthys* stands above the phylogenetic level of †*Prohalecites*, and the topology of all nodes above †*Atacamichthys* is identical (see Fig. 9).

Systematic Paleontology

Infraclass Teleosteomorpha Arratia, 2001
Family †Atacamichthyidae, new family

Diagnosis (based on a combination of characters; uniquely derived characters are identified with an asterisk). Elongate fish of about 30 cm total length, with dorsal fin slightly in front of the pelvic fin and anal fin origin just behind posterior margin of dorsal fin (Fig. 4). Long parietal [= frontal] bone – its length is about 89 % of the skull roof [*]. Parietals fused posteriorly. Main postorbital expansion of the skull roof due to the enlargement of the dermopterotic [*]. Small and square-shaped postparietals, about 15 % of the parietal length. Extrascapulae with crenulate posterior margin [*]. One long supraorbital bone. Crescentic preopercle slightly expanded ventrally. Supraorbital, otic, and preopercular canals ramified open onto the surface by numerous, small pores. Abdominal neural arches without epineural processes. Epipleural bones absent. Caudal region with diplospondylous chordacentral vertebrae. Supracleithrum and postcleithra with crenulated posterior borders [*]. Pectoral propterygium fused with first pectoral fin ray. Pelvic basipterygium elongate and plate-like. Dorsal and anal fins acuminate. With four or more epurals. Elasmoid scales of amioid-type.

Content. Only one species known, †*Atacamichthys greeni* Arratia & Schultze, 1987.

†*Atacamichthys greeni* Arratia & Schultze, 1987

Diagnosis. Same as family diagnosis.

Geographical distribution and age. Known only from Sierra de Varas, Cordillera de Domeyko, Antofagasta Region, northern Chile (Arratia & Schultze 1987; Fig. 1). Upper Jurassic, Oxfordian (Hillebrandt 1970, Chong 1977).

Discussion and Conclusions

The main goal of this study was to clarify the position of †*Atacamichthys* that is a fish bearing a mixture of characters, some of which belong to amiiforms and others to teleosteomorphs. The two phylogenetic analyses performed here support the hypothesis that the fish is a primitive teleosteomorph, just above the phylogenetic level of the Triassic fish †*Prohalecites*. This result is important because it adds new information concerning stem teleosts, especially at the most basal nodes of Teleosteomorpha and consequently, its combination of characters influences our understanding of characters distribution. The addition of two new taxa to the ingroup, †*Atacamichthys* and †*Longileptolepis*, change the interpretation and distribution of characters, increasing their level of homoplasy.

Because of the combination of characters (amiiform and teleostean) that †*Atacamichthys* presents, the phylogenetic analysis performed here includes more distant outgroup taxa (†*Birgeria* and *Polypterus*) than ones used in previous analyses by Arratia (e.g., 1999, 2013) that lead to a re-coding of characters. The impact of using different outgroups in relation to changes in character polarization and the phylogenetic position

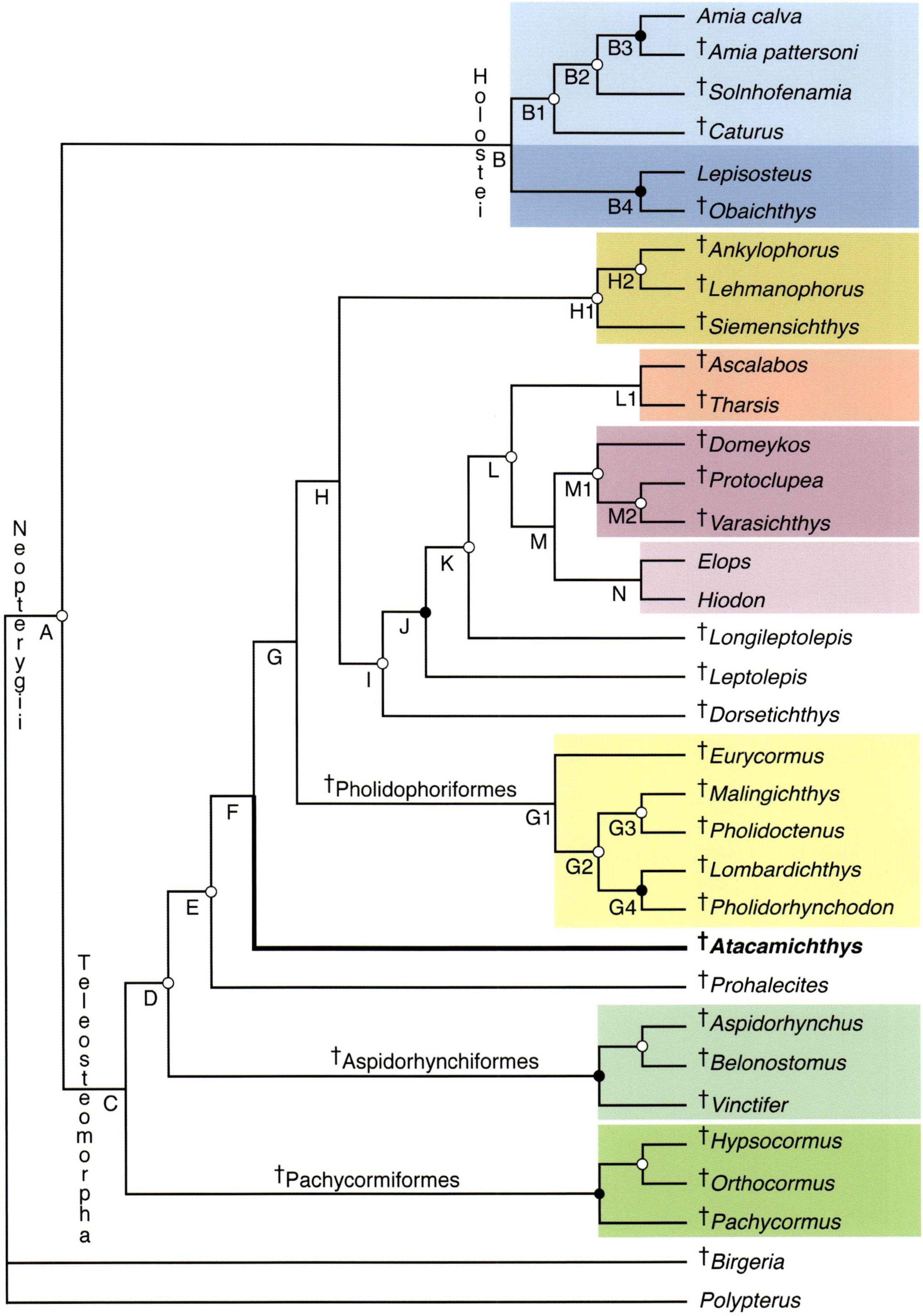

Fig. 9. Phylogenetic hypothesis of relationships of advanced actinopterygians showing the position of the marine Oxfordian teleosteomorph †*Atacamichthys greeni*. The phylogenetic hypothesis is based on 35 taxa and 130 characters; †*Birgeria* and *Polypterus* are the outgroup taxa. For characters supporting nodes see text and Appendix 1. A uniquely derived character is identified with an asterisk [*]. Node A: Ch. 25[1], 34[1]*, 57[1], 83[1], 114[1], 126[1]*, 128[1]*, and 129[1]*. Node B (holosteans): Ch. 13[1]*, 15[1], 28[0], 29[1], 36[2], 39[1]*, 58[0], 83[1], 102[1], and 113[1]. Node B1 (amiiforms): Ch. 25[0], 55[1], 74[0], 82[1], 87[0], 120[1], 127[1]. Node B4 (lepisosteiforms): Ch. 4[1], 13[1]*, 14[1]*, 15[1], 28[0], 29[1], 33[1], 36[2], 39[1]*, 45[0], 53[2], 55[3], 58[0], 59[3], 76[1]*, 79[1]*, 85[1]*, 97[1], 99[1], 102[1], 104[1], 113[1], and 121[1]*. Node C (Teleosteomorpha): Ch. 23[2], 48[2], 73[1], 93[1]*, 110[1], 122[1]*, 123[1]*, 124[1]*, 125[1]*, and 127[1]. Node D: Ch. 15[1], 53[1], 63[1], 91[2], 102[1], 111[1], and 116[1]. Node E: Ch. 3[1], 50[1]*, 52[1]*, 58[0], 75[1], 82[1], 87[0], 88[1], 97[1], 99[1], 104[1], 106[1], 110[0], and 115[1]. Node F: Ch. 22[2], 35[1], 60[1], and 109[1]. Node G: 57[2], 59[1], 62[1], 66[1], 82[0], and 95[1]. Node H: Ch. 16[1]*, 53[0], 72[1]*, 73[2], 77[1]*, 92[1], 115[0], and 116[0]. Node I: Ch. 27[0], 29[1], 40[1], 61[1], 65[1]*, 90[1], 91[0], 94[1], and 100[1]. Node J: Ch. 7[2], 8[2]*, 27[1], 31[1]*, 32[1]*, 35[0], 40[3], 53[1], 64[1], 78[1]*, 87[1], 97[0], 99[0], 102[0], 104[0], 105[1], 107[1], 108[1], 117[1]*, 118[1], and 12[2]. Node K: Ch. 29[0], 32[2]*, 34[2]*, 37[0], 45[2], and 47[2]. Node L: Ch. 66[0], 86[1]*, and 89[1]. Node M: Ch. 33[1], 96[1], and 114[0]. Node N (crown-group Teleostei): Ch. 29[1], 41[1], 62[0], and 118[0].

of pachycormiforms will require future work. Depending on the outgroup, pachycormiforms are included within teleosteomorphs (Fig. 9), or are just outside of neopterygians (when †*Australosomus* and †*Birgeria* are outgroup taxa). The phylogenetic position of †Pachycormiformes is outside of the scope of this study and is being investigated by Gouiric-Cavalli and Arratia.

The monotypic Oxfordian family †Atacamichthyidae from the Jurassic of Chile is endemic to the southern region of the Paleo-Pacific Ocean. No fish with a similar combination of characters have been mentioned for any other regions across the globe. This is not surprising, because El Profeta Formation in the Cordillera de Domeyko, northern Chile, is characterized by other endemic teleosteomorph taxa, such as the family †Chongichthyidae (with one genus) and †Varasichthyidae (with five genera). The evidence, as presented above, supports the uniqueness of †*Atacamichthys greeni* and its interpretation in a new family.

Acknowledgements

For loan of valuable specimens, the authors thank W. Saul (ANSP); the late C. H. von Daniels (BGHan); P. Wellnhofer, O. Rauhut, and M. Moser (SNSB-BSPG); W. Eschmeyer and D. Catania (CAS); D. Berman (CM); L. Grande, W. Simpson, M. Westneat, and M. A. Rogers (FMNH); W. Westphal and the late R. Reif (GPIT); H. Jahnke (GZG); W. Mette and W. Resch (Innsb.); J.-Y. Zhang (IVVP); A. Simons and V. Hirt (JFBM); G. Viohl, M. Kölbl-Ebert and M. Ebert (JME); E. O. Wiley and A. Bentley (KU:KUI); the late L. Martin and D. Miao (KUVP); F. Witzman (MB); A. Paganoni (MCSNB); A. Tintori (MCSNIO); the late K. Liem, K. Hartel, the late F. Jenkins and J. Cundiff (MCZ); the late C. Patterson and A. Longbottom (NHM); H. Bjerring (SMNH); R. Böttcher and E. Maxwell (SMNS); and J. McEacharan and M. Retzer (TCWC). SG-C thanks the support of PICT 2015-0253 and CQ-R thanks the support of CONICYT Chile, Proyecto Fondecyt 11181259. We thank T. J. Meehan (Lawrence, Kansas, USA) for revising the style of the manuscript. We acknowledge the comments of two anonymous reviewers that improved the presentation of a prior version.

We especially thank to the editors for the invitation to contribute to the Festschrift for our colleague and friend John Maisey, whom we wish an active and enjoyable retirement.

Literature cited

Allis, Jr., E. P. 1922. The cranial anatomy of *Polypterus*, with special reference to *Polypterus bichir*. Journal of Anatomy, 56: 189–294.

Arratia, G. 1981. *Varasichthys ariasi* n. gen. et sp. from the Upper Jurassic from Chile (Pisces, Teleostei, Varasichthyidae n. fam.). Palaeontographica Abt. A 175: 107–139.

Arratia, G. 1987. Jurassic fishes from Chile and critical comments. – Pp. 257–286 in: W. Volkheimer & E. A. Musacchio (eds.), Biogeografía de los Sistemas Regionales del Jurásico y Cretácico en América del Sur, vol. 1: Jurásico anterior a los movimientos Málmicos. Editorial Inca, Mendoza.

Arratia, G. 1991. The caudal skeleton of Jurassic teleosts. A phylogenetic analysis. – Pp. 249–334 in: M.-M. Chang, H. Hai & G.-R. Zhang, (eds.), Early Vertebrates and Related Problems in Evolutionary Biology. Science Press, Beijing.

Arratia, G. 1997. Basal teleosts and teleostean phylogeny. Palaeo Ichthyologica, 7: 1–168.

Arratia, G. 1999. The monophyly of Teleostei and stem group teleosts. – Pp. 265–334 in: G. Arratia & H.-P. Schultze (eds.), Mesozoic Fishes – Systematics and the Fossil Record. Verlag Dr. F. Pfeil, München.

Arratia, G. 2000. New teleostean fishes from southern Germany and the systematic problems concerning the 'pholidophoriforms.' Paläontologische Zeitschrift, 74: 113–143.

Arratia, G. 2001. The Sister-group of Teleostei: Consensus and Disagreements. Journal of Vertebrate Paleontology, 21: 767–773.

Arratia, G. 2003. *Leptolepis*, *Paraleptolepis* (Teleostei) and a new fish name. Mitteilungen aus dem Museum für Naturkunde Berlin, Geowissenschaftliche Reihe (= Fossil Record), 6: 157–159.

Arratia, G. 2008. Actinopterygian postcranial skeleton with special reference to the diversity of fin ray elements, and the problem of identifying homologies. – Pp. 40–101 in: G. Arratia, H.-P. Schultze & M. V. H. Wilson, (eds.), Mesozoic Fishes 4 – Homology and Phylogeny. Verlag Dr. F. Pfeil, München.

Arratia, G. 2009. Identifying patterns of diversity of the actinopterygian fulcra. Acta Zoologica, Stockholm, 90 (Suppl. 1): 220–235.

Arratia, G. 2013. Morphology, taxonomy, and phylogeny of Triassic pholidophorid fishes (Actinopterygii, Teleostei). Journal of Vertebrate Paleontology, vol. 33, Suppl. to Nr. 6, Memoir, 13: 1–138.

Arratia, G. 2015a. Los peces osteíctios fósiles de Chile y su importancia en los contextos paleobiogeográfico y evolutivo. – Pp. 35–83 in: R. Rubilar-Rogers, A. Otero, A. Vargas & M. Sallaberry (eds.), Vertebrados Fósiles de Chile. Publicación Ocasional del Museo Nacional de Historia Natural N° 63, Santiago, Chile.

Arratia, G. 2015b. Complexities of early Teleostei and the evolution of particular morphological structures through time. Copeia, 103: 999–1025.

Arratia, G. 2016. New remarkable Jurassic teleosts from Southern Germany: Ascalaboidae n. fam., its content, morphology and its phylogenetic relationships. Fossil Record, 19: 31–59.

Arratia, G. 2017. New Triassic teleosts (Actinopterygii, Teleosteomorpha) from northern Italy and their phylogenetic relationships among the most basal teleosts. Journal of Vertebrate Paleontology, 37. DOI: 10.1080/02724634.2017.1312690.

Arratia, G. & P. Lambers. 1996. The caudal skeleton of pachycormiform fishes: Parallel evolution? – Pp. 219–241 in: G. Arratia & G. Viohl (eds.), Mesozoic Fishes – Systematics and Paleoecology. Verlag Dr. F. Pfeil, München.

Arratia, G. & H.-P. Schultze. 1985. Late Jurassic teleosts (Actinopterygii, Pisces) from northern Chile and Cuba. Palaeontographica Abt. A, 189: 29–61.

Arratia, G. & H.-P. Schultze. 1987. A new halecostome fish (Actinopterygii, Osteichthyes) from the Late Jurassic of Chile and its relationships. Dakoterra, 3: 1–13.

Arratia, G. & H.-P. Schultze. 1990. The urohyal: Development and homology within osteichthyans. Journal of Morphology, 203: 247–282.

Arratia, G. & H.-P. Schultze. 1991. The palatoquadrate and its ossifications: Development and homology within osteichthyans. Journal of Morphology, 208: 1–81.

Arratia, G. & H.-P. Schultze. 1992. Reevaluation of the caudal skeleton of certain actinopterygian fishes. III. Salmonidae. Homologization of caudal skeletal structures. Journal of Morphology, 214: 1–63.

Arratia, G. & H.-P. Schultze. 1999. The Mesozoic fishes from Chile. – Pp. 565–594 in: G. Arratia & H.-P. Schultze (eds.), Mesozoic Fishes 2 – Systematics and the Fossil Record. Verlag Dr. F. Pfeil, München.

Arratia, G. & H.-P. Schultze. 2007. *Eurycormus-Eurypoma*, two Jurassic actinopterygian genera with mixed identity. Fossil Record, 10: 17–37.

Arratia, G. & H.-P. Schultze. 2013. Outstanding features of a new Late Jurassic pachycormiform fish from the Kimmeridgian of Brunn, Germany and comments on current understanding of pachycormiforms. – Pp. 87–120 in: G. Arratia, H.-P. Schultze & M. V. H. Wilson (eds.), Mesozoic Fishes 5 – Global Diversity and Evolution. Verlag Dr. F. Pfeil, München.

Arratia, G. & D. Thies. 2001. A new teleost (Osteichthyes, Actinopterygii) from the Early Jurassic Posidonia shale of Northern Germany. Fossil Record, 4: 167–187.

Arratia, G. & A. Tintori. 1999. The caudal skeleton of the Triassic actinopterygian *Prohalecites* and its phylogenetic relationships. – Pp. 121–142 in: G. Arratia & H.-P. Schultze (eds.), Mesozoic Fishes – Systematics and Fossil Record. Verlag Dr. F. Pfeil, München.

Arratia, G., H.-P. Schultze & J. R. Casciotta. 2001. Vertebral column and associated elements in dipnoans and comparison with other fishes. Development and homology. Journal of Morphology, 250: 101–172.

Arratia, G., H.-P. Schultze & H. Tischlinger. 2019. On a remarkable new species of *Tharsis*, a Late Jurassic teleostean fish from southern Germany: Its morphology and phylogenetic relationships. Fossil Record, 22: 1–23. DOI: 10.5194/fr-22-1-2019

Bean, L. & G. Arratia. 2019. Anatomical revision of the Australian teleosts *Cavenderichthys talbragarensis* and *Waldmanichthys koonwarri* impacting on previous phylogenetic interpretations of teleostean relationships. Alcheringa: An Australasian Journal of Paleontology. DOI: 10.1080/03115518.2019.1666921

Betancur-R. R., E. O. Wiley, G. Arratia, A. Acero, A., N. Bailly, M. Miya, G. Lecointre & G. Orti. 2017. Phylogenetic classification of bony fishes. BMC Evolutionary Biology. 42 pages. DOI 10.1186/s12862-017-0958-3

Brito, P. 1997. Révision des Aspidorhynchidae (Pisces, Actinopterygii) du Mésozoïque: ostéologie et relations phylogénétiques, données environnementales et biogéographique. Geodiversitas, 19: 681-772.

Cawley, J. J., J. Kriwet, S. Klug & M. J. Benton. 2019. The stem group teleost *Pachycormus* (Pachycormiformes: Pachycormidae) from the Upper Lias (Lower Jurassic) of Strawberry Bank, UK. Paläontologische Zeitschrift, 93(2): 285-302.

Chen, X.-Y. & G. Arratia. 1994. Olfactory organ in Acipenseriformes and comparisons with other actinopterygians: Patterns of diversity. Journal of Morphology, 222: 242-267.

Chong, G. 1977. Contribution to the knowledge to the Domeyko Range in the Andes of northern Chile. Geologische Rundschau, 66: 374-404.

Ebert, M. 2014. The genus *Belonostomus* Agassiz, 1834 (Neopterygii, Aspidorhynchiformes) in the late Jurassic of the Solnhofen Archipelago, with a focus on *Belonostomus kochii* Münster, 1836 from Ettling (Germany). Archaeopteryx, 32: 15-43.

Friedman, M., K. Shimada, L. D. Martin, M. J. Everhart, J. Liston, A. Maltese & M. Triebold. 2010. 100-million-year dynasty of giant planktivorous bony fishes in the Mesozoic seas. Science, 327: 900-933.

Gaudant, J. 1978. Essai de révision taxonomique des *"Pholidophorus"* (Poissons actinoptérygiens) du Jurassique supérieur de Cerin (Ain). Nouvelles Archives du Muséum d'Histoire naturelle, Lyon, 16: 101-121.

Giles, S., G. H. Xu, T. J. Near & M. Friedman. 2017. Early members of 'living fossil' lineage imply later origin of modern ray-finned fishes. Nature, 549: 261-268.

Grande, L. 2010. An empirical synthetic pattern study of gars (Lepisosteiformes) and closely related species, based mostly on skeletal anatomy. The resurrection of Holostei. American Society of Ichthyologists and Herpetologists, Special Publication, 6, Suppl. Copeia, 10(2A): X + 871 pp.

Grande, L. & W. L. Bemis. 1998. A comprehensive phylogenetic study of amiid fishes (Amiidae) based on comparative skeletal anatomy. An empirical search for interconnected patterns of natural history. Society of Vertebrate Paleontology, Memoir, 4, Suppl. Journal of Vertebrate Paleontology, 18(1): X + 690 pp.

Hillebrandt, A. v. 1970. Zur Biostratigraphie und Ammoniten-Fauna des südamerikanischen Jura (insbes.) Chile. Neues Jahrbuch für Geologie und Paläontologie, Abhandlungen, 136: 166-211.

Jessen, H. 1972. Schultergürtel und Pectoralflosse bei Actinopterygiern. Fossil and Strata, 1: 1-101.

Jollie, M. 1962. Chordate Morphology. XIV + 478 pp. Reinhold Publishing Co., New York.

Lambers, P. 1992. On the ichthyofauna of the Solnhofen Lithographic Limestone (Upper Jurassic), Germany. Unpublished Doctoral Thesis, Rijksuniversiteit Groningen, The Netherlands, 1-366.

Lehman, J.-P. 1949. Étude d'un *Pachycormus* du Lias de Normandie. Kungliga Svenska VetenskapsAkademiens Handlingar, 4. Serien, 1: 1-44.

Lehman, J.-P. 1952. Étude complémentaire des poissons de l'Eotrias de Madagascar. Kungliga Svenska VetenskapsAkademiens Handlingar, 4. Serien, 2: 1-201.

Li, G.-Q. & M. V. H. Wilson. 1996. Phylogeny of Osteoglossomorpha. – Pp. 167-174 in: M. L. J. Stiassny, L. R. Parenti & G. D. Johnson (eds.), Interrelationships of Fishes. Academic Press, San Diego.

Li, G.-Q. & M. V. H. Wilson. 1999. Early divergence of Hiodontiformes sensu stricto in East Asia and phylogeny of some Late Mesozoic teleosts from China. – Pp. 369-384 in: G. Arratia & H.-P. Schultze (eds.), Mesozoic Fishes 2 – Systematics and Fossil Record. Verlag Dr. F. Pfeil, München.

Mainwaring, A. J. 1978. Anatomical and systematic revision of the Pachycormidae, a family of Mesozoic fossil fishes. Unpublished Doctoral Thesis, Westfield College, London, 1-127.

Müller, J. 1845. Über den Bau und die Grenzen der Ganoiden, und über das natürliche System der Fische. Physikalisch-Mathematische Abhandlungen der königlichen Akademie der Wissenschaften zu Berlin, 1844: 117-216.

Nelson, G. J. 1973. Relationships of the clupeomorphs, with remarks on the structure of the lower jaw in fishes. – Pp. 333-350 in: P. H. Greenwood, R. S. Miles & C. Patterson (eds.), Interrelationships of Fishes. Journal Linnean Society (Zoology), 53, Suppl. 1. Academic Press, London.

Nelson, J. S., T. C. Grande & M. V. H. Wilson. 2016. Fishes of the World. Fifth edition. J. Wiley & Sons, Hoboken, New Jersey. XLI + 707 pp.

Nielsen, E. 1949. Studies on Triassic fishes from East Greenland. II. *Australosomus* and *Birgeria*. Meddelelser om Grønland, 146: 1-309.

Nybelin, O. 1963. Zur Morphologie und Terminologie des Schwanzskelettes der Actinopterygier. Arkiv för Zoologi (Uppsala), (2) 15: 485-516.

Nybelin, O. 1966. On certain Triassic and Liassic representatives of the family Pholidophoridae s. str. Bulletin of the British Museum Natural History, Geology, 11: 351-432.

Patterson, C. 1968. The caudal skeleton in Lower Liassic pholidophorid fishes. Bulletin of the British Museum of Natural History, Geology, 16: 202-239.

Patterson, C. 1973. Interrelationships of holosteans. – Pp. 233-305 in: P. H. Greenwood, R. S. Miles & C. Patterson (eds.), Interrelationships of Fishes. Journal Linnean Society (Zoology), 53, Suppl. 1. Academic Press, London.

Patterson, C. 1975. The braincase of pholidophorid and leptolepid fishes, with a review of the actinopterygian braincase. Philosophical Transactions of the Royal Society of London B, 269: 275-579.

Patterson, C. 1977. The contribution of paleontology to teleostean phylogeny. – Pp. 579-643 in: M. P. C. Hecht, P. C. Goody & B. Hecht, Major Patterns in Vertebrate Evolution. NATO Advanced Study Institute, Series 14. Plenum Press, New York.

Patterson, C. & D. E. Rosen. 1977. Review of ichthyodectiform and other Mesozoic teleost fishes and the theory and practice of classifying fossils, Bulletin of the American Museum of Natural History, 158: 81-172.

Pinna, M. de. 1996. Teleostean monophyly. – Pp. 147-162 in: M. L. J. Stiassny, L. R. Parenti & G. D. Johnson (eds.), Interrelationships of Fishes. Academic Press, San Diego.

Saint-Seine, P. de. 1949. Les poissons des calcaires lithographiques de Cerin (Ain). Nouvelle Archive du Muséum d'Histoire naturelle, Lyon, 2: 1-357.

Schultze, H.-P. 1966. Morphologische und histologische Untersuchungen an den Schuppen mesozoischer Actinopterygier (Übergang von Ganoid- zu Rundschuppen). Neues Jahrbuch für Geologie und Paläontologie, Abhandlungen, 126: 232-312.

Schultze, H.-P. 1989. Three-dimensional muscle preservation in Jurassic fishes of Chile. Revista Geológica de Chile, 16: 183-215.

Schultze, H.-P. 1996. The scales of Mesozoic actinopterygians. – Pp. 83-93 in: G. Arratia & G. Viohl, Mesozoic Fishes – Systematics and Paleoecology, Verlag Dr. F. Pfeil, München.

Schultze, H.-P. 2008. Nomenclature and homologization of cranial bones in actinopterygians. – Pp. 23-48 in: G. Arratia, H.-P. Schultze & M. V. H. Wilson (eds.), Mesozoic Fishes 4 – Homology and Phylogeny. Verlag Dr. F. Pfeil, München.

Schultze, H.-P. 2015. Scales, enamel, cosmine, ganoine, and early osteichthyans. Comptes Rendus Palevol, 15: 83-102. http://DOI: 10.1016/j.crpv.2015.04.001

Schultze, H.-P. 2018. Hard tissues in fish evolution: history and current issues. Cybium, 42: 29-39.

Schultze, H.-P. & G. Arratia. 1986. Reevaluation of the caudal skeleton of actinopterygian fishes. I. *Lepisosteus* and *Amia*. Journal of Morphology, 190: 215–241.

Schultze, H.-P. & G. Arratia. 1988. Reevaluation of the caudal skeleton of actinopterygian fishes. II. *Hiodon*, *Elops* and *Albula*. Journal of Morphology, 195: 257–303.

Schultze, H.-P. & G. Arratia. 1989. Reevaluation of the caudal skeleton of teleosts (Actinopterygii, Osteichthyes). Zoological Journal of the Linnean Society, 97: 189–231.

Schultze, H.-P. & G. Arratia. 2013. The caudal skeleton of basal teleosts, its conventions, and some of its major evolutionary novelties in a temporal dimension. – Pp. 187-246 in: G. Arratia, H.-P. Schultze & M. V. H. Wilson (eds.), Mesozoic Fishes 5 – Global Diversity and Evolution. Verlag Dr. F. Pfeil, München.

Schultze, H.-P. & G. Arratia. 2015. Knochenfische im weiteren Sinne (Osteichthyes oder Osteognathostomata). – Pp. 360-409 in: G. Arratia, H.-P. Schultze, H. Tischlinger & G. Viohl (eds.), Solnhofen. Ein Fenster in die Jurazeit. 2. Verlag Dr. F. Pfeil, München.

Schultze, H.-P. & D. Stöhr. 1996. *Vinctifer* (Pisces, Aspidorhynchidae) aus der Unterkreide (oberes Aptium) von Kolumbien. Neues Jahrbuch für Geologie und Paläontologie, Abhandlungen, 119: 395-415.

Stensiö, E. A. 1932. Triassic fishes from East Greenland. Collected by the Danish Expeditions in 1920-1931. Meddelelser om Grønland, 83: 1-305.

Stiassny, M. L. J. 1986. The limits and relationships of the acanthomorph teleosts. Journal of Zoology (B), 1: 411–460.

Swofford, D. L. 2000. PAUP*: Phylogenetic Analysis Using Parsimony (* and other methods). Version 4.0 Beta, Sinauer Associates, Sunderland, Massachusetts.

Taverne, L. 1981. Les Actinoptérygiens de l'Aptien inférieur (Töck) d'Helgoland. Mitteilungen aus dem Geologisch-Paläontologischen Institut der Universität Hamburg, 51: 43-82.

Taverne, L. 2011. Ostéologie et relations de *Catervariolus* (Teleostei, "Pholidophoriformes") du Jurassique moyen de Kisangani (Formation de Stanleyville) en République Démocratique du Congo. Bulletin de l'Institut Royal des Sciences Naturelles de Belgique, Sciences de la Terre, 81: 175–212.

Taverne, L. 2014. Osteology and relationships of *Kisanganichthys casieri* gen. and sp. nov. (Teleostei, Catervariolidae) from the Middle Jurassic (Stanleyville Formation) of Kisangani (Congo R.D.). Comments on the systematic position of Catervarioliformes. Geo-Eco-Trop: Revue Internationale de Géologie, de Géographie et d'Écologie Tropicales, 38: 241–258.

Thies, D. & J. Waschkewitz. 2015. Redescription of *Dapedium pholidotum* (Agassiz, 1832) (Actinopterygii, Neopterygii) from the Lower Jurassic Posidinia Shale, with comments on the phylogenetic position of *Dapedium* Leach, 1822. Journal of Systematic Paleontology. DOI: 10.1080/14772019.2015.1043361

Tintori, A. 1990. The actinopterygian fish *Prohalecites* from the Triassic of northern Italy. Palaeontology, 33: 155–174.

Tintori, A., Z. Sun, P.-G. Ni, C. Lombardo, D. Jiang & R. Motani. 2015. Oldest stem teleost from the Late Ladinian (Middle Triassic) of Southern China. Revista Italiana di Paleontologia e Stratigrafia, 121: 285–296.

Westoll, T. S. 1943. The origin of the tetrapods. Biological Reviews, 18: 78–98.

Wiley, E. O. 2008. Homology, identity and transformation. – Pp. 9–21 in: G. Arratia, H.-P. Schultze & M. V. H. Wilson (eds.), Mesozoic Fishes 4 – Homology and Phylogeny. Verlag Dr. F. Pfeil, München.

Wiley, E. O., A. M. Fuiten, H. M. Dosey, B. K. Looman, C. Merker & M. Azuma. 2015. The caudal skeleton of the zebra fish, *Danio rerio*, from a phylogenetic perspective: A polyural interpretation of homologous structures. Copeia, 103: 740–750.

Xu, G.-H. & X.-Y. Ma. 2018. Redescription and phylogenetic reassessment of *Asialepidotus shingyiensis* (Holostei: Halecomorphi) from the Middle Triassic (Ladinian) of China. Zoological Journal of Linnean Society, 184: 95–114. DOI: 10.1093/zoolinnean/zlx105

Zambelli, R. 1977. Note sui Pholidophoriformes. II. *Pholidoctenus serianus* gen. n. sp. n. Rendiconti della Accademia Nazionale dei 40 [Quaranta], Roma, 5: 101–124.

Zambelli, R. 1980a. Note sui Pholidophoriformes. III contributo: *Pholidophorus gervasuttii* sp. n. Rivista del Museo Civico di Scienze Naturali "E. Caffi", Bergamo, 1: 5–44.

Zambelli, R. 1980b. Note sui Pholidophoriformes. IV contributo: *Pholidorynchodon malzannii* gen. n. sp. n. Rivista del Museo Civico di Scienze Naturali "E. Caffi", Bergamo, 2: 129–159.

Zambelli, R. 1989. Note sui Pholidophoriformes. VII contributo. *Eopholidophorus forojuliensis* n. g., n. sp. Gortania. Atti del Museo Friulano di Storia Naturale, 11: 63–76.

Appendix 1. List of characters

Character states are represented by 0, 1, 2, 3, and 4. The dash [–] represents an inapplicable character state. A question mark [?] represents unknown information or incomplete information or ambiguous information concerning a specific character state. All characters are unordered and unweighted. The list of characters is the result of successive lists from Arratia (2013, 2016, 2017) and Arratia et al. (2019) with additions of the Jurassic fishes †*Atacamichthys* and †*Longileptolepis* and the holosteans †*Caturus* and †*Solnhofenamia*; however, most characters were re-polarized following the new outgroups used for this study. Outgroup taxa used to polarize characters now include †*Birgeria*, and *Polypterus* (first phylogenetic analysis) and †*Australosomus* and †*Birgeria* (second phylogenetic analysis). For systematic position of *Polypterus* see Giles et al. (2017) and Schultze (2018). The coding of †*Australosomus* is mainly based on Nielsen (1949) and Lehman (1952). The coding of †*Birgeria* is mainly based on Stensiö (1932), Nielsen (1949) and Lehman (1952) and that of *Polypterus* on direct observation on specimens and Allis (1922).

In addition to the authors's own studies, interpretations of the following literature were considered in the coding of the following taxa:

†*Ankylophorus* (Saint-Seine 1949, Gaudant 1978); †Aspidorhynchiformes (Brito 1997, Schultze & Stöhr 1996, Ebert 2014); †*Dorsetichthys* = †*Pholidophorus bechei* (Nybelin 1966, Patterson 1973, 1975, 1977); †*Eurycormus* (Patterson 1973, Arratia & Schultze 2007, Schultze & Arratia 2013); †*Lehmanophorus* (Saint-Seine 1949, Gaudant 1978); †*Longileptolepis* (Arratia & Thies 2001, Arratia 2003); †*Lombardichthys* = †*Pholidophorus gervasuttii* (Zambelli 1980a, Arratia 2017); †*Malingichthys* (Tintori et al. 2015); †Pachycormiformes (Lehman 1949, Patterson 1973, Lambers 1992, Friedman et al. 2010, Arratia & Schultze 2013; †*Pholidoctenus* (Zambelli 1977); †*Pholidorhynchodon* (Zambelli 1980b); †*Prohalecites porroi* (Tintori 1990, Arratia & Tintori 1999); and †*Siemensichthys macrocephalus* (Schultze 1966, Patterson 1975, Arratia 2000).

Neurocranium and skull roof

1. Skull roof bones (parietals, postparietals, autosphenotics, and dermopterotics): unfused [0]; fused into a bony plate in large specimens [1].

Comments. All pholidophorids possess this feature that is interpreted as a synapomorphy of the family †Pholidophoridae (e.g., Arratia, 2013, figs. 97A, 98A–D; Fig. 9, node G2). Character 2[1] is interpreted as a synapomorphy of †Pholidophoriformes, but it is also present in *Amia*, with further transformations in [†*Aspidorhynchus* + †*Belonostomus*] (Ch. 2[2]; a homoplastic character).

2. Postparietal [= parietal] bones: independent [0]; fused to each other [1]; fused to each other plus other skull bones in large specimens [2]; fused with dermopterotic [3].

Comments. Character-state 3 was added because it represents the condition observed in *Polypterus bichir* (Allis 1922).

3. Skull roof: no distinct broadening between orbital and postorbital regions [0]; orbital region slightly narrower than postorbital one [1]; narrow orbital region versus a broad postorbital region – about four times broader or more [2].

Comments. See Figure 4 for interpretation of this character.

4. Ratio of preorbital snout length to postorbital head length: less than 1.2 [0]; 1.5 or greater [1]).

Comments. See Grande (2010) for an explanation of this character.

5. Contact between acutely sharp (anteriorly) parietal bones and rostral bone: absent [0]; present [1]; non-applicable, e.g., numerous rostrals [–].

6. Antero-lateral margin of parietal bone [= frontal]: smooth, without process [0]; with tongue-like process [1].

7. Independent rostral bone: roughly rounded or rectangular-shaped without distinct lateral processes [0]; roughly rhomboidal with well-developed, distinct lateral processes [1]; rostral bone absent [2]; non-applicable, e.g. numerous rostral elements or fused to other bone [–].

Comments. The rostral is rhomboidal-shaped and with distinct lateral processes in †*Eurycormus* (e.g., SMNS 95445/12). This feature is shared with most Triassic pholidophorids (see Arratia 2013).

8. Independent rostral bone: without teeth [0]; bearing teeth [1]; bone absent [2]; non-applicable e.g. numerous rostral elements or fused to other bone [–].

Comments. A rostral bone armed with a few, large conic teeth is present in the Triassic pholidophorids †*Pholidorhynchodon* (see Arratia 2013, fig. 14B, Arratia 2017, fig. 8A, B) and †*Neopholidophorus* (Zambelli 1989, contra Taverne 2011 and 2014).

9. Large compound rostrodermethmoid: absent [0]; present [1].

Comments. Character and interpretation of rostrodermethmoid follows Mainwaring (1978) and Lambers (1992).

10. Teeth on rostrodermethmoid: uniform in size [0]; combination of large tusk-like tooth and small dentition [1]; non-applicable [–].

Comments. Interpretation of teeth on rostrodermethmoid follows Lambers (1992) and Arratia & Schultze (2013). See also Cawley et al. (2019).

11. Laterodermethmoids: bones absent or without teeth [0]; bearing teeth [1].

12. Premaxillae: fused or joined medially [0]; displaced by toothed laterodermethmoids that join medially [1]. Non-applicable, premaxillae absent [–].
Comments. See Arratia (2013 and 2017, figs. 6, 8) for interpretation of the laterodermethmoid and this character.

13. Autosphenotic with small dermal component absent [0]; present [1].
Comments. See Grande (2010) for an interpretation of this character.

14. Braincase: including basisphenoid [0]; basisphenoid absent [1].
Comments. Character slightly modified from Arratia (1997). The so-called sphenoid region of *Polypterus* contains the orbitosphenoid, alisphenoid and median basisphenoid of holosteans and teleosts according to Allis (1922).

15. Orbitosphenoid bone: large, and separating completely both orbital cavities [0]; absent or reduced in size [1].
Comments. See explanation of this character in Arratia (1997), but the polarity is changed according to the new outgroups. The orbitosphenoid region is the so-called sphenoid region in *Polypterus bichir* after Allis (1922).

16. Supraoccipital bone: absent [0]; present [1].
Comments. The teleostean supraoccipital is understood here in the sense of Patterson (1975, p. 297). See text for comments on †*Atacamichthys*.

17. Occipital process formed by fusion of intercalar and autopterotic: absent [0]; present [1].
Comments. This character is after Brito (1997), who interpreted it as a synapomorphy of aspidorhynchiforms.

18. Temporal boss: absent [0]; present [1].
Comments. Interpretation of temporal boss (enlargement of temporal-supratemporal region of the cranium) after Mainwaring (1978), Lambers (1992), and Arratia & Schultze (2013, figs. 3B, 4B). A temporal boss is a synapomorphy shared by some pachycormiforms (see Lambers 1992, Arratia & Schultze 2013, Friedman et al. 2010).

19. Temporal boss: poorly developed, scarcely rising above the surface of the skull roof or absent [0]; well developed, extending above and in front of postparietal region [1]; non-applicable [–].
Comments. Interpretation of temporal boss and its development among certain pachycormiforms follows Lambers (1992) and Arratia & Schultze (2013, figs 3B, 4B).

20. Extrascapula: with smooth, thin anterior margin [0]; with a large rollover bony layer at anterior region of the bone present [1].
Comments. See Arratia (2013, 2017). State character 1 is a synapomorphy of Pholidophoridae (Fig. 9, Node G2).

21. Extrascapula(e) with posterior margin: smooth [0]; serrated [1]; crenulated [2].
Comments. See Arratia (2013, 2017).

22. Interparietal [= interfrontal] suture: smooth (or harmonica) [0]; serrated (serrata) or dentated (dentata) [1]; undulated or wavy in its middle section [2]; no suture [–].

23. Nasal bones: separated from each other by the rostral [0]; joined in midline [1]; separated from each other by parietal bones [= frontals] [2]; separated from each other by the rostrodermethmoid [3].

24. Nasal bone with a large foramen forming the wall of posterior nostril: absent [0]; present [1].
Comments. The presence of the posterior nostril enclosed by the nasal bone is a synapomorphy of certain pholidophorids (Arratia 2017, fig. 6; Arratia et al. 2019, fig. 10: node C5a).

25. Lateral margin of nasal bone: joined with the whole margin of antorbital or partially sutured with [0]; not sutured with antorbital [1]
Comments. Change of polarity in comparison to previous phylogenetic analysis (e.g., Arratia 2013, 2017, Arratia et al. 2019) because of the condition present in the outgroup †*Birgeria*.

26. Nasal bone: included in the orbital margin [0]; not forming part of orbital margin [1].
Comments. Change of polarity in comparison to previous analysis (e.g., Arratia 2013, 2017, Arratia et al. 2019) because the nasal bone is included in the orbital margin in †*Australosomus* and †*Birgeria*.

27. Nasal bone: broad, rectangular-like and moderately large [0]; elongate and narrow (mainly bearing anterior portion of supraorbital canal) [1]; very small, square-like [2].
Comments. A nasal bone reduced in size is a synapomorphy of †*Leptolepis coryphaenoides* plus most advanced teleosts (Arratia et al. 2019, fig. 12, node G; Fig. 9, Node J herein).

28. Vomer (in adult individuals): paired [0]; unpaired [1]; none or unknown condition [–].
Comments. Although the presence of a median vomer was proposed first by Patterson (1977) for primitive teleosts, most stem teleosts present different conditions of this character. As far as is known, pachycormiforms have a paired vomer. In contrast, aspidorhynchiforms have a median, narrow vomer (Brito 1997); the condition is unknown in †*Belonostomus*, as well as in pholidophorids (Arratia 2013). *Australosomus* has two vomers; the condition is unknown in †*Birgeria;* and *Polypterus* has one single bone, so that the polarization of this character has been changed in comparison to Arratia's previous matrices, e.g., 1999, 2013, 2017, and Arratia et al. 2019.

29. Parasphenoid: toothless [0]; with small teeth [1]; large teeth [2].
Comments. For an explanation of this character and its coding see Arratia (1999, p. 286). The coding was modified following the new outgroups, †*Australosomus*, †*Birgeria*, and/or *Polypterus*.

30. Ossified aortic canal: present [0]; absent [1].
Comments. From Patterson & Rosen (1977) and Arratia (1997).

31. Canals for occipital arteries in basi-exoccipital region or basioccipital bone: present [0]; absent [1].
Comments. From Patterson & Rosen (1977) and Arratia (1997).

32. Spiracular canal: developed [0]; greatly reduced [1]; absent [2].
Comments. From Arratia (1997), slightly modified from Patterson & Rosen (1977).

33. Anterior myodome: present [0]; absent [1].
Comments. Modified from Patterson & Rosen (1977) and Arratia (1997). The anterior myodome is present in †*Australosomus* and †*Birgeria*.

34. Foramen for glossopharyngeal nerve position: in basi-exoccipital [0]; in prootic or prootic-exoccipital suture [1]; in exoccipital [2]; other position, e.g., between prootic and opisthotic [3].
Comments. Slightly modified Patterson & Rosen (1977). Occipital region is formed by the basi-exoccipital bone in †*Australosomus*, †*Birgeria*, and *Polypterus*.

35. Supraorbital canal with: simple tubules [0]; branched tubules [1].
Comments. For explanation on this character see Arratia (1997, 2013, p. 122).

36. Supraorbital and infraorbital canal: do not join [0]; join within parietal [= frontal] bone [1]; within dermosphenotic [2].
Comments. The canals do not join in †*Birgeria* and †*Australosomus*. The canals do not join within the dermosphenotic in †*Tharsis* plus †*Ascalabos* and more advanced teleosts (Fig. 9, Node L).

37. Middle pit-line groove: not crossing dermopterotic (or pterotic) [0]; crossing dermopterotic (or pterotic) [1].
Comments. The middle pit-line does not cross the dermopterotic in †*Birgeria* and *Polypterus* (Grande's 2010 illustration), but it crosses in Allis' illustration and in specimens examined here.

Circumorbital bones

38. Antorbital bone: present [0]; absent [1].
Comments. †*Australosomus* does not have an antorbital, whereas †*Birgeria* has the bone.

39. Tube-like canal bearing anterior arm of antorbital: absent [0]; present [1]; non-applicable, antorbital absent [–].
Comments. Character from Grande (2010).

40. Posterior region of infraorbital 3 or jugal: not extending below suborbital bone [0]; extending below suborbital bone reaching anterior margin of preopercle [1]; non-applicable (e.g., no suborbital bone) [–] or independent infraorbital 3.

41. Fourth infraorbital bone: small and narrow [0]; expanded, broad bone [1]; bone is fused with another infraorbital (e.g., infraorbital 3) [2]; Non-applicable [–].
Comments. The fourth infraorbital bone is small in part of the outgroup and also in most teleosts. However, an expanded broad bone is present in some teleosts, e.g. osteoglossomorphs (Li & Wilson 1996, 1999) and some ankylophorids (see Arratia 2000), except †*Lehmanophorus* with infraorbitals 3 and 4 fused (Gaudant 1978).

42. Infraorbital 4 with: smooth posterior margin [0]; serrated posterior margin [1]; non-applicable, e.g., fusion of bones or absence of bone [–].

43. Infraorbital 5: smooth posterior margin [0]; with serrated posterior margin [1]; non-applicable, e.g., absent of bone or fusion of infraorbitals [–].

44. Postero-dorsal infraorbital bones: less than six [0]; more than six [1]. Non-applicable, e.g., other condition, presence of ossicles [–].

45. Suborbital bones (without accessory suborbitals): many [0]; one to three [1]; none [2].

Comments. For an explanation of suborbital bones see Arratia (1997, p. 117). †*Birgeria* has many suborbitals.

46. Accessory suborbital bones positioned ventrolateral to postorbital region of skull roof, posterior to dermosphenotic: none [0]; one or two [1].

Comments. Accessory suborbitals are known only in a few fossil teleosts (e.g., †*Eurycormus* and some pholidophorids; see Arratia 2013).

47. Supraorbital bones: none [0]; two or more [1]; one [2].

Comments. †*Australosomus* has no supraorbital, but †*Birgeria* has three.

48. Sclerotic ring: three or more sclerotic bones enclosing the eye [0]; absence of sclerotic bones [1]; complete ring of two sclerotic bones (oriented anterior and posterior to eye) [2]; incomplete ring of two sclerotic bones (oriented anterior and posterior to eye) [3].

Comments. Sclerotic bones are absent in *Polypterus*. Two or three bones are present in †*Australosomus*, but the condition is unclear (Nielsen 1949, Lehman 1952). Four bones are present in †*Birgeria*. Remnants of sclerotic bones are preserved in †*Atacamichthys* (Arratia & Schultze 1987, fig. 7), but after the incomplete preservation this character is coded with a question mark for this taxon.

Jaws

49. Premaxillary teeth: placed at an angle of about 90 degrees to premaxilla [0]; placed obliquely to premaxilla (posteriorly-directed teeth) [1]; non-applicable (absence of teeth or premaxilla) [–].

Comments. Character from Lambers (1992).

50. Premaxilla: without ascending process [0]; with ascending process [1].

51. Premaxilla forming a rostral tube that projects into the ethmoidal region: absent [0]; present [1].

Comments. From Brito (1997). This character is interpreted as a synapomorphy of †Aspidorhynchiformes.

52. Mobility of premaxilla: fixed premaxilla, sutured with other cranial bones [0]; mobile [1].

Comments. From Patterson (1977) who interpreted this character as a synapomorphy of Teleostei. This character is an unchallenged synapomorphy of the stem teleost †*Prohalecites* plus more advanced teleosts (Arratia 2013, fig. 95, Arratia et al. 2019, fig. 12: node B; Fig. 9, Node E, herein).

53. Maxilla: elongate, extending behind orbit [0]; moderately long, extending below orbit [1]; very short, anterior to orbit [2]; non-applicable, e.g., fusion of bones (e.g., *Polypterus*) or absence [–].

Comments. From Arratia (1997, p. 121, 1999). The elongation of the maxilla is a feature with taxonomic and also phylogenetic value and is a synapomorphy of certain teleosteomorph clades.

54. Maxilla, with dorsal and ventral margins: with different degrees of curvature [0]; almost straight [1]; non-applicable, e.g., fusion of bones (e.g., *Polypterus*) [–].

Comments. From Arratia et al. (2019).

55. Posterior margin of maxilla: rounded or straight [0]; notched or concave [1]; acute, markedly oblique [2]; irregularly shaped [3]; non-applicable, e.g., fusion of bones (e.g., *Polypterus*) [–].

Appendix 2. Character-taxon data matrix for 130 characters used in the phylogenetic analysis. Polymorphic states are represented by A [0/1] and B [1/2]. State [-] represents the character is not applicable. Missing or ambiguous information is represented with [?].

	1–10	11–20	21–30	31–40	41–50	51–60
Amia calva	0 A 0 0 0 0 0 0 0 -	0 0 1 0 1 0 0 0 - 0	0 0 1 0 0 1 0 0 1 0	0 0 0 1 1 1 1 1 0 1 -	1 0 - 0 2 0 0 1 0 0	0 0 0 0 1 1 1 0 0 0
Amia pattersoni	0 0 0 0 0 0 0 0 0 -	0 0 1 0 1 0 0 0 - 0	0 0 1 0 0 1 0 0 1 ?	? 0 0 1 1 1 1 1 0 1 -	1 0 - 0 2 0 0 1 0 0	0 0 0 0 1 1 1 0 0 0
Ankylophorus	0 0 1 0 1 1 0 0 0 -	1 ? ? ? 0 ? ? 0 - 0	0 2 2 0 0 1 0 ? ? ?	? ? ? ? 1 0 1 0 0 -	2 - - 0 2 0 1 ? 0 1	0 1 0 1 2 ? ? 0 1 ?
Ascalabos	0 0 1 0 0 0 2 2 0 -	0 0 0 ? 1 1 ? 0 - 0	0 0 2 0 1 1 ? 1 0 ?	? 2 ? 2 0 ? 0 0 0 -	0 0 0 0 2 0 2 3 0 1	0 1 1 0 0 0 2 0 1 1
Atacamichthys	0 0 1 0 ? 0 ? ? 0 -	? 0 ? ? ? ? ? 0 - 0	2 2 2 ? 1 1 0 ? ? ?	? ? ? ? 1 ? 1 0 0 0	0 0 ? 0 1 0 2 ? 0 1	0 1 1 0 ? ? 1 0 0 ?
Australosomus	0 0 0 0 0 ? ? ? 0 -	0 0 0 0 0 0 0 0 - 0	0 0 0 0 ? 0 0 0 0 0	0 0 0 0 0 0 ? 1 - -	- - - 0 2 0 0 ? 0 0	0 0 0 0 0 0 0 - 0 0
Aspidorhynchus	0 2 0 0 0 0 - - 0 -	0 0 0 0 1 0 1 0 - 0	0 1 2 1 1 1 2 1 1 ?	? ? 0 1 1 ? 1 1 - 0	0 0 0 0 1 0 B 2 0 0	1 0 0 0 0 0 1 1 1 3
Belonostomus	0 2 0 1 0 0 - - 0 -	0 0 0 0 1 0 1 0 - 0	0 1 2 1 1 1 2 ? 1 ?	? ? 0 1 0 ? 1 1 - 0	0 0 0 0 1 0 1 2 0 0	1 0 1 0 0 0 1 1 1 3
Birgeria	0 0 0 0 0 ? 0 0 0 -	0 0 0 0 0 0 0 0 - 0	0 2 0 0 0 0 0 ? 0 0	0 0 0 0 0 0 0 0 0 0	0 0 0 0 0 0 1 0 0 0	0 0 0 0 0 0 0 - 0 0
Caturus	0 0 0 0 0 0 0 0 0 -	? 0 1 ? 1 0 0 0 - 0	0 2 0 0 0 1 0 0 1 ?	? 0 ? ? ? 2 ? 0 1 0	0 0 - 1 1 0 1 2 0 0	0 0 0 0 1 ? 1 0 0 0
Domeykos	0 0 1 0 0 0 ? ? 0 -	0 0 0 0 1 1 ? 0 - 0	2 0 2 ? ? 1 ? 1 0 ?	? 2 ? 2 0 ? ? ? ? -	? ? ? 0 2 0 ? 3 0 1	0 ? 1 ? 0 0 2 0 1 1
Dorsetichthys	0 0 1 0 1 0 0 0 0 -	0 0 0 0 1 1 0 0 - 0	0 0 2 0 1 1 0 1 1 0	0 0 0 1 1 0 1 0 0 1	0 0 0 0 1 0 1 2 0 1	0 1 0 0 0 0 2 0 1 1
Elops	0 0 1 0 0 0 2 2 0 -	0 0 0 0 1 1 0 0 - 0	0 0 2 0 1 1 1 1 1 1	1 2 1 2 0 0 0 0 0 -	1 0 0 0 2 0 2 3 0 1	0 1 0 0 0 0 2 0 0 1
Eurycormus	0 2 A 0 0 0 1 0 0 -	1 1 0 ? 1 0 0 0 - 0	0 1 2 0 0 1 1 1 0 ?	? ? ? ? 1 0 1 0 0 0	0 0 0 0 1 1 1 2 0 1	0 1 1 0 0 0 2 0 1 ?
Hiodon	0 0 1 0 0 0 2 2 0 -	0 0 0 0 1 1 0 0 - 0	0 0 2 0 1 1 1 1 2 1	1 2 0 2 0 0 0 1 - -	1 - 0 0 2 0 0 1 0 1	0 1 1 0 0 0 0 - 1 0
Hypsocormus	0 ? 0 0 0 0 - - 1 1	0 0 0 0 0 0 0 1 1 0	0 0 3 0 ? ? ? ? 0 ?	? 0 0 3 1 ? ? ? ? 0	0 0 0 1 1 0 0 2 1 0	0 0 0 ? 0 0 ? 1 0 0
Lehmanophorus	0 0 1 0 0 1 ? 0 0 -	1 ? ? ? 0 ? ? 0 - 0	0 2 2 0 0 1 0 ? ? ?	? ? ? ? 1 0 1 0 0 -	2 - - 0 2 0 1 ? 0 1	0 1 0 1 2 ? 2 0 1 ?
Leptolepis coryphaenoides	0 0 1 0 0 0 2 2 0 -	0 0 0 0 1 1 0 0 - 0	0 0 2 0 1 1 1 1 1 1	1 1 0 1 0 0 1 0 0 1	0 0 0 0 1 0 1 3 0 1	0 1 1 0 0 0 2 0 1 1
Lepisosteus	0 0 0 1 0 0 0 0 0 -	0 0 1 1 1 0 0 0 - 0	0 0 1 0 1 1 0 0 1 0	0 0 1 1 0 2 1 0 1 -	0 - - 0 0 1 1 1 0 0	0 0 2 0 3 0 A 0 3 0
Lombardichthys	1 - 2 0 1 0 1 0 0 -	0 0 0 ? 1 0 ? 0 - 1	0 - 2 1 1 0 0 ? 0 ?	? ? ? ? 0 ? 1 0 0 0	0 0 0 0 1 1 1 2 0 1	0 1 0 0 B 0 2 0 0 ?
Longileptolepis	0 ? 1 0 0 0 2 2 0 -	0 0 ? ? 1 ? ? 0 - 0	0 ? ? ? ? ? ? ? ? ?	? ? ? ? 0 ? ? ? 0 1	0 0 ? 0 2 0 2 3 0 1	0 1 1 0 0 ? 2 0 1 ?
Malingichthys	A 1 2 0 0 0 0 0 0 -	0 0 ? ? 1 0 ? 0 - ?	1 - 1 0 ? 0 0 ? ? ?	? ? ? ? 0 ? 1 0 0 1	0 1 1 0 1 0 A ? 0 ?	0 1 A 0 ? 1 2 0 A ?
Obaichthys	0 0 0 1 ? 0 ? ? 0 -	0 0 1 1 1 0 0 0 - 0	0 0 ? ? ? ? ? ? 1 ?	? ? 1 ? ? 2 ? ? ? -	0 - - 0 ? 0 1 1 0 0	0 0 2 0 3 0 1 0 3 0
Orthocormus	0 ? 0 0 0 0 - - 1 1	0 0 0 0 0 0 0 1 A 0	0 0 3 0 ? ? 0 ? ? ?	? 0 ? ? ? ? ? ? ? 0	0 0 0 1 1 0 0 2 1 0	0 0 0 0 0 0 1 1 0 ?
Pachycormus	0 0 0 0 0 0 - - 1 0	0 0 0 0 0 0 0 1 0 0	0 0 3 0 0 0 0 0 1 ?	? 0 0 3 0 0 1 0 ? 0	0 0 0 1 1 0 0 2 0 0	0 ? 0 0 0 0 1 1 0 ?
Polypterus	0 3 0 0 0 0 0 0 0 -	0 0 0 0 0 0 0 0 - 0	0 0 1 0 1 1 0 1 1 0	0 0 0 0 0 1 1 0 0 -	0 0 - 0 1 1 0 1 0 0	0 0 - - - - 0 - 0 0
Pholidoctenus	1 - 2 0 0 0 0 0 0 -	0 0 0 ? 1 0 ? 0 - 1	1 - 1 0 1 0 0 ? 0 ?	? ? ? ? 0 0 1 0 0 1	0 1 1 0 1 1 1 2 0 1	0 1 1 0 A A 2 0 1 ?
Pholidorhynchodon	1 - 2 0 0 0 1 1 0 -	1 1 0 ? 1 0 ? 0 - 1	0 2 2 1 0 0 0 ? 0 0	? ? ? ? 0 0 1 0 0 1	0 0 0 0 1 1 1 2 0 1	0 1 0 0 0 0 2 0 0 1
Prohalecites	0 A 1 0 0 0 0 0 0 -	0 0 ? ? 1 0 0 0 - 0	0 0 1 0 0 1 0 ? 0 ?	? ? ? ? 0 0 A 0 0 -	0 1 1 0 1 0 1 ? 0 1	0 1 1 0 0 0 0 - 0 0
Protoclupea	0 0 1 0 ? 0 ? ? 0 -	? ? 0 ? 1 1 ? 0 - 0	2 0 2 0 ? 1 ? ? ? ?	? 2 ? ? 0 0 1 0 ? -	0 0 0 0 2 0 ? 3 0 ?	0 ? ? 0 ? 0 2 0 1 1
Siemensichthys	0 2 1 0 0 1 0 0 0 -	1 1 0 0 1 1 0 0 - 0	0 2 2 1 ? 1 0 1 0 0	0 ? 0 1 1 0 1 0 0 0	1 0 0 0 2 0 1 2 0 1	0 1 0 1 2 0 1 0 0 1
Solnhofenamia	0 0 1 0 0 0 1 0 0 -	0 0 1 ? 1 0 0 0 - 0	0 0 1 0 0 1 0 0 0 ?	? ? ? ? ? 2 ? 0 1 0	0 0 - 0 1 0 1 3 0 0	0 0 0 0 1 1 1 0 0 ?
Tharsis	0 0 1 0 0 0 2 2 0 -	0 0 0 0 1 1 0 0 - 0	0 0 2 0 1 1 1 1 0 1	1 A 0 2 0 0 0 0 0 -	1 0 0 0 2 0 2 3 0 1	0 1 1 0 0 0 2 0 1 1
Varasichthys	0 0 1 0 0 0 2 2 0 -	0 0 0 0 1 1 0 0 - 0	2 0 2 0 1 1 1 ? 0 1	1 2 1 2 0 0 1 0 0 -	0 0 0 0 2 0 2 3 0 1	0 1 1 0 1 0 2 0 1 2
Vinctifer	0 0 0 0 0 2 - 0 -	0 0 0 0 1 0 1 0 - 0	0 1 2 1 1 1 2 1 0 ?	0 ? 0 1 ? 2 1 1 - 0	0 0 0 0 1 0 2 2 0 0	1 0 1 0 0 0 0 - 1 3

Comments. Character modified from Grande & Bemis (1998) to include other character states.

56. Dorsal margin of maxilla: without supramaxillary process [0]; with supramaxillary process [1]; non-applicable, e.g., fusion of bones (e.g., *Polypterus*) or absence [–].

Comments. †*Pholidoctenus* is coded with state 1, because the fishes have a supramaxillary process just in front of the anterior tip of supramaxilla 1 (Arratia 2013, fig. 76, Arratia 2017, fig. 2A, B).

57. Supramaxillary bone(s): none [0]; one [1]; two [2].

Comments. The presence of two supramaxillae is a synapomorphy of the apomorphy-based Teleostei (Arratia 2017, fig. 9, node C, Arratia et al. 2019, fig. 12: node F; Fig. 9: Node G) with further transformation in more advanced teleosts.

58. Supramaxillary bone or most posterior supramaxilla: dorsal to maxilla [0]; posterior or dorso-posterior to maxilla [1]; non-applicable (absence of bone) [–].

Comments. Character from Arratia (1997). For an explanation see Arratia (1997, p. 121, 1999, p. 290).

59. Quadrate-mandibular articulation: posterior to orbit [0]; below the posterior half of orbit [1]; below anterior half of orbit [2]; anterior to orbit [3].

Comments. Character from Arratia (1999).

60. Articular bone: not fused with angular and/or retroarticular bones [0]; fused with angular and retroarticular [1]; fused with angular forming an anguloarticular [2]; fused with retroarticular [3].

Comments. Modified from Patterson (1977) to include stem teleosts and other primitive fossil teleosts studied herein.

61. Postarticular process of lower jaw: developed or not developed [0]; well developed [1].

Comments. Modified from Arratia (1997) with change in the polarity of the character depending on the outgroups used here.

62. Oral margin of dentalosplenial or dentary: without a notch [0]; with a notch (= "leptolepid" notch) [1].

Comments. Character from Arratia (1997).

63. Lower jaw: with coronoid bone(s) [0]; without coronoid bones [1].

Comments. Character from Nelson (1973). The loss of coronoid bones is a synapomorphy of teleosts (Arratia 2017, fig. 9: node D; Arratia et al. 2019, fig. 12: node F; Fig. 9, Node D herein).

64. Lower jaw: with surangular bone [0]; without surangular bone [1].

Comments. Character from Patterson (1977). The loss of the surangular bone stands as a synapomorphy of †*Leptolepis coryphaenoides* plus more advanced teleosts (Arratia 2017, fig. 9: node F, Arratia et al. 2019, fig. 12: node G; Fig. 9, Node J herein). The condition in the outgroup taxa varies: †*Australosomus* and †*Birgeria* present a surangular bone, whereas the bone is absent in *Polypterus*.

65. Lower jaw: with prearticular bone [0]; without prearticular bone [1].

Comments. Character from Patterson (1977). This character stands as a synapomorphy of †*Dorsetichthys* plus more advanced teleosts (Fig. 9, Node I herein).

66. Dentary: with no separation between dental and splenial regions [0]; with a well-developed, protruding lateral bony ridge separating dental and splenial regions [1].

Comments. This character is interpreted as a synapomorphy at the basal level of the apomorphy-based teleosts (Arratia 2017, fig. 9: node C, Arratia et al. 2019, fig. 12: node C; Fig. 9, Node G herein), with a reversal in more advanced nodes.

67. Toothed predentary in lower jaw: absent [0]; present [1].

Comments. Character from Brito (1997). A predentary bone is present only in aspidorhynchiforms among the studied fishes. This character is interpreted as a synapomorphy of †Aspidorhynchiformes (Brito 1997).

61–70	71–80	81–90	91–100	101–110	111–120	121–130
0000000-00	1000000000	0110000001	1000000000	0000000010	-010000011	0000011110
0000000-00	1000000000	011000000?	1000000000	0000000010	-010000011	0????11110
0????10-00	????10??00	001?0????0	??????101?	?10???????	???1??0000	0??????110
1111100-00	1121101100	0011011111	0111?00000	10001A1110	1001101102	0????11110
?01?000-00	1????0A000	01?000000?	20?1001010	01010?0010	-0?????0?1	0??????110
0000000?00	000--0?000	0000000000	0000000000	0000000000	00000---00	0????00000
0011001000	300A000001	00A1001000	?110?01010	0100-00031	1001010000	0????11110
0011001100	3001100001	00A1001000	?11??00000	0100-00031	1001000000	0????11110
0000000?00	000--00000	0000001000	0000000000	00000?0??0	00000---00	0????0?000
0?0000?-0?	10000?0000	0110000000	??0000?0?0	010???0?10	-011000001	0????11110
1111100-00	112110??00	001?01111?	01???10001	0000101110	1000101?02	0????11111
1110110-10	112110?000	101000010?	?111101011	01010100A0	1001110000	0????11110
1011100-00	1121101100	0010011111	0111110001	0000010110	1000001202	0111111110
0100010-00	1011100000	0010000100	20101110?0	0101010001	0001110001	0????11110
1010100-00	1121101100	0011011101	0111010001	0000101120	1000001002	0111111110
0000000-00	2?????0?00	00100--000	??11?0?100	0?1?000201	01210?0001	0??????110
0????10-?0	1?2?????00	001???????	??????101?	?100??????	???1??0000	0??????110
1111110-00	1121101100	0010001101	3111100001	0000111110	1001001102	0????11110
0000000-00	1001010010	0000101000	0000001010	0101000000	-011000000	1000010110
01?0?10-11	201AA00000	1010000000	20?0111011	?101??????	???1110000	0????11110
?????10-00	?????0??00	101?001101	?1??100001	00001011?0	1001001?02	0??????110
0????10-11	1???????00	001??0?0??	?????01?11	?101??????	???11?0000	0????????0
0000000-00	1001010010	0010101000	0000001010	?1010?0000	-011000000	1????101?0
00???00-00	????????00	00100--000	??11000100	0010000201	0121000001	0????1?110
0000000-00	2AB1A?0000	00100--000	0?A11011-0	0010000201	01210?0001	0????11110
1001000-0-	100--00000	0000?01000	0000000000	00000000?0	-010----10	0000000000
0100?10-11	101??00000	0110000000	20?0111011	?1010000?0	?0?1110000	0????11110
0100010-11	20???00000	1010000000	2010111011	0101??????	???11?0000	0?????1110
00?0?00-00	1011100000	011?000100	??1??01010	0101010000	100111000-	0????11110
1111100-00	1121101?00	001?011111	?1??110011	0000010110	1000101?02	0????11111
01?0010-00	112110?000	001000010?	21?0101010	01010??0??	??01000000	0????11110
?000000-00	1000000000	011000000?	??00?0000?	0000??0000	-010000011	0????11110
0111100-00	1121101100	0010011111	0111100000	0000010110	1001001102	0????11110
0111100-00	1121101100	0010011111	0111110001	00000?01?0	1000?01102	0????11111
001100100-	3010000001	0001001000	2110000000	0000-00231	10010?0000	0????11110

68. Predentary: short, less than half length of premaxillary tube [0]; almost as long as premaxillary tube [1]; non-applicable, predentary absent [–].

Comments. The length of the predentary bone is a character separating members of the †Aspidorhynchiformes (e.g., Brito 1997, Ebert 2014, Arratia & Schultze 2013).

69. Maxilla: without or scarce ganoine ornamentation [0]; heavily ornamented with characteristic longitudinal ridges of ganoine [1].

Comments. This feature stands as a synapomorphy shared by some pholidophorids (Arratia 2017, fig. 9, node C3).

70. Supramaxilla(e): without or scarce ganoine ornamentation [0]; heavily ornamented with characteristic longitudinal or concentric ridges of ganoine: [1]; non-applicable, no supramaxilla present [–].

Palatoquadrate, hyoid arch, and urohyal

71. Suspensorium: short, with an antero-dorsal inclination of hyomandibula [0]; short, with a vertical alignment of the hyomandibula [1]; elongate, with a postero-ventral inclination [2]; elongate, with an antero-ventral inclination [3].

72. Elongated postero-dorsal or postero-ventral process of quadrate (depending on the angle of the quadrate): absent [0]; present [1].

Comments. Arratia (2013, fig. 99B) illustrated a SEM photograph of the quadrate in †*Leptolepis coryphaenoides* to illustrate what Arratia described as an elongate process. This elongate process as in †*L. coryphaenoides* is also present in †*Siemensichthys*, †*Dorsetichthys*, †*Ichthyokentema* and more advanced teleosts, including primitive living forms (e.g., *Elops*; Arratia & Schultze 1991, figs. 15B, 21A–D). †*Eurycormus* was coded with a question mark in Arratia's (2013, 2016) matrices. However, the parsimony analysis suggested that the process was present. New specimens (e.g., SMNH 88987/3) with the complete quadrate exposed (Arratia et al. 2019, fig. S2A) show that an elongate process is not present in †*Eurycormus*, but a tiny process is, as the one illustrated for some pholidophorids (Arratia 2013, figs. 13, 18).

According to Taverne (2014, p. 252) "in Aspidorhynchidae, the ventral margin of the quadrate often is swollen, forming a sort of bony pad (Brito 1997, fig. 15A) and a true bony quadratic process could even be individualized (Taverne 1981, fig. 5)." Brito (1997, figs. 15A, B) did not illustrate nor describe a postero-ventral process (quadratic process) of the quadrate for any aspidorhynchiform (see Arratia 2017, fig. S3, Arratia et al. 2019, fig. S3).

73. Posterior or postero-ventral margin of quadrate: not especially thickened ossification [0]; markedly thickened and with a diminutive process distally [1]; markedly ossified into a postero-dorsal (or ventral depending on angle of quadrate) process partially separated by a deep notch from main body of the bone [2].

Comments. See Arratia (2013, figs. 15, 16, 18), Arratia (2017, fig. S2A, B), and Arratia et al. (2019, fig. S2A, B).

74. Symplectic: articulates with lower jaw [0]; does not articulate with lower jaw [1]; non-applicable, e.g., absence of bone [–].

Comments. See comments on this character in Patterson (1973), Brito (1997), and Arratia (1999). A symplectic is absent in the outgroup taxa, e.g., †*Australosomus*, †*Birgeria*, and *Polypterus*. *Lepisosteus* and †*Obaichthys* were coded with 0 in Arratia (2013, 2016). However, *Lepisosteus* presents a short symplectic, as illustrated by Arratia & Schultze (1990, fig. 13A, B) and Grande (2010). The coding of *Lepisosteus* and †*Obaichthys* is replaced with state 1.

75. Position of symplectic: posterior to posterior margin of quadrate [0]; medial to posterior margin of quadrate [1]. Non-applicable, e.g., absence of symplectic [–].

Comments. A symplectic is absent in the outgroup taxa, e.g., †*Australosomus*, †*Birgeria*, and *Polypterus*. For information on this character see Pinna (1996), Brito (1997), and Arratia (1999).

76. Pars quadrata or quadrate and pars metapterygoidea or metapterygoid: close together and articulated with each other [0]; distant to each other [1].

Comments. Character state 1 is only known from lepisosteids among the studied fishes (see also Grande 2010).

77. Hyoid arch: with one hypohyal articulated with anterior ceratohyal [0]; with dorsal and ventral hypohyals articulated with anterior ceratohyal [1].

Comments. This character is from Patterson (1977). For explanation on this character, its homology and its importance as a teleostean character see Arratia & Schultze (1990), Pinna (1996), and Arratia (2015b).

78. Urohyal formed as an unpaired tendon bone of the sternohyoideus muscle: absent [0]; present [1].

Comments. This character is from Arratia & Schultze (1990). For explanation of the tendon-bone urohyal, its homology and its importance as a teleostean character see Arratia & Schultze (1990), Pinna (1996), and Wiley (2008).

Opercular bones, branchiostegals, and gular plate

79. Ventral or anterior extent of preopercle: not extending below anterior region of orbit [0]; extending below anterior region of orbit [1].

80. Posterior position of preopercular sensory canal in a peculiarly-shaped preopercle: absent [0]; present [1].

Comments. Character from Brito (1997). This feature is interpreted as a synapomorphy of †Aspidorhynchiformes.

81. Postero-ventral or posterior margin of preopercle: smooth [0]; notched [1].

Comments. A notch at the posterior margin of the preopercle was described first by Nybelin (1966) for certain Triassic and Liassic pholidophorids.

82. Crescent-shaped preopercle: absent [0]; present [1].

Comments. From Grande & Bemis (1998).

83. Interopercle: absent [0]; present [1].

Comments. †*Aspidorhynchus* and †*Belonostomus* are coded with states 0/1 because specimens from Ettling (Bavaria, Germany) show an interopercle, whereas specimens from other localities in the Solnhofen limestones seem not to have the bone (see also Ebert 2014). †*Vinctifer* does not have an interopercle (after Brito 1997). An interopercle is absent in outgroup taxa, †*Australosomus*, †*Birgeria*, and *Polypterus*.

84. Cranial ventral region: with median gular plate [0]; without gular plate [1].

Vertebrae and intermuscular bones

85. Opisthocoelous centra with convex anterior articular surface and concave posterior surface: absent [0]; present [1].

Comments. Character from Arratia (1999). Opisthoceolous vertebrae are only known from lepisosteids.

86. Notochord of midcaudal centra: unconstricted or moderately constricted [0]; strongly constricted by the walls of vertebral centra (hourglass-shaped centrum) [1]; non-applicable, e.g., no centra, functional notochord present [–].

Comments. Character from Arratia (1991). For information on structure of vertebrae see Arratia (1991, 1997) and Arratia et al. (2001).

87. Mid-caudal centra (adults): diplospondylous centra [0]; monospondylous centra [1]; non-applicable, e.g., no centra, functional notochord present [–].

Comments. Character from Arratia (1991). For information on structure of vertebrae see Arratia (1991, 1997) and Arratia et al. (2001). Among the outgroup, †*Australosomus* presents diplospondyly (Nielsen 1949).

88. Abdominal neural arches: without epineural processes [0]; with epineural processes [1].

Comments. From Arratia (1977). For an explanation of epineural process and epineural bones see Arratia (1997, 1999, 2015b).

89. Epipleural intermuscular bones: not developed [0]; developed in the middle part of body [1].

Comments. Character from Patterson (1977).

Pectoral and pelvic girdles, and axillary processes

90. Posttemporal: without a dorsal process [0]; with a distinct dorsal process to articulate with cranium [1].

91. Serrated appendages: absent [0]; anterior and posterior toothed elements present on the ventral arm of cleithrum [1]; one long, toothed element covering the whole medial surface of cleithrum [2]; one long, narrow element covering the medial border of cleithrum [3].

Comments. See Arratia & Schultze (1987) and Arratia (2013, p. 123, 125) for explanation on serrate appendage versus clavicle.

92. Pectoral girdle: with clavicle(s) [0]; without clavicle [1].
Comments. Among the outgroup taxa, †*Australosomus* does not have a clavicle (Lehman 1952). The teleosts have been traditionally described without a clavicle. The present results also suggest this, because the absence of a clavicle stands as a synapomorphy of teleosteomorphs (e.g., Arratia et al. 2019, fig. 12: node A). However, a clavicle is present in pholidophorids (Arratia 2013, p. 125, fig. 20) and †*Eurycormus*, and this character is interpreted here as a synapomorphy of pholidophoriforms (a reversal at node C1 in fig. 9 in Arratia 2017; Fig. 9, Node H herein).

93. Pectoral proximal radials: more than four [0]; four [1].
Comments. †*Vinctifer* has four proximal radials after Brito (1997). Taverne (2014) criticized Arratia's coding [0] of †*Pachycormus* as wrong based on Jessen's figure (1972, pl. 25, fig. 1) and citing Mainwaring (1978), who also referred to the same figure. Jessen's plate 25, fig. 1 does not show radials. This is a misreading of Jessen's illustration.

Furthermore, Jessen (1972, p. 75) wrote the following sentence about †*Pachycormus*: "Zum Endoskelett der Brustflosse gehören das Propterygium, fünf Radien und eine Anzahl Radialia", which translates as, "The propterygium, five radii (proximal pectoral radial) and a number of radialia (distal pectoral radials) belong to the endoskeleton of the pectoral fin" (translation by H.-P. Schultze). Furthermore, Jessen (1972, p. 89, fig. 10) concluded that the endoskeleton of the pectoral girdle of †*Pachycormus* is similar to that of †*Pteronisculus* (with five proximal radials). Independently of this, we have not seen four pectoral radials in any of the studied pachycormiforms.

94. Pectoral propterygium: free [0]; fused with first pectoral-fin ray [1].
Comments. Patterson (1977) proposed this character as a synapomorphy of Teleostei, at the level of pachycormiforms and more advanced teleosts. But see Arratia (1999, 2013, and herein). A pectoral propterygium fused with the first pectoral ray is present in †*Atacamichthys* (Arratia & Schultze 1987, herein).

95. First pectoral fin-ray: simple ray, not fused to fulcra [0]; compound ray fused with basal fulcra [1].
Comments. For an explanation on this character see Arratia (2008, 2013).

96. Pectoral fin: without pectoral axillary process [0]; with pectoral axillary process [1].
Comments. For an explanation on this character see Arratia (1997, 1999, 2013).

97. Leading margin of pectoral fin: without fringing fulcra [0]; with fringing fulcra [1].
Comments. Fringing fulcra associated with paired and unpaired fins are treated independently for each fin and not as a group, e.g., "paired fins bearing fringing fulcra", because the pectoral fin is homologous to other pectoral fins, not to the pelvic or other fins. Thus, each is treated independently to keep the homology concept. The polarity of this character was revised following the fact that the outgroup taxa do not have fringing fulcra associated to the pectoral fin.

98. Pectoral-fin shape: rounded or acuminate or lanceolate [0]; scythe-like [1].
Comments. A scythe-like pectoral fin has only been described in pachycormiforms, and it is considered a synapomorphy of the group (e.g., Mainwaring 1978, Lambers 1992, Friedman et al. 2010).

99. Leading margin of pelvic fin: without fringing fulcra [0]; with fringing fulcra [1].
Comments. Fringing fulcra associated with paired and unpaired fins are treated independently for each fin and not as a group, e.g., "paired fins bearing fringing fulcra", because the pectoral fin is homologous to other pectoral fins, not to the pelvic or other fins. Thus, each is treated independently to keep the homology concept. The polarity of this character was revised following the fact that the outgroup taxa do not have fringing fulcra associated to the pelvic fin.

100. Pelvic fin: without pelvic axillary process [0]; with pelvic axillary process [1].
Comments. Character from Arratia (1997). For an explanation on pelvic axillary processes see Arratia (1997, p. 134).

Dorsal and anal fins

101. First dorsal pterygiophore: without processes or with anteroventral processes of similar size [0]; with three or more anteroventral processes, the first one expanded [1] (see Arratia et al. 2019, fig. 8).
Comments. This character stands as a synapomorphy of ascalaboids (Arratia 2017, fig. 9: node G1, Arratia et al. 2019, fig. 12).

102. Leading margin of dorsal fin: without fringing fulcra [0]; with fringing fulcra [1].
Comments. Change of polarity with respect to previous phylogenetic analyses because of the outgroup taxa now used to polarize characters.

103. Dorsal fin carrying: few, short basal fulcra or no fulcra [0]; many, long basal fulcra [1].
Comments. This kind of basal fulcra has only been observed in the studied pachycormiforms (see also Arratia & Schultze 2013).

104. Leading margin of anal fin: without fringing fulcra [0]; with fringing fulcra [1].
Comments. Change of polarity with respect to previous phylogenetic analyses because of the outgroup taxa now used to polarize characters.

Caudal skeleton

105. Preural vertebrae 3 and 2 of adult individuals: with haemal arches autogenous [0]; with haemal arches laterally fused to their respective centra [1].

106. Neural spine of preural centrum 2: as long as neural spine of preural centrum 3 [0]; shorter than neural spine of preural centrum 3 [1].
Comments. Character from Arratia (1991).

107. Arch of parhypural in adults: autogenous [0]; laterally fused with its centrum [1].
Comments. Modified from Arratia (1991).

108. Number of ural centra (in adults): more than two (polyural condition) [0]; commonly two (diural condition) [1]; no ural centra [2].
Comments. The diural caudal skeleton was proposed as a synapomorphy of teleosts by Patterson (1977). For further explanation on this character and the homology problems involved see Schultze & Arratia (1988, 1989, 2013) and Wiley et al. (2015).

109. Number of epural(s): five or more [0]; three or four [1]; two or one [2]; none [3].
Comments. Modified from Patterson (1977). The number of epurals is a character not affected by growth. It has been consistently used in taxonomic and systematic studies of teleosteomorphs, as well as other neopterygians. See for instance: Patterson (1968, 1977), Pinna (1996), Brito (1997), Grande & Bemis (1998), Schultze & Arratia (1986, 1987, 1989, 2013), Arratia & Schultze (1992), Arratia & Lambers (1996), Arratia (1997, 1999, 2000, 2013), and Grande (2010). The polarity of the character was changed in relation to previous phylogenetic analysis considering the new outgroup taxa used.

110. All or a few preural neural spines: unmodified [0]; enlarged and expanded laterally (uroneural-like bones) [1].
Comments. For information concerning uroneural-like bones, see Patterson (1973), Arratia & Lambers (1996), Arratia & Schultze (2013), and Schultze & Arratia (2013). Uroneural-like bones are only known in pachycormiforms, in some aspidorhynchiforms, and in †*Eurycormus*. In contrast, uroneurals are modified ural neural arches, in other words, are modifications of the ural region, not preural region.

111. Ural neural arches: unmodified [0]; modified as uroneurals [1]; non-applicable [–].
Comments. The interpretation of ural neural arches modified as uroneurals is from Patterson (1968). The presence of uroneurals as a synapomorphy of teleosts was suggested by Patterson (1973, 1977), an interpretation that have changed after study in different teleosteomorphs and extant teleosts (e.g., Arratia & Lambers 1996, Arratia & Schultze 2013, Schultze & Arratia 2013). For information concerning presence and/or absence of uroneurals and their meaning as teleostean characters see for instance: Patterson (1968, 1977), Brito (1997), Schultze & Arratia (1986, 1988, 1989, 2013), Arratia & Schultze (1992, 2013), Arratia (1997, 1999, 2000, 2013). For origin and homologization of uroneurals see Schultze and Arratia (1989, 2013), and Arratia and Schultze (1992).

112. Hypurals: independent [0]; fused into a plate [1].
Comments. Character state 1 stands as a synapomorphy of the studied pachycormiforms (Arratia 2017, fig. 9, node A3).

113. Arrangement of hypurals and caudal fin rays: each hypural normally articulated with a few caudal rays [0]; each hypural normally articulated with one caudal ray [1]; a hypural plate articulated with many rays [2].
Comments. Character from Grande & Bemis (1998), slightly modified.

114. Dorsal or epaxial leading margin of caudal fin: without epaxial basal fulcra [0]; with basal fulcra [1].

115. Epaxial lobe of the caudal fin: without rudimentary ray [0]; with rudimentary ray [1].
Comments. Interpretation of epaxial rudimentary ray follows Patterson (1968) and Arratia (1991, 2008).

116. Ventral hypaxial leading margin of caudal fin: without accessory fulcra [0]; with accessory fulcra irregularly intercalated between hypaxial fringing fulcra [1].
Comments. For explanation of accessory fulcra see Arratia (2008).

117. Epaxial and hypaxial leading margins of the caudal fin: including two or three principal rays, respectively [0]; including only the first and last principal caudal ray, respectively [1]. Non-applicable [–]
Comments. Character from Arratia (2008).

118. Tendon-bone 'urodermals' associated with first principal caudal rays: none [0]; two [1]; one [2].
Comments. For an explanation on tendon-bone 'urodermals' and urodermals (and their homology) see Arratia & Schultze (1992). In †*Tharsis*, as well as in other teleosts (Arratia et al. 2019, fig. 12, node G), 'urodermal' tendon-bones are external bones lying lateral to the first or first and second principal caudal rays. They should not be confused with the dorsal series of short uroneurals present in †*Tharsis*, that are medial to the bases of the rays.

119. Posterior margin of caudal fin convexly rounded: absent [0]; present [1].
Comments. Character slightly modified from Grande & Bemis (1998).

Scales

120. Scales: ganoid of lepisosteoid-type [0]; elasmoid of amioid-type [1]; elasmoid of cycloid-type [2]; non-applicable, e.g., absence of scales [–].
Comments. For type of scales see Schultze (1966, 1996, 2015). For the special type of scales of pachycormiforms, see Schultze (1966) and Lambers (1992).

Miscellaneous characters

121. Separation between olfactory organ and eye: close together [0]; broadly separated from each other [1].
Comments. This feature stands as a synapomorphy of lepisosteiforms (Fig. 9, Node B4).

122. Olfactory organ: without accessory nasal sacs [0]; with accessory nasal sacs [1].
Comments. This feature was proposed as a synapomorphy of crown teleosts by Chen and Arratia (1994), Pinna (1996), and Arratia (1999). This feature is known only in extant teleosts, so that all fossil taxa studied here are coded with a question mark. The parsimony analysis predicts this as a feature of teleosteomorphs (Arratia 2017, fig. 9: node A, Arratia et al. 2019, fig. 12: node A; Fig. 9, Node C herein).

123. Craniotemporal muscle: absent [0]; present [1].
Comments. This feature was proposed as a synapomorphy of crown teleosts by Stiassny (1986), Pinna (1996), and Arratia (1999). This feature is known only in extant teleosts. The parsimony analysis predicts this as a feature of teleosteomorphs (Arratia 2017, fig. 9: node A, Arratia et al. 2019, fig. 12: node A; Fig. 9, Node C herein).

124. Heart with two arterial valves (in the conus arteriosus): absent [0]; present [1].
Comments. Character is from Müller (1845) who used it for the first time to diagnose his subclass Teleostei. For additional comments see Arratia (2015b).

125. Muscles at the basal arteria (ventral aorta): present [0]; absent [1].
Comments. Character is from Müller (1845) who used it for the first time to diagnose his subclass Teleostei. For additional comments see Arratia (2015b).

127. Quadratojugal: present [0]; absent [1].

128. Most dorsal fin rays equal in number than their support: absent [0]; present [1].

129. Most anal fin rays equal in number to their support: absent [0]; present [1].

130. Scales: without transverse lines in the middle field [0]; with circuli crossed by transverse lines in the middle field [1]; non-applicable, e.g., other type of scale.

Note

For references cited in Appendix 1, see Literature cited.

Ancient Fishes and their Living Relatives: a Tribute to John G. Maisey. Alan Pradel, John S. S. Denton & Philippe Janvier (eds.): pp. 37–47, 7 figs.
© 2021 by Verlag Dr. Friedrich Pfeil, München, Germany – ISBN 978-3-89937-269-4

Paleontological treasures among commonplace fossils: a paradigm to study evolutionary innovation

Jose Xavier-Neto* and Ismar de Souza Carvalho**

Abstract

In the last fifteen years, preserved remnants from a number of internal organs have been described. This type of soft, non-mineralized, tissue preservation opens exciting new possibilities to understand the biology of extinct species. Information obtained from fossil organs can indicate the occupation of certain ecological niches and suggest behaviors. From a developmental standpoint, fossil organs may reveal novel ontogenetic sequences and morphogenetic events previously unexpected from phylogenetic analyses dominated by extant species. Recently, we described the first known fossilized heart. The fossil heart was found in *Rhacolepis buccalis*, an extinct basal teleost from the Santana Formation, a Cretaceous Konservat Lagerstätte in the Araripe Basin in the Northeast of Brazil (119–113 Ma). Finding the fossilized heart to be intermediate between the hearts of vigorous predators such as tunas and sedentary bottom feeders such as codfish helped to further clarify the classically-defined role of *R. buccalis* as an avid predator of shrimps – such as *Paleomattea deliciosa* – and as a small fry cannibal. The cardiac outflow tract of *R. buccalis* displayed unforeseen degrees of complexity, with multiple associated individual valves, in contrast with the single outflow tract valves of derived teleosts. These findings reaffirm that evolution does not always associate with complexity, as younger species may display morphologies that are substantially less sophisticated than those displayed by more primitive forms.

Introduction

The primary goal of paleontology since its inception as a discipline with a clear scientific plan has been to study all aspects of past life on Earth. Within this broad scope, the focus of the discipline has changed many times since the eighteenth century. Nevertheless, one of the strongest elements of paleontological fascination has always been the anticipated elation of bringing unknown species back to life, at least to our intellectual lives.

As the study of past life, paleontology is a historical science. The paleontological narrative on Earth results from continued interactions between evolving species and their changing environments. What is the relationship of paleontology with biology? If all extant life forms are destined to be past life sometime in the future, then biology, the study of present life is both a science of the actuality, and a direct product of the previous history of living beings in this Earth. Biology is thus as historical a discipline as paleontology. If biology is a historical science, where is it written? Biology is the substance of all living beings, but many of its current historical secrets can only be assessed through the inspection of two very special domains, the realm of fossils (the province of paleontology) and a very short window of early development (the domain of comparative developmental biology).

Fossils provide direct evidence of past life owing to the resilience of some living materials, chiefly mineralized tissues, comprising hard protective or structural elements. In special circumstances associated with quick insulation of remains within anoxic or hypoxic environments, these materials stand a good chance of avoiding the disaggregating actions of scavengers, of bacteria, and of other microorganisms (e.g. Allisson & Briggs 1991, Behrensmeyer & Kidwell 1985). Under special circumstances, fragile, non-mineralized tissues are also preserved, often in association with mass mortality events caused by, or linked to, catastrophes such flash floods, or tectonic events. The quality of the information obtained from paleontological materials is, therefore, dependent on the degree of post-mortem modification (taphonomy), as well as contingent on a particular chemistry at the microenvironment of burial sites, including preservation of biologic materials, fossil diagenesis, and their consolidation into sedimentary rocks (Martin 1999).

Although the morphology of metazoans and plants may change after sexual maturity, the most relevant differences between species are already established during their early development. Therefore, the origin of morphologic and functional diversity during evolution is intrinsically linked to the processes that control development (Erwin & Davidson 2012). To study evolution is to study descent with modification of development. In fact, the study of comparative early development is also a valuable complement to fossils, when the goal is to understand evolution of animal forms.

After the embryo grows from a cellular agglomerate to a group of cells organized in orthogonal axes of the three-dimensional space, the developing individual undergoes morphogenetic movements that place tissue and organ progenitors closer to their mature dispositions in an ontogenetic phase classically known as gastrulation (Stern 2004, and references therein). From this period on, the events of patterning, cell migration, and morphogenesis that collectively shape embryonic structures maintain relationship with the order of acquisition of structural complexity through evolutionary time. Thus, the primitive attributes appear first, while the most specialized, sometimes diagnostic, attributes of individual phyla appear only later. It is always important to state that the relationship between ontology and embryology does not follow Haeckel's long superseded view of recapitulation, in which the sequential embryonic phenotypes of derived animals represented sequential adult forms in the phylogeny. Rather, the modern view is that early embryonic forms of derived animals do recapitulate the phenotypes from embryos of more primitive animals (i.e. those that branched earlier in the phylogenetic tree). Therefore, ontogeny does provide valuable clues about the relative novelty of characters. For instance, elephant trunks appear much later

* Universidade Federal do Ceará (UFC), Departamento de Morfologia, Ceará, Brazil; josexavierneto@gmail.com (corresponding author)
** Universidade Federal do Rio de Janeiro, Departamento de Geologia, CCMN/IGEO 21.949-900 Cidade Universitária – Ilha do Fundão, Rio de Janeiro, Brazil; Universidade de Coimbra, Centro de Geociências. Rua Sílvio Lima, 3030-790, Coimbra, Portugal

in ontogeny than elephant limb buds, reaffirming the notion that trunks appeared much later than limbs in the phylogeny (Hautier et al. 2012). Likewise, bat wings, which are formed by a thin membrane, the patagium, appear so late in bat development, that sometimes it is difficult for the untrained eye to tell bat from mouse embryos, for example (Cretekos et al. 2008). Often, however, relationships between ontology and phylogeny are scrambled by heterochrony, when some structures may form before, or after, they would be expected to, based on the timing of the appearance of that structure in an ancestor (Gould 1977, Hall 2003).

Thus, embryonic development can be similar to a quick movie trailer, a video clip of the first steps needed to build a multicellular animal. Too much emphasis on a single movie genre can severely limit one's ability as a film critic. Likewise, excessive devotion to only a few species, or phyla, will easily mislead a paleontologist, or a developmental biologist. This is because individual developmental video clips can be quite deceiving. Sometimes, the sequence order of developmental events may be shuffled, the duration of individual events may be altered, or some combination of these two through heterochrony (Gould 1977, Hall 2003). Thus, the key to interpret the clues that development offers about the evolution of form and function is to weed out the heterochronic events from the narrative and to build a representative account that can work as an initial, generic script. With this basic script at hand, it is possible to dissect the eccentricities that often compound the establishment of form and function during ontogeny. The latter, by their own peculiarity, constitute shared-derived characters, which will become valuable tools to label specific evolutionary lineages (i.e. a chronological series of organisms linked by descent).

Updating the synthesis: Enter developmental biology, gene regulatory networks, genomes, and the environment. The last 30 years witnessed enormous successes in studying the commonalities and similarities of development across metazoans as diverse as arthropods, chordates, mollusks, corals and comb jellies. This collective effort produced at least three major realizations. First, developmental processes display the same mechanistic bases in all multicellular life forms. Second, morphological (and functional) evolution depends on the interaction among a limited, recurrent set of chemical signals and gene products as proximate causes of developmental innovation (Held 2014). Ultimately, however, developmental changes depend on the orchestration of gene regulatory networks (Erwin & Davidson 2012) that are often in perpetual transformation due to the metastable nature of genomes. Third, genome innovation is fostered by the dynamic accumulation of changes induced by mutation, gene duplication, as well as by the agency of mobile elements such as transposons, retrotransposons and viruses (e.g. retroviruses, endogenous retroviruses), which act as continuous generators of diversity (Jacobs et al. 2014). If genomic changes influence the organism's overall survivability and/or fertility, they will be selected for, or against. Often however, genomic changes may randomly accumulate, generating specific allelic patterns in small populations, especially when gene flow is at a minimum (genetic drift). These changes are often neutral in that they do not significantly modify the fitness associated with the individual. However, these so-called neutral changes may have consequences for both morphology and function. With time, genomic changes, selectable and neutral, can accumulate to the point of leading to events of speciation.

To some, the above-mentioned achievements in our efforts to understand biology signify that crucial developmental and genomic paradigms are already quite reasonably established. This is a recognition that so much has been achieved since the beginning of the scientific revolution brought on by the discovery of DNA as an agent of heredity, that the time is ripe for efforts focused at broader levels of organization. One possible direction for this suggested program is to study the mechanisms of anatomic and functional innovation in the inclusive historic and ecologic scenarios (Jablonka & Lamb 2008, Moczek 2008, Held 2014). What we and others with similar objectives want is to work together with other pertinent disciplines to establish the relevant physical and ecological backgrounds associated with important morpho-functional transitions and, when possible, examine the extent to which the environment may have worked as a selector, or agent, of morpho-functional innovation.

Is it possible for the environment to drive permanent changes in living beings? Indeed, there is the eloquent example of polyphenisms, when distinct phenotypes arise from the same genotype in response to changes in climate and habitat, for instance, sex determination by temperature in alligators and other reptiles (Woodward and Murray, 1993) and changes of worker to queen caste by ingestion of royal jelly (Haydak 1970). There are other proposed ways through which the environment is thought to produce permanent changes in the individual, such as the Baldwin effect and genetic assimilation, which depend on the environment selecting developmental programs that are already present in the genome as part of the genetic variability associated with one given species. Thus, the Baldwin effect and genetic assimilation are fundamentally different from the more familiar emergence of mutations and further purifying selection of mutants (Crispo 2007, Hall 1992).

In truth, it is less than firmly established that the environment operates as a proximate cause of evolution, but multiple recent lines of evidence indicate that inheritance of characters from parents (heredity) is not limited to the transmission of genes from parents to offspring. Epigenetic marks on the DNA are what first come to mind in this regard, because these changes are passed along from progenitors to progeny, but there are other relevant forms of environmental character transmission such as maternal transmission of environmentally-sensitive cellular contents, learning, and cultural inheritance (Creanza & Feldman 2016, Horsthemke 2018, Whiten 2019).

Refining our views about the lives of fossil animals. Because most of the information we get from extinct animals comes from preserved hard mineralized tissues, we often turn a blind eye to opportunities for a more inclusive reconstruction of past lives based on soft tissue counterparts. There are many examples of possible inferences that we can make for the behavior of extinct animal species from even partially preserved soft tissue organs (Sansom et al. 2011, Sansom 2016). Two organs come quickly to our minds when we think about behavior: the brain and the heart. It is common to see inferences about overall intelligence, balance, visual, olfactory, and auditory acuities extracted from digital reconstructions of the brain from endocasts (Alonso et al. 2004, Zelenitsky et al. 2009, Balanoff et al. 2013, Watanabe et al. 2018) or, less often, by direct inspection of fossilized cerebral remnants (Brasier et al. 2016). These observations can provide essential detail regarding the ecological roles and behavioral as well as physiological abilities of these extinct animals. However, it is worth considering that, sometimes, brain endocasts may actually obscure, rather than illuminate functional and behavioral inferences. Chimeroid fishes, for instance, display a bizarre size discrepancy between a very large endocranial capacity and minute brain (Pradel et al. 2009), raising suspicions about the validity of cranial endocasts as surrogates for brain size in fishes (e.g. Watanabe et al. 2018).

Morphological information about hearts, or other kinds of circulatory pumps, are also highly valuable when assessing ecological niche occupation. Fishes are one example of the correlation between cardiac anatomy and niche, as it is well documented that the hearts of active predatory fishes such as tuna display robust and thickened V-shaped, pyramidal ventricles, while sedentary fishes such cod possess thin-walled, tubular ventricles (Santer et al. 1983).

Expanding a core paleontological paradigm. Until quite recently, there was little opportunity to study soft-tissue evolution with the fossil record. Indeed, for most soft-tissue organs, there was no such a thing as a fossil record. Today, however, there is good evidence that organs and tissues such as vertebrate gut and gut contents (Brown et al. 2020, Martill 1990, Wilby & Martill 1992), both arthropod and vertebrate brains (Xiaoya et al. 2015, Pradel et al. 2009), umbilical cord (Long et al. 2008), coelacanth lungs (Cupello et al. 2015), and the heart (Maldanis et al. 2016) are preserved in high quality fossils. Given this emerging data, we could in theory begin to establish ontogenies (and phylogenies) for selected, well-preserved, organs, if such taxa were preserved in high-enough numbers.

Frozen in time by rare combinations of mass mortality events and conducive chemical conditions, entire populations, including adults, juveniles, larvae and embryos, or at least some of these developmental stages, could be accessed. Some geological sites would be more appropriate to find specific ontological stages than others, but it should be possible to collect at least a fragmented view of organ development through the study of highly productive sites of exceptional preservation.

A case for cardiovascular paleontology

Chambered hearts as an evolutionary novelty. A major part of the reconstructed narrative associated to an extinct animal species must include, if possible, what it could, or could not do. We confront questions such as: "does this fossil suggests a sessile filter feeder, or a microbial mat grazer?" "Was this animal an ambush or a pursuit predator?" "Was this extinct fish species an aquatic, or an air breather?" "Was this dinosaur warm blooded, or cold blooded?" As discussed above, many of these questions have been traditionally addressed using extrapolation from surrogates, such as endocasts. However, the capabilities of animal circulatory pumps can provide primary data about animal lifestyles. Therefore, information from preserved heart tissues can directly suggest, corroborate, or reject hypotheses about the ecologic niche of an extinct animal.

Our hearts are sophisticated positive displacement pumps that employ two different types of chambers – the atrium and the ventricle – to segregate the functions of fluid inflow and outflow, respectively. In simpler terms, such a positive biological displacement pump generates flow by capturing a given amount of fluid and squeezing it through an outlet. The cardiac atrium receives the blood that returns to the heart from the veins and operates both as a variable capacity reservoir and as filler to the ventricle. In turn, the ventricle is a powerful variable stroke pump that rhythmically propels blood into the aorta, thus generating flow to the organs and tissues (Xavier-Neto et al. 2010). The structures of all vertebrate and mollusk hearts preserve this simple organization (Fig. 1).

From where did this mechanical arrangement come? Here it is appropriate to clarify that, in evolutionary terms, the chambered heart is an amazing oddity. This is because, as currently accepted, true chambered hearts are present in only two groups of animals: vertebrate and mollusks (Brusca & Brusca 2003). The origins of chambered hearts in these two highly interesting groups have been previously discussed (Xavier-Neto et al. 2007). From the standpoint of comparative anatomy, it is clear that the presence of chambered hearts in vertebrates and mollusks cannot be traced back to a common origin in a more primitive clade of animals (i.e. a homology due to the presence of a chambered heart in a common ancestor). Thus, there are arguments in favor of a completely independent evolution for vertebrate and mollusk hearts (evolutionary convergence), or in support of a parallel origin in a distant ancestor (parallel evolution; Xavier-Neto et al. 2007, fig. 2).

Regardless of whether the chambered hearts of vertebrates and mollusks are the result of convergent or parallel evolution, there is strong indication that they both evolved from peristaltic pumps not unlike our own guts, which are primitive (in the sense of basal) displacement pumps that equip most bilaterians. Peristaltic pumps push fluids through the action of sequential waves of contraction that pass along the walls of their hollow muscular structures, driving their contents forwards and backwards (reviewed in Xavier-Neto 2010, Fig. 3).

While currently there are no reasons to doubt that chambered hearts originated from peristaltic pumps, the actual sequence of events is not clear. There are at least three hypotheses for this crucial morpho-functional transition: the sequential hypothesis, the recruitment hypothesis, and the patterning hypothesis (Xavier-Neto et al. 2010). The sequential hypothesis finds support in comparative adult anatomy, while recruitment and patterning hypotheses derive from comparative developmental biology. These ideas are discussed below in condensed and simplified fashion. The reader is directed to the original materials for more comprehensive treatments (Fishman & Olson 1997, Davidson et al. 2008, Hochgreb et al. 2003, Simões-Costa et al. 2005) (Fig. 3).

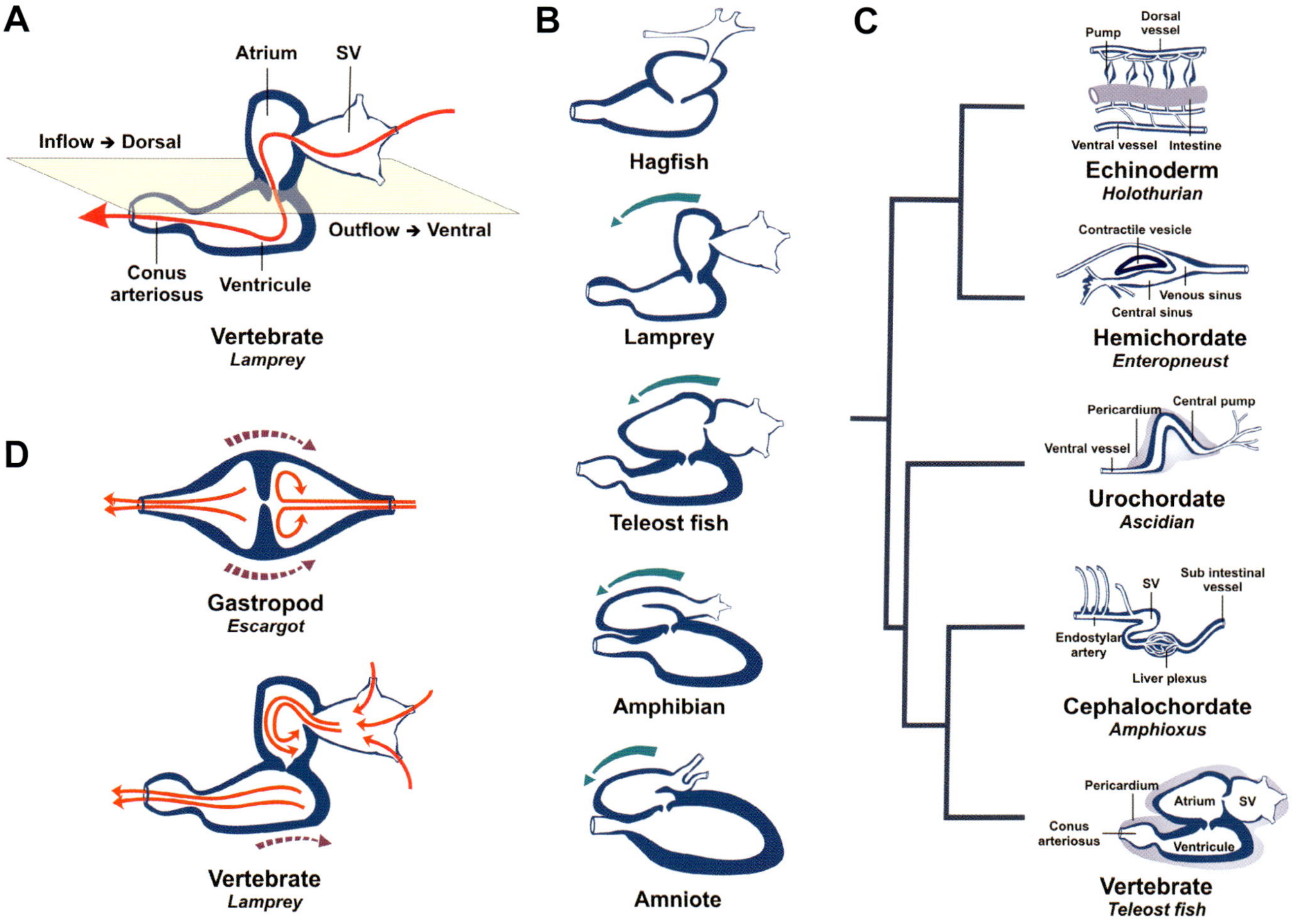

Fig. 1. The organization of chambered hearts. **A.** S-shaped, vertebrate heart. Inflow chambers are dorsal, outflow chambers, ventral. Red arrows indicate flow direction. **B.** In amphibians and amniotes, atrium and sinus venosus (SV) are displaced rostrally (green arrow), obscuring the original dorso/ventral disposition. **C.** Deuterostome circulatory pumps. The four-chambered vertebrate heart contrasts with the simpler peristaltic pumps of cephalochordates and tunicates (urochordates). Enteropneust hemichordates display a vesicle squeezing blood from a central sinus. Multiple pumps of the hemal system in holothuroid echinoderms. **D.** Among mollusks, the chambered hearts of gastropods are linear. Hemolymph enters the atrium symmetrically, reflecting away from the atrioventricular valve. Hemolymph ejection in the aorta produces ventricular recoil (dotted arrows) against the atrium, preventing efficient filling. The dorsal/ventral arrangement of inflow/outflow chambers in vertebrate hearts produces asymmetrical streams directed to the valve. Systolic ventricular recoil is away from the atrium and assists in flow. Figure and legends reproduced from Simões-Costa et al. 2005, with permission.

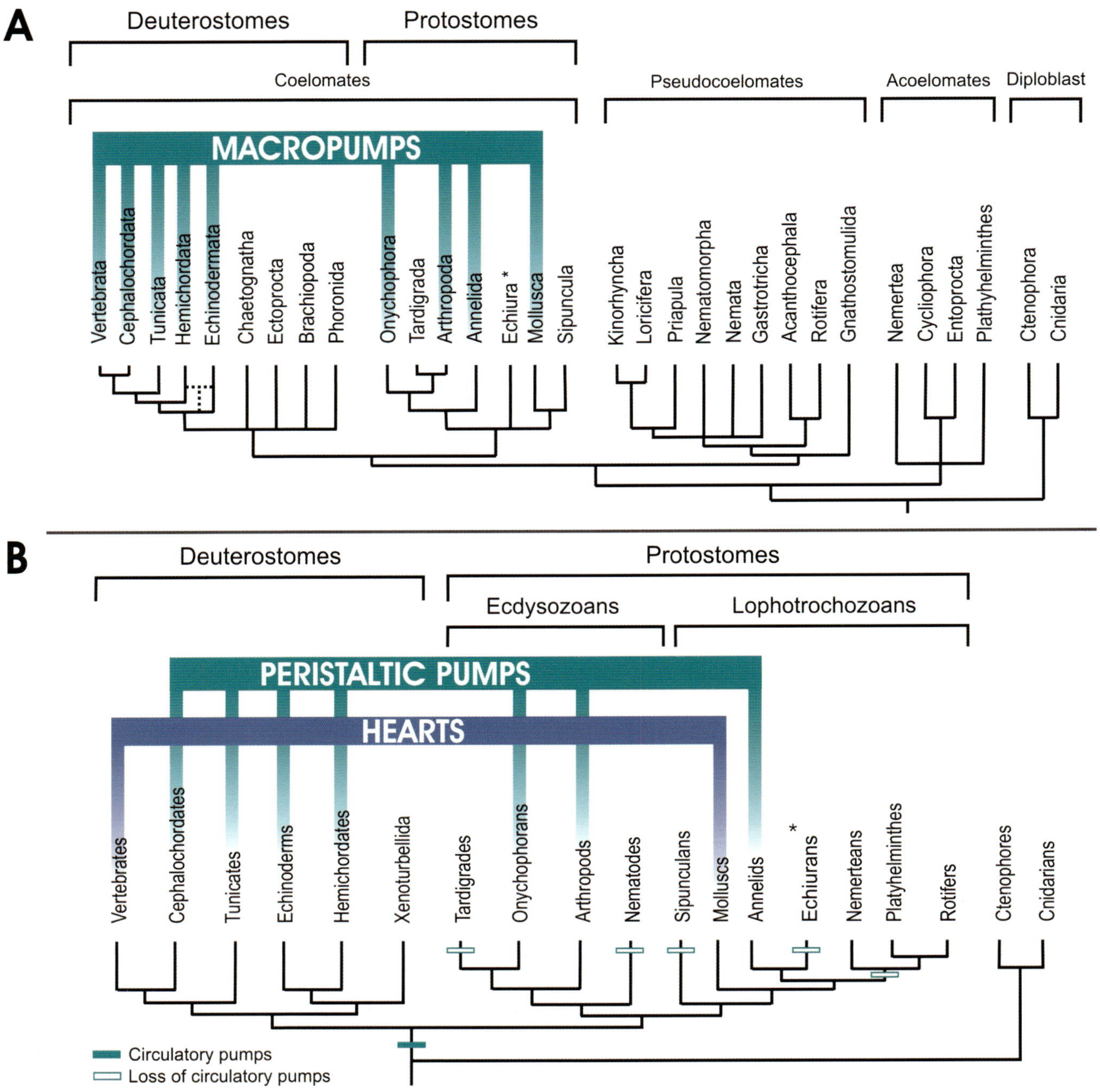

Fig. 2. The origins of animal pumping organs. A. Morphological metazoan phylogenies classify animals according to the presence of characters such as: true presence of tissues, embryonic layers (diploblastic and triploblastic), types of body symmetry (bilateral versus radial), and absence (acoelomate) or presence of a fluid-filled space derived from the blastocoel (pseudocoelomates) or from the mesoderm (coelomates). Triploblastic bilaterians are further divided into protostomes and deuterostomes. In these phylogenies, acoelomates and pseudocoelomates (mostly pumpless, heartless animals) are located at the base of bilaterians, suggesting that they are intermediates between diploblastic animals and coelomate triploblastic bilaterians (protostomes and deuterostomes), which display specialized pumping organs. B. Molecular phylogenies placed acoelomates and pseudocoelomates together with protostomes, displacing them from the basal positions they occupied in morphological phylogenies. Protostomes were divided between ecdysozoans and lophotrochozoans. Deuterostomes were reaffirmed in a natural group, but lost a few phyla to protostomes. Molecular phylogenies suggest that all bilaterian pumping organs descend from an ancestral bilaterian peristaltic pump, and that acoelomate and pseudocoelomate animals without pumping organs must have lost them secondarily. *Despite the absence of circulatory pumps in most echiurans, a specialized pumping organ is present in the family Ikedae. Figure and legends reproduced from Xavier-Neto et al. 2010, with permission.

The sequential hypothesis. The sequential hypothesis simply states that the four types of recognized cardiac chambers, the sinus venosus, the atrium, the ventricle, and the conus arteriosus arose one by one during evolution (Fishman & Olson 1997, Xavier-Neto et al. 2010). One logical consequence of this hypothesis is that the transition from peristaltic pump to chambered heart necessitated the existence of a one-chambered heart. One-chambered hearts, however, are not found among the central pumps of either extant vertebrates or mollusks. However, among arthropods, there is evidence for the evolution of something akin to a possible one-chambered heart from peristaltic vessels. During their embryonic and early juvenile periods, crustaceans display a lengthy peristaltic pump – the dorsal vessel – at their dorsal sides. The dorsal vessel is formed by the repetition of virtually the same units, in line with the segmental principle of arthropod body organization. However, in their adult stages, some decapods such as lobsters clearly depart from the peristaltic organization to follow a model of concentrated pumping function in the anterior part of their dorsal vessels (Wilkens 1999). How is this concentration anatomically expressed? In lobsters, it appears that multiple anterior segments of the dorsal vessel coalesced into a single compartment, in a process perhaps analogous to the collapsing of serial thin segments of a telescope into a single thicker unit to produce a much more robust pump. It has been suggested that this conjectural "one-chambered heart" is one of the evolutionary novelties that may have allowed the emergence of the largest extant crustaceans that we know, the Japanese spider crab (*Macrocheira kaempferi*), the American lobster (*Homarus americanus*), and the coconut crab (*Birgus latro*) (Wilkens 1999). Perhaps in the near future it will be possible to do more than speculate about the types of circulatory pumps that equipped even larger, extinct arthropods such as the sea scorpions (e.g. *Jaekelopterus rhenaniae*) the giant millipede *Arthropleura*, or the giant dragonfly *Meganeuropsis permiana* (Fig. 4).

The recruitment hypothesis. The recruitment hypothesis evolved from studies of circulatory pump development in the tunicate model species *Ciona intestinalis* (Davidson et al. 2008). The organ that moves circulation in *Ciona intestinalis* operates as a classic peristaltic pump, albeit an advanced one, displaying some features such as morphological specialization (i.e. V-shaped) and encasing into a pericardial membrane (Robb 1965). During *Ciona intestinalis* development, the pump forms from a small group of four founder cells. Normally, founder cells divide asymmetrically into rostral and caudal groups. Under the force of Fibroblast Growth Factor (FGF) signaling, only the rostral daughter cells are instructed to give rise to the lineage of pump cells, which eventually organize themselves in the 3D space to form a single rounded compartment, while caudal cells will give rise to skeletal muscles. However, if the strength of FGF signaling increases through transgenesis, even caudal cells are recruited to the pump phenotype. Significantly, in a small population of transgenic individuals, the artificially stronger FGF signal induces the formation of an additional, smaller, rounded compartment. In a few individuals, these two compartments were connected and recorded pumping fluid in concert. The experiment described above is a beautiful demonstration of how, in principle, two-chambered hearts may have evolved from a peristaltic pump (Davidson et al. 2008). Moreover, the experiment demonstrates the power of developmental biology to provide testable hypothesis concerning the feasibility of crucial transitions in morphology (Fig. 3).

The patterning hypothesis. The patterning hypothesis is based on the normal mechanisms of vertebrate chambered-heart development and on the power displayed by retinoic acid signaling to modify the chamber composition of the developing vertebrate heart (Fig. 3). Indeed, by increasing the force of retinoid signaling, it is possible to induce the formation of hearts almost completely dominated by atrial chambers. Conversely, inhibiting retinoic acid leads to hearts formed essentially by ventricles (Moss et al. 1998, Rosenthal & Xavier-Neto 2000, Xavier-Neto et al. 2001, Hochgreb et al. 2003). The ability of custom retinoid signals to produce these atrialized, or ventriculized hearts is at the core of the normal developmental processes employed during cardiac chamber development. During normal vertebrate development, retinoic signaling is localized to the posterior half of the cardiac precursor domain. This well-defined location is not a fortuitous coincidence, because posterior precursors are exactly the ones destined to form the atrial chambers (which occupy a posterior position in early embryonic hearts, but curiously migrate to the cardiac top later on). Likewise, anterior cardiac precursors are distanced from retinoic acid signaling and, again, far from being a coincidence, anterior cardiac precursors will form the ventricular chambers, which occupy an anterior position in early embryonic hearts (Hochgreb et al. 2003). During early vertebrate cardiac development, both anterior and posterior cardiac precursors first organize themselves as a primordial cardiac tube (more correctly a cardiac trough, because its dorsal part is formed by the anterior gut wall) (De Ruiter et al. 1992).

The primitive cardiac trough aligns to the cranio-caudal axis, with ventricular precursors at the anterior pole and atrial precursors at the posterior pole. At this early stage of development, the primordial vertebrate pump works as a peristaltic device. This tiny peristaltic pump provides for all needs of the vertebrate embryo during a stage of exponential growth. Because of its cranio-caudal patterning, the vertebrate peristaltic pump is adjusted to propel blood from veins to the arteries. This is due to a decreasing gradient of excitability from its posterior (later atrial) to its anterior poles (later ventricular) (i.e. a higher frequency of excitation and contraction in the posterior pole, the atria, dominates all ventricular pacemakers, resulting in a net transfer of fluid to the arteries) (De Jong et al. 1992). After a series of complex tube twisting and differential growth of its myocytes, the primordial peristaltic vertebrate pump becomes a chambered heart. These developmental changes represent the basic script in all known vertebrates, and beautifully illustrate the classic idea that the study of ontogenesis does provide important clues about the origins of phylogenetically important morpho-functional transitions.

In summary, all that was learned from early vertebrate cardiac development suggests that vertebrate chambered hearts evolved directly from a peristaltic pump precursor organ without passing through a necessary one-chambered heart stage. Instead, what seems to be the norm for vertebrates is a mechanism whereby an apparent homogeneous field of cardiac precursors is first patterned in the cranio-caudal axis to establish an intermediate state pump (the patterned primordial peristaltic cardiac trough). This intermediate pump is later reshaped into an advanced pumping organ with specific chambers dedicated to inflow (atria) and outflow (ventricle) (Xavier-Neto et al. 2010) (Fig. 3).

The role of paleontology in illuminating the origin and evolution of chambered hearts. The sequential, recruitment, and patterning hypotheses for the evolution of chambered hearts illustrate well the powers of comparative anatomy and comparative developmental biology for generating scenarios with explanatory potential for important morpho-functional transitions. However, it is clear that these scenarios do not provide direct evidence for what has happened historically, but rather for what may have happened.

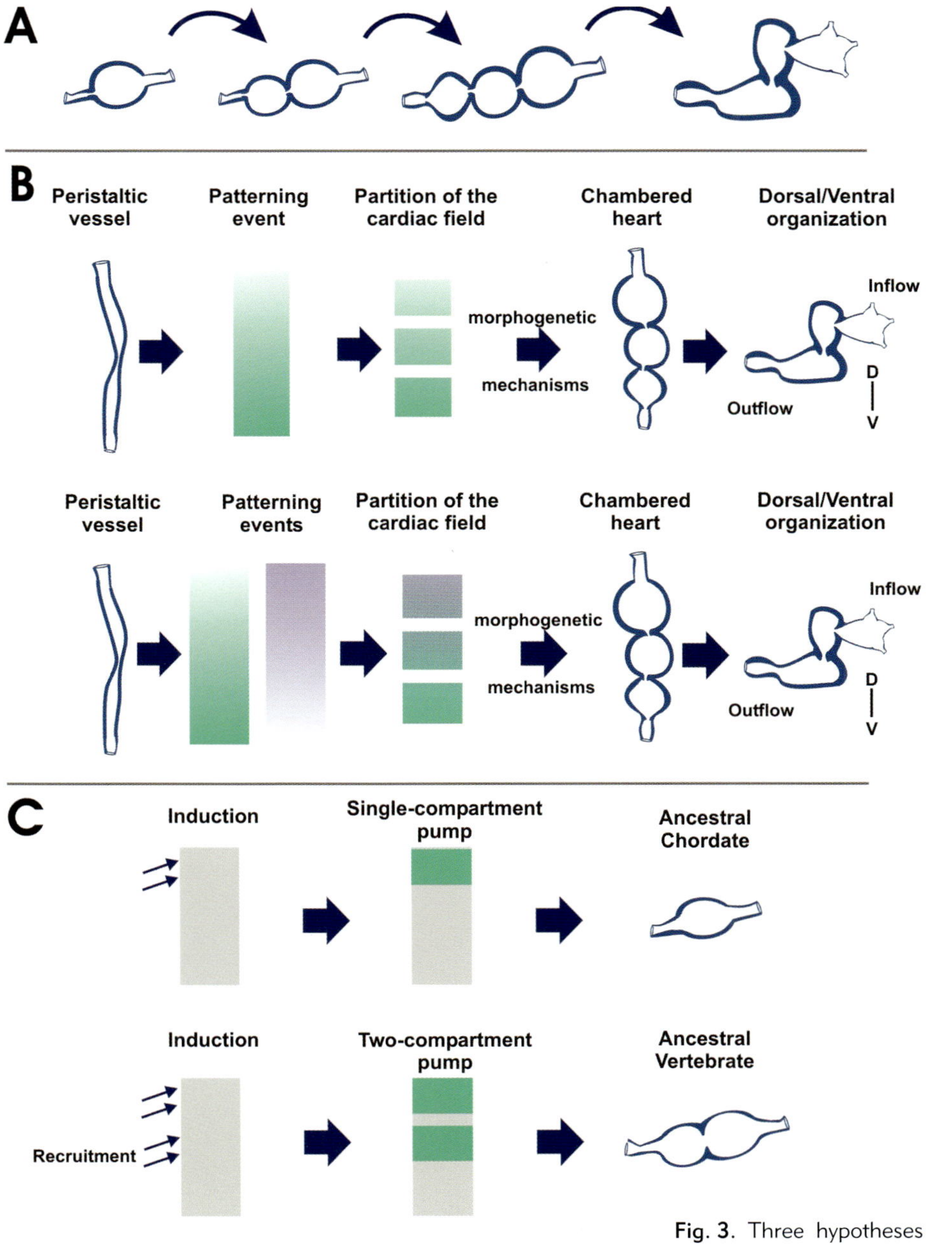

Fig. 3. Three hypotheses for the evolution of vertebrate cardiac chambers. **A.** The sequential hypothesis: cardiac chambers arose successively on top of a single-chambered heart from the ancestral chordate. **B.** The recruitment hypothesis: cardiac chambers originated when an additional progenitor domain was recruited to a pump fate, creating a two-compartment peristaltic pump that was later molded by morphogenetic and other mechanisms into inflow and outflow chambers. **C.** The patterning hypothesis: two or more cardiac chambers appeared simultaneously when patterning mechanism(s) divided the progenitor area of a peristaltic pump into two or more domains that were subsequently fashioned into inflow and outflow cardiac chambers by morphogenetic and other mechanisms. Figure and legends reproduced from Xavier-Neto et al. 2010, with permission.

In the end, only paleontology can provide the primary data to ascertain the true course of the few tractable ontogenetic and phylogenetic problems that we painfully identify and select for investigation. In this sense, one of the best-known examples, a critical problem in cardiac evolution, and a longstanding problem in teleost evolution since the nineteenth century, concerns the present state of the outflow tract of basal actinopterygian hearts. Derived teleosts display a minimalist arrangement of a single valve in their outflow tracts. By contrast, basal actinopterygian hearts possess an outflow decorated with a highly complicated set of one-way valves. The contrast is impressive. In one extreme there are hearts displaying dozens of outflow tract valves, as in *Polypterus*, a primitive ray-finned fish (i.e. actinopterygian, Duran et al. 2014), while at the other extreme there are hearts showing a single valve row in the cardiac outflow tract, as in the derived actinopterygian model species *Danio rerio* (the zebrafish). This large cardiovascular anatomy spectrum constitutes an evolutionary conundrum noted by the celebrated German anatomist Karl Gegenbauer, or even before, by other earlier anatomists (see Maldanis 2016, and references therein).

Cardiac outflow tract evolution in teleosts is important because it provides evidence for the counterintuitive realization that younger species often display morphologies that are substantially less sophisticated than those displayed by earlier-diverging species. How might these drastic changes have evolved? Was the outflow tract simplification in teleosts the result of a radical developmental event of macroevolutionary scale, or yet another garden-variety example of phyletic gradualism? Furthermore, did the changes associated with outflow tract simplification arise because of genetic drift (i.e.

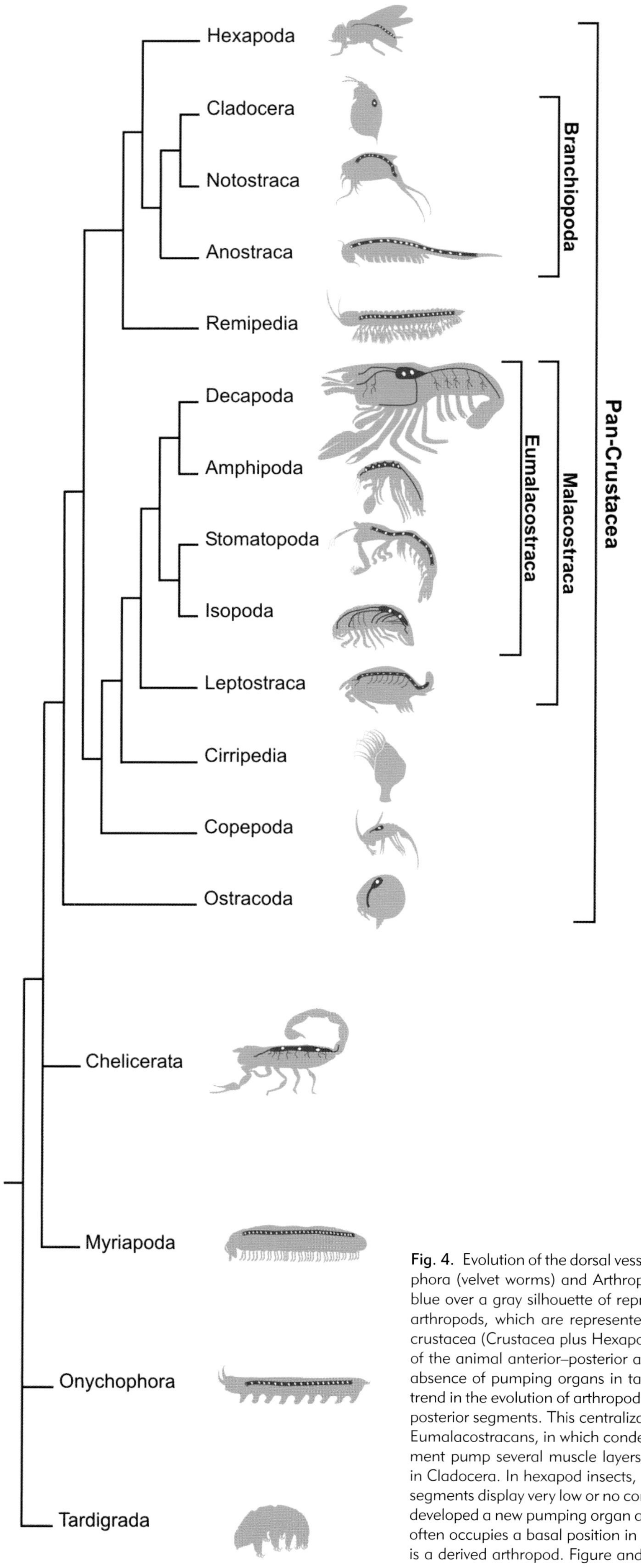

neutral evolution), or did they reflect a progressive adjustment in a selective environment (discussed in Maldanis et al. 2016)? If so, what were the ecologic and environmental selectors, and what were the mechanical reasons associated? Until recently, we had no idea of how such radical anatomical changes may have evolved, chiefly because we were missing suitable intermediate morphologies from a very large period of time. In summary, it is clear that the answers to the problems of cardiac chamber origins and cardiac outflow tract evolution in actinopterygians lie with the relevant fossils, because they can provide us with two invaluable clues: evidence for transitional morphologies and for the time intervals between them (Maldanis et al. 2016).

The paleontological context of the first-known fossilized heart. The pursuit of the origins of important morpho-functional transitions outlined above requires that we identify examples of tractable investigative scenarios. Here, we have argued that evolution of chambered hearts is a worthy case. Why is this so? The answer is simple: because we can find fossilized hearts.

The first indication that superabundant vertebrate fossil species can give us clues about cardiac evolution came from the famous Cretaceous site of Brazil, the Santana Formation in the Araripe Basin (Maisey 1991, Maldanis et al. 2016). The Araripe Basin contains one of the larger outcrop areas of Cretaceous rocks among the intracratonic basins of the Brazilian northeast (cratons are very old pieces of the earth's lithosphere that survived all the cycles of destruction and reformation by the merging and rifting of the continents, from their formation until now). The sedimentary area of the Araripe Basin spreads over 12 200 km^2, as a result of a tectonic event that led to formation of the Equatorial and Eastern Atlantic opening. The Araripe basement is composed of magmatic and metamorphic rocks, filled with clastic and chemical rocks (Carvalho et al. 2000). During Early Cretaceous times, there were multiple ecologic and environmental changes in the Southern hemisphere, mostly due to the rupture of Gondwanaland, which allowed the emergence of new ecological spaces (Carvalho et al. 2000). These geological events are registered in the interior of Northeastern region of Brazil, especially in the Araripe Basin, in two distinct Konservat Lagerstätten (sedimentary deposits containing numerous specimens of fossils with exceptional preservation): the Crato and Romualdo fossiliferous deposits (Selden & Nudds 2012).

The Crato Lagerstätten is in the Crato Member of the Santana Formation, which presents exquisitely preserved fossils that are ideal for detailed morphological analysis and reconstructions of past morphologies. The deposits record evidence of a mildly saline, evaporitic environment, displaying freshwater fossils and terrestrial species that presumably died near the shores of multiple interconnected lakes (Maisey 1991). Why is the Crato Member such a fine preservational space? Recent petrographic and chemical studies conducted by Catto (2015) and Catto et al. (2016) on the carbonates of the Crato Member (Aptian, Lower Cretaceous) displayed a population of autotrophic prokaryotes such as filamentous cyanobacteria, coccoids, and *Spirulina*. These bacteria are found inside microbial mats and are associated with well-preserved extracellular polymeric substance (EPS), and also preserved as carbonate, consistent with the suggested role of EPS in carbonate precipitation. Fossils from

Fig. 4. Evolution of the dorsal vessel in Panarthropoda, which includes Tardigrada (water bears), Onychophora (velvet worms) and Arthropoda. Dorsal vessels and some major efferent vessels are depicted in blue over a gray silhouette of representative species. Tardigrades and onychophorans are outgroups to arthropods, which are represented by an unresolved trichotomy of Chelicerata, Myriapoda, and Pancrustacea (Crustacea plus Hexapoda). The onychophoran dorsal vessel, a tubular organ spanning most of the animal anterior–posterior axis, is thought to be similar to the ancestral arthropod condition. The absence of pumping organs in tardigrades is a potential consequence of small body sizes. One major trend in the evolution of arthropod pumping organs was the concentration of pumping work in anterior or posterior segments. This centralization may be a consequence of the fusion of body segments as seen in Eumalacostracans, in which condensation of anterior body segments created a powerful single compartment pump several muscle layers thick (decapods), or secondary to a loss of body segments, as seen in Cladocera. In hexapod insects, pumping work was concentrated in posterior segments, while anterior segments display very low or no contractility. Sessile cirripeds lost the dorsal vessel, but the class Thoracica developed a new pumping organ actuated by skeletal muscles. Remipedia, a primitive-looking arthropod, often occupies a basal position in morphological phylogenies, but recent molecular evidence suggests it is a derived arthropod. Figure and legends reproduced from Xavier-Neto et al. 2010, with permission.

the Crato Member show micro- to nanospherical, rhombohedric, and acicular geometries. This diminutive quality of crystals associated with fossils from the Crato Member is probably one of the reasons explaining the exceptionally preserved fossils of this lithostratigraphic unit, much as the small size of grains in a photographic emulsion is responsible for its high definition. The Romualdo Lagerstätten comprises the fossils famously found as carbonate concretions in the Romualdo Member of the Santana Formation. The age of these fossils also seems to be Aptian, and they have been related to a great Aptian transgression as suggested by Arai (2014) and to further intermittent entrance of water from the Central Atlantic. These transgressive events allowed the deposition of evaporites, in a context of a restricted saltwater environment. While these observations are more or less supported by direct evidence, there are far more proposed scenarios than effectively tested hypotheses for the actual paleoenvironments associated with Romualdo concretions and fossils. Martill suggested that fossils from the Romualdo Member occurred in a context of either a lagoon environment, or as an area with restricted connections to waters of normal marine salinity (Martill 1993). On the other hand, Maisey (1991) suggested that the environment was a shallow embayment nearside a coastal region, in which episodic marine incursions allowed the mixing of salt waters and less saline waters (Fig. 5).

The literature dealing with the paleoenvironments associated with concretions from the Romualdo Member is prolific, full of sweeping syntheses and of contradictory hypotheses. No one has yet presented a strong case for any specific environment proposed, perhaps because there was never such a thing as one uniform scenario across the heterogeneous Araripe Basin at the geologic times represented in the Romualdo strata. Because it is impossible to project the enormous diversity of environments of the Araripe through time and space into a single Romualdo paleoenvironment, perhaps it is better to start by stating what it was not. The depositional environment associated with the Romualdo Member was not a stratified fresh water lake. It was neither a hypersaline body of water, nor a frankly marine environment. This is because the Romualdo Member is very poor on marine markers such as the dinoflagellate *Subtilisphaera* and echinoids (echinoderms) which, excepting some brackish species, are strictly marine (Beurlen 1966, Brusca & Brusca 2003).

Recent and detailed work on the eastern side of the Araripe Basin (Serra do Mãozinha) has been shining light on many aspects of the depositional environments associated with the Romualdo Member. The Romualdo succession has been interpreted as the deposits of two cycles of marine transgression and regression (Custódio et al. 2018). The predominance of ostracods, small bivalves, and fewer gastropods is associated with conditions ranging from anoxia to dysoxia (Maisey 1991). Quite interestingly, the study of Custódio et al. (2018) suggests that high-energy events are overrepresented in the Romualdo Member. In effect, it appears that much less sediment was deposited during calm, low-energy conditions than during disruptive episodes of transgression and regression. This reasoning, if correct, may indicate that the various peaceful ecological scenarios that we traditionally imagined as associated to the Romualdo Member may not be factual (Fig. 5). Indeed, there is growing evidence that the waters in the Araripe basin were connected to an internal extension of the central Atlantic for at least one single, long-lasting interval. There are clues that, at this particular time, the Araripe was invaded by the sea through a North-West-South-East axis, sequentially involving the states of Maranhão, Piauí, and Ceará (Arai 2014).

Going back to our goal of finding direct evidence for relevant vertebrate cardiac morpho-functional transitions in the fossil record, we argued for the study of well-known and prolific species as an appropriate strategy to find fossil hearts. In this regard, what better resources to explore in the Araripe basin than *Rhacolepis*, a basal teleost fish from the Pachyrhizodontoidei clade (related to Elopomorpha) (Fig. 5D), the most common fossil fish from the Romualdo Member, Santana Formation?

Fig. 5. A. A modern lagoon environment close to the mouth of river Piranji, in Eastern Ceará, Brazil 4°23'12" S 37°50'14" W. Drone picture provided by Prof. Odorico de Moraes Filho, Universidade Federal do Ceará. The context is reminiscent of the transitional environment suggested for the Romualdo region, 125–113 million years ago, perhaps without the lush vegetation represented. **B.** Geologic cross section displaying the stratigraphic profile at a Gypsum Mine in the Araripe Basin, in Nova Olinda County, Ceará, Brazil. The Romualdo Member is the litostratigraphic unit above the white Gypsum layers at the bottom. Two humans as scale. **C.** Carbonatic nodules with fossil fishes from the Romualdo Member, with a human as scale. Locality of Pedra Branca, Nova Olinda County, Ceará, Brazil. **D.** Exquisitely preserved *Rhacolepis buccalis* fossil totally embedded in the concretion matrix.

Rhacolepis

Rhacolepis is preserved in three dimensions, usually displaying a fusiform body enveloped by a robust-looking mesh of cycloid scales (Voltani 2016) (Fig. 5D). The fine preservation of the *Rhacolepis* exterior frequently tricks the examiner into hopes of similar internal preservation. However, there is no such correlation. When opened, the best-looking fossils more often than not display only a few scattered vertebrae in the midst of amorphous crystalline materials. In fact, our experience indicates that internal preservation is linked not to external appearances, but to one of the three lithologies described by Maisey for the Romualdo Member concretions (Maisey 1991). In Maisey's view, the variance associated with Romualdo Member concretions could be tentatively described with at least three categories loosely associated with fossil provenance: Santana concretions, Jardim concretions, and Missão Velha concretions. Santana concretions are ovoid and elongated, formed by a poorly laminated, coarsely granular, beige limestone matrix essentially devoid of clay and pyrite (Maisey 1991). In the original description, Santana concretions do not harbor *Rhacolepis* fossils, a notion that was recently supported by controlled excavations of the Romualdo Member southwest of the city of Santana do Cariri (Faraa et al. 2005). The most promising *Rhacolepis* fossils as far as internal preservation is concerned are those classified by Maisey as typical from the Jardim lithology. The *Rhacolepis* fossils from the Jardim lithology range from those laterally compressed, or platy, to three-dimensional forms. The latter include fusiform types, highly bloated fish, and some specimens with a curious triangular cross section. As originally described,

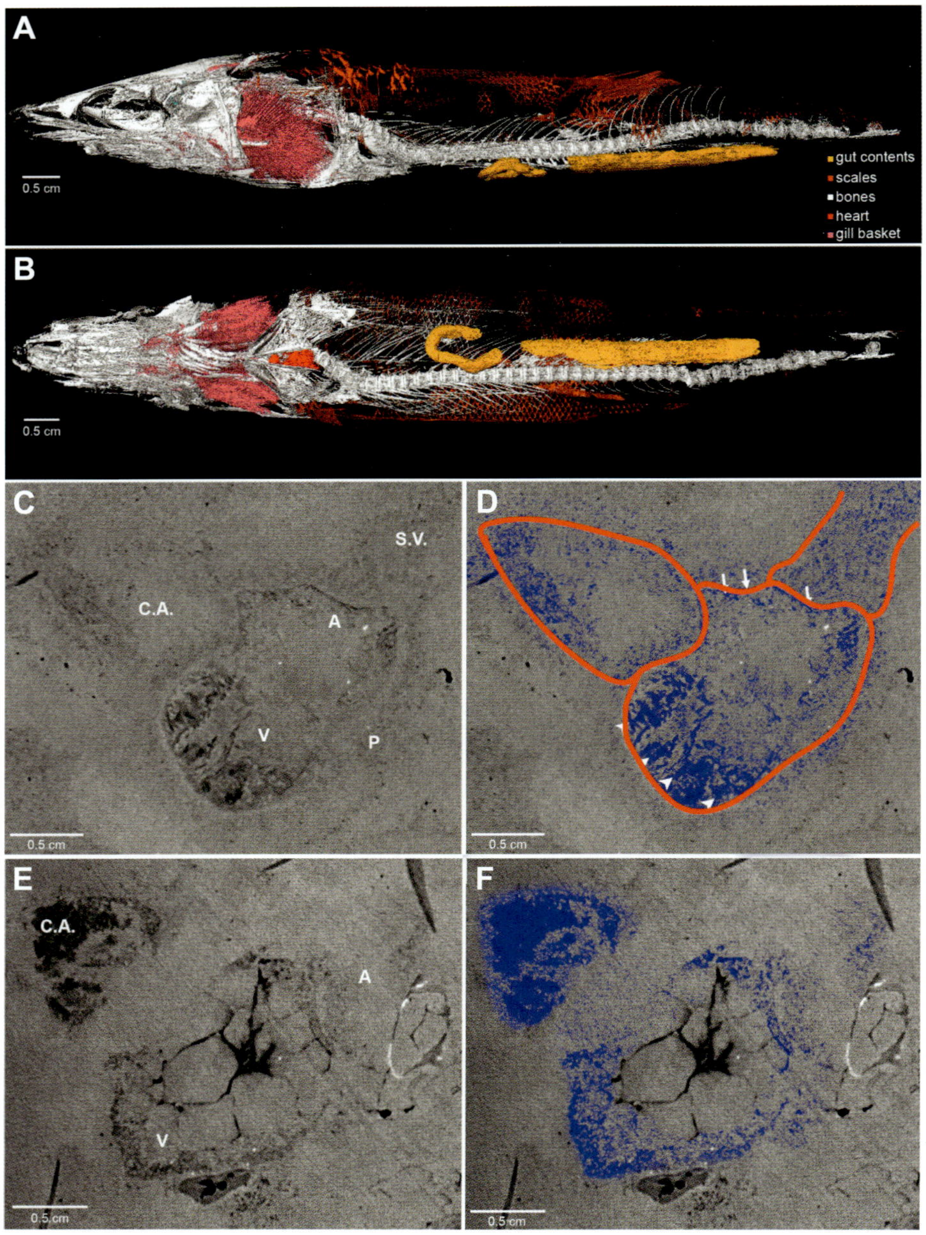

Fig. 6. Phase contrast synchrotron microtomography of teleost fossil hearts. 3D reconstructions. **A.** Left lateral view. **B.** Ventral view. **C–F.** Sagittal sections of two specimens. Blue masks in **D** and **F** highlight fossil cardiac chambers and pericardium in the two specimens. Note that thin trabeculae are associated to the atrium (arrows) and that thick trabeculae are typical of the ventricle (arrowheads). Abbreviations: A, atrium; C.A., conus arteriosus; P, pericardium; S.V., sinus venosus; V, ventricle. Figure and legends reproduced from Maldanis et al. 2016, with permission.

the fossil matrix is well laminated and contains observable amounts of clay. Iron oxides (oxihydroxides) are common and pyrite is present in minute quantities, if at all, in the matrix. Recrystallized calcite is often found inside the fish associated with large body cavities, which may have been produced by necrosis and gas production. Septarian crack formations are also common in these fossils (Maisey 1991). *Rhacolepis* fossils from the Missão Velha lithology are perhaps the most beautiful externally, displaying finely preserved scales. The matrix is typically dark green and abundant, so that relatively small fossils are associated with brick-size rocks. Importantly, the pyrite content of Missão Velha fossils is high and they are rich in clay (Maisey 1991). However, *Rhacolepis* fossils from the Missão Velha lithology often disappoint the examiner hoping for good internal preservation in specimens with finely preserved scales, as they frequently contain very few conserved internal features among large amounts of chalky materials. In sum, from the perspective of internal conservation, *Rhacolepis* fossils with a Jardim lithology (Fig. 5D) are the ones to be tried first.

The significance of concretion lithologies for the preservation of *Rhacolepis* structures. Maisey's 1991 description of the three Romualdo Member lithologies spanning West to East region of the Araripe basin from Santana do Cariri, Jardim and Missão Velha represented a useful overview for subsequent studies on the mechanisms of fossilization from this rich horizon. At that time, it was already clear that the three lithologies represented only the most common fossils arising from a given locality across the Occidental-Oriental axis of the Araripe basin, rather than the only possible fossilization mode there. In effect, it is not difficult to find the typical pyrite-poor Jardim fossils in the municipality of Jardim less than 10 km apart from fossils that reek intensely of petrol in the Romualdo Member. That such different fossilization environments can coexist in the same acknowledged stratum is, no doubt, a reflection of the four-dimensional nature of the diverse taphonomic and diagenetic processes that acted upon the remains of fishes from the region in a given time period.

Enter the stone heart. The discovery of the fossilized heart was largely the product of a developmental biology program initiated in the late 1990's (Moss et al. 1998), when one of us (J.X.N.) established that signaling by the morphogen retinoic acid was key to giving rise to atrial, rather than to ventricular, cardiac chambers in amniotes (Hochgreb et al. 2003). From then on, we set out to inquire whether retinoid signaling was a potential ancestral mechanism of chamber formation in vertebrates (Simões-Costa et al. 2005) and, if so, whether its operations could suggest how chambered hearts evolved from peristaltic pumps. Ideally, this was a matter to be decided by paleontology, by close examination of the fossil record of hearts, more precisely with a focus on samples from Cambrian, or Pre-Cambrian sites of exceptional preservation, perhaps such as the Burgess Shale in Canada 520–512 million years ago (Ma), or the Doushantuo Formation in China (635–551 Ma).

Regrettably, there was no such a thing as a fossil record for chambered hearts. Nonetheless, rather than a completely void area, the field of fossil hearts had a few highly interesting non-conclusive findings and at least a high-profile disappointment in Willo, the dinosaur. Willo was a *Thescelosaurus* and its bones were found preserved in sandstone with a presumptive cardiac remnant lying in a very provocative region, i.e. protected by the pectoral girdle and behind the sternum (Fischer et al. 2000), which is the typical position of a bona fide amniote heart (Kardong 2009). The 3D analysis of Willo's putative heart revealed some fascinating similarities with a vertebrate heart, prompting the authors to suggest that the dinosaur heart was perhaps closest in structure to an avian than to a lizard's heart (Fischer et al. 2000). These conclusions were received with strong skepticism, chiefly because sandstone is not a documented vehicle for large masses of soft tissue preservation outside the protective influence of bone structure (Rowe et al. 2000). Indeed, when reassessed several years later, Willo's putative heart appeared to be composed of cell-like material of vegetal, rather than of animal, origin, suggesting that Willo's putative heart was actually an iron-rich concretion, perhaps formed by sand aggregates and some organic material under water (Cleland et al. 2011).

Given the state of affairs with the cardiac fossil record, in 2005 we were convinced that it was appropriate to take a few steps back from the initial goal to find clues about the origins of chambers in Cambrian and Pre-Cambrian fossils and instead try to find evidence that hearts fossilize. The exquisitely preserved fossils from the Santana formation proved to be more than equal to this task. After a brief stint with the magnificent fish fossils of the Crato Member, we quickly concentrated on selecting suitable specimens of *Rhacolepis* for invasive analyses using classic acid preparations (Maisey 1991). In a stroke of luck, we found a suitable heart candidate (yet to be described) in the second specimen analyzed. The distinguishing feature that called our attention was a cone-shaped structure lying squeezed between the bones of the pectoral girdle. Further acid dissolution revealed the outlines of a complete fish heart, from its typical conus arteriosus anterior (the cone-shaped structure that drew our attention), to a succession of chambers such as ventricle, atrium, sinus venosus and Cuvier ducts. We spent the next 11 years in search of further fossilized hearts and in an investigation of the taphonomic conditions associated with cardiac fossilization.

The unfortunate experience with the putative Willo heart told us that we needed to amass unequivocal evidence for cardiac fossilization before publication of our original findings. For that, we reasoned that we would require multiple fossil hearts, preferably displayed through non-invasive methods. At first, we tried conventional computed tomography (CT). However, CT proved of limited value for direct visualization of cardiac remains. In our hands, CT was only useful for selecting specimens with

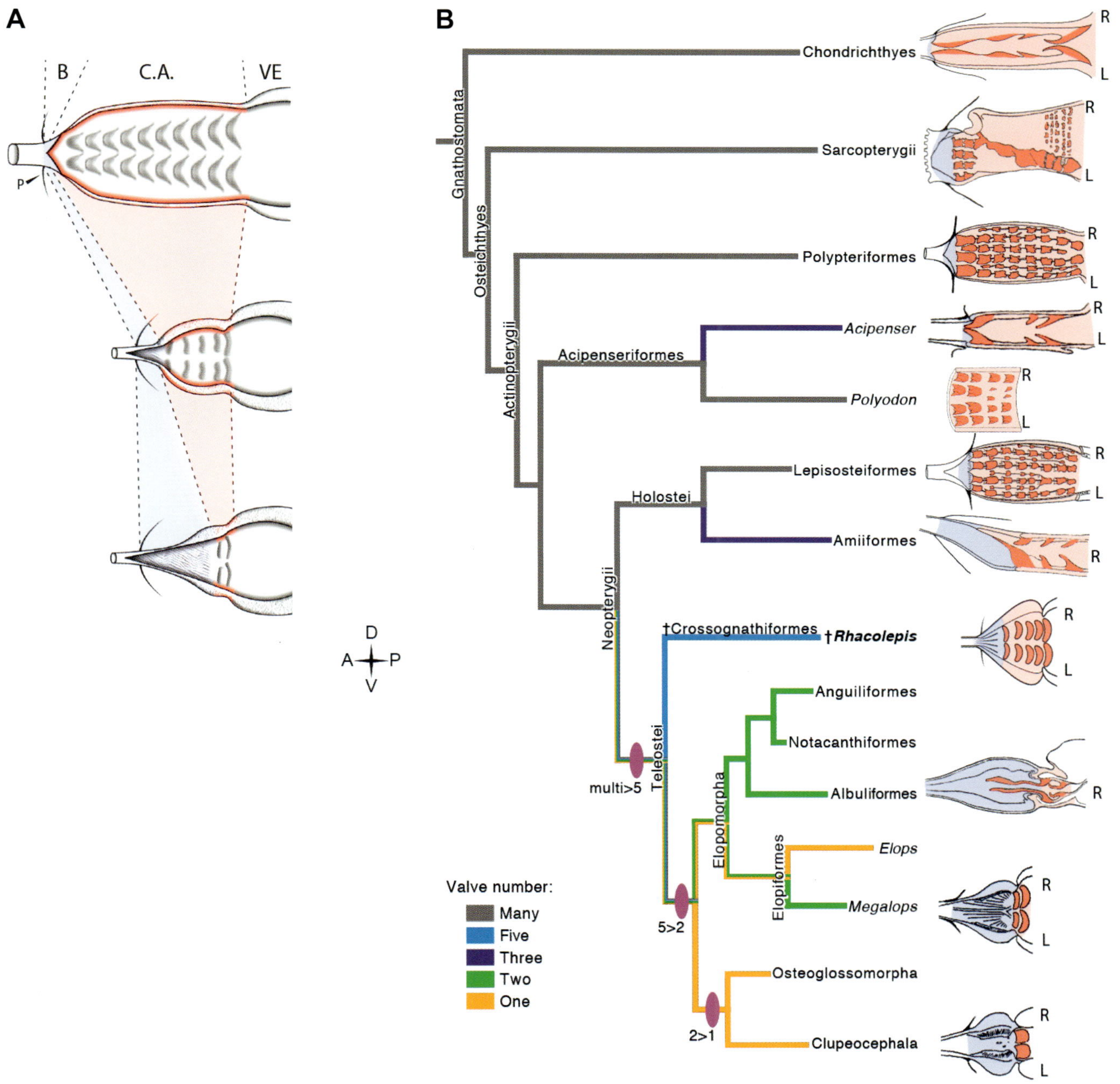

Fig. 7. The *Rhacolepis buccalis* conus arteriosus is morphologically intermediate in actinopterygian cardiac outflow tract evolution. **A.** Hypothetical transition from a character state composed of an array of multiple valve rows in the conus arteriosus of basal actinopterygians, such as Polypteriformes (top), to a derived state characterized by the dominance of the valveless bulbus arteriosus in living teleosts (here represented by a generalized elopomorph at the bottom), through an intermediate state represented in the conus arteriosus of fossilized *R. buccalis* hearts (middle). Anterior to left. **B.** Cladogram depicting phylogenetic relationships among early and derived gnathostomes and their corresponding morphologies of the cardiac outflow region. Drawings represent either the inner sides of right (R) and left (L) counterparts, or only the inner right side of the cardiac outflow tract. Blue and pink coloring highlight, respectively, bulbus and conus arteriosus (and respective valves) in extant species. A parsimony ancestral character state reconstruction was performed for the number of conal valves, following the color code in terminals. Genera illustrating the conal condition in each branch are: *Squalus* for Chondrichthyes; *Neoceratodus* for Sarcopterygii; *Polypterus* for Polypteriformes; *Lepisosteus* for Lepisosteiformes; *Amia* for Amiiformes; *Pterothrissus* for Albuliformes; *Gadus* for Clupeocephala. Abbreviations: B, bulbus; C.A., conus arteriosus; L, left side; P, pericardium; R, right side; VE, ventricle. Figure and legends reproduced from Maldanis et al. 2016, with permission.

a better likelihood of displaying internal conservation. This is probably due to the small radiographic contrast between fossil and matrix and to the limited resolution of conventional CT.

High quality, non-invasive data displaying in situ *Rhacolepis* hearts were eventually obtained through collaboration with Vincent Fernandez and Paul Tafforeau from the ESRF in Grenoble, France. Using phase-contrast X-ray synchrotron microtomography, we showed hearts from three *Rhacolepis* specimens which, collectively, revealed convincing evidence that exceptional cardiac preservation is possible (Fig. 6).

What we learned from the first-known fossilized heart. The fossil heart of *Rhacolepis* highlights some important concepts. The most fundamental concept is easy enough to understand: today it is perfectly possible, but not necessarily simple, to search for the origins of vertebrate organs using a combination of paleontology, developmental biology and evolutionary approaches. Moreover, finding a heart for *Rhacolepis* helped to refine our views about the life of this iconic Cretaceous species.

The place of *Rhacolepis buccalis* in the trophic chain. The *Rhacolepis* heart is of the so-called saccular nature, intermediate between the athletic hearts of vigorous predators, which display very thick ventricles and typical pyramidal forms, and the hearts of sedentary fishes, which exhibit thin ventricles shaped as elongated tubes (Santer 1983). The fossilized *Rhacolepis* heart's intermediate, saccular morphology was found to be in agreement with the ecological role and position in the trophic chain as hypothesized twenty five years ago, when Maisey and de Carvalho (1995) used acid preparations to investigate gut contents in fossils and found within *Rhacolepis* juveniles substantial numbers of the shrimp *Paleomattea deliciosa* – described in holotype from these gut contents (Maisey & de Carvalho 1995). Incidentally, there are few species so aptly named as the sergestid shrimp *Paleomattea deliciosa*. This small shrimp was probably a favorite delicacy (*mattea* in Latin) for *R. buccalis*, who ate it with voracity, and perhaps gusto, hence the name "deliciosa" (delicious, in Portuguese). Curiously, and unbeknownst to Maisey and de Carvalho (1995) at the time, mattea also stands for "a gift of God" (more specifically from Yahweh) in Greek, so, after all, perhaps *Paleomattea deliciosa* was indeed a God-given delicacy for *Rhacolepis buccalis*, who devoured it to the point of filling its entire gut, as revealed by inspection of preserved acid-prepared gut contents (Maisey & de Carvalho 1995), or by three-dimensional reconstitution of phase contrast synchrotron radiation microtomographic sections (unpublished). Therefore, the fossil heart of *Rhacolepis* supports the ecological role of this species as mid-level in the trophic chain Maisey 1991, Maisey & de Carvalho 1995).

The heart of *Rhacolepis* illuminates the process of outflow tract simplification in teleosts. Finding a fossilized heart was fortunate enough, but in a second strike of luck, we found that *Rhacolepis* hearts were exquisitely preserved and allowed a detailed examination of their anatomic structure, including a highly detailed depiction of the valvar apparatus associated with its outflow tract. The truncus arteriosus of *Rhacolepis* contains five rows of valves, while more basal actinopterygians such as *Polypterus* display nine valve rows, and higher actinopterygians sport only one valve. Clearly, the valvar apparatus of *R. buccalis* represents a long-lost intermediate morphology, a snapshot of evolution in the making taken at the Aptian/Albian border, 113–119 Ma in the Cretaceous.

With the long-lost morphology of the *R. buccalis* heart recovered, it was possible to tentatively answer at least one of the questions raised. Was the outflow tract simplification in teleosts the result of a more punctuational macroevolutionary event, or the outcome of phyletic gradualism? The provisional answer is that the scenario is more compatible with phyletic gradualism. This is because there were at least three steps of valvar simplification in teleosts: multiple valve rows to five (e.g. from *Polypterus* to *R. buccalis*), five valve rows to two (from *R. buccalis* to living elopomorphs), and two valve rows to one single valve row retained in all other teleosts (Maldanis et al. 2016).

Because there are basal extant actinopterygian species still living with morphologies that changed very little for millions of years when compared to their fossils, it was possible for us to suggest that valvar reduction in actinopterygians was more likely the result of gradual phyletic changes, than the result of a punctuational event (Fig. 7).

In a review of our findings, Long (2016) warns us that we have but one data point, meaning that we have the very first cardiac fossil and, therefore, that it may be premature to exclude any hypothesis. The point is well taken. However, our reasoning was that, as stated before, most of the morphologic transitions in the actinopterygian cardiac outflow are documented in very old animals that are still living and, therefore, are accessible for detailed study of their organs (our interpretation assumes that cardiac anatomy in these animals did not change much). Long's argument would thus be particularly relevant if, sometime in the future, we or others could find cardiac fossils from extant derived teleosts displaying the ancestral multivalvar arrangement, rather than their current derived, monovalvar, condition (e.g. cypriniformes with multiple valves rows in the conus arteriosus). This is an interesting proposition, which illustrates how far we are from clarifying many relevant patterns of evolution. In any event, it is great that we live in a fascinating time in which we can pose these ambitious questions that may potentially be answered using a paleontological perspective.

Acknowledgements

The authors are indebted to John Denton for editing the text. We thank Klaus Hartfelder and Marcos Simões Costa for critical reading, as well as Odorico de Moraes Filho for providing us with a drone picture of a modern lagoon environment. This work was supported by grants from CNPq to J.X.N. and I.S.C. and from FUNCAP to J.X.N.

Literature cited

Allison, P. A. & D. E. G. Briggs. 1991. Taphonomy of non-mineralized tissues. – Pp. 25–69 in: Allison, P. A. & D. E. G. Briggs (eds.). Taphonomy: Releasing the data locked in the fossil record: New York, Plenum Press.

Allison, P. A. & D. E. G. Briggs, 1993. Exceptional fossil record: Distribution of soft-tissue preservation through the Phanerozoic. Geology 21, 527–530.

Alonso, P. D., A. C. Milner, R. A. Ketcham, M. J. Cookson & T. B. Rowe. 2004. The avian nature of the brain and inner ear of Archaeopteryx. Nature 430: 666–669.

Arai, M. 2014. Aptian/Albian (Early Cretaceous) paleogeography of the South Atlantic: a paleontological perspective. Brazilian Journal of Geology 44: 339–350.

Balanoff, A. M., G. S. Bever, T. B. Rowe & M. A. Norell. 2013. Evolutionary origins of the avian brain. Nature 501: 93–96.

Behrensmeyer, A. K. & S. Kidwell. 1985. Taphonomy's contributions to paleobiology. Paleobiology 11: 105–119.

Beurlen, K. 1966. Novos equinóides no Cretaceo do Nordeste do Brasil. Anais da Academia Brasileira de Ciências 38, 455–464.

Brasier, M. D., D. B. Norman, A. G. Liu, L. J. Cotton, J. E. H. Hiscocks, R. J. Garwood, J. B. Antcliffe & D. Wacey. 2016. Remarkable preservation of brain tissues in an Early Cretaceous iguanodontian dinosaur. Geological Society, London, Special Publications 448: 1–16.

Brown, C. M., D. R. Greenwood, J. E. Kalyniuk, Braman, D. M. Henderson, C. L. Greenwood & J. F. Basinger. 2020. Dietary palaeoecology of an Early Cretaceous armoured dinosaur (Ornithischia; Nodosauridae) based on floral analysis of stomach contents. Royal Society Open Science, 7: p.200305.

Brusca, R. C. & G. J. Brusca. 2003. Invertebrates. 2nd ed. Sunderland, Mass: Sinauer Associates, 936 pp.

Carvalho, I. S. 2000. Geological environments of dinosaur footprints in the intracratonic basins from Northeast Brazil during the South Atlantic opening (Early Cretaceous), Cretaceous Research 21: 255–267.

Catto, B. 2015. Laminitos microbiais no Membro Crato (Neoaptiano), Bacia do Araripe, Nordeste do Brasil. Master of Sciences Dissertation. Instituto de Geociências e Ciências Exatas, Universidade Estadual Paulista Júlio de Mesquita Filho, 93 pp.

Catto, B., R. J. Jahnert, L. Warren, F. G. Varejão & M. L Assine. 2016. The microbial nature of laminated limestones: lessons from the Upper Aptian, Araripe Basin, Brazil. Sedimentary Geology 341: 304–315.

Cleland, T. P., M. K. Stoskopf & M. H. Schweitzer. 2011. Histological, chemical, and morphological reexamination of the "heart" of a small Late Cretaceous *Thescelosaurus*. Naturwissenschaften, 98: 203–211.

Creanza, N. & M. W. Feldman. 2016. Worldwide genetic and cultural change in human evolution. Current Opinion in Genetics & Development 41: 85–92.

Cretekos, C. J., Y. Wang, E. D. Green, J. F. Martin, J. J. Rasweiler, 4th & R. R. Behringer. 2008. Regulatory divergence modifies limb length between mammals. Genes & Development, 22: 141–151. DOI: 10.1101/gad.1620408.

Crispo, E. 2007. The Baldwin effect and genetic assimilation: Revisiting two mechanisms of evolutionary change mediated by phenotypic plasticity. Evolution 61: 2469–2479.

Cupello, C., P. Brito, M. Herbin, F. J. Meunier, P. Janvier, H. Dutel & G. Clement. 2015. Allometric growth in the extant coelacanth lung during ontogenetic development. Nature Communications, 6: 1–5. DOI: 10.1038/ncomms9222.

Custódio, M. A., F. Quaglio, L. Warren, M. Simões, F. Fürsich, A. Perinotto & M. Assine. 2017. The transgressive-regressive cycle of the Romualdo Formation (Araripe Basin): Sedimentary archive of the Early Cretaceous marine ingression in the interior of Northeast Brazil. Sedimentary Geology, 359. DOI: 10.1016/j.sedgeo.2017.07.010.

Davidson B. W., Shi, J. Beh, L. Christiaen & M. Levine. 2006. FGF signaling delineates the cardiac progenitor field in the simple chordate, *Ciona intestinalis*. Genes & Development 20: 2728–2738.

De Jong, F., T. Opthof, A. A. Wilde, M. J. Janse, R. Charles, W. H. Lamers & A. F. Moorman. 1992. Persisting zones of slow impulse conduction in developing chicken hearts. Circulation Research 71: 240–250.

De Ruiter, M. C., R. E. Poelmann, I. Vander Plas-de Vries, M. M. Mentink & A. C. Gittenberger-de Groot. 1992. The development of the myocardium and endocardium in mouse embryos. Fusion of two heart tubes? Anatomy and Embryology, 185: 461–473.

Duran, A. C., I. Reyes-Moya, B. Fernandez, C. Rodriguez, V. Sans-Coma & A. C. Grimes. 2014. The anatomical components of the cardiac outflow tract of the gray bichir, *Polypterus senegalus*: their evolutionary significance. Zoology 117: 370–376. DOI: 10.1016/j.zool.2014.05.003.

Erwin, D. H. & E. H. Davidson. 2012. The last common bilaterian ancestor. Development 129: 3021–3032.

Faraa, E., A. A. F. Saraiva, D. A. Campos, J. K. R. Moreira, D. C. Siebrab & A. W. A. Kellner. 2005. Controlled excavations in the Romualdo Member of the Santana Formation (Early Cretaceous, Araripe Basin, northeastern Brazil): stratigraphic, palaeoenvironmental and palaeoecological implications. Palaeogeography, Palaeoclimatology, Palaeoecology 218: 145–160.

Fishman, M. C. & E. N. Olson. 1997. Parsing the heart: genetic modules for organ assembly. Cell 91: 153–156.

Fisher, P. E., D. A. Russell, M. K. Stoskopf, R. E. Barrick, M. Hammer & A. A. Kuzmitz. 2000. Cardiovascular evidence for an intermediate or higher metabolic rate in an ornithischian dinosaur. Science 288: 503–505.

Gould, S. J. 1977. Ontogeny and phylogeny. Cambridge, Mass.: Belknap Press. ISBN 978-0674639416. OCLC 2508336.

Hall, B. K.1992. Waddington's Legacy in Development and Evolution American Zoologist 32: 113–122.

Hall, B. K. 2003. Evo-Devo: evolutionary developmental mechanisms. International Journal of Developmental Biology 47: 491–495.

Haydak, M. H. 1970. Honey bee nutrition. Annual Review of Entomology 15: 143–156.

Hautier, L., F. J. Stansfield, W. T. Allen & R. J. Asher. 2012. Skeletal development in the African elephant and ossification timing in placental mammals. Proceedings of the Royal Society B, 279: 2188–2195.

Held, L. I. 2014. How the Snake Lost its Legs. Curious Tales from the Frontier of Evo-Devo. Cambridge University Press, pp. 126–127.

Hochgreb, T., V. L. Linhares, D. C. Menezes, A. C. Sampaio, C. Y. Yan, W. V. Cardoso, N. Rosenthal & J. Xavier-Neto. 2003. A caudorostral wave of RALDH2 conveys anteroposterior information to the cardiac field. Development 130: 5363–5374.

Horsthemke, B. 2018. A critical view on transgenerational epigenetic inheritance in humans. Nature Communications 9: 1–4.

Jablonka, E. & M. J. Lamb. 2008. Soft inheritance: challenging the modern synthesis. Genetics and Molecular Biology 31: 389–395.

Jacobs F. M., D. Greenberg, N. Nguyen, M. Haussler, A. D. Ewing, S. Katzman, B. Paten, S. R. Salama & D. Haussler. 2014. An evolutionary arms race between KRAB zinc-finger genes ZNF91/93 and SVA/L1 retrotransposons. Nature, 516: 242–245.

Kardong, K. V. 2009. Vertebrates: comparative anatomy, function, evolution. 18th ed. Boston: McGraw-Hill Higher Education. Harvard, 800 pp.

Long, J. A. 2016. Getting to the heart of a good fossil. Elife, 5:e16207. DOI: 10.7554/eLife.16207.

Long J. A., K. Trinajstic, G. C. Young & T. Senden. 2008. Live birth in the Devonian period. Nature 453: 650–652.

Maisey, J. G. 1991. Santana Fossils. An illustrated atlas. Neptune, NJ, TFH Publications, 459 pp.

Maisey, J. G. & M. G. P. Carvalho. 1995. First records of fossil sergestid decapods and fossil brachyuran crab larvae (Arthropoda, Crustacea), with remarks on some supposed palaemonid fossils, from the Santana Formation (Aptian–Albian, NE Brazil). American Museum Novitates 3132: 1–20.

Maldanis L., M. Carvalho, M. R. Almeida, F. I. Freitas, J. A. Andrade, R. S. Nunes, C. E. Rochitte R. J., Poppi, R. O. Freitas, F. Rodrigues, S. Siljeström, F. A. Lima, D. Galante, I. S. Carvalho, C. A. Perez, M. R. Carvalho, J. Bettini, V. Fernandez & J. Xavier-Neto. 2016. Heart fossilization is possible and informs the evolution of cardiac outflow tract in vertebrates. Elife, 5:e14698. DOI: 10.7554/eLife.14698.

Martill, D. M. 1990. Macromolecular resolution of fossilized muscle tissue from an elopomorph fish. Nature 346: 171–172.

Martill, D. M. 1993. Fossils of the Santana and Crato Formations, Brazil. London, The Paleontological Association. Field Guides to Fossils, 159 pp.

Martin, R. E. 1999. Taphonomy: a process approach. Cambridge Paleobiology Series. Cambridge University Press, 508 pp.

Moss, J. B., J. Xavier-Neto, M. D. Shapiro, S. M. Nayeem, P. McCaffery, U. C. Dräger & N. Rosenthal. 1998. Dynamic patterns of retinoic acid synthesis and response in the developing mammalian heart. Developmental Biology 199: 55–71.

Moczek, A. P. 2008. On the origins of novelty in development and evolution. BioEssays 30: 432–447.

Pradel, A., M. Langer, J. G. Maisey, D. Geffard-Kuriyama, P. Cloetens, P. Janvier, P. Tafforeau. 2009. Skull and brain of a 300-million-year-old chimaeroid fish revealed by synchrotron holotomography. Proceedings of the National Academy of Sciences of the USA, 106: 5224–5228.

Robb, J. S. 1965. Comparative Basic Cardiology. Grune & Stratton, New York, 602 pp.

Rosenthal, N. & J. Xavier-Neto. 2000. From the bottom of the heart: anteroposterior decisions in cardiac muscle differentiation. Current Opinion in Cell Biology 12: 742–746.

Rowe, T., E. F. McBride & P. C. Sereno. 2001. Dinosaur with a heart of stone. Science 291: 783.

Sansom, R. S. 2016. Preservation and phylogeny of Cambrian ecdysozoans tested by experimental decay of *Priapulus*. Scientific Reports 6: 1–12.

Sansom, R. S., S. E. Gabbott & M. A. Purnell. 2011. Decay of vertebrate characters in hagfish and lamprey (Cyclostomata) and the implications for the vertebrate fossil record. Proceedings of the Royal Society B, 278: 1150–1157.

Santer, R. M., M. Greer-Walker, L. Emerson & P. R. Witthames. 1983. On the morphology of the heart ventricle in marine teleost fish (Teleostei). Comparative Biochemistry and Physiology 76: 453–457.

Selden, P. & J. Nudds. 2012. Evolution of Fossil Ecosystems. 2nd Edition in: Waltham, MA, Academic Press, 288 pp.

Simões-Costa, M. S., M. Vasconcelos, A. C. Sampaio, R. M. Cravo, V. L. Linhares, T. Hochgreb, C. Y. Yan, B. Davidson & J. Xavier-Neto. 2005. The evolutionary origin of cardiac chambers. Developmental Biology 277: 1–15.

Voltani, C. G. 2016. Redescrição osteológica comentada e aspectos paleontológicos dos gêneros *Notelops* e *Rhacolepis* (Pachyrhizodontoidei). Ph.D. Thesis. Universidade Estadual Paulista, Instituto de Geociências e Ciências Exatas do Campus de Rio Claro, 178 pp.

Wilby, P. R. & D. M. Martill. 1992. Fossil fish stomachs: A microenvironment for exceptional preservation, Historical Biology 6: 25–36.

Wilkens, J. L. 1999. Evolution of the cardiovascular system in Crustacea. American Zoologist 39: 199–214.

Whiten, A. 2019. Cultural Evolution in Animals. Annual Review of Ecology, Evolution, and Systematics 50: 27–48.

Woodward, D. E. & J. D. Murray. 1993. On the Effect of Temperature-Dependent Sex Determination on Sex Ratio and Survivorship in Crocodilians. Proceedings of the Royal Society B, 252: 149–155.

Xavier-Neto, J., N. Rosenthal, F. A. Silva, T. G. Matos, T. Hochgreb, & V. L. Linhares. 2001. Retinoid signaling and cardiac anteroposterior segmentation. Genesis 31: 97–104.

Xavier-Neto, J., R. A. Castro, A. C. Sampaio, A. P. Azambuja, H. A. Castillo, R. M. Cravo & M. S. Simões-Costa. 2007. Parallel avenues in the evolution of hearts and pumping organs. Cellular and Molecular Life Sciences, 64: 719–734.

Xavier-Neto, J., B. Davidson, M. S. Simões-Costa, R. A. Castro, H. A. Castillo & A. C. Sampaio. 2010. Evolutionary Origins of Hearts. – Pp. 3–45 in: Rosenthal, N. & R. P. Harvey (eds.). Heart Development and Regeneration. London, UK, Elsevier Inc.

Xavier-Neto, J., S. S. Trueba, A. Stolfi, H. M. Souza, T. J. Sobreira, M. Schubert & H. A. Castillo. 2012. An unauthorized biography of the second heart field and a pioneer/scaffold model for cardiac development. Current Topics in Developmental Biology 100: 67–105.

Xiaoya, M., G. D. Edgecombe, X-g. Hou, T. K. Goral & N. J. Strausfeld. 2015. Preservational Pathways of Corresponding Brains of a Cambrian Euarthropod. Current Biology 25: 2969–2975.

Watanabe, A., P. M. Gignac, A. M Balanoff, et al. 2018. Are endocasts good proxies for brain size and shape in archosaurs throughout ontogeny? Journal of Anatomy 234, 291–305.

Zelenitsky, D. K., F. Therrien & Y. Kobayashi. 2009. Olfactory acuity in theropods: palaeobiological and evolutionary implications. Proceedings of the Royal Society B, 276: 667–673.

Ancient Fishes and their Living Relatives: a Tribute to John G. Maisey. Alan Pradel, John S. S. Denton & Philippe Janvier (eds.): pp. 49–57, 5 figs.
© 2021 by Verlag Dr. Friedrich Pfeil, München, Germany – ISBN 978-3-89937-269-4

Influence analysis of fossil chondrichthyan taxa

John S. S. Denton* and Eric W. Goolsby**

Abstract

The question of a taxon's effect on our ability to infer the shape of evolutionary history is a fundamental question in systematics, and one that becomes more complicated when considering the fossil record. Although some (predominantly parsimony-based) studies predict certain features of influential taxa, model-based studies have not yet been brought to bear on the question of taxon influence. Here, we analyze a dataset of fossil chondrichthyan taxa using Bayesian methods and a new implementation of a measurement (taxon influence index; TII) using full tree posterior distributions. We show that the resampling-based methods are a good approximation to the prohibitively intensive calculation based on the full posterior distribution. Agreement between the resampling-based and full-posterior approaches to TII suggests that the inferred taxon rank order in this case is robust. Analysis identifies a mixture of Paleozoic and Recent taxa as having the highest influence, including two specimens (*Pucapampella*, *Doliodus*) worked on extensively by John. The emergence of this largely consistent ranking from a modest-sized character matrix suggests that there may be a great deal of information of use to paleontologists in posterior distributions of trees.

Introduction

Debate on the relative merits of fossil and Recent taxa has a longstanding history in phylogenetics. Regarding cladistic topology inference, some authors (e.g. Patterson & Rosen [1977], Patterson [1981]) have suggested fossils play a secondary role, arguing in favor of using only Recent taxa to infer cladograms, and of assigning fossils to groups based on these inferred relationships among Recent taxa, as "fossils overturning theories of relationship based on Recent organisms are very rare, and may be nonexistent (Patterson 1981: 218). Similarly, Ax and Jeffries (1987) argue from an implicitly analytical standpoint (see, for example, Kearney [2002], Kearney & Clark [2003], Wiens [2003]) that because fossils are incomplete, cladograms should be constructed based on Recent taxa alone.

By contrast, other authors, using parsimony-based computational approaches (e.g. Gauthier, Kluge & Rowe 1988, Donoghue et al. 1989, Novacek 1992), have shown across many taxonomic groups that fossils may affect the shape of inferred cladograms and that, as a source of new characters and character states, fossils may significantly affect inferred character polarity and ancestral state inferences (e.g. Doyle & Donoghue 1987, Huelsenbeck 1991, Wilson 1992), and that it might be expected that a group's earliest-diverging taxa (and/or most 'surprising' taxa) may have the greatest phylogenetic contribution (e.g. Huelsenbeck 1991, Wilson 1992). However, see Gauthier, Kluge & Rowe (1988) for an example in which intermediate-age taxa exhibit the greatest effect. Other authors have argued for the importance of fossils for informing the evolutionary elaboration of physiological processes (Maisey 1986). Still other authors have argued that inconsistent treatment of fossil taxa in a cladistic framework makes it fail as a paradigm (Van Valen 1978). The literature on debate regarding the relative importance of fossils and Recent taxa is rich and only summarized briefly here.

Despite some degree of conflicting and/or contingent results regarding the importance of fossils in cladistic analysis, there is general consensus that fossil taxa pose a unique analytical problem. In the context of cladistic inference, patterns of missing ("?") data are distinct from inapplicable ("–") data, and may alter inference in unpredictable ways (e.g. Kearney [2002]; Kearney & Clark [2003], Wiens [2003]). Furthermore, for fossil taxa, the degree of intraspecific polymorphism is frequently unknown, but potentially significant for inference (e.g. Watanabe [2016]), and phylogenetic software programs infrequently distinguish among these three conditions by default (but see, for example, Goloboff [2014]), despite their ontological distinctions (Platnick, Griswold & Coddington 1991). Although parsimony-based methods have predominated in the cladistic analysis of fossil taxa (and in analyses of their importance), recent debate has centered on the validity of a suggested improvement in phylogenetic accuracy of model-based analyses (e.g. the Mk model (Lewis 2001) and its different ascertainment bias corrections in likelihood and Bayesian analysis) on morphological datasets (Goloboff 2014, Wright & Hillis 2014, O'Reilly et al. 2016, Wright, Lloyd & Hillis 2016, Goloboff, Torres & Arias 2018, Sansom et al. 2018, Goloboff & Arias 2019)).

In contrast with this recent debate on methodology itself, the subject of influential taxa (as distinct from rogue and unstable taxa; see Denton & Goolsby (2018) for a brief review) has been revisited only infrequently with model-based methods (but see Mariadassou, Bar-Hen & Kishino (2012)for a likelihood-based taxon jackknifing approach, Denton & Goolsby (2018) for an approximation to the Mariadassou et al. Bayesian formulation of taxon influence, and Koch & Parry (2020) for an examination of the properties and effects of fossil taxa). Model-based taxon influence estimation, and especially Bayesian approaches, is worth revisiting more broadly on paleontological datasets, given the emphasis of such methods on accounting for sources of uncertainty.

It is worth noting that Patterson (1981), in his oft-quoted publication on the significance of fossils, opens the work not with the famous quotation cited earlier (which instead appears in his concluding section), but rather with reference to the paucity of chondrichthyan fossils (p. 195). The chondrichthyan fossil record, which records the evolution of forms leading to modern-day sharks and rays, is notoriously incomplete relative to other taxonomic groups, with largely separate taxonomies (and estimated completeness values) for Paleozoic and Mesozoic forms based on their differing representations by body fossils and teeth, respectively (e.g. Foote & Sepkoski Jr. 1999,

* Florida Museum of Natural History, Dickinson Hall, 1659 Museum Road, Gainesville, FL 32611; jsdenton@ufl.edu (corresponding author)

** University of Central Florida, Department of Biology, 4110 Libra Dr, Orlando, FL 32816

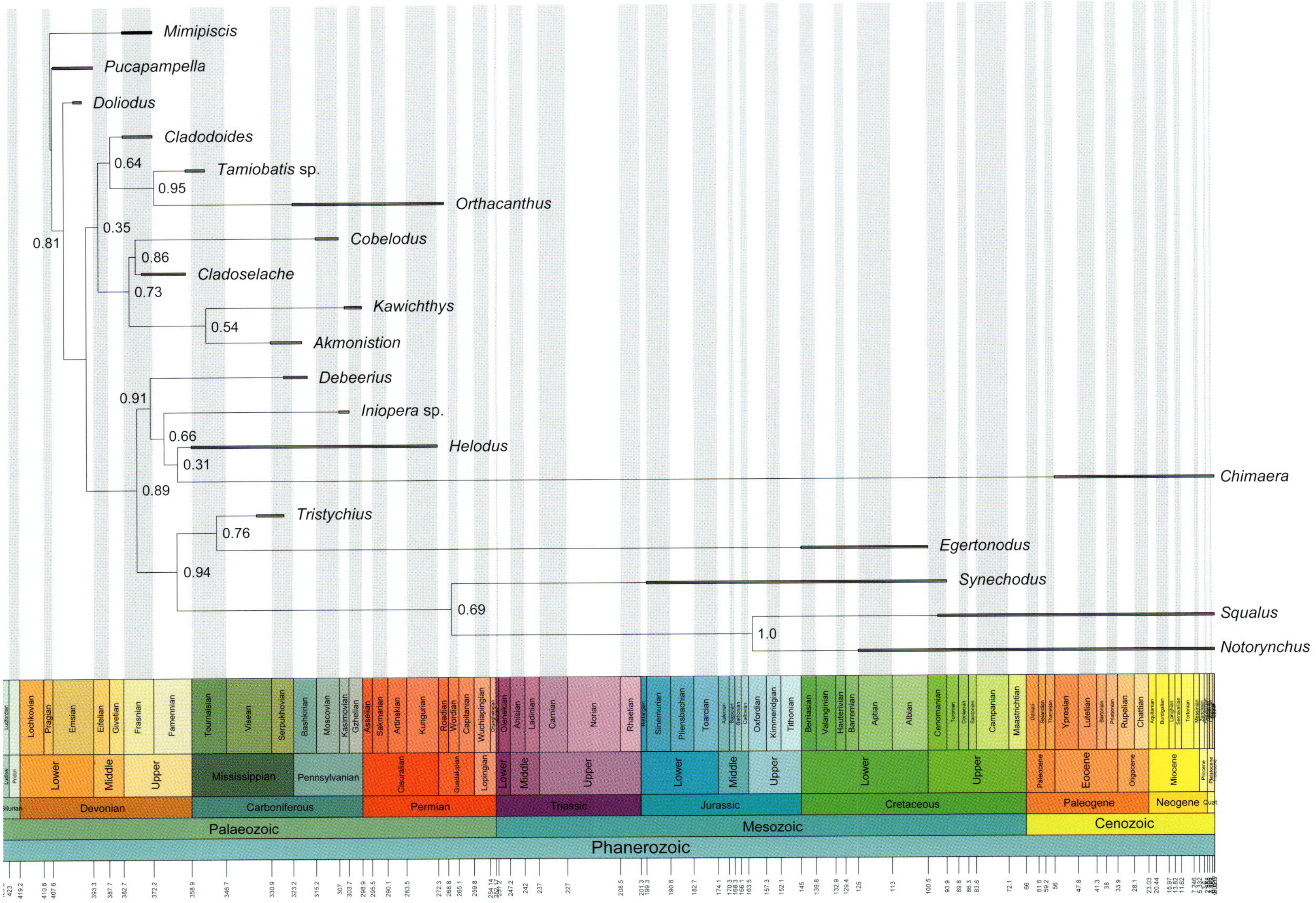

Fig. 1. Maximum clade credibility (MCC) tree derived from post-burnin Bayesian tree posterior of the P2011 dataset. Clade credibility values presented at nodes. Stratigraphic ranges for taxa represented by bars. The MCC tree was the same topology as that derived from an implied weights parsimony analysis, which recovered a single tree. The fit of the MCC topology to the stratigraphic record was significantly better than random for the majority of the metrics used to assess fit (SCI = 0.529, $p > 0.01$; RCI = –64.69, $p < 0.01$; GER = 0.588, $p < 0.01$; MSM* = 0.316, $p < 0.01$). Fit of topology to stratigraphic data, and congruence analysis, was conducted in the R package strap.

Fig. 1–2). The intervening years since Patterson's acknowledgement have seen significant advances in understanding of the chondrichthyan fossil record, due significantly to John's work on Paleozoic taxa and his early embrace of tomographic imaging methods (e.g. Maisey 2001c, Maisey 2001b), which has led to significant insight on chondrichthyan character evolution (e.g. Pradel et al. 2014). However, the placement of many fossil taxa at even higher levels within Chondrichthyes remains uncertain, and so it is worth asking what model-based influence analysis accounting for such uncertainty might reveal. To that end, this contribution focuses on a specific published dataset spanning the complete stratigraphic age of Chondrichthyes, with representation by both fossil and Recent forms, and with reduced proportions of missing and inapplicable data. We infer a best-fit Mk model of evolution to the data, develop a new implementation for the full Mariadassou et al. Bayesian influence calculation, and compare its results to those of the heuristic presented in an earlier work. Finally, we correlate the inferred influence estimates with hypothesized predictors of influence, including stratigraphic first occurrence and character mosaic uniqueness.

Materials and Methods

Dataset. Although many taxon-dense datasets spanning both gnathostome diversity (e.g. Davis, Finarelli & Coates 2012, Zhu et al. 2013, Giles, Friedman & Brazeau 2015) and the acanthodian-chondrichthyan divide (e.g. Burrow et al. 2016, King et al. 2017, Coates et al. 2018) have been published in recent years, the dataset of Pradel et al. (2011) was selected for analysis based on its comparatively small proportion of missing data (states coded as "?"), its absence of inapplicables (states coded as "–", treated the same way as "?" by existing software implementations), and its restriction to a single anatomical module (the chondrocranium). The Pradel et al. dataset (referred to subsequently as P2011) consisted of 19 taxa and 35 chondrocranial characters, with a median of 5.7 % missing data per taxon. The outgroup taxon referred to as *Mimia* in Pradel et al. (2011) was updated to *Mimipiscis*, following the revision of Choo (2012).

Model selection. Model selection for taxon influence analysis was carried out by comparing different models in a Bayesian framework. Bayesian analyses of morphological character data have been suggested to outperform parsimony approaches under certain conditions (Wright & Hillis 2014, O'Reilly et al. 2016, but see Goloboff, Torres & Arias 2018, Sansom et al. 2018, and Goloboff & Arias 2019 for recent debate), and variation in the values of the parameters of the beta distribution for (hyper-) priors applied to character transition rate asymmetries (Wright, Lloyd & Hillis 2016) has been shown in some cases to lead to model fits that outperform the traditional Mk model variants for morphology (Lewis 2001).

Because morphological character matrices are selected to reflect character variation and so lack constant sites, model-based analysis was conducted using an ascertainment bias correction for the presence of only parsimony-informative sites (Mki). Three models with varying character state asymmetries were assessed, following the terminology of Wright, Lloyd and Hillis (2016): the standard Mki model (corresponding to a prior of $\alpha = \infty$), which prohibits transition rate asymmetry among characters; a 'relaxed' model ($\alpha = 10$), which allows some transition rate asymmetry among characters; and an asymmetric model ($\alpha = 0.05$), which constrains character transition rates to be strongly asymmetrical. All models were additionally parameterized with a discrete gamma distribution (Γ4) for rate heterogeneity among characters. All analyses were conducted in mrbayes 3.2.6 (Ronquist et al. 2012), each using a single run of three chains, sampling every 1000 generations, and run for 40 million generations with a 10 % burn-in. Convergence in non-topology model parameter estimates, as determined by parameter ESS values >200 was confirmed in Tracer 1.6. Model comparisons were conducted using the AICM (Baele et al. 2012). Corrected approximate topological ESS values for the post-burn-in best-fit model were assessed in the R package rwty (Warren, Geneva & Lanfear 2017) by dividing the estimated approximate topological ESS by the approximate plateau value in the topological autocorrelation plot. Stratigraphic consistency measures (SCI, RCI, GER, and MSM*) were calculated for the trees in the posterior distribution, for the MCC tree, and for the unweighted parsimony tree from the original P2011 publication, in the R package strap (Bell & Lloyd 2015), using 1000 random and 1000 sample trees.

Taxon influence analysis. To assess the degree of "importance" of each taxon for inferring evolutionary relationships, and to generate a ranked list of taxon importance based on influence values, a taxon influence analysis was conducted, following the procedure of Denton & Goolsby (2018)'s approximation of the Mariadassou, Bar-Hen & Kishino (2012) equation for the Bayesian version of the taxon influence index (TII)

$$TII(i) = med_p\left[\sum_{T^*,T'} w_i(T^*)w_i(T')d(T^*,T')\right]$$

where T* is the posterior distribution of trees from analysis using all taxa, T′ is the posterior distribution of trees in which a focal taxon *i* is dropped before analysis, T'_i is a phylogenetic tree from posterior T′ for which taxon *i* was dropped before analysis, T^*_i is a tree from the posterior T* in which taxon *i* was dropped after analysis for comparison with T'_i, w_i is the posterior probability of tree *i*, and d(·,·) is the topological distance between the two trees, calculated as the ratio of the SPR distance (e.g. Goloboff [2008]) between the two trees to the size of their maximum agreement subtree (MAST; Gordon [1979]; Finden & Gordon [1985]). Here, the TII is calculated as the median (med) value over *p* resampling iterations, where trees T* and T′ are drawn according to their posterior probabilities to generate pseudoposteriors with length equal to min(|T*|, |T′|).

Denton & Goolsby (2018) estimated TII values under the assumption of sampling trees without replacement. We therefore assessed the effect of resampling scheme by comparing sampling with and without replacement. Additionally, because the resampling-based TII is based predominantly on high-probability trees, and so may represent a biased sample when tree distributions are strongly skewed, we also compared these heuristic sampling approaches with a new implementation of the original (full) Mariadassou equation comparing all |T*| × |T′| trees from unresampled and untruncated posterior distributions

$$TII(i) = E_{T_i*\otimes T_i}[d(T^*,T')] = \sum_{T^*,T'} w_i(T^*)w_i(T')d(T^*,T')$$

where the $E_{T_i*\otimes T_i}$ indicates the expected value over all pairwise distances derived from all trees in the two post-burnin posterior distributions.

Using these approaches, we inferred a rank list of taxon influence for all taxa in the P2011 dataset, examined whether the Recent taxa (*Chimaera*, *Notorynchus*, *Squalus*) exhibited any clustering in TII values at either the upper or lower end of taxon ranking (relative to the fossil taxa included), and examined how effective the heuristic methods were in recovering a ranking similar to that based on the exhaustive pairwise comparisons of the full equation. *Mimipiscis* was fixed as the outgroup, and rooted all jackknifed analyses. Resampling for both the with and without sampling schemes was done 1000 times. Differences between rankings were calculated using rank-biased overlap (rbo; Webber, Moffat & Zobel [2010]) from the R package gespeR (Schmich et al. 2015) as a test statistic, with the top ten rankings having 86 % of the overall ranking weight ($p = 0.9$). Rank-biased overlap was selected as the statistic for list comparison based on its desirable properties relative to several other existing ranking comparison methods (Okada, Pedronette & da S Torres 2015). A one-tailed permutation test in which values in one ranking were fixed, and values in the comparison list permuted, was used to assess the significance of ranking differences against the null hypothesis of dissimilar rankings (H_0: rbo = 0; H_a: rbo > 0), in a manner similar to the statistical assessment of significance for Kendall's τ. Finally, to assess whether influential taxa may be rogue taxa, a rogue taxon analysis (Aberer, Krompass & Stamatakis 2013) was conducted in RAxML 8.2.9 (Stamatakis 2014).

Correlates of taxon influence. To assess potential correlates of taxon influence, we conducted a regression of earliest stratigraphic record, stratigraphic range, proportion missing data per taxon, and taxon 'uniqueness' on the TII values derived from the full analysis. Taxon uniqueness was calculated, following Koch & Parry (2020), as the mean of each taxon's MORD (maximum observable rescaled distance) to the other taxa, in the R package Claddis (Lloyd 2016), with equally-weighted and unordered characters. Regression was carried out using the following equation:

$$Y^{(\lambda)} = X_1 + X_2 + X_1X_2 + X_3 + X_2X_3 + X_4$$

where Y is taxon influence, X_1 is earliest stratigraphic record, X_2 is stratigraphic range, X_3 is proportion of missing data, X_4 is taxon uniqueness, and λ is the likelihood-optimized value of a Box-Cox transformation (Box & Cox 1964), such that Y^λ is a normalized response variable. Stratigraphic ages used for regression, with references, are presented in Table 1. All analyses were carried out in R 3.5.1 (R Core Team 2016).

Table 1. Stratigraphic ages of taxa from the P2011 dataset.

Taxon	Stratigraphic Range (Ma)	Source
Mimipiscis	382.7-372.2	Choo (2012)
Akmonistion	331-320	Coates & Sequeira (2001b)
Chimaera	56-present	Stahl (1999)
Cladodoides	382.7-372.2	Maisey (2001a); Maisey (2005)
Cladoselache	376.1-360.7	Williams (2001)
Cobelodus	315.2-307	Ginter, Hampe & Duffin (2010)
Debeerius	326.4-318.1	Grogan & Lund (2000)
Doliodus	400-397	Maisey et al. (2019)
Egertonodus	145-100.5	Maisey (1987)
Helodus	358.9-272.5	Maisey (pers. comm.)
Iniopera sp.	306.95-303.4	Zangerl & Case (1973)
Kawichthys	305-299	Pradel et al. (2011)
Notorynchus	125-present	Cappetta (1987)
Orthacanthus	323.2-270	Ginter, Hampe & Duffin (2010)
Pucapampella	407-393	Maisey et al. (2019)
Squalus	97-present	Adnet & Cappetta (2001)
Synechodus	199.3-93.9	Ginter, Hampe & Duffin (2010)
Tamiobatis sp.	360.7-353.8	Ginter, Hampe & Duffin (2010)
Tristychius	336-326.4	Coates & Tietjen (2018)

Results

Model selection and phylogenetic analysis. The AICM selected the standard Mki + Γ model (AICM = 530.159 ± 0.044) over the "nearly symmetrical" model (AICM = 532.262 ± 0.138) and the strongly asymmetrical model (AICM = 562.471 ± 0.383). Bayes Factor comparisons of the Mki and "nearly symmetrical" model did not strongly favor the Mki model (BF = 1.408), but it was selected for computational expediency. A similar pattern of fit was seen among the "mismatched" ascertainment bias correction models, with the Mkv + Γ model (AICM = 545.149 ± 0.064) outperforming the "nearly symmetrical" model (AICM = 545.784 ± 0.074) and the strongly asymmetrical model (AICM = 582.764 ± 0.08). In all cases, the Mki models outperformed their "mismatched" counterparts with strong Bayes Factor supports.

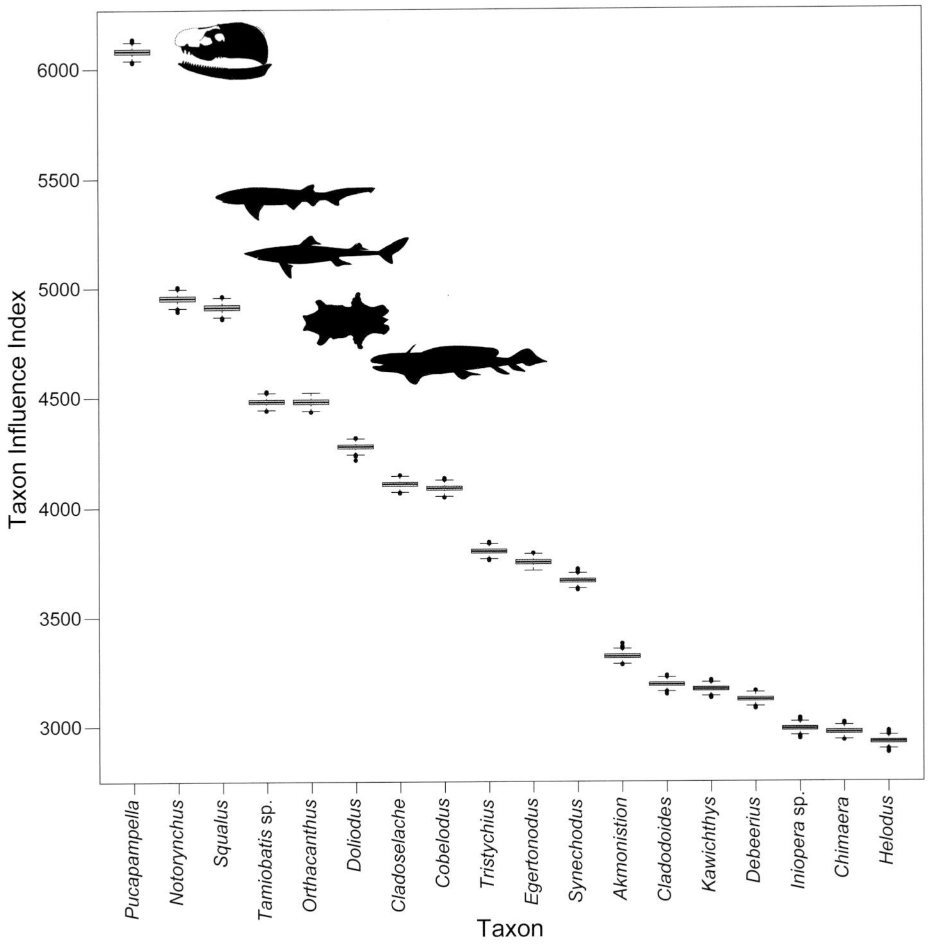

Fig. 2. Resampling-with-replacement-based taxon influence estimates, ranked by the Denton & Goolsby median estimate. A mixture of Paleozoic and Recent taxa (*Pucapampella*, *Notorynchus*, *Squalus*, *Tamiobatis*, *Orthacanthus*), identified in silhouette, occupy the upper quartile of influence estimate, and holocephalan taxa occupy the lower quartile by this estimate. Identification of both rootward-diverging and early-within-clade-diverging taxa in the highest quartile of influence estimates somewhat reiterates previous hypotheses regarding the role of fossils and of early-diverging taxa in general. Resampled values appear to exhibit mostly balanced, and well-separated distributions, with mostly unskewed outliers. Silhouette attributions provided in acknowledgements.

The post-burnin topological ESS under the Mki + Γ model was ~2546. Bayesian analysis (Fig. 1, maximum clade credibility [MCC] tree) recovered *Doliodus* and *Pucapampella* at the base of "conventionally-defined chondrichthyans (sensu Zhu et al. [2013])," followed by a clade of ctenacanths and xenacanths (*Cladodoides*, *Orthacanthus*, *Tamiobatis*), and symmoriiforms + *Cladoselache* (*Akmonistion*, *Kawichthys*, *Cladoselache*, *Cobelodus*), and a clade of euchondrocephalans (*Debeerius*, *Iniopera*, *Chimaera*, *Helodus*) plus hybodontiform and neoselachian chondrichthyan taxa (*Egertonodus*, *Tristychius*, *Synechodus*, *Notorynchus*, *Squalus*). Clade supports were generally low, with the exception of some sister-taxon pairings. Implied-weights parsimony (e.g. Goloboff et al. 2008) with k = 3 recovered the same topology as the MCC tree. These trees were different from the single unweighted parsimony tree. Stratigraphic consistency of trees in the post-burnin posterior distribution based on four common consistency measures (stratigraphic consistency index [SCI], relative completeness index [RCI], gap excess ratio [GER], and modified Manhattan stratigraphic measure [MSM*]; Sansom et al. (2018)) were largely significantly different from random (median ± IQR [proportion *p*-values > 0.01]: SCI = 0.529 ± 0.658 [0.47]; RCI = -67.255 ± 19.86 [0.12]; GER = 0.579 ± 0.072 [0.203]; MSM* = 0.311 ± 0.036 [0.043]). Consistency measures for the MCC and unweighted parsimony topologies fell within the ranges of the values in the posterior distribution.

Taxon influence analysis. Resampling-based taxon influence analysis produced estimates of similar overall magnitudes, with little overlap in the interquartile ranges, and with slight differences in ranking depending on whether resampling was conducted with or without replacement. When sampling with replacement (Fig. 2), the early-diverging taxon *Pucapampella*, two Recent taxa *Squalus* and *Notorynchus*, the ctenacanth *Tamiobatis*, and xenacanth *Orthacanthus* occupied the upper quartile of taxon influence. The Recent taxa (*Squalus*, *Notorynchus*, *Chimaera*) were distributed at both tails of the ranking. The holocephalan taxa (*Iniopera*, *Debeerius*, *Helodus*, *Chimaera*), and one symmoriiform taxon (*Kawichthys*), occupied the lower quartile of taxon influence. *Helodus* exhibited the lowest taxon influence estimate.

Ranking patterns when sampling without replacement were similar to those sampled with replacement (Fig. 3), with most of the holocephalan taxa (*Chimaera*, *Helodus*, *Iniopera*), one symmoriiform taxon (*Kawichthys*), and one ctenacanth (*Cladodoides*), occupying the lower quartile of taxon influence. *Helodus* exhibited the lowest taxon influence estimate. Recent taxa were again distributed bimodally in the ranking, with minor differences in other taxon rank positions. The null hypothesis of dissimilarity between the two resampling-based rankings was rejected (rbo = 0.955; $p < 0.001$).

The full |T*| × |T′| taxon influence analysis exhibited wider separation among values in the tails of the rank estimates than was recovered by the resampling-based approaches (Fig. 4). In the full ranking, the difference in influence estimate between *Pucapampella* and the next-highest-ranked taxon (*Notorynchus*) was larger than the range of influence values separating the rest of the taxa. The holocephalan and symmoriiform taxa occupied the lower quartile of taxon influence, with the taxon *Iniopera* exhibiting the lowest influence estimate. The null hypothesis of dissimilarity between this ranking and the resampling-based rankings was rejected (rbo = 0.897; $p < 0.001$). The rbo values and *p*-values in comparisons between the full and resampled rankings were nearly identical. Graphical summaries of the

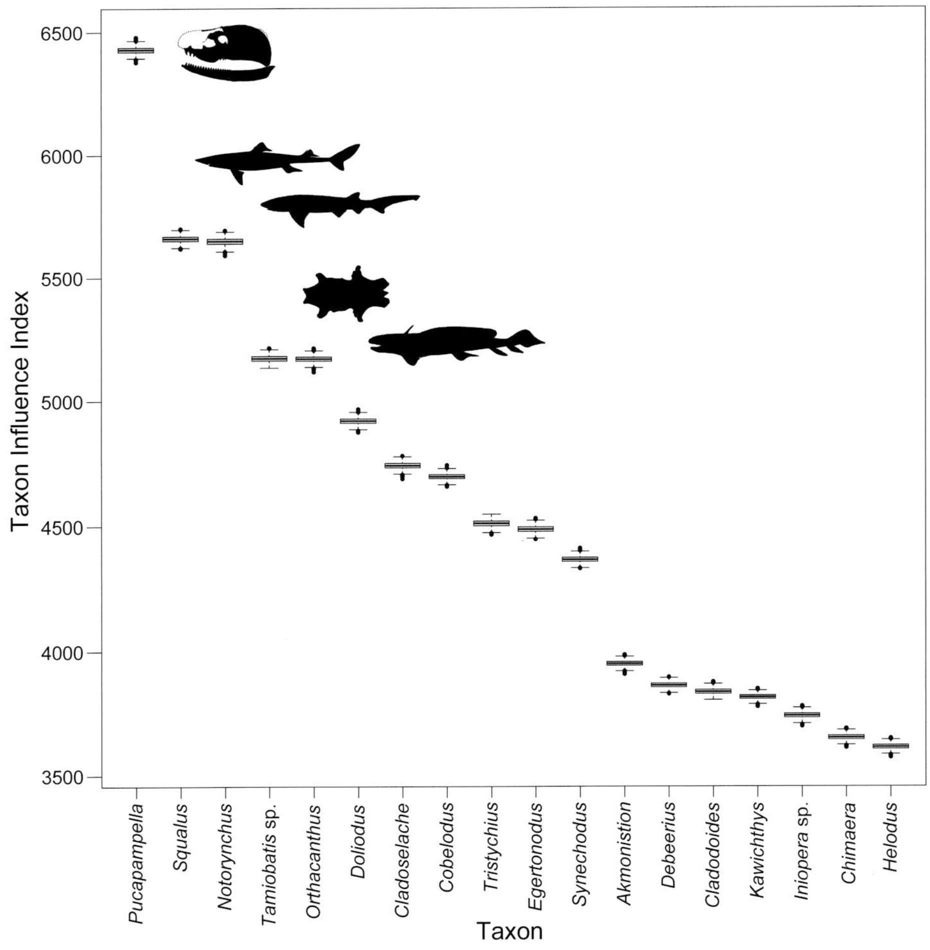

◁ **Fig. 3.** Resampling-without-replacement-based taxon influence estimates, ranked by the Denton & Goolsby median estimate. Similar to the ranking sampled with replacement, a mixture of Paleozoic and Recent taxa (*Pucapampella*, *Notorynchus*, *Squalus*, *Tamiobatis*, *Orthacanthus*), identified in silhouette, occupy the upper quartile of influence estimate, and holocephalan taxa occupy the lower quartile by this estimate. Resampled values appear to exhibit mostly balanced, and well-separated distributions, with mostly unskewed outliers. The hypothesis of dissimilarity in resampling-based rankings was rejected (rbo = 0.955; $p < 0.001$).

relationships between the resampling-based and full influence values are presented in Figure 5. Rogue taxon analysis did not identify any taxa to drop.

Regression. The optimized exponent $\lambda = -2.323$ suggested a non-standard Box-Cox transformation. Quantile-quantile plots did not reveal heavy tails, although *Akmonistion*, *Iniopera*, and *Pucapampella* were identified as slight outliers.

There was a significant fit of the overall model to the data ($BIC = 116.929$; $F = 41.65$; $df = 12$; $R^2 = 0.931$; $p < 0.001$). Taxon uniqueness ($t = 2.545$; $p = 0.026$) was a slightly significant main predictor of taxon influence. Neither of the stratigraphic parameters (taxon first occurrence: $t = -0.838$; $p = 0.419$; stratigraphic range: $t = -1.367$; $p = 0.197$) nor the proportion of missing data ($t = 0.193$; $p = 0.85$) were significant main predictors of taxon influence. There were no significant interactions.

Discussion

Discussion of the inferred phylogeny and the results of taxon influence analysis focuses herein on structural differences and analytical findings, rather than on cladistic, stratigraphic, or character-related implications.

Phylogeny. The results of phylogenetic analysis presented here (Fig. 1), which recover the 'traditional' placement of *Doliodus* and *Pucapampella* at or near the base of "conventionally-defined" chondrichthyans (e.g., Davis, Finarelli & Coates 2012, Zhu et al. 2013, Giles, Friedman & Brazeau 2015; extended Data Fig. 8; King et al. 2017, Fig. 3; Coates et al. 2018, Fig. 3b), differ strongly from those presented in the original Pradel et al. (2011) paper, in which *Doliodus* and *Pucapampella* were recovered on the stem of a clade of "cladodont" sharks (*Cladodoides*, *Tamiobatis*, *Akmonistion*, *Cobelodus*, and *Cladoselache*), sister to a clade of crown chondrichthyans (stem- and crown-holocephalans and hybodontiformes plus neoselachians), based on a new technology equal-weights parsimony analysis in TnT (Goloboff, Farris & Nixon 2008).

Taxon influence analysis. The approach to taxon influence calculation in this study follows that outlined in Denton and Goolsby (2018), using the ratio of the SPR distance to the size of the maximum agreement subtree (MAST). Because the distance between trees is divided by the size of the subtree in common, the measurement gives higher weight to "size-corrected" changes among internal branches that move clades. Inferred influence values reflect either a stabilizing or destabilizing effect. Although a resampling-based approximation to the Bayesian taxon influence index was presented in Denton and Goolsby (2018), the present analysis is the first implementation of the full calculation as outlined by Mariadassou, Bar-Hen & Kishino (2012).

The significant correlation of the resampling-based TII rankings (Figs. 2–3) both with one another and with the full ranking (Fig. 4) has two significant implications. First, it suggests robustness of the inferred results to the method of influence calculation. Across all ranking methods, *Pucapampella*, ctenacanth *Tamiobatis*, xenacanth *Orthacanthus*, and Recent taxa *Notorynchus* and *Squalus* were ranked in the upper quartile, with *Doliodus* ranking in the upper quartile in the full ranking.

The upper-quartile ranking of the early-diverging taxa *Pucapampella* and *Doliodus* seems intuitive, given recent analyses have suggested the acanthodian-chondrichthyan transition to be a significant evolutionary gradation, both in terms of the evolution of body plans (e.g., Maisey et al. 2017, Coates et al. 2018) and in terms of a diverse mosaic of cartilage mineralization patterns (e.g. Maisey 2013) leading to the assembly of the complex modern tessellated calcified cartilage system ('intertesseral joint system', sensu Seidel et al. 2016) that exists as a supposedly definitive chondrichthyan feature (e.g. Brazeau 2009, character 1; Davis, Finarelli & Coates 2012, character 1; Zhu et al. 2013, character 1). Both *Pucapampella* and *Doliodus* straddle this transition, and *Doliodus* has been suggested to have acanthodian affinities, reflected conspicuously in the presence of fin spines (Maisey et al. 2017). It is worth noting that the dataset analyzed in the present study was restricted to a published matrix of chondrocranial characters, and thus the present analysis has not utilized any potentially informative postcranial anatomy that might affect the TII estimate for *Doliodus*. Furthermore, recently-(re-)described taxa occupying the same general topological zone as *Pucapampella* (e.g. *Gydoselache* Maisey et al. [2019]; *Gladbachus* [Coates et al. 2018]) were not sampled in the analyzed character matrix.

High rank influence for rootward extinct taxa such as *Pucapampella* and *Doliodus* is in agreement with previous simulation results based on ancestral state reconstruction (Huelsenbeck 1991), but it is also an encouraging result, given John's history of work on Paleozoic taxa. The large separation of *Pucapampella* from the remaining rank estimates is notable, given the taxon's combinations of potentially neotenic and autapomorphic features that "do not provide a shred of evidence that chondrichthyans and 'acanthodians' are closely related (Maisey et al. 2019: 104–105)." However, the explanation for the magnitude of difference in ranking is unclear and warrants attempted replication with larger datasets.

The upper quartile ranking of *Notorynchus* and *Squalus*, both extant taxa, is unusual given some established context for importance based on simulation suggests that, in the absence of skewed preservational artifacts, and with increasing evolutionary rates and wider timescales, older taxa and taxa near the roots of clades are predicted to be more influential for reconstructing relationships (e.g., Huelsenbeck 1991, Wilson 1992). *Notorynchus* is a member of an order of neoselachian taxa (Hexanchiformes) that is either inferred to be early-diverging or assumed as outgroups (e.g. Adnet & Cappetta 2001, Fig. 3; Maisey, Naylor & Ward 2004, Fig. 3; Vélez-Zuazo & Agnarsson 2011, Fig. 2, previous trees). It is worth noting that the order Hexanchiformes presents a stratigraphic puzzle, as the tooth fossil genus *Mcmurdodus* (Mcmurdontidae White, 1968) from the Devonian of Australia and Antarctica (Burrow et al. 2008) suggests a very long ghost lineage for this particular order. By contrast, *Squalus* (Squaliformes) is comparatively tipward and, although the lineage is early-diverging within the order (Straube

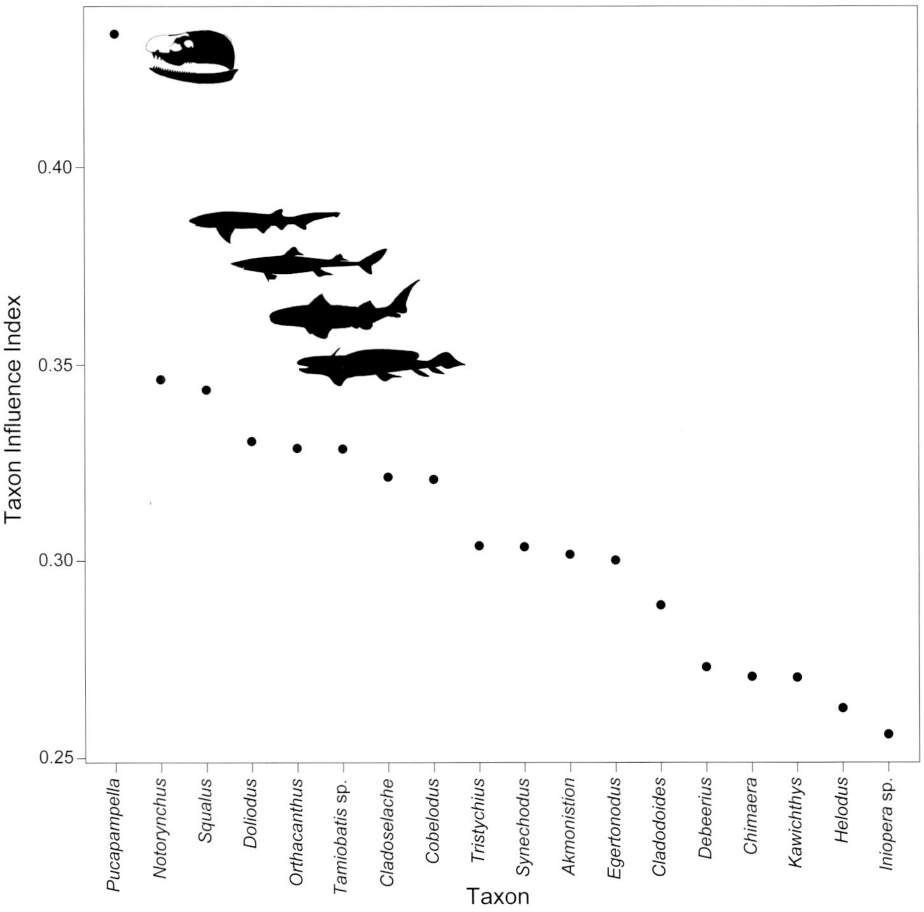

Fig. 4. Resampling based on full post-burnin posterior distributions of trees. A mixture of Paleozoic and Recent taxa (*Pucapampella*, *Notorynchus*, *Squalus*, *Doliodus*, *Orthacanthus*), identified in silhouette, occupy the upper quartile of influence estimate, and holocephalan taxa occupy the lower quartile by this estimate. The hypothesis of dissimilarity between the ranked lists and the full list was rejected ($rbo = 0.897$; $p < 0.001$). Silhouette attributions provided in acknowledgements.

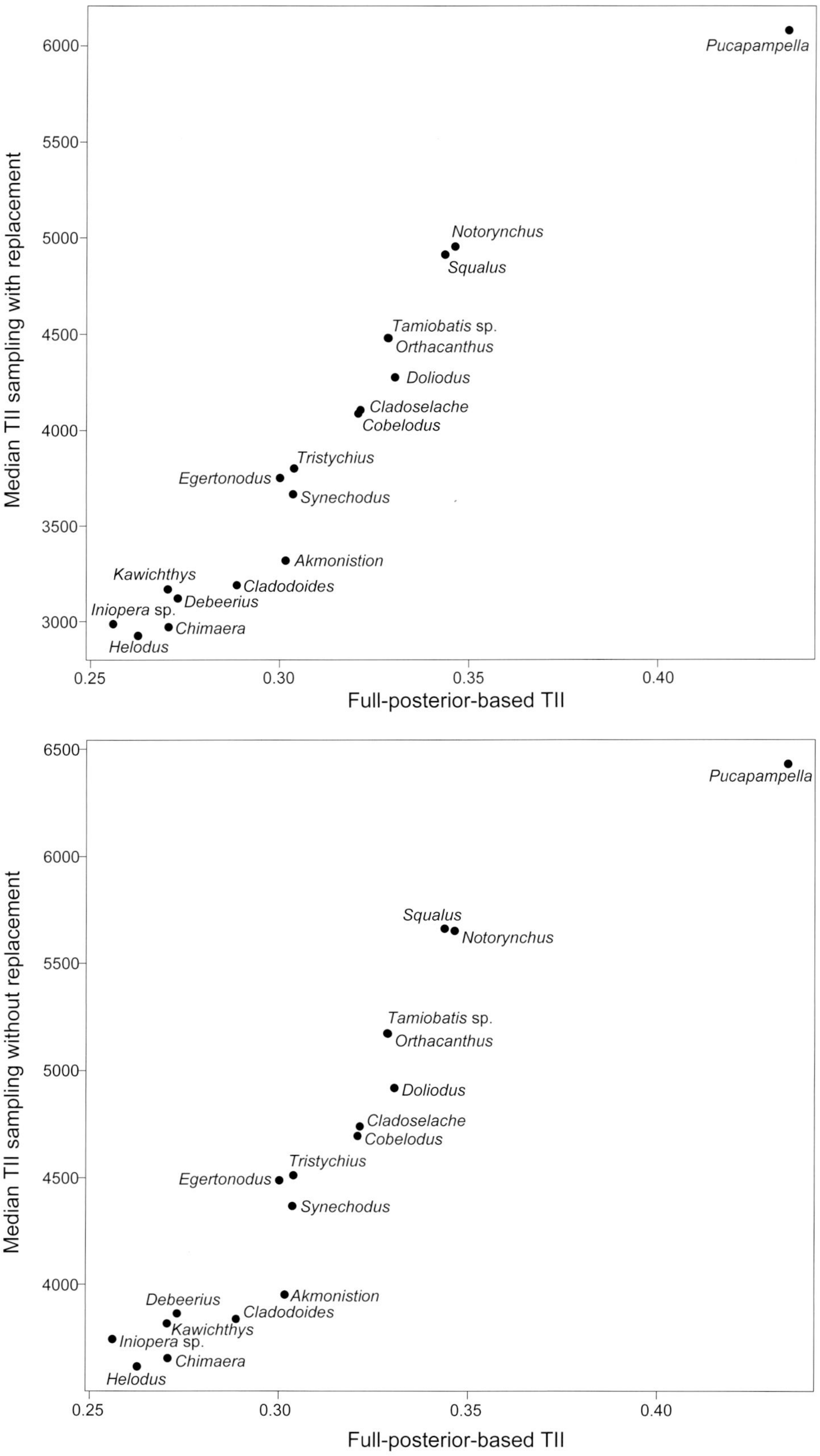

Fig. 5. Correlations between the resampling-based and full-posterior-based TII estimates. Rankings both sampling with ($F = 259.8$; $R^2 = 0.938$; $p < 0.001$) and without ($F = 140.3$; $R^2 = 0.891$; $p < 0.001$) replacement were strongly correlated with the rank values estimated from the full posterior, suggesting resampling-based approximation of the TII is an effective heuristic for the computationally intensive calculation from the full posterior.

et al. 2015), it does not appear to exhibit any stratigraphic mystery, as fossils for the genus appear to align with ages inferred from molecular data (Adnet & Cappetta 2001, Fig. 4; Straube et al. 2015, Fig. 2). The rank position for *Squalus* may potentially be a sampling artifact, but other explanations are worth investigating.

The consistent recovery across analysis methods of holocephalan taxa and one symmoriiform taxon (*Helodus*, *Kawichthys*, *Iniopera*, *Chimaera*, *Debeerius*) in the lower quartile of taxon influence belies the fact that these taxa are the focus of a major debate in chondrichthyan paleontology. This debate centers on the phylogenetic placement of symmoriiform taxa, which has varied based on the results of differential character weighting of individual datasets (e. g. Coates & Sequeira 2001b, Fig. 17A–B) and on unweighted parsimony analyses with different primary homologies and codings of several, mostly chondrocranial, characters (e. g., Coates & Sequeira 2001a, Pradel et al. 2011, Coates et al. 2017), reviewed in Maisey et al. (2019, p. 109). It is worth noting that although the holocephalan clade presents many unusual anatomies (e. g. Janvier 1996, Fig. 4.36B–C; Grogan & Lund 2000, Fig. 19), comparatively less discussion and amended taxonomic and/or anatomical (re-)description has centered on these taxa, relative to other taxa in this analysis (e. g. *Kawichthys* (Pradel et al. 2011), *Akmonistion* (Coates & Sequeira 2001b), *Tristychius* (Coates & Tietjen 2018), *Doliodus* (Maisey, Miller & Turner 2009, Burrow et al. 2017, Maisey et al. 2017, Maisey et al. 2019), and *Pucapampella* (Maisey et al. 2019). Holocephalan taxa are in need of broader taxonomic work.

Methods. The strong correlation between, and similar rbo-based rankings (Fig. 5) of, the resampling-based and full-posterior-based TII estimates suggest the resampling-based approaches proposed in Denton & Goolsby (2018) are both an effective and computationally feasible approximation to the prohibitively intensive full analysis. Resampling-based methods can run on laptop computers, making them accessible for researchers without cluster accounts.

It is worth noting that although the present analysis relies on the Mk family of models, alternatives to Mk models may be useful for future research on taxon influence, especially for large-scale morphological character matrices. During review of this chapter, a reader pointed out that, under likelihood approximations (Goloboff & Arias 2019), unweighted parsimony methods exhibited better AIC-based fit to the present dataset than did either the Mki + Γ model or implied weights parsimony. This observation highlights the need for exploration of the fit of models to morphological data (Goloboff et al. 2018), especially with regard to Bayesian TII approaches, as the methods employed stand to significantly influence the inferred posterior distributions of trees, regardless of how they are summarized. One possible way forward in this vein using existing methods would be to conduct Bayesian TII analysis using the No Common Mechanism model of Tuffley and Steel (Tuffley & Steel 1997) to generate a posterior distribution of trees with ranks equivalent to those obtained under unweighted parsimony, and conduct TII using this distribution of trees.

The inferred relationship between the Koch & Parry (2020) taxon uniqueness measure and the full-posterior taxon influence index is in agreement with the authors of the study, who observe a significant correlation between their measure and topological impact of taxa (Koch & Parry 2020, Fig. 6B). In the present analysis, the relationship was weak, but somewhat supported, and this relationship raises the question of why more analytically involved approaches to taxon-jackknifing-based influence assessment should be used, given the MORD-based uniqueness value is far quicker to compute than even the resampling-based TII estimates. However, given the differences in the analytical approach and study goal – we are interested in incorporating uncertainty into estimates of the structure of ranked lists – attempts to compare the performance of the uniqueness value with TII estimates in terms of capturing features of interest is premature.

A major analytical question that follows from our analysis is how significance values for taxon-specific TII estimates might be derived. A procedure for future work in this direction is outlined, as follows: given that the original Bayesian TII estimate is a point value obtained as the sum over all $|T'| \times |T^*|$ pairings of topologies from the T' (taxa pruned from alignment before tree inference) and T^* (taxa dropped from topology after tree inference) posterior distributions of trees, this sum, $\mathrm{med}_p[d(T^*, T')]$, may serve as a test statistic (TII*) against which a one-tailed test is applied, using pseudoreplicates. Preliminary investigations in this direction suggest that an approach using a basic bootstrapping procedure similar to that used for generating clade support values (Felsenstein 1985) fails to produce appropriate test distributions. An alternative approach lies in generating simulated datasets under a model of morphological evolution, using the majority-rule consensus or MCC tree and parameter estimates derived from the full analysis used to generate T^*, and running taxon influence analyses on each simulated dataset. Significance for each taxon i would then be assessed as the proportion of simulation-based TII values

(TII_i') greater than TII_i^*. However, because the computational complexity of TII in this implementation depends on the shape of the posterior distribution, the size of the individual trees to compare, and the number of taxa to remove, significant work must be undertaken to make such an approach practical for many-taxon datasets and for datasets producing flattened or long-tail posterior distributions, such as those with either low or conflicting phylogenetic signal.

Perhaps a more fundamental question, related to the one of taxon-specific significance, is how to assess the significance of the overall ranking itself. Such a question might be implied in, for example, asking whether the overall taxon influence ranking had significant structure. For this test, it seems plausible to suggest the same procedure for simulated datasets used to generate taxon-specific TII significance estimates. However, both the test statistic and significance test in the case of these ranked lists is unclear. Measures such as RBO, discussed above, or weighted Kendall's τ_w (Kendall 1938, Shieh 1998), another measure which allows for differential weighting at the tails of the lists, may be a potential way to generate a descriptive statistic, albeit without a significance value, by taking the ratio of the 'between-group' rank correlations between the TII* and TII' rankings and the 'within-group' rank correlations among TII rankings. This is an area in need of significant theory development.

Conclusions

In conclusion, we show that the results of model-based taxon influence analysis, incorporating a new implementation of a method to leverage a full posterior distribution of trees, produces a measurement that is not simply reducible to indices previously hypothesized to predict fossil taxon importance, including stratigraphic first occurrence. Resampling-based inference of taxon influence appears to be a good approximation of the computationally prohibitive TII based on the full posterior distribution. Agreement between resampling-based and full-posterior-based TII rankings suggests the inferred rank order to be robust. In the inferred rankings, holocephalan and symmoriiform taxa rank lowest, and a mixture of Paleozoic and Recent taxa rank highest, a somewhat surprising result that nonetheless underscores the need for continued work on early chondrichthyans, as pioneered by John. That these results emerge from a small dataset with limited taxon sampling relative to contemporary character matrices suggests, somewhat tantalizingly, that there may be significantly more information of use to paleontologists in posterior distributions of trees than is currently leveraged by existing methods.

Acknowledgements

The first author would like to thank John Maisey for his years of collegiality, friendship, and postdoctoral support while the first author was an Axelrod Postdoctoral Fellow at the AMNH, and for introducing the author to the conceptual problem of the value of fossils. We thank both Cécile Ané and Pablo Goloboff for their input on an earlier version of this chapter, which greatly improved both the presentation and computational elements. The first author also thanks Gloria Arratia and Allison Bronson for discussion related to what would become this volume, and Philippe Janvier and Alan Pradel for their editorial work, collaboration, and friendship. Specimen silhouettes presented in Figures 2–4 were generated from non-silhouette original line drawings and illustrations, as follows: *Orthacanthus*, modified from original artwork of *O. senckenbergianus* (© Nobu Tamura) on Wikimedia Commons, and shared by a Creative Commons Attribution ShareAlike license; *Tamiobatis*, modified from Romer (1964, Fig. 1), after original artwork (Museum of Comparative Zoology, Harvard University and ©President and Fellows of Harvard College); *Notorynchus*, modified from original artwork of *N. cepedianus* by Dr. Tony Ayling from a drawing in Ayling & Cox (1982), and shared on Wikimedia Commons by a Creative Commons ShareAlike 1.0 Generic license (https://creativecommons.org/licenses/by-sa/1.0/deed.en); *Squalus*, modified from artwork of *S. acanthias* originally by H. L. Todd, in the Marine and Freshwater Image Bank at the University of Washington; *Pucapampella*, modified from Janvier & Racheboeuf (2018, Fig. 3); *Doliodus*, modified from original artwork by Dr. Susan Turner © 2003 (www.paleodeadfish.com). The second author would like to acknowledge the University of Central Florida Advanced Research Computing Center.

Literature cited

Aberer, A. J., D. Krompass & A. Stamatakis. 2013. Pruning rogue taxa improves phylogenetic accuracy: an efficient algorithm and webservice. Systematic Biology, 62: 162–166.

Adnet, S. & H. Cappetta. 2001. A palaeontological and phylogenetical analysis of squaliform sharks (Chondrichthyes: Squaliformes) based on dental characters. Lethaia, 34: 234–248.

Ax, P. & R. Jeffries. 1987. The phylogenetic system: the systematization of organisms on the basis of their phylogenesis. John Wiley & Sons, Chichester, NY, 340 pp.

Ayling, T. & G. J. Cox. 1982. Guide to the Sea Fishes of New Zealand. International Specialized Book Service Incorporated. 343 pp.

Baele, G., P. Lemey, T. Bedford, A. Rambaut, M. A. Suchard & A. V. Alekseyenko. 2012. Improving the accuracy of demographic and molecular clock model comparison while accommodating phylogenetic uncertainty. Molecular Biology and Evolution, 29: 2157–2167.

Bell, M. A. & G. T. Lloyd. 2015. strap: an R package for plotting phylogenies against stratigraphy and assessing their stratigraphic congruence. Palaeontology, 58: 379–389.

Box, G. E. & D. R. Cox. 1964. An analysis of transformations. Journal of the Royal Statistical Society: Series B (Methodological), 26: 211–243.

Brazeau, M. D. 2009. The braincase and jaws of a Devonian 'acanthodian' and modern gnathostome origins. Nature, 457: 305–308.

Burrow, C., J. den Blaauwen, M. Newman & R. Davidson. 2016. The diplacanthid fishes (Acanthodii, Diplacanthiformes, Diplacanthidae) from the Middle Devonian of Scotland. Palaeontologia Electronica, 19: 1–83.

Burrow, C. J., D. C. Hovestadt, M. Hovestadt-Euler, S. Turner & G. C. Young. 2008. New information on the Devonian shark *Mcmurdodus*, based on material from western Queensland, Australia. Acta Geologica Polonica, 58: 155–163.

Burrow, C. J., S. Turner, J. G. Maisey, S. Desbiens & R. F. Miller. 2017. Spines of the stem chondrichthyan *Doliodus latispinosus* (Whiteaves) comb. nov. from the Lower Devonian of eastern Canada. Canadian Journal of Earth Sciences, 54: 1248–1262.

Cappetta, H. 1987. Handbook of Paleoichthyology, Vol. 3B: Chondrichthyes II. Gustav Fischer Verlag, Stuttgart, Germany and New York, NY, 193 pp.

Choo, B. 2012. Revision of the actinopterygian genus *Mimipiscis* (= *Mimia*) from the Upper Devonian Gogo Formation of Western Australia and the interrelationships of the early Actinopterygii. Earth and Environmental Science Transactions of the Royal Society of Edinburgh, 102: 77–104.

Coates, M. I., J. A. Finarelli, I. J. Sansom, P. S. Andreev, K. E. Criswell, K. Tietjen, M. L. Rivers & P. J. La Riviere 2018. An early chondrichthyan and the evolutionary assembly of a shark body plan. Proceedings of the Royal Society B, 285: 20172418.

Coates, M. I., R. W.Gess, J. A. Finarelli, K. E. Criswell & K. Tietjen. 2017. A symmoriiform chondrichthyan braincase and the origin of chimaeroid fishes. Nature, 541: 208–211.

Coates, M. I. & S. E. K. Sequeira. 2001a. Early sharks and primitive gnathostome interrelationships. Major Events in Early Vertebrate Evolution (ed. P. E. Ahlberg), pp. 241–262. Taylor & Francis, London, UK.

Coates, M. I. & S. E. K. Sequeira. 2001b. A new stethacanthid chondrichthyan from the Lower Carboniferous of Bearsden, Scotland. Journal of Vertebrate Paleontology, 21: 438–459.

Coates, M. I. & K. Tietjen. 2018. The neurocranium of the Lower Carboniferous shark *Tristychius arcuatus* (Agassiz, 1837). Earth and Environmental Science Transactions of the Royal Society of Edinburgh, 108: 19–35.

Davis, S. P., J. A. Finarelli & M. I. Coates. 2012. *Acanthodes* and shark-like conditions in the last common ancestor of modern gnathostomes. Nature, 486: 247–250.

Denton, J. S. S. & E. W. Goolsby. 2018. Measuring inferential importance of taxa using taxon influence indices. Ecology and Evolution, 8: 4484–4494.
Donoghue, M. J., J. A. Doyle, J. Gauthier, A. G. Kluge & T. Rowe. 1989. The importance of fossils in phylogeny reconstruction. Annual Review of Ecology and Systematics, 20: 431–460.
Doyle, J. A. & M. J. Donoghue. 1987. The importance of fossils in elucidating seed plant phylogeny and macroevolution. Review of Palaeobotany and Palynology, 50: 63–95.
Felsenstein, J. 1985. Confidence limits on phylogenies: an approach using the bootstrap. Evolution, 39: 783–791.
Finden, C. & A. Gordon. 1985. Obtaining common pruned trees. Journal of Classification, 2: 255–276.
Foote, M. & J. J. Sepkoski, Jr. 1999. Absolute measures of the completeness of the fossil record. Nature, 398: 415–417.
Gauthier, J., A. G. Kluge & T. Rowe. 1988. Amniote phylogeny and the importance of fossils. Cladistics, 4: 105–209.
Giles, S., M. Friedman & M. D. Brazeau. 2015. Osteichthyan-like cranial conditions in an Early Devonian stem gnathostome. Nature, 520: 82–85.
Ginter, M., O. Hampe & C. J. Duffin. 2010. Handbook of Paleoichthyology, Volume 3D: Chondrichthyes: Paleozoic Elasmobranchii: Teeth. Verlag Dr. Friedrich Pfeil, München, Germany, 168 pp.
Goloboff, P. A. 2008. Calculating SPR distances between trees. Cladistics, 24: 591–597.
Goloboff, P. A. 2014. Extended implied weighting. Cladistics, 30: 260–272.
Goloboff, P. A. & J. S. Arias. 2019. Likelihood approximations of implied weights parsimony can be selected over the Mk model by the Akaike information criterion. Cladistics, 35: 695–716.
Goloboff, P. A., J. M. Carpenter, J. S. Arias & D. R. M. Esquivel. 2008. Weighting against homoplasy improves phylogenetic analysis of morphological data sets. Cladistics, 24: 758–773.
Goloboff, P. A., J. S. Farris & K. C. Nixon. 2008. TNT, a free program for phylogenetic analysis. Cladistics, 24: 774–786.
Goloboff, P. A., M. Pittman, D. Pol & X. Xu. 2018. Morphological data sets fit a common mechanism much more poorly than DNA sequences and call into question the Mkv model. Systematic Biology, 68: 494–504.
Goloboff, P. A., A. Torres & J. S. Arias. 2018. Weighted parsimony outperforms other methods of phylogenetic inference under models appropriate for morphology. Cladistics, 34: 407–437.
Gordon, A. 1979. A measure of the agreement between rankings. Biometrika, 66: 7–15.
Grogan, E. D. & R. Lund. 2000. *Debeerius ellefseni* (fam. nov., gen. nov., spec. nov.), an autodiastylic chondrichthyan from the Mississippian Bear Gulch Limestone of Montana (USA), the relationships of the Chondrichthyes, and comments on gnathostome evolution. Journal of Morphology, 243: 219–245.
Huelsenbeck, J. P. 1991. When are fossils better than extant taxa in phylogenetic analysis? Systematic Biology, 40: 458–469.
Janvier, P. 1996. Early Vertebrates. Oxford University Press, Oxford, UK, 393 pp.
Janvier, P. & P. Racheboeuf. 2018. The Paleozoic vertebrates of Bolivia, with comments on the faunal and environmental context of the 'Malvinokaffric Realm'. Fósiles y Facies de Bolivia (eds. Riglos, M. S., A. D. Farjat & M. Á. P. Leyton), pp. 22–35. Sincronía Diseño & Publicidad, Santa Cruz de la Sierra, Bolivia.
Kearney, M. 2002. Fragmentary taxa, missing data, and ambiguity: mistaken assumptions and conclusions. Systematic Biology, 51: 369–381.
Kearney, M. & J. M. Clark. 2003. Problems due to missing data in phylogenetic analyses including fossils: a critical review. Journal of Vertebrate Paleontology, 23: 263–274.
Kendall, M. G. 1938. A new measure of rank correlation. Biometrika, 30: 81–93.
King, B., T. Qiao, M. S. Lee, M. Zhu & J. A. Long. 2017. Bayesian morphological clock methods resurrect placoderm monophyly and reveal rapid early evolution in jawed vertebrates. Systematic Biology, 66: 499–516.
Koch, N. M. & L. A. Parry. 2020. Death is on our side: paleontological data drastically modify phylogenetic hypotheses. Systematic Biology, 69: 1052–1067.
Lewis, P. O. 2001. A likelihood approach to estimating phylogeny from discrete morphological character data. Systematic Biology, 50: 913–925.
Lloyd, G. T. 2016. Estimating morphological diversity and tempo with discrete character-taxon matrices: implementation, challenges, progress, and future directions. Biological Journal of the Linnean Society, 118: 131–151.
Maisey, J. G. 1986. Heads and tails: a chordate phylogeny. Cladistics, 2: 210–256.
Maisey, J. G. 1987. Cranial anatomy of the Lower Jurassic shark *Hybodus reticulatus* (Chondrichthyes, Elasmobranchii): with comments on hybodontid systematics. American Museum Novitates, 2878: 1–39.
Maisey, J. G. 2001a. CT-scan reveals new cranial features in Devonian chondrichthyan *"Cladodus" wildungensis*. Journal of Vertebrate Paleontology, 21: 807–810.
Maisey, J. G. 2001b. A primitive chondrichthyan braincase from the Middle Devonian of Bolivia. Major Events in Early Vertebrate evolution (ed. P. E. Ahlberg), pp. 263–288. Taylor & Francis, London, UK.
Maisey, J. G. 2001c. Remarks on the inner ear of elasmobranchs and its interpretation from skeletal labyrinth morphology. Journal of Morphology, 250: 236–264.
Maisey, J. G. 2005. Braincase of the Upper Devonian shark *Cladodoides wildungensis* (Chondrichthyes, Elasmobranchii), with observations on the braincase in early chondrichthyans. Bulletin of the American Museum of Natural History, 288: 1–103.
Maisey, J. G. 2013. The diversity of tessellated calcification in modern and extinct chondrichthyan. Revue de Paleobiologie, 32: 355–371.
Maisey, J. G., P. Janvier, A. Pradel, J. S. S. Denton, A. W. Bronson, R. Miller & C. J. Burrow. 2019. *Doliodus* and Pucapampellids: Contrasting Perspectives on Stem Chondrichthyan Morphology. Evolution and Development of Fishes (eds. Johanson, Z., C. Underwood & M. Richter), pp. 87–109. Cambridge University Press, Cambridge, United Kingdon.
Maisey, J. G., R. Miller, A. Pradel, J. S. Denton, A. Bronson & P. Janvier. 2017. Pectoral morphology in *Doliodus*: bridging the 'acanthodian'-chondrichthyan divide. American Museum Novitates, 2017: 1–16.
Maisey, J. G., R. Miller & S. Turner. 2009. The braincase of the chondrichthyan *Doliodus* from the Lower Devonian Campbellton Formation of New Brunswick, Canada. Acta Zoologica, 90: 109–122.
Maisey, J. G., G. J. P. Naylor & D. Ward. 2004. Mesozoic elasmobranchs, neoselachian phylogeny and the rise of modern elasmobranch diversity. – Pp. 17–56 in Mesozoic Fishes 3: Systematics, Paleoenvironments, and Biodiversity. Verlag Dr. Friedrich Pfeil, München, Germany.
Mariadassou, M., A. Bar-Hen & H. Kishino. 2012. Taxon influence index: assessing taxon-induced incongruities in phylogenetic inference. Systematic Biology, 61: 337–345.
Novacek, M. J. 1992. Fossils, topologies, missing data, and the higher level phylogeny of eutherian mammals. Systematic Biology, 41: 58–73.
O'Reilly, J. E., M. N. Puttick, L. Parry, A. R. Tanner, J. E. Tarver, J. Fleming, D. Pisani & P. C. Donoghue. 2016. Bayesian methods outperform parsimony but at the expense of precision in the estimation of phylogeny from discrete morphological data. Biology Letters, 12: 20160081.
Okada, C. Y., D. C. G. Pedronette & R. da S Torres. 2015. Unsupervised distance learning by rank correlation measures for image retrieval. Proceedings of the 5th ACM on International Conference on Multimedia Retrieval, pp. 331–338. ACM, Shainghai, China.
Patterson, C. 1981. Significance of fossils in determining evolutionary relationships. Annual Review of Ecology and Systematics, 12: 195–223.
Patterson, C. & D. E. Rosen. 1977. Review of ichthyodectiform and other Mesozoic teleost fishes, and the theory and practice of classifying fossils. Bulletin of the American Museum of Natural History, 158: 81–172.
Platnick, N. I., C. E. Griswold & J. A. Coddington. 1991. On missing entries in cladistic analysis. Cladistics, 7: 337–343.
Pradel, A., J. G. Maisey, P. Tafforeau, R. H. Mapes & J. Mallatt. 2014. A Palaeozoic shark with osteichthyan-like branchial arches. Nature, 509: 608–611.
Pradel, A., P. Tafforeau, J. G. Maisey & P. Janvier. 2011. A New Paleozoic Symmoriiformes (Chondrichthyes) from the Late Carboniferous of Kansas (USA) and cladistic analysis of early chondrichthyans. PLoS ONE, 6: e24938.
R Core Team. 2016. R: A language and environment for statistical computing. R Foundation for Statistical Computing, Vienna, Austria.
Romer, A. S. 1964. The braincase of the Paleozoic elasmobranch *Tamiobatis*. Bulletin of the Museum of Comparative Zoology, 131: 89–106.

Ronquist, F., M. Teslenko, P. van der Mark, D. L. Ayres, A. Darling, S. Höhna, B. Larget, L. Liu, M. A. Suchard & J. P. Huelsenbeck. 2012. MrBayes 3.2: efficient Bayesian phylogenetic inference and model choice across a large model space. Systematic Biology, 61: 539–542.

Sansom, R. S., P. G. Choate, J. N. Keating & E. Randle. 2018. Parsimony, not Bayesian analysis, recovers more stratigraphically congruent phylogenetic trees. Biology Letters, 14: 20180263.

Schmich, F., E. Szczurek, S. Kreibich, S. Dilling, D. Andritschke, A. Casanova, S. H. Low, S. Eicher, S. Muntwiler & M. Emmenlauer. 2015. gespeR: a statistical model for deconvoluting off-target-confounded RNA interference screens. Genome Biology, 16: 220.

Seidel, R., K. Lyons, M. Blumer, P. Zaslansky, P. Fratzl, J. C. Weaver, & M. N. Dean. 2016. Ultrastructural and developmental features of the tessellated endoskeleton of elasmobranchs (sharks and rays). Journal of Anatomy, 229: 681–702.

Shieh, G. S. 1998. A weighted Kendall's tau statistic. Statistics & Probability Letters, 39: 17–24.

Stahl, B. J. 1999. Handbook of Paleoichthyology Volume 4: Chondrichthyes III: Holocephali. Verlag Dr. Friedrich Pfeil, München, Germany, 164 pp.

Stamatakis, A. 2014. RAxML version 8: a tool for phylogenetic analysis and post-analysis of large phylogenies. Bioinformatics, 30: 1312–1313.

Straube, N., C. Li, J. M. Claes, S. Corrigan & G. J. P. Naylor. 2015. Molecular phylogeny of Squaliformes and first occurrence of bioluminescence in sharks. BMC Evolutionary Biology, 15: 162.

Tuffley, C. & M. Steel. 1997. Links between maximum likelihood and maximum parsimony under a simple model of site substitution. Bulletin of Mathematical Biology, 59: 581–607.

Van Valen, L. M. 1978. Why not to be a cladist. Evolutionary Theory, 3: 285–299.

Vélez-Zuazo, X. & I. Agnarsson. 2011. Shark tales: a molecular species-level phylogeny of sharks (Selachimorpha, Chondrichthyes). Molecular Phylogenetics and Evolution, 58: 207–217.

Warren, D. L., A. J. Geneva & R. Lanfear. 2017. RWTY (R We There Yet): An R package for examining convergence of Bayesian phylogenetic analyses. Molecular Biology and Evolution, 34: 1016–1020.

Watanabe, A. 2016. The impact of poor sampling of polymorphism on cladistic analysis. Cladistics, 32: 317–334.

Webber, W., A. Moffat & J. Zobel. 2010. A similarity measure for indefinite rankings. ACM Transactions on Information Systems (TOIS), 28: 1–38.

Wiens, J. J. 2003. Missing data, incomplete taxa, and phylogenetic accuracy. Systematic Biology, 52: 528–538.

Williams, M. E. 2001. Tooth retention in cladodont sharks: with a comparison between primitive grasping and swallowing, and modern cutting and gouging feeding mechanisms. Journal of Vertebrate Paleontology, 21: 214–226.

Wilson, M. V. 1992. Importance for phylogeny of single and multiple stem-group fossil species with examples from freshwater fishes. Systematic Biology, 41: 462–470.

Wright, A. M. & D. M. Hillis. 2014. Bayesian analysis using a simple likelihood model outperforms parsimony for estimation of phylogeny from discrete morphological data. PLoS ONE, 9: e109210.

Wright, A. M., G. T. Lloyd & D. M. Hillis. 2016. Modeling character change heterogeneity in phylogenetic analyses of morphology through the use of priors. Systematic Biology, 65: 602–611.

Zangerl, R. & G. R. Case. 1973. Iniopterygia: a new order of Chondrichthyan fishes from the Pennsylvanian of North America. Field Museum of Natural History. Fieldiana Geology Memoirs, 1167: 1–67.

Zhu, M., X. Yu, P. E. Ahlberg, B. Choo, J. Lu, T. Qiao, Q. Qu, W. Zhao, L. Jia & H. Blom. 2013. A Silurian placoderm with osteichthyan-like marginal jaw bones. Nature, 502: 188–193.

Ancient Fishes and their Living Relatives: a Tribute to John G. Maisey. Alan Pradel, John S. S. Denton & Philippe Janvier (eds.): pp. 59–70, 6 figs., 6 tabs.
© 2021 by Verlag Dr. Friedrich Pfeil, München, Germany – ISBN 978-3-89937-269-4

Modularity of the Weberian apparatus in the zebrafish using micro-CT technology and 3-D geometric morphometrics

Kevin K. Duclos*, Terry C. Grande** and Richard Cloutier***

Abstract

Complex morphological systems are subject to constraints that divide the phenotype into modules (i.e., discrete units of variation), which influence their developmental and evolutionary trajectories. Some constraints are intrinsically developmental, and some arise due to mechanical and functional restrictions. The covariance structure of objects can be used to investigate the modularity of a system and test hypotheses based on functional, ecological, developmental or other factors. This approach is often used in geometric morphometrics and allows for in-depth investigation of developmental and evolutionary constraints. The Weberian apparatus, a crucial system for the auditory function of otophysan fishes (e.g., minnows, suckers, catfishes), is known for its complexity and its disparity across taxa. The ecological success of otophysans has occasionally been attributed to the evolution of the Weberian apparatus. Despite this, little is known about the processes that shape the evolution of the Weberian apparatus and how variable it is within taxa. The main objective of this study is to test the modular partitioning of the Weberian apparatus, and its modularity as distinct from cranial and trunk skeleton. Here we assess phenotypic variation and modularity of the Weberian apparatus and associated elements in the zebrafish, *Danio rerio*, using micro-computed tomography and three-dimensional geometric morphometrics. Our results show high levels of phenotypic variation in the Weberian apparatus. Maximum-likelihood analyses suggest that the Weberian apparatus is modular with respect to the basicranium and post-Weberian vertebrae. Our results further suggest patterning of the Weberian apparatus into three functional modules: a module including centra and modified pleural ribs, a second module including neurals and supraneural with the Weberian ossicles sensu stricto forming their own modules. This modular patterning also involves strong covariational tendencies between modules, which appear to be dependent to some extent on allometry and other developmental processes, that constrain their variability. This approach provides new insights into the evolution of complex structures such as the Weberian apparatus and how the space of possible phenotypes is constrained during evolution.

Introduction

Adult phenotypes result from dynamic interactions among developmental mechanisms and their sensitivity to external factors. Complex morphological systems are subjected to a variety of intrinsic and extrinsic constraints (e.g., mechanical constraints, ecological context) during development (Gould & Lewontin 1979, Maynard-Smith 1985, Arnold 1992, Revell et al. 2007, Hallgrímsson et al. 2009, Sanger et al. 2011). These intrinsic and extrinsic factors limit the amount of phenotypic variability, the ability to produce variation, for a given species or population. Many intrinsic factors have to do with developmental stability and instability that affect the capacity of a given genotype to produce a stable, repeatable, phenotypic outcome across environments (Waddington & Robertson 1966, Siegal & Bergman 2002, Félix & Wagner 2008).

An important developmental property that enables traits to vary coordinately or independently from one another is modularity. Modularity is broadly defined as the capacity for systems to behave as quasi-independent units (Wagner et al. 2007, Goswami 2007, Klingenberg 2008, Mitteroecker 2009). This translates at different levels of biological organization (Wagner et al. 2007, Klingenberg 2008, Mitteroecker 2009). For instance, developmental modularity often refers to the capacity of a trait or a group of traits to develop consistently independent of their context (Raff & Raff 2000, Schlosser & Wagner 2004, Wagner et al. 2007, Klingenberg 2008). It has also been used to describe traits that share developmental timing and/or pathways. More commonly, modularity is mentioned in the context of variational modularity where modules are networks of interacting elements behaving as highly integrated traits (i.e., traits that are strongly correlated) which covary strongly together but weakly with other sets of elements, acting as discrete units of variation (Klingenberg et al. 2001, Chernoff & Magwene 1999, Zelditch et al. 2008, Larouche et al. 2015, 2018). Complex phenotypes are thus decomposed into modules, and this modular partitioning influences developmental and evolutionary trajectories. Modularity, regardless of the level of biological organization investigated, allows systems to vary or evolve independently from other systems, allowing for phenotypic space to be explored while reducing the risk that adapted systems become maladapted as a result of covariation (Hansen 2003, Hansen & Houle 2008, Pavlicev & Hansen 2011, Vilmoare 2013). Modularity has thus been targeted as a primary evolutionary developmental component facilitating the evolvability of complex systems (Raff & Raff 2000, Hansen 2003, Schlosser & Wagner 2004, Wagner et al. 2007, Klingenberg 2008, Hansen & Houle 2008, Pavlicev & Hansen 2011, Vilmoare 2013).

The Weberian apparatus, originally described by Ernst Heinrich Weber (1820), and considered diagnostic of the ostariophysan subgroup Otophysi (e.g., carp, minnows, catfishes; Rosen & Greenwood 1970) is one such complex system. It consists of a series of modifications of the first four vertebral centra and associated neural arches, supraneurals and pleural ribs that lie in a linear sequence forming a mechanical linkage between the wall of a modified gas bladder to the sinus impar of the inner ear. This linkage enables otophysans to hear at frequencies up to several thousand Hertz thereby increasing auditory sensitivity throughout the audible frequency range

* Department of Cell Biology and Anatomy, The University of Calgary, 2500 University Dr NW, Calgary, Alberta, T2N 1N4, Canada; Kevin.duclos@ucalgary.ca (corresponding author)
** Department of Biology, Loyola University Chicago, 1032 W Sheridan Rd, Chicago, Illinois, 60660, USA
*** Département de biologie, chimie et géographie, Université du Québec à Rimouski, 300, allée des Ursulines, C.P. 3300, succ. A, Rimouski. Québec, G5L 3A1, Canada

(Fay 1988, Higgs et al. 2004, Braun & Grande 2008). It is not surprising that some researchers have attributed the presence of the Weberian apparatus in these hearing specialists with the group's widespread ecological and evolutionary success (Grande & de Pinna 2004). Despite this potentially crucial role in the evolution of otophysans, little is known concerning the developmental constraints and modular partitioning which has impacted the evolution of the Weberian apparatus across taxa. Among otophysan taxa, Cypriniformes exhibit the most generalized Weberian apparatus morphology (Watson 1939, Ramaswami 1955, Chardon & Vandewalle 1997); Cyprinidae in turn, most resembles the ancestral condition (Bird & Hernandez 2007). As described for the cyprinid *Danio rerio* (Sanger & McCune 2002, Bird & Mabee 2003, Grande & Young 2004), the Weberian apparatus is divided into two sections. First, the pars auditum consists of paired auditory ossicles (i. e., scaphium intercalarium and tripus), that are directly involved in sound transmission. The tip of the tripus (i. e., transformator process) articulates with the tunica externa, or anterior surface of the gas bladder, and is connected rostrally to the intercalarium by the ligamentum tripe. The concha scaphium forms the posterolateral wall of the sinus impar perilymphaticus. Second, the pars sustentaculum serves a more structural role supporting the auditory ossicles of the pars auditum and is not directly involved in sound transmission. Elements of the pars sustentaculum include the centra themselves as well as neural arches, supraneurals, parapophyses of the first, second and fourth vertebrae, the fourth pleural rib and the os suspensorium. As a fish swims through a high frequency sound field, displacement waves cause the gas bladder to pulsate. Vibrations are transmitted from it to the tripus and then to the intercalarium. From the intercalarium, vibrations are transmitted to the scaphium and then finally into the inner ear (Grande & Young 2004).

Significant disparity in Weberian apparatus structure exists among otophysan subgroups. For example, within Siluriphysi (e. g., catfishes and knifefishes), the scaphium is missing as a separate ossicle in callichthyids (Coburn & Grubach 1998), and in many gymnotiforms, the intercalarium is reduced to an intraligamentous ossification (Fink & Fink 1981). The claustrum is secondarily lost in gymnotids (Fink & Fink 1981). Chardon & Vendewalle (1997) suggest that the reduction in size and ossicle number might serve to increase the efficiency of sound transmission from the gas bladder to the inner ear by decreasing the inertia of the ossicles. According to Bird & Hernandez (2007), among cypriniforms, disparity of the Weberian apparatus is mostly restricted to the pars sustentaculum. Variation among ossicles is minor and restricted to subtle changes in shape, length, or the loss of processes of the tripus, intercalarium and scaphium. Nevertheless, the morphology of the Weberian apparatus in cypriniforms appears to be diagnostic for each family, with Cyprinidae exhibiting the generalized morphology as seen in the zebrafish (Bird & Hernandez 2007). Bird and Hernandez (2007) suggested that the conservatism in ossicle shape within cypriniform families might be the result of structural constraints associated with the functionality of the Weberian ossicles.

Although a body of work regarding developmental pathways and timing in the Weberian apparatus is slowly emerging (Bogutskaya 1991, Ichiyanagi et al. 1996, de Pinna & Grande 2003, Britz & Hoffmann 2006), few studies have looked at within-taxa developmental properties and variability. Conway & Britz (2007) have observed important sexual dimorphism in the Weberian apparatus as well as adjacent morphological systems such as the basioccipital bone and post-Weberian vertebrae in *Sundadanio axelrodi*. The extreme morphological differences in the Weberian apparatus between males and females, and the fact that they are not evenly distributed across Weberian elements suggest that elements of the Weberian apparatus could act as modules and vary independently from one another. Thus, as suggested by Patterson (1984), the evolution of the Weberian apparatus is much more complex than once thought and likely evolved in subunits that together function as a sound transducer in otophysans.

Patterns of modularity and integration within the Weberian apparatus and interacting elements are therefore likely to affect variability both at the intra- and interspecific scales. Furthermore, these patterns could play a crucial role in the evolution and tailoring of the acoustic sensitivity of different groups to their environments. Higgs et al. (2003) have observed that acoustic sensitivity in the zebrafish, *Danio rerio*, changed ontogenetically (from 10 mm to 45 mm of total length) between 200 Hz and 4000 Hz. During that phase of development, Weberian ossicles form and become ossified. Plastic developmental responses targeting modules within the Weberian apparatus could therefore have a significant bearing on the auditory function and its variability across environments within species and populations. The objectives of this study are twofold. First, we establish framework for the study of variation and variational constraints within the Weberian apparatus and offer new testable hypotheses regarding variation and modularity in the Weberian apparatus. We use micro-computed tomography and three-dimensional geometric morphometrics to describe and quantify shape variation within the Weberian apparatus and associated elements. Secondly, we test the modular independence of the Weberian vertebrae and the Weberian apparatus with regards to non-Weberian vertebrae and the cranium as well as test four hypotheses of modular partitioning of the Weberian apparatus itself, based on (1) the ossification sequence of *Danio rerio* (developmental groupings), (2) structural groupings and (3) functional groupings.

Material and methods

Model system and specimen collection. The zebrafish is commonly used as a model for studies of developmental variation (Meyer et al. 1993, Guryev et al. 2006, Westerfield 2007). Because of its widespread use as a model, a wide array of information concerning its skeletal development, including chondrification and ossification sequences and genetic pathways involved in morphogenesis, has been collected (Meyer et al. 1993, Sprague et al. 2006, 2007, Guryev et al. 2006, Westerfield 2007). Incidentally, the developmental sequence of the Weberian apparatus and its anatomy are well described in zebrafish, making it an ideal model to investigate modularity and variability (Grande & Young 2004).

A total of 42 zebrafish specimens were obtained from a private collection and catalogued into the Loyola University fish collection (LUC 7362). Specimens were preserved in 10 % buffered formalin and transferred to 70 % ethyl alcohol within three days for long term storage. All specimens used for this study were wild-type, obtained from a wild-caught zebrafish line, not laboratory reared lineages. Covariation structure and patterns of modularity in laboratory reared models are often different from what is encountered in natural populations (Jamniczky & Hallgrímsson 2009, Parsons et al. 2018). Because of this, wild specimens are generally a better indicator of developmental properties of natural populations.

Microcomputed tomography and data processing. X-ray microtomography (μ-CT) enables in-depth studies of the skeleton and its in situ organization. By offering high spatial resolution and a means to obtain and process information digitally without dissection and the risk of breaking skeletal elements, μ-CT allows for systems such as the Weberian apparatus to be described and measured with unprecedented depth (Maisey 2001, 2004, Metscher 2009, Cunningham et al. 2014, Jamniczky et al. 2015).

Specimens of *Danio rerio* were scanned within seven days of initial preservation using a Quantum GX microCT imaging system microtomographer (PerkinElmer). All specimens were scanned with the same settings (Voxel size: 26.5 μm, source current: 88 mA, source voltage: 90 V, exposure time: 840 ms). Projection slices were exported as Dicom files, which were then imported within the Stratovan Checkpoint software for subsequent analyses (Version 2018.09.07.0325, Stratovan Corporation, 2018, U.S.).

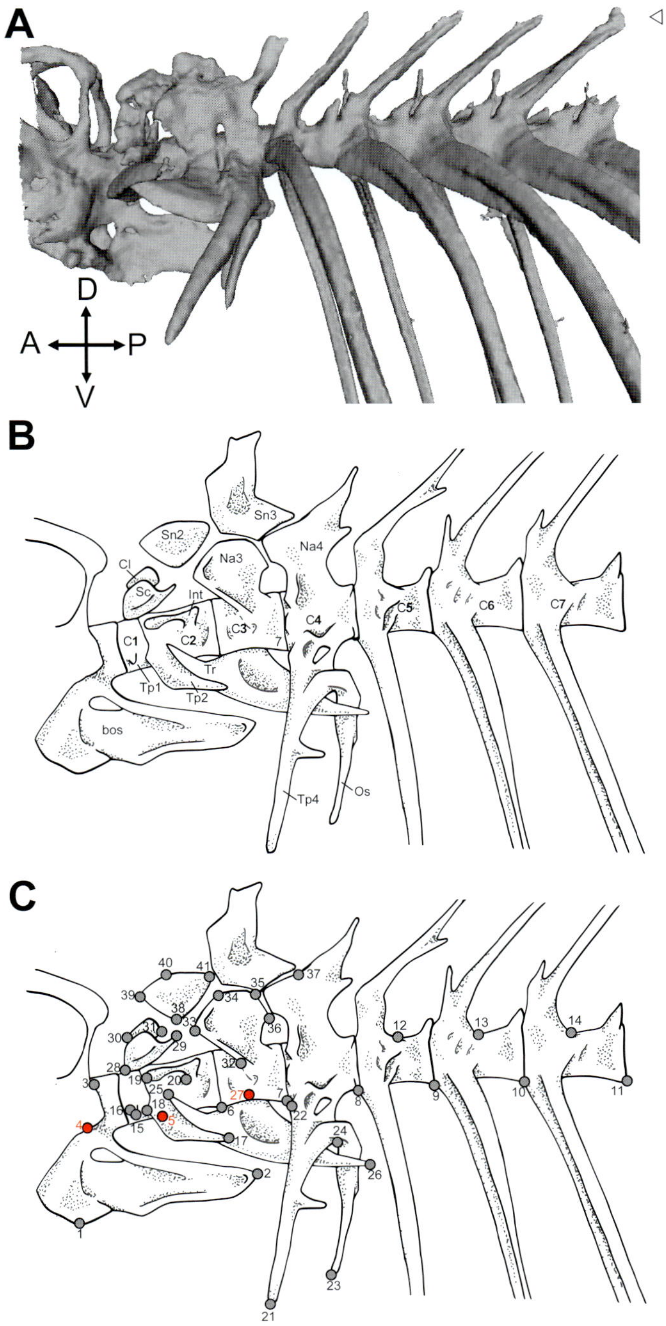

◁ **Fig. 1.** Anatomy of the Weberian apparatus and landmark digitization scheme. **A.** 3D render of the Weberian apparatus of *Danio rerio*. **B.** Annotated anatomy of the Weberian apparatus. **C.** Landmark digitization scheme (landmarks 5–8 and 15–41 correspond to the Weberian apparatus sensu stricto). Landmarks are numbered following their digitization order. Red landmarks are beneath structures and are not directly visible in this illustration. Landmark configuration and illustration are based on mean shape for the dataset studied. bos, basioccipital; C1–7, centra 1 to 7; Cl, claustrum; Int, intercalarium; Na3–4, neural arches 3 and 4; Os, os suspensorium; Sc, scaphium; Sn2–3, supraneurals 2 and 3; Tp1, 2 and 4, transverse processes 1, 2 and 4; Tr, tripus. The axes shown are the dorso-ventral axis (D–V) and the antero-posterior axis (A–P).

Geometric morphometrics. Recent progress in the implementation of three-dimensional geometric morphometrics (3DGM) combined with μ-CT technology, which produces high-quality and high-fidelity models, are powerful tools to quantify subtle shape variation in three-dimensional space (Adams et al. 2004, Mitteroecker & Gunz 2009, Parr et al. 2012, Pomidor et al. 2016). Landmark-based 3DGM are an increasingly popular tool to describe and quantify morphological variation in both ecology and evolutionary developmental biology to investigate ecological, functional and developmental constraints such as modularity and integration (Klingenberg 2009, Klingenberg & Marugán-Lobón 2013, Aguirre et al. 2016, Bardua et al. 2019, Gray et al. 2019). While the use of 3DGM has tended to be limited to simple systems, it is progressively being used to describe more and more complex shapes (Adams et al. 2013). Current and future research on the Weberian apparatus and its evolution can benefit greatly from geometric morphometrics and quantitative approaches which focus on variation and how it is structured within populations.

Landmark digitization was performed in the Stratovan Checkpoint software. Specimens were isosurfaced using a density threshold within the software and a total of 70 landmarks were digitized (41 different landmarks, 29 being bilateral landmarks and 12 being midline landmarks, see Table 1 and Figure 1). Low pixel resolution, or individual variation in bone density and size may cause artefactual variation in microtomography image (Abel et al. 2012). Because of this artefactual variation, landmarks could not be placed reliably on the posterior edge or the neural spine of neural arch 4 and no landmark was placed on supraneural 3. The basioccipital and centra of vertebrae 5 through 7 were included to test whether Weberian vertebrae covaried with adjacent cranial elements and/or with posterior vertebrae. Landmark digitization was performed twice on the entirety of specimens to assess landmark placement repeatability (Klingenberg & McIntyre 1998).

After landmark digitization, landmarks for each specimen were exported as nts files and loaded within the statistical programming environment R (Version 3.5.0, R Core Team 2018) where all subsequent analyses were conducted. Analyses were performed using the geomorph (version 3.0.6., Adams et al. 2018), paleomorph (version 0.1.4., Lucas & Goswami 2017), smatr (3.4.8., Warton et al. 2012) and EMMLi (version 0.0.3., Goswami et al. 2017) packages. Landmarks were separated

Table 1. Anatomical correspondence of landmarks. Landmarks were placed on isosurfaced models. Landmarks 2 and 4 through 14 are median landmarks, the remaining ones are bilateral and were digitized on the left and right sides of specimens. Bilateral landmarks are indicated in boldface. Transverse processes are also referred to as lateral processes and parapophysis in the literature.

Structure	Landmarks	Anatomical correspondence
Basioccipital (Bos)	**1**	Lateral edge of the masticatory plate of the basioccipital
	2	Tip of the posterior process of the basioccipital
	3	Dorsal base of the posterior process of the basioccipital
	4	Base of the exoccipital
Centra 2 to 4 (C2–4)	5	Ventral midpoint between centra 1 and 2
	6	Ventral midpoint between centra 2 and 3
	7	Ventral midpoint between centra 3 and 4
	8	Ventral midpoint between centra 4 and 5
Centra 5 to 7 (C5–7)	9	Ventral midpoint between centra 5 and 6
	10	Ventral midpoint between centra 6 and 7
	11	Ventral midpoint between centra 7 and 8
	12	Dorsal midpoint of centrum 5
	13	Dorsal midpoint of centrum 6
	14	Dorsal midpoint of centrum 7
Transverse process 1 (Tp1)	**15**	Tip of transverse process 1
	16	Anterior base of transverse process 1
Transverse process 2 (Tp2)	**17**	Tip of transverse process 2
	18	Anterior base of transverse process 2
Intercalarium (Int)	**19**	Tip of the ascending process of the intercalarium
	20	Anterior base of the manubrium of the intercalarium
Transverse process 4 (Tp4)	**21**	Tip of transverse process 4
	22	Ridge on anterior edge of transverse process 4
Os suspensorium (Os)	**23**	Ventral tip of descending process of the os suspensorium
	24	Midpoint on the ventral curve between rib 4 and os suspensorium
Tripus (Tr)	**25**	Tip of the anterior process of the tripus
	26	Tip of the transformator process of the tripus
	27	Tip of the articular process of the tripus
Scaphium (Sc)	**28**	Anterior edge of the concha scaphium
	29	Tip of the ascending process of the scaphium
Claustrum (Cl)	**30**	Anterior edge of the claustrum
	31	Posterior edge of the claustrum
Neural arch 3 (Na3)	**32**	Anterior edge of the articulation with centrum 3
	33	Anterodorsal apex of neural arch 3
	34	Anteroventral apex of neural arch 3
	35	Posterodorsal apex of neural arch 3
	36	Posteroventral apex of neural arch 3
Neural arch 4 (Na4)	**37**	Base of the neural spine
Supraneural 2 (Sn2)	**38**	Curvature maximum on the ventral edge of supraneural 2
	39	Anterior tip of supraneural 2
	40	Anterior limit of medial edge of supraneural 2
	41	Posterior limit of medial edge of supraneural 2

Table 2. Modularity hypotheses for the Weberian apparatus and adjacent structures. Modules within hypotheses are non-nested. The order of modules in each hypothesis is arbitrary.

Hypothesis		Modules	# of modules
1	Developmental groupings	[c2–4+c5–7] / [int+sc] / [tp4+os+tr] / [tp1+cl] / [sn2] / [tp2+na3+na4] / [bos]	7
2	Structural groupings	[bos] / [c2–4+c5–7+tp1+tp2+int+tp4+os] / [tr] / [sc] / [cl] / [na3+na4+sn2]	6
3	Functional groupings	[int+tr+sc+cl] / [c2–4+tp1+tp2+tp4+os] / [na3+na4+sn2] / [c5–7] / [bos]	5
4	Integrated Weberian module	[c2–4+tp1+tp2+int+tp4+os+tr+sc+cl+na3+na4+sn2] / [c5–7] / [bos]	3

into left and right sides for each specimen and right-side configurations were reflected onto the left side by multiplying the x coordinate by "−1". This allowed for the right side to be superimposable on the left side. Landmark configurations were then subjected to a Generalised Procrustes Analysis (GPA). Performing GPA removes scale and rotation differences among specimens and allows the comparison of geometric shape (Rohlf & Slice 1990). Following GPA, coordinates were averaged between left and right sides to obtain one configuration per specimen per replicate. Averaging between sides removes information about asymmetry in our dataset; however, we opted to remove this information to have a more focused dataset to test for modularity and integration. In addition, landmarks 2 and 4–14 are midline landmarks, relative positioning of landmarks to the midline is therefore not lost. Replication error was tested using a Procrustes ANOVA analysis using 999 permutations (Klingenberg & McIntyre 1998). The effect of replication on shape variation was not significant (*p*-value: 0.056, R^2: 0.019) indicating that placement of landmarks was similar between replicates. Coordinates of landmarks were thus average between replicates to obtain a single configuration per specimen.

Allometry. The effect of size, measured as the logarithm of centroid size, on shape variation (Gouldian allometry, as described in Gould 1966, Gould 1971, Klingenberg 2016) was assessed using a Procrustes ANOVA with 999 permutations. Because the effect of size was found to be significant, shape data were thus regressed against size to obtain size-free residuals that were added to the consensus configuration of landmarks to obtain size-adjusted shapes for the entirety of the dataset. All subsequent analyses were performed on both the original shape data and size-adjusted shapes. The regression was used to obtain predicted configurations for minimum and maximum values of size. The difference between minimum and maximum size configurations were reported on consensus shapes to show shape change due to allometry.

Principal component analysis and landmark variation. Principal component analysis (PCA) was performed to decompose phenotypic variation within the datasets (Mitteroecker & Gunz 2009). The first three principal components (PC) were plotted. Shape changes from minimum to maximum shapes along each PC were reported on consensus shape to show shape variation along PCs. Individual landmark variation was computed by measuring the distance between individual landmarks for each specimen and the corresponding landmarks on the consensus shape.

Analysis of modularity hypotheses. Four hypotheses of modularity were tested (see Table 2). Hypothesis 1 describes inferred developmental modularity obtained from ossification sequence data in *Danio rerio* (Grande & Young 2004). Here, elements that developed concurrently are hypothesised to form modules. Hypothesis 2 is based on structural contiguity between structures. Hypothesis 3 is based on functional groupings related to functional roles of structures reflecting among other modules the pars auditum and the pars sustentaculum. The Weberian apparatus has been identified as a center of development in the axial skeleton, suggesting it could act as a module, independent from the craniofacial skeleton and the rest of the axial skeleton (Grande & de Pinna 2004, Grande & Young 2004, Bird & Hernandez 2009). We thus tested hypothesis 4 where the Weberian apparatus could act as an integrated module separated from the basioccipital and centra 5 through 7 (centra 5–7 are used as representative of the remainder of the vertebral column).

To test these hypotheses, we used a combined approach. Firstly, we tested hypotheses using the EMMLi maximum-likelihood test developed by Goswami & Finarelli (2016). This method compares log-likelihoods of correlation matrices using the corrected Akaike Information Criterion (AICc), an indicator of the relative quality of models in given sets of data. Additionally, this method compares all proposed models to a null model of no modularity and subdivides each model into four hypotheses with either equal or different strengths of correlation within modules and equal or different strengths of correlation between modules. For this analysis, we computed vector correlation matrices for landmarks using the vector congruence coefficient following Goswami & Finarelli (2016). This produced 41×41 matrices showing correlations between individual landmarks rather than between individual landmark coordinates (i.e., x, y, and z coordinates for each landmark) (Goswami & Finarrelli 2016, Lucas & Goswami 2017).

We then calculated covariance ratios based on each hypothesis. The covariance ratio measure is a measure of modularity between hypothesized modules (Adams 2016). This measure should be equal to 1 for landmarks randomly assigned to subsets; values lower than 1 indicate modularity while values higher than 1 indicate stronger association between subsets than expected.

We assessed covariation between modules for the supported hypothesis using pairwise partial least squares correlation between modules. We then used fitted values of small major axis regression of the first singular value decomposition vectors for each module of the supported hypothesis to obtain shape predictors of phenotypic variation due to association between modules. Phenotypic variation due to association with all other modules was computed for all modules and reported on consensus shape to show how covariation affected phenotypic variation.

Results

Allometry and phenotypic variation. A Procrustes ANOVA was used to test the effect of size on shape, using the natural logarithm of centroid size. The effect of size on shape was significant (*p*-value = 0.001, Table 3), meaning that changes in size directly incurred variation in shape. A second dataset containing size-adjusted shapes (i.e., with the effect of allometry removed) was created for subsequent analyses. All subsequent analyses were run on both the original dataset and the size-adjusted dataset to assess how allometry affected variational trends in the data.

Shape changes due to allometry (Figure 2) were mostly evident at the level of the basioccipital where the posterior process seemed to elongate, and the masticatory plate moved posteriorly due to size. Centra 5 through 7 appear to elongate as well, as would be expected through growth, but centra 2 through 4 appear to change more subtly. Similarly, vectors in Figure 2 suggest neural

Table 3. Procrustes ANOVA for size effect on shape for the complete dataset. Df, degrees of freedom; SS, sum of squares; MS, mean sums of squares; R^2, determination coefficient; F, F value; Z, Z-score, effect size.

	Df	SS	MS	R^2	F	Z	*p*-value
Log (Centroid size)	1	0.018	0.018	0.119	5.539	5.066	0.001
Residuals	40	0.130	0.003	0.881			
Total	41	0.148					

arches and supraneural 2 exhibit reduced growth compared to the rest. Weberian ossicles exhibit relatively small changes with the exception of transverse process 4 and the os suspensorium that appear to change angles to extend more anteriorly.

A PCA was used to assess phenotypic variation and its components within the dataset. For both the original and size-adjusted datasets, only PC 1 through 3 were considered (Fig. 3). For the original dataset, PCs 1–3 account for 43.33 % of overall variation. The first 20 PCs contain 90 % of overall variation. For the size-adjusted data, PCs 1–3 regroup 40.973 % of overall variation, while 90 % of the overall variation is spread across the first 21 PCs. Visually, both datasets appeared to behave similarly with specimens being distributed evenly and no groupings of individuals were evidently apparent. The shape of the Weberian apparatus and associated elements appear variable with and without the effects of allometry, indicating that the variation observed is not simply a consequence of allometric growth.

Shape changes associated with each PC were also computed and reported per landmark changes in the form of vectors on the consensus shape (Fig. 4). Landmarks that appear to be the most subject to shape changes along each PC were the same for both the original and size-adjusted datasets. Interestingly, while the landmarks subject to change shape were the same, shape changes varied depending on whether allometry was included or excluded. For PC 1, the magnitude of shape changes appears to be stronger when allometry is included but direction seemed to remain the same (Fig. 4A, D). Along PC 1, there appear to be mostly variation in the length of centra 6 and 7 as well as transverse process 4 and the os suspensorium. Transverse process 4 and the os suspensorium angled more anteriorly for positive values of PC 1. The os suspensorium also appears shorter for positive values of PC 1. This tendency is the same with or without allometric effects.

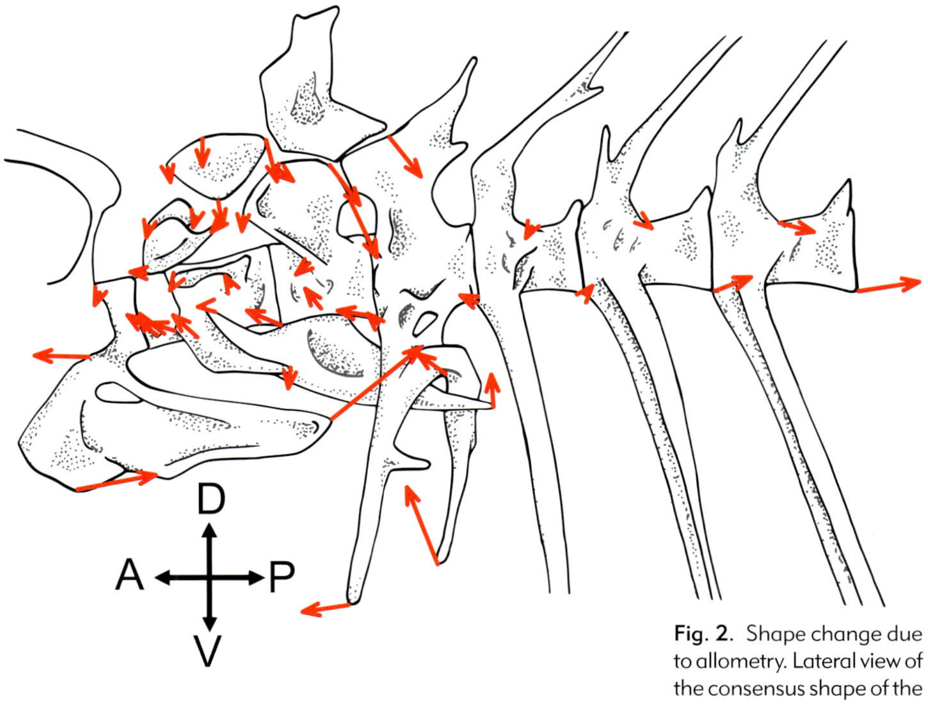

Fig. 2. Shape change due to allometry. Lateral view of the consensus shape of the Weberian apparatus and adjacent elements in *Danio rerio*. Vectors illustrate the magnitude and direction of shape changes associated with overall size changes for each landmark. Vectors were amplified by a factor of 2 for ease of visualisation. Arrowheads indicate landmark displacement in the largest specimens with regards to smallest specimens.

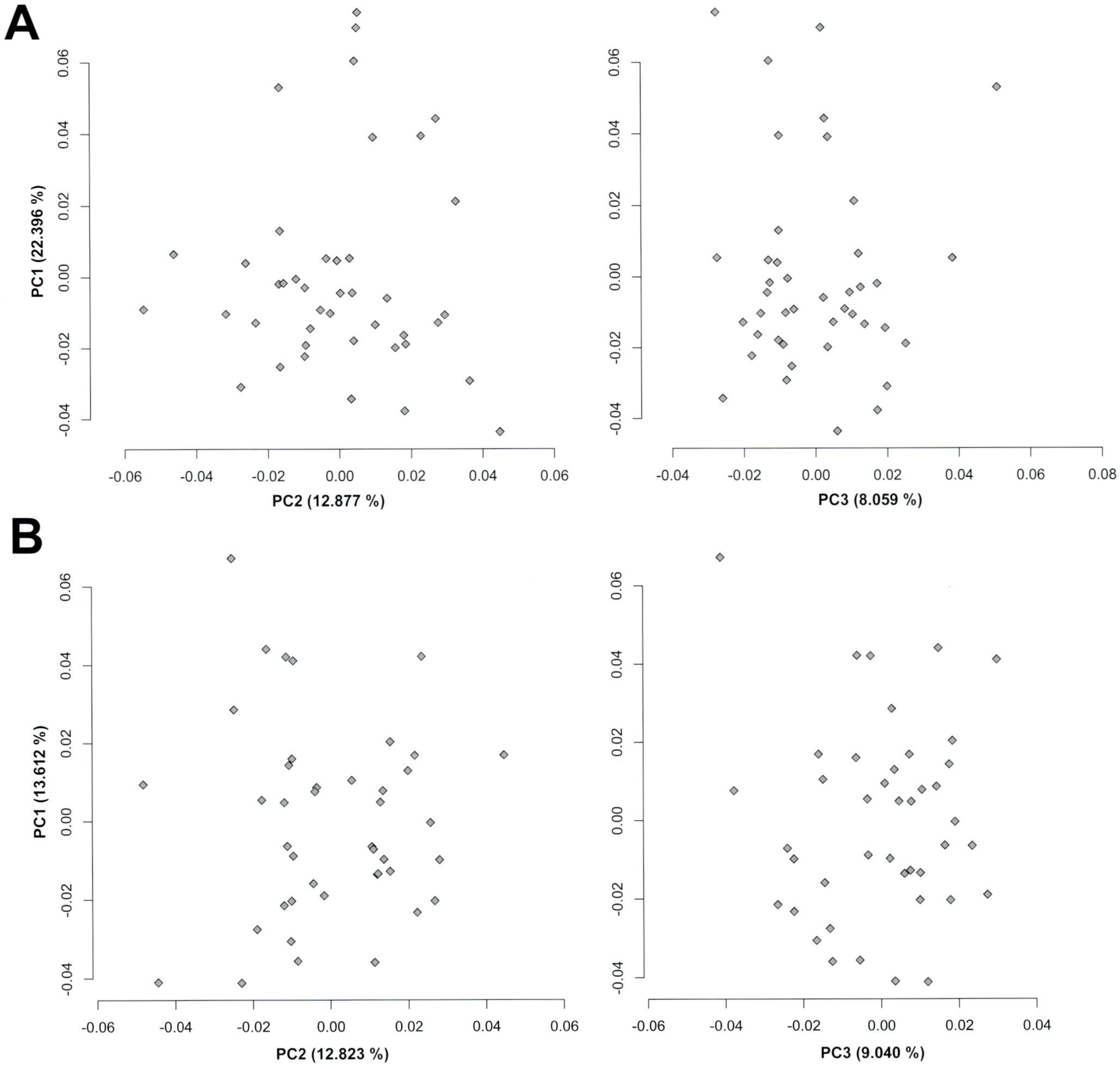

Fig. 3. Prinicipal Components Analysis of shape in the Weberian apparatus and associated elements in *Danio rerio*. The first three principal components (PCs 1–3) for shape in both the original dataset (**A**) and the size-adjusted dataset (**B**) are displayed.

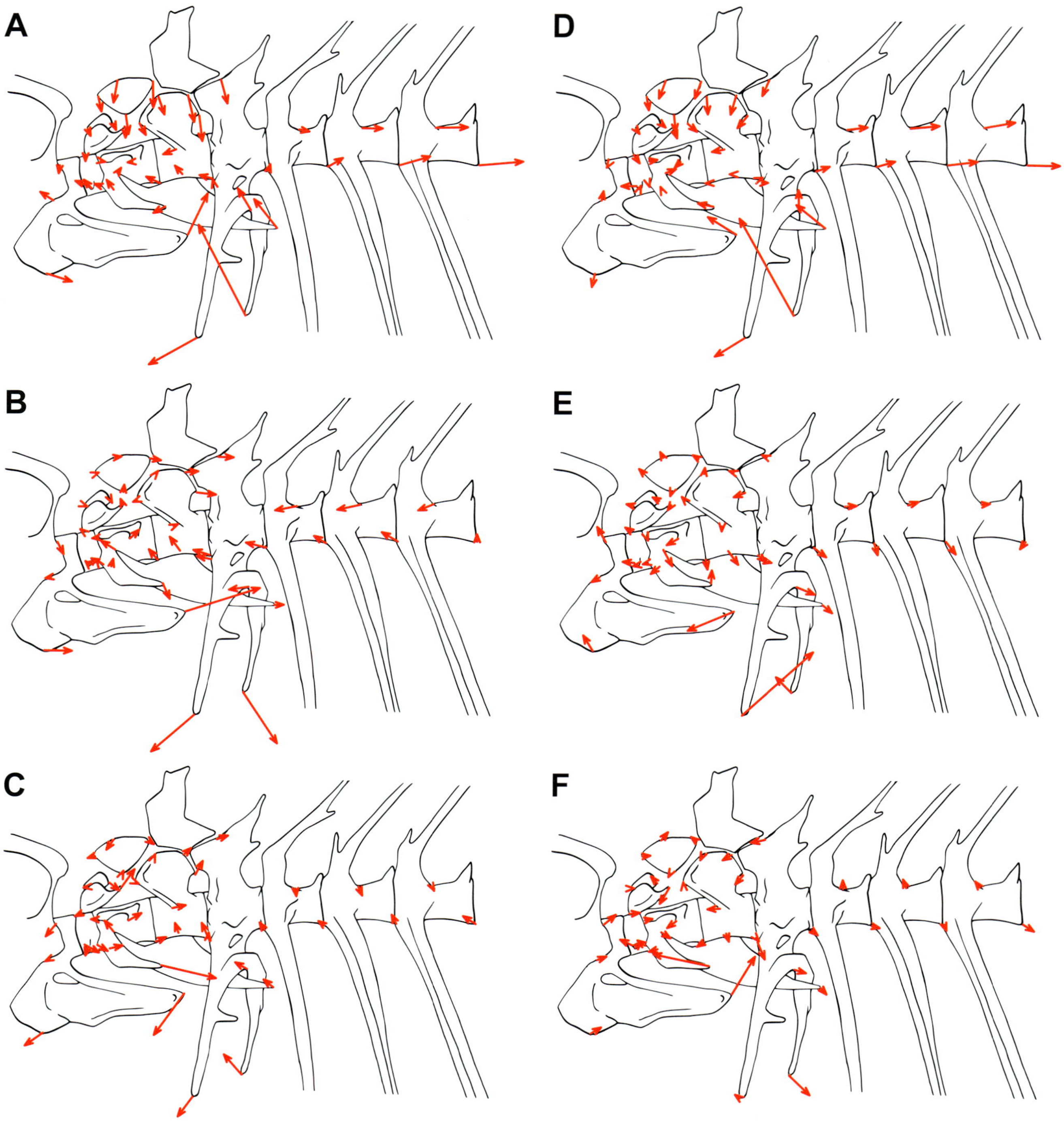

Fig. 4. Shape change along principal components in the Weberian apparatus and associated elements in *Danio rerio*. The first three principal components (PCs 1–3) for shape in both the original dataset (PC1: **A**, PC2: **B**, PC3: **C**) and the size-adjusted dataset (PC1: **D**, PC2: **E**, PC3: **F**) are displayed. Vectors illustrate the magnitude and direction of shape change associated with overall size changes for each landmark along PCs. Vectors were amplified by a factor of 2 for ease of visualisation. Arrowheads indicate landmark displacement from the specimens with the lowest score along each PC with regards to specimens with the largest score.

For PCs 2–3, the direction of shape change appears to differ whether allometry is included or not (Fig. 4B,C,E,F). For the original dataset, there appears to be a lengthening of both transverse process 4 and the os suspensorium and the base of transverse process 4 appears to move anteriorly. Shape change vectors along PC2 are in the opposite direction for the size-adjusted dataset.

Table 4. Maximum-likelihood of modularity hypotheses for the Weberian apparatus and associated elements in *Danio rerio*. dAICc, difference in corrected Akaike Information Criterion; MaxL, Maximum-likelihood score; Post-prob, posterior probability. Boldface characters indicate supported hypotheses.

Hypothesis	Degree of correlation within and between modules	With allometry			Without allometry		
		MaxL	dAICc	Post-prob	MaxL	dAICc	Post-prob
Null	Absence of modularity	1234.373	96.429	1.14E-21	1281.954	81.312	2.20E-18
1	Same within modules, same between modules	1257.931	51.328	7.09E-12	1298.746	49.743	1.58E-11
1	Different within modules, same between modules	1270.165	39.053	3.28E-09	1307.079	45.269	1.48E-10
1	Same within modules, different between modules	1272.824	62.899	2.18E-14	1307.099	74.394	7.00E-17
1	Different within modules, different between modules	1285.058	51.247	7.38E-12	1315.432	70.543	4.80E-16
2	Same within modules, same between modules	1248.775	69.639	7.49E-16	1292.358	62.519	2.65E-14
2	Different within modules, same between modules	1260.783	55.771	7.69E-13	1296.996	63.392	1.72E-14
2	Same within modules, different between modules	1254.310	87.302	1.09E-19	1296.785	82.398	1.28E-18
2	Different within modules, different between modules	1266.319	73.792	9.39E-17	1301.423	83.630	6.91E-19
3	Same within modules, same between modules	1255.046	57.097	3.96E-13	1298.423		1.14E-11
3	**Different within modules, same between modules**	**1287.649**	**0**	**0.991798**	**1327.672**	**0**	**0.999319**
3	Same within modules, different between modules	1259.521	66.505	3.59E-15	1299.812	65.967	4.73E-15
3	Different within modules, different between modules	1292.124	9.590	0.008202	1329.061	15.760	0.000378
4	Same within modules, same between modules	1243.038	81.112	2.42E-18	1284.659	77.916	1.20E-17
4	Different within modules, same between modules	1269.245	32.743	7.70E-08	1317.352	16.574	0.000252
4	Same within modules, different between modules	1243.110	85.013	3.44E-19	1285.112	81.056	2.50E-18
4	Different within modules, different between modules	1269.317	36.663	1.08E-08	1317.805	19.735	5.18E-05

For the original dataset, along PC3, transverse process 2 elongates and its tip moves posteriorly (Fig. 4C,F). The tips of transverse process 4 and the os supsensorium move anteriorly. The process of the basioccipital and the masticatory plate appear to move ventrally. For the size-adjusted dataset, the tip of the os suspensorium appears to move more posteriorly while the tip of transverse process 2 appears to move anteriorly.

The variation display by individual landmarks was calculated and displayed in Figure 5. Most landmarks display fairly low levels of variation across both datasets and variation appears to be mostly evenly distributed. However, landmarks 1, 2, 17, 21 and 23, corresponding to the basioccipital, the tip of lateral processes 2 and 4 and the tip of the descending process of the os suspensorium respectively, appeared to be more variable.

Modularity. The maximum-likelihood analysis (Table 4, EMMLi, Goswami & Finarelli 2016) supports hypothesis 3 with different within-module correlations but similar between-module correlations for both the original dataset and size-adjusted shapes. Maximum-likelihood scores for this functional hypothesis are 1287.649 and 1327.672 for the original and the size-adjusted dataset, respectively (Table 4). The maximum-likelihood analysis thus supports a partitioning into five modules. Module 1 contains the intercalarium, the tripus, the scaphium and the claustrum, Module 2 contains centra 1 through 4, transverse process 1, 2 and 4 and the os suspensorium, Module 3 contains neural arches 3 and 4 and supraneural 2, Module 4 consists of centra 5 through 7, and Module 5 contains solely the basioccipital.

Covariance ratios were calculated for all hypotheses and summarised in Table 5. When including the effects of allometric shape variation, hypotheses 1, 2 and 3 are supported with a covariance ratio of 0.799, 0.800 and 0.875, respectively and a *p*-value of 0.001, 0.004 and 0.023, respectively (Table 5). The covariance ratio for hypothesis 3 was the highest. When the effects of allometric shape variation are excluded, only hypotheses 1 and 2 are significantly supported (CR of 0.804 and 0.828, respectively; P-value of 0.002 and 0.019, respectively; Table 5).

A partial least square analysis was run to assess correlations between modules for hypothesis 3. Partial least square correlation scores (r.pls) are 0.751 for the original dataset (*p*-value = 0.001) and 0.738 for the size-adjusted dataset (*p*-value = 0.001). Pairwise covariance ratios for modules range from 0.709 to 1.083 for the original dataset and from 0.747 to 1.049 for the size adjusted dataset. All pairwise covariance ratios are lower than 1, except for the covariance ratio between Modules 2 and 3 which are 1.083 and 1.049 for the original and size-adjusted datasets, respectively. R.pls scores range from 0.561 to 0.856 for the original dataset and from 0.551 to 0.843 for the size-adjusted dataset. The highest r.pls scores are between Modules 2 and 4 for both datasets. R.pls scores between Modules 2 and 3 are 0.848 and 0.817 for the original and size-adjusted datasets, respectively. R.pls scores between Module 5, which contains only the basioccipital, and the other modules are the lowest scores ranging from 0.561 to 0.658 for the original dataset and from 0.551 to 0.654 for the size-adjusted dataset. Pairwise covariance ratios and partial least squares scores are shown in Table 6. In summary, most modules display low levels of covariation with other modules with the basioccipital bone displaying the lowest amount of covariation with all other modules. However, Modules 2 and 4 are the most covarying, meaning that centra 2–4 covary more importantly with centra 5–7 than all other modules do between themselves.

Phenotypic variational constraints owing to covariation between modules are shown in Figure 6. This figure shows lines of direction of shape change due to correlation between modules. These lines do not represent directional shape change but indicate the direction along which variation of a given landmark tends to be distributed as a result of covariation. It is worth noting that variation outside of the lines drawn here is not precluded but that variation along those lines is more likely as a result of the covariation with other modules. Longer lines indicate a tendency to vary more along that direction. For Module 1, variation of the claustrum, scaphium and intercalarium is small in magnitude and spread out evenly around landmarks. These effects are similar in both the original and size-adjusted datasets. The transformator process of the tripus shows more variation. Indeed, in both original and size-adjusted shapes,

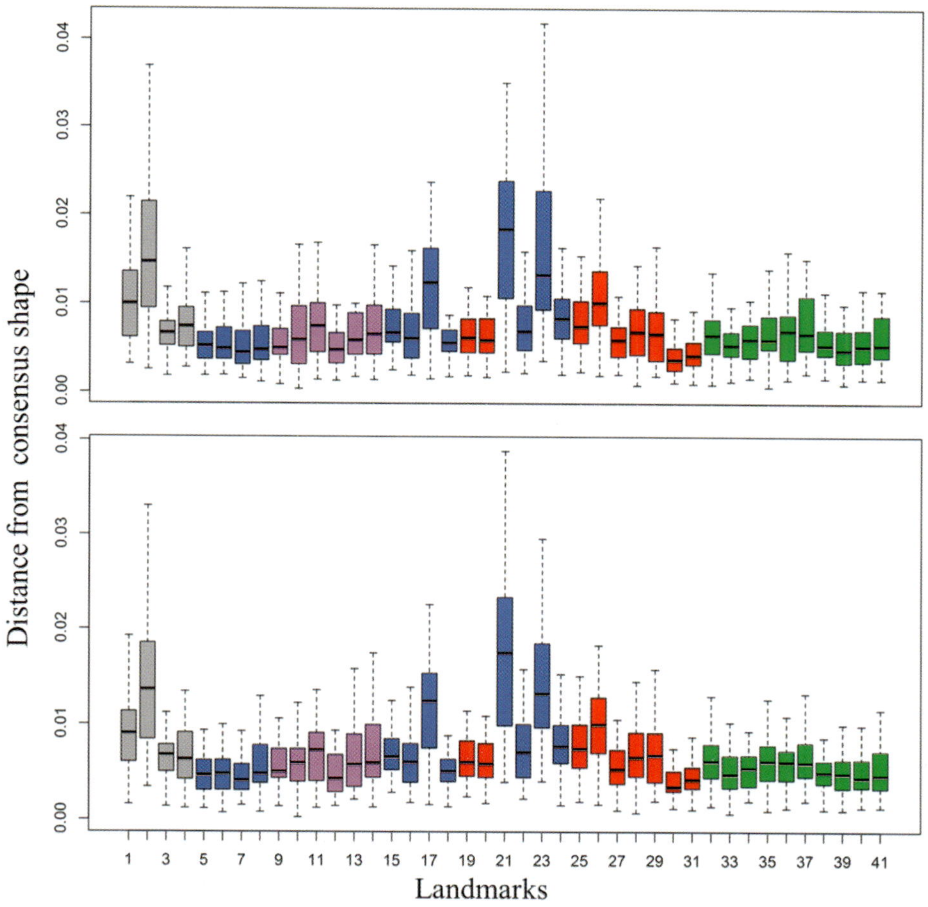

Fig. 5. Variation of individual landmarks. Boxplots of Euclidean distance between landmarks and the corresponding landmarks from the consensus shape in shape space are shown for the original dataset (above) and the size-adjusted dataset (below). The x-axis corresponds to landmarks indexed according to their digitising order, as seen in Figure 1 and Table 1. Because replicate error was not significant, this figure shows where variation is localised and where it seems to be more constrained. Colors indicate to which module a given landmark belongs: Module 1, red; Module 2, blue; Module 3, green; Module 4, purple; Module 5, grey.

Table 5. Covariance ratio for modularity hypotheses for the Weberian apparatus and associated elements in *Danio rerio*. CR, covariance ratio. Boldface characters indicate hypotheses supported by a significant P-value.

Hypothesis	With allometric effect		Without allometric effect	
	CR	P-value	CR	P-value
1	**0.799**	**0.001**	**0.804**	**0.002**
2	**0.800**	**0.004**	**0.828**	**0.019**
3	**0.875**	**0.023**	0.895	0.165
4	0.852	0.061	0.886	0.265

Table 6. Covariance ratio and partial least squares correlation matrices between modules defined in hypothesis 3. Below diagonal: original dataset, above diagonal: size-adjusted dataset. CR, covariance ratio; R.PLS, partial least squares correlation scores. Modules are denoted 1 through 5 and correspond to the modules defined in hypothesis 3. Module 1: int, tr, sc and cl; Module 2: c2–4, tp1,2,4 and os; Module 3: na3 and 4 and sn2; Module 4: c5–7; Module 5: bos.

	1	2	3	4	5
CR matrix					
1		0.945	0.747	0.851	0.794
2	0.843		1.049	0.955	0.922
3	0.709	1.083		0.874	0.940
4	0.729	0.955	0.862		0.871
5	0.811	0.966	0.956	0.833	
R.PLS matrix					
1		0.808	0.806	0.827	0.614
2	0.841		0.817	0.843	0.654
3	0.803	0.848		0.833	0.627
4	0.843	0.856	0.844		0.551
5	0.635	0.658	0.625	0.561	

covariation with Modules 2, 3 and 4 appears to allow variation from a shortened transformator process in a more dorsal position towards a longer transformator process with a more postero-ventral position. The variation shown by the transformator process does not affect this structure alone and would likely result in a variation of the angle of the curve between the tip of the anterior process and the transformator process of the tripus. Overall, covariation appears to be structured similarly in the size-adjusted dataset apart from covariation with Module 5, which allowed the transformator process to move towards an antero-ventral position. In addition, there is a slight increase in the magnitude of variation along all directions for the anterior tip of the tripus. For Module 2, the tips of transverse process 4 and the os suspensorium display important covariation with Modules 1 and 3 to 4. This covariation results in the tip of the transverse process moving along the antero-posterior axis and the tip of the os suspensorium moving along the dorso-ventral axis. When the effect of allometry is removed, the direction of covariation with Modules 1, 3 and 4 remains the same but the effect of covariation with Module 5 change to variation skewed towards a more ventral position. For Module 3, variation for neural arches 3 and 4 and supraneural 2 appears to be constrained along the dorso-ventral axis allowing change in their height relative to associated centra. With the effects of allometry removed, there is a slight increase in lateral variation.

Discussion

The Weberian apparatus has garnered a lot of interest for its putative role in the evolutionary and ecological success of otophysans as well as its apparent morphological and developmental complexity (Chardon & Vandewalle 1997, Grande & de Pinna 2004, Braun & Grande 2008, Bird & Hernandez 2009). It has been extensively described in classical anatomical description works with some studies drawing attention to the impressive disparity displayed across taxa (Bird & Hernandez 2007, 2009, Braun & Grande 2008). However, with a few notable exceptions (Conway & Britz 2007), little attention has been given to phenotypic variation of the Weberian apparatus at the specific and populational levels and the developmental constraints underlying this variation. Here, we apply geometric morphometrics tools to the study of the Weberian apparatus and its modularity and offer new hypotheses regarding its evolution and the developmental constraints directing it. To this end, we tested the modular independence of the Weberian vertebrae and the Weberian apparatus with regards to non-Weberian vertebrae and the cranium as well as to test hypotheses of modular partitioning of the Weberian apparatus itself within the zebrafish using μ-CT and 3DGM. Our results show that there is intraspecific variation in the shape of the Weberian apparatus of the zebrafish and that phenotypic variation was not evenly distributed across morphological landmarks. We identified factors affecting and constraining the variability of the Weberian apparatus in zebrafish, more specifically, we identified allometry as being a significant factor contributing to overall shape variation as well as a modular partitioning reflecting functional groupings. Finally, our results support the hypothesis that the Weberian apparatus sensu stricto is modular with regards to cranial and axial elements.

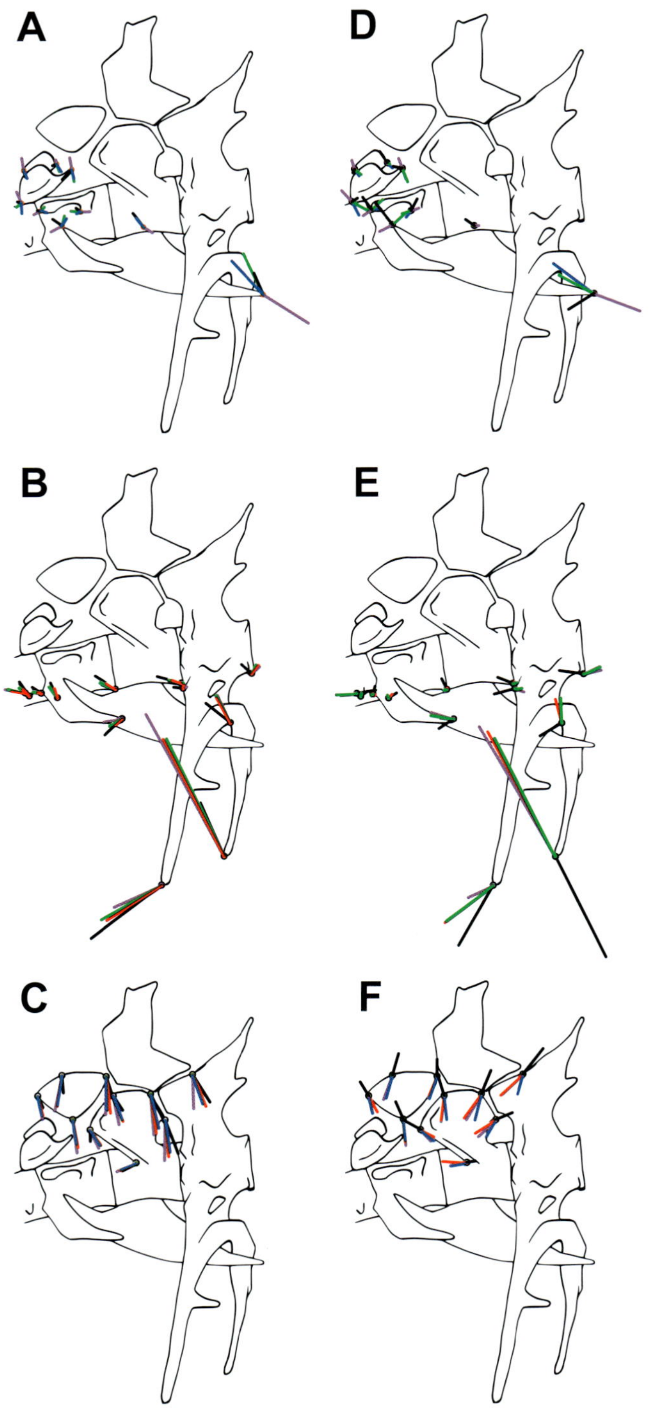

Fig. 6. Shape change in Weberian apparatus modules due to inter-module covariation. **A, B, C** show phenotypic variation in the original dataset and **D, E, F** show phenotypic variation in the size-adjusted dataset. **A, D** show phenotypic variation for Module 1, **B, E** for Module 2 and **C, F** for Module 3. Red lines indicate phenotypic changes due to covariation with Module 1, blue lines with Module 2, green lines with Module 3, purple lines with Module 4 and black lines with modules 5. Vectors were amplified by a factor of 2 for ease of visualisation.

Variability in the Weberian apparatus is affected by developmental and functional factors. Variation within the Weberian apparatus of *Danio rerio* was not distributed evenly; some elements [i. e., basioccipital, Weberian apparatus sensu stricto (transverse processes 2 and 4, tripus and os suspensorium)] were more variable while others appeared more constrained [i. e., centra 5-7, Weberian apparatus sensu stricto (centra 2-4, transverse process 1, intercalarium, scaphium, claustrum, neural arches 3 and 4 and supraneural 2)]. This was evident in the first module of the functional groupings hypothesis (Module 1) that contained the ossicles involved in sound transmission (i. e., intercalarium, tripus, scaphium and claustrum). Indeed, most of the visible variation in Module 1 was small in magnitude and spread somewhat evenly around landmarks 19–20, 25, 27 and 28–31. Strong functional pressures on this system are likely to limit the variation it displays. The transformator process of the tripus (landmark 26) displayed more variation than the other elements composing that module. Covariation with Modules 2 (centra 2–4, transverse processes 1, 2 and 4 and the os suspensorium), 3 (neural arches 3 and 4 and supraneural 2) and 4 (centra 5-7) could shift the positioning of the transformator process as well as the overall angle for the tripus. This variation could be crucial in ensuring proper transmission of sound from the gas bladder or could affect acoustic sensitivity. The functional effects of the variation shown here will require further functional analyses.

Allometry was a significant factor in the shape variation of the Weberian apparatus and associated elements in the zebrafish. Although the R^2 coefficient for the allometric effect was small (0.119, Table 3), there were marked shape changes due to increases in size. Allometry is not the only driver of phenotypic variation but had a visible effect on overall shape variation as well as on the amount and direction of covariation across modules (Table 5, Fig. 6). The most evident effects of allometry on covariation were in the basioccipital, more specifically the masticatory plate. However, these effects might be owing to a change in trophic niche as much as due to growth-induced changes. Indeed, size-change in fishes is predicted to change their trophic niche and, consequently, change the type of food ingested, potentially changing mechanical load in systems such as the masticatory plate (Wainwright et al. 1991, Sharf et al. 2000, Wainwright 2005). Similarly, centra tended to elongate. Phenotypic changes owing to allometry in the Weberian apparatus sensu stricto were more restricted, with effects on the relative height of neural arches and supraneural relative to centra, and small effects on transverse process 4 and the os suspensorium.

It is worth noting that removing allometry did not seem to affect the magnitude of phenotypic variation per landmark (Fig. 5, 6); however, it appeared to change how that variation was distributed in terms of covariation with other systems. As seen in Figure 6, covariation between Modules 1–4 with Module 5, the basioccipital, changed direction. Without allometry, the direction of covariation with the basioccipital for the tip of the transformator process shifted toward a more antero-posterior axis (Fig. 6D). The variation displayed by the tip of os supsensorium due to covariation with the basioccipital kept the same orientation but extended towards a more ventral position (Fig. 6E), and covariation with the basioccipital extended the range of variation for all landmarks in Module 3 towards a more dorsal position. Again, the basioccipital tended to covary weakly with other modules (Table 5) and this change could simply reflect the effects of an ecological change, for instance in trophic niche, rather than pure allometric effects. Nonetheless, allometry does have a role in the phenotypic variation of the Weberian apparatus while it may not affect the structure of covariation within it.

Modularity and integration were also key factors in shaping variation within the Weberian apparatus. The maximum-likelihood approach used to assess modularity within the Weberian apparatus and associated elements supported a subdivision into modules based on functional interactions. This subdivision was subsequently confirmed using the CR values although the same hypothesis was non-significant when allometry was removed. As mentioned above, the EMMLi approach subdivides hypotheses into four hypotheses testing for equal correlation strengths within and between modules (Goswami & Finarelli 2016). Our results supported similar strengths of association across modules but variable strengths within modules. This result suggests that the supported modules do indeed behave modularly from one another, but that elements within them could be further subdivided into smaller modular units.

In addition, significant CR for hypotheses 1 and 2 supports the idea that our modules could be further subdivided as these hypotheses decomposed our dataset into 7 and 6 modules, respectively, while hypothesis 3 divides it into 5. It is thus likely that parts of supported modules could form independent modules, and that parts of different modules might even form a module together. As proposed by Raff (1996), one possibility is that each centra could act as an independent module since they are iterative elements. Scores from the CR and PLS analyses show slightly stronger association between Modules 2 and 3 and 2 and 4, respectively, but these tendencies were not strong enough as to suggest that these modules are integrated between each other. Elements within Modules 2, 3 and 4 are possibly structured along a different pattern.

Overall, our results do confirm that the Weberian apparatus sensu stricto behaves modularly from adjacent cranial and axial skeletal elements. Furthermore, we have support that the Weberian apparatus is further subdivided into modules that reflect functional constraints more than developmental constraints with the intercalarium, tripus, scaphium and claustrum (i.e., the sound transmission chain between the gas bladder and the otoliths) forming an independent module. Nevertheless, the interaction between allometric effects and modularity suggest strong developmental associations underlying covariation structure. These associations might be hidden in our dataset by functional pressures. This is also supported by the fact that some modules from the functional groupings hypothesis within the Weberian apparatus show unexpectedly strong covariation between each other.

Using geometric morphometrics to study the development of the Weberian apparatus. Geometric morphometrics, as a field and as a set of tools, is becoming ubiquitous across ecology and evolutionary developmental biology (Adams et al. 2004, 2013). Nevertheless, there are still too few studies that apply these tools to the study of non-classical model systems, amorphous systems or structures that were too small to be adequately described using accessible technologies. Imaging technologies such as μ-CT scanners have improved access to these systems allowing them to be observed and quantified non-destructively and with unprecedented details (Metscher 2009, Abel et al. 2012, Cunningham et al. 2014).

Using a landmarking scheme describing elements of the Weberian apparatus simply, we have shown intraspecific variation in the shape of the Weberian apparatus of the zebrafish. This variation can be used to explore developmental properties and trajectories affecting the variability within the Weberian apparatus. More expansive landmarking may yet reveal variational properties and interactions driving individual elements. Indeed, some of the more variable landmarks were situated on transverse process 4 and the os suspensorium, as well as the tripus. These structures interact directly with the gas bladder, either via simple physical contact or via mechanical interactions, and a landmarking scheme with improved resolution on these structures could reveal localized variation which may affect function in this system. This could easily be extended to physical or functional associations of the Weberian apparatus with systems like the pectoral girdle, known to play a role in sound production in some taxa (Ladich & Fine 1994, Ladich 2004).

The landmark digitization scheme used here could be improved and bolstered with the inclusion of more specimens and with higher imaging resolution, but we believe the present scheme to be adequate for the exploration of variability in the Weberian apparatus considering what little is known about it. Depending on the taxonomic breadth envisioned for study, this landmarking scheme could be modified to include more or less landmarks. For instance, we opted not to include landmarks on structures such as the flange of the os suspensorium because this structure is not ubiquitous across cypriniforms (Bird & Hernandez 2007, Conway & Britz 2007), but it could be added easily. As a generalized landmarking scheme, it should therefore be applicable across the Cypriniformes; however, a landmarking scheme extending outside of this order to more taxa may not allow for a precise description of biological variation across taxa because landmarks may not be placed in homologous positions and will not account for non-homologous variation (i.e., presence of absence of ossicles and extreme differences in ossicle shape).

Using geometric morphometric approaches to quantify shape and shape variation should not be a replacement for developmental or anatomical work but should be used as a complement to address the same questions (Klingenberg 2002, 2015, Adams et al. 2004, 2013). Some aspects of anatomy, such as intraspecific variation or small-scale phenotypic changes might not be measured or captured by more traditional approaches (Adams et al. 2004, 2013, Zelditch et al. 2012). Novel approaches such as anatomical network analysis (Rasskin-Gutman & Esteve-Altava 2014, Esteve-Altava 2017) are also emerging as new tools to address topology and structural modularity in morphological systems and are being applied to an increasing number of systems and taxa, such as the axial and appendicular skeleton.

The zebrafish is a popular model organism (Westerfield 2007) for which a wide bank of genetic and developmental information is available (see ZFIN database as an example, Sprague et al. 2006, 2007). In addition, the zebrafish displays the complete configuration for the Weberian apparatus (i.e., containing all known ossicles and vertebral modifications) (Bird & Hernandez 2007). A better understanding of the constraints affecting the development and evolution of the Weberian apparatus at higher levels of complexity for that system may allow for the building of a theoretical morphospace for all possible Weberian apparatus configurations across otophysans. Using the zebrafish, we can thus begin to emit more specific hypotheses regarding variation and developmental constraints within this system before extrapolating them to a broader taxonomic spectrum.

Exploring the evolution of developmental constraints and variability in the Weberian apparatus. The phenotypic variation displayed by the zebrafish's Weberian apparatus was not evenly distributed across morphological landmarks. Some landmarks displayed important variation. Within the Weberian apparatus sensu stricto, transverse processes 2 and 4 as well as the os suspensorium displayed a lot of variation. It could be inferred that these structures may vary because they are under neutral selection and that phenotypic change in these elements does not have strong consequences for the fitness of individuals (Huynen 1996).

Similarly, there could be some developmental plasticity affecting the Weberian apparatus (de Pinna & Grande 2003, Bird 2009). The Weberian apparatus is involved in the hearing function, across different groups and populations, this function can be involved in environmental awareness and sensing predators (Ladich 1999, 2004, 2015, Chardon et al. 2003). It could also be involved in communication between conspecifics and play a role in group cohesion or schooling behavior which are behaviors sensitive to environmental fluctuations (Ladich 1999, 2004, 2015, Chardon et al. 2003). Vertebrae in fish are known to display developmental plasticity, it is possible that modified vertebral elements composing the Weberian apparatus could retain their sensitivity to external cues (Sfakianakis et al. 2004, Bird 2009, Witten & Hall 2015). There could therefore be some plasticity within the Weberian apparatus triggered by changes in the environments.

Conversely, some landmarks displayed very little variation; this is most apparent in Module 1. The intercalarium, the scaphium and the claustrum displayed little variation while variation in the tripus was mostly limited to the transformator process. It is possible that there are strong selective pressures on these elements because of the crucial role they play or that these elements are under strong developmental canalization limiting the amount of variation in the auditory chain past the transformator process of the tripus. The intercalarium, the tripus and the scaphium seem to be conserved across most lineages. However, while these elements displayed little variation in the population studied, they might be variable in shape across taxa and contribute to the disparity in form of the Weberian apparatus across taxa.

Plasticity and canalization, along with allometry, modularity and integration are important factors directing morphological development and its evolution (West-Eberhard 2003, Lenski et al. 2006, Hallgrímsson et al. 2009, Von Dassow & Davidson 2011). Assessing these properties in the Weberian apparatus at the scales of genera or families would be a crucial step in understanding what constraints are important and have indeed directed the evolution of the impressive disparity display by that system across taxa. Here we have used covariance ratio and the EMMLi approach to assess modularity in the Weberian apparatus of the zebrafish with successful results. However, these methods do not allow for analyses of nested modular partitions nor do they offer a competent means to compare patterns of modularity among multiple groups (Goswami & Finarelli 2016, Adams 2016, Adams & Collyer 2016, Esteve-Altava 2017). These methods also do not account for the possibility of varying complexity in patterning across a dataset, as could be expected in a phylogenetic context (Goswami & Finarelli 2016, Adams 2016, Adams & Collyer 2016, Esteve-Altava 2017). In a phylogenetic perspective, other analytical tools, such as the minimum deviance method, could be applied to test whether modular partitioning remains the same across groups (Marquez 2008, Larouche et al. 2015, 2018). This method, which fits models to the covariance matrix of landmark coordinates, has been successfully used to test modularity in the zebrafish, among other models, and can handle a high number of alternative hypotheses with varying complexities (Marquez 2008, Larouche et al. 2015, 2018, Parsons et al. 2018).

Conclusion

The Weberian apparatus is a system of considerable importance in the evolution of otophysan fishes and could be a driver of their ecological and evolutionary success. Despite this, this system has not been capitalized upon the current evo-devo research and the breadth of methods used by researchers in that field. We propose that current and further research on the Weberian apparatus and its evolution could benefit greatly from modern evo-devo approaches and a solid framework to investigate developmental and evolutionary constraints to the variability of this system.

Here, we have explored variational modularity within the Weberian apparatus of the zebrafish. Our results have shown that there is abundant intraspecific variation in the shape of the Weberian apparatus and that Weberian ossicles are modular amongst themselves and with adjacent cranial and axial skeletal elements. While the patterns of modularity appeared to be determined along functional groups, it is likely that other developmental properties such as allometry, plasticity and canalization are at play and influence the capacity for variation to arise within the Weberian apparatus both at the scale of populations and at broader taxonomic scales. Going forwards, quantitative methods to describe shape variation, such a landmark-based geometric morphometrics, applied to the study of Weberian apparatus across taxa will elucidate how developmental mechanisms and constraints have directed the evolution of the impressive diversity of forms seen across otophysans.

Acknowledgements

We thank Jeffrey Peters for scanning the specimens and Thomas Grünbaum for his insights and ideas. We also thank two anonymous reviewers for their comments and suggestions to improve this manuscript. Finally, we thank the editors, Alan Pradel, John Denton and Philippe Janvier, for their commitment and hard work in making this volume.

KD is supported by fellowships from the Fonds de Recherche du Québec – Nature et Technologies, the Richard J. Schmeelk Foundation and the Eyes-High Doctoral Recruitment Scholarship of The University of Calgary. TCG is funded by the Loyola University Chicago Foundation Fund LUC 211155. RC is funded by a Discovery Grant from the Natural Sciences and Engineering Research Council of Canada.

Literature cited

Abel, R. L., C. R. Laurini & M. Richter 2012. A palaeobiologist's guide to 'virtual' micro-CT preparation. Palaeontologia Electronica, 15: 1–17.

Adams, D. C., F. J. Rohlf & D. E. Slice. 2004. Geometric morphometrics: ten years of progress following the 'revolution'. Italian Journal of Zoology, 71: 5–16.

Adams, D. C., F. J. Rohlf & D. E. Slice. 2013. A field comes of age: geometric morphometrics in the 21st century. Hystrix, 24: 7–14.

Adams, D. C. 2016. Evaluating modularity in morphometric data: challenges with the RV coefficient and a new test measure. Methods in Ecology and Evolution, 7: 565–572.

Adams, D. C. & M. L. Collyer. 2016. On the comparison of the strength of morphological integration across morphometric datasets. Evolution, 70: 2623–2631.

Adams, D. C., M. L. Collyer & A. Kaliontzopoulou. 2018. Geomorph: Software for geometric morphometric analyses. R package version 3.0.6. https://cran.r-project.org/package=geomorph.

Aguirre, W. E., S. E. Contreras, K. M. Carlson, A. J. Jagla & L. Arellano. 2016. Evolutionary diversification of body form and the axial skeleton in the Gasterosteoidei – the sticklebacks and their closest relatives. Evolutionary Ecology Research, 17: 373–393.

Arnold, S. J. 1992. Constraints on phenotypic evolution. The American Naturalist, 140: S85–S107.

Bardua, C., M. Wilkinson, D. J. Gower, E. Sherratt & A. Goswami. 2019. Morphological evolution and modularity of the caecilian skull. BMC Evolutionary Biology. 19: 1–30.

Bird, N. C. & P. M. Mabee. 2003. Developmental morphology of the axial skeleton of the zebrafish, *Danio rerio* (Ostariophysi: Cyprinidae). Developmental Dynamics, 228: 337–357.

Bird, N. C. & L. P. Hernandez. 2007. Morphological variation in the Weberian apparatus of Cypriniformes. Journal of Morphology, 268: 739–757.

Bird, N. C. & L. P. Hernandez. 2009. Building an evolutionary innovation: differential growth in the modified vertebral elements of the zebrafish Weberian apparatus. Zoology, 112: 97–112.

Bird, N. C. 2009. Development, growth, and evolution of the cypriniform Weberian apparatus (Doctoral dissertation, The George Washington University).

Bogutskaya, N. G. 1991. Development of the Weberian apparatus during ontogeny of some species of Cyprinidae. Journal of Ichthyology, 31: 363–372.

Braun, C. & T. Grande. 2008. Evolution of peripheral mechanisms for the enhancement of sound reception. – In: Webb, J., R. Fay & A. Popper (eds.). Fish Bioacoustics, 32: 99–144.

Britz, R. & M. Hoffmann. 2006. Ontogeny and homology of the claustra in otophysan Ostariophysi (Teleostei). Journal of Morphology, 267: 909–923.

Chardon, M. & P. Vandewalle. 1997. Evolutionary trends and possible origin of the Weberian apparatus. Netherland Journal of Zoology, 47: 383–403.

Chardon, M., E. Parmentier & P. Vandewalle. 2003. Morphology, development and evolution of the Weberian apparatus in catfish. Catfishes, 1: 71–120.

Chernoff, B. & P. M. Magwene. 1999. Afterword – Morphological integration: Forty years later. – Pp. 319–353 in Olson,E. C. & C. G. Miller (eds.), Morphological Integration. Chicago, IL: The University of Chicago Press.

Coburn, M. M. & P. G. Grubach. 1998) Ontogeny of the armored catfish *Corydoras paleatus* (Siluriformes: Callichthyidae). Copeia, 1998: 301–311.

Conway, K. W. & R. Britz. 2007. Sexual dimorphism of the Weberian apparatus and pectoral girdle in *Sundadanio axelrodi* (Ostariophysi: Cyprinidae). Journal of Fish Biology, 71: 1562–1570.

Cunningham, J. A., I. A. Rahman, S. Lautenschlager, E. J. Rayfield, & P. C. Donoghue. 2014. A virtual world of paleontology. Trends in Ecology & Evolution, 29: 347–357.

de Pinna, M. & T. Grande. 2003. Ontogeny of the accessory neural arch in pristigasteroid clupeomorphs and its bearing on the homology of the otophysan claustrum (Teleostei). Copeia, 2003: 838–845.

Esteve-Altava, B. 2017. In search of morphological modules: a systematic review. Biological Reviews, 92: 1332–1347.

Fay, R. R. 1988. Hearing in Vertebrates: A Psychophysics Databook. Winnetka, IL: Hill-Fay Associates.

Félix, M. A. & A. Wagner. 2008. Robustness and evolution: concepts, insights and challenges from a developmental model system. Heredity, 100: 132–140.

Fink, S. V. & W. L. Fink. 1981. Interrelationships of the ostariophysan fishes (Teleostei). Zoological Journal of the Linnean Society, 72: 297–353.

Goswami, A. 2007. Cranial modularity and sequence heterochrony in mammals. Evolution and Development, 9: 290–298.

Goswami, A., V. Weisbecker & M. R. Sanchez-Villagra. 2009. Journal of Experimental Zoology (Molecular and Developmental Evolution), 312B: 186–195.

Goswami, A. & J. A. Finarelli. 2016. EMMLi: A maximum likelihood approach to the analysis of modularity. Evolution, 70: 1622–1637.

Gould, S. J. 1966. Allometry and size in ontogeny and phylogeny. Biological Reviews, 41: 587–638.

Gould, S. J. 1971. Geometric similarity in allometric growth: a contribution to the problem of scaling in the evolution of size. The American Naturalist, 105: 113–136.

Gould, S. J. & R. C. Lewontin. 1979. The spandrels of San Marco and the Panglossian paradigm – a critique of the adaptationist program. Proceedings of the Royal Society of London B, Biological Sciences, 205: 581–598.

Grande, T. & M. de Pinna. 2004. The evolution of the Weberian apparatus: A phylogenetic perspective. – Pp 429–448 in: Arratia G. & A. Tintori (eds.). Mesozoic Fishes 3 – Systematics, Paleoenvironments and Biodiversity. Verlag Dr. Friedrich Pfeil, Munich, Germany.

Grande, T. & B. Young. 2004. The ontogeny and homology of the Weberian apparatus in the zebrafish *Danio rerio* (Ostariophysi: Cypriniformes). Zoological Journal of the Linnean Society, 140: 241–254.

Gray, J. A., E. Sherratt, M. N. Hutchinson & M. E. H. Jones. 2019. Changes in ontogenetic patterns facilitate diversification in skull shape of Australian agamid lizards. BMC Evolutionary Biology, 19: 7.

Guryev, V., M. J. Koudijs, E. Berezikov, S. L. Johnson, R. H. Plasterk, F. J. Van Eeden & E. Cuppen. 2006. Genetic variation in the zebrafish. Genome Research, 16: 491–497.

Hallgrímsson, B., H. Jamniczky, N. M. Young, C. Rolian, T. E. Parsons, J. C. Boughner & R. S. Marcucio. 2009. Deciphering the palimpsest: studying the relationship between morphological integration and phenotypic covariation. Evolutionary Biology, 36: 355–376.

Hansen, T. F. 2003. Is modularity necessary for evolvability?: Remarks on the relationship between pleiotropy and evolvability. Biosystems, 69: 83–94.

Hansen, T. F. & D. Houle. 2008. Measuring and comparing evolvability and constraint in multivariate characters. Journal of Evolutionary Biology, 21: 1201–1219.

Higgs, D. M., A. K. Rollo, M. J. Souza & A. N. Popper. 2003. Development of form and function in peripheral auditory structures of the zebrafish (*Danio rerio*). Journal of the Acoustical Society of America, 113: 1145–1154.

Higgs, D. M., D. T. T. Plachta, A. K. Rollo, M. Singheiser, M. C. Hastings & A. N. Popper. 2004. Development of ultrasound detection in American shad (*Alosa sapidissima*). Journal of Experimental Biology, 207: 155–163.

Huynen, M. A. 1996. Exploring phenotype space through neutral evolution. Journal of Molecular Evolution, 43: 165–169.

Ichiyanagi, T., H. Kohno, K. Fujita & Y. Taki. 1996. Ontogenetic development of the Weberian ossicles in two cyprinids, *Tribolodon hakonensis* and *Zacco platypus*. Journal of the Tokyo University of Fisheries, 82: 119–124.

Jamniczky, H. A. & B. Hallgrímsson. 2009. A comparison of covariance structure in wild and laboratory muroid crania. Evolution, 63: 1540–1556.

Jamniczky, H. A., T. N. Barry & S. M. Rogers. 2015. Eco-evo-devo in the study of adaptive divergence: examples from threespine stickleback (*Gasterosteus aculeatus*). Integrative and Comparative Biology, 55: 166–178.

Krumholz, L. A. 1943. A comparative study of the Weberian ossicles in North American Ostariophysine fishes. Copeia, 1943: 33–40.

Klingenberg, C. P. & G. S. McIntyre. 1998. Geometric morphometrics of developmental instability: analyzing patterns of fluctuating asymmetry with Procrustes methods. Evolution, 52: 1363–1375.

Klingenberg, C. P. 2002. Morphometrics and the role of the phenotype in studies of the evolution of developmental mechanisms. Gene, 287: 3–10.

Klingenberg, C. P. 2008. Morphological integration and developmental modularity. Annual Review of Ecology, Evolution, and Systematics, 39: 115–132.

Klingenberg, C. P. 2009. Morphometric integration and modularity in configurations of landmarks: tools for evaluating a priori hypotheses. Evolution and Development, 11: 405–421.

Klingenberg, C. P. 2015. Analyzing fluctuating asymmetry with geometric morphometrics: concepts, methods, and applications. Symmetry, 7: 843–934.

Klingenberg, C. P. 2016. Size, shape, and form: concepts of allometry in geometric morphometrics. Development Genes and Evolution, 226: 113–137.

Klingenberg, C. P., A. V. Badyaev, S. M. Sowry & N. J. Beckwith. 2001. Inferring developmental modularity from morphological integration: analysis of individual variation and asymmetry in bumblebee wings. The American Naturalist, 157: 11–23.

Klingenberg, C. P. & J. Marugán-Lobón. 2013. Evolutionary covariation in geometric morphometric data: analyzing integration, modularity, and allometry in a phylogenetic context. Systematic Biology, 62: 591–610.

Ladich, F. & M. L. Fine. 1994. Localization of swimbladder and pectoral motoneurons involved in sound production in pimelodid catfish. Brain, Behavior and Evolution, 44: 86–100.

Ladich, F. 1999. Did auditory sensitivity and vocalization evolve independently in otophysan fishes? Brain, Behavior and Evolution, 53: 288–304.
Ladich, F. 2004. Sound production and acoustic communication. – Pp. 210–230 in: von dem Emde, G., J. Mogdans & B. G. Kapoor (eds.). The Senses of Fish. Springer, Dordrecht.
Ladich, F. 2015. Ontogenetic development of sound communication in fishes. – Pp. 127–148 in Ladich, F. (ed.). Sound Communication in Fishes. Springer, Vienna.
Larouche, O., R. Cloutier & M. L. Zelditch. 2015. Head, body and fins: Patterns of morphological integration and modularity in fishes. Evolutionary Biology, 42: 296–311.
Larouche, O., M. L. Zelditch & R. Cloutier. 2018. Modularity promotes morphological divergence in ray-finned fishes. Scientific Reports, 8: e7278.
Lenski, R. E., J. E. Barrick & C. Ofria. 2006. Balancing robustness and evolvability. PLoS Biology, 4: e428.
Lucas, T. & A. Goswami. 2017. Paleomorph: Geometric Morphometric Tools for Paleobiology. R package version 0.1.4. https://CRAN.R-project.org/package=paleomorph
Maisey, J. G. 2001. CT-scan reveals new cranial features in Devonian chondrichthyan *"Cladodus" wildungensis*. Journal of Vertebrate Paleontology, 21: 807–810.
Maisey, J. G. 2004. Morphology of the braincase in the broadnose sevengill shark *Notorynchus* (Elasmobranchii, Hexanchiformes), based on CT scanning. American Museum Novitates, 44: 1–52.
Márquez, E. J. 2008. A statistical framework for testing modularity in multidimensional data. Evolution, 62: 2688–2708.
Maynard-Smith, J., R. Burian, S. Kauffman, P. Alberch, J. Campbell, B. Goodwin, R. Lande, D. Raup & L. Wolpert. 1985. Developmental constraints and evolution. Quarterly Review of Biology, 6: 265–287.
Metscher, B. D. 2009. MicroCT for comparative morphology: simple staining methods allow high-contrast 3D imaging of diverse non-mineralized animal tissues. BMC Physiology, 9: e11.
Meyer, A., C. H. Biermann & G. Orti. 1993. The phylogenetic position of the zebrafish (*Danio rerio*), a model system in developmental biology: an invitation to the comparative method. Proceedings of the Royal Society B, 252: 231–236.
Mitteroecker, P. 2009. The developmental basis of variational modularity: insights from quantitative genetics, morphometrics, and developmental biology. Evolutionary Biology, 36: 377–385.
Mitteroecker, P. & P. Gunz. 2009. Advances in geometric morphometrics. Evolutionary Biology, 36: 235–247.
Patterson, C. 1984. *Chanoides*, a marine Eocene otophysan fish (Teleostei: Ostariophysi). Journal of Vertebrate Paleontology, 4: 430–456.
Parr, W. C. H., S. Wroe, U. Chamoli, H. S. Richards, M. R. McCurry, P. D. Clausen & C. McHenry. 2012. Toward integration of geometric morphometrics and computational biomechanics: new methods for 3D virtual reconstruction and quantitative analysis of Finite Element Models. Journal of Theoretical Biology, 301: 1–14.
Parsons, K. J., Y. H. Son, A. Crespel, D. Thambithurai, S. Killen, M. P. Harris & R. C. Albertson, 2018. Conserved but flexible modularity in the zebrafish skull: implications for craniofacial evolvability. Proceedings of the Royal Society B, 285: e20172671.
Pavlicev, M. & T. F. Hansen. 2011. Genotype-phenotype maps maximizing evolvability: Modularity revisited. Evolutionary Biology, 38 371–389.
Pomidor, B. J., J. Makedonska & D. E. Slice. 2016. A landmark-free method for three-dimensional shape analysis. PLoS One, 11: e0150368.
Raff, R. A. 1996. The Shape of Life: Genes. Development, and the Evolution of Animal Form: University of Chicago Press.
Raff, E. C. & R. A. Raff. 2000. Dissociability, modularity, evolvability. Evolution and Development, 2: 235–237.
Ramaswami, L. S. 1955. Skeleton of cyprinoid fishes in relation to phylogenetic studies: VII. The skull and Weberian apparatus of Cyprininae (Cyprinidae). Acta Zoologica, 36: 1–44.
Rasskin-Gutman, D. & B. Esteve-Altava. 2014. Connecting the dots: anatomical network analysis in morphological EvoDevo. Biological Theory, 9: 178–193.
R Core Team 2018. R: A language and environment for statistical computing. R Foundation for Statistical Computing, Vienna, Austria. URL https://www.R-project.org/.
Revell, L. J., L. J. Harmon, R. B. Langerhans & J. J. Kolbe. 2007. A phylogenetic approach to determining the importance of constraint on phenotypic evolution in the neotropical lizard *Anolis cristatellus*. Evolutionary Ecology Research, 9: 261–282.
Rosen, D. E. & P. H. Greenwood. 1970. Origin of the Weberian apparatus and the relationships of the ostariophysan and gonorynchiform fishes. American Museum Novitates, 2428: 1–25.
Rohlf, F. J. & D. Slice. 1990. Extensions of the Procrustes method for the optimal superimposition of landmarks. Systematic Biology, 39: 40–59.
Sanger, T. J. & A. R. McCune. 2002. Comparative osteology of the *Danio* (Cyprinidae: Ostariophysi) axial skeleton with comments on *Danio* relationships based on molecules and morphology. Zoological Journal of the Linnean Society, 135: 529–546.
Sanger, T. J., L. J. Revell, J. J. Gibson-Brown & J. B. Losos. 2011. Repeated modification of early limb morphogenesis programmes underlies the convergence of relative limb length in *Anolis* lizards. Proceedings of the Royal Society B, 279: 739–748.
Schlosser, G. & G. P. Wagner (eds.). 2004. Modularity in Development and Evolution. University of Chicago Press.
Sfakianakis, D. G., G. Koumoundouros, P. Divanach & M. Kentouri. 2004. Osteological development of the vertebral column and of the fins in *Pagellus erythrinus* (L. 1758). Temperature effect on the developmental plasticity and morpho-anatomical abnormalities. Aquaculture, 232: 407–424.
Siegal, M. L. & A. Bergman. 2002. Waddington's canalization revisited: developmental stability and evolution. Proceedings of the National Academy of Sciences of the USA, 99: 10528–10532.
Sprague, J., L. Bayraktaroglu, D. Clements, T. Conlin, D. Fashena, K. Frazer, M. Haendel, G. H. Douglas, P. Mani, S. Ramachandran, K. Schaper, E. Segerdell, P. Song, B. Sprunger, S. Taylor, C. E. Van Slyke & M. Westerfield. 2006. The Zebrafish Information Network: the zebrafish model organism database. Nucleic Acids Research, 34 (suppl. 1): D581–D585.
Sprague, J., L. Bayraktaroglu, Y. Bradford, T. Conlin, N. Dunn, D. Fashena, K. Frazer, M. Haendel, D. G. Howe, J. Knight, P. Mani, S. A. T. Moxon, C. Pich, S. Ramachandran, K. Schaper, E. Segerdell, X. Shao, A. Singer, P. Song, B. Sprunger, C. E. Van Slyke & M. Westerfield. 2007. The Zebrafish Information Network: the zebrafish model organism database provides expanded support for genotypes and phenotypes. Nucleic Acids Research, 36(suppl. 1): D768–D772.
Stratovan Corporation. Stratovan Checkpoint [Software]. Version 2018.09.07. Sep 07, 2018. URL: https://www.stratovan.com/products/checkpoint
Villmoare, B. 2013. Morphological integration, evolutionary constraints, and extinction: a computer simulation-based study. Evolutionary Biology, 40: 76–83.
Von Dassow, M. & L. A. Davidson. 2011. Physics and the canalization of morphogenesis: a grand challenge in organismal biology. Physical Biology, 8: 045002.
Waddington, C. H. & E. Robertson. 1966. Selection for developmental canalisation. Genetics Research, 7: 303–312.
Wainwright, P. C., C. W. Osenberg & G. G. Mittelbach. 1991. Trophic polymorphism in the pumpkinseed sunfish (*Lepomis gibbosus* Linnaeus): effects of environment on ontogeny. Functional Ecology, 5: 40–55.
Wainwright, P. C. 2005. Functional morphology of the pharyngeal jaw apparatus. Fish Physiology, 23: 77–101.
Wagner, G. P., M. Pavlicev & J. M. Cheverud. 2007. The road to modularity. Nature Reviews Genetics, 8: 921–931.
Watson, J. M. 1939. The development of the Weberian ossicles and anterior vertebrae in the goldfish. Proceedings of the Royal Society B, 127: 452–472.
Warton, D. I., R. A. Duursma, D. S. Falster & S. Taskinen. 2012. smatr 3 – an R package for estimation and inference about allometric lines. Methods in Ecology and Evolution, 3: 257–259.
Weber, E. H. 1820. De aure et auditu hominis et animalium, Pars I. De Aure Animalium. Aquatilium. Leipzig.
Westerfield, M. 2007. The Zebrafish Book: A Guide for the Laboratory Use of Zebrafish *Danio* (*"Brachydanio Rerio"*). University of Oregon.
Witten, P. E. & B. K. Hall. 2015. Teleost skeletal plasticity: modulation, adaptation, and remodelling. Copeia, 103: 727–739.
Zelditch, M. L., A. R. Wood, R. M. Bonett & D. L. Swiderski. 2008. Modularity of the rodent mandible: Integrating bones, muscles, and teeth. Evolution and Development, 10: 756–768.
Zelditch, M. L., D. L. Swiderski & H. D. Sheets. 2012. Geometric Morphometrics for Biologists: A Primer. Academic Press.

Ancient Fishes and their Living Relatives: a Tribute to John G. Maisey. Alan Pradel, John S. S. Denton & Philippe Janvier (eds.): pp. 71–79, 5 figs., 3 supp.
© 2021 by Verlag Dr. Friedrich Pfeil, München, Germany – ISBN 978-3-89937-269-4

You are how you look: potential utility of quantitative body shape analysis in classification of Eocene cypriniforms

Juan Liu*

Abstract

The oldest skeleton-based fossil cypriniforms (Teleost: Ostariophysi: Cypriniformes) were recently recovered from the Eocene sediments of North America and East Asia. These fossils belong to three families: Cyprinidae, Catostomidae, and Jianghanichthyidae, and represent the principal sources for understanding the early diversity and divergence of the Cypriniformes. However, relatively conserved overall body plan (low morphological disparity) among Eocene cypriniforms is a major problem for confident taxonomic assignment. To better understand the early morphological disparity of cypriniforms, I conducted geometric morphometric shape analysis on all available Eocene cypriniform specimens following a systematic taxonomic review. I selected complete and undeformed two-dimensional specimens preserved laterally on slabs for image capture using a surface scanner. Fourteen homologous anatomical landmarks which represent body shape were digitized. A sample of 90 specimens from three families, four genera, and ten species were studied, consisting of 54 individuals of catostomids, 31 jianghanichthyids, and five cyprinids. First, in the morphospace of PC1 (34.1 %) and PC2 (26.6 %) from Principal Component Analysis (PCA) on Procrustes shape, members of the three families and taxa from North America and East Asia occupy different regions with minimum overlap. Second, PCA on Procrustes shape and appended size found a similar distribution pattern but more overlap between each pair of three families, and between taxa from North America and East Asia (PC1 99.66 %, PC2 0.11 %). Third, allometry visualization using the common allometric component suggested that growth trajectories of the three families are divergent. In a more focused allometric test, significant common allometric signal is inferred in species of *Amyzon*. After removal of size effect (size correction) using regression residuals in performing PCA, body shape variations among species of *Amyzon* show a random pattern which confirms the homogeneity of shape-size regression of *Amyzon*. Fourth, in the body shape morphospace accounting for phylogeny created by PC1 and PC2, projection of the phylogenetic tree shows species of each family are clustered without crossing over in the morphospace using mean shape of each species and the phylogenetic signal is statistically significant ($K = 0.743$, $p = 0.006$). Last but not least, Procrustes ANOVA on the two closely related *Amyzon aggregatum* and *Amyzon gosiutense* show no significant Procrustes distance difference on shape only variables, but shows significant differences after accounting for covariation of body shape and size. This finding explains the confusion that resulted in their synonymization by using linear measurement and meristic counts only, and supports the position that both should be unique species. Also, there are significant body shape differences between the three families of Eocene cypriniforms, and between North American and East Asian Eocene cypriniforms with and without accounting for covariation of body size. The distinctiveness of these three families in overall body shape and size correlated with phylogeny suggests that there was minimum homoplasy in body shape characteristics during the early divergence of cypriniforms. More research is needed to understand morphospace occupation of extant cypriniforms, and the relative positions of fossils in that space in order to better understand evolutionary trajectory and patterns of body shape changes through time during Cypriniformes' critical early divergence and diversification periods.

Introduction

Cypriniform fishes include suckers, carps, minnows, loaches, algae eaters, and other common and economically important species; they are globally distributed except in Australia and South America (Nelson et al. 2016). The order Cypriniformes is the most diversified and largest freshwater fish group with over 4500 species with new species still being rapidly discovered (Eschmeyer et al. 2018). The number of the families within this order continues to grow because of widespread interest and continued efforts to better understand their biology (Kottelat 2004, Chen & Mayden 2009, Conway 2011, Kottelat 2012, Schönhuth et al. 2018, Tan & Armbruster 2018). The extinct family Jianghanichthyidae from Eocene deposits of Asia has recently been described, as the first and only known extinct cypriniform family (Liu et al. 2015).

The oldest cypriniforms were long believed to occur in Cretaceous (100.5–66 Ma). However, only one unambiguous skeleton-based fossil record is known prior to the Eocene (33.9–56 Ma). That single record is from the Paleocene (56–66 Ma) Paskapoo Formation, represented by disarticulated bones (Wilson 1980). Following the sparse and often questionable fossil record in Cretaceous and Paleocene (Liu et al. 2015), cypriniforms underwent their first burst of evolutionary radiation during early to late Eocene. Sixteen unambiguous species belonging to 11 genera and three families (Catostomidae, Cyprinidae, and Jianghanichthyidae) have been described based on complete or nearly complete articulated skeletons (Cope 1872, Cope 1875, Cope 1893, Tang 1959, Wilson 1977, Wang et al. 1981, Grande et al. 1982, Sytchevskaya 1986, Chang et al. 2001, Liu & Chang 2009, Su 2011, Chen et al. 2015, Liu et al. 2015, Liu et al. 2016). Additionally, two cyprinid species were described based on pharyngeal teeth (Sytchevskaya 1986, Böhme et al. 2013), whereas seven cyprinid species (taxonomy pending clarification) were described from Sumatra based on complete skeletons (Sanders 1934). Two of the Sumatra species were very recently re-described and their taxonomic position reassessed (Murray 2019). These records suggest an apparent burst of taxonomic diversity recorded in a relatively short time; in contrast, morphological disparity appears to be relatively low.

Morphological differences among species of Cypriniformes are substantial, yet subtle. Their morphological disparity is relatively low compared to their extremely high taxonomic diversity.

* Current address: Department of Integrative Biology and University of California Museum of Paleontology, University of California, Berkeley. Berkeley, CA 94720, USA; liujuan@berkeley.edu
Division of Paleontology, American Museum of Natural History. New York, NY 10024, USA.

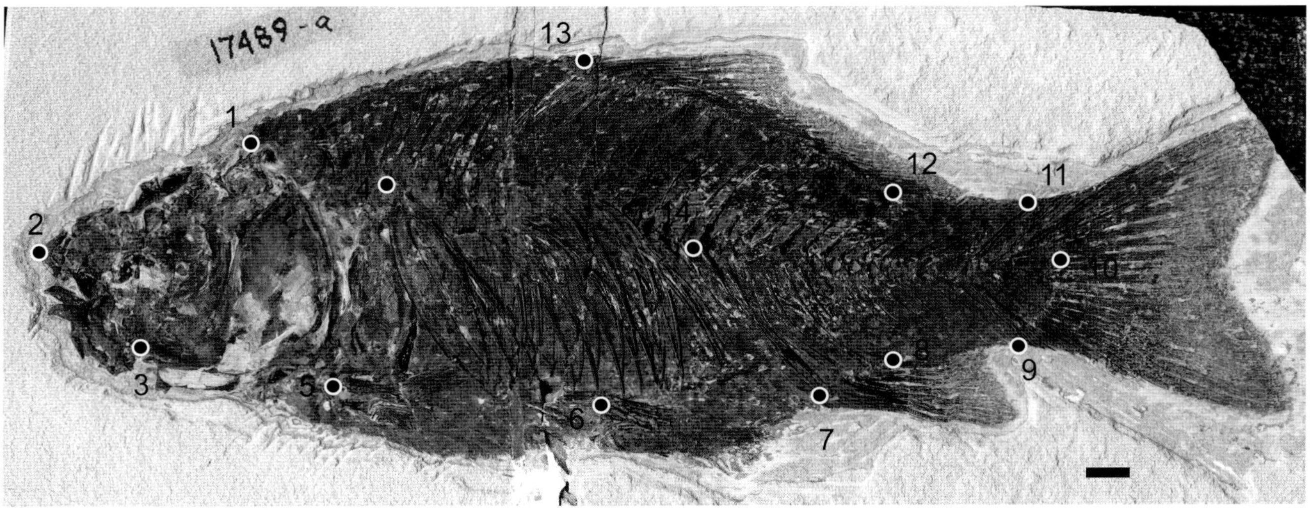

Fig. 1. Anatomical landmarks used in geometric morphometrics analysis using example of left-side preserved specimen of *Amyzon aggregatum* (UALVP 17489a). See Material and Methods for explanation of the landmarks. Scale bar = 10 mm.

Classification is made even more difficult for fossil cyprinids, as only osteological characters are normally available. Closely related species which share high similarity on osteological characters, especially in meristic and morphometric characters, had caused problematic taxonomic assignment in some cases. Based on highly overlapping meristic and metric characters, Bruner (1991) designated the specimen-rich *Amyzon gosiutense* as a junior synonym of *Amyzon aggregatum*.

The jianghanichthyid *Jianghanichthys hubeiensis* was once considered to be an Eocene catostomid because of similarity in general appearance (Chang & Chen 2008). Although the taxonomic status of above species were clarified by osteological comparison (Liu et al. 2015, Liu et al. 2016), the taxonomic confusion raised the question of whether the body shape (containing features of general appearance, meristics, and metric characters) of Eocene cypriniforms are significantly different among groups or species. If there are significant differences among groups and species, then body shape data may be informative for assisting taxonomic assignment when new fossil cypriniforms are collected in the future.

Eocene cypriniforms, when present in a fish fauna, were often dominant in number of specimens. Large numbers of specimens, subtle differences in meristics and metrics, and usually lateral preservation on a slab make Eocene cypriniforms amenable to a morphological differentiation study using quantitative shape analysis. Geometric morphometrics can be used to quantify body shape variation and allows multivariate analyses and statistical tests of potential shape differences between groups (Zelditch et al. 2004). This study aims to quantify body shape in Eocene cypriniforms for comparison among groups and species, through ontogeny, and across phylogeny, with the goal of clarifying shared and unique body shape traits that characterize putative species and groups.

Materials and Methods

Study sample. All complete skeleton specimens of known Eocene cypriniforms were digitized for this study. Deformed and incomplete (i. e., any designated landmark missing) specimens were removed from the dataset (see Methods below for landmarks). In total, 90 specimens/individuals from three families, four genera, and ten species were studied (numbers in brackets indicate specimen sample sizes used in this study): **Catostomidae**: *Amyzon aggregatum* (19), *Amyzon gosiutense* (8), "UndescribedGenus" *brevipinne* (2), *Amyzon commune* (2), *Amyzon hunanense* (9), and *Amyzon kishenehnicum* (14); **Jianghanichthyidae:** *Jianghanichthys hubeiensis* (29), *Jianghanichthys sanshuiensis* (2); **Cyprinidae**: "UndescribedSpecies" (3); *Tianshanicus_liui* (2). See appendix for a full list of specimens.

All known Eocene cypriniform species have been sampled for this study, except *Amyzon mentale*, *Amyzon fusiforme*, *Amyzon pandatum*, *Eoprocypris maomingensis*, and *Vasnetzovia artemica*. No known specimens of above taxa qualified for the completeness and non-deformation requirements for geometric morphometric analysis, and thus were excluded. Among the sampled species, "UndescribedGenus" *brevipinne* was originally described as *Amyzon brevipinne*, which is currently undergoing a taxonomic revision. "UndescribedSpecies" represents undescribed new materials from Xiangxiang Basin, Hunan, China, that were collected by the author and colleagues.

Image capture. Eocene cypriniforms were laterally compressed fishes in life resembling their extant relatives, and each is preserved laterally on a single plane of a slab within a shallow depth. This type of preservation is ideal for flatbed scanning to capture body shape and minimizing the deformation from imaging. The scanners Epson Perfection 3590 in University of Alberta (UALVP) and Uniscan M800U in the Institute of Vertebrate Paleontology and Paleoanthropology, Chinese Academy of Sciences (IVPP) were used. A few specimens from Royal Ontario Museum (ROM) and Canadian Museum of Nature (CMN) were photographed using a Canon 70D with camera stand and manually adjusting the specimen's surface and camera to be horizontal using a bubble level.

Anatomical landmarks. To define body shape, 14 homologous and anatomical landmarks were recorded in the following order (Fig. 1): 1, posterior border of parietal; 2, anteromedial corner of premaxilla; 3, anterior tip of preopercle; 4, center of first centrum excluding Weberian apparatus; 5, origin of pectoral fin; 6, origin of pelvic fin; 7, origin of anal fin; 8, posterior insertion of anal fin; 9, ventral insertion of caudal fin; 10, distal end of hypural 3; 11, dorsal insertion of caudal fin; 12, posterior insertion of dorsal fin; 13, origin of dorsal fin; 14, center of first caudal centrum. Most of these landmarks represent points that serve as bony endpoints in common linear measurements of bony fish. For example, the distance from 2 (anteromedial corner of premaxilla) to 10 (distal end of hypural 3) represents the distance between snout and caudal fin base which is approximately the standard length of a fish.

The origins of dorsal and anal fins (landmarks 7 and 13) were recorded at the base of the first rudimentary fin ray. The insertion of paired fins (landmarks 5 and 7) are recognized on the left fin if the fish is left-side preserved, and vice versa.

Landmark digitization was done in the TPS software package: The TPS file for each taxon represented by image stacks was created using TPS utility v. 1.68 (Rohlf 2016). Landmarks were digitized based on selected images using tpsDig2 (Rohlf 2010). The 14 landmarks were recorded in numerical order on left-side preserved specimens and horizontally flipped right-side preserved specimens. Scales for each image were digitized at the same time.

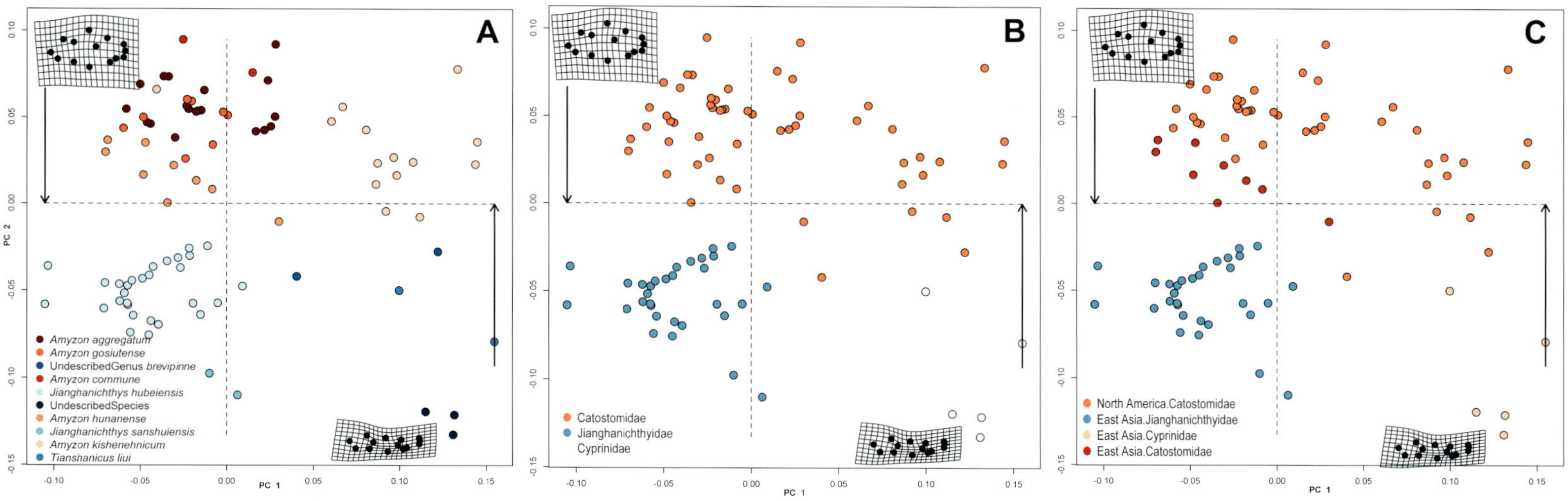

Fig. 2. Scatter plot of PC scores from principal component analysis (PCA) of landmark based body shape changes on species of Eocene catostomids, cyprinids, and jianghanichthyids. **A.** PC plot by species; **B.** PC plot by family; **C.** PC plot by interaction of family and geographic distribution. First two principal components (PC1 and PC2) are displayed. Body shape deformations along axis of PC1 and PC2 are mapped with transformation grids to visualize shape changes. PC1 and PC2 together account for 60.7 % of the total variance (PC1 34.2 %, PC2 26.5 %).

Geometric morphometric analyses. The following analysis series were performed in R 3.3.3 for Mac OS (R Core Team 2019) with a customized script (https://github.com/liujuan-leaf/Body-shape-analysis-of-Eocene-cypriniform-fishes.git). In the script, I used the abind package (Plate & Heiberger 2016) to assemble landmark dataset, geomorph package (Adams & Otárola-Castillo 2013, Adams et al. 2020) with depended RRPP (Collyer & Adams 2018, 2020) for shape analysis, ape package (Paradis & Schliep 2018) to incorporate phylogeny in shape analysis, and RColorBrewer (Neuwirth 2014) to plot results from analysis using colorblind friendly colors.

After the dataset had been exported from tpsDig2, the individual dataset of each species was imported and combined in R. Procrustes fit was performed to align the data. Procrustes superimposition was used to remove the position and orientation effect. Subsequent analyses were based on Procrustes coordinates for shape and centroid size for size (Zelditch et al. 2004).

Principal component analysis (PCA) and Procrustes ANOVA were performed to describe and assess body shape variations and disparity among species, families, and between North American and Asian taxa. PCA identified the major axes of variation in the body shape data, which plotted data with PCA scores and visually provided insights into patterns in the data. I applied Procrustes Analysis of Variance (ANOVA) with a permutation test using Procrustes distance to assess the statistical hypothesis of variation between species and between groups in terms of body shape.

To test the covariation of shape and size, I conducted regression analysis on Procrustes coordinates and log-transformed centroid size, which represent body shape and size, respectively. The null hypothesis tested was that size was not a significant predictor of body shape variations and difference among groups. Three linear models for simple allometry, common allometry, and unique allometry were tested for all sampled species (Mitteroecker et al. 2004). In addition to the body shape only analysis, PCA and pairwise comparison of Procrustes ANOVA were repeated using body shape and appended body size. Then, in a more focused allometry analysis on species of *Amyzon*, further examination of body shape variations after size-correction was performed. To remove size effect, regression residuals were extracted from the regression analysis between centroid size and shape (Klingenberg 2016). Another series of PCA were conducted on the regression residuals to test whether allometry significantly affect body shape variation among groups.

A phylogeny based on the phylogenetic tree from Liu (2016), with the addition of cyprinids, was mapped in morphospace to examine phylogenetic patterns of morphological variation. The composite tree was composed in Mesquite v.3.51 (Maddison & Maddison 2011). The composited phylogeny was considered ultrametric and constructed with Grafen's computation (Grafen & Hamilton 1989) of branch lengths in R. Phylogenetic signals were tested using both mean shape (Fig. 5) and largest individual (Supplement III) of each species. The phylomorphospace (projecting phylogeny into a multivariate morphospace (Sidlauskas 2008)) was constructed using the first two PCA axes with mapped phylogeny. A phylogenetic signal test using similar methods as in Armbruster et al. (2015) was performed on the null hypothesis of no phylogenetic signal to the distribution of taxa in the morphospace.

Results

Body shape variations using PCA and Procrustes ANOVA. PCA on body shape data of Eocene cypriniforms was performed and displayed in the morphospace defined by PC1 and PC2 (Fig. 2A). The first two PC axes together account for 60.7 % of the total variance, of which the first axis accounts for 34.2 % and second axis 26.5 % (PC3, 9.7 %). Additional PC axes displaying less than 10 % of variance explained were not analyzed. Specimens located on the positive end of PC1 axis exhibit a deeper body, larger head, more ventrally placed snout, and more posteriorly placed anal fin, whereas specimens with more positive PC2 values show a larger dorsal fin, shorter caudal peduncle, more dorsally placed first vertebra (excluding Weberian apparatus), more anteroventrally placed pectoral fin, and more anterodorsal placed anal fin (wireframe graph along PC axis of Fig. 2). Taxa from three families (Catostomidae, Cyprinidae, and Jianghanichthyidae) are apparently clustered in three non-overlapping regions in the morphospace (Fig. 2B). Meanwhile, North American and Asian taxa occupy minimally overlapping morphospace regions (Fig. 2C).

Procrustes ANOVA indicates significant correlation between body shape and taxonomic assignment (species and family respectively) and geographic distribution ($p = 0.001$ in all three tests). The pairwise comparison of Procrustes distance between mean shape compared each pair of species, each pair of families, and North American and Asian taxa. The results indicate significant differences between families and between American and Asian taxa (Table 1). All the p values of compared geographical groups and families are < 0.05 from permutation tests using Procrustes distance, and permit rejection of the null

Table 1. Pairwise Procrustes ANOVA with permutation test using Procrustes distance on body shape data only and covariation of shape-size of Eocene cypriniforms. Higher level groups and closely related *Amyzon* species are summarized here. EA-NA cypriniforms, East Asian and North American Eocene cypriniforms.

Groups	Compared taxa	Shape only Procrustes distance/p-value	Shape-size Procrustes distance/p-value
geographical groups	EA – NA cypriniforms	0.100/0.001	0.098/0.001
families	catostomids – cyprinids	0.194/0.001	0.192/0.001
	catostomids – jianghanichthyids	0.109/0.001	0.100/0.001
	cyprinids – jianghanichthyids	0.193/0.001	0.180/0.001
species	*Amyzon aggregatum* – *A. gosiutense*	0.054/0.189	4.260/0.024
	A. aggregatum – *A. kishenehnicum*	0.119/0.001	2.798/0.970
	A. gosiutense – *A. kishenehnicum*	0.127/0.001	5.018/0.093

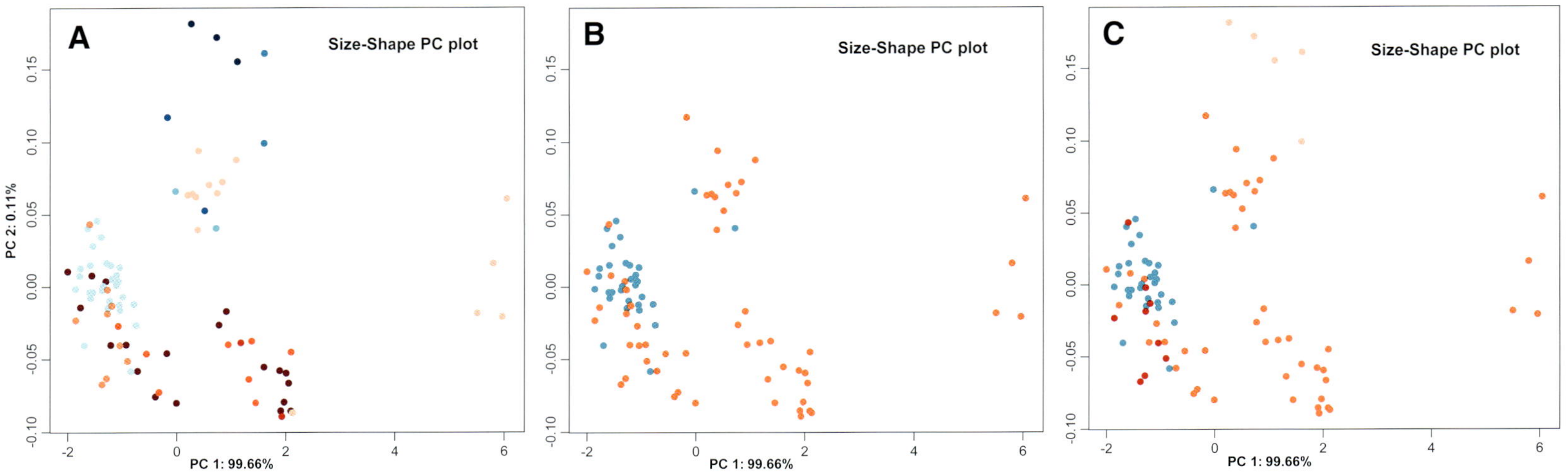

Fig. 3. Size-shape PCA of landmark based body shape changes on species of Eocene catostomids, cyprinids, and jianghanichthyids. **A.** PC plot by species; **B.** PC plot by family; **C.** PC plot by interaction of family and geographic distribution. Group legends are the same as of Figure 2A, B, and C respectively. The PCA was repeated using same shape data as shown in Figure 2 and appended with centroid size. PC1 and PC2 together account for 99.77 % of the total variance (PC1 99.66, PC2 0.11 %). The PC1 is highly correlated with size which is typical for an allometric data set.

hypothesis of no significant differences at the $p = 0.05$ level. Among the pairwise comparison of three closely related species, body shape of *Amyzon kinshenehnicum* is significantly different from that of *A. gosiutense* and *A. aggregatum*, but the latter two are not significantly different from each other.

Covariation of body shape and size. Simple allometry fit of all sampled species using Procrustes coordinates and log-transformed centroid size suggested significant covariation of body shape and size. To characterize the observed allometry, linear models of common allometry of species (parallel slopes) and species-specific allometry (unique) were fitted (Mitteroecker et al. 2004). ANOVA using Procrustes distance with 1000 permutations shows significant covariation of body shape and species-specific size ($p < 0.05$). Also, the Residual Sum-of-Squares (RSS) of the three linear models are 0.94, 0.36, and 0.30. Therefore, species of the complete sample display unique allometry among species, and this simple allometry is the best descriptor of the allometry of the sample of all species. To examine the covariation of body shape and size, and understand the effect of size on body shape variation, PCA on body shape was repeated and supplemented by size (Fig. 3). There are overlapping data points of species in the size-shape morphospace of PC (Fig. 3A). However, taxa from the three families (Catostomidae, Cyprinidae, and Jianghanichthyidae) (Fig. 3B), and from NA and EA (Fig. 3C) are apparently clustered in three and two regions respectively in the morphospace. The pairwise comparison using Procrustes ANOVA accounting for simple allometry (1000 permutations) shows significant differences between species of families and between geographical areas ($p = 0.001$, the same result as from the same analysis using shape data only. Among the pairwise comparison of three closely related species, contrary to previous shape only analysis, body shape of *Amyzon kinshenehnicum* is not significantly different from either that of *A. gosiutense* or *A. aggregatum*, but the latter two are significantly different from each other ($p < 0.05$). In other words, size resulting from growth contributes significantly to body shape differences among closely related species.

Size-standardized body shape variation of *Amyzon*. For the speciose and specimen-rich genus *Amyzon*, linear models of common allometry and unique allometry were fitted. In the subsample of *Amyzon* species, body shape is significantly correlated with body size and species status ($p = 0.001$ respectively), but not significantly correlated with species-specific sizes ($p = 0.67$). RSS of common allometry and unique allometry are 0.22 and 0.20, so that common allometry better describes the allometry of *Amyzon* species. In this case, the "common allometry" is estimated by a pooled regression of the shape variables corrected on size (Mitteroecker et al. 2004). To remove the allometric effect and "correct size", regression residuals are extracted from pooled regression. PCA using the covariance matrix of the residuals produced a size-corrected morphospace (Fig. 4C). The first two PCs account for 44.4 % of the variance (PC1 30.4 %, PC2 14.0 %, and PC3 11.5 %). Different from the well-defined clusters from shape only PCA (Fig. 4A; PC1 49.91 %, PC2 8.99 %) and size-shape PCA (Fig. 4B; PC1 99.81 %, PC2 0.09 %), data points of each species are randomly distributed in the size-corrected PC morphospace. Representing the relationship of allometric shape to size itself, the plot (Fig. 4D, E) of the common allometric component (CAC) and PC1 of the residuals of shape coordinates after size regressed (RSC1) by log centroid size shows the strong linear correlation of scale with the common allometric component within all five species. Also, the growth trajectory in the CAC-RSC1 plot suggests parallel allometric slope (Fig. 4E).

Mapping phylogeny to PC morphospace. A phylogenetic tree including all the sampled species is projected into morphospace created by the PC1 and PC2 of mean shape (consensus shape of each species) (Fig. 5). The ultrametric phylogeny has limited crossover in the phylomorphospace. Under a Brownian motion model of evolution, the phylogenetic signal test using Procrustes shape variables and the projected phylogeny with permutation test show significant phylogenetic signal ($K = 0.743$, $p = 0.006$), and the Procrustes distances are not significantly different among species after accounting for phylogeny. This result allows me to partially reject the null hypothesis of no phylogenetic signal.

Discussion

Body shape of Eocene Cypriniformes and differences among species and groups. The geometric morphometric analyses show that there are significant differences in the cypriniforms examined and between families and between North American and East Asian cypriniforms (Fig. 2). Among the differences found from the PCA, PC1 (34.2 %) is associated with variations in body depth, head length, snout position (terminal to sub-terminal), caudal peduncle length, and length of caudal vertebral column (distance between the 1st caudal vertebra and the distal end of hypural 3, corresponding to the number of caudal vertebrae). PC2 (26.5 %) is mostly associated with the relative positions of dorsal fin rays insertions and ends. PC3 (not showing in figure) is the axis associated with the relative insertion point of the pectoral and the pelvic fin, as well as the origins and ends of the anal fin. PC1, PC2, and PC3 together summarize 60.7 % of the variance in the overall body shape consisting of above morphometric and corresponding meristic differences.

Some of the characters that are associated with PC scores are the most distinguished interspecific differences in descriptive comparison of the sampled species. For example, the relative position of origin and end points of the dorsal fin (along PC2) represents the length of dorsal fin base, and reflects the number of dorsal fin rays. It is the most prominent diagnostic character in new species recognition of Eocene cypriniforms, and main difference between species of *Amyzon*, the speciose genus of Eocene catostomid (Liu et al. 2016). Interestingly, dorsal fin base length and relative position (PC2) are not the

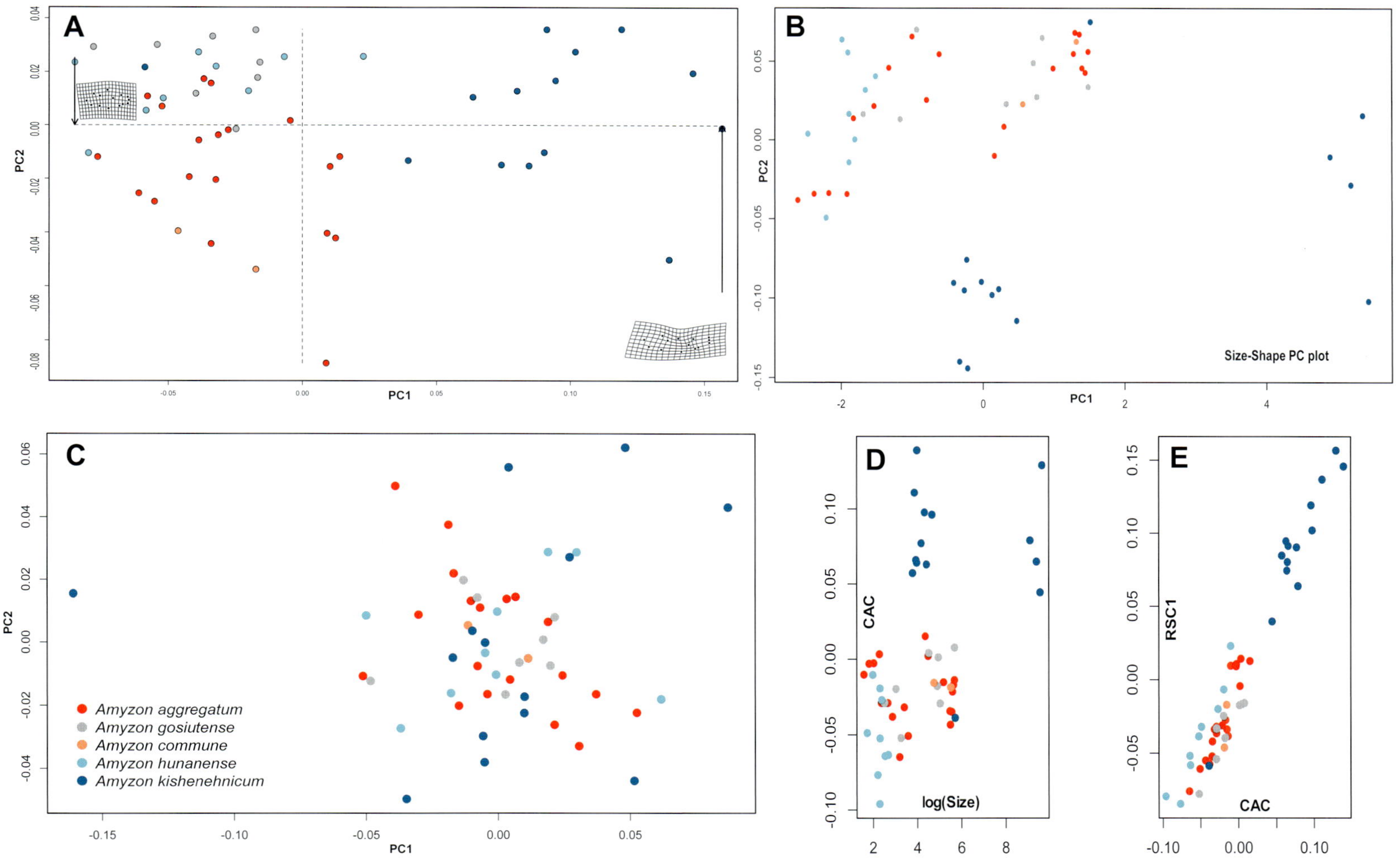

Fig. 4. Scatter plot of PC scores and allometric regression on species of *Amyzon*. **A.** PCA on landmark based Procrustes body shape of *Amyzon* species; **B.** PCA using Procrustes body shape and centroid size; C. Regression residuals are from the linear regression model of common allometry; **D.** plot of common allometric component (CAC) to log transformed centroid size; **E.** plot of CAC and PC1 of the residuals of shape coordinates after size regressed (RSC1).

most prominent differences in this geometric morphometric analysis. A combination of characters, which summarize overall differences in body depth, head length, snout position, caudal peduncle length, and possibly the number of caudal vertebrae, shows higher variance (PC1). The combination of these characters has the potential to evaluate interspecific distinctiveness and to help in defining higher-level phylogenetic relationships in fossil cypriniforms.

Besides distinctive clustering in morphospace and significant difference from Procrustes distance observed between NA and EA taxa, interfamilial and interspecific differences among some species of *Amyzon* are also recognized by additional analyses in Procrustes ANOVA of both shape only and covariation of shape and size (Table 1). Although the meristic and metric characters of *Amyzon* species overlap extensively both in count or measurement range (Bruner 1991), comparisons using quantitative analyses of body shape indicate significant differences among some species. *Amyzon aggregatum* and *A. gosiutense*, which were suggested to be senior and junior synonyms because of highly overlapping meristics and metrics (Bruner 1991), occupy adjacent and moderately overlapping morphospaces in PCA. The additional Procrustes distance analysis (Table 1) suggests they are not significantly different in Procrustes body shape data, but significantly different in body shape appended with body size, in accordance with the osteological study of Liu et al. (2016). This later finding contradicts the statistical comparison using linear measurement (metrics) and meristics but supports the qualitative comparisons in osteological morphology. It is indicated that growth characteristics are essential in taxonomical evaluation and geometric morphometric methods can more comprehensively summarize the overall body plan of cypriniforms, providing an informative source of distinguishing characteristics in taxonomic and phylogenetic study of fossil cypriniforms.

Allometric and size-standardized body shape. Maximum body size of Eocene catostomid species ranges from 80 mm in "UndescribedGenus" *brevipinne* to about 450 mm in *Amyzon kishenehnicum* (Liu et al. 2016) in standard length. Even with different maximum body size, allometry has been noticed and often described as shallow body in juvenile and deeper in the adults of Eocene catostomids and jianghanichthyids (Wilson 1974, Liu et al. 2015, Liu et al. 2016). Specifically, the ratio of body depth to standard length is usually positively correlated with body size. However, one should not assume all groups in the studied sample share parallel allometries, meaning that the ontogenetic shape change is the same across all taxonomic levels (Mipounga et al. 2020). The covariation analysis of body shape and size in this study tested three linear regression models using Procrustes coordinates of 14 anatomical landmarks and centroid size, which represent a number of measurable parameters (morphometric characters) and stands for overall body shape of the fish in lateral view and their covariation with overall size. The results suggest overall body shape of the tested species is significantly correlated with size, but Eocene species of the three cypriniform families display unique allometry. In other words, their allometric slopes are not parallel. An additional explorative analysis using CAC-RSC1 plot (Supplementary II) was performed and clearly showed that the growth trajectory of three families were divergent in the Eocene.

Moreover, interspecific variation among tested species is evident in both descriptive morphology comparison and morphospace analysis. The body depth of Eocene catostomids and jianghanichthyids ranges from extremely deep in *Plesiomyxocyprinus arratiae* (Liu & Chang 2009) to relatively shallow body in "UndescribedGenus" *brevipinne* (Wilson 1977). However, the maximum standard sizes of these two species also differ dramatically, with the former being ~300 mm and later ~80 mm. To detect whether the interspecific body shape difference is a result of size allometry, size should be corrected or standardized when comparing the species with large range of body sizes. Three linear models were tested for the allometric type of *Amyzon* species and the results indicate that a common allometry is shared by *Amyzon* species, which means homogeneity of their allometric slopes. The method using regression residual to remove size effect on morphological comparison (Sidlauskas et al. 2011, Klingenberg 2016) is then performed in this study. After size-correction, specimens of each species became essentially randomly distributed, rather than occupying the same region as that in PC morphospace without size correction. This

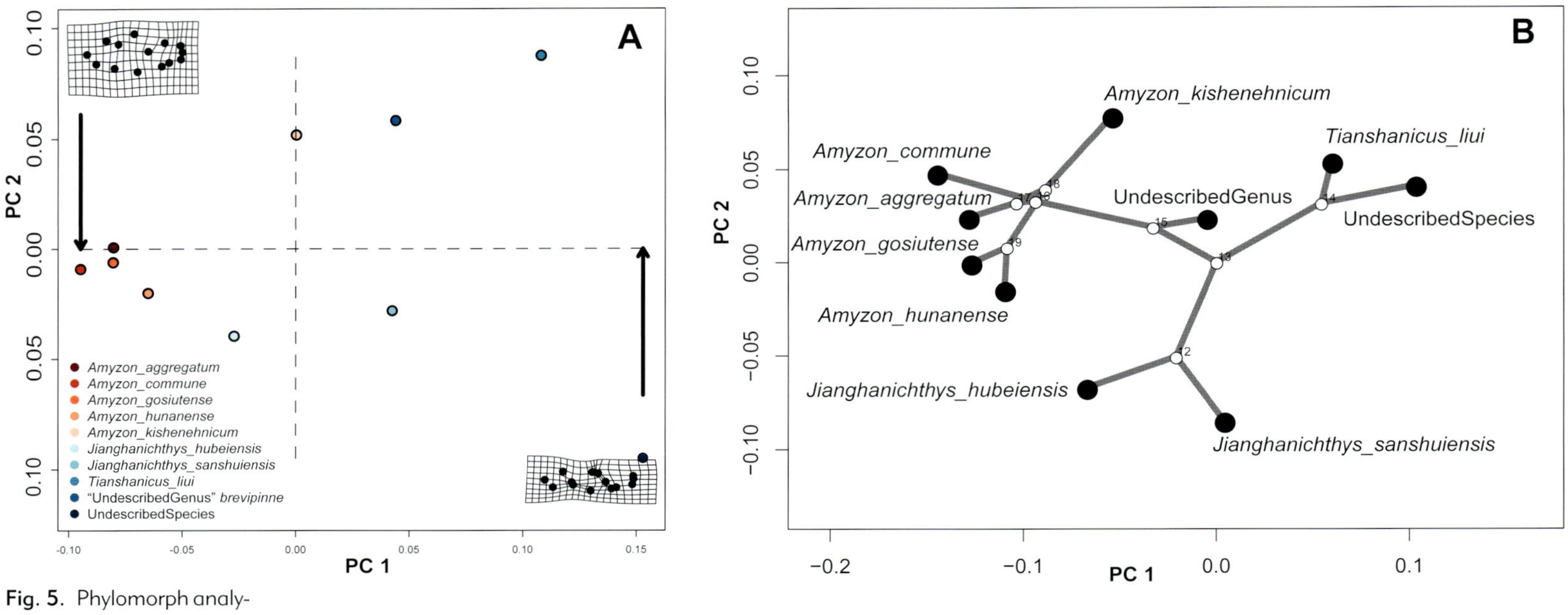

Fig. 5. Phylomorph analysis using mean shape of each species. **A.** PCA of mean shape of each species. **B.** Phylogeny mapped over PC1 versus PC2 scores. Phylogeny is modified and composited from Liu (2016) computed with ultrametric branch length using Grafen's method. PC scores are a consensus from each species. PC1 and PC2 together account for 69.24 % of the total variance (PC1 49.81, PC2 19.43 %). Body shape deformation along axis of PC1 and PC2 are mapped with transformation grids to visualize shape changes. Observed Phylogenetic Signal (K): 0.7431, *p* value: 0.006.

result demonstrates that allometry plays an important role in differentiating species of *Amyzon*.

Furthermore, data points of the three families are less clustered in the size-shape morphospace (Fig. 3B) than they are in Procrustes-shape-only PC space. Geographically, North American and East Asian species show more overlap under allometric influences than shape only analyses, as well (Fig. 3C). Despite the pattern changes in the morphospace, the pairwise comparison in Procrustes distance and permutation shows significant differences between families and between geographical distribution with or without appended size (Table 1). To sum up the above findings, in terms of body shape under the influence of allometry, the interfamilial differences are greater than those of interspecific variation, whereas the interspecific variation of overall body shape of Eocene cypriniforms is greater than their intraspecific variation even though maximum body size varies largely. This has important implications for qualitative comparisons, because the results suggest that morphological differences could be described from different sized specimens from different taxa with consideration of allometric bias. For instance, in direct comparison of two groups of fish consisting of individuals displaying ontogenetic series, in addition to providing mean and range of measurements, one could detect the difference or similarity among groups by comparing the growth trajectory estimated by CAC-RSC1. This will complement the osteological study towards taxonomy, especially with closely related fossil fish.

Evolutionary trajectories through morphospace. The geometric morphometric analyses of this study show that body shape variation among the fossil cypriniforms analyzed is significantly affected by phylogeny using the consensus shape of each species, and the Procrustes distances are not significantly different among species after accounting for phylogeny. In the consensus mean body shape morphospace created by PC1 and PC2, projection of the phylogenetic tree shows species of each family (Catostomidae, Cyprinidae, and Jianghanichthyidae) are clustered without crossing over (Fig. 5). It is apparent that there are strong phylogenetic patterns in the data as these clades occupy specific regions of the body shape space. In other words, speciation and diversification in these fishes appears to have occurred via explorations of new morphospace regions, rather than convergence or constraint within the same morphotypes.

Most fish exhibit indeterminate growth, such that morphological comparisons between species are often performed on the largest available specimen to most accurately represent their adult morphology. As an exploratory analysis, I also used the body shape data of the largest specimen (centroid size) of each species to plot the phylogeny in morphospace, and tested for phylogeny signal. Species of each family are still clustered but display extensive cross-over with other families. The phylogenetic signal test using Procrustes shape variables and the projected phylogeny with permutation test show insignificant influence of phylogeny ($K = 0.479$, $p = 0.928$). This contradictory result raises the question of whether phylogeny signal or allometric influence more strongly affects the body shape variations of Eocene cypriniforms. The mean shape approach seems inclusive for intraspecific variations and some ontogenetic changes. However, despite exhaustive sampling, specimen sample sizes of Eocene cypriniforms are not balanced and equally abundant, instead ranging from two (*Tianshanicus liui*) to dozens (species of *Amyzon*). Thus, the mean shape of current datasets may not adequately represent the true consensus shape of each species. On the other hand, the largest specimen of each species is an acceptable representation of adult body shape in those cypriniform species that exhibit metamorphic stages. If a single specimen were selected to represent the species, the intraspecific and allometric data would be lost inevitably. In this case, allometry will not be evaluated when accounting for phylogeny, with less than ideal results. Therefore, I performed this analysis on both datasets, and preferred using mean shape for this project to account for allometry.

To sum up the divergence in body shape and growth trajectory when cypriniforms diversified and geographically expanded in the Eocene, body shape of catostomids differs from that of cyprinids and jianghanichthyids in the length of the dorsal fin base (PC2), which normally corresponds to the number of dorsal fin rays. Body shape of cyprinids differs from that of jianghanichthyids mostly in body depth, head length, snout position, and caudal peduncle length (PC1). The distinctiveness of these three families in overall body shape and divergence of growth trajectory suggests that there was minimum homoplasy in body shape characteristics during the early divergence of cypriniforms.

Conclusion

Using geometric morphometrics, significant body shape differences were found among the three families of Eocene cypriniforms, between North American and East Asian Eocene cypriniforms, and among some closely related species of the Eocene genus *Amyzon*. The differences and patterns of distribution in morphospace remain the same for higher level groups, but species of *Amyzon* cannot be distinguished morphometrically after size correction. Whereas growth trajectory was shown to diverge among families, species of *Amyzon* share common allometry and exhibit parallel growth trajectory. Allometry also affects the body shape comparison of previous synonymized *A. aggregatum* and *A. gosiutense*; these species exhibit no significant differences in Procrustes shape, but do show a significant difference in shape appended with size. The overall body shape and distribution patterns in morphospace represented by mean shape of each species are also significantly associated with phylogenetic signal. Further research should expand the analyses to extant cypriniforms to contextualize allometric characteristics, evolutionary trajectory and patterns of Cypriniformes body shapes over their 50-million-year divergence and diversification into one of the most widespread freshwater vertebrate groups living today. Having such data also will permit secondary analyses, such as inference of macroevolutionary rates and modes, to be conducted, and will eventually help integrate understanding about cypriniforms evolution into a global environmental change framework.

Acknowledgements

Part of the analysis and text in this study were prepared in the Division of Paleontology, American Museum of Natural History (AMNH). The author is truly grateful to John Maisey and Gloria Carvalho for their help, support, and encouragement. The author is also thankful to Brian Sidlauskas, Jonathan W. Armbruster, and an anonymous reviewer for constructive comments, to Z. Jack Tseng for reading this manuscript and suggesting stylistic improvement, to Meemann Chang, Mark V. H. Wilson, and Alison Murray for their guidance at the early stage of this research, to collection staff for specimen access and data collecting at UALVP, IVPP, AMNH, CMN, ROM, and Field Museum of Natural History (FMNH), and to Jiang-yong Zhang for fieldwork assistance and scan of IVPP cyprinid specimens. The research is supported by Open Grant from State Key Laboratory of Palaeobiology and Stratigraphy (Nanjing Institute of Geology and Palaeontology, CAS) (173116), National Natural Science Foundation of China (41862001), the International Partnership Program of Chinese Academy of Sciences (Grant No. 132311KYSB20170022), and Natural Science Foundation of Guangxi, China (2017GXNSFAA198291).

Literature cited

Adams, D. C. & E. Otárola-Castillo. 2013. Geomorph: an r package for the collection and analysis of geometric morphometric shape data. Methods in Ecology and Evolution, 4: 393–399.

Adams, D. C., M. L. Collyer & A. Kaliontzopoulou. 2020. Geomorph: Software for geometric morphometric analyses. R package version 3.2.1.

Armbruster, J., C. Stout & M. Hayes. 2015. An empirical test for convergence using African barbs (Cypriniformes: Cyprinidae). Evolutionary Ecology, 30: 435–450.

Böhme, M., M. Aiglstorfer, P.-O. Antoine, E. Appel, P. Havlik, G. Métais, L. The Phuc, S. Schneider, F. Setzer, R. Tappert, D. Ngoc Tran, D. Uhl & J. Prieto. 2013. Na Duong (northern Vietnam) – an exceptional window into Eocene ecosystems from Southeast Asia. Zitteliana, A 53: 121–167.

Bruner, J. C. 1991. Comments on the genus *Amyzon* family Catostomidae. Journal of Paleontology, 65: 678–686.

Chang, M.-M., D. S. Miao, Y. Y. Chen, J. J. Zhou & P. F. Chen. 2001. Suckers (Fish, Catostomidae) from the Eocene of China account for the family's current disjunct distributions. Science in China Series D-Earth Sciences, 44: 577–586.

Chang, M.-M. & G. Chen. 2008. Fossil Cypriniformes from China and its adjacent areas and their palaeobiogeographical implications. Geological Society Special Publication, 295: 337–350.

Chen, G., M.-M. Chang & H. Liu. 2015. Revision of *Cyprinus maomingensis* Liu 1957 and the first discovery of *Procypris*-like cyprinid (Teleostei, Pisces) from the late Eocene of South China. Science China-Earth Sciences, 58: 1123–1132.

Chen, W.-J. & R. L. Mayden. 2009. Molecular systematics of the Cyprinoidea (Teleostei: Cypriniformes), the world's largest clade of freshwater fishes: Further evidence from six nuclear genes. Molecular Phylogenetics and Evolution, 52: 544–549.

Collyer, M. L. & D. C. Adams. 2018. RRPP: An R package for fitting linear models to high-dimensional data using residual randomization. Methods in Ecology and Evolution, 9: 1772–1779.

Collyer, M. L. & D. C. Adams. 2020. RRPP: Linear Model Evaluation with Randomized Residuals in a Permutation Procedure, R package version 0.5.2.

Conway, K. W. 2011. Osteology of the South Asian Genus *Psilorhynchus* McClelland, 1839 (Teleostei: Ostariophysi: Psilorhynchidae), with investigation of its phylogenetic relationships within the order Cypriniformes. Zoological Journal of the Linnean Society, 163: 50–154.

Cope, E. D. 1872. On the Tertiary coal and fossils of Osino, Nevada. Proceedings of the American Philosophical Society, 12: 478–481.

Cope, E. D. 1875. On the fishes of the Tertiary shales of the South Park. Bulletin of the US Geological and Geographical Survey of the Territories, 2: 4–5.

Cope, E. D. 1893. Fossil fishes from British Columbia. Proceedings of the Academy of Natural Sciences of Philadelphia, 1893: 401–402.

Eschmeyer, W., R. Fricke & R. Van der Laan. 2018. Catalog of Fishes: Genera, Species, References.

Grafen, A. & W. D. Hamilton. 1989. The phylogenetic regression. Philosophical Transactions of the Royal Society of London B, Biological Sciences, 326: 119–157.

Grande, L., J. T. Eastman & T. M. Cavender. 1982. *Amyzon gosiutensis*, a new catostomid fish from the Green River Formation Copeia, 1982: 523–532.

Klingenberg, C. P. 2016. Size, shape, and form: concepts of allometry in geometric morphometrics. Development Genes and Evolution, 226: 113–137.

Kottelat, M. 2004. *Botia kubotai*, a new species of loach (Teleostei: Cobitidae) from the Ataran River Basin (Myanmar), with comments on botiine nomenclature and diagnosis of a new genus. Zootaxa, 401: 1–18.

Kottelat, M. 2012. Conspectus cobitidum: an inventory of the loaches of the world (Teleostei: Cypriniformes: Cobitoidei). Raffles Bulletin of Zoology, Supplement 26: 1–199.

Liu, J. & M.-M. Chang. 2009. A new Eocene catostomid (Teleostei: Cypriniformes) from Northeastern China and early divergence of Catostomidae. Science in China Series D Earth Sciences, 52: 189–202.

Liu, J., M.-M. Chang, M. V. H. Wilson & A. M. Murray. 2015. A new family of Cypriniformes (Teleostei, Ostariophysi) based on a redescription of †*Jianghanichthys hubeiensis* (Lei, 1977) from the Eocene Yangxi Formation of China. Journal of Vertebrate Paleontology, 35: 1–23.

Liu, J., M. V. H. Wilson & A. M. Murray. 2016. A new Catostomid fish (Ostariophysi, Cypriniformes) from the Eocene Kishenehn Formation and remarks on the North American species of †*Amyzon* Cope. Journal of Paleontology, 90: 288–304.

Maddison, W. P. & D. R. Maddison. 2011. Mesquite: a modular system for evolutionary analysis. 2.75 ed.

Mipounga, H. K., J. Cutler, J. H. Mve Beh, B. Adam & B. L. Sidlauskas. 2020. *Enteromius pinnimaculatus* sp. nov. (Cypriniformes: Cyprinidae) from southern Gabon. Journal of Fish Biology, 96: 1218–1233.

Mitteroecker, P., P. Gunz, M. Bernhard, K. Schaefer & F. L. Bookstein. 2004. Comparison of cranial ontogenetic trajectories among great apes and humans. Journal of Human Evolution, 46: 679–698.

Murray, A. M. 2019. Redescription of *Barbus megacephalus* Günther, 1876 and *Thynnichthys amblyostoma* von der Marck, 1876 (Cypriniformes: Cyprinidae) from probable Eocene deposits of

Southeast Asia, and an assessment of their taxonomic positions. Journal of Systematic Palaeontology, 17: 1433–1455.

Nelson, J. S., T. C. Grande & M. V. H. Wilson. 2016. Fishes of the World. John Wiley & Sons, Hoboken, 707 pp.

Neuwirth, E. 2014. RColorBrewer: ColorBrewer Palettes. R package version 1.1–2.

Paradis, E. & K. Schliep. 2018. ape 5.0: an environment for modern phylogenetics and evolutionary analyses in R. Bioinformatics, 35: 526–528.

Plate, T. & R. Heiberger. 2016. abind: Combine Multidimensional Arrays. R package version 1.4–5.

R Core Team. 2019. R: A Language and Environment for Statistical Computing. Vienna, Austria.: R Foundation for Statistical Computing.

Rohlf, F. J. 2010. tpsDig (computer program). Department of Ecology and Evolution, State University of New York, Stony Brook.

Rohlf, F. J. 2016. tpsUtil32 (computer program). Department of Ecology and Evolution, State University of New York, Stony Brook.

Sanders, M. 1934. Die fossilen Fische der alttertiären Süsswasserablagerungen aus Mittel-Sumatra. Verhandelingen van het geologisch-mijnbouwkundig genoostschap voor Nedeland en kolonien. Geologische Serie, 11: 1–44.

Schönhuth, S., J. Vukic, R. Sanda, L. Yang & R. L. Mayden. 2018. Phylogenetic relationships and classification of the Holarctic family Leuciscidae (Cypriniformes: Cyprinoidei). Molecular Phylogenetics and Evolution, 127: 781–799.

Sidlauskas, B. 2008. Continuous and arrested morphological diversification in sister clades of characiform fishes: a phylomorphospace approach. Evolution, 62: 3135–3156.

Sidlauskas, B. L., J. H. Mol & R. P. Vari. 2011. Dealing with allometry in linear and geometric morphometrics: a taxonomic case study in the *Leporinus cylindriformis* group (Characiformes: Anostomidae) with description of a new species from Suriname. Zoological Journal of the Linnean Society, 162: 103–130.

Su, D.-Z. 2011. A new cyprinid fish from Paleogene of northern Xinjiang, China. Vertebrata PalAsiatica, 49: 141–154.

Sytchevskaya, E. K. 1986. Palaeogene freshwater fish fauna of the USSR and Mongolia. Hayka Publishing House, Moscow, 154 pp.

Tan, M. & J. W. Armbruster. 2018. Phylogenetic classification of extant genera of fishes of the order Cypriniformes (Teleostei: Ostariophysi). Zootaxa, 4476: 6–39.

Tang, X. 1959. A new cyprnid species from Linli, Hunan. Paleovertebrata Et Paleoanthropologia, 1: 211–213.

Wang, J., G. Li & J. Wang. 1981. The early Tertiary fossil fishes from Sanshui and its adjacent basin, Guangdong., 90 pp.

Wilson, M. V. H. 1974. Fossil fishes of the Tertiary of British Columbia. Toronto, Canada: University of Toronto, 680 pp.

Wilson, M. V. H. 1977. Middle Eocene freshwater fishes from British Columbia. Life Science Contribution Royal Ontario Museum, 113: 1–61.

Wilson, M. V. H. 1980. Oldest known *Esox* (Pisces: Esocidae), part of a new Paleocene Teleost fauna from Western Canada. Canadian Journal of Earth Sciences, 17: 307–312.

Zelditch, M. L., D. L. Swiderski, H. D. Sheets & W. L. Fink. 2004. Geometric Morphometrics for Biologists: A Primer. Elsevier Academic Press, New York and London, 437 pp.

Supplement I: List of specimens used in geometric morphometric analysis

Amyzon aggregatum
UALVP 15825, UALVP 15833, UALVP 15835, UALVP 17456a, UALVP 17462, UALVP 17465, UALVP 17466, UALVP 17489a, UALVP 22875a, UALVP 32265, UALVP 33114, CMN 6190, ROM 11034b, ROM 11042a, ROM 11041a, ROM 11051a, ROM 11094b, ROM 11067a, ROM 11070a, and ROM 11019a.

Amyzon commune
AMNH FF 2579 and FMNH PF 116.

Amyzon gosiutense
UALVP 14723, UALVP 14725, AMNH FF 10460, FMNH PF 10425, FMNH PF 10427, FMNH PF 10577, FMNH PF 10579, and FMNH PF 10580.

Amyzon hunanense
IVPP V 12571.1–6, IVPP V 17906.6a, IVPP V 17906.7a, IVPP V 17906.9, IVPP V 17906.15b, IVPP V 17906.23, IVPP V 17906.31a, IVPP V 17906.35a, IVPP V 17906.38, IVPP V 17906.47a, IVPP V 17906.48, IVPP V 17906.49, IVPP V 17908.1a, and IVPP V 17908.2a.

Amyzon kishenehnicum
UALVP 55260, UALVP 23947, UALVP 24137, UALVP 24139, UALVP 24140, UALVP 24142, UALVP 24144, UALVP 24145, UALVP 24149, UALVP 24150, UALVP 24151, UALVP 24157, UALVP 38967, and UALVP 39031.

"UndescribedGenus" *brevipinne*
CMN 2042 and ROM 11163a.

Jianghanichthys hubeiensis
IVPP V 12163.1a, IVPP V 12163.4–7, IVPP V 12163.9–12, IVPP V 12163.14a, IVPP V 12163.15, IVPP V 12163.17–23, IVPP V 12163.25, IVPP V 12163.26, IVPP V 12163.28–31, IVPP V 12163.33, IVPP V 12163.334, IVPP V 12163.36b, IVPP V 12163.38–40, IVPP V 12163.49, IVPP V 12163.52, IVPP V 12163.53, IVPP V 12163.56, IVPP V 12163.57, IVPP V 12163.62, IVPP V 15712.2a, IVPP V 15712.3, GMC V1810-1 (V 2504 in Lei, 1977; 1987).

Jianghanichthys sanshuiensis
IVPP uncatalogued specimens.

Tianshanicus liui
IVPP V 12172.1a, IVPP V 12172.5a.

UndescribedSpecies_cyprinid
IVPP undescribed Eocene cyprinid specimens from Hunan

Supplement II

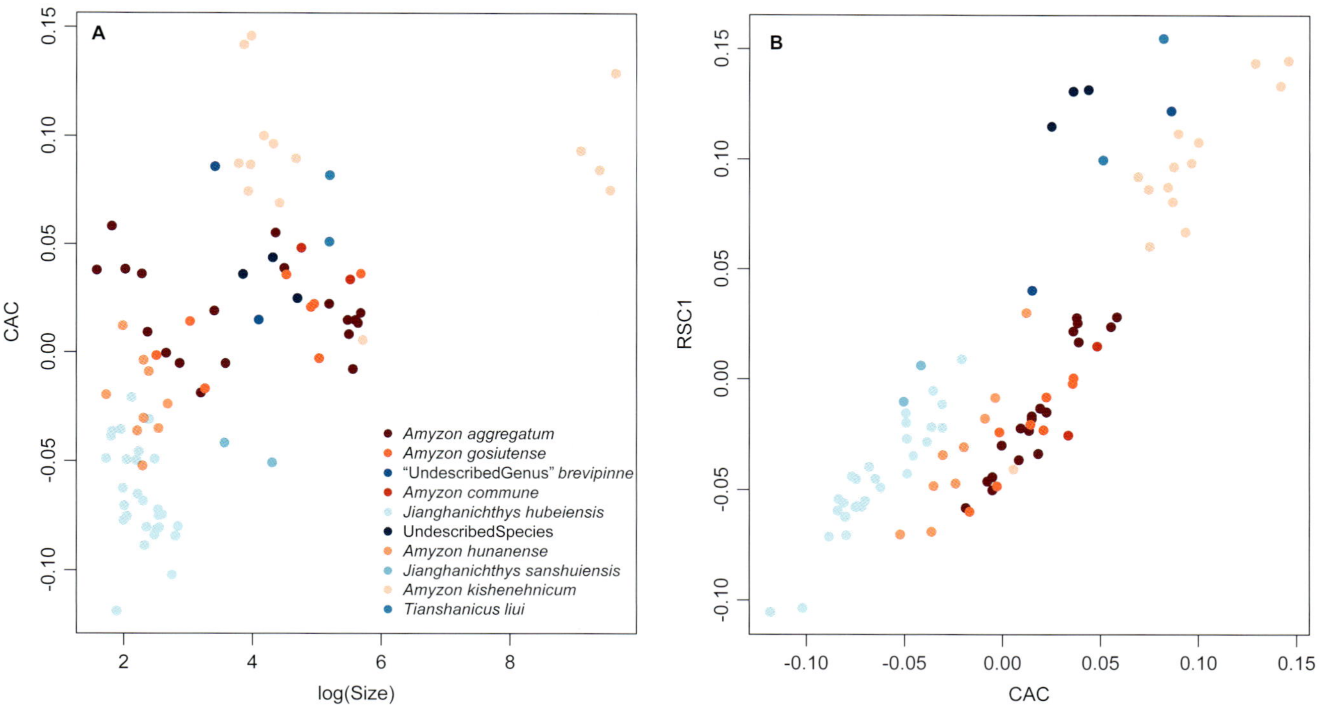

Fig. S1. Correlations of common allometric component (CAC) and the residuals of shape coordinates after size regressed (RSC1) to log transformed centroid size (logCS) in all sampled species. **A.** plot of CAC and logCS; **B.** plot of CAC and RSC1.

Supplement III

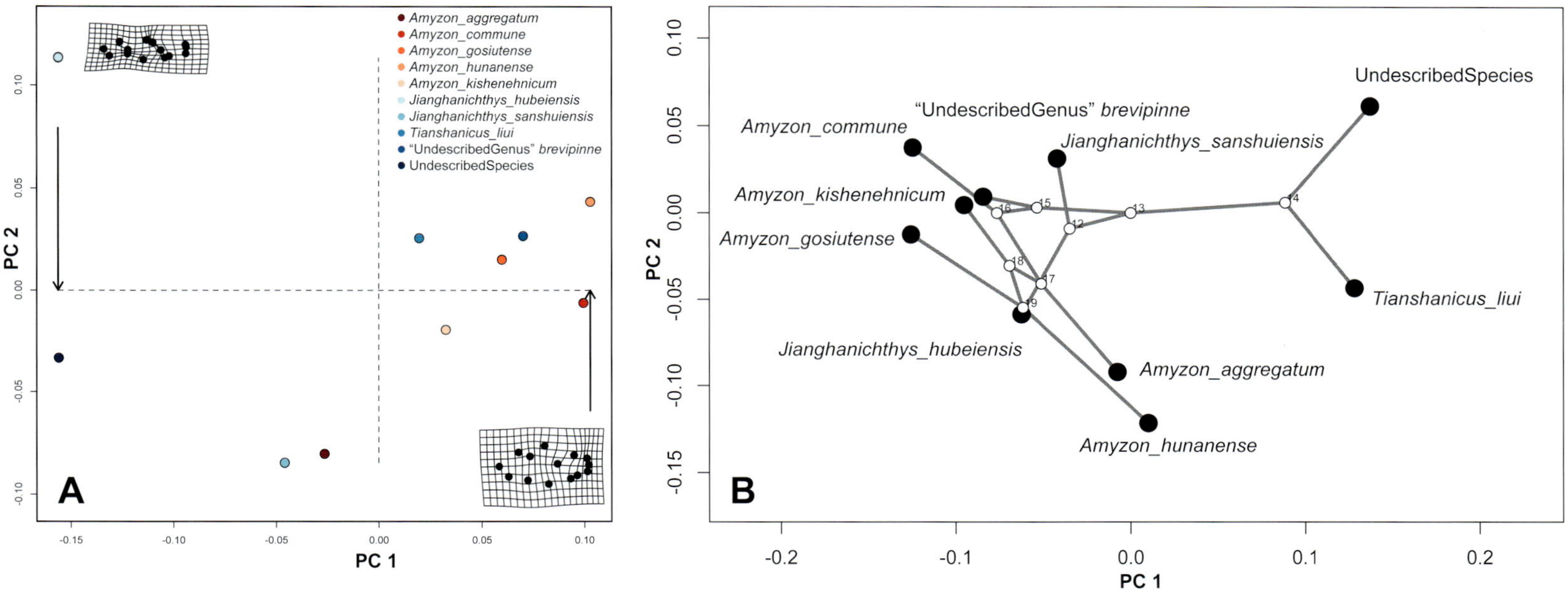

Fig. S2. Phylomorph analysis using largest specimen of each species. A phylogenetic tree including all the sampled species is projected into morphospace created by the PC1 and PC2 of the largest specimen of each species. Under a Brownian motion model of evolution, phylogenetic signal test using Procrustes shape variables and the projected phylogeny with permutation test shows insignificant signal ($K = 0.4798$, $p = 0.928$).

Ancient Fishes and their Living Relatives: a Tribute to John G. Maisey. Alan Pradel, John S. S. Denton & Philippe Janvier (eds.): pp. 81–91, 10 figs.
© 2021 by Verlag Dr. Friedrich Pfeil, München, Germany – ISBN 978-3-89937-269-4

Endoskeletal tissues of acanthodians (stem Chondrichthyes)

Carole J. Burrow* and Jan L. den Blaauwen**

Abstract

Consensus is emerging that the acanthodian grade of stem gnathostomes populates the chondrichthyan stem. One of the main defining features of crown chondrichthyans is traditionally considered to be prismatic and tessellated, or only tessellated, calcified cartilage, so it is of interest to determine the distribution of potential precursors of this tissue in acanthodians. The structure of the head and branchial endoskeleton in acanthodians is poorly known, other than for the highly derived acanthodiform *Acanthodes*. Here, we detail the distribution and type of mineralised tissues in representatives of the acanthodian families Euthacanthidae, Climatiidae, Gyracanthidae, Ischnacanthidae, Diplacanthidae, Mesacanthidae, Cheiracanthidae and Acanthodidae, and the 'climatiiform' *Nostolepis scotica*. We find that this distribution tends to be family-specific, although showing no recognisable phylogenetic trend relative to crown Chondrichthyes.

Introduction

As noted by Ørvig (1951, p. 329) regarding vertebrate hard tissues, "From a phylogenetic point of view the mesodermal hard tissues, dentine, bone and calcified cartilage, are equally old". He based this statement on the earliest known occurrence of all these mineralised tissues being in the upper Middle Ordovician Harding Sandstone, Colorado, having been the first worker (Ørvig 1951, fig. 18A) to note the presence of globular calcified cartilage (gcc) in the vertebrate assemblages from that deposit. Subsequently, Denison (1967) deduced that the globular calcified cartilage associated with vertebrate microremains from that formation derived from the pteraspidomorph agnathan *Eryptichius americanus* Walcott, 1892, based on its presence in a partially articulated head shield. Calcified cartilage has not yet been found in any younger pteraspidomorphs, but is, however, found in Silurian-Devonian osteostracan and galeaspid agnathans (e.g. Ørvig 1951, Wang et al. 2005), and in all the major groups of jawed fishes (Ørvig 1951, plate 4.2 [Placodermi], plate 5.2 [Chondrichthyes], plate 6.2 [Acanthodii], plate 8.3 [Actinopterygii]).

The next oldest vertebrate endoskeletal mineralised tissue known is perichondral bone, found in osteostracan and galeaspid agnathans from the Silurian of the Northern Hemisphere (Stensiö 1927, Janvier 1996). This tissue is generally regarded as the only type of endoskeletal bone found in placoderms and acanthodians, although several authors have reported endochondral bone in placoderms (Denison 1978, Brazeau et al. 2020). Otherwise, endochondral bone is only certainly known in the Osteichthyes, with the oldest taxon in which it is preserved being *Psarolepis romeri* Yu, 1998, found in the uppermost Silurian (Pridoli) to lowermost Devonian (Lochkovian) of Yunnan, China (Zhu et al. 2006).

The acanthodians

A consensus is emerging that all acanthodians are stem chondrichthyans (Coates et al. 2018), rather than being a group including both stem chondrichthyans and stem osteichthyans (Janvier 1996, Brazeau 2009). Coates et al. (2018, p. 6) proposed "resurrecting the term Acanthodii to define the diplacanthid-ischnacanthid-acanthodid clade". Their phylogenetic analysis omitted at least one taxon included in previous analyses, *Euthacanthus*, which presumably caused problems in determining relationships; for now, we continue to use the general term acanthodians, in the traditional sense, to describe the whole grade. A number of Devonian taxa other than acanthodians *sensu lato* are also regarded as being more or less derived stem chondrichthyans, despite lacking diagnostic features of crown chondrichthyans. For example, Coates et al. (2018) recovered *Gladbachus* Heidtke & Krätschmer, 2001 as a more basal taxon than those in the acanthodian grade, *Pucapampella* Janvier & Suarez-Riglos, 1986 in a derived group within the acanthodian grade, and *Doliodus* Traquair, 1893 as a basal member of the 'conventionally defined chondrichthyans' (sensu Maisey et al. 2017). Given the current focus on relationships of taxa within the chondrichthyan stem, it seems appropriate to better define the composition of the endoskeletal tissues within the acanthodian grade, as a prelude to further 'dissecting' the types of calcified cartilage found higher in the chondrichthyan stem.

The history of taxonomic descriptions of acanthodian fishes extends back for nearly 200 years, since the seminal work by Louis Agassiz (1833–1843). Although many new taxa were erected in the nineteenth century, those descriptions were almost exclusively focused on morphology, although Hugh Miller made thin sections of some fishes to investigate histology. During the twentieth century, detailed work on the histology of hard tissues in acanthodians focussed on the structure of the dermal skeleton, in particular the scales, with the main contributors being Walter Gross in the mid-20th century and Juozas Valiukevičius in the late 20th and early 21st centuries. Only in recent years has close attention been paid to the structure of endoskeletal tissues, which were previously assessed as being either unmineralised, or comprising globular calcified cartilage, or perichondral bone (e.g. Denison 1979). Endochondral bone is not present, contra Donoghue & Sansom (2002, p. 356). The oldest known occurrence of mineralised endoskeletal tissues in stem chondrichthyans is globular calcified cartilage in the fin spines of the Sinacanthidae (Sansom et al. 2005) from the late Llandovery of the Tarim Basin, northwest China. In this case, the globular calcified cartilage forms a thin layer on the internal surface of the spine. This arrangement differs from that in Devonian species (e.g. cheiracanthids and diplacanthids from Tynet Burn,

* Geosciences, Queensland Museum, 122 Gerler Rd, Hendra, Queensland 4011; carole.burrow@gmail.com (corresponding author)
** University of Amsterdam, Science Park 904, 1098 XH, Amsterdam, The Netherlands

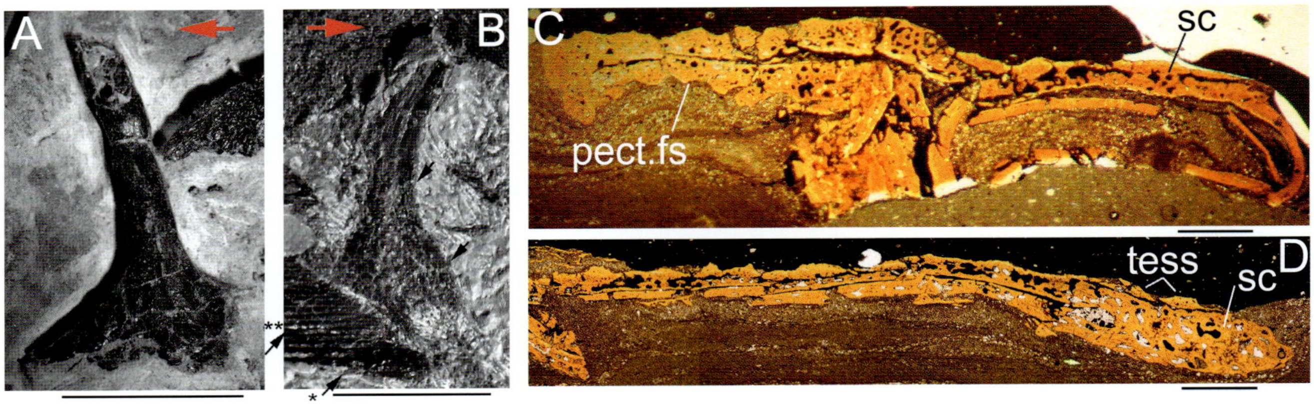

Fig. 1. Acanthodian pectoral girdle structure. **A.** *Euthacanthus macnicoli* NMS G.2010.7.39 left scapulocoracoid, lateral view. **B–D.** *Climatius reticulatus*: **B.** QM F57178, right scapulocoracoid articulated with pectoral fin spine, lateral view; **C. D.** thin sections of NMS G.2013.8.2, lateral surface uppermost: **C.** LD150 = NMS G.2013.8.2.4, through base of fin spine and scapula, approximate line of section indicated by single asterisked arrows on B; **D.** LD154 = NMS G.2013.8.2.8, through shaft and base of scapula, approximate line of section indicated by double asterisked arrows on B. Scale bars = 10 mm in A, B; 1 mm in C, D. Red arrows indicate anterior direction. Abbreviations: pect.fs, pectoral fin spine; sc, scapulocoracoid; tess, tesserae.

Scotland), where the whole of the spine central cavity can be infilled with a calcite which, by analogy with extant chondrichthyans (Maisey 1979), represents the mineralisation of the presumed hyaline cartilage that originally extended into this space.

Contributions by the present authors and colleagues on Scottish Devonian acanthodians over the last decade have incorporated descriptions and illustrations of the histological structure of the endoskeleton in the climatiid *Climatius reticulatus* Agassiz, 1844 (Burrow et al. 2015), the ischnacanthid *Ischnacanthus gracilis* (Egerton, 1861) (Burrow et al. 2018), and most recently, cheiracanthids from the Middle Devonian of Scotland (den Blaauwen et al. 2019, Burrow et al. 2020, Newman et al. 2020). Based on the investigations undertaken for the present study and previous works, we document here the basic tissue forms, and variations within those forms, found in different acanthodian families, and compare them with the composition of the endoskeletal tissues in other stem chondrichthyans.

Material and methods

Several specimens were newly sectioned for this project. *Diplacanthus crassisimus* specimen NMS G.2019.3.5 was revealed in slicing a nodule from Gamrie, also containing a *Cheiracanthus murchisoni* specimen which was being sacrificed for sectioning. The rest of the *D. crassisimus* specimen was subsequently exposed, and the head region and original slice were serially sectioned employing the techniques used by JdB in previous projects (e.g. Burrow et al. 2018). Other specimens serially sectioned for this project include *Mesacanthus mitchelli* NMS G.2019.3.2, *Mesacanthus pusillus* NMS G.2019.3.4 and 11, *Cheiracanthus latus* NMS G.2019.3.3, and *Homalacanthus concinnus* NMS G.2019.3.1.

Description

Mineralisation of the scapulocoracoid. Almost universally in acanthodians, the endoskeletal scapulocoracoid is preserved as a hollow, thick-walled bone structure, often having only the scapula ossified (Fig. 1). The core of the structure is presumed to have been unmineralised cartilage. Indeed, having a 'pork-chop' shaped, perichondrally ossified scapulocoracoid is one of the few general features currently recognized for the acanthodians as a grade or clade.

Histological structure of the neurocranium and visceral skeleton. During ontogeny, the neurocranium (braincase) derives mainly from the somitic and lateral-plate mesoderm, and the visceral skeleton or splanchnocranium (jaws and gill arches) derives from the ectomesenchyme (Janvier 1996). These differing origins, however, do not necessarily lead to different tissues forming the individual elements (e.g. Hirasawa & Kuratani 2015). Here we consider the type of tissues found to occur in representatives of most recognised acanthodian families (Table 1).

Euthacanthidae Berg, 1940. *Euthacanthus macnicoli* Powrie, 1864, the type and best known euthacanthid, from the Lochkovian (Lower Devonian) Midland Valley of Scotland, was considered by Denison (1979) to be the most primitive acanthodian, based on his and previous workers' (e.g. Watson, 1937) assessment that the visceral arches were completely unmineralised. However, Newman et al. (2014, fig. 6A, B) showed that the jaw cartilages are preserved in rare cases as a granular mineralisation, comprising globular calcified cartilage globules. *Euthacanthus* lacked teeth. None of the sections made by JdB through the head of an articulated *E. macnicoli* (NMS G.2013.8.1) showed any endoskeletal mineralisation. We have found no morphological evidence in articulated specimens that the neurocranium or the branchial arches were mineralised, and surmise that the loose calcified cartilage globules found in several disarticulated specimens, e.g. NMS G.2013.10.1 (Fig. 2) are highly likely to be from the jaws. These spherical globules show the rings of Liesegang that typify globular calcified cartilage, but show no evidence of compression or a tessellate organisation.

Climatiidae Berg, 1940. Burrow et al. (2015) described and figured the structure of the visceral skeleton in the Lower Devonian type species *Climatius reticulatus* Agassiz, 1844 (Fig. 3A), also from the Midland Valley of Scotland. The jaws were shown to be formed of a granular mineralisation, superficially similar to that observed on rare specimens of *Euthacanthus macnicoli*, and presumed to encase an unmineralised cartilage core. We have found no evidence for perichondral bone in the head region of *C. reticulatus*, with calcified cartilage being the only mineralised endoskeletal tissue preserved (Fig. 3B–D). Sections through the jaw cartilages of NMS G.2013.8.2 (Fig. 3C, D) show that the calcified cartilage globules form a subtessellate structure of two or more layers, with the individual units shaped like irregular blocks rather than like the spheres observed in the neurocranium of *C. reticulatus*, and in the other major groups

Table 1. Composition of the neurocranium and splanchnocranium in acanthodians. Abbreviations: cc, calcified cartilage; cgcc, consolidated globular calcified cartilage; gcc, globular calcified cartilage; mcc, multilayered subtessellate calcified cartilage; pcb, perichondral bone; scc, one or two-layered subtessellate calcified cartilage; tmt, tessellate mineralised tissue; x, not present; –, not mineralised; ?, present, but histological structure not known; * only known in rare specimens.

Taxon / endoskeletal element	*Euthacanthus*	*Climatius*	*Nostolepis*	*Diplacanthus*	*Ischnacanthus*	*Mesacanthus*	*Cheiracanthus peachi*	*Cheiracanthus grandispinus*	*Homalacanthus*	*Acanthodes*
neurocranium	–	–	gcc	* ?pcb+gcc	–	–	–	?pcb+gcc	–	pcb
jaws	gcc	mcc	gcc	*cgcc	mcc	tmt	gcc	scc	scc	pcb+gcc
hyoid arch	–	?	gcc	*cgcc	–	–	?layered cgcc	?	pcb+cc	pcb+gcc
other branchial bars	–	?	gcc	*	–	–	cgcc	cgcc	cgcc	?pcb+gcc
mandibular bone	x	x	x	x	x	x	x	x	x	?pcb+cc

of jawed fishes. It seems reasonable to infer that the subtessellate structure provided greater strength to the jaws, which bore robust tooth whorls. This type of cartilage is also preserved under the branchiostegal plates (Fig. 3E); it is unclear if this is a displaced jaw cartilage fragment, or if the branchiostegal plates are from one side of the fish and the cartilage from the other side. The least likely scenario is that the cartilage is from the branchial arches, as it appears to have the same sheet-like structure as seen in the jaw cartilages.

Family indet., *Nostolepis scotica* (Newton, 1892). This taxon, also from the Scottish Midland Valley, was originally assigned by Newton (1892) to the sarcopterygian *Onychodus*, subsequently referred by Traquair (1898) to the chondrichthyan *Protodus*, and only relatively recently assigned by Burrow & Turner (2010) to the acanthodian *Nostolepis*. The latter authors tentatively referred the species to the family Brochoadmonidae Bernacsek & Dineley, 1977, but the type and only species in that family, *Brochoadmones milesi* (from the Lochkovian Delorme Formation, Northwest Territories, Canada), has a totally unmineralised endoskeleton, thus lacking the key diagnostic feature for the acanthodians, a scapulocoracoid formed of perichondral bone. It seems likely that this lack is a derived feature however, associated with its deep body and small, cone-shaped pectoral spine, and possibly also with the geochemistry of the environment in which the fish lived. Recent cladistic analyses, which have not included *N. scotica*, recover *Brochoadmones* in various positions amongst the acanthodian 'climatiiforms' and other stem chondrichthyans (e.g. Burrow et al. 2016, figs. 39–41; Lu et al. 2017, suppl. fig. 7), and this is no doubt where *N. scotica* also belongs.

The mineralised endoskeletal tissues preserved in *N. scotica* are parts of the neurocranium, jaws and branchial skeleton, with all formed of globular calcified cartilage (Fig. 4). No thin sections have been made of the only partial articulated specimen known, NMS G.1891.92.696, from the Lochkovian Tillywhandland Quarry, Scotland.

Gyracanthidae Woodward, 1906. Mineralisation of the visceral skeleton has not been reported in the gyracanthids, of which the only species known from articulated fish is *Gyracanthides murrayi* Woodward, 1906. However, a specimen labelled as *Mitrodus* in the MCZ collection preserves a set of branchial arches apparently composed of tessellate calcified cartilage, and squamation comprising *Mitrodus*-type denticles. *Mitrodus quadricornis* Owen, 1867 was erected for polyodontode scales from the British Carboniferous coal measures, which were first identified as gyracanthid scales by Hancock & Atthey (1868), and were subsequently confirmed as such by their presence on articulated (*Gyracanthides murrayi*: Warren et al. 2000) and disarticulated ("*Gyracanthus*" *sherwoodi*: Snyder et al. 2017) gyracanthids. This specimen is under investigation by D. Snyder, to determine whether the scales might be ctenacanthid rather than gyracanthid, given the structure of the branchial arches.

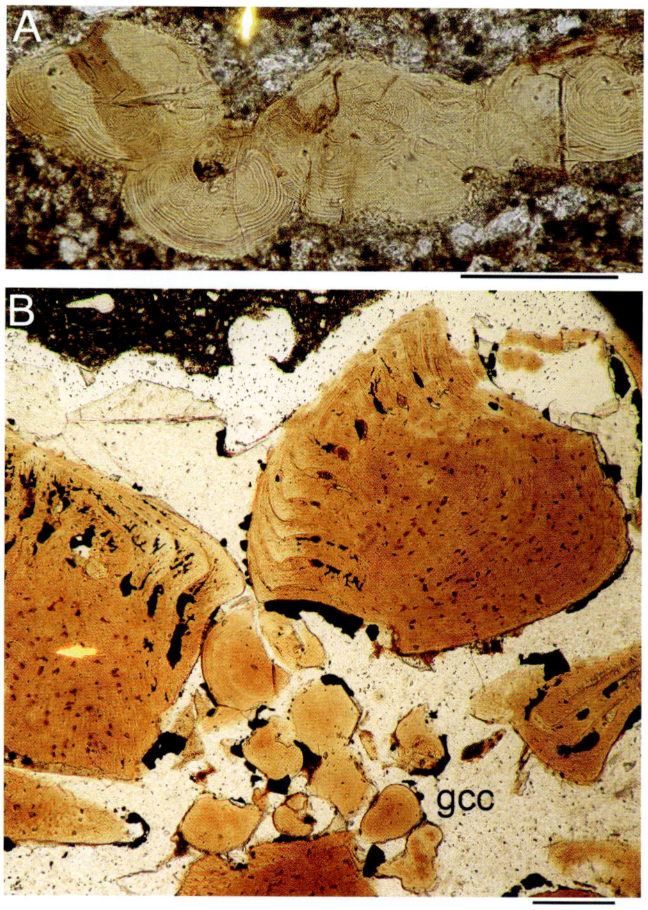

Fig. 2. *Euthacanthus macnicoli* endoskeletal mineralisation. **A, B.** NMS G. 2013.10.1 disarticulated specimen. **A,** thin section LD120 (= NMS G.2013. 10.1.11; **B,** thin section LD122 (= NMS G.2013. 10.1.12): globular calcified cartilage (gcc). Scale bars = 1 mm.

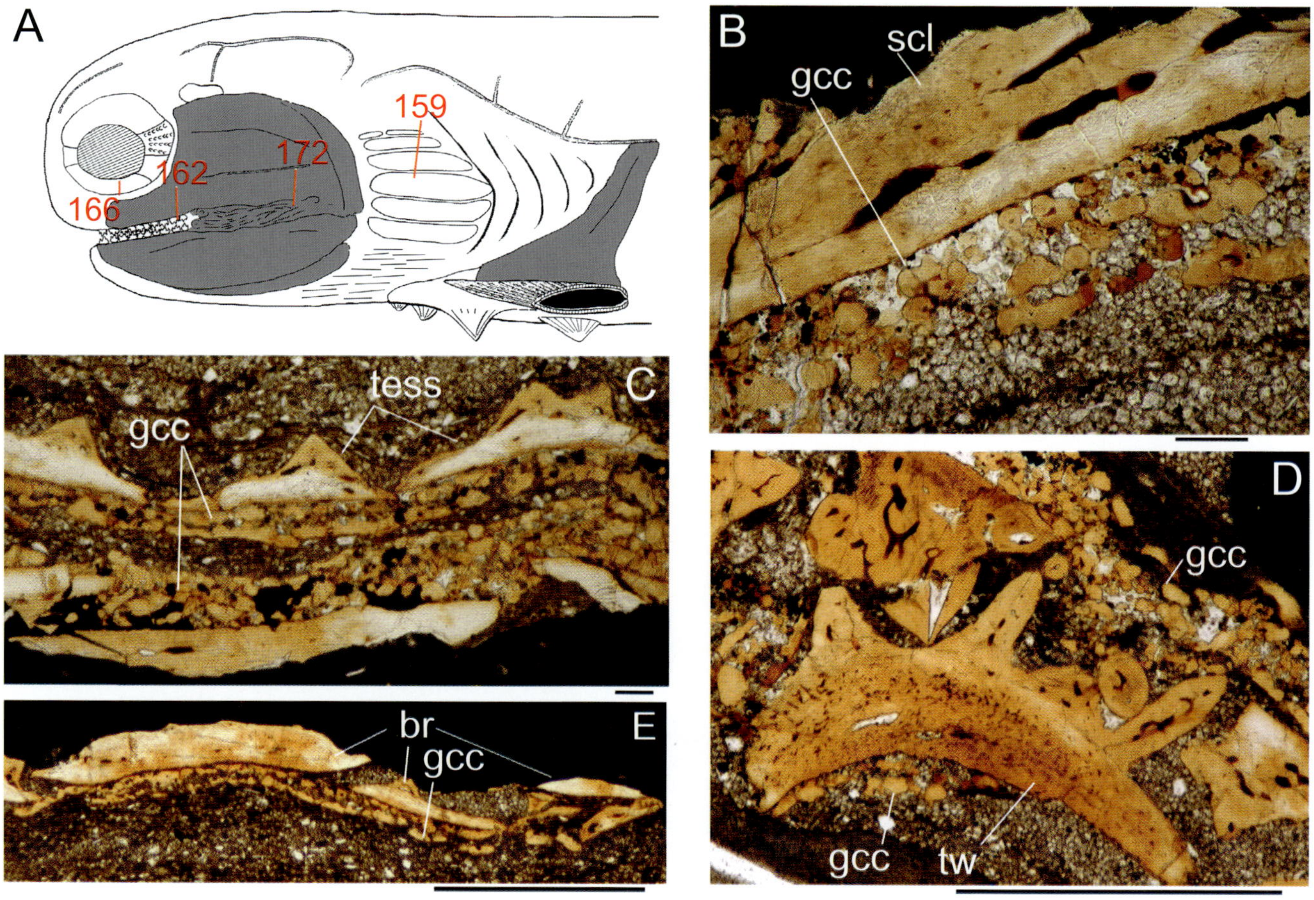

Fig. 3. *Climatius reticulatus* visceral skeleton mineralisation. **A.** reconstruction of head to pectoral region, showing approximate position of LD sections made of specimen NMS G.2013.8.2. **B.** LD166, through sclerotic plate and globular calcified cartilage. **C.** LD172, through jaw cartilages and head plates. **D.** LD162, through tooth whorls and jaw cartilages. **E.** LD159, through branchiostegal plates and underlying globular calcified cartilage (gcc). Scale bars = 1 mm in D, E; 0.1 mm in B, C. Abbreviations: br, branchiostegal plate; gcc, globular calcified cartilage; scl, sclerotic plate; tess, tesserae; tw, tooth whorl.

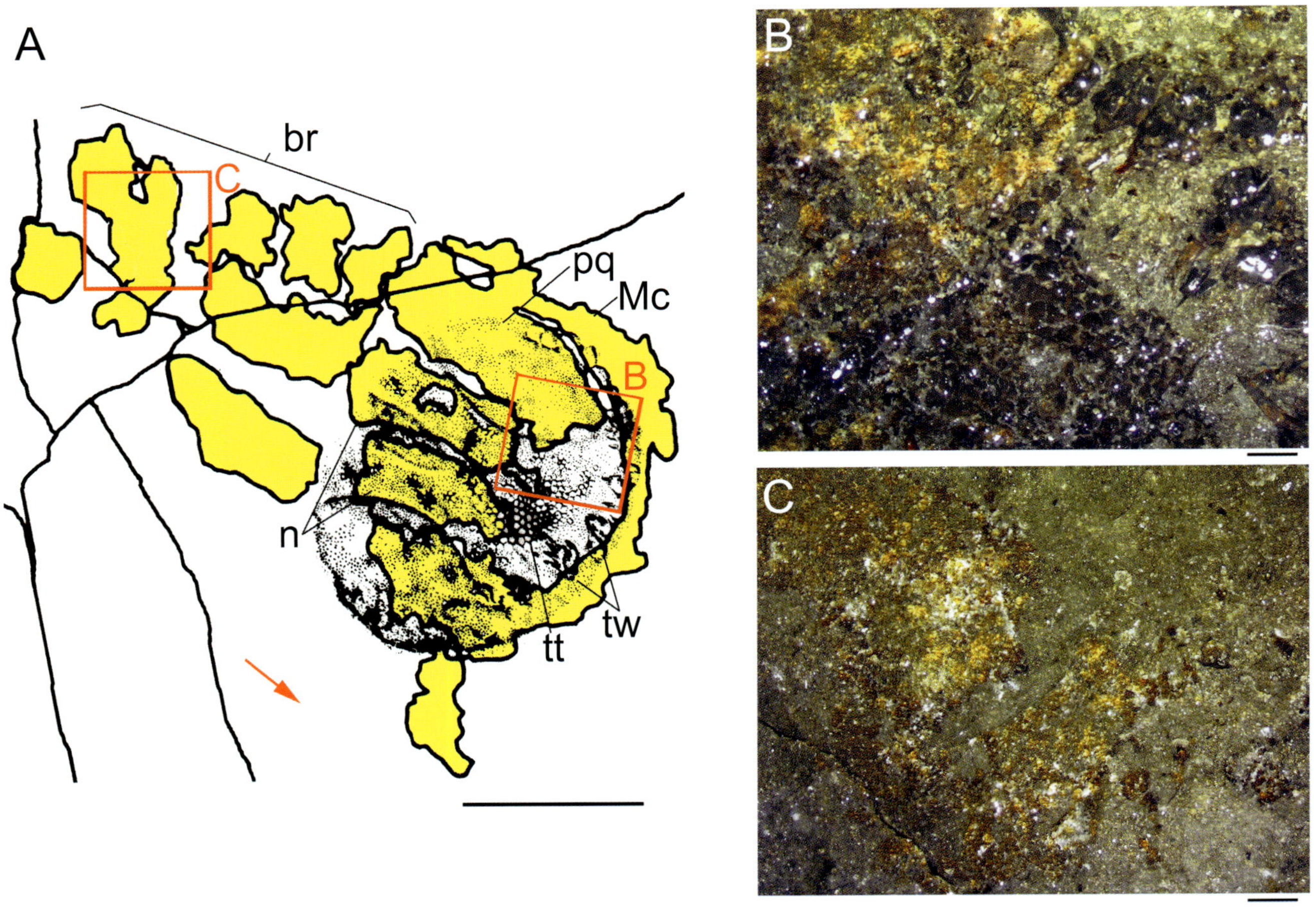

Fig. 4. *Nostolepis scotica* endoskeletal mineralisation in NMS G.1891.92.696, dorsoventrally compressed head and branchial region of fish, from Tillywhandland Quarry, Midland Valley, Scotland. **A.** sketch of specimen showing distribution of globular calcified cartilage (gcc), coloured yellow; **B.** box B region; **C.** box C, in the branchial region. Scale bars = 10 mm in A; 1 mm in B, C. Red arrow indicates anterior direction. Abbreviations: br, branchial endoskeletal elements; Mc, Meckel's cartilage; n, neurocranium; pq, palatoquadrate; tt, tectal tesserae; tw, tooth whorls.

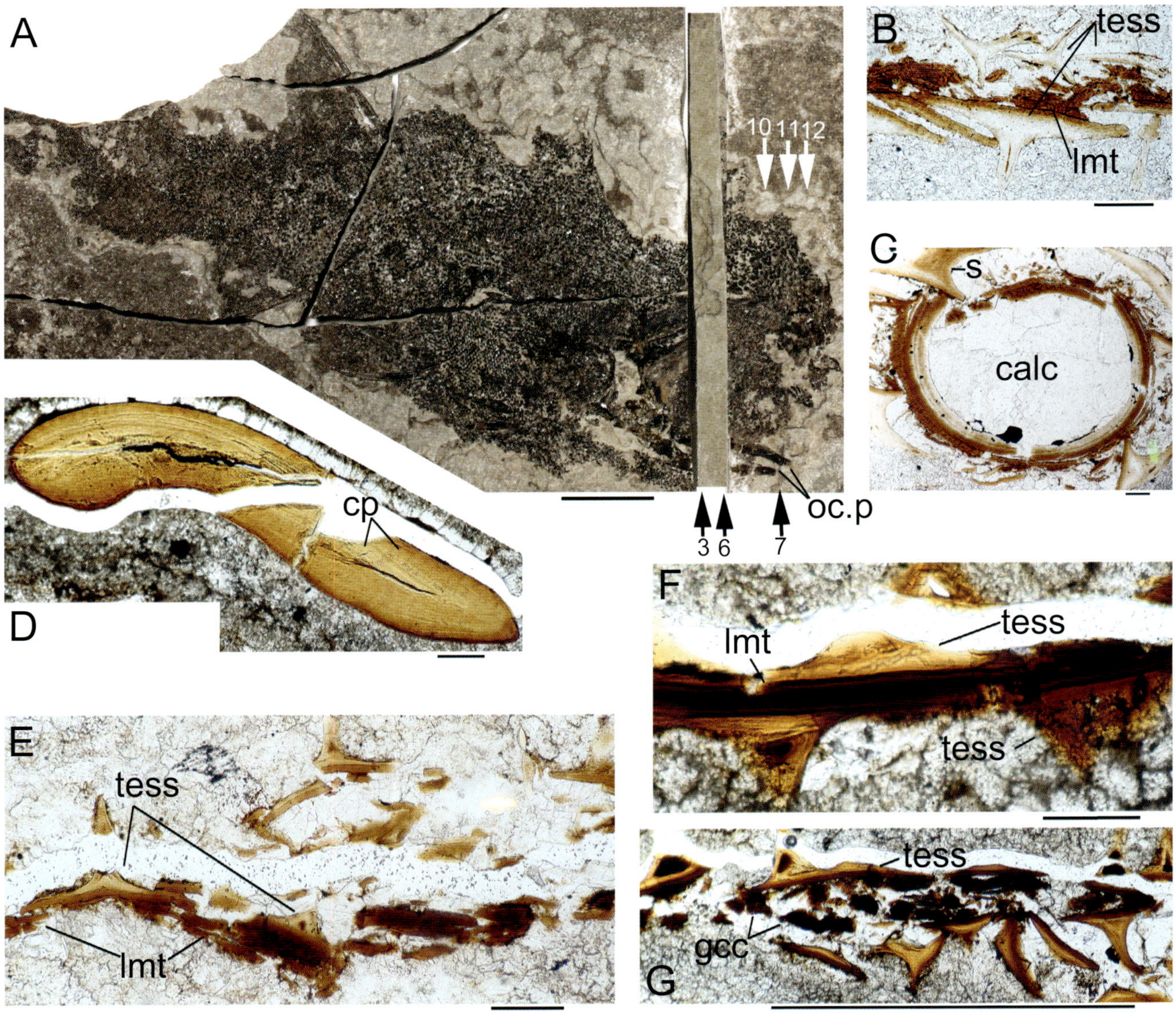

Fig. 5. *Diplacanthus crassisimus* visceral skeleton mineralisation. **A.** NMS G.2019.3.5 from Gamrie, part, exposed surface after one cut, before sectioning; arrows showing orientation of numbered sections, bottom arrows indicate sections through mouth region, top arrows indicate sections through tectal region; anterior to right. **B, C.** NMS G.2019.3.5.6 (6 in A): **B.** tesserae and denticles overlying dense layered tissue in the branchial region; **C.** ?ceratohyal transverse section. **D.** NMS G.2019.3.5.7 (7 in A), occlusal plate transverse section. **E.** NMS G.2019.3.5.10 (10 in A), head tesserae and scales overlying dense layered tissue. **F.** NMS G.2019.3.5.12 (12 in A), tesserae from both sides of head pressed together, showing thin lamellar layer underlying them. **G.** NMS G.2019.3.5.11 (11 in A), head tesserae and scales overlying globular calcified cartilage. Scale bars = 10 mm in A; 0.1 mm in B–G. Abbreviations: calc, calcitic infill; cp, ?cell processes/canaliculi; gcc, globular calcified cartilage; lmt, layered mineralised tissue; s, scales; tess, tessera.

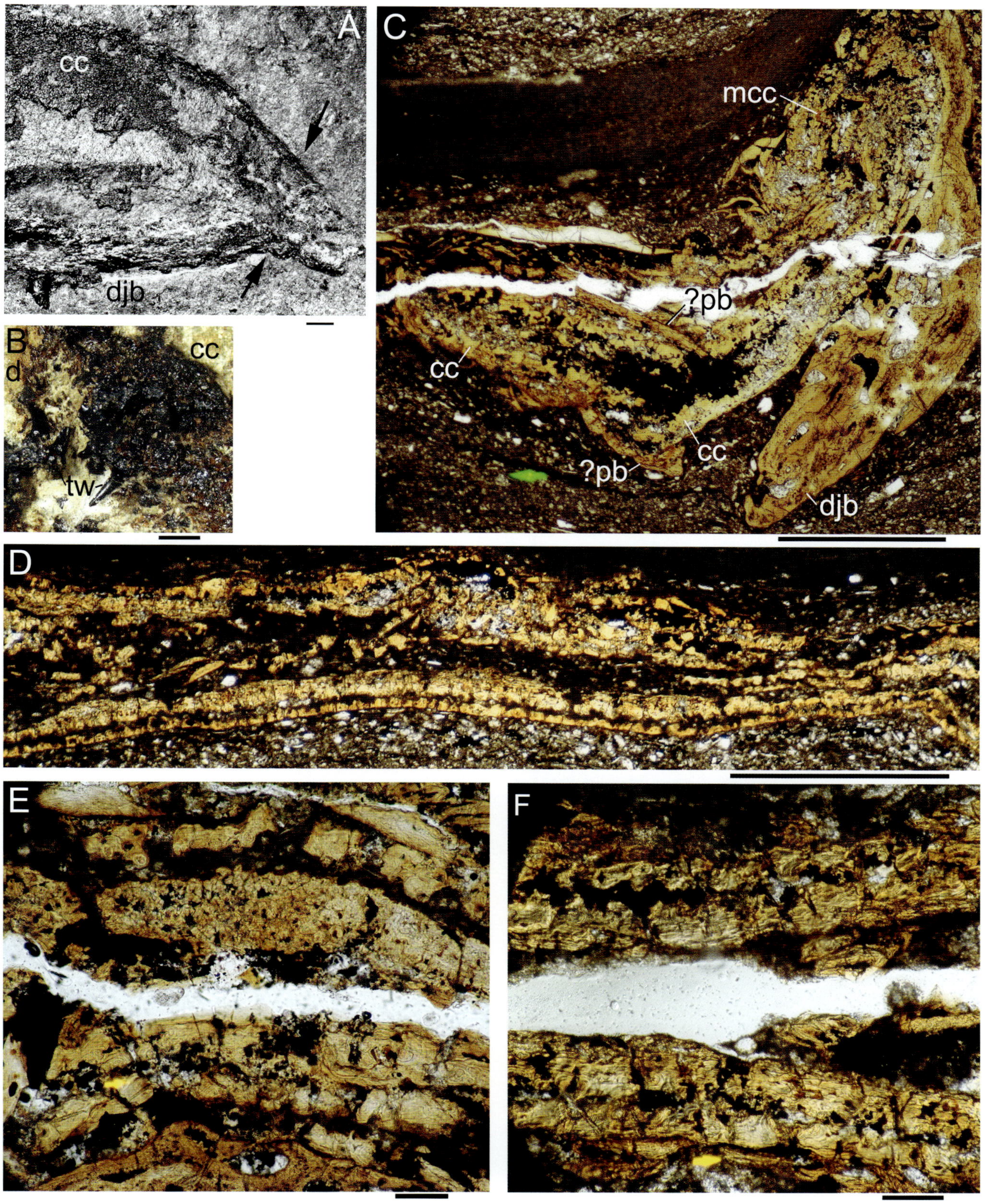

Fig. 6. *Ischnacanthus gracilis* visceral skeleton mineralisation. **A.** QMF 58848 from Tillywhandland, palatoquadrate, showing granular surface texture; arrows indicate approximate position of section shown in C. **B.** QMF 58851, tooth whorls, extraoral denticles, and jaw cartilage globules. **C.** NMS G.2016.22.3 TS44 oblique section, through posterior end of palatoquadrate cartilage and flange of dentigerous jaw bone. **D.** NMS G.2016.22.3 TS50, vertical section through compressed palatoquadrates (dorsal to left). **E.** NMS G.2016.22.3 TS50x, closeup of cartilage structure. **F.** NMS G.2016.22.3 TS51A, closeup of cartilage structure. Scale bars = 1 mm in A–D; 0.1 mm in E, F. Abbreviations: cc, calcified cartilage; djb, dentigerous jaw bone; mcc, multilayered calcified cartilage; pb, perichondral bone.

Diplacanthidae Woodward, 1891. Watson (1937, plate 10.1) noted that branchial arches, jaws and other endoskeletal elements are only mineralised in rare specimens of the type species, *Diplacanthus crassisimus* (Duff, 1842), and these are all from the Tynet Burn locality in Moray, Scotland. The specimen figured by Watson, NMS G.1891.92.334, differs from all other known specimens in having "massive ossifications in the head skeleton, vertebral column and basals of the fins, and possesses well-preserved dermal fin rays. Other specimens show slight traces of ossification of the same nature in the branchial arches" (Watson 1937, p. 88). No such specimens have been serially sectioned, so until now the histological structure could only be inferred from the appearance of exposed structures (e.g. Burrow et al. 2016, fig. 4.8-10: NHM OR.36582). NMS G.1891.92.334 was recently scanned in a synchrotron (Dearden 2019), revealing the layout of the endoskeletal elements, which Dearden recognised as chondrifications rather than ossifications. The synchrotron scans showed that the jaws and ceratohyal have the same type of mineralisation, comprising layers described by Dearden (2019, fig. 5.1) as inner and outer layers of globular calcified cartilage. New sections of a specimen of *D. crassisimus* from Gamrie, NMS G.2019.3.5 (Fig. 5), show that despite no endoskeletal tissues being visible on the exposed surfaces of the specimen prior to sectioning, some were preserved. In the branchial region, thin layers of dense mineralised tissue underlie tesserae and denticles (Fig. 5B). A hollow cylindrical element, presumed to be a ceratohyal by its position close to the occlusal plate (now recognised as a dermal bone, see Burrow et al. 2016), is formed of thin layers of consolidated globular calcified cartilage (Fig. 5C). As expected, the dermal occlusal plates are composed of a dense lamellar bone (Fig. 5D), lacking bone cell lacunae but with fine branching canaliculi extending through the layers, and more or less perpendicular to the outer surface. In the head region above the jaws, a thin layer of dense mineralised tissue, possibly perichondral bone rather than globular calcified cartilage, is preserved under some of

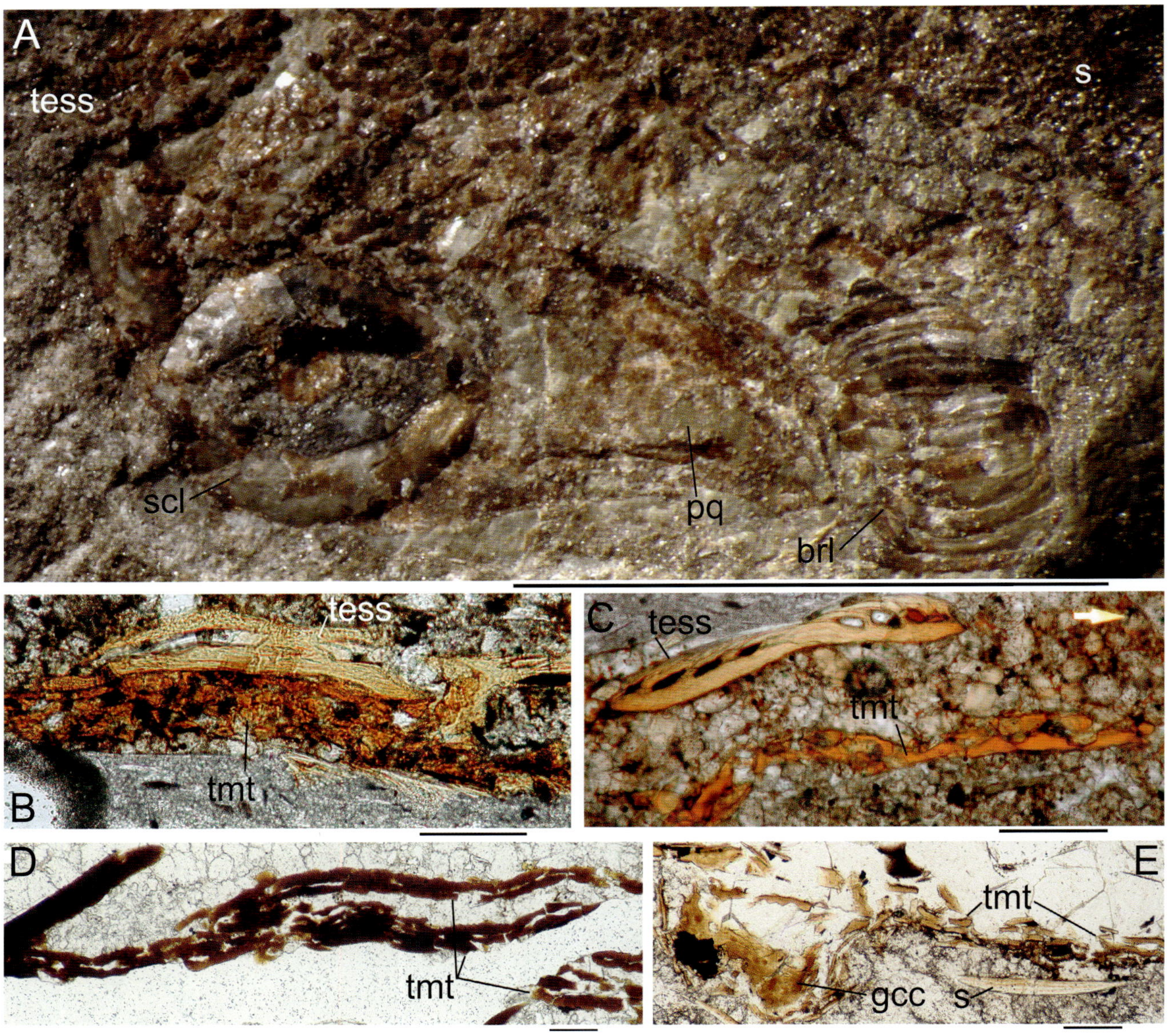

Fig. 7. *Mesacanthus* jaw mineralisation. **A–C.** *Mesacanthus mitchelli* from Tillywhandland Quarry: **A.** NMS G.2009.36.1, showing granular mineralisation of palatoquadrate; **B, C.** specimen NMS G. 2019.3.2, sacrificed for thin sectioning, transverse sections through upper jaws: **B.** NMS G.2019.3.2.3; **C.** NMS G.2019.3.2.2. **D, E.** *Mesacanthus pusillus*: **D.** NMS G.2019.3.4 from Cromarty, Sutors, NMS G.2019.3.4.15, lower jaw transverse section, dorsal to left; **E.** NMS G.2019. 3.11 from Corbie Den, Banffshire, NMS G.2019.3. 11.4, lower jaw transverse section, ventral edge to left. Scale bars = 10 mm in A; 0.1 mm in B–E. Abbreviations: brl, branchiostegal plates; gcc, globular calcified cartilage; pq, palatoquadrate; s, scales; tess, head tesserae; tmt, tessellate mineralised tissue.

the dermal tesserae (Fig. 5E, F), whereas in other areas globular calcified cartilage appears to be present (Fig. 5G).

No endoskeletal tissues have been observed in the head and branchial regions of the other diplacanthids from Scotland, *Diplacanthus tenuistriatus* Traquair, 1894 and *Rhadinacanthus longispinus* (Agassiz, 1844). The only record elsewhere of endoskeletal structures in these regions in other diplacanthiforms is of purported ceratobranchials in *Rhadinacanthus horridus* (Woodward, 1892) specimen MNHM 03-737 from the Frasnian Escuminac Formation, Miguasha, Canada (Gagnier 1996).

Ischnacanthidae Woodward, 1891. As described by Watson (1937) and confirmed by Burrow et al. (2018), the jaws are the only mineralized endoskeletal structures recorded to date in the head-branchial region of the type species *Ischnacanthus gracilis* (Egerton, 1861) from the Lower Devonian of Scotland. Burrow et al. (2018, figs. 5A, B, H, 8K, 14) showed that the jaws are formed of one or more layers of subtessellate cartilage blocks, overlying an uncalcified cartilage core. The surface of the mineralised cartilage has a granular appearance (Fig. 6A, B), identical to that of the same elements in *Climatius reticulatus*. The thickness (and number of layers) of the cartilage varies along the jaw, being thickest posterior to the dentigerous jaw bones, at and near the jaw articulation (Fig. 6C), and thin, comprising just one or two layers, under the robust occlusal jaw bones (Fig. 6D–F). The structure of the thinnest regions of cartilage appears to comprise two different layers. A thin layer of perichondral bone possibly also is developed in the region of the jaw articulation surfaces (Fig. 6C). *Xylacanthus* appears to have had perichondral bone lining the Meckel's cartilage (Ørvig 1967, Hanke et al. 2001).

Acanthodiformes Berg, 1940. Ørvig (1951, p. 414) studied the structure of the visceral skeleton of an "undetermined Acanthodian from Scaumenac Bay". Unfortunately, it is still unknown to which taxon the latter material belongs; two acanthodiforms, the cheiracanthid *Homalacanthus concinnus* (Whiteaves, 1887) and the mesacanthid *Triazeugacanthus affinis* (Whiteaves, 1889) have been described from the Frasnian Escuminac Formation in Scaumenac Bay, Canada, but the histological structure of the endoskeleton of neither species had been investigated since Ørvig's publication.

Mesacanthidae Moy-Thomas, 1939. According to Watson (1937), the jaws and a mandibular splint supporting the lower jaw are the only mineralised endoskeletal elements in the head and branchial regions of the type species *Mesacanthus mitchelli* (Egerton, 1861) from the Lochkovian of the Midland Valley, Scotland. Watson (1937) considered the tissue forming the elements to be bone. However, serial sectioning of the fish shows the jaws to be composed of thin layers of tessellated mineralised tissue (tmt; Fig. 7B, C). We could not identify rings or lines of Liesegang in the separate blocks, which however show no evidence of the lamellae or cell lacunae that would identify the tissue as bone. Watson (1937, p. 76) stated that in the ontogenetic development of *M. mitchelli*, the mandibular bone mineralises before the palatoquadrate and Meckel's cartilages, but we consider that the latter have a thickened ventral edge, with no mandibular bone developed. The Middle Devonian *Mesacanthus pusillus* (Agassiz, 1844) from the north of Scotland shows similar endoskeletal mineralisation as in *M. mitchelli*, with the jaws formed of layered blocks of amorphous mineralised tissue and a thickened ventral edge on the Meckel's cartilage encasing a globular calcified cartilage core (Fig. 7D, E). The ventral edge of the Meckel's cartilage is thickened by a core of globular calcified cartilage (to the left in Fig. 7E).

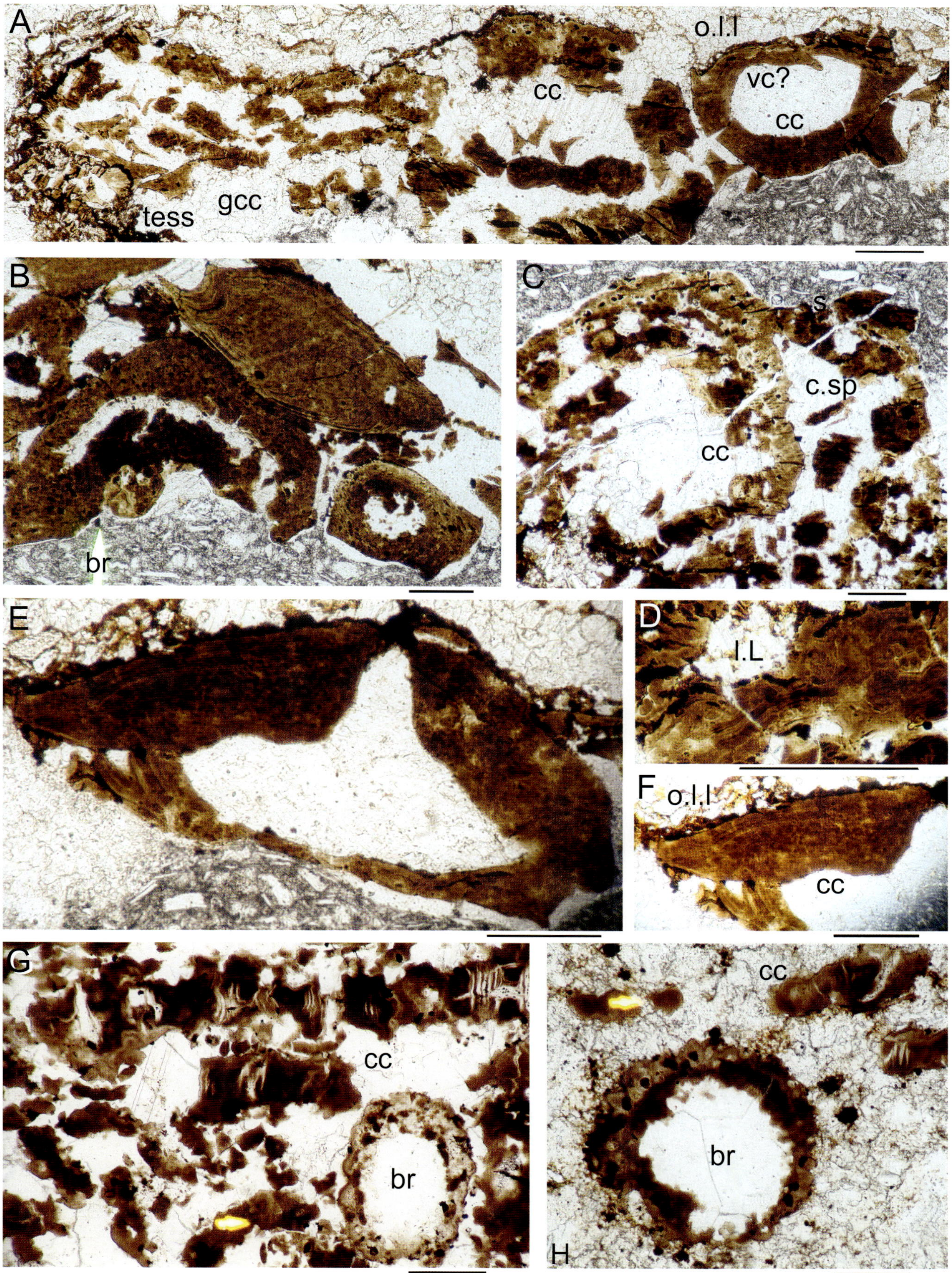

Fig. 8. *Cheiracanthus* visceral skeleton mineralisation. **A–F.** thin sections through the jaws and branchial arches of *Cheiracanthus peachi*, NMS G.2018.28.25: **A.** NMS G.2018.28.25.10, globular calcified cartilage of jaw, and ?two-layered consolidated calcified cartilage of ceratohyal; **B.** NMS G.2018.28.25.14, cross sections of three consolidated cartilage branchial bars; **C.** NMS G.2018.28.25.12, anterior end of a branchial bar, with partly consolidated calcified cartilage and large cell spaces; **D.** NMS G.2018.28.25.11, close-up of less consoliated cartilage with lines and rings of Liesegang. **E, F.** NMS G.2018.28.25.8, anterior ends of left and right ceratohyal junction, with lamellation of outer layer visible in F. **G, H.** *Cheiracanthus latus* NMS G.2019.3.3 from Balintore, thin section through the jaws and branchial bars NMS G.2019.3.3.5. Scale bars = 0.1 mm. Abbreviations: br, branchial bars; cc, calcified cartilage; c.sp, cell spaces; o.l.l, outer lamellar layer; l.L, lines of Liesegang; tess, head tessera; vc, vascular canal.

Two other younger mesacanthids *Lodeacanthus gaujensis* Upeniece, 1996 (Givetian, Lode Quarry, Latvia) and *Triazeugacanthus affinis* are represented by abundant populations showing ontogenetic series, from which it has been possible to determine the sequence of development of the endoskeleton. Chevrinais et al. (2015) recorded that in the latter species, traces of the notochord are the first endoskeletal elements to develop in the smallest specimens (about 4–5 mm long). However, Chevrinais et al. (2016) stated that these elements showed no sign of mineralisation. At around 25 mm, the palatoquadrates and Meckel's cartilages were mineralised. The histological structure of the elements was not investigated. Chevrinais et al. (2016) analysed the biomineralisation of the skeletal elements in this species, noting differences in the chemical composition of bone and cartilage, but this difference was only based on the composition of the mineralised tissues developed during ontogeny which those authors presumed to be bone rather than calcified cartilage, and the unmineralised cartilage present in the earliest ontogenetic stage. As noted by Ørvig (1951, p. 351), the main distinction between calcified cartilage and bone in early vertebrates is the relative richness of 'chondromucoid' vs. collagen in the tissue; the chemical composition of bone and mineralised cartilage are very similar.

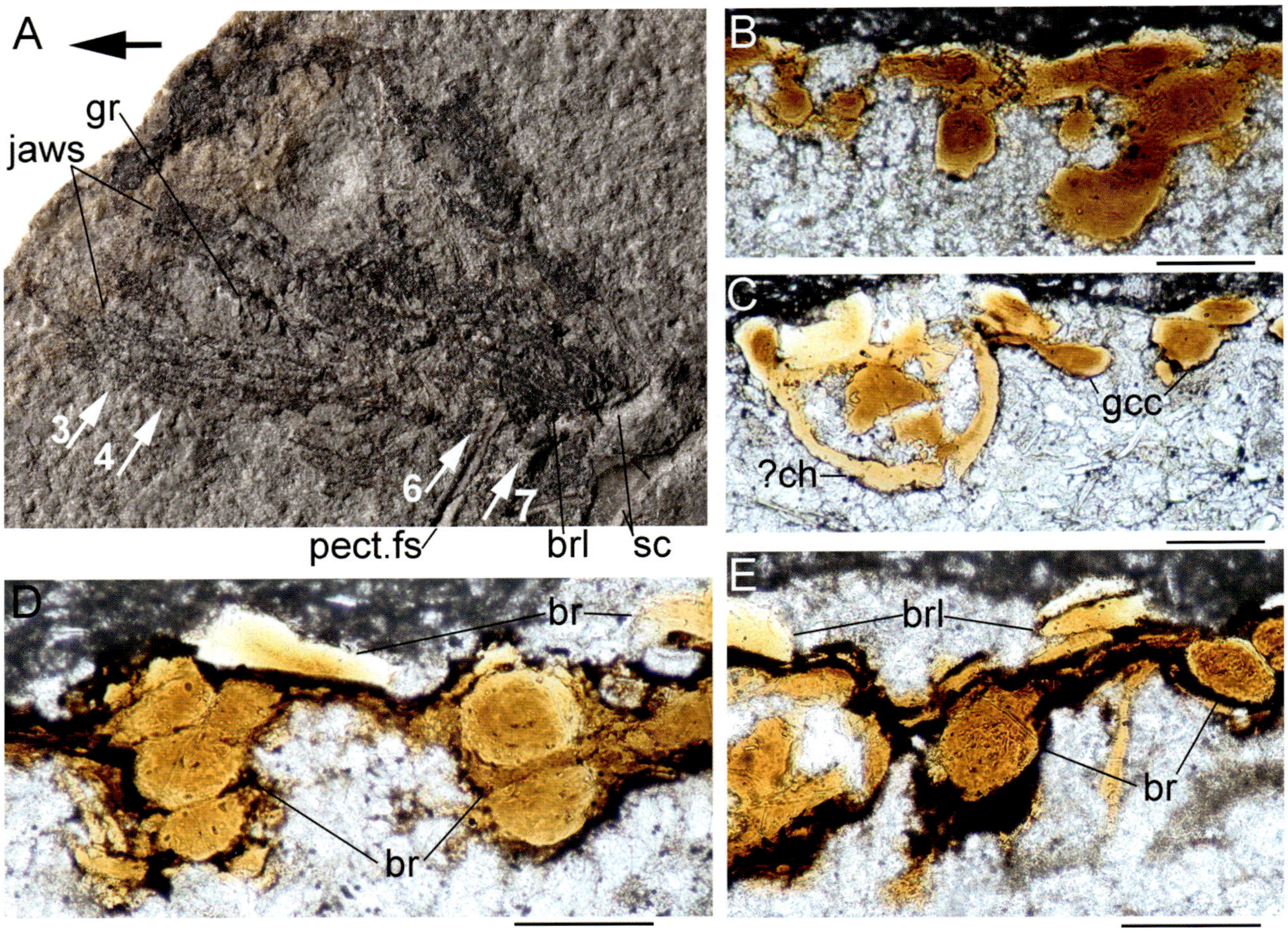

Fig. 9. Jaw mineralisation in *Homalacanthus concinnus* specimen NMS G. 2019.3.1 from the Escuminac Formation, Canada. **A.** head and branchial region of the specimen as preserved, numbered arrows show the orientation of thin sections figured in B–E; **B.** NMS G.2019.3.1.3 (3 in A), vertical section through the jaw; **C.** NMS G.2019.3.1.4 (4 in A), section through the jaw and ceratohyal?; **D, E.** NMS G.2019.3.1.6, .7, sections through the branchial bars and branchiostegal plates. Scale bars = 0.1 mm. Arrow indicates anterior direction. Abbreviations: br, branchial bars; brl, branchiostegal plates; gcc, calcified cartilage; gr, gill rakers; ch, ceratohyal; pectfs, pectoral fin spine; sc, scapulocoracoid.

In *Lodeacanthus gaujensis*, Upeniece (2011) noted that in the youngest stage, before squamation was developed, the only endoskeletal element developed was the scapulocoracoid; the neurocranium mineralisation commenced when scale cover extended to about a third of the body, and the jaw and branchial arch cartilages only mineralised after the scale covering of the body and head was complete. Mandibular bones are not present in either *L. gaujensis* or *T. affinis*. Again, the histological structure of the elements was not described.

Another mesacanthid, *Teneracanthus toombaensis* Burrow & Young, 2005, is known only from disarticulated elements in the vertebrate assemblage from the ?upper Emsian or lower Eifelian Cravens Peak Beds, western Queensland, Australia. Other than the scapulocoracoid, the only endoskeletal tissue recognised is a very thin tissue, presumed to be perichondral bone, preserved in the cleft between the scapulocoracoid and the pectoral fin spine, and also forming the palatoquadrate (Burrow & Young 2005, fig. 4K–M, R–T). This material was prepared by acetic acid treatment of the limestone matrix, which might have dissolved any calcite-replacement of unmineralised cartilages.

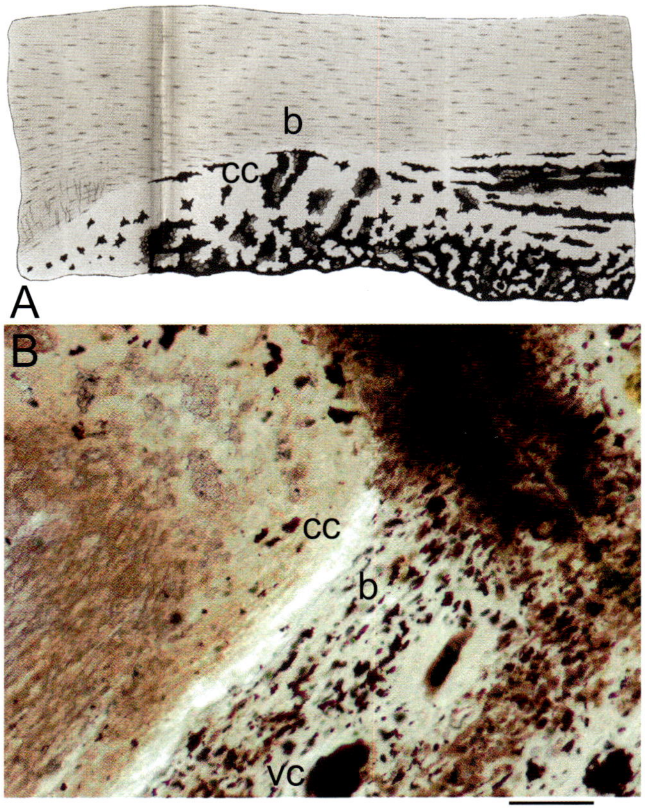

Fig. 10. Acanthodid jaw mineralisation. **A.** *Acanthodes bronni*, thin section of a visceral element, after Reis (1896, pl. 7.1; scale not specified), image rotated 90° so that inner cartilage is below the outer bone layer. **B.** *Acanthodopsis russelli*, thin section through the mandibular splint of the lower jaw QMF 51266. Scale bar = 0.1 mm in B. Abbreviations: b, bone; cc, calcified cartilage; vc, vascular canal.

Cheiracanthidae Berg, 1940. Ørvig (1951) was the first to investigate the histological structure of the cheiracanthid endoskeleton, working on the visceral skeleton of specimens of *Cheiracanthus murchisoni* Agassiz, 1835 from Gamrie, Scotland. Ørvig found no traces of perichondral bone, only globular calcified cartilage, which he considered corresponded to the subperichondral form in *Acanthodes*. He also noted that the rings of Liesegang in the cartilage globules tended to be somewhat polygonal rather than the circular shape found in the arthrodires.

However, the serial sections made by JdB through the head and branchial regions of several *Cheiracanthus* spp. from the Middle Devonian of Scotland reveal that several different types of tissue form the endoskeletal elements (Fig. 8). In a serially sectioned specimen of *Cheiracanthus peachi* from the flagstones of the Orcadian Basin, northern Scotland, we found that the jaw cartilages were only formed of thin layers of globular calcified cartilage (Fig. 8A). The branchial bars were formed of a more consolidated globular calcified cartilage with relatively large cell spaces, with the posterior ends of the elements being dense and mostly lacking the rings and lines of Liesegang (Fig. 8B), and the anterior portions being less dense and showing some rings and lines of Liesegang (Fig. 8C, D). None of these elements showed evidence of an outer layer of perichondral bone. The ceratohyal, however, was described by den Blaauwen et al. (2019) as formed of perichondral bone overlying an inner calcified cartilage layer. This interpretation was based on the lamellar structure of the outer layer (Fig. 8A, E, F), but it seems possible that the lamellar layer is rather formed of dense calcified cartilage layers similar to those in the ceratohyal of *Diplacanthus*

(Fig. 5C). The tissues identified in sections of *Cheiracanthus latus* Egerton, 1861 from Balintore, Eastern Ross differed slightly, with the jaws formed of blocks rather than globules of calcified cartilage, no doubt of a similar structure to that observed by Ørvig (1951) in *C. murchisoni*.

Serial sections of a specimen of *Homalacanthus concinnus* from the Escuminac Formation in eastern Canada (Fig. 9A–E) show that its jaws have a similar histological structure to that of *Cheiracanthus*, mineralised as a thin layer of calcified cartilage (Fig. 9B). The ceratohyal appears to be preserved as a thin hollow perichondral bone (Fig. 9C), and the branchial arches are formed of the same type of cartilage as those of *Cheiracanthus* (Fig. 9D,E).

Acanthodidae Huxley, 1861. Reis (1890, 1896, pl. 7.1) described and illustrated the histological structure of the visceral skeleton of *Acanthodes bronni* Agassiz, 1833 from the Lower Permian (Autunian) of Germany, noting that the individual elements are lined with a superficial layer of mineralised tissue overlying calcified cartilage (Fig. 10A). Reis maintained that the superficial layer was not bone, as it was not laminated, and he did not identify processes emanating from the cell spaces in the tissue. However, when Ørvig (1951) studied sections of the skeleton of *Acanthodes* cf. *bronni* from Lebach under polarised light, he recognised such cell processes, as well as Sharpey's fibres, establishing that this tissue is bone. He also noted that the tissue under this perichondral bone is globular calcified cartilage, and that there is a gradual transition between the two tissues, unlike their relationship in arthrodire placoderms.

Ontogenetic growth series are known in *Acanthodes lopatini* Rohon, 1889 from the Lower Carboniferous (Tournaisian) of Siberia. Beznosov (2009) noted that the neurocranium was mineralised, and then only partially, in only one large specimen of the 100+ specimens he examined; it was presumed to be cartilaginous in all other specimens. The Meckelian cartilage was observed to ossify in specimens more than 60 mm long, with the gap between the two ossifications (anterior mentomandibular and posterior articular) gradually closing during ontogeny. The palatoquadrate was ossified later and more incompletely than the lower jaw. The hyomandibula was rarely preserved, whereas the ceratohyal was preserved as a thin perichondral ossification, or (like the gill arches) as calcifed cartilage without a perichondral bone layer in some large individuals.

Another acanthodid, *Acanthodopsis russelli* Burrow, 2004 from the Lower Carboniferous of Queensland, Australia, was found to also have perichondral bone forming the superficial layer on the denticulated jaws, but the cartilage must have been unmineralised (or dissolved during diagenesis), as the jaws were hollow shells. Mineralised cartilage was however preserved in or in contact with the mandibular splint, formed by a thick layer of cellular bone pierced by vascular canals (Fig. 10B).

Comparison

Clearly there is quite large variation in the degree and form of mineralisation of the endoskeleton in the head and branchial regions of acanthodians, even within the same taxon, but some consistent family-level characters are supported. The cheiracanthids in particular show the same distribution of mineralised tissue forms independent of preservation conditions and locality, as evidenced by Scottish *Cheiracanthus* spp. from the Eifelian lacustrine nodule beds of the Moray Firth (e.g. Gamrie) and varved flagstones of the Orcadian Basin, and Canadian *Homalacanthus* from the Frasnian brackish to marine Escuminac Formation (Matton et al. 2012). The family-specific occurrence of the different types of mineralised tissue is also supported by one nodule from Gamrie preserving both a diplacanthid (*Diplacanthus crassisimus*: Fig. 5), and a cheiracanthid (*Cheiracanthus murchisoni*), with each showing the endoskeletal tissue forms deemed typical for their family.

Although the diplacanthiform–ischnacanthiform–acanthodiform clade is well supported in phylogenetic analyses (e.g. Coates et al. 2018), it is perhaps surprising that ischnacanthids and cheiracanthids have a calcified cartilage structure that appears intermediate between the globular calcified cartilage typical of more basal vertebrates and the tessellated calcified cartilage which characterises crown chondrichthyans. As noted above, *Climatius*, which is nested within the 'climatiiform' stem chondrichthyans, also has calcified cartilage similar to that seen in the ischnacanthid *Ischnacanthus*. It is notable that the tissue structure in the Cheiracanthidae shows no evidence to support this family being intermediate between the Mesacanthidae and Acanthodidae, as has previously been proposed (e.g. Denison 1979). As well as their endoskeletal structure as detailed here, they have ornamented scales whereas the other two families have smooth scales; the main character uniting the three families is having one dorsal fin spine.

Unfortunately, because the long pre-Devonian history of acanthodians is mostly only represented by isolated elements, there is no certainty about when the trend to tessellation arose, but it would seem it must have occurred before the split between the 'climatiiform' grade and other acanthodians. In this regard, it is interesting that the only mineralised tissue forming the endoskeleton of *Euthacanthus* appears to have been globular calcified cartilage of the form found in placoderms and agnathans, perhaps supporting Denison's (1979) view that *Euthacanthus* was the most basal acanthodian (based on articulated fish) known at that time. A large suite of taxa populate the 'climatiiform' grade of stem chondrichthyans – i.e. acanthodians *sensu lato*, exclusive of the Ischnacanthiformes, Diplacanthiformes and Acanthodiformes. The histological structure of endoskeletal tissues in 'climatiiforms' is only known for a few species. *Ptomacanthus anglicus* Miles, 1973 is the only 'climatiiform' in which neurocranial features are known (Brazeau 2009, 2012), but these are only inferred from casts of impressions, with no preservation of the hard tissues. Brazeau (2009, p. 306) noted that these impressions showed "... a rough crystalline surface comparable to the mineralized jaws known from *Climatius*". If we assume that this indicates that the endoskeleton of *Ptomacanthus* was calcified cartilage, then that is the only mineralised tissue known to form the neurocranium and splanchnocranium in the 'climatiiforms', albeit in several different manifestations (Table 1). Calcified cartilage is also the only endoskeletal tissue of the head-branchial region in the ischnacanthiforms (Ischnacanthidae + Poracanthodidae + Acritolepidae), based on the surface appearance of the jaws (the only endoskeletal elements usually mineralized). The diplacanthiforms mostly show no morphologically visible mineralisation of the endoskeleton of the head-branchial region, except in the rare instances of some specimens of *Diplacanthus crassisimus* from Tynet Burn. In those fish, the endoskeleton shows a similar distribution of tissue types as seen in the cheiracanthids (Table 1), but possibly with the whole visceral skeleton formed of layered consolidated calcified cartilage. This distribution differs from that in the acanthodids, which appear to have had perichondral bone encasing globular calcified cartilage forming all endoskeletal elements. The mesacanthid acanthodiforms *Mesacanthus mitchelli* and *M. pusillus*, however, had an amorphous tessellate mineralised tissue forming the jaws.

Looking crownward from the acanthodians, *Doliodus* from the Emsian of New Brunswick, Canada is often recovered as the sister taxon to crown Chondrichthyes (e.g. Coates et al. 2018). Maisey et al. (2017) first described the acanthodian-type arrangement of fin spines in *Doliodus*, and Maisey et al. (2019) noted that it has "... crystalline tesserae-like perichondral blocks, although it is unclear whether these are arrayed in layers". Those authors noted that the characterisation of calcified cartilage in phylogenetic analyses as either tessellated or nontessellated (or in older analyses, prismatic or nonprismatic), is an oversimplification. Investigations on the detailed structure of the tessellated calcified cartilage of crown chondrichthyans were compared with that known for stem chondrichthyans by Maisey et al. (2021).

Acknowledgments

We thank the reviewers for their advice. CJB also thanks the Queensland Museum for provision of basic facilities.

Literature cited

Agassiz, L. 1833–1843. Recherches sur les poissons fossiles. 5 vols. and atlas. Imprimerie de Petitpierre et Prince, Neuchâtel, 1420 pp.

Agassiz, L. 1844–1845. Monographie de poissons fossiles des Vieux Grès Rouges ou Système Dévonien (Old Red Sandstone) des Îles Britanniques et de Russie. Imprimerie de H. Wolfrath, Neuchâtel, 171 pp.

Berg, L. S. 1940. Classification of fishes, both recent and fossil. Trudy Instituta Zoologicheskikh Akademiia Nauk, 5 (2): 85–517.

Bernacsek, G. M. & D. L. Dineley. 1977. New acanthodians from the Delorme Formation (Lower Devonian) of N.W.T., Canada. Palaeontographica A, 158: 1–25.

Beznosov, P. 2009. A redescription of the Early Carboniferous acanthodian *Acanthodes* lopatini Rohon, 1889. Acta Zoologica (Stockholm) 90 (suppl. 1): 183–193.

Blaauwen, J. L. den, M. J. Newman & C. J. Burrow. 2019. A new cheiracanthid acanthodian from the Middle Devonian (Givetian) Orcadian Basin of Scotland and its biostratigraphic and biogeographical significance. Scottish Journal of Geology: sjg2018-023. DOI: 10.1144/sjg2018-023

Brazeau, M. D. 2009. The braincase and jaws of a Devonian 'acanthodian' and modern gnathostome origins. Nature, 457 (7227): 305–308. DOI: nature07436[pii]10.1038/nature07436

Brazeau, M. D. 2012. A revision of the anatomy of the Early Devonian jawed vertebrate *Ptomacanthus anglicus* Miles. Palaeontology, 55: 355–367. DOI: 10.1111/j.1475-4983.2012.01130.x

Brazeau, M. D., S. Giles, R. P. Dearden, A. Jerve, Y. A. Ariunchimeg, E. Zorig, R. Sansom, T. Guillerme & M. Castiello. 2020. Endochondral bone in an Early Devonian 'placoderm' from Mongolia. Nature Ecology & Evolution, 4: 1477–1484. DOI: 10.1101/2020.06.09.132027

Burrow, C. J. 2004. Acanthodians with dentigerous jaws: the Ischnacanthiformes and *Acanthodopsis*. Fossils and Strata, 50: 8–22.

Burrow, C. J. & S. Turner. 2010. Reassessment of "*Protodus*" *scoticus* from the Early Devonian of Scotland. – Pp. 123–144 in D. K. Elliott, J. G. Maisey, X. Yu & D. Miao (eds.), Morphology, Phylogeny and Paleobiogeography of Fossil Fishes. Verlag Dr Friedrich Pfeil, Munich.

Burrow, C. J. & G. C. Young. 2005. The acanthodian fauna of the Craven Peaks Beds (Early to Middle Devonian), western Queensland. Memoirs of the Queensland Museum, 51: 3–25.

Burrow, C. J., R. G. Davidson, J. L.den Blaauwen & , M. J. Newman. 2015. Revision of *Climatius reticulatus* Agassiz, 1844 (Acanthodii, Climatiidae), from the Lower Devonian of Scotland, based on new histological and morphological data. Journal of Vertebrate Paleontology, 35: e913421, 1–15. DOI: 10.1080/02724634.2014.913421

Burrow, C., J. den Blaauwen, M. Newman & R. Davidson. 2016. The diplacanthid fishes (Acanthodii, Diplacanthiformes, Diplacanthidae) from the Middle Devonian of Scotland. Palaeontologia Electronica, 19.1.10A: 1–83. DOI: palaeo-electronica.org/content/2016/1398-scottish-diplacanthid-fishes

Burrow, C. J., M. Newman, J. den Blaauwen, R. Jones & R. Davidson. 2018. The Early Devonian ischnacanthiform acanthodian *Ischnacanthus gracilis* (Egerton, 1861) from the Midland Valley of Scotland. Acta Geologica Polonica, 68: 335–362. DOI: 10.1515/agp-2018-0008

Burrow, C. J., M. J. Newman & J. L. den Blaauwen. 2020. Cheiracanthid acanthodians from the lower fossil fish-bearing horizons (Eifelian, Middle Devonian) of the Orcadian Basin, Scotland. Scottish Journal of Geology, 57: sjg2020-006. DOI: 10.1144/sjg2020-006

Chevrinais, M., R. Cloutier & J.-Y. Sire. 2015. The revival of a so-called rotten fish: the ontogeny of the Devonian acanthodian *Triazeugacanthus*. Biology Letters, 11: 20140950. DOI: 10.1098/rsbl.2014.0950

Chevrinais, M., E. Balan & R. Cloutier. 2016. New insights in the ontogeny and taphonomy of the Devonian acanthodian *Triazeugacanthus affinis* from the Miguasha fossil-Lagerstätte, eastern Canada. Minerals, 6: 1–17. DOI: 10.3390/min6010001

Coates, M. I., J. A. Finarelli, I. J. Sansom, P. S. Andreev, K. E. Criswell, K. Tietjen, M. L. Rivers & P. J. La Riviere. 2018. An early chondrichthyan and the evolutionary assembly of a shark body plan. Proceedings of the Royal Society B: 285: 20172418, 10 pp. DOI: 10.1098/rspb.2017.2418

Dearden, R. 2019. The anatomy and evolution of "acanthodian" stem-chondrichthyans. Ph.D. thesis (unpublished). Imperial College, London.

Denison, R. H. 1967. Ordovician vertebrates from western United States. Fieldiana Geology, 16: 131–192.

Denison, R. H. 1978. Placodermi. Volume 2 in H.-P. Schultze (ed.) Handbook of Paleoichthology. 122 pp. Gustav Fischer Verlag, Stuttgart and New York.

Denison, R. H. 1979. Acanthodii. Volume 5 in H.-P. Schultze (ed.) Handbook of Paleoichthyology. 62 pp., Gustav Fischer Verlag, Stuttgart and New York.

Donoghue, P. C. J. & I. J. Sansom. 2002. Origin and early evolution of vertebrate skeletonization. Microscopy Research and Technique, 59: 352–372.

Duff, P. 1842. Sketch of the geology of Moray. 72 pp. Forsyth and Young, Elgin.

Egerton, P. G. 1861. British fossils. Memoirs of the Geological Survey of the United Kingdom (British Organic Remains), Decade 10: 51–75.

Gagnier, P.-Y. 1996. Acanthodii. – Pp. 149–164 in H.-P. Schultze & R. Cloutier (eds.), Devonian fishes and plants of Miguasha, Quebec, Canada. Pfeil, Munich.

Hancock, A. & T. Atthey. 1868. Notes on the remains of some reptiles and fishes from the shales of the Northumberland coal field. Annals and Magazine of Natural History, 1: 266–278, 346–378.

Hanke, G. F., M. V. H. Wilson, L. A. Lindoe. 2001. New species of Silurian acanthodians from the Mackenzie Mountains, Canada. Canadian Journal of Earth Sciences, 38: 1517–1529.

Heidtke, U. H. J. & K. Krätschmer. 2001. *Gladbachus adentatus* nov. gen. et sp., ein primitiver Hai aus dem oberen Givetium (Oberes Mitteldevon) der Bergisch Gladbach-Paffrath-Mulde (Rheinisches Schieffergebirge). Mainzer geowissenschaftliche Mitteilungen, 30: 105–122.

Hirasawa, T. & S. Kuratani. 2015. Evolution of the vertebrate skeleton: morphology, embryology, and development. Zoological Letters, 1: 17 pp. DOI: 10.1186/s40851-014-0007-7

Huxley, T. H. 1861. Preliminary essay upon the systematic arrangement of the fishes of the Devonian epoch. Memoirs of the Geological Survey of the United Kingdom, Decade 10: 1–40.

Janvier, P. 1996. The dawn of the vertebrates: characters versus common ascent in the rise of current vertebrate phylogenies. Palaeontology, 39: 259–287.

Janvier, P. & M. Suarez-Riglos. 1986. The Silurian and Devonian vertebrates of Bolivia. Bulletin de l'Institut Français d'Études Andines, 15: 73–114.

Lu, J., S. Giles, M. Friedman & M. Zhu. 2017. A new stem sarcopterygian illuminates patterns of character evolution in early bony fishes. Nature Communications, 8: e1932. DOI: 10.1038/s41467-017-01801-z

Maisey, J. G. 1979. Finspine morphogenesis in squalid and heterodontid sharks. Zoological Journal of the Linnean Society, 66: 161–183.

Maisey, J. G., R. Miller, A. Pradel, J. S. S. Denton, A. Bronson & P. Janvier. 2017. Pectoral morphology in *Doliodus*: bridging the 'acanthodian'-chondrichthyan divide. American Museum Novitates, 3875: 1–15. DOI: 10.1206/3875.1

Maisey, J. G., P. Janvier, A. Pradel, J. S. S. Denton, A. Bronson, R. Miller & C. J. Burrow. 2019. *Doliodus* and pucapampellids: contrasting perspectives on stem chondrichthyan morphology. – Pp. 87–109 in C. Underwood, M. Richter & Z. Johanson (eds.), Evolution and Development of Fishes. Cambridge University Press, Cambridge.

Maisey, J. G., J. S. S. Denton, C. Burrow & A. Pradel. 2021. Architectural and ultrastructural features of tessellated calcified cartilage in modern and extinct chondrichthyan fishes. Journal of Fish Biology, 98: 919-941. DOI: 10.1111/jfb.14376

Matton, O., R. Cloutier & R. Stevenson. 2012. Apatite for destruction: Isotopic and geochemical analyses of bioapatites and sediments from the Upper Devonian Escuminac Formation (Miguasha, Québec). Palaeogeography, Palaeoclimatology, Palaeoecology, 361–362: 73–83. DOI: 10.1016/j.palaeo.2012.08.004

Miles, R. S. 1973. Articulated acanthodian fishes from the Old Red Sandstone of England, with a review of the structure and evolution of the acanthodian shoulder-girdle. Bulletin of the British Museum (Natural History), Geology 24: 113–213.

Moy-Thomas, J. A. 1939. Palaeozoic Fishes, Methuen and Co. Ltd, London, 149 pp.

Newman, M. J., C. J. Burrow, J. L. den Blaauwen & R. G. Davidson. 2014. The Early Devonian acanthodian *Euthacanthus macnicoli* Powrie, 1864 from the Midland Valley of Scotland. Geodiversitas, 36: 321–348. DOI: 10.5252/g2014n3a1

Newman, M. J., J. L. den Blaauwen & C. J. Burrow. 2020. Two newly identified cheiracanthid acanthodians from the Mey Flagstone Formation (Givetian, Middle Devonian) of the Orcadian Basin, Scotland. Scottish Journal of Geology, 57: sjg2020-009. DOI: 10.1144/sjg2020-009

Newton, E. T. 1892. Note on a new species of *Onychodus* from the Lower Old Red Sandstone of Forfar. Geological Magazine, Decade 3, 9: 51–52.

Ørvig, T. 1951. Histological studies of placoderms and fossil elasmobranchs. 1. The endoskeleton, with remarks on the hard tissues of lower vertebrates in general. Arkiv för Zoologi, 2: 321–454.

Ørvig, T. 1967. Some new acanthodian material from the Lower Devonian of Europe. Journal of the Linnean Society (Zoology) 47: 131–153.

Owen, R. 1867. On the dental characters of genera and species, chiefly of fishes from the Lower Main seam and shales of coal, Northumberland. Transactions of the Odontological Society of Great Britain, 5: 323–376.

Powrie, J. 1864. On the fossiliferous rocks of Forfarshire and their contents. Quarterly Journal of the Geological Society of London, 20: 413–429.

Reis, O. M. 1890. Zur Kenntniss des Skelets der Acanthodinen. Geognostische Jahrbücher, 3: 1–43.

Reis, O. M. 1896. Über *Acanthodes bronnii* Agassiz. Morphologische Arbeiten, 6: 143–220.

Rohon, J. V. 1889. Über fossile Fische vom oberen Jenissei. Memoires de L'Académie Impériale des Sciences de St-Pétersbourg, VII Séri 36: 1–17.

Sansom, I. J., N. Z. Wang & M. Smith. 2005. The histology and affinities of sinacanthid fishes: primitive gnathostomes from the Silurian of China. Zoological Journal of the Linnean Society, 144: 379–386.

Snyder, D., S. Turner, C. J. Burrow & E. B. Daeschler. 2017. *"Gyracanthus" sherwoodi* (Gnathostomata, Gyracanthidae) from the Late Devonian of North America. Proceedings of the Academy of Natural Sciences of Philadelphia, 165: 195–219.

Stensiö, E. A. 1927. The Downtonian and Devonian vertebrates of Spitsbergen. Part 1. Family Cephalaspidae. Skrifter om Svalbard og Nordishavet, 12: 1–391.

Traquair, R. H. 1893. Notes on the Devonian fishes of Campbelltown and Scaumenac Bay in Canada. Parts 1 and 2. Proceedings of the Royal Physical Society of Edinburgh, 12: 111–125.

Traquair, R. H. 1894. On a new species of *Diplacanthus*, with remarks on the acanthodian shoulder-girdle. Geological Magazine, (Decade 4) 1: 254–257.

Traquair, R. H. 1898. Notes on Palaeozoic fishes. No. II. The Annals and Magazine of Natural History, 2, ser. 7: 67–70.

Upeniece, I. 1996. *Lodeacanthus gaugicus* n. g. et sp. (Acanthodii: Mesacanthidae) from the Late Devonian of Latvia. Modern Geology, 20: 383–398.

Upeniece, I. 2011. Palaeoecology and juvenile individuals of the Devonian placoderm and acanthodian fishes from Lode site, Latvia. 221 pp. Doctoral thesis, University of Latvia, Riga. http://www.lu.lv/fileadmin/user_upload/lu_portal/zinas/Upeniece_Ieva_-_2011_-_promocijas_darbs.pdf

Walcott, C. D. 1892. Notes on the discovery of a vertebrate fauna in Silurian (Ordovician) strata. Bulletin of the Geological Society of America, 3: 153–172.

Wang, N. Z., P. C. J. Donoghue, M. M. Smith & I. J. Sansom. 2005. Histology of the galeaspid dermoskeleton and endoskeleton, and the origin and early evolution of the vertebrate cranial endoskeleton. Journal of Vertebrate Paleontology, 25: 745–756.

Warren, A., B. P. Currie, C. J. Burrow & S. Turner. 2000. A redescription and reinterpretation of *Gyracanthides murrayi* Woodward 1906 (Acanthodii, Gyracanthidae) from the Lower Carboniferous of the Mansfield Basin, Victoria, Australia. Journal of Vertebrate Paleontology, 20: 225–242.

Watson, D. M. S. 1937. The acanthodian fishes. Philosophical Transactions of the Royal Society (B), 228: 49–146.

Whiteaves, J. 1887. Illustrations of the fossil fishes of the Devonian rocks of Canada. Proceedings and Transactions of the Royal Society of Canada, 4: 101–110.

Whiteaves, J. F. 1889. Illustrations of the fossil fishes of the Devonian rocks of Canada. Part II. Transactions of the Royal Society of Canada, 6: 77–96.

Woodward, A. S. 1891. Catalogue of the fossil fishes in the British Museum (Natural History). Part II. British Museum (Natural History), London, 567 pp.

Woodward, A. S. 1892. Further contributions to knowledge of the Devonian fishes of Canada. The Geological Magazine, 9 (new series Decade 3): 481–485.

Woodward, A. S. 1906. On a Carboniferous fish fauna from the Mansfield district, Victoria. Memoirs of the National Museum, Melbourne, 1: 1–32.

Yu, X. 1998. A new porolepiform-like fish, *Psarolepis romeri*, gen.et sp. nov. (Sarcopterygii, Osteichthyes) from the Lower Devonian of Yunnan, China. Journal of Vertebrate Paleontology, 18: 261–274.

Zhu, M., X. Yu, W. Wang, W. Zhao & L. Jia. 2006. A primitive fish provides key characters bearing on deep osteichthyan phylogeny. Nature, 441: 77–80.

Ancient Fishes and their Living Relatives: a Tribute to John G. Maisey. Alan Pradel, John S. S. Denton & Philippe Janvier (eds.): pp. 93–100, 12 figs.
© 2021 by Verlag Dr. Friedrich Pfeil, München, Germany – ISBN 978-3-89937-269-4

A three-dimensionally preserved stethacanthid cranium and endocast from the Late Mississippian Fayetteville Shale (Arkansas, USA)

Allison W. Bronson*

Abstract

Fossilized cartilage is rare, but three-dimensionally preserved cartilage is even scarcer. The Fayetteville Shale of northwestern Arkansas (USA) has yielded numerous well-preserved fossil chondrichthyans over the last several decades, due to a reducing, anoxic depositional environment. One such fossil is a three-dimensional partial cranium of a symmoriiform chondrichthyan, revealed here by computed tomography (CT) scanning. This cranium is preserved alongside fragmentary branchial elements, mandibular arch fragments, teeth, and "*Lambdodus*" cranial cap denticles. A partial endocast can also be reconstructed from the CT scan, including three-dimensionally preserved skeletal labyrinths of the inner ears. The cranium shows some similarities with *Stethacanthus* and *Akmonistion*, but with a more dorsoventrally expanded otic roof and a large, upwardly angled foramen magnum. The inner ear has a clear crus commune, and there is no chondrified median capsular wall. Thus, the labyrinth is confluent with the cranial cavity (as is the case in most early chondrichthyans). This new taxon adds another chondrichthyan to the diversity described from the Fayetteville Shale assemblage, in addition to *Ozarcus*, *Carcharopsis*, and '*Cobelodus*'.

Introduction

This study describes a new symmoriiform chondrichthyan from the Fayetteville Shale of Arkansas (USA) which, due to the unique depositional environment of the formation, preserves remarkably three-dimensional skeletal labyrinths of the inner ear that typify a primitive chondrichthyan inner ear morphology. Symmoriiforms are somewhat phylogenetically controversial, and the combination of characters in this fossil is unique and could provide new characters for future phylogenetic analyses.

Cartilage fossils and the Fayetteville Shale

Historically, complete chondrichthyan skeletal material has been sparse in the fossil record. However, well-preserved skeletons of cartilaginous fishes have the potential to reveal vitally informative characters for understanding early gnathostome relationships (Pradel et al. 2014, Coates et al. 2017, 2018, Maisey et al. 2019). The Fayetteville Shale of northwestern Arkansas (USA) represents a late Mississippian (326 million year old) post-extinction recovered community, following the catastrophic Hangenberg Event at the end of the Devonian (Sallan & Coates 2010, Friedman & Sallan 2012). The Fayetteville Shale is an oil-bearing black shale, well-studied by geochemists and sedimentologists (Zangerl et al. 1969, Handford 1986, Murthy et al. 2004, Alase 2012), invertebrate paleontologists (Girty 1910, Manger & Quinn 1972, Doguzhaeva et al. 1997, 2002), paleobotanists (Mapes 1966, Tomescu et al. 2001, Dunn 2002, 2004, Dunn et al. 2002), and more recently, paleoichthyologists (Maisey 2007, Pradel et al. 2014, Bronson et al. 2018). The anoxic, reducing depositional environment of the Fayetteville Shale preferentially preserves prismatic calcified cartilage rather than bone (Bronson et al. in prep), resulting in numerous symmoriiform skeletal remains (Maisey 2007, Pradel et al. 2014) and fairly three-dimensionally preserved elasmobranch crania (Bronson et al. 2018) from deposits close to Fayetteville, Arkansas (Fig. 1).

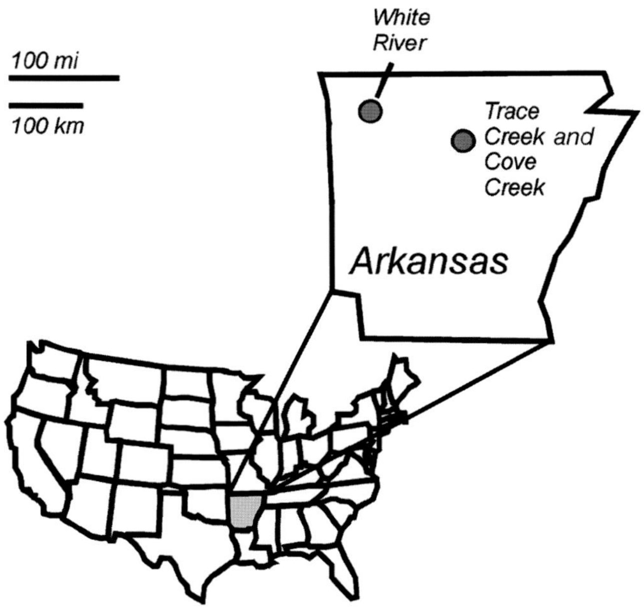

Fig. 1. Map of Arkansas showing the location of Fayetteville Shale exposures that have previously yielded chondrichthyan fossils (modified from Bronson et al. 2018, after Dunn et al. 2002).

Utility of CT scanning for making the most of rare cartilage specimens

Computed tomography (CT) scanning, originally used for materials science and for medical diagnostics and research, has revolutionized morphology and paleontology (Cunningham et al. 2014). John Maisey pioneered the use of this technology for paleoichthyology in his work on fossil and extant chondrichthyan crania and endocasts (Maisey 2001a,b). Forgoing traditional preparation techniques in favor of using scanning (where possible) to elucidate morphology within a matrix avoids the potential issues of over-preparation and removal of potentially informative features, and reveals anatomy in a relatively short period of time compared with traditional techniques. Tomographic reconstruction can be taken one step further using synchrotron tomography, which produces extremely high

* Humboldt State University, Department of Biological Sciences, 1 Harpst Street, Arcata, California, 95521 USA; awb18@humboldt.edu

resolution anatomical data (Tafforeau et al. 2006, Pradel et al. 2014) or, for extant specimens, using contrast enhancing stains for visualizing soft tissue (Metscher 2009, Gignac & Kley 2014).

Symmoriiform morphology

Fishes in the order Symmoriida have been recorded from the Upper Devonian to the Upper Carboniferous of Europe and North America, often from black shales (Zangerl 1981). They generally have no denticles other than those along the lateral line, cranial cap denticles, and the spine-brush complex (Zangerl 1981). Morphologically, symmoriiforms are unlike any elasmobranchs alive today, having very different cranial proportions and foramina and unusual appendages and denticles on the head and dorsum of several taxa (Zangerl 1981).

The taxonomy of the group has changed over the past several decades. Formerly, symmoriiforms were grouped into three families based on dermal ornamentation: Family Symmoriidae lacking any spine-like structures, versus the families Falcatidae and Stethacanthidae that possess dermal ornamentation, with Stethacanthidae further defined as possessing a spine-brush complex (Lund 1984). The classification was later revised to focus on skeletal features, leaving just the Symmoriidae and Falcatidae (Zangerl 1981), but issues remained surrounding the differentiation of symmoriiforms between the "symmoriids" and "stethacanthids." It is still unclear whether these are monophyletic sister groups, and whether some taxa may be synonyms – some specimens may be the males of a particular taxon (with a spine-brush complex) while other 'taxa' could instead be female members of the same species that lack a spine-brush complex due to sexual dimorphism. Some genera such as *Damocles* and *Falcatus* are different from all other symmoriiforms and can be more confidently classified within Family Falcatidae (Zangerl 1990), but most symmoriiforms are now classified as Symmoriiformes incertae sedis (Maisey 2007). Most recently, phylogenetic analyses by Coates and colleagues resolve a monophyletic Symmoriida inclusive of holocephalans, based on cranial characters (Coates et al. 2017), in contrast to prior phylogenetic analyses (Pradel et al. 2011).

Many symmoriiform specimens are flattened due to taphonomic deformation (Lund 1985, Maisey 1989, 2008). Despite the challenges presented by taphonomy, the cranial structure has been thoroughly described in several genera, including *Falcatus* (Lund 1985), *Stethacanthus* (Lund 1974), *Akmonistion* (Coates & Sequeira 1998), *Kawichthys* (Pradel et al. 2011) and '*Cobelodus*' (likely *Ozarcus*) (Maisey 2007). All of these apart from '*Cobelodus*' and *Kawichthys* are significantly taphonomically deformed. *Dwykaselachus*, described from the Karoo Basin of South Africa, is exceedingly well preserved and three-dimensional, but no denticles are known (Coates et al. 2017). These exceptionally preserved specimens probably make symmoriiforms (Order Symmoriiformes (Maisey 2007)) the best-known Paleozoic chondrichthyans.

Chondrichthyan ears

The specimen described in this chapter has three-dimensionally preserved skeletal labyrinths of the inner ear. The inner ear of elasmobranchs functions in both proprioception (via otolith-type inner ear mechanoreceptors) and sound detection (Corwin 1977, Maisey 2001b). The vestibular labyrinths within the otic capsules are composed of three semicircular canals (anterior, posterior, and lateral/horizontal/external) and the utriculus, sacculus, and lagena (otolith organs). In fossils, inferences related to the function of the ear must be restricted to the shape and size of the skeletal labyrinth, as the membranous labyrinth is not generally preserved. However, features of the skeletal labyrinth, such as the morphology and position of the semicircular canals, may be informative for systematics and for inferring the ecology of fossil taxa (Yi & Norell 2015, Giles et al. 2018). Gravity receptors are present within the utriculus, sacculus, and lagena, with the utriculus and sacculus likely responding to lateral and fore-aft tilting of the shark's body (Bleckman & Hofmann 1999). The maculae and gravity receptors within the lagena are also sensitive to changes in position, but would not be represented by corresponding shape change of the skeletal labyrinth.

Chondrichthyan vestibular systems function as teloreceptors (sense organs which perceive faraway stimuli), and in extant elasmobranchs, this system can determine the direction of a low-frequency sound source at great distance (Bleckman & Hofmann 1999, Myrberg Jr. 2001). This is due to a combination of anatomical features in elasmobranchs, reviewed by Maisey & Lane (Maisey & Lane 2010), which, in modern sharks (but not in chimaeroids or many Paleozoic sharks), include the separation of the posterior semicircular canal from the anterior and lateral semicircular canals, and the closing of the otico-occipital fissure during ontogeny. Presence or absence of these features appears to indicate whether a chondrichthyan would have been able to detect low frequency sound similar to the way extant elasmobranchs do today.

Material and Methods

Material Examined

AMNH FF 20531: a three-dimensional braincase with the most dorsal and most ventral surfaces lost to taphonomic forces, with associated head denticles and isolated fragmentary mandibular and pharyngeal arch elements, in a pyritized phosphatic concretion (Figs. 2–3), Fayetteville Shale (Serpukhovian, Upper Mississippian). Collected by Dr. Royal Mapes and Dr. Gene Mapes, housed at the American Museum of Natural History (AMNH).

CT Scanning Methods

AMNH FF 20531 was CT scanned on September 10, 2017 at the AMNH Microscopy and Imaging Facility, using a GE Phoenix v tome x|s microCT scanner. Scan Data: 1800 images, 180 kV, 180 µA, one 0.5 mm copper filter, voxel size 0.09 mm. CT data was reconstructed using Datos/x software (GE) and relevant anatomical structures were segmented using Mimics software (Materialise, Belgium). The braincase is preserved in fairly three-dimensional condition, though the specimen is beginning to show signs of damage due to pyrite disease. The high density of the phosphatic concretion provided a challenge in CT scanning, and so certain regions of the cranium were easier to reconstruct than other regions, due to differing attenuation of x-rays and issues with beam hardening (Abel et al., 2012). Despite a high wattage and amperage (and using appropriate filters), the scanner had trouble penetrating some denser portions of the specimen.

The resulting image stack and renderings of the cranium and inner ear (as STL files) are available for viewing on MorphoSource (www.morphosource.org), under Project Number 752.

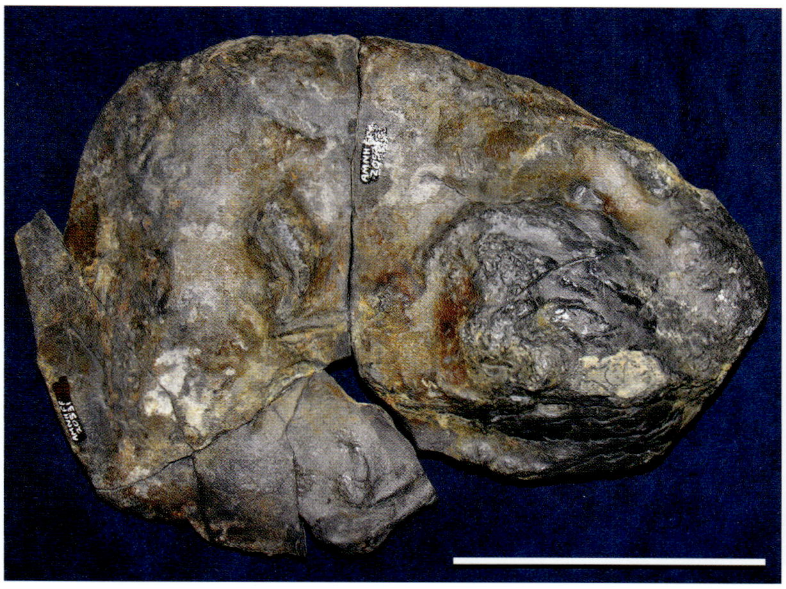

Fig. 2. Dorsal view of the entire specimen AMNH FF 20531. Note exposure of cartilage in cross section, in the lighter area toward the lower right of the specimen. Scale = 10 cm.

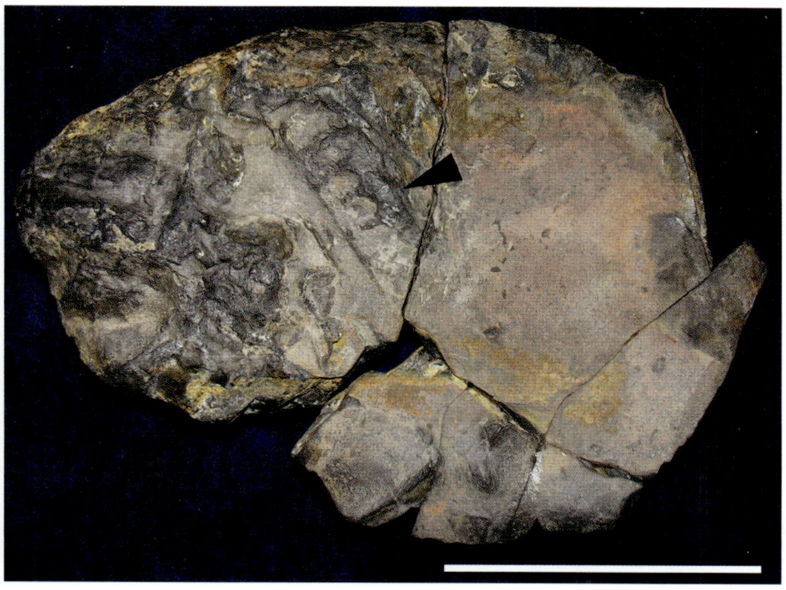

Fig. 3. Ventral view of the specimen, AMNH FF 20531. Note the scalloped depressions (arrowhead) in a fragment of the mandibular (Meckel's) cartilage. Scale bar = 10 cm.

Systematic Paleontology

Class Chondrichthyes Huxley, 1880
Order Symmoriiformes Maisey, 2007
Genus *Maiseyacanthus* gen. nov. Bronson, 2021
Species *Maiseyacanthus ozarkanum* sp. nov. Bronson, 2021

Type Species. *Maiseyacanthus ozarkanum* sp. nov.

Diagnosis. According to the present description, *Maiseyacanthus* may be diagnosed as follows:

Chondrichthyan features: Tessellated calcified cartilage, placoid scales (cranial cap denticles)

Symmoriiform features: Enclosed, median dorsal aorta; single-crowned cranial cap denticles, cladodont teeth with small lateral cusps, no serrations, and a reniform tooth base.

Autapomorphies of genus and species: platybasic skull, cranium with expanded roof in the otic region, and with a very large foramen magnum (approximately 1/5 of the width of the cranium), foramen magnum steeply angled in relation to the rest of the cranial cavity (appx. 100° relative to the angle of the external semicircular canal), four spino-occipital foramina present, lateral otic shelf upwardly curved and scooplike, large fossa present just medial to lateral otic process, periotic shelf present but very slight.

Etymology. "Maisey" after John G. Maisey, one of the first paleoichthyologists to address Fayetteville Shale vertebrates and use Computed Tomography for fossil fishes, and "acanthus" (Greek ἄκανθος for spine or thorn) for the head denticles associated with the specimen. The epithet "ozarkanum" is "of the Ozarks" for the region of the United States in which the holotype was discovered.

Holotype. AMNH FF 20531, Figure 2 and 3, partial cranium and disarticulated elements of gill arches, teeth, and cranial cap denticles, all preserved within a single broken phosphatic concretion.

Locality and horizons. Fayetteville Formation (lower shale member), Chesterian, Upper Mississippian. Precise locality is unknown, but the holotype is very likely from the ARC-07 locality 24 (section (sec.) 22, township (T.) 14N, range (R.) 15W) near Leslie, Searcy County, Arkansas, United States.

Description

Braincase

The braincase (Fig. 4) is covered by tessellated calcified cartilage (Fig. 5), which is visible on the surface of the specimen. Remarkably, the braincase appears to be hardly crushed by taphonomic processes. However, not only is the portion of the braincase anterior to the orbits missing, but the lower and upper surfaces of the braincase have also been abraded away taphonomically. This braincase lacks the multiple layers of calcification seen in some other Paleozoic chondrichthyan braincases (such

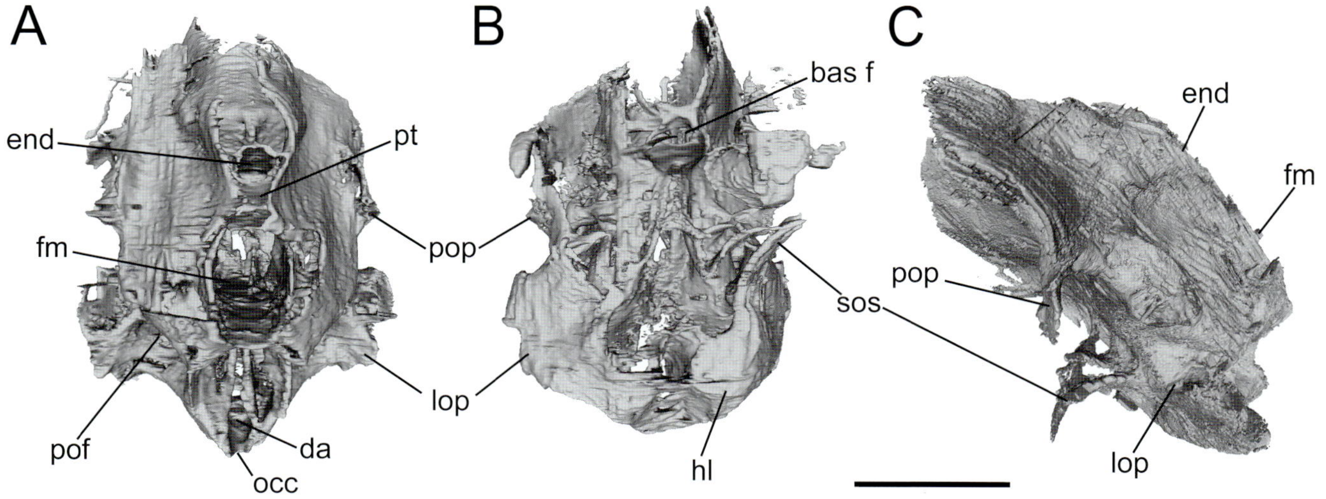

Fig. 4. 3D dorsal (**A**), ventral (**B**), and (**C**) left lateral renderings of the braincase of *Maiseyacanthus*, produced in Mimics. Anterior to top in A and B. Abbreviations: bas f, basicranial fenestra; da, foramen for median dorsal aorta; end, endolymphatic foramen; fm, foramen magnum; hl, hypotic lamina; lop, lateral otic process; occ, occipital cotylus; pof, preoccipital fossa; pop, postorbital process; pt, posterior tectum; sos, suborbital shelf. Scale bar = 5 cm.

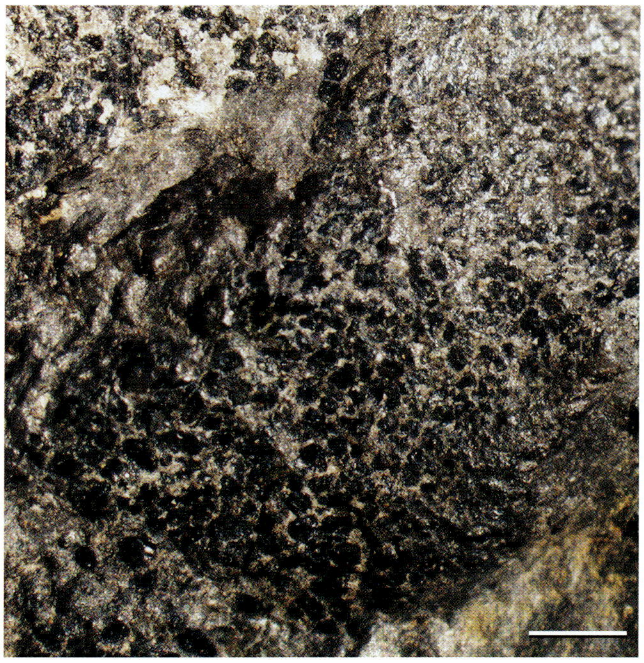

Fig. 5. Tessellated calcified cartilage on ventral surface of the specimen (AMNH FF 20531). Scale bar = 1 mm.

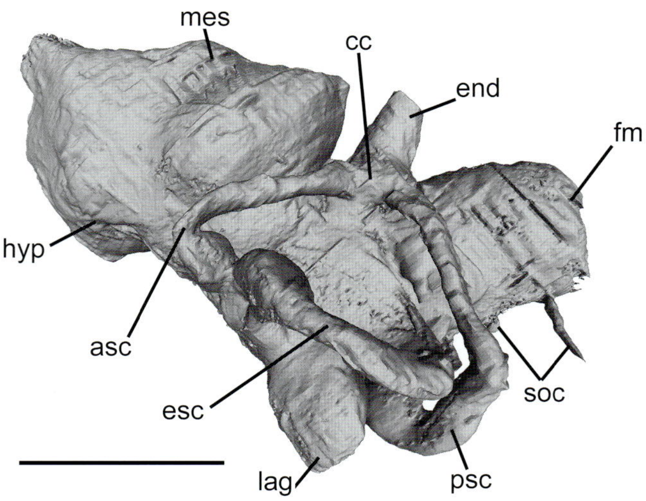

Fig. 6. Endocast of *Maiseyacanthus* in left lateral view. Abbreviations: asc, anterior semicircular canal; cc, crus commune; end, endolymphatic duct; esc, external semicircular canal; fm, foramen magnum; hyp, hypophyseal chamber; lag, lagena; mes, mesencephalic chamber; psc, posterior semicircular canal; soc, passages for spino-occipital nerves. Scale bar = 5 cm.

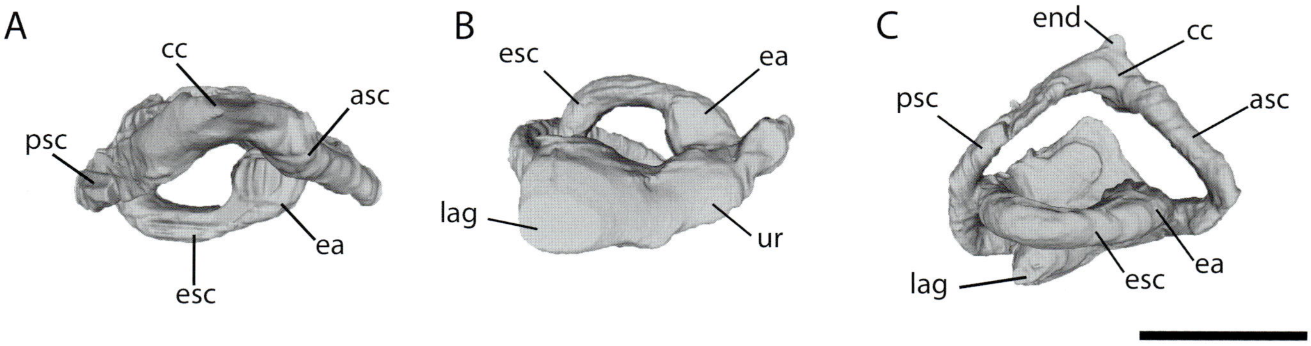

Fig. 7. Three-dimensional rendering (produced in Mimics from a CT scan) of the right inner ear of AMNH FF 20531 in dorsal (**A**), ventral (**B**), and lateral (**C**) view. Abbreviations: asc, anterior semicircular canal; cc, crus commune; ea, external ampulla; end, endolymphatic duct; esc, external semicircular canal; lag, lagena; psc, posterior semicircular canal; ur, utriculus. Scale bar = 5 cm.

as xenacanths), and yet it is in three-dimensional condition, suggesting little postmortem transport and rapid deposition and diagenesis.

The braincase roof is broad, and slightly tapers anteriorly and posteriorly. It may taper more dramatically closer to the anterior, but that part of the braincase is not preserved intact. The braincase is approximately 101.32 mm long, and is 74.61 mm wide, measured at widest preserved part of the chondrocranium. The most dorsal and ventral parts of the cranium have been worn away and are not preserved, leaving large holes in the roof and floor of the skull – the most anterior of which represents the endolymphatic foramen. The foramen magnum is large and oriented at a steep angle (of approximately 100°) relative to the lateral semicircular canal (Figs. 4, 6). An endolymphatic fossa is present and is separated from the occipital block by a posterior tectum (Fig. 4). The otic capsule is separated from the occipital block, there is no evidence of a persistent otico-occipital fissure, and there is no "wedging" of the occipital region between the otic capsules (Fig. 4).

The cranial cavity appears to project between the orbits, as in a platybasic cranium (Maisey 2007). The foramen magnum is very large (approximately 13.33 mm wide). However, its cartilage is worn away at the dorsal surface of the specimen, and it is possible that the roof of the braincase could have extended farther dorsally and the opening for the foramen magnum was not as large in life. The foramen magnum is flanked by four pairs of openings for the occipito-spinal nerves (Fig. 6).

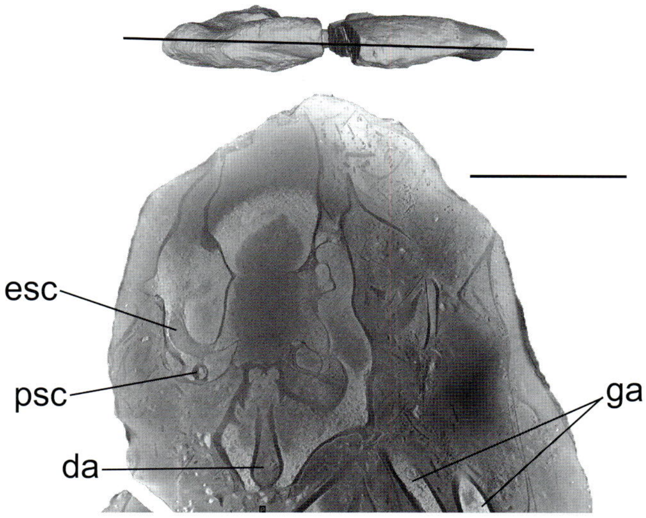

Fig. 8. Axial section through the cranium at the level shown by the inset 3D reconstruction of the entire fossil, showing the inner ear. Note partial gill arches present at the bottom right of the image. Abbreviations: esc, external semicircular canal; da, foramen for median dorsal aorta; ga, gill arch fragments; psc, posterior semicircular canal. Scale bar = 5 cm.

Endocast and Inner Ear

The endocast (infilling of the cranial cavity) is only partial, truncated by taphonomy (Fig. 6), but what is present is three-dimensionally preserved. The cranial cavity shows a hypophyseal chamber and a chamber for the mesencephalon. Openings for the spino-occipital nerves can be traced along the path of the spinal cord through the foramen magnum (Fig. 6).

Three semicircular canals are present, each approximately 4.65 mm in diameter (Fig. 7). The external semicircular canal spans approximately 26.5 mm in length (Figs. 7, 8). The skeletal labyrinth is confluent with the cranial cavity, due to the absence of a chondrified medial capsule wall. The anterior and posterior canals are joined at a crus commune, connected to an endolymphatic duct that opens into the endolymphatic fossa (also known as the posterior dorsal fontanelle) that is approximately 1/10 of the width of the braincase (Fig. 4). No glossopharyngeal canal is apparent, but the lagenar chamber is large (Fig. 6).

Denticles

"Lambdodus" type cranial cap denticles are visible along the dorsal surface of the specimen (Fig. 9). The location of these denticles makes it impossible to properly reconstruct them in 3D from the CT scan (due to imaging problems at the margins

of the fossil, potentially due to beam hardening). These denticles are characteristic of symmoriiforms in the "stethacanthid" group, but are not diagnostic to species, particularly because no discernable organization (e.g. medial vs. lateral rows of denticles) is apparent in this fossil.

Teeth

Two teeth are visible on the external surface of the specimen (one appears to be fairly complete (Fig. 10) and the other is only recognizable by its base). However, scanning reveals a small clump of teeth in the matrix below the more complete exposed tooth (Fig. 11). These are diagnosable as cladodont teeth, based on the reniform shape of the base in all teeth, and on the cusp morphology of the single complete tooth as well as fragmentary bases and cusps. Resolution of the scan, and embedding of much of the externally visible teeth in the matrix, is such that neither the presence or absence of nutritive foramina nor the number of lateral cusps can be resolved. No serrations are present, based on the limited resolution of the scan and surface observations, and the overall shape appears to consist of a small lingual torus and a large central cusp that appears to be flanked by a single pair of small lateral cusps. However, they cannot be reconstructed in detail at this scan resolution.

Mandibular cartilage and gill arches

The ventral surface of the specimen reveals a partial Meckel's (mandibular) cartilage, with a scalloped lingual margin that, in life, would have housed tooth families (Fig. 3). The teeth have all been lost from the cartilage; none are in situ. In addition to the mandibular cartilage, there are several broken structures within the matrix that likely represent fragmentary gill arch elements (Fig. 8). They are no longer articulated and, though still three-dimensional, do not possess features that can assign them to particular arches.

Discussion

Symmoriiform identification is supported by the presence of "*Lambdodus*" cranial cap denticles, and tooth morphology: cladodont teeth with a reniform tooth base, with a central cusp and two small lateral cusps, all of which are non-serrated. Symmoriiform diagnosis is further supported by the lack of a chondrified medial capsule wall separating the inner ear from the cranial cavity, and the presence of an enclosed, median dorsal aorta.

Cranium

The somewhat bulbous shape of the dorsal roof of the braincase (Fig. 4) is unusual in comparison with other symmoriiform cranial material, with the exception of the slightly domed braincase of *Kawichthys* (Pradel et al. 2011). However, comparison of this shape with other taxa is hindered by the taphonomic deformation of other fossils compared with the preservation of Fayetteville material. Conclusions can nevertheless be drawn regarding the fossil's systematic affinities based on the proportions of the braincase, platybasic arrangement, and presence of a large and somewhat dorsally-directed foramen magnum.

Compared with the more obviously tropibasic *Ozarcus* from this formation (Pradel et al. 2014), *Maiseyacanthus* appears to have a platybasic condition, more similar to the condition in *Akmonistion* or *Stethacanthus* (Coates & Sequeira 1998, Maisey 2007), however this fossil bears some differences from those taxa, including a small suborbital shelf (Fig. 4, sos) and enlarged foramen magnum, positioned close to the endolymphatic foramen (Fig. 12). Additionally, the cranial roof is not well preserved, and there is subsequently no evidence of an otico-occipital fissure. The braincase is proportionally fairly similar to that of *Akmonistion* (Coates & Sequeira 1998, 2001) but, like the braincase of '*Cobelodus*' and *Dwykaselachus*, the braincase of *Maiseyacanthus* possesses an elevated region anterior to the foramen magnum that could correspond with the location of

Fig. 9. Detail of "*Lambdodus*" denticles on dorsal surface of specimen, AMNH FF 20531. Scale bar = 2 mm.

Fig. 10. A single visible cladodont tooth embedded in the matrix surrounding the cranium, AMNH FF 20531. Scale bar = 1 mm.

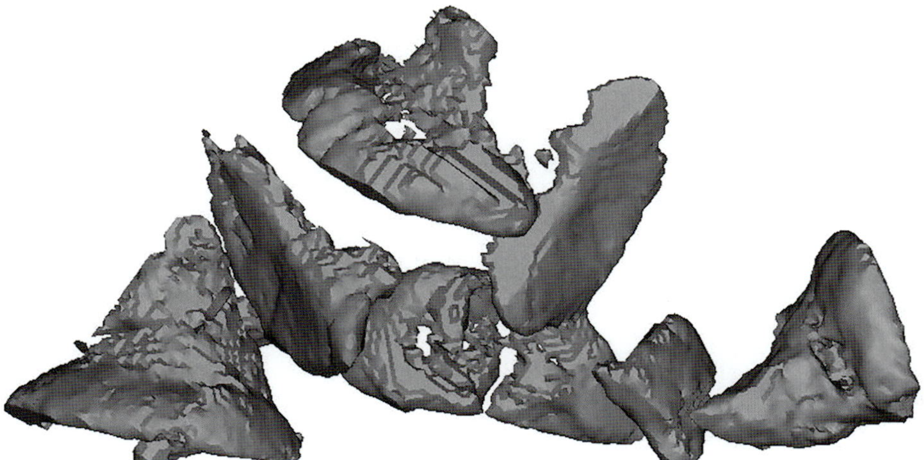

Fig. 11. 3D rendering of a group of teeth embedded in the matrix surrounding the cranium of AMNH FF 20531, revealed by CT scanning. Scale bar = 6 mm.

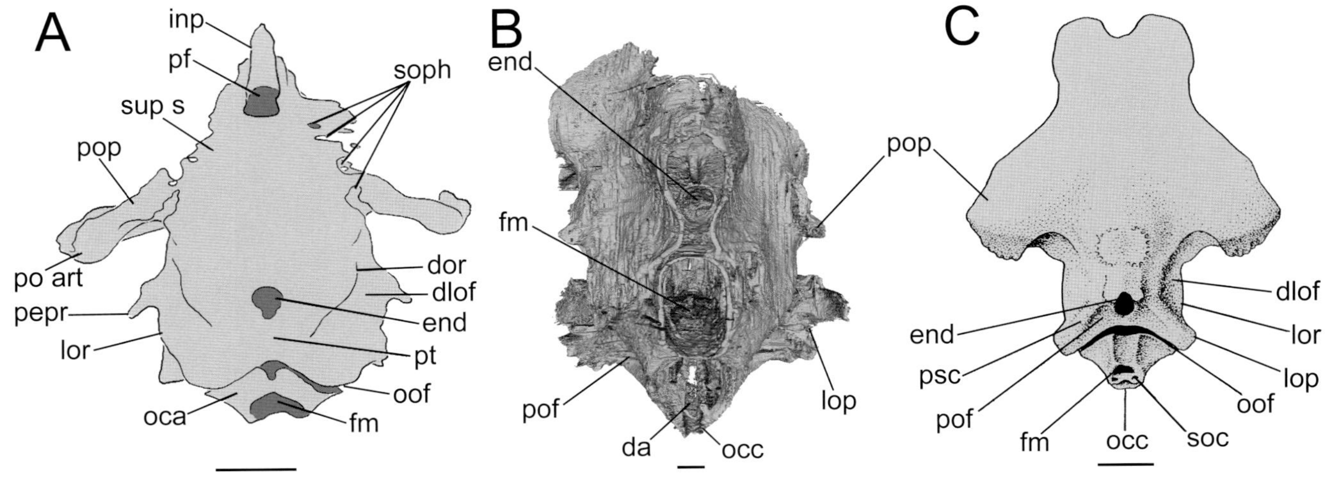

Fig. 12. Comparison of crania in dorsal view. **A.** *'Cobelodus'* modified from Maisey 2007; **B.** *Maiseyacanthus* AMNH FF 20531; **C.** *Akmonistion*, modified from Coates & Sequeira 1998. Anterior to top. Abbreviations: da, foramen for dorsal aorta; dlof, dorsolateral otic fossa; dor, dorsal otic ridge; end, endolymphatic duct foramen; fm, foramen magnum; inp, internasal plate; lop, lateral otic process; lor, lateral otic ridge; oca, occipital arch; occ, occipital cotylus; oof, otico-occipital/metotic fissure; pepr, periotic process; pf, precerebral fontanelle; po art, postorbital articulation for palatoquadrate; pof, preoccipital fossa; pop, postorbital process; psc, posterior semicircular canal; pt, posterior tectum; soc, passages for spino-occipital nerves; soph, course of the superficial ophthalmic complex; sup s, supraorbital shelf. Scale bars = 1 cm.

the midbrain (Maisey 2007, Coates et al. 2017), which is also characteristic of chimaeroids.

More detail regarding the path of nerves and vessels through the cranium may be obtained with more targeted scanning efforts. The concretion is simply too dense to be scanned at a higher resolution using the scanner at the AMNH, but as an alternative to traditional preparation, neutron or synchrotron scanning might yield higher resolution images of cranial foramina and pathways of nerves or vessels leaving the cranial cavity, such as the location of the lateral head vein or glossopharyngeal nerve. The internal carotid arteries of *Kawichthys* for example, are entirely enclosed within basicranial canals, and *Kawichthys* (Pradel et al. 2011), *Akmonistion* and *Stethacanthus* all have distinct foramina for the hyoidean artery (Coates & Sequeira 1998, 2001), but these features cannot be distinguished in this fossil.

A posterior dorsal fontanelle or endolymphatic foramen (the term for this opening in modern elasmobranchs) is present, and similar to the condition in *'Cobelodus'* (Maisey 2007), the endolymphatic foramen is surrounded by a slight "chimney" that elevates its aperture above the level of the cranial roof (Fig. 4). The cranial roof itself is also broadly sloping, giving a somewhat bulbous appearance to the cranium (Fig. 4). Just medial to the broad lateral otic shelf (Fig. 4, los), there is a broad fossa (Fig. 4, pof), similar in appearance to that of *Hybodus basanus* (Maisey 1983), and larger than the preoccipital fossa adjacent to the posterior tectum of *Akmonistion* (Coates & Sequeira 1998). The occipital region is broadly expanded, showing no "wedging" of the occipital region between the otic capsules (Maisey 2007), not even to the minimal degree of *Akmonistion* (Coates & Sequeira 2001). Instead, the occipital region extends back behind the otic capsules and tapers toward the occipital cotylus (Fig. 12).

There could have been a short median dorsal ridge present ahead of the endolymphatic fossa, but that region is not preserved. Unlike the condition in *Akmonistion*, the braincase in the Fayetteville stethacanthid is broad ahead of the otic region, with part of the postorbital process preserved (Figs. 4, 12). A superficial ophthalmic complex is not visible in the specimen, and whether an internasal plate was present also cannot be determined due to the preservation of the braincase. This is probably because the cartilage in the anterior half of the cranium is thinner than the calcified cartilage of the occipital region. Taphonomic processes have destroyed some of the orbit, but a small suborbital shelf appears to be present, as well as a possible interorbital fenestra (or opening for the palatine ramus of the trigeminal nerve) (Figs. 4, 12).

Inner ear

The inner ear is similar to those described from other symmoriiforms such as *'Cobelodus'*, in possessing a crus commune and a posterior semicircular canal that is not separated posteriorly from other semicircular canals, and in having no chondrified medial capsular wall (Maisey 2005, 2007, Maisey & Lane 2010).

The inner ear has been described from *'Cobelodus'* and *Dwykaselachus*, and these ears are similar to that of *Maiseyacanthus*. The ears of *Dwykaselachus*, *'Cobelodus,'* and *Maiseyacanthus* all have a crus commune, however the endolymphatic fenestra (posterior dorsal fontanelle) of *Dwykaselachus* is much smaller, similar to the condition in *'Cobelodus,'* while the endolymphatic fenestra of *Maiseyacanthus* is a bit larger relative to the cranium and is positioned closer to the foramen magnum. The density of the otic region compared with surrounding matrix makes segmentation challenging, but the ear also appears to have a very large lagena, which has not been described in other Paleozoic chondrichthyans.

The posterior semicircular canal is connected with the anterior canal via a crus commune, and therefore this stethacanthid was not specialized for detection of low frequency sound in the same manner of modern elasmobranchs (Maisey 2001b). Development of a system for detection of low frequency sound is well known in extant selachians, but is absent from more primitive taxa such as symmoriiforms. The low frequency sound detection system (LFSDS) of extant elasmobranchs enables responses to auditory cues from struggling aquatic prey. Symmoriiforms are known for their large orbits, and probably relied on visual rather than auditory cues for identifying potential prey items.

Denticles

"Lambdodus" denticles are common within 'stethacanthidae' and Falcatidae, present in *Falcatus*, *Stethacanthus*, *Akmonistion* (Coates & Sequeira 2001, Maisey 2007), and *Kawichthys* (Pradel et al. 2011). The presence of such denticles therefore cannot place *Maiseyacanthus* beyond the level of Symmoriiformes incertae sedis. Isolated denticles of this type have also been identified from the Fayetteville Shale (Bronson in prep).

All denticles appear to be the same size, as is the case in *Stethacanthus* & *Akmonistion* (Coates & Sequeira 2001). Due to taphonomy, it is impossible to discern whether the denticles would have been, in life, arranged in the pattern seen in *Kawichthys*, with median denticles sandwiched by a pair of lateral denticles (Pradel et al. 2011).

Systematics

The phylogenetic position of some symmoriiform groups is under debate: either they are stem chondrichthyans (Pradel et al. 2011, Maisey et al. 2019) or possibly closely related to holocephalans (Coates & Sequeira 2001, Coates et al. 2017). Cranial specializations that unite some symmoriiforms with holocephalans include large orbits, similar morphology of the endolymphatic ducts, and ventral displacement of the forebrain endocranial cavity (Coates et al. 2017), but holocephalans have numerous apomorphic features such as the fusion of the cranium and palatoquadrate and a specialized dentition. Careful attention to character coding, and discovery of additional well-preserved specimens may help to resolve this conflict.

Conclusions

The presence of Lambdodus head denticles and cladodont teeth, along with the closed dorsal portion of the otico-occipital fissure and median dorsal aorta support a stethacanthid affinity for this fossil. The braincase shows some superficial similarity to that of *Akmonistion* (Coates & Sequeira 1998), *Dwykaselachus* (Coates et al. 2017), and *Stethacanthus* (Coates & Sequeira 1998), but these taxa do not have a domed braincase and *Akmonistion* has a very broad supraorbital shelf and wide postorbital arcade, which is not apparent in *Maiseyacanthus* (Figs. 4, 12). It is possible, however, that the difference in braincase roof shape is due mainly to the exceptional lack of deformation in this specimen, and that the more flattened specimens of *Falcatus*, *Stethacanthus*, and *Akmonistion* could have had a rounded roof to the braincase, prior to taphonomic deformation. *Kawichthys* has a domed cranial roof like *Maiseyacanthus* (Pradel et al. 2011), but *Kawichthys* has a tropibasic cranium and the cranial cavity of *Maiseyacanthus* appears to be platybasic. Particularly unusual in *Maiseyacanthus* is the presence of an upwardly inflected foramen magnum, more extreme than has been described in any other symmoriiform.

The dentition scattered within this specimen is not sufficiently reconstructed to support any placement within a particular family of symmoriiform chondrichthyans. Further work either by mechanical preparation or targeted scanning using a neutron scanner or synchrotron scanner may yield more detail. Regardless of whether further investigation reveals these teeth to be like those of *Akmonistion* or markedly different from described stethacanthiform genera, this specimen adds a new member to the Fayetteville Shale chondrichthyan fauna. This stethacanthid joins the existing symmoriiform taxa from the Shale (*Ozarcus* (Pradel et al. 2014), *'Cobelodus'* (Maisey 2007), and a new taxon with elongate hyoid rays (Bronson et al. in prep) as well as the euselachian *Carcharopsis wortheni* (Bronson et al. 2018). The Fayetteville Shale continues to yield well-preserved fossils that inform our understanding of Carboniferous chondrichthyan morphology, with more discoveries likely in the future.

Acknowledgments

After six years of studying fishes, I am only beginning to scratch the surface of the depths of information aggregated by my PhD advisor and mentor John G. Maisey. I am immensely grateful for his guidance and expertise. Thanks also to John Denton, Alan Pradel, and Philippe Janvier for organizing and editing this volume, to Morgan Hill and Andrew Smith in the American Museum of Natural History (AMNH) Microscopy and Imaging Facility for assistance with CT scanning, to Dave Baston at Humboldt State University for use of the Core Research Facility, to Alana Gishlick (AMNH) for specimen loans, to Royal and Gene Mapes for sharing their Fayetteville Shale discoveries, and to three reviewers for their very helpful comments.

References

Abel, R. L., C. R. Laurini & M. Richter. 2012. A palaeobiologist's guide to 'virtual' micro-CT preparation. Palaeontologica Electronica, 15: 1–16.

Alase, A. O. 2012. Gamma spectrometry and geochemical investigation of the Mississippian (Chesterian) Fayetteville Shale and Imo Shale, Arkoma Basin, Arkansas. Oklahoma State University, 1–138 pp.

Bleckman, H. & M. H. Hofmann. 1999. Special Senses. – Pp. 300–328 in W. C. Hamlett (ed.), Sharks, Skates, and Rays: The Biology of Elasmobranch Fishes. The Johns Hopkins University Press, Baltimore and London.

Bronson, A. W., R. H. Mapes & J. G. Maisey. 2018. Chondrocranial morphology of *Carcharopsis wortheni* (Chondrichthyes, Euselachii incertae sedis) based on new material from the Fayetteville Shale (Upper Mississippian, Middle Chesterian). Papers in Palaeontology, 4: 1–14.

Coates, M. I. & S. E. K. Sequeira. 1998. The braincase of a primitive shark. Transactions of the Royal Society of Edinburgh: Earth Sciences, 89: 63–85.

Coates, M. I. & S. E. K. Sequeira. 2001. A new stethacanthid chondrichthyan from the Lower Carboniferous of Bearsden, Scotland. Journal of Vertebrate Paleontology, 21: 438–459.

Coates, M. I., R. W. Gess, J. A. Finarelli, K. E. Criswell & K. Tietjen. 2017. A symmoriiform chondrichthyan braincase and the origin of chimaeroid fishes. Nature, 541: 208–211.

Coates, M. I., J. A. Finarelli, I. J. Sansom, P. S. Andreev, K. E. Criswell, K. Tietjen, M. L. Rivers & P. J. La Riviere. 2018. An early chondrichthyan and the evolutionary assembly of a shark body plan. Proceedings of the Royal Society B, 285: 20172418.

Corwin, J. T. 1977. Morphology of the macula neglecta in sharks of the genus Carcharhinus. Journal of Morphology, 152: 341–362.

Cunningham, J. A., I. A. Rahman, S. Lautenschlager, E. J. Rayfield & P. C. J. Donoghue. 2014. A virtual world of paleontology. Trends in Ecology & Evolution, 29: 347–357.

Doguzhaeva, L., R. H. Mapes & H. Mutvei. 1997. Beaks and radulae of Early Carboniferous goniatites. Lethaia, 30: 305–313.

Doguzhaeva, L., H. Mutvei & R. H. Mapes. 2002. Chaetognath grasping spines from the Upper Mississippian of Arkansas (USA). Acta Palaeontologica Polonica , 47: 421–430.

Dunn, M. T. 2002. The Fayetteville Flora of Arkansas (USA): Structurally Modern Wetland Communities from the Upper Mississippian (Middle Chesterian/Lower Namurian A). Geological Society of America Annual Meeting. Session No. 91 Wetlands Paleoecology Through Time: A101/103.

Dunn, M. T. 2004. The Fayetteville Flora I: Upper Mississippian (middle Chesterian/lower Namurian A) plant assemblages of permineralized and compression remains from Arkansas, USA. Review of Palaeobotany and Palynology, 132: 79–102.

Dunn, M. T., G. W. Rothwell & G. Mapes. 2002. Additional observations on *Rhynchosperma quinnii* (Medullosaceae): A permineralized ovule from the Chesterian (Upper Mississippian) Fayetteville Formation of Arkansas. American Journal of Botany, 89: 1799–1808.

Friedman, M. & L. C. Sallan. 2012. Five hundred million years of extinction and recovery: A phanerozoic survey of large-scale diversity patterns in fishes. Palaeontology, 55: 707–742.

Gignac, P. M. & N. J. Kley. 2014. Iodine-enhanced micro-CT imaging: Methodological refinements for the study of the soft-tissue anatomy of post-embryonic vertebrates. Journal of Experimental Zoology Part B: Molecular and Developmental Evolution, 322: 166–176.

Giles, S., M. Rogers & M. Friedman. 2018. Bony labyrinth morphology in early neopterygian fishes (Actinopterygii: Neopterygii). Journal of Morphology, 279: 426–440.

Girty, G. H. 1910. New genera and species of Carboniferous fossils from the Fayetteville Shale of Arkansas. Annals of the New York Academy of Sciences, 20: 189–238.

Handford, C. R. 1986. Facies and bedding sequences in shelf-storm-deposited carbonates – Fayetteville Shale and Pitkin Limestone (Mississippian), Arkansas. Journal of Sedimentary Petrology, 56: 123–137.

Huxley, T. 1880. On the application of the laws of evolution to the arrangement of the Vertebrata and more particularly of the Mammalia. Proceedings of the Scientific Meetings of the Zoological Society of London, 1880: 649–652.

Lund, R. 1974. *Stethacanthus altonensis* (Elasmobranchii) from the Bear Gulch limestone of Montana. Annals of the Carnegie Museum, 45: 161–178.

Lund, R. 1984. On the spines of the Stethacanthidae (Chondrichthyes), with a description of a new genus from the Mississippian Bear Gulch Limestone. Geobios,17: 281–295.

Lund, R. 1985. The morphology of *Falcatus Falcatus* (St. John and Worthen), a Mississippian stethacanthid chondrichthyan from the Bear Gulch Limestone of Montana. Journal of Vertebrate Paleontology, 5: 1–19.

Maisey, J. G. 1983. Cranial anatomy of *Hybodus basanus* Egerton from the Lower Cretaceous of England. American Museum Novitates, 2758: 1–64.

Maisey, J. G. 1989. Visceral skeleton and musculature of a Late Devonian shark. Journal of Vertebrate Paleontology, 9: 174–190.

Maisey, J. G. 2001a. CT-Scan Reveals New Cranial Features in Devonian chondrichthyan "*Cladodus*" *wildungensis*. Journal of Vertebrate Paleontology, 21: 807–810.

Maisey, J. G. 2001b. Remarks on the inner ear of elasmobranchs and its interpretation from skeletal labyrinth morphology. Journal of Morphology, 264: 236–264.

Maisey, J. G. 2005. Braincase of the Upper Devonian shark *Cladodoides wildungensis* (Chondrichthyes, Elasmobranchii), with observations on the braincase in early chondrichthyans. Bulletin of the American Museum of Natural History, 288: 1–103.

Maisey, J. G. 2007. The braincase in Paleozoic symmoriiform and cladoselachian sharks. Bulletin of the American Museum of Natural History, 307: 1–122.

Maisey, J. G. 2008. Some observations on *Denaea fournieri* (Chondrichthyes, Symmoriiformes) from the Lower Carboniferous of Belgium. Acta Geologica Polonica, 58: 185–190.

Maisey, J. G. & J. A. Lane. 2010. Morphologie du labyrinthe et évolution de la phonoréception à basse fréquence chez les élasmobranches. Comptes Rendus – Palevol, 9: 289–309.

Maisey, J. G., P. Janvier, A. Pradel, J. S. S. Denton, A. Bronson, R. Miller & C. J. Burrow. 2019. *Doliodus* and Pucapampellids: Contrasting perspectives on stem chondrichthyan morphology. – Pp. 87–109 in Z. Johanson (ed.), Evolution and Development of Fishes.

Manger, W. L. & J. H. Quinn. 1972. Carboniferous dimorphoceratid ammonoids from Northern Arkansas. Journal of Paleontology, 46: 303–314.

Mapes, R. H. 1966. Late mississippian lycopod branch from Arkansas. Oklahoma Geology Notes, 26: 117–120.

Metscher, B. D. 2009. MicroCT for comparative morphology: simple staining methods allow high-contrast 3D imaging of diverse non-mineralized animal tissues. BMC Physiology, 11: 1–14.

Murthy, R., D. Kidder, R. Mapes & R. Hannigan. 2004. Rare-earth element chemistry of Mississippian-age phosphate nodules in the Fayetteville Shale of Oklahoma and Arkansas. Environmental Geosciences, 11: 99–111.

Myrberg, Jr., A. 2001. The acoustical biology of elasmobranchs. Environmental Biology of Fishes, 60: 31–44.

Pradel, A., P. Tafforeau, J. G. Maisey & P. Janvier. 2011. A new Paleozoic symmoriiformes (chondrichthyes) from the late Carboniferous of Kansas (USA) and cladistic analysis of early chondrichthyans. PLoS ONE, 6: e24938.

Pradel, A., J. G. Maisey, P. Tafforeau, R. H. Mapes & J. Mallatt. 2014. A Palaeozoic shark with osteichthyan-like branchial arches. Nature, 509: 608–611.

Sallan, L. C. & M. I. Coates. 2010. End-Devonian extinction and a bottleneck in the early evolution of modern jawed vertebrates. Proceedings of the National Academy of Sciences of the United States of America, 107: 10131–10135.

Tafforeau, P., R. Boistel, E. Boller, A. Bravin, M. Brunet, Y. Chaimanee, P. Cloetens, M. Feist, J. Hoszowska, J. J. Jaeger, R. F. Kay, V. Lazzari, L. Marivaux, A. Nel, C. Nemoz, X. Thibault, P. Vignaud & S. Zabler. 2006. Applications of X-ray synchrotron microtomography for non-destructive 3D studies of paleontological specimens. Applied Physics A: Materials Science and Processing, 83: 195–202.

Tomescu, A. M. F., G. W. Rothwell & G. Mapes. 2001. *Lyginopteris royalii* sp. nov. from the Upper Mississippian of North America. Review of Palaeobotany and Palynology, 116: 159–173.

Yi, H. & M. A. Norell. 2015. The burrowing origin of modern snakes. Science Advances, 1: e1500743–e1500743.

Zangerl, R. 1981. Handbook of Paleoichthyology Volume 3A: Chondrichthyes I, Paleozoic Elasmobranchii (H.-P. Schultze (ed.)). Gustav Fischer Verlag, Stuttgart.

Zangerl, R. 1990. Two new stethacanthid sharks (Stethacanthidae, Symmoriida) from the Pennsylvanian of Indiana, U.S.A. Palaeontographica Abteilung A, Palaeozoologie – Stratigraphie, 213: 115–141.

Zangerl, R., B. G. Woodland, E. S. Richardson Jr. & D. L. Zachry Jr. 1969. Early diagenetic phenomena in the Fayetteville Black Shale (Mississippian) of Arkansas. Sedimentary Geology, 3: 87–119.

Ancient Fishes and their Living Relatives: a Tribute to John G. Maisey. Alan Pradel, John S. S. Denton & Philippe Janvier (eds.): pp. 101–182, 44 figs., 3 tabs., 29 pls.
© 2021 by Verlag Dr. Friedrich Pfeil, München, Germany – ISBN 978-3-89937-269-4

The new family Mesiteiidae (Chondrichthyes, Orectolobiformes), based on *Mesiteia emiliae* Kramberger, 1884.

A contribution to the Upper Cretaceous (early Cenomanian) shark fauna from Lebanon

Friedrich H. Pfeil*

"It is likely that this tooth pattern is primitive for hexanchoids, however (see ...), and species having this pattern are not necessarily congeneric. To avoid confusion and facilitate comparison in any future studies, I propose placing the relatively better known *Notidanus muensteri* in a new genus rather than assuming (as in Pfeil, 1983) it is congeneric with the poorly-founded genus *Eonotidanus*." Maisey 1986.

Yes John, I agree, and *Mesiteia* is a good sample to prove it.

Contents

Abstract

Mesiteia emiliae Kramberger, 1884, is one of the most enigmatic and rarest species of fossil sharks. For more than a hundred years after its first description in 1884 it was thought that the only known specimen was found in the Eocene limestones of Monte Bolca in Italy. In 1980 Cappetta corrected this error and described *Mesiteia* as belonging to the family Hemiscylliidae, together with *Acanthoscyllium* and *Almascyllium*, all three from the Upper Cretaceous limestones from Lebanon. *Mesiteia* is known only from Hakel (Haqil) (early Cenomanian).

New finds of specimens combined with some excellent preparations and an overview on all 26 known specimens, included those from private collections, provides astonishing insights and allows anatomical descriptions of many parts of the skeleton and a review of the teeth. Following Goto (2001), 41 characters were found on *Mesiteia* that were used to analyze the interrelationships of *Mesiteia* to Recent Orectolobiformes. Many unique anatomical characters separate *Mesiteia* from Hemiscylliidae and make it necessary to erect for *Mesiteia* a new family Mesiteiidae nov. fam. This new family is supported by 26 autapomorphic characters such as an elongated low 1st and 2nd dorsal fin, moved in posterior direction, close to the caudal fin; a very long anal fin, wide overlapping the caudal fin; up to 97 series of radials in the extremely long anal fin; differentiated caudal fin, with up to 22 series of preepichordal cartilages, up to 50 series of ciliate epichordal radials, numerous series of prehypochordal cartilages and up to 61 series of sickle-shaped hypochordal processes; and a naked skin with absent, or almost completely reduced dermal denticles. The close relationship of Mesiteiidae nov. fam. to Hemiscylliidae is supported by two synapomorhic characters, the origin of 1st dorsal fin above the end of the pelvic fin, and prehypochordal cartilages fused with the hemal arches. More synapomorphic and plesiomorphic characters support the taxonomic position of *Mesiteia* within the Orectolobiformes.

* Verlag Dr. Friedrich Pfeil, Wolfratshauser Str. 27, 81379 München, Germany; fritz.pfeil@pfeil-verlag.de

In contrast to the idea that fossil sharks could be classified solely on the basis of their tooth morphology, anterior teeth of *Mesiteia* show great similarity with teeth of both the fossil genus *Eostegostoma* and the Recent genus *Pseudoginglymostoma*, which belong to different families, *Eostegostoma* currently to fam. Brachaeluridae and *Pseudoginglymostoma* to fam. Ginglymostomatidae. Posterior teeth of *Mesiteia* rather remind of teeth of the Recent genus *Brachaelurus*. Only a combination of characters allows the determination of teeth of *Mesiteia*, taking into account the differences between teeth of anterior and posterior files. For *Eostegostoma*, I propose that it should be re-allocated together with *Pseudoginglymostoma* to the subfamily Pseudoginglymostomatinae.

1. Introduction

On the eastern edge of the Mediterranean Sea, in the mountains of Lebanon, not far from Byblos, fossils have been discovered and excavated in small limestone quarries for centuries, probably since Herodotus (5th century BC). The first written quotation of the world famous Lebanese fossil quarries was due to Eusebius of Caesarea (4th century AD; Capasso 2017). A huge number of incredibly well-preserved Late Cretaceous plants and animals of a tropical shallow water sea have been collected by local families. Finds that I saw from the fossil deposits of Hakel (Haqil) and Hajula (Hgula) (early Cenomanian; Gayet et al. 2003), of En Nammoura (middle Cenomanian; e.g. Forey et al. 2003) and Sahel Alma (late Santonian; Cappetta 1980) are among the best preserved fossils in the world. Like fossils of the Jurassic limestones of Bavaria, they can be found in all museums around the world, but unfortunately many groups, despite their fame, have so far been poorly scientifically researched.

The brothers Pierre and Albert Abi Saad have excavated fossils in their own quarries and have been exhibiting their carefully prepared finds at the Fossil Shows of Tucson, Sainte-Marie-aux-Mines and Munich for several years. At the Mineral Show 2011 in Munich a specimen of *Mesiteia emiliae* was purchased by the "Friends of the Bavarian State Collection" for the Bavarian State Collection of Palaeontology and Geology (Pfeil 2012). It is excellently preserved, much better than the holotype and the few other specimens kept in museums and, in addition, the history of the scientific description of *Mesiteia* is an exciting story.

During the past eight years I had the chance to collect and study some more specimens of *Mesiteia* and I found some strange characters that are unique within not only the family Hemiscylliidae but all sharks; this led me to the question of whether *Mesiteia* can remain within the family Hemiscylliidae.

2. The first descriptions of *Mesiteia*

The genus and only species of *Mesiteia* was first described by Dragutin Gorjanović-Kramberger (1856–1936) (Fig. 1), a Croatian geologist, palaeontologist and archaeologist who began his studies in palaeontology in Zurich, then moved to Munich to Karl Zittel and received his doctorate in 1879 in Tübingen with a thesis on fossil fishes. Beginning in 1880 Kramberger was curator of the Mineralogical Collection of the Croatian National Museum in Zagreb. In 1884 he published – in the Croatian language – a paper entitled "Palaeoichtyolozki prilozi" on fish "from the island Lesina in Dalmatia, Kutschlin and Warnsdorf in Bohemia, Sagor in Krain, Trifail in Styria, Szakadát in Transylvania, as well as fish from Monte Bolca in Italy, Valjevo in Serbia, and finally in the appendix – those from Podsused in Croatia" (Gorjanović-Kramberger 1886). In this conglomeration of different fishes and localities is the description of a new genus *Mesiteia*, its only species *Mesiteia emiliae*, represented only by its holotype from the Eocene limestones of Monte Bolca (Fig. 2). It is the only shark described in Kramberger's publication (pp. 44–47, plate III, fig. 2, with a general overview, and some details in figs. 2a,c). Kramberger attributed it to the family "Scylliida", to the cat sharks (Scyliorhinidae), because of the generally slender shape, the shape and length of the tail and the low, elongated shape and position of the two dorsal fins. It is unknown how this fossil reached Zagreb, and the etymology of the species and genus name remains hidden.

Fig. 1. Dragutin Gorjanović-Kramberger at the Krapina Museum *Homo neanderthalensis* excavation site. From Wikimedia Commons: Divna Jaksic – https://commons.wikimedia.org/w/index.php?curid=3172086.

Two years later, in 1886, it was possible for the first time in Croatia to publish foreign-language articles in the journal of the Croatian "Naturforscher-Vereines" Zagreb. Kramberger took the opportunity and summarized his "Palaeoichtyologische Beiträge" published in 1884 in German language a second time, on pp. 136–137 (and without figures) in a highly abbreviated translation of his former description in the Croatian language. The first description was already very detailed, as far as it was possible at that time and considering that the holotype is only 16 cm long. The rear part of the tail is missing after the 2nd dorsal fin. Otherwise the specimen is well preserved and Kramberger could recognize details, e.g., 93 + x vertebrae, 24 rays on the 1st, 27 + x rays on the 2nd dorsal fin, over 60 rays on the anal fin, 19 each on the ventral fins and 16 on the pectoral fins. Remarkably, he also mentioned "At the spinal column there are still elongated, cylindrical arches and on top of these again smaller intermediate arches". Later I will show these structures in detail on better preserved specimens (Pls. 13–14). About the shape of the teeth, which are so important for the determination of fossil (and Recent) sharks, he could only report "Teeth small, pointed, in several rows (3 [?])". With a head width of only 21 mm one can imagine how tiny the teeth are and it was impossible at that time to see and describe such small teeth properly. He could not really prove the presence of dermal denticles but he did not see any; therefore, he wrote "body probably naked". Kramberger named the geological collection of Agram as the place of storage.

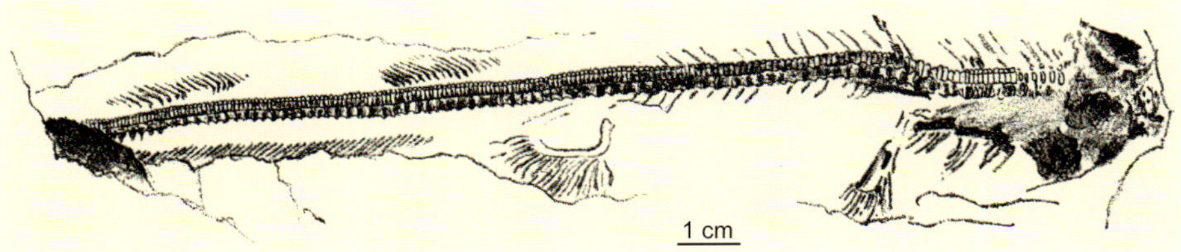

Fig. 2. The holotype of *Mesiteia emiliae* Kramberger, 1884, copied from Gorjanović (1884), from the Monte Bolca monograph by Jaekel (1894, fig. 39).

Shortly afterwards, in 1889, Arthur Smith Woodward's standard work "Catalogue of the fossil fishes in the British Museum, Pt. I Elasmobranchii" was published. Woodward added another species to Kramberger's *Mesiteia*, the catshark (as thought at that time) species "*Scyllium sahel-almae*", now as "*Mesiteia* sahel-alm" (Pictet & Humbert 1866). Pictet & Humbert had described this species in 1866 from the Upper Cretaceous of "Sahel Alma, Mount Lebanon, Syria". The "ae" in "sahel-almae" as correctly spelled in Pictet & Humbert 1866 was probably lost during printing. Woodward cited Kramberger's description of the year 1885 and also the page number is indicated as "p. 54". He did obviously not know the first description of 1884. Woodward believed the middle Eocene age of *Mesiteia emiliae* and the type locality as Monte Bolca and he knew no other specimen. Within his generic description of *Mesiteia* Woodward (1889, p. 345) also mentioned the absence of dermal denticles. One year earlier, Woodward had described in a short article the lateral line of "*Scyllium sahel-almae*" (1888, p. 127), and so he noted (1889, p. 345) as a generic character for *Mesiteia*: "... lateral line supported by a series of calcified incomplete rings." Probably he thought the calcified "long, cylindrical arches" that Kramberger had noticed above the vertebrae, and which are the dorsal arcualia, would be the "calcified incomplete rings", that he found in "*Scyllium sahel-almae*", and considered to be supporting rings of a lateral line organ. Otherwise it is not understandable how this character became part of the generic diagnose of *Mesiteia*. Woodward described the teeth of *Mesiteia emiliae* as "minute, numerous, tricuspid".

Otto Jaekel's monograph "Die eozänen Selachier vom Monte Bolca. Ein Beitrag zur Morphologie der Wirbelthiere" was published in 1894. Of course, Jaekel needed to describe

Mesiteia emiliae Kramberger in this monograph because he believed it was found at Monte Bolca in Italy. As he wrote, he had not seen the holotype. Jaekel did not mention the year of its first description (1884), but as page number of the first description he mentioned "pag. 54 (45)". The page number "(45)" in brackets proves that he knew the first publication of 1884. Since Jaekel knew no other specimen of *Mesiteia* either – neither from Monte Bolca nor from elsewhere – he actually only cited the complete text from Kramberger's German translation of 1886 and also copied the overview illustration from the original paper of Kramberger (1884 or 1885). He added the remark "that the genus *Mesiteia* represents a type of sharks of the Upper Cretaceous and the Lower Tertiary which occupies an isolated position in several respects. The complete lack of dermal denticles and the occurrence of a lateral line, which is – following the description of Woodward – recognizable through closely spaced calcified rings, separates this kind of sharks from all younger relatives. On the other hand, the skeletal anatomy, namely the shape of the tail and the position of the two dorsal fins, so much reminds of the corresponding conditions of the Scyllidae that according to the present state of our knowledge it is probably justified that this type of sharks should be added to it." (Jaekel 1894: pp. 175–176; translated from German). Jaekel had accepted the outstanding position of both "*Scyllium sahel-almae*" and *Mesiteia emiliae* united in the genus *Mesiteia*, based mainly on the complete lack of dermal denticles and the (supposed) occurrence of a lateral line also in *Mesiteia emiliae*, and argued, based on the shape of the tail and the position of the two dorsal fins, that the genus *Mesiteia* should belong to catsharks (Scyllidae). The extremely elongated anal fin in *Mesiteia emiliae* was not seen as an important difference, and the caudal fin with its striking anatomical differences to all other sharks was missing on the holotype of *Mesiteia emiliae*.

Mesiteia was forgotten during the following years. It was too rare to appear in collections and finds of fossils from Monte Bolca were considered suspect anyway because of too many professional forgeries.

Only Jacques Herman (1977) worked in detail on the genus *Mesiteia* in his monograph "Les Sélaciens des terrains néocrétacés & paléocènes de Belgique & des contrées limitrophes Eléments d'une biostratigraphie intercontinentale". Herman discovered that Kramberger's specimen was kept in the Geological Museum in Zagreb and received a photo of the holotype. Even if no details of the teeth could be seen on this photograph, Herman was able now to confirm the affiliation of Woodward's (1889) Upper Cretaceous species "*Mesiteia sahel-almae*" to the genus *Mesiteia*. He continued to accept Monte Bolca as the type locality of the genotype *Mesiteia emiliae* Kramberger, 1885. The tooth morphological examination of a further specimens of "*Mesiteia sahel-almae*", which were kept in the museum in Paris, led him to place *Mesiteia* not in the family Scyliorhinidae (catsharks) as before, but in the Orectolobidae sensu lato (nurse sharks). The easily recognizable characteristics of orectolobid teeth, which I cannot go into here in detail, were sufficient to justify this significant change of the previous classification of the genus. Herman saw the reason for the rarity of evidence of the genus *Mesiteia* in the small size of its teeth: teeth of *M. emiliae* reach only 0.5 mm, those of "*M. sahel-almae*" hardly exceed 1 mm. From bulk samples of Belgian sites, teeth of several species from 1 to 2.5 mm in size were known, all of which had previously been described as belonging to the genus *Ginglymostoma*: *G. daimeriesi*, *G. ypresiensis* and *G. bruxelliensis*. Herman now included them all in the genus *Mesiteia*, with "?", because the teeth of the genotype *M. emiliae* were still not known in detail at that time. *Scyllium humboldtii*, described by Reuss (1845) from the Cretaceous of Bohemia, which had been placed until then with the cat sharks, was now also switched by Herman to *Mesiteia*. The same applies to the nurse shark *Brachaelurus greeni* described by Cappetta (1973) from the USA and a species described by Estes (1964) as *Lonchidion selachos*, also from the USA, which Herman (1977) renamed to the new species *Mesiteia? estesi*. Thus, according to Herman, the distribution area of *Mesiteia* expanded from Lebanon via Italy, Bohemia, Belgium, and France to the USA and stratigraphically from the lowest Upper Cretaceous (Cenomanian) to the lower middle Eocene.

Fig. 3. Above (**a**) the specimen (MNHN.F.HAK513; ex coll. Arambourg 1946-17-331) of a *Mesiteia emiliae* from Cappetta (1980, plate 14, fig. 1), which Cappetta discovered at the Muséum national d'Histoire naturelle in Paris. Below (**b**) the holotype of *Mesiteia emiliae* Kramberger, 1884 on the photo of the Museum Zagreb, together with *Diplomystus brevissimus* (Blainville), from Cappetta (1980, plate 14, fig. 2). – Scale 5 cm each. Reprint with kind permission of Dr. A. Nägele, E. Schweizerbart Verlag.

But then in 1980, Henri Cappetta's monograph "Les Sélaciens du Crétacé supérieur du Liban. I. Requins" was published and everything that had been written before about *Mesiteia* changed. At that time, palaeoichthyologists in Belgium (Herman), England (Ward), France (Cappetta) and Germany (Pfeil) began to pay more attention to the teeth of small species, by washing, sieving and selecting even the smallest fractions from 0.3 mm upwards. At the same time palaeontologists began to understand that without a better knowledge of the dental morphology of comparable extant species, it would not be possible to progress. Cappetta compared for the first time in his Lebanon monograph teeth of extant species, whose names hardly anyone had heard before, with the teeth of the small and smallest fossil sharks found in Lebanon. A single tooth, cleaned and drawn at 30× magnification and/or photographed with a SEM, allowed the reliable identification of a genus and species. Thus, for the first time, the families of the order Orectolobiformes could be distinguished based on their tooth morphological characters. *Mesiteia* was according to Cappetta (1980) no longer a genus of the family Orectolobidae but was now placed by him in the family Hemiscylliidae, the bamboo sharks with the two recent genera *Hemiscyllium* and *Chiloscyllium*. The fossil sites in Lebanon proved to be highly interesting for the detection of early forms of Hemiscylliidae. Cappetta created a new genus for "*Mesiteia sahel-almae*" Woodward, 1889, so that this species was now called "*Acanthoscyllium sahel almae*" (Pictet & Humbert 1866). He also added another new genus *Almascyllium* with "*Almascyllium cheikh-eliasi*" (Signeux, 1949) to Hemiscylliidae ... and then there was a surprise. In the palaeontological collection of the Muséum national d'Histoire naturelle in Paris Cappetta found a plate with a small shark from the Cenomanian of Hakel in Lebanon, which seemed to be very similar to *Mesiteia emiliae* (from the middle Eocene of Monte Bolca). On request to the museum in Zagreb, Cappetta received a photograph and a rock sample of the original plate of the holotype (Fig. 3b). Surprisingly, the photo showed the holotype of *Mesiteia emiliae* as well as a teleostean fish identified by Patterson as *Diplomystus brevissimus* (Blainville), a species frequently found at Hakel but never at Bolca. The microfossils found in the rock sample of the holotype's original

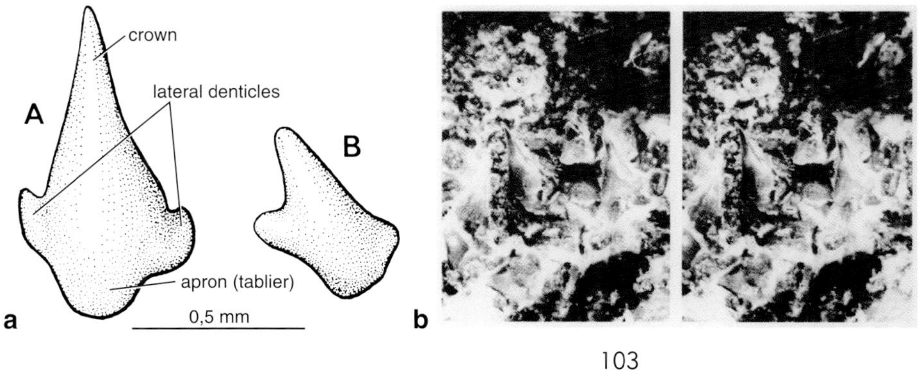

Fig. 4. a. A, anterior and **b.** lateral tooth of the lower jaw, both from the labial side, of the specimen (MNHN.F.HAK513; ex coll. Arambourg 1946-17-331) of *Mesiteia emiliae* from Cappetta (1980: p. 122, fig. 25) which was found at the Muséum national d'Histoire naturelle in Paris. **b.** stereo-pair photos of anterior teeth of the same specimen from Cappetta (1980: pl. 13, fig. 2). – Reprint with kind permission of Dr. A. Nägele, E. Schweizerbart Verlag.

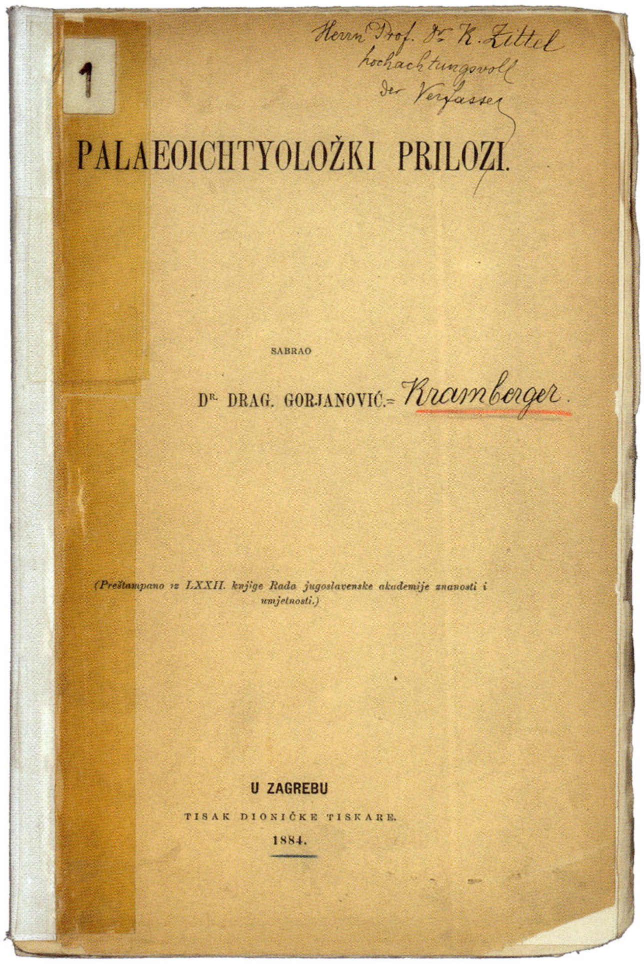

page 44 44 → D. GORJANOVIĆ,

• • •

V. Kučak iz M. Bolke u Italiji.

U našoj geologičkoj zbirci čuva se liep malen kučak, koji potiče iz M. Bolke, bogatoga nalazišta terciarnih riba eocenske stvore.

S razloga, što ova riba reprezentira do sada nepoznati rod, bijaše nuždno ju opisati, tim više, što je do danas još malo tako dobro sačuvanih srodnih riba nadjeno. Nadam se s toga, da će ovaj makar i maleni prinesak pružiti dosta zanimivih podataka tačnijemu poznavanju izumrlih kučaka.

Squalidea.

Fam. **Scylliida.**

→ Nov. gen. Mesiteia Kramberger.

Sravnimo li našega kučka sa živućimi i izumrlimi rodovi familije *Scylliida*, tad ćemo naći, da je on još najsrodniji genusu *Scyllium*,

• • •

page 45 PALAEOICHTYOLOŽKI PRILOZI. 45

• • •

kami po jedna krepka, granulirana bodlja (?). — Nosiljke tračica neparnih splovaka sa tračićem spojene. — Tielo valjda golo.

→ Mesiteia Emiliae Kramb.
(Tab. III., Sl. 2., a, b, c).

Duljina tiela		160 $^{m}/_{m}$
Širina glave	c.	21,$_{10}$ „
Udaljenost prve hrbtene splovke od kraja glave		95,$_{15}$ „
„ druge „ „ „ prve „		36,$_{10}$ „
„ trbušnih splovka od kraja glave		80,$_{15}$ „
„ analne splovke „ „ „		106,$_{15}$ „
Od prsnih splovka do trbušnih		41,$_{10}$ „
Od trbušnih splovaka do analne		25,$_{15}$ „
Širina prve dorsalne splovke		17,$_{18}$ „
„ druge „ „		28+x $^{m}/_{m}$
„ analne splovke		42+x „

• • •

page 10

Palaeoichtyoložki prilozi.

→ SABRAO DR. DRAG. GORJANOVIĆ-KRAMBERGER.

Priobćeno u sjednici matematičko-prirodoslovnoga razreda jugoslavenske akademije znanosti i umjetnosti dne 1. ožujka 1884.

Godinē 1882. poslalo me je ravnateljstvo mineralogičko-geologičkoga muzeja u *Beč*, da u tamošnjih zbirkah pokupim potrebni

• • •

page 53 PALAEOICHTYOLOŽKI PRILOZI. 53

• • •

V. Kučak iz M. Bolke u Italiji.

U našoj geologičkoj zbirci čuva se liep malen kučak, koji potiče iz M. Bolke, bogatoga nalazišta terciarnih riba eocenske stvore.

S razloga, što ova riba reprezentira do sada nepoznati rod, bijaše nuždno ju opisati, tim više, što je do danas još malo tako dobro sačuvanih srodnih riba nadjeno. Nadam se s toga, da će ovaj makar i maleni prinesak pružiti dosta zanimivih podataka tačnijemu poznavanju izumrlih kučaka.

Squalidea.

Fam. **Scylliida.**

→ Nov. gen. Mesiteia Kramberger.

Sravnimo li našega kučka sa živućimi i izumrlimi rodovi familije *Scylliida*, tad ćemo naći, da je on još najsrodniji genusu *Scyllium*,

• • •

page 54 54 → D. GORJANOVIĆ,

• • •

→ Mesiteia Emiliae Kramb.
(Tab. III., Sl. 2., a, b, c).

Duljina tiela		160 $^{m}/_{m}$
Širina glave	c.	21,$_{10}$ „
Udaljenost prve hrbtene splovke od kraja glave		95,$_{15}$ „

page 123

Palaeoichtyologische Beiträge.*

von

→ **Dr. Drag. Gorjanović-Kramberger.**

Es möge mir gestattet sein, in Kürze meinen verehrten Herren Fachgenossen, die der kroatischen Sprache nicht mächtig sind, einen

• • •

* »Rad« der südslavischen Akademie der Künste und Wissenschaften — Band LXXII. 1884. pag. 10—66. mit 5. Tafeln.

page 136 136

• • •

→ n. gen. **Mesiteia** Kramb. (1884.)

→ Gorjanović-Kramberger: „Palaeoichtyoložki prilozi" „Rad" etc. pag. 53 [44.]

Diese Gattung habe ich für einen kleinen Haifisch, welcher aus den eocenen Bildungen des Mt. Bolca in Italien herrührt, aufgestellt.

• • •

page 137 137

• • •

→ **Mesiteia Emiliae** Kramb.
l. cit. pag. 54. [45.]., Tab. III. Fig. 2. 2. a. b. c.

Dieser Fisch ist blos 160$^{m}/_{m}$ lang (es fehlt etwas vom Schwanzstiele); die Kopfbreite beträgt c. 21$^{m}/_{m}$. Die Wirbelsäule zählt 93 + *x*

• • •

Fig. 5. The different use of the author's surname in the three publications of Gorjanović-Kramberger with the description of *Mesiteia* and *Mesiteia emiliae*.
a. On the cover page of the first publication of Gorjanović (1884) the author's surname was printed as "Dr. Drag. Gorjanović" but someone added by hand "- Kramberger". This copy in the library of the BSPG in Munich has a personal handwritten dedication of the author to Prof. Dr. K. Zittel and therefore it must have been the author himself who changed his name on that copy to "Gorjanović-Kramberger". Karl Kramberger had changed his name officially in 1882 to Dragutin Gorjanović-Kramberger.
b. On page 44 of the first publication is a header "D. Gorjanović" according to the author's name printed on the cover page. As usually such headers are added by the publishers; the author might not even had seen this before printing. It is much more interesting that the author wrote (on this same page) the name "Kramberger" behind the name *Mesiteia* when he first mentioned it in his title line. And later, on page 45 with the description of the new species *Mesiteia emiliae* one can see "Kramb." as an abbreviation for "Kramberger".
c. On the first page of the print of the magazine in 1885, there is on page 10 in the line with the authors name "Sabrao Dr. Drag. Gorjanović-Kramberger", and on page 53 the name "Kramberger", following the generic name *Mesiteia* and on page 54 again the abbreviation "Kramb.".
d. In the German translation of 1886, there is on page 123 in the line with the authors name "Dr. Drag. Gorjanović-Kramberger" and on page 136 the abbreviation "Kramb." for the name "Kramberger" following the name *Mesiteia*. Here the author added the year of the first publication "(1884)". In the following line the author cited his first publication and used the name "Gorjanović-Kramberger" and mentioned the page number 53 (of the publication in 1885) with page number 44 (of the first publication in 1884) only in parentheses. Similarly on page 137 he used the abbreviation "Kramb." and in the following line, he referred to his first publication and mentioned the page number 54 (of the publication in 1885) with page number 45 (of the first publication in 1884) only in parentheses.

plate showed an age of Albian or Cenomanian, not Eocene. Finally, it was proven beyond doubt that *Mesiteia emiliae* did not originate in the middle Eocene lagoon of Monte Bolca, as had been thought for a hundred years, but from the Upper Cretaceous limestones of Hakel, in the mountains of Lebanon. Kramberger had studied fishes from Bolca and Lebanon at the same time and presumably the labels were exchanged. For the sake of completeness, it should be mentioned that Cappetta in his publication on sharks from Lebanon also described another family of Orectolobiformes, the Parascylliidae with the genus

Pararhincodon Herman in Cappetta, 1976 and a new species *Pararhincodon lehmani* Cappetta, 1980.

So, there was a second specimen of *Mesiteia emiliae* in Paris (Fig. 3a). Cappetta described it in detail, but unfortunately the illustrations are not very meaningful. It is somewhat more complete than the holotype. On plate 14 in Cappetta (1980) it is illustrated together with the photo from Zagreb and plate 13, fig. 2 shows a stereo image of a 23× enlarged section of the anterior part of the dentition, in – as Cappetta himself wrote – very bad condition (Fig. 4b). I cannot go into the many anatomical details here, so I will confine myself to a few remarks about the teeth that are the key for the determination and assignment to related species. Cappetta (1980) showed drawings of teeth of the two Recent genera of Hemiscylliidae: *Hemiscyllium ocellatum*, *Chiloscyllium punctatum* and *Chiloscyllium indicum*. The teeth of the fossil genera *Acanthoscyllium*, *Almascyllium* and *Mesiteia* described by him are so different between and among them that I cannot understand that these three fossil genera should belong to the family Hemiscylliidae. Cappetta also suggested the possibility that they might be placed in different families. Whereas the teeth of *Acanthoscyllium* and *Almascyllium* are very well preserved, unfortunately there is not much to see of the shape of the *Mesiteia* teeth of both the holotype and the specimen in Paris. Only the labial sides (outer sides) of some 1 mm high, sharply pointed anterior teeth are visible, as well as a few about half as high lateral teeth inclined towards the corner of the mouth. Nothing can be seen from the lingual side or the root with its many features. It can be observed that the teeth are labially smooth, sharply pointed, drop-shaped, and convex in cross-section, with a well-developed rounded apron (tablier) at the base of the crown (Fig. 4a) and a very low lateral cusplet on both sides of the crown. However, these characteristics alone clearly distinguish them from *Acanthoscyllium sahelalmae*, whose crowns show labially a high base, with a very wide apron that is not set off as in *Mesiteia*, with an extremely pointed low main principal cusp and extremely high and pointed lateral cusplets on both sides. The teeth of *Almascyllium cheikheliasi* are low, with a wide apron and a concave labial surface, which above all show clear ornamentation in the form of several folds running from tip to base. It must be mentioned at that point that all the other species that were placed by Herman (1977) in *Mesiteia*, now had to be re-placed again in other genera:

Mesiteia daimeriesi	to *Chiloscyllium daimeriesi* (Herman, 1973)
Mesiteia ypresiensis	to *Protoginglymostoma ypresiensis* (Casier, 1946)
Mesiteia bruxelliensis	to *Hemiscyllium bruxelliensis* (Herman, 1977)
Mesiteia humboldtii	to *Chiloscyllium humboldti* (Reuss, 1845)
Mesiteia greeni	to *Chiloscyllium greeni* (Cappetta, 1973)
Mesiteia estesi	to *Protoginglymostoma estesi* (Herman, 1977)

This list is completed by *Mesiteia* ? sp., described by Thies (1989) from the Middle Jurassic (Bajocian) of northwestern Germany. Its assignment to *Mesiteia* seems to be questionable because of its age compared to the finds from the Upper Cretaceous of Lebanon and because of a strong striation of the crown surface visible in the upper part of the crown on the labial side. Further findings would be necessary to confirm the assignment to *Mesiteia*. But it should be mentioned that this fauna of Early and Middle Jurassic Orectolobidae described by Thies (1989) with *Palaeobrachaelurus*, *Paraginglymostoma*, *Palaeorhincodon*, *Annea* and other Orectolobiformes incertae sedis could include relatives of the ancestors of *Mesiteia*, simply because of the high diversity of small Orectolobiformes with teeth "similar" to *Mesiteia*. The relationships of these small Orectolobiformes with a long apron and a principal cusp flanked by one or two pairs of lateral denticles to each other and to Brachaeluridae but also *Mesiteia* are at the moment not resolved.

In summary, after Cappetta's Lebanon monograph one can state:

- *Mesiteia emiliae* Kramberger, 1884 does not originate from the middle Eocene of Monte Bolca, but from the early Cenomanian of Hakel in Lebanon.
- *Mesiteia emiliae* Kramberger, 1884 is the only species of the genus *Mesiteia*. It does not belong to the cat sharks, family Scyliorhinidae, but to the bamboo sharks, family Hemiscylliidae. The latter assignment, however, is with reservation, as will be explained below.

The authorship of *Mesiteia emiliae* Kramberger, 1884 has been cited differently by the relatively few revisors as: Kramb., 1884, Kramberger, 1884, Kramberger, 1885 but also Gorjanović-Kramberger, 1885. In Kramberger's three publications the author's name is printed differently but the name of the new genus *Mesiteia* is always connected with the author's name "Kramberger", and the abbreviation "Kramb." always follows the name of the new species *Mesiteia emiliae* (Fig. 5). Of course, the original work can only have been published in one of these years. My colleague Markus Moser (SNSB-BSPG München) has taken the trouble to view the relevant publications of Gorjanović-Kramberger and helped solving this puzzle:

- In 1884 a pre-print for the author with its own separate pagination was published in the Croatian language: pp. 1–56, plates 1–5. The beginning of the description of the new genus *Mesiteia* falls on page 44, and the description of the new species *Mesiteia emiliae* on pages 45–47. The illustrations are in plate III, fig. 2a–c.
- In 1885 the actual print of the magazine for the Society with new pagination was published in the Croatian language: pp. 10–65, plates 1–5. The beginning of the description of the new genus *Mesiteia* falls here on page 53, and the description of the new species *Mesiteia emiliae* on pages 54–56. The illustrations are on plate III, fig. 2a–c.
- In 1886 a shortened translation was published in the German language with the description of the new genus *Mesiteia* on page 136 and the species *Mesiteia emiliae* on page 137 and without illustration.

According to the International Code of Zoological Nomenclature, the pre-print from 1884 is a valid publication with its own publication date, separate from the actual print. See ICZN Art. 21 No. 8 (http://www.nhm.ac.uk/hosted-sites/iczn/code/index.jsp?article=21&nfv=true#8). Therefore, 1884 is the correct "date of birth" of the genus *Mesiteia* Kramberger and the species *Mesiteia emiliae* Kramberger.

Hans-Peter Schultze helped me to find the correct use of the genus and species name and draw my attention to an article published by Frayer & Radovçic (2008) in the "New Dictionary of Scientific Biography". According to this, in 1882 Kramberger, of German heritage in the Austro-Hungarian Empire, dropped "Karl" from his name and added "Gorjanovic" to reflect Croatian solidarity and generally signed his scientific papers Dragutin Gorjanovic-Kramberger afterward.

3. The fossil record of *Mesiteia*

Mesiteia emiliae is only known from Hakel (Haqil) in Lebanon, which is located in North Lebanon between Tripoli and Beirut, about 40 km N of Beirut and about 10 km NE of the old Phoenician settlement Byblos or in Arabic Jbeil (Figs. 6a, b).

Two families are excavating fossils in two excavation sites on the north and south flanks of a small valley E of Hakel. The family Abi-Saad runs during the summer months a small quarry on the north slope (Fig. 6c) and the family Nohra on the south side (Fig. 6d). For three generations both families have sold fossils to tourists and fossil collectors and the incredibly rich and diverse fauna and flora is represented in many museums worldwide but still almost unknown to many scientists.

The situation in Lebanon is similar to what happened with the fossils from the Plattenkalk localities of Solnhofen and Eichstaett. In Lebanon, there is no official state collection where type material could be preserved and the old fossil collection at the American University of Beirut is in a very bad situation. The little local museum at Byblos is closed. The same as in

Fig. 6. a. Geographic and stratigraphic position of the main Cretaceous fossils bearing localities in Lebanon; Hakel > early Cenomanian; En Nammoura, Hgula and Kousba > middle Cenomanian; Sahel Alma > late Santonian. **b.** View from the Abi-Saad excavation site down to the village Hakel. **c.** The Abi-Saad quarry on the north slope of the valley. Some of the fossil layers are incredibly rich and can e.g. contain several specimens of different sharks and rays. **d.** The Nohra excavation site on the south side. **e.** folded and relatively thick limestone layer from the Abi-Saad quarry. **f.** paper-thin "Plattenkalk" from the same site.

Bavaria happens with the fossils from Lebanon: the most beautiful, most interesting and of course some of the most expensive fossils were sold to rich dealers and collectors everywhere in the world and many of them are lost for scientific studies. The local families support scientists wherever they can, but these families deserve professional training in how to excavate fossils, how to collect exact stratigraphic data, and how to do the preparation in a professional way. In Lebanon the public interest in fossils is extremely high as can be seen every year when Pierre Abi Saad organizes an exhibition at the MIM museum at Saint Joseph University in Beirut and nine thousand visitors come to see the fossils during the one "night of museums".

The age of the limestone (Plattenkalk) of Hakel is early Cenomanian (Late Cretaceous), a little bit older than the nearby middle Cenomanian localities En Nammoura, Hgula and Kousba or Sahel Alma, which is late Santonian (Fig. 6a).

The preservation of all fossils from Hakel is excellent and the fossils are usually complete and not disarticulated. But unfortunately, the fossils break apart to a left and a right side of the plate and right in the middle of the vertebrae of fishes. Much damage is done when the plates are split, and the best preparation would not help when to many fragile parts of the fossil get lost in the quarry.

Most shark fossils from Lebanon are extremely interesting but it is also frustrating to see the damage due to not always the best preparation, especially of specimens in the older collections. Taking into account that every specimen is unique in its position of fossilisation, in its age and sex, one can imagine that for an understanding of a species more than just one or two specimens should be available for studies. Every specimen helps to give a few more insights to a fossil species.

I had the chance to study a specimen of *Mesiteia emiliae* (*Mesiteia* #04) that was bought at the Munich fossil show in 2011 from Pierre Abi-Saad by the "Friends of the BSPG" for the BSPG in Munich. Thanks to the careful micro-preparation by Renate Liebreich, details of teeth, details of the spinal column, and size and position of all fins, especially the unusually long anal fin and the sickle-shaped rays of the caudal fin became visible. Helmut Tischlinger helped generously and patiently with dozens of long-wave UV pictures, a technique that is indispensable to study such fossils. These first results extended the relatively poor knowledge based before only on the holotype and a second specimen at the MNHN in Paris (Cappetta 1980) and were published in Pfeil (2012).

After purchasing and studying (Pfeil 2013, 2014) two *Cantioscyllium decipiens* Woodward, 1889 specimens for the BSPG at the Mineral Show in Munich (a male in 2012, and a female in 2013), I visited the fossil localities and collections in Lebanon in November 2013. Altogether approx. 300 originals and in addition some pictures of previously sold sharks from Hakel and Hjula could be documented. With the help of Pierre Abi-Saad and Roy Nohra it was possible to purchase some excellently preserved specimens of *Mesiteia emiliae* for my private collection (FHP).

This study would not have been possible if only the few specimens in state collections were figured and described. The holotype in Zagreb and the two specimens in Paris and Munich are incomplete and most of the characters described here are not preserved on these specimens. Therefore, I used mainly the specimens of my own private collection. They are available for further research by contacting my publishing house at any time until they are deposited in an official state collection. The specimens of the collections Abi Saad and Nohra can be seen at the private museums of these families in Lebanon, as can the specimen that is kept at the private museum of René Kindlimann in Switzerland. It seemed to me very important also to figure those other specimens in private collections that are currently not available for study. Otherwise even the few existing pictures could get lost. The pictures document that most of these few specimens whose owner is currently unknown are not as completely preserved that there would be an urgent need to find them. Most likely my list of specimens is not complete, and I hope that due to this paper, one or more specimens in an old collection of fossils from Lebanon may be found. Indeed, two specimens, *Mesiteia* #25 and #26, were discovered in old collections of the Abi Saad family in Lebanon during this study. I checked many of the official collections; some kept specimens that were incorrectly identified, but I found no further specimens. *Mesiteia* always was and still is very rare and it may take many years until a new specimen will be found.

Fossils in private collections matter. Not only would it be ignorant not to study fossils that are kept in private collections – often better than in state collections – it would be highly unscientific. Without exaggeration, centuries of experience with fossils from the Jurassic limestones of Bavaria show that without the private initiative of enthusiastic collectors, only a few and most likely insignificant specimens would have become known. What good is it to work on poorly and incompletely preserved specimens if better preserved ones are available in private collections and can be studied scientifically? Experienced curators know that sooner or later almost every important specimen will end up in any state collection. Calls, such as those currently circulated by institutions like the SVP, that paleontologists should not work on fossils that are in private collections, or that publishers should not publish manuscripts in which fossils from private collections have been utilized, or the strict requirement to publish only in certain magazines, are unacceptable censorship. Such paternalism and attempted repressions are not in the spirit of free independent research. The argument that published fossils must be available to everyone at any times is only a pretext. Why should they not be available when kept in a private collection? I am more concerned about access to study if they are kept in a state collection. I have been collecting fossils for more than 65 years, and I have not yet met a collector who would not make his finds available for serious scientific studies at any time. Unfortunately, I have sometimes experienced that scientists have been so arrogant towards collectors that they have refused access to their collections. Certainly, there are exceptions, but problems can be solved through respectful treatment. It must be the common concern of collectors and scientists to protect fossils and fossil deposits. This will not work through bans and controls.

All 26 specimens of *Mesiteia emiliae* known to date are summarized here (Plates 1 to 6), together with pictures of a few specimens that were previously sold and of which the current whereabouts are unknown, one Internetfind, and a specimen of the collection Kurt Henne in Stuttgart, figured in Frickhinger 1991. This overview is one of the main concerns of this study. It allows later researchers to quickly identify which specimens should be used for further investigation and which ones can be omitted because certain details are not preserved. Of some specimens that were previously sold from the collections in Lebanon to unknown private collectors, only the pictures presented here have survived, but they do show details that would be worthy of further study, as I will explain later. The pictures are a documentation that cannot be renounced. They also serve to pay respect to the former and present owners of these fossils who kindly allowed the use of the pictures.

Table 1. List of all *Mesiteia* specimens known to date.

specimen number	place of collection, inventory number	comments	here figured in
specimens kept in official collections			
#01	Zagreb, Croatian Natural History Museum, inventory number 9354.	holotype	Figs. 2, 3b, 31c, 38a–b; Pls. 1(1), 10a, 14a
#02	Paris, MNHN.F.HAK513	coll. Arambourg 1946-17-331	Figs. 3a, 4a–b; Pls. 1(2), 14b
#03	Paris, MNHN.F.HAK559	coll. Beyrouthy 1990-4-4	Pls. 1(3), 18f
#04	Munich, SNSB-BSPG 2011 I 124	ex coll. Abi Saad	Fig. 30; Pls. 1(4a–b), 7(1a–c), 11e, 14c–d, 15e–f, 18d–e
#05	Munich, SNSB-BSPG 2013 XXVII 1	ex coll. Ulrich Leonhardt (ex coll. Abi Saad)	Figs. 33b–c; Pls. 1(5a–b), 7(2a–d), 15c–d, 18a–b
specimens kept in private collections			
#06	Munich, coll. FHP *Mesiteia* #06	ex coll. Abi Saad – dermal denticles preserved	Pls. 1(6a–b), 7(3a–c), 16a–d
#07	Munich, coll. FHP *Mesiteia* #07	ex coll. Roy Nohra – male with claspers, teeth	Figs. 33h, 34e,h, 40; Pls. 2(1), 7(4a–c), 9(2a–b)
#08	Munich, coll. FHP *Mesiteia* #08	ex coll. Abi Saad – best preserved specimen prep. by Pino & Paul Voelkl	Figs. 27a, 28a, 29a, 31a, 32a, 33a, 34a,f–g, 35c, 36a–b, 37g; Pls. 2(2a–c), 7(5a–c), 9(3a–b), 10c, 11d, 12a,d–e, 13f–g, 14f–g
#09	Munich, coll. FHP *Mesiteia* #09	ex coll. Roy Nohra – biggest specimen, muscle structures and lateral line	Figs. 33i, 41; Pls. 3(1a–b), 8(1a–c), 17a–b
#10	Munich, coll. FHP *Mesiteia* #10	ex coll. Roy Nohra – best preserved teeth	Figs. 28b, 31d; Pls. 3(2), 8(2a–b), 9(1a–c), 10e
#11	Munich, coll. FHP *Mesiteia* #11	ex coll. Abi Saad – spiral valvular intestine; probably preservation of heart, coronary arteries, ventral aorta and other blood vessels	Figs. 7, 33e; Pls. 3(3), 8(3a–c), 10g, 18c
#12	Munich, coll. FHP *Mesiteia* #12	ex coll. Abi Saad	Fig. 35d; Pls. 3(4), 11g, 14e
#13	Switzerland, coll. René Kindlimann – RK/7-711	ex coll. Abi Saad	Figs. 27b, 33j; Pls. 4(1), 8(4a–c), 10f
#14	Lebanon, coll. Expo Hakel (Roy Nohra)	branchial arches	Figs. 31f, 33d; Pls. 4(2), 10d
#15	Lebanon, unknown new owner	ex coll., and picture from Roy Nohra	Fig. 33f; Pl. 4(3), 11f
#16	Lebanon, coll. Expo Hakel (Roy Nohra)	cluspers	Fig. 33g; Pl. 4(4), 18g
#17	Lebanon, unknown new owner	ex coll., and picture from Roy Nohra	Pl. 4(5a–b)
#18	Lebanon, unknown new owner	ex coll., and picture from Roy Nohra – multiple narial barbels?, branchial arches	Figs. 31e, 32b; Pl. 5(1), 10b
#19	Lebanon, unknown new owner	ex coll., and picture from Abi Saad	Pl. 5(2)
#20	Lebanon, coll. Abi Saad	figured in Gayet, Abi Saad & Gaudant 2012	Pl. 5(3), 11h
#21	Lebanon, coll. Abi Saad	figured in Gayet, Abi Saad & Gaudant 2012	Fig. 31b; Pl. 5(5)
#22	Munich, coll. FHP *Mesiteia* #22	from Greenstone Fine Mineralia (ex coll. Abi Saad)	Pl. 5(6)
#23	Stuttgart, coll. Kurt Henne, current whereabouts unknown	figured in Frickhinger, 1991	Pl. 6(1)
#24	Internetfind, unknown owner	cranium and branchial region, blood vessels (?)	Pl. 6(2a–b)
#25	Munich, coll. FHP *Mesiteia* #25	ex coll. Abi Saad	Pls. 6(3), 8(5a–b)
#26	Munich, coll. FHP *Mesiteia* #26	ex coll. Abi Saad	Pl. 6(4a–b)

MNHN, Muséum national d'Histoire naturelle, Paris; SNSB-BSPG, Bayerische Staatssammlung für Paläontologie und Geologie, Munich; FHP, Pfeil, Munich; contact info@pfeil-verlag.de.

Plates 1–6

The 26 specimens of *Mesiteia emiliae*. For an easier correlation of the illustrations in this study to the different specimens, all *Mesiteia* specimens are marked as "#01, #02 ...". This correlation is explained in Table 1. It was also used to organize the many thousands of photographs used to prepare this study. For better comparison, all specimens on Plates 1 to 6 are figured at the same scale, with a scale bar of 10 cm.

Plate 1. Overall view of specimens #01 to #06 of *Mesiteia emiliae* Kramberger, 1884.

1. The holotype of *Mesiteia emiliae* Kramberger, 1884, *Mesiteia* specimen #01 (Zagreb, Croatian Natural History Museum (CNHM), inventory number 9354); cranium, branchial region and paired fins in dorsal view, than spine and unpaired fins in lateral view from right side; skull and branchial region are poorly preserved but branchial rays are nicely seen; there is not much to see from pectoral and pelvic girdles; the two dorsal fins can be recognized but don't show much details; only part of the anal fin is preserved; the tail with all details of the caudal fin is missing; some of the vertebra with arcualia and ribs are well preserved; photo Sanja Japundžič.

2. *Mesiteia* specimen #02 (Paris, MNHN.F.HAK513, ex coll. Arambourg 1946-17-331), the specimen that Cappetta (1980) used for his description; anterior parts in dorsal, then in lateral view, as in the holotype; besides the shape of a short and broad cranium, very few details are visible, the tail is missing; the vertebral ribs on monospondylous vertebrae were emphasized in preparation; posterior to the pectoral fins, the spine is twisted by 180°, probably due to a fatal injury; similar injuries can also be seen in other specimens, e. g. specimen #06 (this Plate, Fig. 6); photo Martin Pickford.

3. *Mesiteia* specimen #03 (Paris, MNHN.F.HAK559; ex coll. Beyrouthy 1990-4-4); in this specimen, the cranium is compressed laterally, and it is complicated to interpret any details but the 1st and 2nd dorsal fin elements are relatively well preserved; only a few anal fin rays are preserved, the caudal fin is missing; photo Martin Pickford.

4. *Mesiteia* specimen #04 (Munich, SNSB-BSPG 2011 I 124, ex coll. Abi Saad), described by Pfeil (2012); a complete specimen in lateral view from right side, typically elongated S-shaped with tail turned up, as most other specimens; the cranium is slightly turned to right and compressed, it looks a bit longer and less broad than specimen #02 (MNHN.F.HAK513, this Plate, Fig. 2); unfortunately, as in most specimens, many of the calcified anatomical elements are broken, split in the middle, or otherwise damaged and only the shape of it can be recognized; however, many autapomorphic characters that led to the establishment of a new family for *Mesiteia* in this study, such as the extremely elongated anal fin, the special shape of the caudal fin elements, the loss of dermal denticles, and the morphology of the teeth, very different to Hemiscylliidae, were first visible on this specimen; the use of long-wave UV-light was a great advance for the investigation of *Mesiteia*. **a,** in normal light; **b,** with long-wave UV-light; photo Helmut Tischlinger.

5. *Mesiteia* specimen #05 (Munich, SNSB-BSPG 2013 XXVII 1), originally from the coll. Abi Saad, the specimen was donated from Ulrich Leonhardt to the BSPG; this specimen is slightly thicker than specimen #04 (SNSB-BSPG 2011 I 124), it is preserved in lateral view from left side, squeezed a little dorsally; much is fragmented but there are interesting details on pectoral and pelvic fins. **a,** in normal light; **b,** with long-wave UV-light; photo Helmut Tischlinger.
6. *Mesiteia* specimen #06 (Munich, coll. FHP, ex coll. Abi Saad); this specimen, in lateral view from the right, skull more in dorsal view, is deformed in a manner similar to specimen #02 (MNHN.F.HAK513, this Plate, Fig. 2), the spine is broken posterior to the 1st dorsal fin, and a second time at the level of the pectoral fin; between the pelvic fins there is a small area with very small reduced dermal denticles, described and figured in detail on Plate 16. **a,** in normal light; **b,** with long-wave UV-light; photo Helmut Tischlinger.

Plate 2. Overall view of specimens #07 to #08 of *Mesiteia emiliae* Kramberger, 1884.
1. *Mesiteia* specimen #07 (Munich, coll. FHP, ex coll. Roy Nohra); male in dorso-lateral view from left side; with smashed cranium but many details of teeth of all series (Plate 7, Figs. 4a–c), details of dorsal fins, and exceptionally well preserved claspers.
2. *Mesiteia* specimen #08 (Munich, coll. FHP, ex coll. Abi Saad), by far the best preserved, most complete, and – by Paul & Pino Voelkl – professionally preparated specimen; cranium and anterior region in ventral, body in lateral view from right side; it is used in this study for the description of most of the characters. **a,** the plate with the fossil, as I received it from Pierre Abi Saad; photo Pierre Abi Saad; **b,** after preparation, showing the incredibly nice preservation; **c,** same, with long-wave UV-light; photo Helmut Tischlinger.

Plate 3. Overall view of specimens #09 to #12 of *Mesiteia emiliae* Kramberger, 1884.
1. *Mesiteia* specimen #09 (Munich, coll. FHP, ex coll. Roy Nohra); biggest specimen with a total length of 47 cm; female in dorso-lateral view from left side; in that specimen only, myomeres, myosepta and traces of both lateral lines are preserved (Plate 17 a,b; Fig. 41). **a,** in normal light; **b,** with long-wave UV-light; photo Helmut Tischlinger.
2. *Mesiteia* specimen #10 (Munich, coll. FHP, ex coll. Roy Nohra); cranium and branchial area in ventral view, trunk in lateral view from left side; many interesting details on skull, branchial region and pectoral girdle and fins, posterior part of caudal fin is missing; this specimen is most important because of its extraordinary fine preservation of teeth in all series. This specimen was used for the description of teeth of *Mesiteia emiliae* and for its comparison with other teeth of Orectolobiformes, e.g. Plate 9, Fig. 1a–c.
3. *Mesiteia* specimen #11 (Munich, coll. FHP, ex coll. Abi Saad); in this specimen, in dorso-lateral view from left side, the branchial area is very special; it seems that there is a whitish muscle mass of a two-chambered heart preserved, with dark traces of blood vessels, small coronary arteries, a strong ventral aorta with branchings to the gills and dorsally more, partly smaller blood vessels (Plate 18 c), but this is still subject to further investigation; in that specimen just behind the pectoral girdle, there is also a round coprolite, showing the spiral pattern.
4. *Mesiteia* specimen #12 (Munich, coll. FHP, ex coll. Abi Saad); a specimen in dorso-lateral view from left side with missing cranium but details of the fins.

Plate 4. Overall view of specimens #13 to #17 of *Mesiteia emiliae* Kramberger, 1884.
1. *Mesiteia* specimen #13 (Switzerland, coll. René Kindlimann – RK/7-711); male in ventro-lateral view from right side, partially crushed during fossilization; cranium with some interesting details, e.g. well preserved medial rostral rod and many teeth; right pectoral fin and part of caudal fin with details, claspers are damaged but the spurs are well preserved.
2. *Mesiteia* specimen #14 (Lebanon, coll. Expo Hakel, Roy Nohra); female in ventral view from cranium up to the end of the 1st dorsal fin, than twisted and in lateral view from left side; the most interesting specimen of those that could not yet be studied; it still needs some professional preparation but note the details on branchial arches, e.g. the well preserved branchial rays, also the pectoral and pelvic girdles and fins.
3. *Mesiteia* specimen #15 (Lebanon, ex coll. Roy Nohra, sold to unknown new owner); female in dorso-lateral view from left side; needs some preparation but it has a well preserved left pectoral fin, 1st and 2nd dorsal fins, anal and caudal fin; also the skull could reveal interesting details, but this cannot be judged from this single photo; photo Roy Nohra.
4. *Mesiteia* specimen #16 (Lebanon, coll. Expo Hakel, Roy Nohra); male in ventro-lateral view, most of caudal fin missing; many interesting details on pectoral and pelvic girdle and fins, especially the big claspers are very well preserved, but also the 1st and 2nd dorsal fin, anal and the beginning of the caudal fin should be studied more in detail; the stomach contents (coprolite) in this specimen has an interesting fine structure, different from the other specimens.
5. *Mesiteia* specimen #17 (Lebanon, ex coll. Roy Nohra, sold to Chris Avila); the few available pictures show plate (**a**) and counter plate (**b**) of a female specimen in dorso-lateral view; the posterior part with the caudal fin is missing; skull, pectoral and pelvic girdle and fins are well preserved and could show interesting details; photo Roy Nohra.

Plate 5. Overall view of specimens #18 to #22 of *Mesiteia emiliae* Kramberger, 1884.
1. *Mesiteia* specimen #18 (Lebanon, ex coll. Roy Nohra, sold to unknown new owner); female in dorso-lateral view; the posterior part, posterior to the pelvic fins is missing but cranium, branchial region, pectoral and pelvic girdle and fins are nice preserved and display many interesting details that would be worth studying; in front of the cranium there are many small cartilages that are probably labial cartilages, but this cannot be determined with the few photographs available, it could also belong to remains of some other fossils; the right pectoral and pelvic fins could provide interesting details after more preparation; photo Roy Nohra.
2. *Mesiteia* specimen #19 (Lebanon, ex coll. Abi Saad, sold to unknown new owner); male specimen (?) in dorsal view from cranium up to the pelvic fins, then tail in lateral view from left side; needs preparation but has many interesting details, e.g. on medial rostral rod, pharyngeal arches, pectoral and pelvic fins; this is the only specimen that reflects the outline of this shark in dorsal view in a most realistic way; photo Pierre Abi Saad.
3. *Mesiteia* specimen #20 (Lebanon, coll. Abi Saad); female in dorso-lateral view from left side; this complete specimen has well preserved pectoral and pelvic girdles and fins, 1st and 2nd dorsal fins are complete and show many details; it seems that in the branchial area there are traces of soft part preservation of heart and blood vessels, as seen in *Mesiteia* #14 (Plate 3, Fig. 3), but better pictures would be needed, and only a study of the original fossil could prove this; photo Pierre Abi Saad (also figured in Gayet, Abi Saad & Gaudant 2012, p. 65).
4. This small shark (6 cm total length) is figured on the same plate in Gayet, Abi Saad & Gaudant (2012, p. 65) and identified as *Mesiteia* (**a**), and zoomed (**b**); but this is a juvenile *Pararhincodon* (Lebanon, coll. Abi Saad); photo Pierre Abi Saad.
5. *Mesiteia* specimen #21 (Lebanon, coll. Abi Saad); female in dorso-lateral view from left side; needs preparation, but overall very well preserved specimen with many details of cranium, branchial arches, pectoral and pelvic girdles and fins, 1st and 2nd dorsal fin, anal and caudal fin; photo Pierre Abi Saad (also figured in Gayet, Abi Saad & Gaudant 2012, p. 65).
6. *Mesiteia* specimen #22 (Munich, coll. FHP); female in dorso-lateral view from right side; an attractive plate, *Mesiteia* together with two other fishes, cranium squeezed, unpaired fins with anatomical details.

Plate 6. Overall view of specimens #23 to #26 of *Mesiteia emiliae* Kramberger, 1884.
1. *Mesiteia* specimen #23 (Stuttgart, coll. Kurt Henne, current whereabouts unknown); female in dorso-lateral view from left side; elements of cranium and branchial region missing, but pectoral and pelvic girdles and fins are well exposed and bear interesting details, same as 1st and 2nd dorsal fins; photo from Frickhinger (1991, p. 177).
2. *Mesiteia* specimen #24 (ex coll. Roy Nohra, Internetfind, unknown current owner); male in dorso-lateral view from right side; with interesting details on cranium, branchial region, pectoral and pelvic girdles and fins; the resolution of the internet picture is not good, and the specimen needs professional preparation but it seems there are also traces of heart and blood vessels preserved, as in *Mesiteia* #11 (Plate 3, Fig. 3, Plate 18 c). **a,** specimen composed of two plates that have been incorrectly mounted; **b,** the plates should be fixed like this, showing the typical elongated S-shaped body with tail turned up; photo from Internet: https://i.pinimg.com/originals/5e/85/81/5e85815821b97bee673e238838dd0e09.jpg.
3. *Mesiteia* specimen #25 (Munich, coll. FHP, ex coll. Abi Saad); female, cranium in dorsal, body in lateral view from right side; needs preparation; tail similarly injured as *Mesiteia* #02 (Plate 1, Fig. 2), and *Mesiteia* #06 (Plate 1, Fig. 6).
4. *Mesiteia* specimen #26 (Munich, coll. FHP, ex coll. Abi Saad); plate (**a**) and counter plate (**b**) of a female specimen in dorso-lateral view; needs preparation.

Plate 1

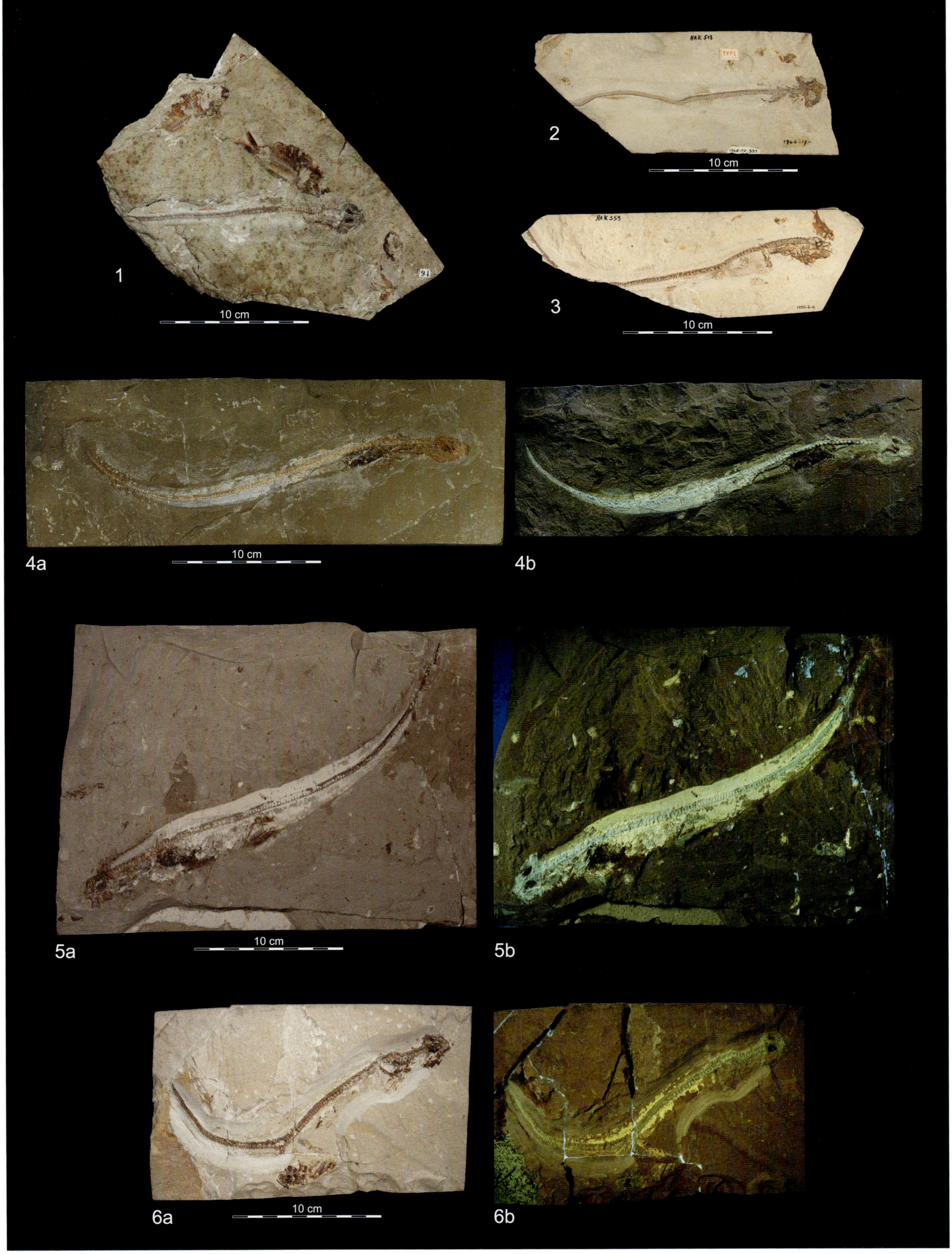

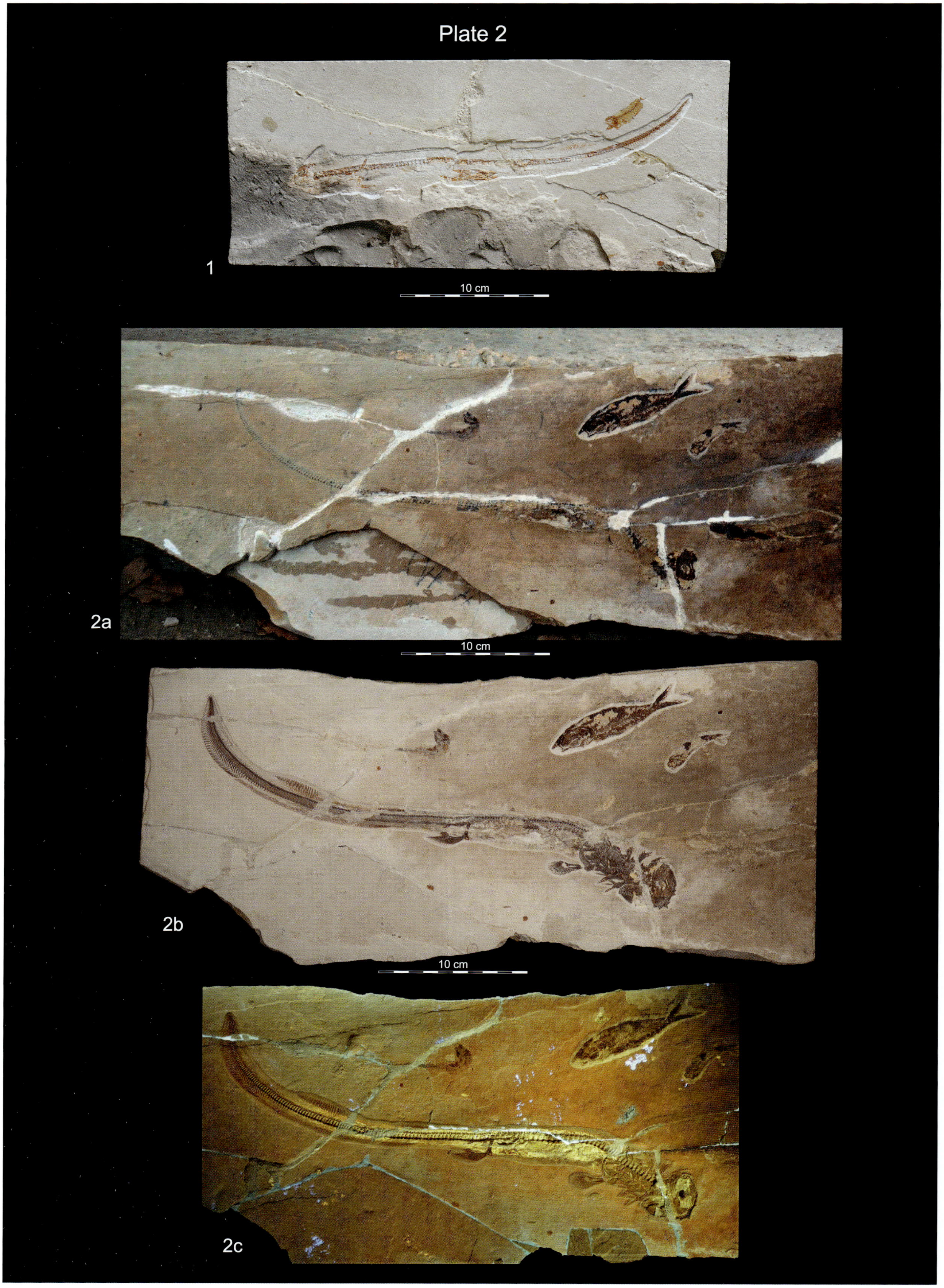
Plate 2
1
10 cm
2a
10 cm
2b
10 cm
2c

Plate 3

Plate 4

Plate 5

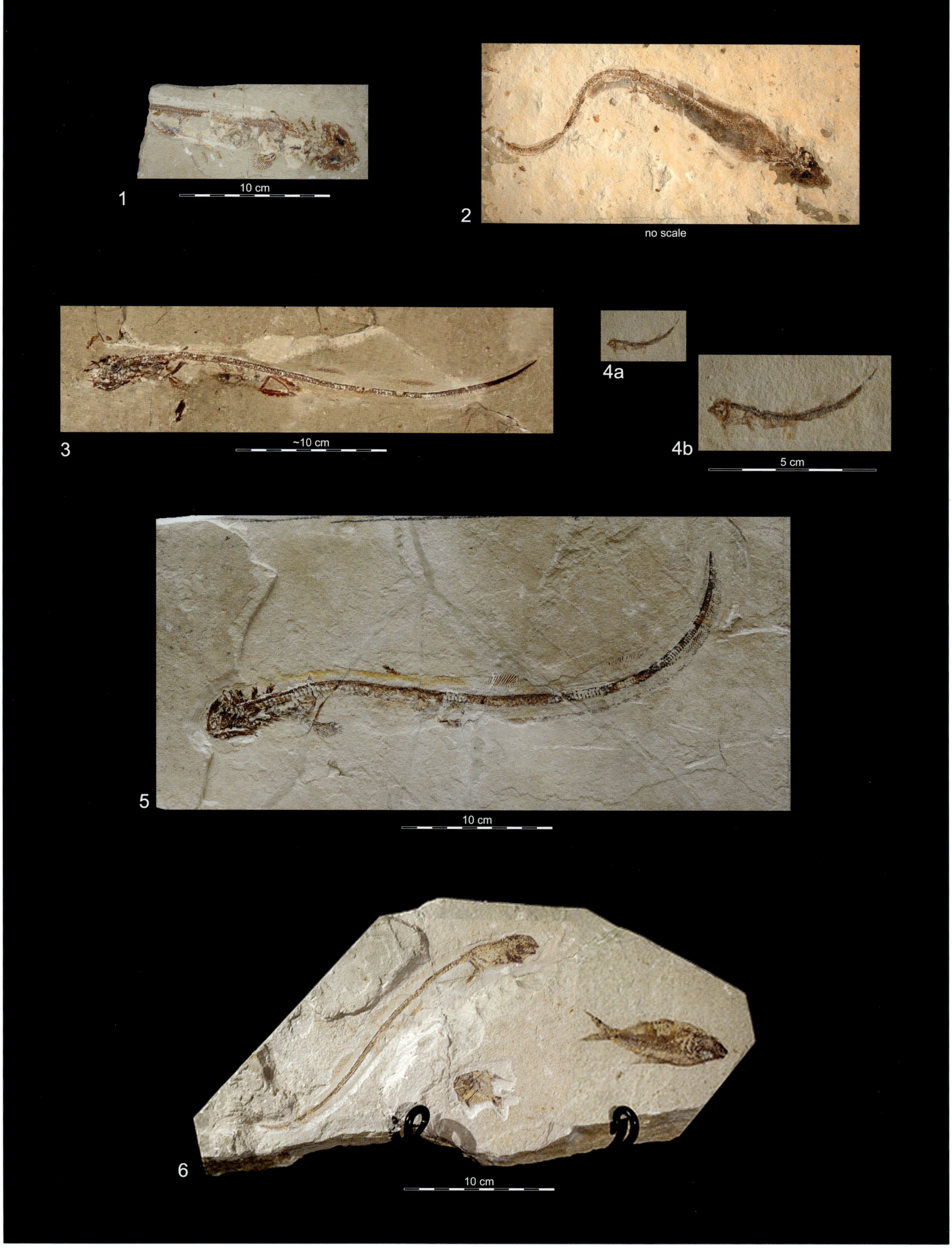

Plate 6

Plate 7
1a
1b
1c
2a
2b
2c
2d
3a
3b
3c
4a
4b
4c
5a
5b
5c

Plate 8

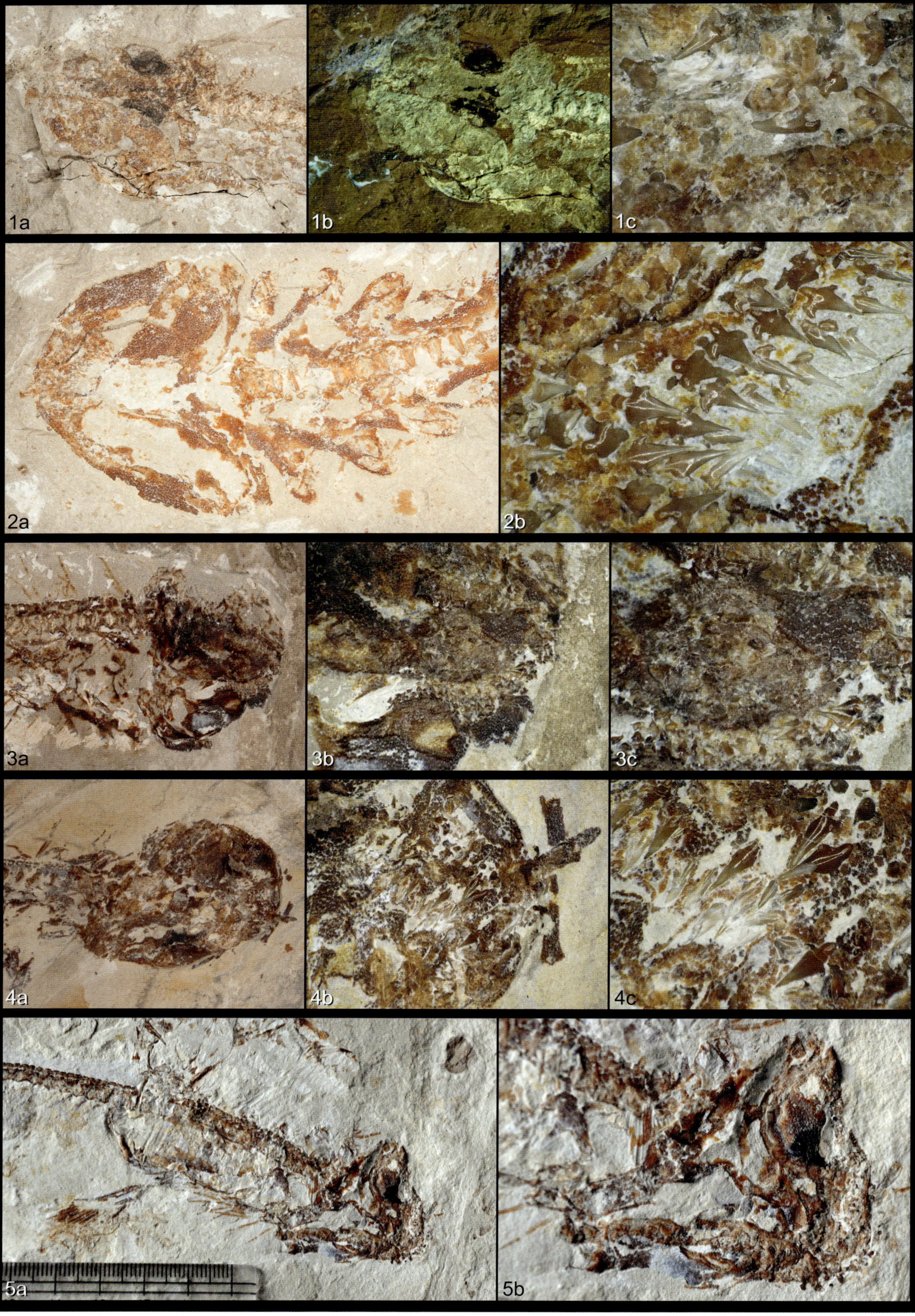

4. Morphological description of *Mesiteia*, and differences from other genera

The purpose of this study is to find autapomorphic characters to test the monophyly of the new family Mesiteiidae, and to establish a classification reflecting the reconstructed genealogical relationships to the family Hemiscylliidae. Phylogenetic relationships of the order Orectolobiformes and all its Recent families were analyzed cladistically on the basis of comparative morphology of external features, skeleton, musculature, and sensory system by previous authors, e. g. Dingerkus & DeFino (1983), Dingerkus (1986), Compagno (1988), Shirai (1992), Shirai (1996), Goto (2001), and Compagno (2001). Fossil sharks cannot provide a comparatively high number of characters but some of the specimens of *Mesiteia* are extraordinary well preserved and it is possible to detect some characters on these fossils that have been used by these authors for their cladistic analysis of Recent sharks.

4.1 Dental characters of *Mesiteia*

Teeth are important parts of a fossil shark for identifying a species. But this is only because, for most fossil species, only the teeth are preserved. To integrate fossil species with genera of Recent sharks or to create new fossil genera or higher categories is always dubious, "... it would be illusory to believe that a group, superorder, order, family, can be defined only on the basis of dental apomorphies. This is particularly true for genera that can be rather defined by a combination of characters, some of which can be apomorphic." (Cappetta 2012, p. 33).

Based on relatively poor knowledge of its tooth morphology – because of the bad preservation of only the holotype (description of teeth by Gorjanović (1884) and the first Paris specimen, here *Mesiteia* #02, see Fig. 3) – Herman & Crochard (1977, p. 145) grouped *Mesiteia* together with many other small Orectolobiformes in the family Orectolobidae. Cappetta (1980) attributed *Mesiteia* to the Recent shark family Hemiscylliidae Gill, 1862, and included two other small fossil Orectolobiformes from Lebanon, *Acanthoscyllium* Cappetta, 1980 and *Almascyllium* Cappetta, 1980. But Cappetta (2012) made clear that the attribution of the three Cretaceous genera *Acanthoscyllium*, *Almascyllium* and *Mesiteia* to Hemiscylliidae was only provisional (Cappetta 1980, p. 113).

The skulls of the different *Mesiteia* specimens are not always well preserved and a preparation without damage to make such small teeth with lateral cusplets visible in a hard and splintering matrix is a bit complicated. It must be mentioned that at the time this study was conducted, the preparation of the teeth was not yet finished. Therefore, I can only describe here what is visible without extracting any teeth. Well preserved teeth are extremely rare and to avoid further damage, until now no isolated teeth were extracted. The teeth are not higher than 0.5 to 1.5 and max. 2 mm. However, at least some of the specimens provide astonishing insights to the tooth morphology of *Mesiteia* (Plates 7-8) and justify the separation of *Mesiteia* from Hemiscylliidae.

Most views of teeth are views of labial faces of teeth from the upper jaws. On most specimens, teeth of anterior and lateral files are much more visible than teeth of posterior files. Specimens #07 (Pl. 7, Figs. 4a–c) and #10 (Pl. 8, Figs. 2a–b) show teeth from all parts of the jaws, but also the other specimens help to get additional views of labial and lateral sides and even a few details of the roots (e. g. specimens #07, Pl. 7, Fig. 4c; #08, Pl. 7, Fig. 5c; #09, Pl. 8, Fig. 1c; and #11, Pl. 8, Fig. 3c). The teeth of the different specimens are very similar and can be described all together, but in specimen #11 (Pl. 8, Fig. 3c) the shape of the teeth is slightly different (this will be discussed under the aspect of heterodonty).

One can recognize teeth from anterior, lateral and even posterior files and deduce from this the following dental characters:

Overall shape

- there are 12 (in *Mesiteia* #07) to 17 (in *Mesiteia* #11) files of teeth on each side of the jaws
- one can count up to 7 rows of teeth, but inner rows are poorly visible
- the teeth sit closely packed behind and next to each other
- teeth of anterior and lateral files are small (up to 2 mm high and 1 mm wide) and relatively high compared to their width
- teeth of anterior and lateral files are usually high-triangular, dagger-shaped in labial view (with apron as grip and lateral cusplets as cross-guard)
- in posterior teeth the height of the crowns decreases, and the principal cusps get wide-triangular
- anterior and lateral teeth are mesio-distally compressed, rather narrow
- anterior and lateral teeth are symmetrical
- posterior teeth get increasingly asymmetrical

Principal cusp

- the principal cusps of anterior files are rather high for their small size
- the principal cusps remain high in lateral files up to the 12th or higher file and get than increasingly lower towards the most posterior files
- the principal cusps of anterior and lateral teeth are not or are only moderately inclined distally
- in posterior files the principal cusps are still straight but more inclined distally
- the basal part of the cusp is rather wide at its base and low
- distinct heels are not developed; laterally the basal part of the cusp originates directly into the lateral cusplets
- the principal cusps are very sharp on top
- the principal cusps are round in cross-section with labial and lingual faces strongly transversely cambered, lingually more than labially
- posterior teeth show a flatter labial face
- the principal cusps are straight and only moderately bent lingually
- the labial face is only moderately convex in mesial view
- the cutting edges are well marked, but not prominent and run mesially and distally from top down to the base and over the lateral cusplets on both sides, separating the labial from the lingual face of the crown

Lateral cusplets

- there is one pair of prominent, relatively low and sharp lateral cusplets on anterior and lateral teeth
- the lateral cusplets are well-separated and broadly united to the base of the principal cusp
- the lateral cusplets are protruding and not close-fitting to the principal cusp

---

◁ **Plate 7**
Selection of specimens of *Mesiteia emiliae* which have at least partially preserved series of teeth in their skulls. These pictures also give a good overall impression of how (poorly) most specimens are preserved in this respect.
1. *Mesiteia* specimen #04 (Munich, SNSB-BSPG 2011 I 124).
2. *Mesiteia* specimen #05 (Munich, SNSB-BSPG 2013 XXVII 1).
3. *Mesiteia* specimen #06 (Munich, coll. FHP).
4. *Mesiteia* specimen #07 (Munich, coll. FHP).
5. *Mesiteia* specimen #08 (Munich, coll. FHP).
Photos 1a–b, 2b–c, 3b–c, 4a–b, 5b–c were taken with long-wave UV-light by Helmut Tischlinger.

◁ **Plate 8**
Further selection of specimens of *Mesiteia emiliae* with partially preserved series of teeth in their skulls.
1. *Mesiteia* specimen #09 (Munich, coll. FHP).
2. *Mesiteia* specimen #10 (Munich, coll. FHP).
3. *Mesiteia* specimen #11 (Munich, coll. FHP).
4. *Mesiteia* specimen #13 (Switzerland, coll. René Kindlimann RK/7-711).
5. *Mesiteia* specimen #25 (Munich, coll. FHP).
Photo 1b was taken with long-wave UV-light by Helmut Tischlinger.

Plate 9

- the lateral cusplets are not inclined towards the principal cusp but divergent and with a hooked apex
- no second pair of lateral cusplets is observed, even not on lateral and posterior teeth
- size of lateral cusplets remains prominent in anterior and antero-lateral files up to the 12th or higher file
- only in posterior and posteriormost files the lateral cusplets get lower, tend to disappear, first mesially, and the heels get reduced to blades without cusplets

Uvula
- the medio-lingual protuberance is well marked and high
- the lingual protuberance is covered with enameloid on its upper part

Apron
- the apron is very salient
- it is well detached from the labial outline of the crown base
- in anterior and first lateral files, the apron is very long and expanded (up to 25 % of the crown height)
- it is rather narrow in antero-lateral files with parallel mesial and distal edges
- it is restricted to the central region of the tooth
- it is narrower than the basal part of the principal cusp
- it is rounded at its basal end
- it is never bifid
- it reaches over the basal face of the root in mesial view
- towards the posterior files the apron becomes shorter, wider, expanded over the base of the principal cusp, and more rounded

Ornamentation
- the enameloid of labial and lingual crown faces is smooth
- the tenuous enameloid folds on the lower edge on each side of the expanded apron, as observed by Cappetta (2012, p. 158) on lateral teeth could not be confirmed

Root
- the root is hemiaulacorhize
- the root is rather bulky and high
- the root is mesio-distally expanded (it is wider than the crown)
- the root exhibits a heart-shaped outline
- there is lingual foramen at the lingual protuberance, sometimes a bit out of axis
- the central foramen is poorly visible, and needs to be better observed
- the basal face of the root is slightly concave
- a pair of margino-lingual foramina seems to be present

Heterodonty
- there is a linear gradient monognathic heterodonty from anterior towards posterior files with teeth changing from …
  high-triangular, narrow, conical principal cusps of anterior and lateral teeth to lower-crowned, wide-triangular posterior teeth;
  symmetrical anterior and lateral crowns to increasingly asymmetrical crowns;
  straight to distally more inclined teeth with flatter labial face, prominent, low and sharp lateral cusplets to heels reduced to blades without cusplets in posteriormost files;
  a long expanded and narrow apron to an apron becoming increasingly shorter, wider expanded and more rounded
- remarkable differences between upper and lower jaw teeth (dignathic heterodonty) or between teeth of males (with claspers) and females (gynandric heterodonty) has not yet been observed

◁ **Plate 9**
Images of dentitions of specimens #07, #08 and #10 that currently show the best views of teeth of *Mesiteia emiliae*.
1. *Mesiteia* specimen #10 (Munich, coll. FHP).
2. *Mesiteia* specimen #07 (Munich, coll. FHP).
3. *Mesiteia* specimen #08 (Munich, coll. FHP).
Photos 1c, 2a, 3a–b were taken with long-wave UV-light by Helmut Tischlinger.

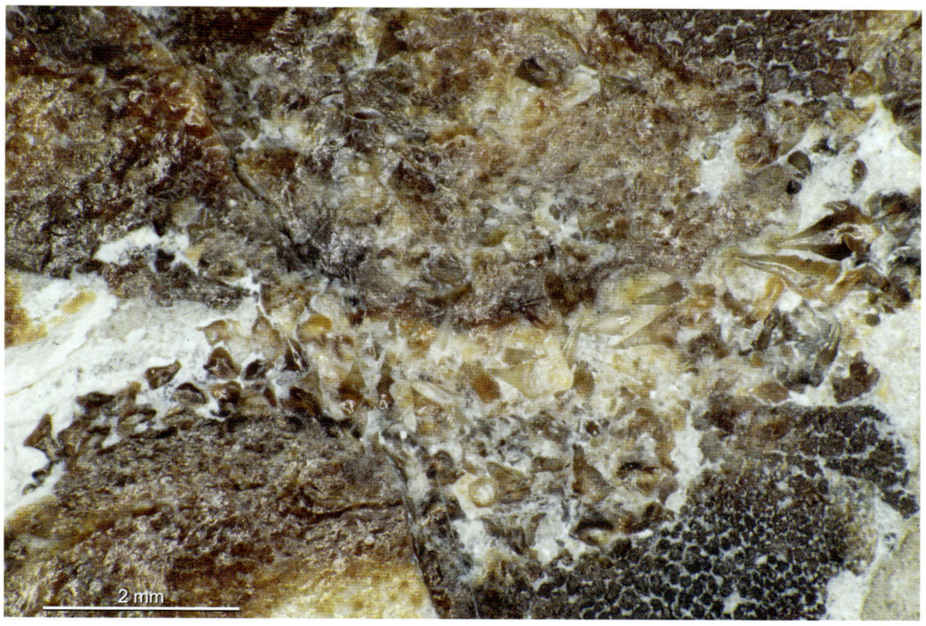

Fig. 7. The complete dentition of 17 files of teeth in the right upper jaw of *Mesiteia* specimen #11. From the anterior teeth (right) to the posterior teeth (left) the height and size decreases very fast and the crown base is relatively wide already in the anterior teeth.

- One female specimen (*Mesiteia* #11) has somehow different principal cusps from anterior to lateral and posterior files (Pl. 8, Fig. 3c, and Fig. 7). The principal cusps of the teeth of the first anterior file are less high and show a wider crown base, compared to specimen #10 (Pl. 8, Fig. 2b, and Pl. 9, Figs. 1a–c), which is also a female. A second different character is the pronounced change from relatively high principal cusps of anterior files to rather low and wide cusps. The height decreases already on the third file and the cusps of most files are relatively low (see Pl. 9, Fig. 1c for comparison). Also, the shape of the apron changes from narrow and elongated to short and wide expanded already within the first lateral files. If these remarkable differences are just a kind of heterodonty or if they characterize a second species of *Mesiteia* must be left for further studies

The gradient monognathic heterodonty from anterior towards posterior files makes it very complicated to find characters that can be used to describe differences from other species or genera. It is necessary to make sure that only teeth from same positions are compared. Isolated fossil teeth always bear some kind of uncertainity. Nevertheless, it is worth comparing the dental characters of *Mesiteia* with those of other species and/or genera.

One most reliable character to distinguish the two Recent genera *Hemiscyllium* and *Chiloscyllium* from *Mesiteia* is the prominent, narrow and rounded apron (= tablier), well set off from the labial outline and restricted to the central region of the tooth. It does reach and overhang the basal plane of the root in profile. This kind of an apron brings the teeth of *Mesiteia* morphologically much closer to those of the fossil genus *Eostegostoma* Herman, 1977 and to those of the Recent genus *Brachaelurus* (both family Brachaeluridae; see Section 4.2.10) than to Hemiscylliidae. A similar apron can be seen on teeth of some Early and Middle Jurassic Orectolobidae described by Thies (1989) as *Annea* and a single tooth described as *Mesiteia ?* sp. but these teeth have all an ornamentation on the labial face of the crown. They will be discussed in the next Section.

4.2 Differences of *Mesiteia* from the other genera of Orectolobiformes

To date 47 genera of Orectolobiformes have been described, 34 as fossils only, and the 13 Recent genera are known by fossils as well. This is not the place to analyze the relationships of all these 47 genera in detail, but I will discuss dental characters, that can be used to separate the clades as described by Goto (2001), including fossil genera and Recent families.

4.2.1 Recent species of *Hemiscyllium*

Very little is known about teeth of the nine Recent (extant) species of *Hemiscyllium*: *H. freycineti* (Quoy & Gaimard, 1824); *H. galei* Allen & Erdmann, 2008; *H. halmahera* Allen

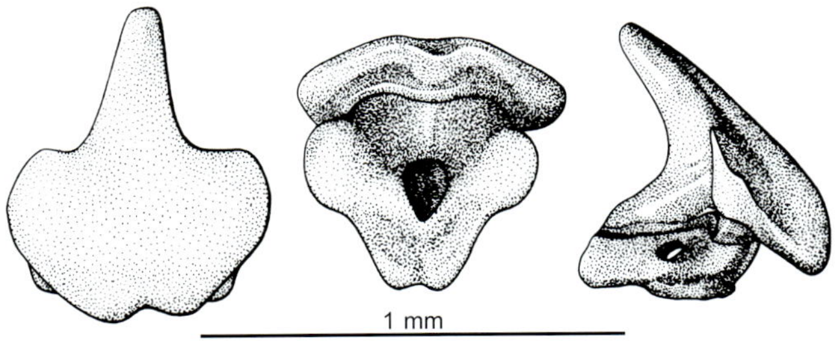

Fig. 8. Anterior tooth of *Hemiscyllium ocellatum* after Cappetta (2012, p. 157, figs. 142 I–K).

& Erdmann, 2013; *H. hallstromi* Whitley, 1967; *H. henryi* Allen & Erdmann, 2008; *H. michaeli* Allen & Dudgeon, 2010; *H. ocellatum* (Bonnaterre, 1788); *H. strahani* Whitley, 1967 and *H. trispeculare* Richardson, 1843. Allen, Erdmann, White, Fahmi & Dudgeon write in their most Recent "Review of the bamboo shark genus *Hemiscyllium* (Orectolobiformes: Hemiscyllidae)" (2016, p. 53): (*Hemiscyllium*) "teeth pavement-like, composed of numerous rows; individual teeth broad-based with single posteriorly-directed cusp, cusps of innermost rows more developed." This general description applies to many families of sharks, e.g. Mustelidae and even rays, e.g. Rajidae, and is less suitable for a determination based on tooth morphological characteristics. Unfortunately also the drawing of a third upper and lower jaw tooth of *H. ocellatum* in Last & Stevens' "Sharks and Rays of Australia" (2nd ed. 2009, p. 148), also re-drawn in Compagno, Dando & Fowler's "A Field Guide to the Sharks of the World" (2005, p. 169) are maybe not sufficiently detailed for palaeontological studies. But at least, in this Field Guide there is a drawing of a tooth of *H. freycineti*. This drawing is schematic, but it clearly shows the typical slender upright crown of a tooth of *Hemiscyllium*.

Herman & Crochard (1977, p. 129) were the first to publish drawings of teeth of three *Hemiscyllium* species: *H. freycineti*, *H. ocellatum* and *H. trispeculare*. From today's perspective, these drawings were simple but very good for showing the dental characters of these three species of *Hemiscyllium* that are summarized here. A very fine drawing of a single tooth of *H. ocellatum* was published in Cappetta (1980, p. 112) and used later as a typical reference for Recent *Hemiscyllium* teeth, e.g. in Cappetta (1987, 2012) and here (Fig. 8). Herman, Hovestadt-Euler & Hovestadt (1992, pl. 12-14) published SEM pictures of *H. ocellatum* clearly showing teeth of all positions in labial, lingual and basal views.

Dental characters of Recent *Hemiscyllium*:

Overall shape

- the teeth of the genus are small (less than 1 mm high and wide)
- a slender cusp is erected on a wide and high semicircular crown base
- teeth are mesio-distally not expanded (as e.g. teeth of *Orectolobus*)

Principal cusp

- in anterior and lateral files very narrow, upright and distinctly cuspidate
- less high or not much higher than the basal part of the crown
- the labial face of the crown is very flat overall
- the lingual face of the principal cusp only (not of the heels) is strongly transversely cambered

Lateral cusplets

- there are no real lateral cusplets but rather two short, subhorizontal, slightly convex heels

Apron

- the apron is very wide, not detached from the labial outline of the crown base, covering the entire crown base
- it is convex, tongue-shaped
- it reaches or almost reaches the basal plane of the root in profile
- it is not very thick
- it is usually distinctly bifid, concave by the pressure of the closely following tooth, except on the posterior teeth where it is mostly convex

Ornamentation

- labial and lingual crown faces are smooth

Heterodonty

- there is a linear gradient monognathic heterodonty from anterior towards posterior files with teeth changing from ... narrow, upright and cuspidate principal cusps to lower-crowned, distally inclined wide-triangular postero-lateral teeth with a wide expanded and still more rounded apron

A slender, erect principal cusp on a wide and high semicircular crown base, without lateral cusplets but heels, a flat labial face of the crown and an apron not detached from the labial outline of the crown base, covering the entire crown base are autapomorphic characters of *Hemiscyllium* and separate it from all other Orectolobiformes and, of course, from *Mesiteia*.

4.2.2 Fossil species of *Hemiscyllium*

Isolated fossil teeth of *Hemiscyllium* are rare, mainly because they are very small, often less than 1 mm in size. The assignment to Recent *Hemiscyllium* seems to be correct but does not correlate with the distribution of the living species, which is restricted to the Australia-New Guinea region with the exception of a single species (*H. halmahera*) from the Indonesian island of Halmahera and another species (*Hemiscyllium* sp.) from the Seychelles.

H. bruxelliensis Herman & Crochard, 1977 – Eocene of Belgium

H. daimeriesi (Herman, 1973) – Paleocene of Belgium

H. hermani Müller, 1989 – from Campanian and Maastrichtian of North Germany (Herman 1982, Müller 1989) to the Early Paleocene in Denmark (Adolffsen & Ward 2013)

H. tailledisensis Adnet, 2006 – Eocene of Southern France

Here added and figured (Fig. 9) is an undescribed species of *Hemiscyllium* from the middle Maastrichtian of Bavaria from the authors collection (FHP). Its overall shape fits very well with Recent *Hemiscyllium* (Fig. 8).

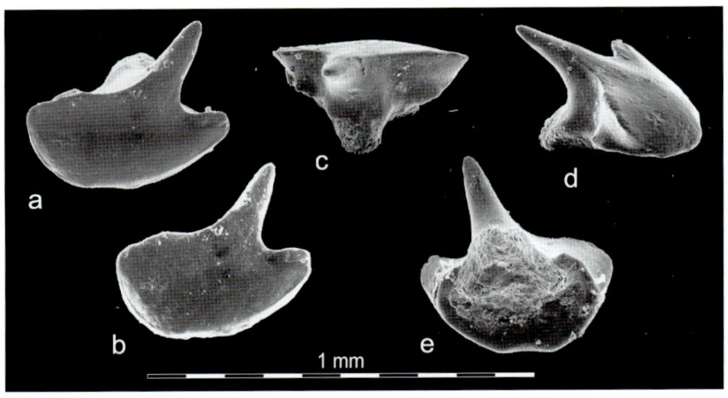

▷▷ **Fig. 9.** Tooth of an undescribed species of *Hemiscyllium* from the middle Maastrichtian of Bavaria (Gerhartsreiter Graben, coll. FHP). **a, b:** labial view; **c.** lingual view; **d.** lateral view; **e.** basal view.

The dental characters of fossil *Hemiscyllium* correspond to those of Recent species of *Hemiscyllium* and are different from those of *Mesiteia*.

4.2.3 *Adnetoscyllium*, *Acanthoscyllium* and *Parahemiscyllium*

The fossil genera *Adnetoscyllium*, *Acanthoscyllium* and *Parahemiscyllium* share at least the following synapomorphies that define a group containing *Hemiscyllium* as a living genus:

Overall shape

- the teeth are small (less than 1 mm wide)
- a slender cusp is erected on a wide and high semicircular crown base
- teeth are mesio-distally not expanded

Principal cusp

- very narrow, upright and distinctly cuspidate
- less high or not much higher than the basal part of the crown
- the labial face of the crown is rather flat overall but the lingual face of only the principal cusp is strongly transversely cambered

Apron

- the apron is very wide, not detached from the labial outline of the crown base, covering the entire crown base
- it is convex, tongue-shaped
- it reaches or almost reaches the basal plane of the root in profile
- it is not very thick

The apomorphies that separate these genera from *Hemiscyllium* are restricted to the existence and appearance of lateral cusplets and the ornamentation of the labial and lingual faces of the crown. These characters are rather different from those of Recent *Hemiscyllium* and *Chiloscyllium*. Having in mind that only *Acanthoscyllium* is known by remains of the whole fish, this is a reminder to be careful with relationships based only on characters that could be homoplastic or reversals (Cappetta 2012, p. 32).

Adnetoscyllium Guinot, Underwood, Cappetta & Ward, 2013 with *Adnetoscyllium angloparisensis* Guinot, Underwood, Cappetta & Ward, 2013 (p. 21, figs. 8V–I') from the late Santonian of the Anglo-Paris Basin is only distinguished from *Hemiscyllium* by a single autapomorphic character, by its distinct ornamentation of the labial crown face:

Ornamentation

- a weak ridge mesio-distally follows the outline of the apron, and an additional discontinuous ridge is present on the lower region of the labial face.

Based on dental characters, *Acanthoscyllium* Cappetta, 1980 from the late Santonian of Sahel Alma, Lebanon seems to be correctly assigned to Hemiscylliidae and separated from *Hemiscyllium*. A thin and sharp principal cusp is erected on a broad and high, semicircular crown base, with a very wide apron. It shares with *Hemiscyllium* the synapomorphies mentioned above and is distinguished by the following autapomorphic characters:

Lateral cusplets

- a pair of prominent and rather high and pointed lateral cusplets

Ornamentation

- a pronounced vertical ridge on the upper region of the lingual face of the crown
- a longer, sometimes divided ridge on the labial face, which does not reach the base of the apron

Acanthoscyllium is currently placed within the family Hemiscylliidae but I agree with Cappetta (2012, p. 155) that it differs from these by several important anatomical characters, and thus might represent a separate family: the head is shorter and broader but the rostrum is very long, the pelvic fins are strikingly more developed than the pectorals and the dorsal fins are smaller than in any Hemiscylliidae. The preservation of the few specimens known is not good enough to prove that an anal fin is really missing but the caudal fin also differs from that in all Hemiscylliidae.

Guinot, Cappetta & Adnet (2014) described from the Valanginian (Lower Cretaceous) of southern France the genus *Parahemiscyllium* with *Parahemiscyllium underwoodwardi* Guinot, Cappetta & Adnet (2014, p. 73-75, fig. 8U–K'). It is distinguished from *Hemiscyllium* by a single autapomorphic character:

Lateral cusplets

- usually two pairs of prominent and rather low triangular lateral cusplets, varying from straight to slightly converging

The very high principal cusps of the anterior teeth and a labial face that seems to be flatter or even slightly concave are characters that can vary from one species to another.

4.2.4 Recent species of *Chiloscyllium*

Seven living species of *Chiloscyllium* are known to date: *C. arabicum* Gubanov, 1980; *C. burmensis* Dingerkus & DeFino, 1983; *C. griseum* Müller & Henle, 1838; *C. hasselti* Bleeker, 1852; *C. indicum* (Gmelin, 1788); *C. plagiosum* (Bennett, 1830) and *C. punctatum* Müller & Henle, 1838.

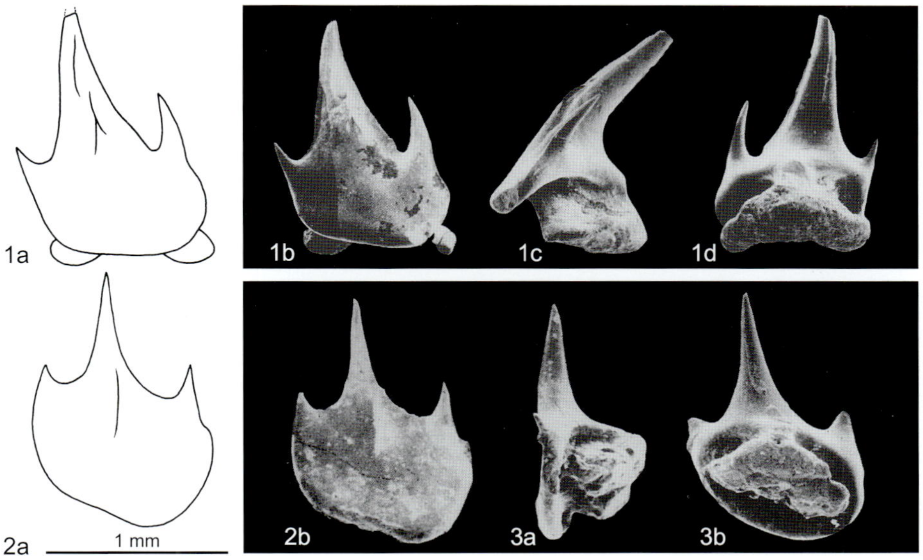

Fig. 10. Teeth of *Acanthoscyllium sahelalmae* after Cappetta (1980, pl. 15, figs. 4–6) and Cappetta (2012, p. 156, figs. 141 C and E). **1.** antero-lateral tooth; a, b. labial view; c. lateral view; d. lingual view; **2.** very lateral tooth; a, b. labial view; **3.** lateral tooth; a. lateral view; b. basal view.

The separation of Recent (and fossil) *Chiloscyllium* species in two dental types, probably suggests heterogeneity of the genus. In a recent phylogenetic mitochondrial study (Naylor et al. 2012) *Hemiscyllium ocellatum* grouped among *Chiloscyllium punctatum*, *C. indicum* and *C. plagiosum*. This is a reminder to not be too optimistic when dental characters seem to support a separation.

Type 1. Here figured are samples of *C. indicum* (Fig. 11a, from Cappetta 2012, p. 157); *C. plagiosum* (Fig. 11b, coll. FHP) and *C. griseum* (Fig. 11c–d, coll. FHP). The crown base with the apron is widely expanded and bears a pair of low but well

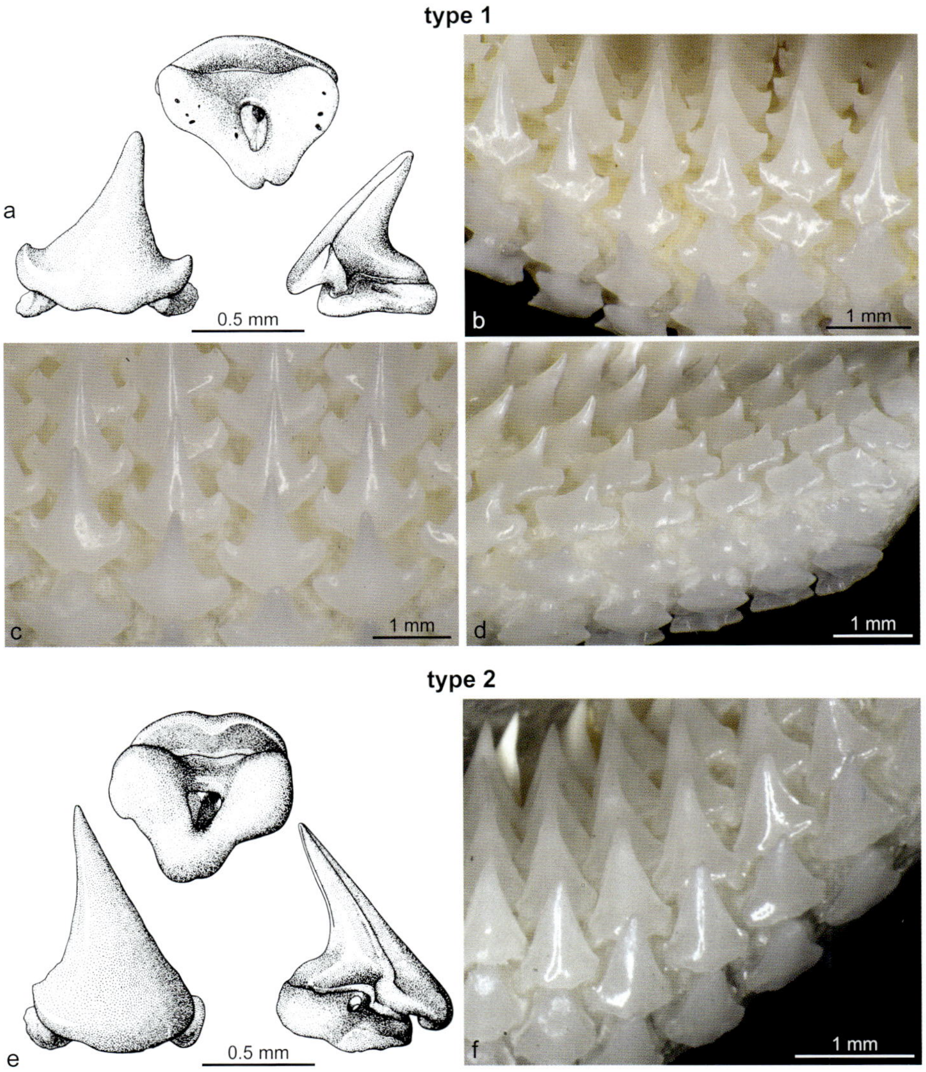

Fig. 11. Teeth of *Chiloscyllium* species can be divided in two types.
Dental type 1: **a.** *C. indicum* in labial, basal and lateral views; after Cappetta (2012, p. 157, figs. A–C); **b.** *C. plagiosum* (coll. FHP-R-2206); **c–d.** *C. griseum* (coll. FHP-R-2235).
Dental type 2: **e.** *C. punctatum* in labial, basal and lateral views; after Cappetta (2012, p. 157, figs. E–G); **f.** *C. punctatum* (coll. FHP-R-238).

developed lateral cusplets, separated from the principal cusp. The apron is usually bifid. In posterior rows the principal cusps of this type get lower and the lateral cusplets are replaced by heels that let these teeth look like teeth of *Hemiscyllium*. In these posterior rows the crowns are more inclined inwards, the principal cusps are less erected, and the narrowly imbricated pavement of the teeth recalls the crushing-type dentitions e.g. of Dasyatidae. Herman, Hovestadt-Euler & Hovestadt (1992, pl. 9–11) published SEM pictures of *C. plagiosum* showing anterior, lateral and posterior teeth in labial, lingual and basal views.

Type 2. The sample here is *C. punctatum* (Fig. 11e, from Cappetta 2012, p. 157 and Fig. 11f, coll. FHP). The crown base with the apron is much less expanded and the mesial and distal cutting edges reach down to a rounded and only sometimes slightly bifid apron. Distinctly separated cusplets are missing.

Dental characters of Recent *Chiloscyllium*

Overall shape

- the teeth of anterior and lateral files are small (up to 1–2 mm high and 1 mm wide)
- a high-triangular cusp originates on a wider, if lateral cusplets are present (type 1), or a narrow, if no lateral cusplets are present (type 2) and very low semicircular crown base
- teeth are mesio-distally slightly (type 1), or not (type 2) expanded

Principal cusp

- high-triangular, upright and towards posterior files distally inclined
- higher than the basal part of the crown
- the labial face of the crown is very flat overall
- the lingual face of the principal cusp only (not of the heels) is strongly transversely cambered

Lateral cusplets

- there is a pair of very low, well-separated lateral cusplets at the lateral extremities of the crown base (type 1), or practically no real lateral cusplets and no heels (type 2)

Apron

- the apron is very wide, slightly detached from the labial outline of the crown base (type 1), or not detached from the labial outline of the crown base, practically covering the entire crown base (type 2)
- it is convex, tongue-shaped
- it is not very high, reaches or almost reaches the basal plane of the root in profile
- it is not very thick
- it is sometimes bifid on the anterior files, concave by the pressure of the closely following tooth, except on the lateral teeth where it is regularly convex

Ornamentation

- labial and lingual crown faces are smooth (type 1), or the lingual or labial faces of the crown are smooth, except that the lingual face of the principal cusp of *C. punctatum* bears a distinct, rectilinear ridge from the apex to almost half-way up the cusp (type 2, Cappetta 2012, p. 156–157)

Heterodonty

- there is a linear gradient monognathic heterodonty from anterior towards posterior files with teeth changing from … high-triangular, upright principal cusps to increasingly lower-crowned, wide-triangular postero-lateral teeth with an almost non-protruding principal cusp, and a still distinct and rounded apron, very similar to that of anterior teeth

Autapomorphic characters of *Chiloscyllium* teeth are the high-triangular labial crown face with a narrower and less semicircular crown base, and a principal cusp that is much higher than the basal part of the crown and not as clearly separated from the crown base as in *Hemiscyllium*. These characters separate it from all other Orectolobiformes and, of course, from *Mesiteia*.

4.2.5 Fossil species of *Chiloscyllium*, and other possibly related genera

Nine fossil species are assigned to *Chiloscyllium* but this is questionable and needs to be corrected: *C. arbizui* Bernardez, 2002 nomen nudum; *C. broennimanni* Casier, 1958, *C. gaemersi* Müller, 1989; *C. greeni* (Cappetta, 1973), *C. humboldti* (Reuss, 1845); *C. meraense* Noubhani & Cappetta, 1997; *C. mereretiae* Werner, 1989; *C. missouriense* Case, 1979; *C. salvani* Noubhani & Cappetta, 1997.

An excellent sample of a fossil *Chiloscyllium* is *Chiloscyllium mereretiae* Werner, 1989 (Fig. 12a) from the upper Cenomanian of Bahariya, Egypt. There is only a distance of about 500 km between Bahariya and the fossil localities in Lebanon and the stratigraphic age of Hakel and Bahariya is Cenomanian. *Chiloscyllium mereretiae* represents a fossil type 1 species as described before with a wide expanded apron and a pair of low but well developed lateral cusplets at the lateral extremities of the crown base. This lingual, rectilinear ridge is present on all the teeth, from the apex to almost half-way up the principal cusp, exactly as described by Cappetta (2012) for the type 1 teeth of Recent *Chiloscyllium*.

▷▷ **Fig. 12.** Fossil *Chiloscyllium* teeth. **a.** dental type 1; *Chiloscyllium mereretiae* Werner, 1989 in labial, lateral and lingual view (from left to right); from Werner (1989, pl. 8, figs. 1a–c); **b.** dental type 2; undescribed fossil *Chiloscyllium* from the middle Maastrichtian of Bavaria (coll. FHP) in labial, lingual, mesial and distal view (from left to right).

Another not-yet-described fossil *Chiloscyllium* from the middle Maastrichtian of Bavaria from my collection (FHP, Fig. 12b) is a representative of the type 2 teeth. Its overall shape fits very well with Recent *Chiloscyllium punctatum*.

Recently three fossil genera were described as Orectolobiformes incertae familiae. They share many synapomorphies with *Chiloscyllium* and its differences are probably of minor weight: *Garrigascyllium* Guinot, Cappetta & Adnet, 2014 (figs. 7H'–M' and 8A–E) from the Valanginian (Lower Cretaceous) of southern France and *Heterophorcynus* Underwood & Ward, 2004 (pl. 6, figs. 1–11) from the Bathonian (Middle Jurassic) of southern England seem to be closer related to *Chiloscyllium* type 1, and *Dorsetoscyllium* Underwood & Ward, 2004 (pl. 7, figs. 1–13) from the Bathonian (Middle Jurassic) of southern England could be closer related to *Chiloscyllium* type 2.

In summary, both Recent and fossil teeth of *Chiloscyllium* share many apomorphic characters and have nothing in common with those of *Mesiteia* (Pl. 9).

4.2.6 *Almascyllium*, and other undescribed orectolobiform sharks from Lebanon

Almascyllium cheikheliasi (Signeux, 1949) from the Santonian of Sahel Alma, Lebanon (Fig. 13) was assigned to Hemiscylliidae by Cappetta (1980, 2012).

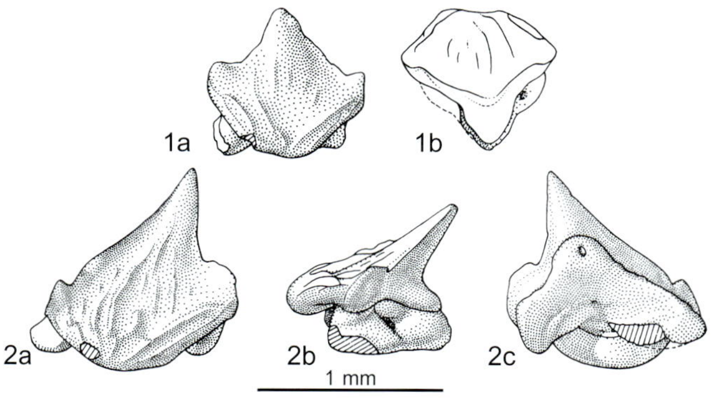

▷▷ **Fig. 13.** Teeth of *Almascyllium cheikheliasi* (Signeux, 1949) after Cappetta (2012, p. 156, figs. 141 F–J). **1.** anterior tooth, in labial (a), occlusal view (b); **2.** lateral tooth, in labial (a), lateral (b), and basal view (c).

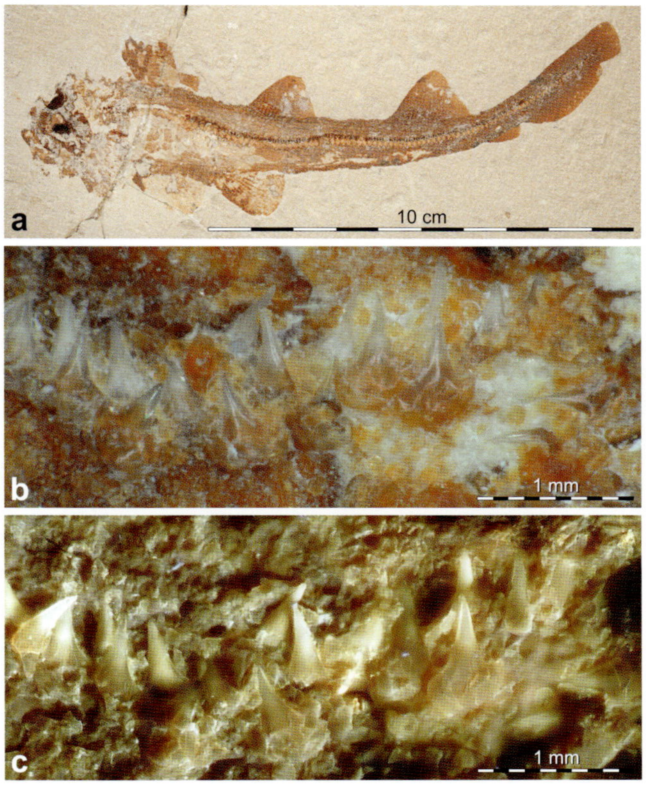

Fig. 14. A new orectolobiform shark from the middle Cenomanian locality Hgula in Lebanon (**a**) and its dentition in normal light (**b**) and long-wave UV-light (**c**). Based on odontological but also some anatomical characters, this could be the first complete fossil shark of the family Hemiscylliidae (coll. FHP); photo (c) with long-wave UV-light Helmut Tischlinger.

Fig. 15. Another new orectolobiform shark from the early Cenomanian locality Hakel in Lebanon (**a**) and a few of its teeth in labial view from anterior, lateral and posterior tooth files (**b**). This complete fossil shark may also belong to the family Hemiscylliidae (coll. FHP).

There are several synapomorphic characters that *Almascyllium* shares with *Hemiscyllium*: the overall shape of the small teeth, a rather low and fine principal cusp, upright in anterior teeth and more inclined towards the rear in lateral and posterior teeth, the very flat labial face of the crown, a broad and high, semicircular crown base, with a wide apron. Also, the pair of tiny, only moderately high lateral cusplets. In posterior files the mesial denticle almost disappeared and the distal denticle is more a heel.

Apomorphic for *Almascyllium* and strikingly different from *Hemiscyllium* is the ornamentation of the labial face of the teeth, bearing strong and flexuous folds in its median region and folds parallel to the edge of the apron under the heels.

At first view the ornamentation of the labial face of the crown of *Almascyllium* could be only a minor difference from *Hemiscyllium*. In all Recent and many fossil Hemiscylliidae the labial face is smooth without an ornamentation but within other Orectelobiformes (Orectolobidae, Ginglymostomatidae) it also happened that fossil genera were ornamented with folds and ridges that were characteristic for the genus but lost in more derived relatives. The differences from *Chiloscyllium* (high-triangular principal cusp, low crown base) are more obvious. It was correct to name a separate genus for *Almascyllium* and to place it under the family Hemiscylliidae.

But this discussion is only about its teeth. *Almascyllium* is known by skeletons, until now only by one (after Cappetta 2012) or two (after Cappetta 1980) specimens but *Almascyllium* could also belong to a different family of small Orectolobiformes with similarly ornamented teeth.

The shark fauna of the Cretaceous from Lebanon is very diverse indeed but very few species have been studied in detail to date. Since I began collecting and studying fossil sharks from Lebanon, I became aware that there are many more surprising finds to expect. As a sample I can show a perfectly preserved 16 cm long orectolobiform shark from the middle Cenomanian locality Hjula of the authors collection (FHP, Fig. 14). Most likely it belongs to the family Hemiscylliidae. The dentition is almost complete. The tiny teeth (less than 0.5 mm wide) – here a detail from the symphyseal and anterior region in normal and long-wave UV-light – share many characters with *Hemiscyllium*. Other anatomical features such as size and position of fins are different but could be used to argue its place in the family Hemiscylliidae.

I found a few more such unknown new sharks when I was searching for juvenile specimens of the much bigger and relatively common *Cantioscyllium decipiens* Woodward, 1889 after two beautiful male and female specimens were purchased for the BSPG in Munich (Pfeil 2012, 2014). Here I can only present a few first impressions (Fig. 15–16). These very small sharks with a total length of less than 10 to 20 cm still need a careful preparation before they can be studied.

Cantioscyllium is known by skeletons up to 1 m and more total length. This genus is wide distributed (Pfeil 2012, 2014) and belongs to the family Ginglymostomatidae. This is supported by synapomorphic dental characters of the adult teeth, such as a wide-triangular and not very high principal cusp, mesio-distally expanded heels bearing two pairs of sharp lateral cusplets and a wide, rounded, not very high and often bifid apron. It is interesting now to see that the anterior teeth (Fig. 16b) have high-triangular principal cusps that are – besides the strong ornamentation of the labial face and the sharp and high lateral cusplets – very similar to teeth of *Chiloscyllium* type 1.

All these new small sharks from Lebanon have nothing in common with *Mesiteia*. But I want to show it not only because of its very different teeth. All these sharks (Figs. 14a, 15a, 16a) are covered with dermal denticles, but – as I will describe later – *Mesiteia* is not.

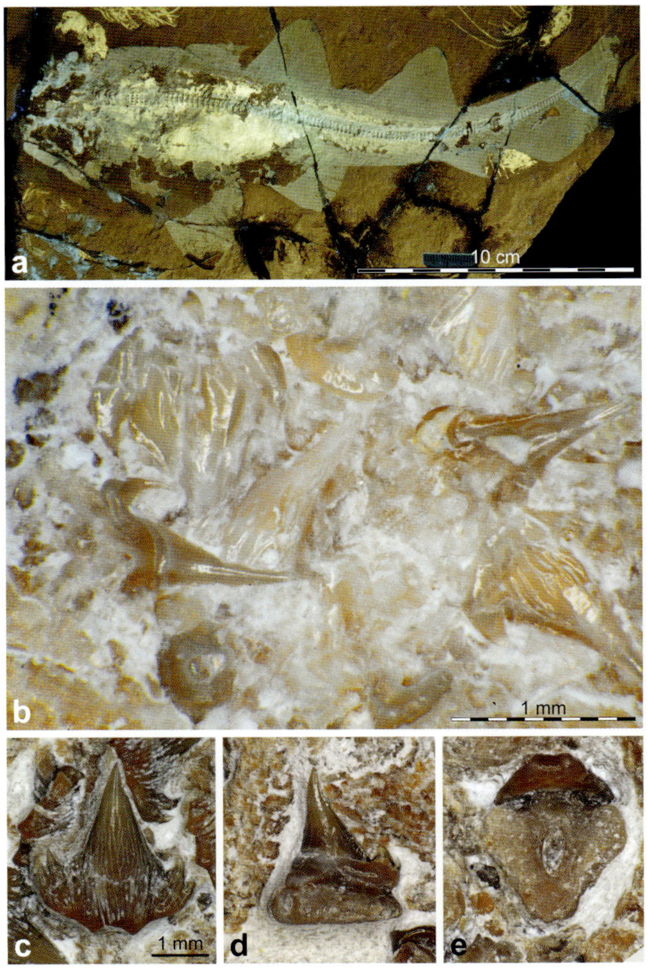

Fig. 16. The smallest and youngest specimen that could be identified as *Cantioscyllium decipiens* Woodward, 1889 (coll. FHP) has a total length of 21 cm (**a**). Its small teeth (**b**) are easy to distinguish from the hemiscylliids of Figure 13 and 14 by their characteristic strongly folded ornamentation of the labial crown face and the sharp lateral cusplets, as seen on teeth from lateral files of the upper jaw of an adult specimen (female, total length 72 cm) of *Cantioscyllium decipiens* Woodward, 1889 in labial (**c**), linguo-lateral (**d**), and basal view (**e**), SNSB-BSPG 2012 I 87; photo a with long-wave UV-light Helmut Tischlinger.

4.2.7 Recent and fossil Brachaeluridae

Based on tooth morphological characters there is no evidence to assign *Mesiteia* to Hemiscylliidae. But there are other Orectolobiformes with teeth that share the striking feature of a prominent, broad and rounded apron. One is the Recent genus *Brachaelurus* (family Brachaeluridae). According to Goto (2001) Brachaeluridae are a monophyletic group supported by eight autapomorphies and are well-separated from Hemiscylliidae.

While the single tooth drawn by Cappetta (1980, 2012) shows a principal cusp less high and with a shorter apron than in *Mesiteia* but otherwise very similar, the numerous SEM pictures in Herman, Hovestadt-Euler & Hovestadt (1992, pl. 1–3) express the differences more clearly.

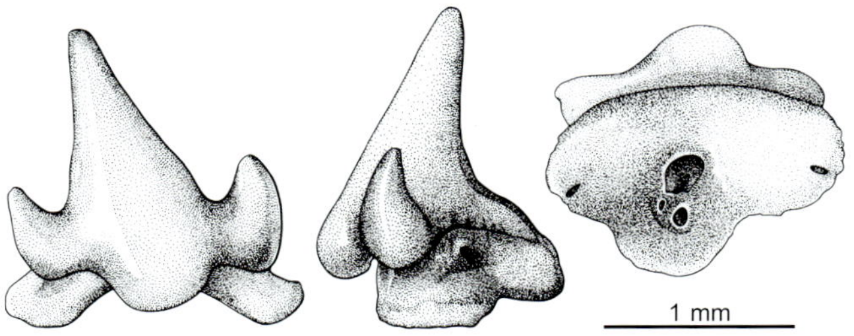

▷▷ **Fig. 17.** Tooth of *Brachaelurus waddi* (Bloch & Schneider, 1801) in labial, lateral, and basal view; after Cappetta (2012, p. 165, figs. 152 A–C).

Dental characters of *Brachaelurus waddi* (Bloch & Schneider, 1801) that are identical to *Mesiteia* (Pl. 9):

Overall shape

- there are 12–16 files of teeth on each side of the jaws
- one can count up to 7 rows of teeth
- the teeth sit closely packed behind and next to each other
- teeth are small (up to 2 mm high and 1 mm wide)
- anterior teeth are high-triangular and relatively high compared to their width
- in lateral and posterior files, the height of the crowns decreases, and the principal cusps get wide-triangular
- anterior teeth are symmetrical, lateral and posterior teeth asymmetrical

Principal cusp

- the principal cusps of anterior and lateral teeth are only moderately inclined distally
- the basal part of the cusp is rather wide at its base and low
- distinct heels are not developed on anterior and lateral teeth; laterally the basal part of the cusp originates directly into the lateral cusplets
- the principal cusps are very sharp on top
- cusps round in cross-section with labial and lingual faces strongly transversely cambered, lingually more than labially
- posterior teeth show a flatter labial face
- the principal cusp is straight and only moderately bent lingually
- the labial face is only moderately convex in mesial view
- the cutting edges are well marked, but not prominent and run mesially and distally from top down to the base and over the lateral cusplets on both sides, separating the labial from the lingual face of the crown

Lateral cusplets

- there is one pair of prominent, relatively low and sharp lateral cusplets on anterior teeth
- lateral cusplets well-separated and broadly united to the base of the principal cusp
- the lateral cusplets are protruding and not close-fitting to the principal cusp
- lateral cusplets divergent and with a hooked apex
- there is no second pair of lateral cusplets, not even on lateral and posterior teeth
- in posterior files the lateral cusplets tend to get very wide and low, in posteriormost files they almost disappear, especially the distal lateral cusplet

Uvula

- the medio-lingual protuberance is well marked and high
- the lingual protuberance is covered with enameloid on its upper part

Apron

- no second pair of lateral cusplets, not even on lateral and posterior teeth
- apron is well detached from the labial outline of the crown base
- apron is restricted to the central region of the tooth
- apron is rounded at its basal end
- on posterior teeth the apron gets wider and shorter

Ornamentation

- the enameloid of labial and lingual crown faces is smooth

Root

- the root is hemiaulacorhize
- the root is rather bulky and high
- the root is mesio-distally expanded (it is wider than the crown)
- the root has a heart-shaped outline
- there is lingual foramen at the lingual protuberance, sometimes a bit out of axis
- the central foramen is broad and elliptical
- the basal face of the root is slightly concave
- a pair of margino-lingual foramina is present

Heterodonty

- there is a linear gradient monognathic heterodonty from anterior towards posterior files with teeth changing from . . .
  high-triangular, narrow, conical principal cusps to lower-crowned, wide-triangular lateral and posterior teeth;
  symmetrical to increasingly asymmetrical crowns;
  straight to distally more inclined teeth with flatter labial face;
  prominent, low and sharp lateral cusplets to heels reduced to blades without prominent cusplets in posteriormost files;
  an apron becoming increasingly shorter, wider and more rounded

Dental characters of *Brachaelurus waddi* (Bloch & Schneider, 1801) that are different to *Mesiteia* (Pl. 9):

Overall shape

- teeth are high-triangular but not as high compared to their width as in *Mesiteia*
- already in lateral teeth the height of the crowns decreases, and the principal cusps get wide-triangular
- teeth are mesio-distally more expanded
- only anterior teeth are symmetrical
- lateral and posterior teeth get soon increasingly asymmetrical

Principal cusp

- the principal cusps are less high
- the principal cusps soon get lower in lateral files and get very low towards the posterior files
- in posterior files the principal cusps are very much inclined distally
- on posterior teeth lateral cusplets are reduced or almost replaced by heels

Lateral cusplets

- size of lateral cusplets soon gets smaller in lateral files
- in posterior files the lateral cusplets become lower and wider, tend to disappear and especially the mesial heels become blades without cusplets

Apron

- by far not as long or as expanded as in *Mesiteia*
- only in anterior files rather narrow and rounded
- not much narrower than the basal part of the principal cusp
- rounded or sometimes flattened at its basal end
- sometimes bifid
- does not extend over the basal face of the root in mesial view
- becomes very short and wide towards the posterior files

Prior to discussing the similarities to *Mesiteia* it is worthwhile to briefly introduce the fossil species of *Brachaelurus*. There are three fossil species accepted as *Brachaelurus* in Cappetta (2012) (*Brachaelurus roklumensis* Thies, 1981, Lower Cretaceous, middle Barremian, N Germany; *B. hornestownensis* Case, Schwimmer, Borodin & Leggett, 2001, Upper Cretaceous, upper Maastrichtian, New Jersey, U.S.A.; *B. templeuvensis* Moreau &

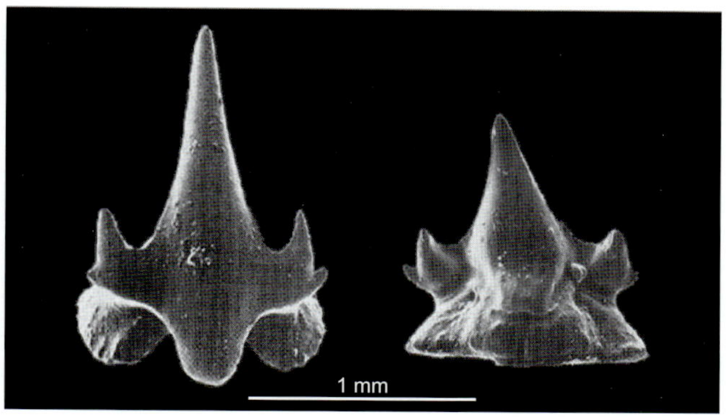

Fig. 18. A fossil tooth of *Brachaelurus* (*Brachaelurus roklumensis* Thies, 1981, in labial and lingual view; Lower Cretaceous, middle Barremian, N Germany, after Thies (1981, p. 478, figs. 5a–b). ◁◁

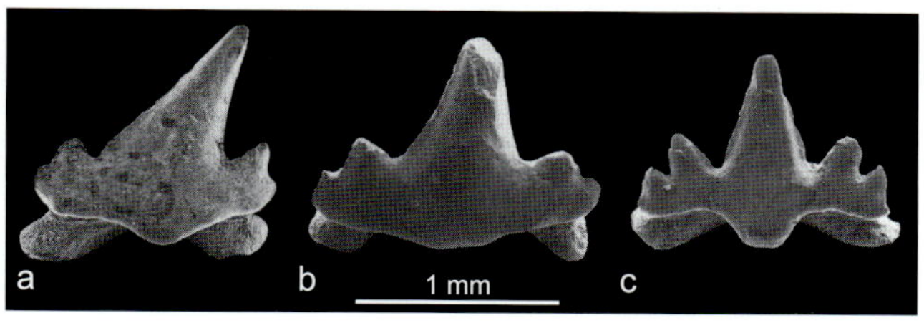

◁◁ **Fig. 19.** Two species of *Palaeobrachaelurus* Thies, 1983. **a–b.** anterior and lateral tooth of *Palaeobrachaelurus aperizostus* Thies, 1983, in labial view, Middle Jurassic, upper Aalenian, northern Germany, after Cappetta (2012, figs. 154A, E); **c.** *Palaeobrachaelurus bedfordensis*. Lateral tooth in labial view, Middle Jurassic, Callovian, England, after Cappetta (2012, fig. 154H).

Mathis, 2000, Upper Paleocene, Thanetian, N France. Based only on very few poorly known isolated teeth, it must remain questionable if these teeth can be attributed to the Recent genus *Brachaelurus*. The teeth of *Brachaelurus roklumensis* Thies, 1981 (Fig. 18) are similar to *Brachaelurus* (Fig. 17) but also to *Mesiteia* (Pl. 9). When Thies (1981, p. 478) argued that his new species should be attributed to *Brachaelurus*, he compared it mainly with *B. greeni* Cappetta, 1973, which has since been transferred to *Chiloscyllium*.

Compared with the illustrations of *Brachaelurus waddi* teeth in Herman, Hovestadt-Euler & Hovestadt (1992, pl. 1–3), the conical crowns of *B. roklumensis* show some interesting differences that make it difficult to accept a close relationship:

Principal cusp

- the principal cusp is higher and more slender

Lateral cusplets

- there is clearly a second pair of lateral cusplets

Apron

- the apron is narrower and shorter

The same applies to *Mesiteia*, which has:

Principal cusp

- a less slender principal cusp with a broader base

Lateral cusplets

- only one pair of lateral cusplets

Apron

- the apron is far longer expanded

The size of the teeth of *B. roklumensis* and *Mesiteia* is the same.

The genus *Palaeobrachaelurus* Thies, 1983 was originally attributed to Brachaeluridae. Batchelor & Ward (1990, p. 196) considered that *Palaeobrachaelurus* was synonymous with *Paraginglymostoma* Herman, 1982 and attributed it to Orectolobidae. The only difference that separates *Palaeobrachaelurus* from *Paraginglymostoma* is the presence in *Paraginglymostoma* of two parallel vertical ridges on the labial face of the principal cusp (Fig. 23). Cappetta (2012, p. 165–166) confirmed the importance of this special ornamentation as a generic character and re-attributed *Palaeobrachaelurus* to Brachaeluridae. But this is problematic, because the mesio-distally wide expansion of the crown base, the relatively low principal cusp, the prominent one or two pairs of lateral cusplets and the variability of the apron from very wide and not separated (Fig. 19a–b) to round and well set off from the labial basal outline (Fig. 19c) are characters not found in Brachaeluridae but in Orectolobidae.

The scheme of a high-conical principal cusp with lateral cusplets and a more or less expanded narrow apron can be found as a homoplastic character between many fossil Orectolobiformes, rather than being homologous. The closer one looks the more differences can be seen, e.g., slenderness of the crown, length and form of the apron etc., and it seems almost impossible to place a fossil species in a genus based only on tooth morphology.

4.2.8 *Eostegostoma*

Cappetta (2012) assigned the fossil genera *Eostegostoma* Herman, 1977, *Palaeobrachaelurus* Thies, 1983, and *Paraginglymostoma* Herman, 1982, to the family Brachaeluridae. As with the Recent *Brachaelurus waddi* (Fig. 17), the teeth of the lower Paleocene to middle Eocene species *Eostegostoma angustum* (Nolf & Taverne in Herman, 1977) share many characters with *Mesiteia* (Pl. 9), but there are also differences:

Principal cusp

- the high and sharp principal cusp of anterior teeth is higher than in *Brachaelurus*, very similar to *Mesiteia*

Lateral cusplets

- there are two pairs of strong lateral cusplets on posterior teeth, whereas in Recent *Brachaelurus* there is only one, in fossil *Brachaelurus* there can be two, and in *Mesiteia* there is always one pair of lateral cusplets
- the lateral cusplets remain well-separated from the principal cusp from anterior to posterior teeth
- on many teeth the mesial lateral cusplet originates slightly deeper than the distal one

Apron

- the apron is similar to *Brachaelurus*, and definitely shorter than in *Mesiteia*
- the apron is sometimes bifid
- the apron remains relatively high in posterior teeth

It remains questionable whether these minor odontological differences can be used to discuss differences between genera belonging to different families. Not only heterodonty but also a simple variation of usually more than one character is complicated to describe. Not knowing its age, it will be almost impossible to distinguish an isolated anterior tooth of *Eostegostoma angustum* from a tooth of *Mesiteia*. However, the posterior teeth of *Eostegostoma* with wider and much lower principal cusps, especially teeth with two pairs of lateral cusplets, can be easily identified. One striking difference is the size: the teeth of *Eostegostoma angustum* (as far as I observed) can be almost three times bigger than the biggest teeth of *Mesiteia*. If *Eostegostoma* belongs to Brachaeluridae, this correlates well with the known size of Recent *Brachaelurus waddi*, which grows up to a length of 1.2 m whereas the biggest *Mesiteia* specimen is only 45 cm long.

Eostegostoma is a good sample to demonstrate how difficult it is to assign fossil teeth to a genus and family of modern sharks. When Herman (1977) created this genus for *Ginglymostoma angustum* Nolf & Taverne in Herman, 1977, he knew both Brachaeluridae and Stegostomatidae and thought

Fig. 20. Teeth of *Eostegostoma angustum* (Nolf & Taverne in Herman, 1977), anterior, lateral and posterior teeth, most in labial view, two in lateral view, from the middle Eocene, Lutetian locality St. Pankraz, Austria (coll. FHP).

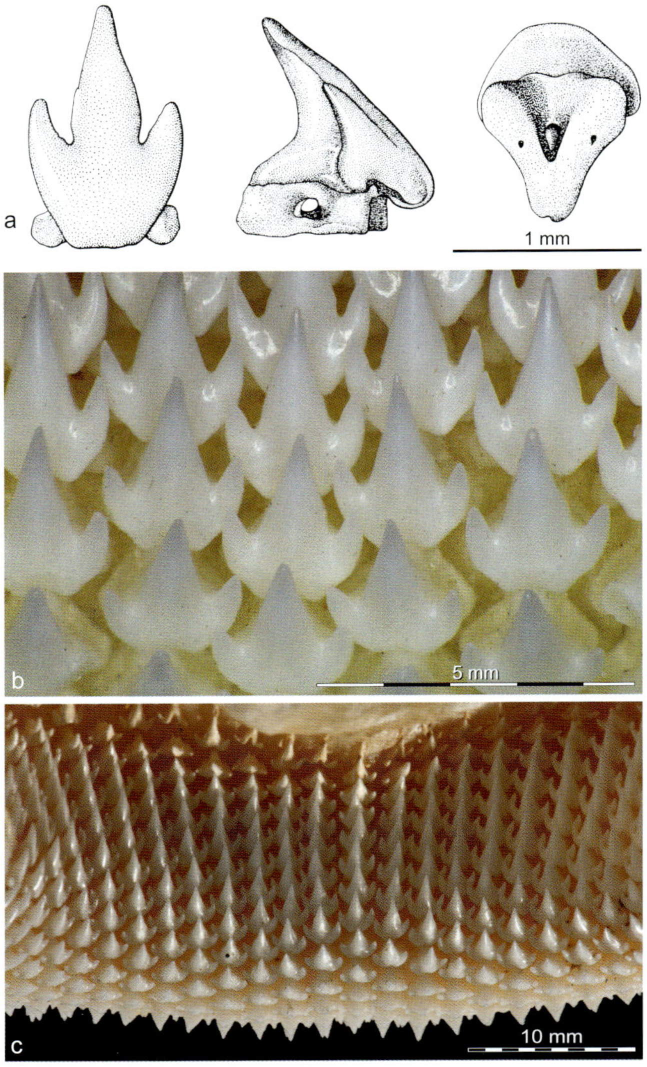

Fig. 21. *Stegostoma fasciatum* (Hermann, 1783). **a.** drawing of a tooth in labial, lateral, and basal view from Cappetta (1980, fig. 17, and 2012, figs. 156 A–C). **b, c.** photographs of a fresh jaw; **b.** detail with teeth of symphysial region; **c.** symphysial region (coll. FHP).

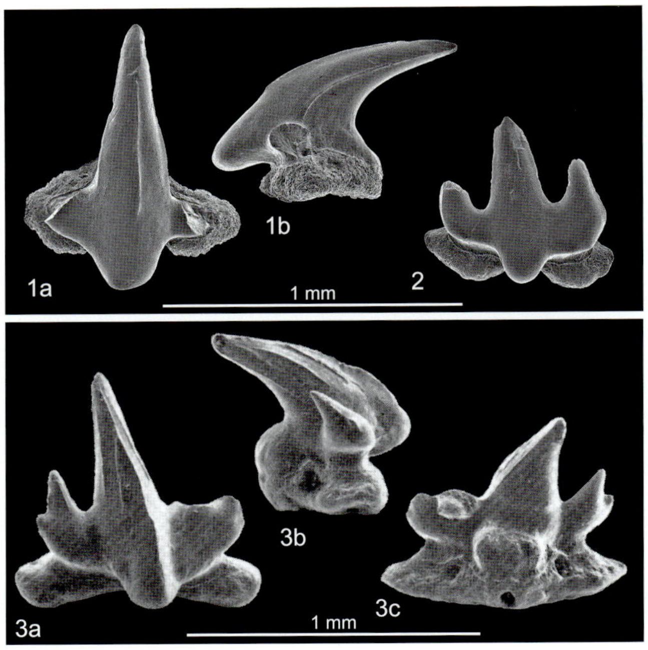

Fig. 22. Lower–Middle Jurassic teeth of *Annea*. **1–2.** *Annea* sp.; anterior tooth (**1**) in labial (a), and lateral view (b); posterior tooth in labial view (**2**); from the Lower Jurassic, Toarcian, SW Belgium, after Cappetta (2012, p. 178, figs. 167 A–B, G); **3.** *Annea carinata* Thies, 1983 in labial (a), lateral (b), and lingual view (c); from the Middle Jurassic, upper Aalenian, N Germany, after Cappetta (2012, p. 178, figs. 166 A–C). ▷▷

the elongated, slender crowns with a subcircular cross-section and with a long tongue-shaped apron would be characteristic of the Stegostomatidae. But he also wrote: "The referral of *Eostegostoma* to the Stegostomatidae is provisional; it could equally be a representative of a family transitional between the Stegostomatidae and Brachaeluridae" (Herman & Crochard 1977, p. 133). Cappetta (1987) could not follow this assessment and wrote: "Teeth of *Eostegostoma* are morphologically much closer to those of the Recent *Brachaelurus* than to *Stegostoma*". Teeth of *Stegostoma* are very special. Cappetta (1980, fig. 17) published a first detailed drawing of a tooth of *Stegostoma* (Fig. 21a), and Herman, Hovestadt-Euler & Hovestadt (1992, pl. 28–31) SEM pictures of teeth from various positions. The principal cusps are symmetrical even in the lateral files and show a constriction at its basal part. A pair of strong and high lateral cusplets, higher than in *Mesiteia* or *Eostegostoma*, sits close to the principal cusp. The apron is wide, less prominent than in *Mesiteia* or *Eostegostoma*, and straight or slightly concave medially.

Knowing only the teeth of Recent Brachaeluridae and Stegostomatidae it is obvious to attribute *Eostegostoma* to Brachaeluridae. Later, in Section 4.2.10 I will show that there is another even more likely option, and that the combination of teeth with a smooth, high-triangular principal cusp and prominent narrow apron (dagger-shaped teeth) has been developed within different groups as a homoplastic character.

4.2.9 *Annea*, *Paraginglymostoma* and *Akaimia*

In searching for possible ancestors of *Mesiteia* one should check the Lower–Middle Jurassic (Toarcian to Callovian) shark fauna described by Thies (1989) with its great variety of Orectolobidae. What there was found and described as *Palaeobrachaelurus*, *Paraginglymostoma*, *Palaeorhincodon*, *Annea* and other Orectolobiformes incertae sedis could lead to the ancestors of *Mesiteia* when more is known about the relationships of these genera to each other.

The teeth of *Annea carinata* Thies, 1983 (Middle Jurassic, upper Aalenian, northern Germany; Fig. 22; 3) are smaller than those of *Mesiteia* (Pl. 9). A dagger-shaped principal cusp (Thies 1989, p. 109) is flanked by one or two pairs of pointed, hook-like lateral cusplets. The apron is relatively expanded. The presence of a prominent strong and sharp vertical ridge with other less distinct divergent folds on the labial face of the crown, as in *Annea*, was thought to be unique among the family Orectolobidae. Teeth of *Annea* sp. (Lower Jurassic, Toarcian, SW Belgium; Fig. 22; 1–2) in Cappetta (2012) have a single axial crest on the labial face of the crown and a single pair of lateral cusplets. A wider crown base, lower principal cusps, relatively higher lateral cusplets and the ornamentation (vertical ridge) on the labial face of the crown separate *Annea* from *Mesiteia*.

It should be mentioned here that more or less prominent vertical ridges on the labial face of the crown seem to be relatively common within Orectolobiformes, and appear also on *Paraginglymostoma*, *Akaimia*, *Parasquatina*, *Restesia*, *Eometlaouia* (Cappetta 2012, p. 160, fig. 145), and on *Acanthoscyllium* (Fig. 10).

▷▷
Fig. 23. Tooth of *Paraginglymostoma* in labial, basal, and lingual view; from the Upper Cretaceous, Maastrichtian, northern Germany, after Cappetta (2012, fig. 155).

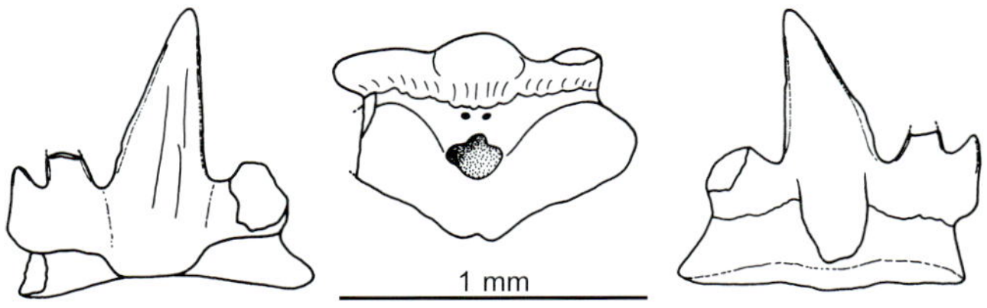

The teeth of *Paraginglymostoma* Herman, 1982 from the Aptian (Lower Cretaceous) of England and from Maastrichtian (Upper Cretaceous) of northern Germany have two parallel vertical ridges on the labial face of the crown. The principal cusp is low compared to the crown base, which is mesio-distally relatively widely expanded. There are two pairs of lateral cusplets, and the apron is short and rounded. These characters separate *Paraginglymostoma* from Brachaeluridae, and from *Mesiteia*.

The teeth of *Akaimia* Rees, 2010 with two species, *Akaimia altucuspis* Rees, 2010 (pl. 1, figs. 1–22) from the Callovian to

▷▷
Fig. 24. *Akaimia myriacuspis* Srdic, Duffin & Martill, 2016. **1.** anterior tooth in labial (a), and lateral view (b); **2.** posterior tooth in labial view; from the Middle Jurassic, Callovian, eastern England, after Srdic Duffin & Martill (2016, figs. 5a–b, d).

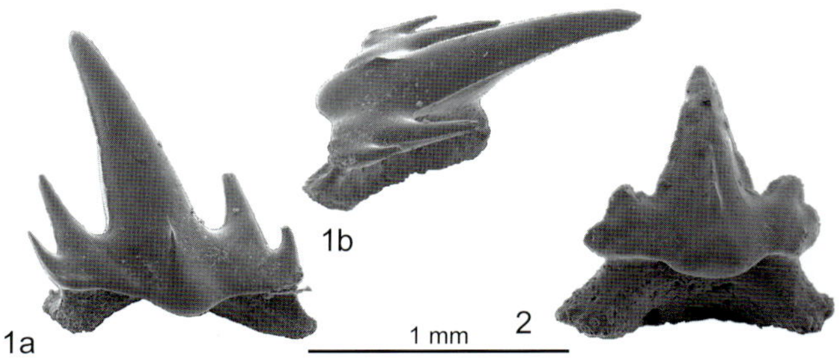

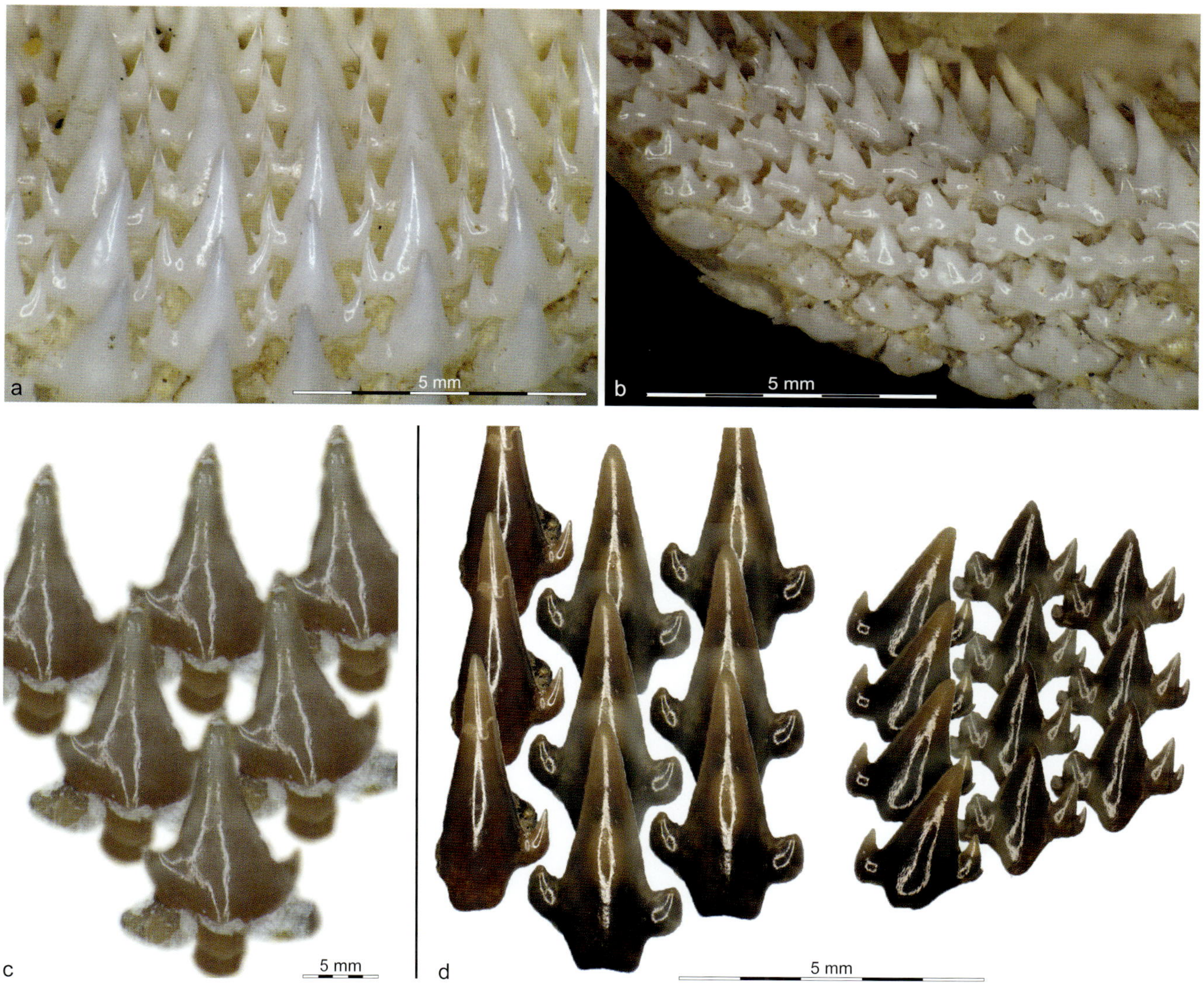

Fig. 25. Teeth of *Pseudoginglymostoma brevicaudatum* (Günther in Playfair & Günther, 1867) (coll. FHP-R-015). **a.** symphysial region; **b.** posterior region; **c.** artificially composed set of anterior teeth of *Mesiteia emiliae* (from specimen #08). The staggered arrangement of the teeth is the reason for the fact that the apron in *Mesiteia* can be so elongated and that it is never bifid, in contrast to the apron in *Pseudoginglymostoma* and *Eostegostoma*; **d.** Artificially composed set of anterior and posterior teeth of *Eostegostoma angustum* (see Fig. 20). The teeth stand one behind the other in a line, same as in *Pseudoginglymostoma*, this way the apron of the following younger tooth is restricted in its growth and tends to get bifid, caused by the pressure of the uvula of the anterior older tooth. The shape of the posterior teeth of *Eostegostoma* with strong and doubled lateral cusplets is indistinguishable from teeth of *Pseudoginglymostoma*. The scale indicates that teeth of *Eostegostoma* have almost the same size as teeth of *Pseudoginglymostoma*, while teeth of *Mesiteia* are much smaller.

Oxfordian (Middle to Upper Jurassic) of southern Poland, and *Akaimia myriacuspis* Srdic, Duffin & Martill, 2016 from the Callovian (Middle Jurassic) of eastern England are very similar to *Annea* and *Paraginglymostoma*, but separated by a higher principal cusp, very sharp and high lateral cusplets and a strong vertical ridge reduced to the central basal part of the labial face of the principal cusp. The apron is round and relatively short.

These three genera, *Annea*, *Paraginglymostoma* and *Akaimia* share the combination of a high principal cusp, a moderately wide expanded crown base with a pair of narrow, sharp lateral cusplets, a short and relatively wide apron, and a characteristic ornamentation with a single (or two) vertical ridge(s) on the labial face of the crown. These characters bring these genera close to Orectolobidae.

4.2.10 *Pseudoginglymostoma*

Now, at the end of my odontological excursion I will discuss the surpising similarity of the teeth of *Mesiteia* to the teeth of the poorly known Recent genus *Pseudoginglymostoma* of the family Ginglymostomatidae.

The smooth and slender crowns with a subcircular cross-section have high and strong principal cusps, closely flanked by a pair of strong lateral cusplets, which have additional and much smaller cusplets at both sides. The principal cusps are slightly inclined and get lower and broader toward the rear in lateral and posterior files. The apron is wide, rounded, set off from the labial outline and restricted to the central region of the tooth. Anterior, lateral and posterior teeth all fit very well with teeth of the Paleo-/Eocene species *Eostegostoma angustum*.

Dingerkus (1986) erected the genus *Pseudoginglymostoma* after Compagno (1984) postulated that *Ginglymostoma brevicaudatum* Günther in Playfair & Günther, 1867 should be placed in its own genus. The teeth of *Pseudoginglymostoma brevicaudatum* (Günther in Playfair & Günther, 1867) (figured in Bass, D'Aubrey & Kistnasamy 1975, pl. 13, and here Fig. 25) share many characters with teeth of *Eostegostoma angustum* Nolf & Taverne in Herman, 1977) (Fig. 20), *Brachaelurus* (Fig. 17), and *Mesiteia* (Pl. 9), but there are also differences if one compares teeth of the same regions within the jaws:

Principal cusp

- in anterior teeth the height of the principal cusps in relation to the width is highest in *Mesiteia* and *Eostegostoma*, and slightly less high in *Brachaelurus* and *Pseudoginglymostoma*
- the height of the principal cusps decreases linearly towards the rear but remains relatively high up to the 10th file and further in *Mesiteia*, whereas the teeth of the 7th file are already reduced to about half the heigth in *Pseudoginglymostoma* and probably in *Eostegostoma*, and even earlier in *Brachaelurus*
- with the reduction of the height of the principal cusps the crown base gets wider and the shape changes from high-triangular to wide-triangular, less visible in *Mesiteia*, clearly recognizable in *Pseudoginglymostoma* and in *Eostegostoma*, and most obvious in *Brachaelurus*

Lateral cusplets

- the number of lateral cusplets seems to be a constant character, but depends on the position of the teeth;
  on all teeth of *Mesiteia* from anterior to posterior files only one pair of strong lateral cusplets could be observed, and in far posterior files the lateral cusplets can be reduced, mesially more than distally;
  anterior and lateral teeth of Recent *Brachaelurus* also have only one pair of lateral cusplets, and the teeth in its posterior files have the cusplet distally reduced to a heel;
  in fossil *Brachaelurus* and *Palaeobrachaelurus* – if these teeth really can be assigned to Brachaeluridae – there can be two pairs of lateral cusplets from anterior to posterior files;
  anterior and lateral teeth of *Eostegostoma* have one pair of lateral cusplets, but the teeth in its posterior files can have two pairs of lateral cusplets;

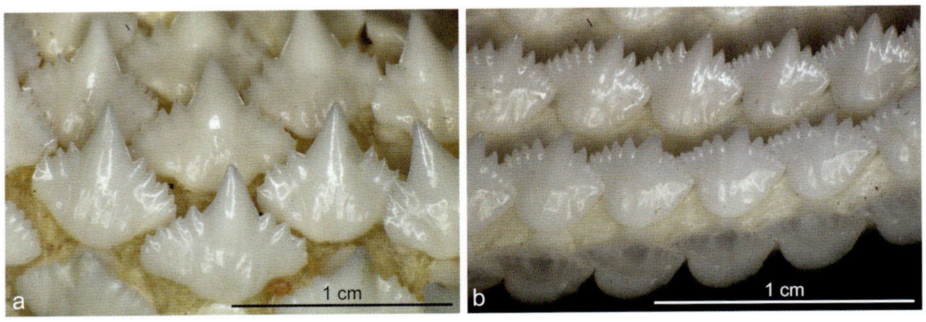

Fig. 26. Teeth of Ginglymostomatidae. **a.** *Ginglymostoma cirratum* (Bonnaterre, 1788) (coll. FHP-R-170); **b.** *Nebrius ferrugineus* (Lesson, 1831) (coll. FHP-R-2118).

and in *Pseudoginglymostoma* there are always two pairs of lateral cusplets on teeth of all files

- the size, shape, and method of their branching from the principal cusp seems also to be a useful constant character; the not very high but strong lateral cusplets of *Mesiteia* are laterally protruding and not close-fitting to the principal cusp, and they remain prominent and well-separated from the principal cusp from anterior up to the far posterior files, reduced only in posteriormost files;
  on anterior and lateral teeth of Recent *Brachaelurus* the lateral cusplets are similar to *Mesiteia*, but even higher in relation to the height of the principal cusp, well-separated from the principal cusp, unless they get reduced and not separated any more in posterior files. In a very different way, the lateral cusplets soon become wider and flatter in labial view, lower, and more inclined towards the rear. Posterior teeth never tend to produce a second pair of lateral cusplets;
  on anterior and lateral teeth of *Eostegostoma* the lateral cusplets are very similar to *Mesiteia*, not very high but strong, laterally protruding well-separated and not close-fitting to the principal cusp. The lateral cusplets remain high and well-separated up to the far posterior files, while its principal cusps get lower and wider and as a remarkable difference from *Mesiteia* and *Brachaelurus* the teeth of posterior files can have a second pair of smaller cusplets, close-fitting to the inner cusplets; the lateral cusplets of teeth of the posterior files never tend to be reduced to heels as in *Mesiteia* and *Brachaelurus*;
  and on anterior and lateral teeth of *Pseudoginglymostoma* the inner of the two pairs of lateral cusplets is in its size similar to *Mesiteia* and *Brachaelurus*, very sharply pointed; the outer pair is very small and narrowly attached to the inner cusplet, not as separated as in *Eostegostoma*; the inner cusplets are very close-fitting to the principal cusp; both cusplets are not laterally protruding; on the wider expanded posterior teeth with lower principal cusps the lateral cusplets are slightly wider, the inner pair still high; there are always two pairs of lateral cusplets on teeth of all files

Apron
The size and shape of the apron, and its extension at the basal rim of the crown are subject to a wide range of variation. This concerns both the mesio-distal extension and the shortening of the apron from anterior teeth to posterior files. But it is especially remarkable that within higher categories, especially within one family or clade, considerable differences can be observed, which can be found in very similar forms in different families as homoplasy; nevertheless the width and height of the apron in relation to a comparable position within the dentition is a reliable and stable feature;

- in anterior and lateral teeth, the height of the apron is highest in *Mesiteia*, and less high in *Brachaelurus*, *Eostegostoma* and *Pseudoginglymostoma*, where it is very similar
- the apron is sometimes bifid in *Brachaelurus*, and often in *Eostegostoma* and *Pseudoginglymostoma*, but it has not been observed in *Mesiteia*; this reflects a different way of overlapping or contact between teeth within a file; in *Brachaelurus*, *Eostegostoma* and *Pseudoginglymostoma* the apron touches the uvula of the earlier-developed tooth and gives it some stability; a close contact between apron and uvula during the developmental phase of the enameloid, when the enameloid of the apron is softer than the more developed uvula of the older tooth, creates a bifid apron; this is different in *Mesiteia*, where the apron fits in the gap between the two teeth of the neighboring earlier-developed teeth of a tooth row
- the apron of *Mesiteia* is narrower than in *Brachaelurus*, *Eostegostoma* and *Pseudoginglymostoma*, and it remains relatively high up to the posterior teeth for the same reason as explained above, the apron of *Mesiteia* fits in the gap between the two teeth of the neighboring earlier-developed teeth of a tooth row

The Recent family Ginglymostomatidae is a paraphyletic group with a subfamily Pseudoginglymostomatinae (with a single genus *Pseudoginglymostoma*) as sister group of the subfamily Ginglymostomatinae (with two genera *Ginglymostoma* and *Nebrius*) (Compagno 2001, Goto 2001). *Eostegostoma* is currently assigned to Brachaeluridae (Cappetta, 2012).

The combination of a high-triangular principal cusp with one or two pairs of prominent lateral cusplets that remain prominent up to the posterior files, and in anterior files with a mesio-distally not wide expanded crown base, are synapomorphic dental characters, that separate it from any other Orectolobiformes. *Pseudoginglymostoma* and *Eostegostoma* are closely related to each other and based on dental characters I propose that *Eostegostoma* should be re-allocated together with *Pseudoginglymostoma* to the subfamily Pseudoginglymostomatinae.

It remains surprising that *Pseudoginglymostoma* (Fig. 25a,b), a Recent genus of the family Ginglymostomatidae has teeth that more closely resemble teeth of the Recent family Brachaeluridae. Therefore, I figure here also for comparison teeth of the only other two genera of Ginglymostomatidae, teeth of *Ginglymostoma* (Fig. 26a) and *Nebrius* (Fig. 26b) with its wide-triangular cusps without separated lateral cusplets, but serrated mesial and distal cutting edges, and a wide expanded apron.

4.2.11 Summarizing dental characters of *Mesiteia*

As I tried to describe in the previous Sections, there are surprising differences, but also surprising similarities between teeth of *Mesiteia* and other genera. The teeth of Recent Hemiscylliidae (*Hemiscyllium* and *Chiloscyllium*) are smaller, wider at their base, have less prominent lateral cusplets, and a short tongue-shaped apron. Based on the morphology of these teeth there is also no close relationship with the two other fossil genera of hemiscylliid sharks from Lebanon, *Acanthoscyllium* and *Almascyllium*. The similarity between teeth of anterior files of *Mesiteia* and *Eostegostoma* is astonishing; both having strong, upright, dagger-shaped principal cusps with one pair of prominent lateral cusplets. This leads to the Recent Brachaeluridae, to which *Eostegostoma* is currently assigned. Teeth of posterior files of *Brachaelurus* are expanded, and have the lateral cusplets partly reduced to heels, similar to the condition in *Mesiteia*, whereas posterior teeth of *Eostegostoma* are laterally expanded, and their lateral cusplets remain high and prominent and are often doubled. This last feature brings teeth of *Eostegostoma* closer to *Pseudoginglymostoma*, to which it should be re-allocated in the subfamily Pseudoginglymostomatinae.

Plate 10 ▷
Selection of specimens of *Mesiteia emiliae* with best preserved cranium and branchial region.
a. *Mesiteia* specimen #01 (holotype, Zagreb, Croatian Natural History Museum, Inv. number 9354); photo Sanja Japundžić.
b. *Mesiteia* specimen #18 (Lebanon, unknown new owner); note the many unknown structures in front of the skull, looking a bit like narial barbels, the branchial arches and the well preserved pectoral fin; photo Roy Nohra.
c. *Mesiteia* specimen #08 (Munich, coll. FHP).
d. *Mesiteia* specimen #14 (Lebanon, coll. Expo Hakel, Roy Nohra), note the details of branchial arches.
e. *Mesiteia* specimen #10 (Munich, coll. FHP).
f. *Mesiteia* specimen #13 (Switzerland, coll. René Kindlimann RK/7-711).
g. *Mesiteia* specimen #11 (Munich, coll. FHP). For explanation see Figure 28. Some specimens were mirrored for better comparison (marked with ↔).

Plate 10
a
b
c
d
e
f
g

Fig. 27. Cranium with rostral region of *Mesiteia emiliae*. **a.** *Mesiteia* specimen #08 (Munich, coll. FHP); **b.** *Mesiteia* specimen #13 (Switzerland, coll. René Kindlimann RK/7-711). It is clearly visible that the mesial rostral rod (rm) consists of two overlapping rods of cartilage. The outer nasal cartilage in specimen #13 is on the right side artificially moved a bit in anterior direction. onc – outer nasal cartilage; rmd – dorsal medial rostral rod; rmv – ventral medial rostral rod; s – symphysis.

There is no single character that would allow identification of a tooth of *Mesiteia*; only the combination of specific characteristics of teeth of anterior and posterior files allows to identify *Mesiteia* by its teeth.

- **Character 01:** Dental characters. In general, teeth of Orectolobiformes, except of *Parascyllium* and *Cirrhoscyllium*, possess an apron, either broadly expanded along the basal part of the labial face of the crown, or narrow and prominent in a median position, and on the opposite lingual face an uvula. Since teeth of some tooth files of *Mesiteia* are practically indistinguishable from teeth of *Brachaelurus* and *Eostegostoma*, and also from *Pseudoginglymostoma*, only a combination of dental characters can be used to distinguish the teeth of *Mesiteia*. In anterior and lateral files, principal cusp upright, smooth, high-triangular, not much laterally expanded, with only one pair of prominent lateral cusplets (dagger-shaped) and a very long, prominent, narrow apron. In posterior files, principal cusp inclined, only slightly expanded, with partly reduced lateral cusplets.

Three states are recognized,
0 apron and uvula absent;
1 combination of dental characters different from the characters described for *Mesiteia*, at least in combination of all tooth files;
2 combination of dental characters as described for *Mesiteia*.

In the plesiomorphic state as seen in *Parascyllium* and *Cirrhoscyllium*, the apron and uvula are not yet developed. For all other Orectolobiformes, the development of apron and uvula is apomorphic but leads to various forms with further apomorphies and reversals. The combination of characters as described above is autapomorphic for *Mesiteia*.

4.3 Anatomical characters of *Mesiteia*

Many details are visible on *Mesiteia*, and each specimen provides special insights, but this might not be the right place to discuss every detail of all 26 specimens. I concentrate on those characters I believe to be important and try to describe the details of those specimens where they are best preserved. This will be especially relevant on specimen #08, and I will try to find characters on *Mesiteia* that were previously used mainly by Goto (2001) for a cladistic analysis of the Recent families of Orectolobiformes.

4.3.1 Cranium

On most specimens, including the holotype (#01), the two specimens in Paris (#02, #03) and the two specimens at the BSPG in Munich (#04, #05) the cephalic region is strongly crushed and does not allow very precise observations. The cephalic regions of the holotype (#01) and specimens #18, #08, #14, #10, #13 and #11 are figured on Plate 10, and with explanations of at least some of the features in Figs. 27-29. Most of these specimens (#01, #18, #08, #14 and #10) provide a ventral view of the skull, #13 is a bit squeezed to a lateral view, and #11 gives us obviously a dorsal view. On many of the other specimens, e.g. #04, the skull is laterally compressed, and the cartilages are split in the middle. This gives the wrong impression that the cranium and the jaws would be narrow and pointed anteriorly.

There is not much to see from the thin cartilages of the chondrocranium of *Mesiteia*. The chondrocranium is shorter than in *Chiloscyllium* and obviously wider than long. Two dark oval areas, the fossilized remains of the eyes, mark the orbit. The mouth is terminal. The wide semicircular anterior margin of the skull corresponds to the palatoquadrate, which is strong but less bulky than the Meckel's cartilages just behind it. The dimensions of the palatoquadrate and Meckel's cartilages seem to be very similar to those of *Chiloscyllium*. Most of the visible teeth belong to the palatoquadrate. The symphysis of the Meckel's cartilages on specimen #08 is well marked (Fig. 27a). The posterior part of the cranium is limited by the hyoid arch.

Labial cartilage elements are displayed on several specimens, e.g. #08, #01, #14. Whether the many small cartilages around the anterior part of the mouth of #18 (Pl. 10b) are all labial cartilages or are part of any other close attached fossil cannot be determined with the few photographs available.

Rostrum

- **Character 02:** Medial rostral rod (rm). The medial rostral rod is depressed and arched dorsally to marginate the ventral profile of the snout. A thin and greatly elongated anteriorly medial rostral rod is plesiomorphic for Galea sensu Shirai (1996) (Goto 2001, p. 11). In most Orectolobiformes, the shape of the rostral rod is thin and feeble, with weak or no expansion in most taxa (Goto 2001, p. 14, figs. 6-12, 14). The rostral rod has a large expansion on its tip in *Stegostoma* (Goto 2001, fig. 13), and in *Rhincodon*, it is greatly broadened and shortened (Goto 2001, fig. 15). In *Mesiteia*, a short and strong medial rostral rod (rm) projects anterior to the nasal capsules, well preserved on specimens #08 and #13 (Fig. 27), and #14 (Pl. 10). The rod consists of two elements, a dorsal (rmd) and a ventral (rmv) cartilage rod, the dorsal longer expanded more anteriorly than the ventral one. The ventral rostral rod originates from the inter-nasal septum (Goto 2001 e.g. fig. 6D), whereas the dorsal rod expands from the antero-mesial margin of the nasal capsule (Goto 2001, e.g. fig. 6C). In *Acanthoscyllium*, the characteristic bifid rostrum also consists of two rostral rod cartilages, but a left and a right one (Cappetta 1980, fig. 22), which form a rostral fenestra with its two elongated rostral rods. The left and right branches of the medial rostral rod in *Acanthoscyllium* are much longer than in *Mesiteia*, consisting of a strong basal part and a thin, feeble ray. In *Pararhincodon*, the medial rostral rod is strong and about twice as long as in *Mesiteia* (Cappetta 1980, fig. 26) but this specimen is not complete. Better preserved specimens have a longer rostrum which becomes also bifid at its anterior end (pers. observation, coll. FHP).

Three states are recognized,
0 medial rostral rod consists of one cartilage;
1 medial rostral rod consists of two short cartilages, a dorsal and a ventral rod;
2 medial rostral rod is bifid and very long.

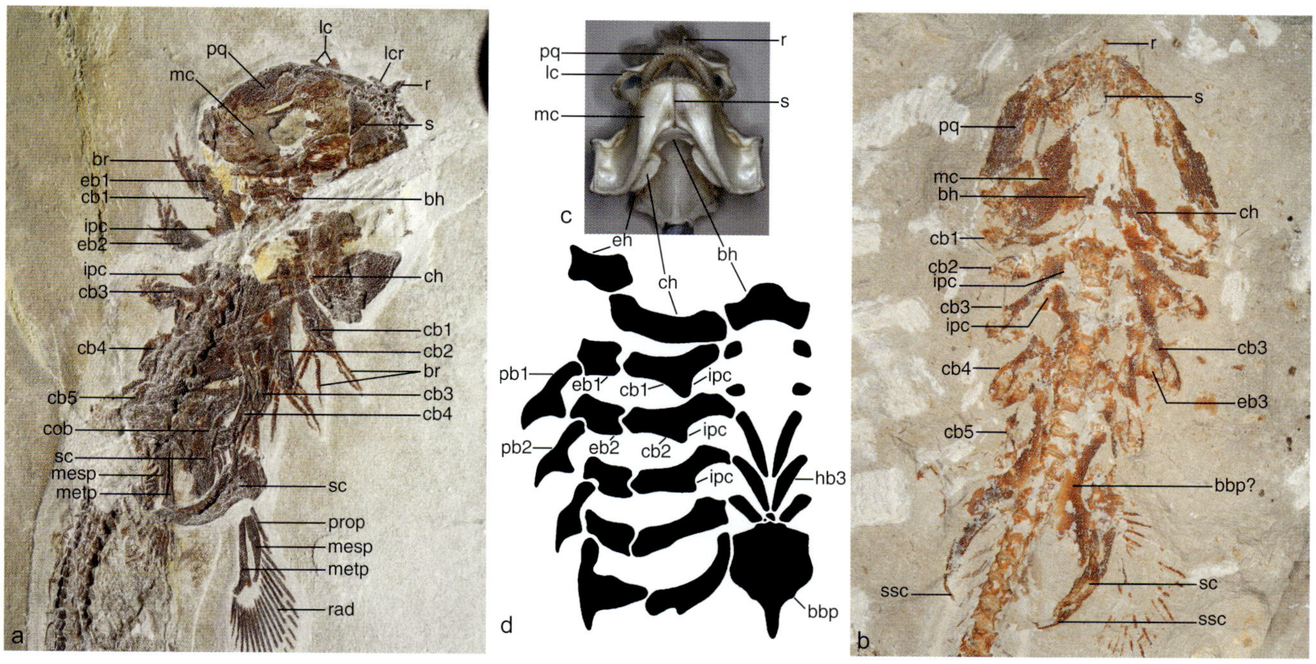

Fig. 28. Hyal and branchial elements of *Mesiteia emiliae*. **a.** *Mesiteia* specimen #08 (Munich, coll. FHP); **b.** *Mesiteia* specimen #10 (Munich, coll. FHP), compared with those of **c.** *Chiloscyllium plagiosum* in ventral view (coll. FHP R-2206), and **d.** *C. punctatum* (after Dingerkus & DeFino 1983, fig. 55).
Cranium: lc – labial cartilages; lcr – lateral cartilages of rostrum; mc – Meckel's cartilages; pq – palatoquadrate; rm – medial rostral rod; s – symphysis.
Pharyngeal arches: bbp – basibranchial plate; bh – basihyal; br – branchial rays (= gill filaments); cb1-5 – ceratobranchial 1-5; ch – ceratohyal; eb1-5 – epibranchial 1-5; eh – epihyal (= hyomandibula); hb – 1-4 hypobranchial 1-4; ipc – inner process of the ceratobranchial 1-3 – pb1-5 – pharyngobranchial 1-5.
Pectoral girdle: cob – coracoid bar; sc – scapular cartilage; ssc – suprascapular cartilage.
Pectoral fin: mesp – mesopterygium; metp – metapterygium; prop – propterygium; rad – radials.

The type of rostrum consisting of a dorsal and ventral rod, as seen in *Mesiteia*, is unique and different from all other Orectolobiformes. It is assumed to be autapomorphic for *Mesiteia*. For *Acanthoscyllium*, the bifid rostrum is also an autapomorphic character.

- **Character 03**: Outer nasal cartilage (onc). The outer nasal cartilage is a ring-like cartilage with an expansion supporting the anterior nasal flap mesially (Goto 2001, p. 12). In most Orectolobiformes, the outer nasal cartilage is arched, opens posteriorly and is completely isolated from the nasal capsule (Goto 2001, figs. 16C–E). In *Rhincodon*, it is circular, and fused with the nasal capsule antero-mesially and laterally (Goto 2001, fig. 16F). *Parascyllium* and *Cirrhoscyllium* have a thin cartilaginous bar extending from the nasal capsule that supports an unbranched nasal barbel (Goto 2001, p. 56, fig. 73B, and for *Parascyllium ferrugineum* fig. 11A). In *Mesiteia*, the medial rostral rod is flanked by two outer nasal cartilages pointing to the sides, giving the rostrum a cross-like shape (Fig. 27). These two cartilages seem to expand proximally from the dorsal medial rostral rod (Fig. 27a) but these cartilages are isolated from the rostral rod and nasal capsule as can be seen on *Mesiteia* #13 (Fig. 27b), where the outer nasal cartilage of the right side is moved a bit anteriorly. The distal edges are chamfered, similar to the drawing of an outer nasal cartilage on a neurocranium of *Hemiscyllium freycineti* (Goto 2001, fig. 7A). The difference from all other taxa, including Orectolobiformes is that in *Mesiteia*, the outer nasal cartilages are not ring-like or arched but straight cartilaginous bars, short, expanding not very widely from the base of the medial rostral rod, and not surrounding the nasal capsule. *Mesiteia* certainly had nasal barbels, and its outer nasal cartilage could have supported laterally the barbel as described for *Parascyllium* and *Cirrhoscyllium* (Goto 2001, p. 56, fig. 73B). In *Acanthoscyllium*, Cappetta (1980, p. 114, fig. 22) described two expansions close to the base of the rostrum as corresponding to the anterior margin of the nasal capsule.

Two states are recognized,
0 outer nasal cartilage is arched, surrounds the anterior nasal capsule;
1 outer nasal cartilage straight, short bar, expanding from rostral rod.

The shape of outer nasal cartilages in *Mesiteia* is autapomorphic.

4.3.2 Gill and hyoid arches

The branchial region is poorly preserved in most specimens, but some of the new specimens allow astonishing insights, e.g. #08, #10, #14, #18 (Plate 10, and Figs. 28–29). Because of the poor preservation of the few specimens that were available for his study, Cappetta (1980) could only mention that on the holotype as "figured by Kramberger (1884), and later by Jaekel (1894), we can see 3 ends of arches, each carrying several rather fine cartilaginous rods, probably corresponding to branchial rays".

I use *Chiloscyllium punctatum* (after Dingerkus & DeFino 1983, fig. 55) for comparison, and I follow the description of characters by Goto (2001). The examination of the new specimens is not yet finished but one can identify a curved basihyal that is wider than long. The ceratohyal is covered by the posterior edge of Meckel's cartilages, but it is long, bent, and robust. The epihyal is also covered by the lateral edge of the Meckel's cartilages. On specimen #10 one can see the rest of a robust plate with long anterior and posterior processes that could be the basibranchial plate.

Ceratobranchials

The five ceratobranchials are strong and long and at least partially visible in many of the specimens. The fifth ceratobranchial is close to the pectoral girdle and somewhat smaller than the others. In smaller (juvenile) specimens it seems that they are more fragile.

- **Character 04:** Inner process on the medial margin of ceratobranchial (ipc). According to Goto (2001, p. 88) the presence of a "prominent inner process on the medial margin of ceratobranchial" (Goto 2001, figs. 4C–D) is one of the two synapomorphies that characterizes the order Orectolobiformes as a monophyletic lineage. In Orectolobiformes except *Rhincodon*, it is prominent, located on the medial margin of the ceratobranchial in the anterior three arches, and can be recognized also on a few specimens of *Mesiteia*, e.g. #08 and #10.

Two states are recognized,
0 inner process on the medial region of ceratobranchial absent;
1 inner process of ceratobranchial present.

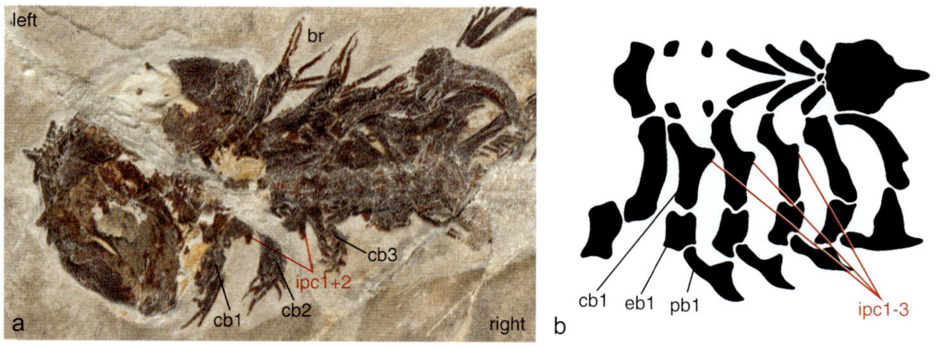

Fig. 29. Pharyngeal arches of *Mesiteia emiliae*. **a.** *Mesiteia* specimen #08 (Munich, coll. FHP), and for comparison; **b.** *Chiloscyllium punctatum* (after Dingerkus & DeFino 1983, fig. 55). The inner process of the ceratobranchial (ipc) is well exposed on the medial margin in the anterior three branchial arches on the right side of this specimen of *Mesiteia*.
Pharyngeal arches: br – branchial rays (= gill filaments); cb1-5 – ceratobranchial 1-5; eb1-5 – epibranchial 1-5; ipc – inner process of the ceratobranchial 1-3; pb1-5 – pharyngobranchial 1-5.

Fig. 30. The comparison with the Munich specimen #04 of *Mesiteia emiliae* (Munich, SNSB-BSPG 2011 I 124) in UV-light shows the "normal stage" of preservation as seen in the few specimens that are kept in "official collections", and demonstrates the dramatic increase of information as much better preserved and professionally prepared specimens from private collections became available.

For *Mesiteia* this is a synapomorphic character, shared with all other Orectolobiformes except *Rhincodon*, for which its absence is a reversal.

Five long, thin and flexible branchial rays or gill filaments articulate distally with both ceratobranchial and epibranchial. The connection of ceratobranchial with epibranchial is not clearly seen, but on specimen #08 it seems that two branchial rays articulate with the ceratobranchial, and three with the epibranchial.

Epi- and pharyngobranchials could not yet be distinguished.

Extrabranchial cartilage could not be identified because of insufficient preservation.

4.3.3 Pectoral girdle and pectoral fins

The pectoral girdle of the earlier known specimens #01 and #04 is poorly preserved and the fragmentary preservation did not allow any exact description. Only the scapula and a few radials of the pectoral fin can be recognized. The pectoral girdle starts at the level of the 11th or 12th vertebra on the holotype #01, the 8th vertebra on the Paris specimen #02, the 11th on the Munich specimen #04, and the 11th on specimen #08. The exact position and the outline of the coracoid bar is virtually impossible to determine at any of the specimens. This is because this area is often too fragmentarily preserved or too many elements lie one over the other. The scapula and pectoral fin of at least one side are rather perfectly preserved, e.g. on specimen #08.

Scapular cartilage

The scapula is strong and more bent on specimens #08 and #14 than e.g. on specimens #02, #04, #10, #18 and #24, where it is thinner and less bent. Compared with *Chiloscyllium* and *Hemiscyllium* (Fig. 32, after Dingerkus & DeFino 1983, figs. 41, 42) the scapula on specimens #08 and #14 is stronger and more bent but on other specimens, e.g. #02, #04, #10, #18 and #24, the scapulae are very similar to those of *Chiloscyllium* and *Hemiscyllium*.

- **Character 05:** Suprascapular cartilage (ssc). The dorsal processes of the scapula of *Mesiteia* have small suprascapular cartilages (ssc) associated with them dorsally (Fig. 28b, 31d,f). According to Goto (2001, p. 9, 31, fig. 32) suprascapular cartilage is present among living taxa only in some Orectolobiformes and *Chlamydoselachus*. It is found in *Chiloscyllium*, *Hemiscyllium*, *Ginglymostoma*, *Stegostoma* and *Rhincodon*, whereas it is absent in the remaining genera. The presence of the suprascapular cartilage is assumed to be an apomorphic character. Goto (2001, p. 74) used it as a character to define its clade I, containing all species of *Chiloscyllium*, *Hemiscyllium*, *Ginglymostoma*, *Stegostoma* and *Rhincodon*.

Two states are recognized,

0 suprascapular cartilage absent;
1 suprascapular cartilage present.

For *Mesiteia* this is a synapomorphic character, shared with *Chiloscyllium*, *Hemiscyllium*, *Ginglymostoma*, *Stegostoma* and *Rhincodon* (clade I in Goto 2001).

Pectoral fins

The pectoral fin cartilages of *Mesiteia* are different from those of Hemiscylliidae (Fig. 32) and one can distinguish two types. In type A, represented in Fig. 31 by specimen #08, #21 and #10, the propterygium, mesopterygium, and metapterygium are all elongated. The propterygium has about half the length of the mesopterygium and the metapterygium is still a bit longer. The propterygium is as thin and straight as the radial elements. The mesopterygium is proximally wider and gets distally thinner without getting grooved, as in *Chiloscyllium*, along its distal leading edge where the associated radial elements fit. The metapterygium is the strongest element, distally forming a rounded expansion, looking Y-shaped, maybe due to its incomplete preservation. There is only one row of radial elements, not three, as in Hemiscylliidae. It is only due to a fragmentary preservation of some parts of the fins that sometimes it looks like there would be two or more rows of radials. The radials are long compared with those of Hemiscylliidae and get thinner and appear flexible distally. There are no fusions of the radials, especially not of the pro- and mesopterygium, as there are in Hemiscylliidae, no thickening of distal radial elements and no paired radials. There is one short radial articulating with the propterygium, six radials are articulating with the mesopterygium, and eleven radials are articulating with the metapterygium, totalling 18 radials in specimens #08 and #21, and 18 or 19? in specimens #10.

In type B, represented in Figure 31 by specimens #01 and #18, the whole pectoral fin is more round than oval, and the rays are all further spread. The pro-, meso-, and metapterygium cartilages are less elongated than in type A but otherwise similar. The propterygium is not as thin and as straight as the metapterygium. It is not very clear to see but the metapterygium is distally also expanded and Y-shaped. There is only one row of radial elements, and the appearance of two or more rows of radials is a preservational artifact. The radials are long, thin, and appear flexible distally. There are no fusions of the radials, no thickening of distal radial elements and no paired radials. There is one radial articulating with the propterygium that is longer than in type A. Eight radials are articulating with the mesopterygium, and seven radials are articulating with the metapterygium. In type B pectoral fins there would be 16 or 17? radials.

Due to the suboptimal preservation of most specimens, but one should not overvalue these two types of pectoral fin structures, but the differences as mentioned between *Mesiteia* and Hemiscylliidae are rather striking.

- **Character 06:** Pectoral basal cartilages. The pectoral basal cartilages, i.e., propterygium (prop), mesopterygium (mesp) and metapterygium (metp) are according to Goto (2001, p. 9, 32, figs. 34A–F) different within Orectolobiformes, and different within Hemiscylliidae. In *Orectolobus*, *Eucrossorhinus*, *Brachaelurus* and *Chiloscyllium* except *C. indicum*, the propterygium, mesopterygium and metapterygium are divided into three isolated cartilages. In *Sutorectus*, *C. indicum*, *Hemiscyllium*, *Ginglymostoma*, *Stegostoma* and *Rhincodon*, the propterygium and mesopterygium are fused into a single cartilage, with a distinct fissure at its distal region. In *Parascyllium* and *Cirrhoscyllium*, there are two cartilages, with the anterior one lacking a fissure at its distal region. There are different interpretations about this "fused" pro- and and metapterygium. Following Compagno (1988) the propterygium is considered lost. Shirai (1992) considered as ambiguous whether the anterior cartilage is the mesopterygium or is a fused cartilage of the propterygium and mesopterygium. Goto (2001) followed Shirai's treatment for this character as ambiguous (?) because of lack of the consistent identification of this element. As most living taxa have the tribasal pectoral basal cartilages (Goto 2001, p. 9), the fused basal cartilages are assumed to be apomorphic. In *Mesiteia* the pectoral basal cartilages are divided into three elements, the propterygium and mesopterygium are widely separated from each other compared to *Chiloscyllium* (Figs. 31–32), and with no tendency to be fused. This condition is estimated to be autapomorphic.

Three states are recognized,

0 separated pro-, meso- and metapterygium;
1 propterygium absent (or completely fused with mesopterygium);

Fig. 31. Pectoral girdle and pectoral fins of *Mesiteia emiliae*. **a.** *Mesiteia* specimen #08 (Munich, coll. FHP); **b.** *Mesiteia* specimen #21 (Lebanon, coll. Abi Saad); photo Pierre Abi Saad; **c.** *Mesiteia* specimen #01 (holotype, Zagreb, Croatian Natural History Museum, Inv. number 9354); photo Sanja Japundžić; **d.** *Mesiteia* specimen #10 (Munich, coll. FHP); **e.** *Mesiteia* specimen #18 (Lebanon, unknown new owner); photo Roy Nohra; **f.** *Mesiteia* specimen #14 (Lebanon, coll. Expo Hakel, Roy Nohra).
Pectoral girdle: cob – coracoid bar; sc – scapular cartilage; ssc – suprascapular cartilage.
Pectoral fins: mesp – mesopterygium; metp – metapterygium; prop – propterygium; rad – radials; in red the number of the radials from anterior to posterior radials.

2 propterygium partly fused with mesopterygium (fissure present).

The fusion of propterygium with mesopterygium could either represent different autapomorphies for the clades including *Sutorectus*, *C. indicum* and *Hemiscyllium*, and *Ginglymostoma*, *Stegostoma* and *Rhincodon*, or a single autapomorphy for clade D (Goto 2001), with reversals to the plesiomorphic state in *Orectolobus*, *Eucrossorhinus*, *Brachaelurus* and *Chiloscyllium* except *C. indicum*. I prefer the possibility of repeated autapomorphies that led to an indistinguishable separation of state 2 to states A2 to A6 in Figure 44. For *Mesiteia*, the widely separated pectoral basal cartilages (pro-, meso- and metapterygium) are a plesiomorphic character, shared with the ancestors of Orectolobiformes.

- **Character 07**: Metapterygial axis (mtx) on the distal tip of the metapterygium. A metapterygial axis on the distal tip of the metapterygium, distinguishable from the adjacent branchial rays is seen in *Orectolobus* except *O. maculatus*, *Eucrossorhinus*, *Sutorectus*, *Brachaelurus* and *Rhincodon*, (Goto 2001, figs. 34A,C). But in most Orectolobiformes the posteriormost radial is indistinguishable from the adjacent radials (Goto 2001, figs. 34B,D–F). The loss of the

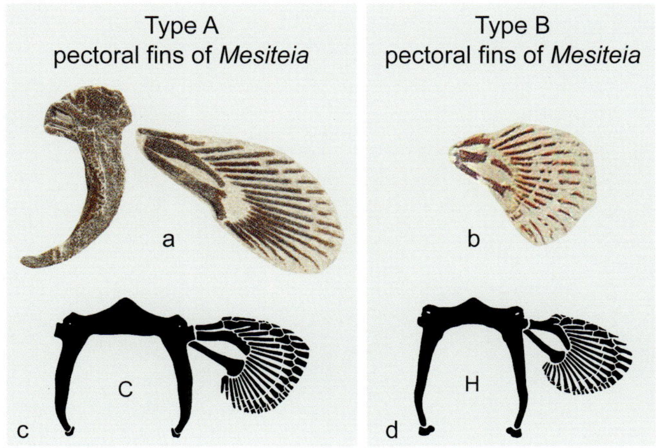

Fig. 32. The pectoral girdle and fin skeletons of *Mesiteia* compared with those of *Chiloscyllium* and *Hemiscyllium*. **a.** scapular cartilage and left pectoral fin of *Mesiteia* specimen #08 (Munich, coll. FHP); **b.** left pectoral fin of *Mesiteia* specimen #18 (Lebanon, unknown new owner; photo Roy Nohra); **c.** pectoral girdle and left pectoral fin of *Chiloscyllium punctatum* (C) (after Dingerkus & DeFino 1983, fig. 41); **d.** pectoral girdle and left pectoral fin of *Hemiscyllium ocellatum* (H) (after Dingerkus & DeFino 1983, fig. 42).

metapteryial axis is assumed to be apomorphic (Goto 2001, p. 62) for Orectolobiformes. In *Mesiteia*, I could not find a metapterygial axis on the metapterygium.

Two states are recognized,
0 metapterygial axis present;
1 metapterygial axis absent.

For *Mesiteia* the absence of a metapterygial axis is a synapomorphic character, shared with *Parascyllium*, *Cirrhoscyllium*, *Hemiscyllium*, *Chiloscyllium*, *Ginglymostoma* and *Stegostoma*. The presence of a metapterygial axis in *Orectolobus* except *O. maculatus*, *Eucrossorhinus*, *Sutorectus*, *Brachaelurus*, and also in *Rhincodon* are two reversals, and the return of a metapterygial axis in *Orectolobus maculatus*, is estimated to be a second reversal back to the state of the first loss of the metapterygial in clade A (Goto 2001).

- **Character 08:** Number of pectoral radial elements. According to Goto (2001, p. 33, fig. 34) in most Orectolobiformes, the pectoral radials are principally composed of three components, and as an apomorphic character (Goto 2001, p. 62), in *Ginglymostoma*, *Rhincodon* and *Stegostoma*, the distal radials are segmented into numerous (up to 10) cartilaginous pieces. In *Mesiteia* the pectoral radial elements are divided into three elements.

Two states are recognized,
0 pectoral radials composed of three components;
1 pectoral radials composed of more than three (up to 10) components.

For *Mesiteia*, the three pectoral radial elements are plesiomorphic, shared with Orectolobiformes, except *Ginglymostoma*, *Rhincodon* and *Stegostoma*. For the latter the more than three radial elements are estimated to be apomorphic.

4.3.4 Pelvic girdle, pelvic fins and claspers

Females: The pelvic girdle is well preserved in specimen #08 and other specimens helping to show the salient features. The pelvic girdle begins at the level of the 39th vertebra on the holotype #01, the 36th vertebra on the Paris specimen #02, and the 43rd on specimen #08.

The puboischiatic bar (pub) is narrow and longer, and the anterior margin is slightly convex, not straight as in Hemiscylliidae. Its posterior edge is concave as in *Hemiscyllium*. The articular condyle and a facet for the pelvic basipterygium on the posterior margin of the puboischiatic bar are not visible on the fossils because they are overlaid by the basipterygium (Fig. 34f).

- **Character 09:** Prepelvic process of the puboischiatic bar. The puboischiatic bar has an anteriorly expanded anterolateral margin, representing the prepelvic process (pvp) on each of the lateral edges. According to Goto (2001, p. 34, figs. 35A, C–D) in most Orectolobiformes, the prepelvic process is reduced, and rounded or bluntly pointed. In *Parascyllium* and *Cirrhoscyllium*, it is projected anteriorly and is strongly pointed distally (Goto 2001, fig. 35B). The prepelvic process in *Mesiteia* is not projected and looks similar to that of *Hemiscyllium* and *Chiloscyllium* (after Dingerkus & DeFino 1983, figs. 43, 44).

Two states are recognized,
0 prepelvic process of the puboischiatic bar rounded or bluntly pointed;
1 prepelvic process of the puboischiatic bar greatly projected anteriorly.

For *Mesiteia*, the condition is estimated to be plesiomorphic, shared with all Orectolobiformes, except *Parascyllium* and *Cirrhoscyllium*, for which the anteriorly projected prepelvic process of the puboischiatic bar is apomorphic.

- **Character 10:** Articular condyle (pac) for the anterior pelvic basal. In most Orectolobiformes (*Parascyllium*, *Cirrhoscyllium*, *Orectolobus*, *Sutorectus*, *Eucrossorhinus*, *Brachaelurus*, *Chiloscyllium* and *Hemiscyllium*) the posterolateral margin of the puboischiatic bar is projected posterolaterally, forming a granulated articular condyle (pac) for the anterior pelvic basal (Goto 2001, fig. 35). In Recent Hemiscylliidae the articular condyle is well developed (Dingerkus & DeFino 1983, figs. 43, 44, and here Fig. 34b–d). In *Mesiteia*, there is no projected articular condyle on the pelvic bar, and the anterior pelvic basal articulates in a groove between the pelvic bar and the first radial, with both the margin of the puboischiadic bar and the basipterygium.

Two states are recognized,
0 articular condyle for the anterior pelvic basal on the puboischiatic bar present;
1 articular condyle for the anterior pelvic basal on the puboischiatic bar absent.

For *Mesiteia*, the absence of an articular condyle for the anterior pelvic basal is an autapomorphic character. It is also an autapomorphy for *Rhincodon*. For all other Orectolobiformes the presence of this condyle is a plesiomorphy.

All together, the pelvic fins of *Mesiteia* are well developed and more elongated compared with Recent Hemiscylliidae.

The pelvic basipterygium (pvb) is rather elongated and arched mesially, with most pelvic radials equally long, articulating on its lateral margin. The proximal end of the pelvic basipterygium has a granulated condyle (cpv, Fig. 34a, f) and a shallow facet (fcpv in Goto 2001, fig. 35F, not visible on the fossils), which articulates with the articular condyle and facet of the puboischiadic bar. The distal end of the pelvic basipterygium looks tapered in *Mesiteia* #08 but this is only because the basipterygium is broken, maybe during fossilization, and the basipterygium lies distally a bit crosswise. To the two fragmented little cartilages there are still radials articulated.

I am not certain about the terminology of the cartilage elements posterior to the pelvic basipterygium, therefore I have tried to describe what I see. At its posterior end, the pelvic basipterygium has two separated basal cartilage elements (b1–2) attached, similar to what Dingerkus & DeFino found on *Chiloscyllium punctatum* (1983, fig. 43, and here Fig. 34c), but not on *Hemiscyllium ocellatum* (fig. 44, and here Fig. 34c). In *C. punctatum* the proximal 1st basal cartilage element is small, triangularly pointed to its distal end, where a 2nd basal cartilage element that looks like a short, strong, laterally curved radial is attached. In *Chiloscyllium* there are no other pelvic fin radials articulated with the first basal cartilage element. This corresponds to another drawing of the pelvic fin of *C. punctatum* in Goto (2001, fig. 35A), and these two basal cartilage elements are also present in *Parascyllium collare* (Goto 2001, fig. 35B) and other sharks. *Mesiteia* #08 (Fig. 34g) has a long triangular 1st basal cartilage (b1), pointed to its distal end, with a fine, flexible terminal radial attached, which represents the 2nd basal cartilage element (b2). Pelvic fin radials are not articulated with the basal cartilage elements of *Mesiteia*.

- **Character 11:** Number of pelvic radial elements. The pelvic radials (rd) are generally divided into elongated, slender proximal elements, the inner row radials (radi), and short, square-shaped or elongated, pointed distal elements, the outer row radials (rado) (Fig. 34, and Goto 2001, fig. 35). In most Orectolobiformes, the pelvic radials consist of these two elements (Goto 2001, figs. 35A–B). In *Ginglymostoma*, *Stegostoma* and *Rhincodon*, the radials are segmented into three elements (Goto 2001, figs. 35C–D). In *Mesiteia* the pelvic radial components are divided into two elements.

Two states are recognized,
0 pectoral radials composed of two elements;
1 pectoral radials composed of three or more elements.

For *Mesiteia*, the two pectoral radial elements are plesiomorphic, shared with all Orectolobiformes, except *Ginglymostoma*, *Rhincodon* and *Stegostoma*. For the latter the three or more radial elements are synapomorphic.

In Recent Hemiscyllidae (Fig. 34b–c) the radials form an arc, and they become increasingly longer from the anterior ones and then shorter again to the posterior ones. This also applies

Fig. 33. Pelvic girdle and pelvic fins of female and male specimens of *Mesiteia emiliae*.
a–f. Females; **a.** *Mesiteia* specimen #08 (Munich, coll. FHP); **b–c.** *Mesiteia* specimen #05 (Munich, SNSB-BSPG 2013 XXVII 1); **b.** in normal light; **c.** with long-wave UV-light; photo Helmut Tischlinger; **d.** *Mesiteia* specimen #14 (Lebanon, coll. Expo Hakel, Roy Nohra); **e.** *Mesiteia* specimen #11 (Munich, coll. FHP); **f.** *Mesiteia* specimen #15 (Lebanon, unknown new owner).
g–j. Males; **g.** *Mesiteia* specimen #16 (Lebanon, coll. Expo Hakel, Roy Nohra); **h.** *Mesiteia* specimen #07 (Munich, coll. FHP); **i.** *Mesiteia* specimen #09 (Munich, coll. FHP); **j.** *Mesiteia* specimen #13 (Switzerland, coll. René Kindlimann RK/7-711).
For description of the elements see Figure 34.

to the outer row radials. In *Parascyllium collare* (Goto 2001, fig. 35b) the inner row radials are equally long, and the short square-shaped outer row radials also remain the same size from anterior to posterior rows. The same with *Mesiteia*.

- **Character 12:** Anterior pelvic basal (apv). I have mentioned above that *Mesiteia* has no projected articular condyle on the pelvic bar, and the cartilage ray that was identified as the anterior pelvic basal articulates in a groove between the pelvic bar and the first radial. It is not quite as long as the following radial, equally strong and not subdivided as in Hemiscylliidae (Fig. 34). According to Goto (2001, p. 34, fig. 35) in most Orectolobiformes this single anteriormost radial is remarkably larger than the adjacent radials, and it articulates directly with the puboischiadic bar. It is slender and simple in *Ginglymostoma* and *Stegostoma* (Goto 2001, fig. 35C), somewhat stout and distally subdivided into a few branches in *Chiloscyllium* and *Hemiscyllium* (Goto 2001, fig. 35A, and here Fig. 34b,c), depressed and distally expanded in *Parascyllium*, *Cirrhoscyllium*, *Orectolobus*, *Sutorectus*, *Eucrossorhinus* and *Brachaelurus* (Goto 2001, fig. 35B), and *Rhincodon* has neither the anterior pelvic basal nor its articular condyle on the puboischiadic bar (Goto 2001, fig. 35D).

Fig. 34. Female and male pelvic girdle elements of *Mesiteia emiliae* compared with those of Hemiscylliidae. **a, f–g**: female *Mesiteia* specimen #08 (Munich, coll. FHP); **b.** female *Chiloscyllium punctatum*; (after Dingerkus & DeFino 1983, fig. 43); **c.** female *Hemiscyllium ocellatum*; (after Dingerkus & DeFino 1983, figs. 44); **d.** male *Chiloscyllium* sp. (reconstructed after Dingerkus & DeFino 1983, fig. 43); **e, h.** male *Mesiteia* specimen #07 (Munich, coll. FHP).
Pelvic girdle: pac – condyle for the anterior pelvic basal; pub – puboischiadic bar; pvp – prepelvic process. Pelvic fin: apv – anterior pelvic basal; b1 – first basal cartilage element; b2 – second basal cartilage element; cpv – condyle, which articulate with the facet of the puboischiadic bar; pvb – pelvic basipterygium; radi – inner rows of radial elements; rado – outer rows of radial elements; in red the number of the radials from anterior to posterior radials.
Clasper: dmc – dorsal marginal cartilage; dtc – dorsal terminal cartilage; vmc – ventral marginal cartilage; vtc – ventral terminal cartilage; spc – spur.

Two states are recognized,
0 anterior pelvic basal present;
1 anterior pelvic basal absent.

For *Mesiteia*, the shape of the anteriormost pelvic radial is indistinguishable from the other radials, as is its articulation with the puboischiadic bar and the basipterygium; the absence of a distinguishable anterior pelvic basal is an autapomorphic character. The absence of an anterior pelvic basal is also an autapomorphy for *Rhincodon*. For all other Orectolobiformes the presence of an anterior pelvic basal is a plesiomorphy.

The inner row pelvic radials (radi) are long, thin, and strong up to their distal articulation with the outer row radials (rado). In *Mesiteia* #08 (Fig. 34), there are thirteen series of rather equally long and strong inner row radials articulated with the pelvic basipterygium, the 13th radial is the longest of all pelvic radials. The inner row radials are articulated with thirteen more or less equally short outer row radial elements. The outer row radials are finer, and one can see that the distal articulation face of the inner row radials and the proximal articulation face of the outer radials has an enlarged process for this articulation and is not the result of any damage. The following five posterior pelvic radials are shorter and finer, get rapidly shorter, and are not articulated with outer row radials. The posteriormost 18th pelvic radial articulates at the posteriolateral margin of the pelvic basipterygium. This radial only is subdivided into two branches. The number of pelvic radial series in *Mesiteia* #08 is 18, and following Dingerkus & DeFino (1983, p. 58) in *Hemiscyllium* there are 17 radial series, and in *Chiloscyllium* there are 16 radial series articulating with the metapterygium (= pelvic basipterygium) (Figs. 34b–c).

Males: In males, the two separated basal cartilage elements (b1-2) at the posterior end of the pelvic basipterygium are modified into the clasper cartilages. According to Goto (2001, p. 33, fig. 35F), the clasper cartilages of Orectolobiformes consist of the joint cartilage, stem cartilage, dorsal and ventral terminal cartilages, spur, and hook. In male specimens of *Mesiteia*, the thick and elongated claspers can easily be recognized but the pelvic girdle with the puboischiadic bar, the right and left pelvic basipterygium and the radials are in all specimens moved sideways and the preservation of these elements is not really good. But one can see a few interesting details of the claspers on *Mesiteia* #16, #07, #09 and #13.

In *Mesiteia* #07 (Figs. 33h, 34e,h) there are one, maybe two, joint cartilages visible, articulated to the distal end of the pelvic basipterygium, composed of one or two small segments (b1–2), with no beta cartilage. The stem cartilage is very large (*Mesiteia* #16, Fig. 33g), its mesial and lateral margins well expanded dorsally, forming the dorsal marginal (dmc) and ventral marginal (vmc) cartilages (*Mesiteia* #07, Fig. 34e). The dorsal terminal cartilage (dtc) or dorsal covering plate (*Mesiteia* #07, Fig. 34e, well preserved on *Mesiteia* #16), originates from the distal tip of the dorsal marginal cartilage. The ventral terminal cartilage (vtc) or ventral sheaths of the claspers originates from the distal tip of the ventral marginal cartilage. The spurs (spc) are relatively big hooks, tapering distally, articulating with the ventral terminal cartilage, one or two on each clasper (Figs. 33 g–j, 34 e,h). The different size and length of the claspers which are otherwise very similar depends on age or maturity of the individual specimen.

4.3.5 1st and 2nd dorsal fins

The 1st and 2nd dorsal fins of *Mesiteia* are rather different in form and position from what one can see on Hemiscylliidae (Plate 11). Unfortunately, both the 1st and 2nd dorsal fins are slightly damaged in most specimens and therefore it is impossible at the moment to count exactly the number of basal radial elements.

Following Goto (2001, p. 34, fig. 36), the dorsal fins of Orectolobiformes are basically supported by the set composed of a basal cartilage and two radials, arranged longitudinally. The basal cartilage (b) is slender and elongated, lying on the midline between the epaxial body muscles. It originates from the completely separated region from the head of the neural arch. The distal end is more or less expanded, with a small facet for the proximal radial on the dorsal margin. The radial consists of a slender, elongated proximal or inner radial element (radi) and a short, distally tapering distal or outer radial element (rado).

1st dorsal fin

- **Character 13**: Position of 1st dorsal fin. In *Mesiteia*, the origin of the 1st dorsal fin is at the end of the pelvic fins at the level of the 47th vertebra and stops at the level of the 58th vertebra on the holotype #01, it starts at the level of the 55th and stops at the level of the 72nd vertebra on the Paris specimen #02, and starts at the level of the 56th and stops at the level of the 66th vertebra on specimen #08. Following Compagno (2001) and Dingerkus & DeFino (1983) the 1st dorsal fin of Recent Orectolobiformes originates in relation to the pelvic fin, well anterior in *Ginglymostoma*, *Nebrius*, *Pseudoginglymostoma*, *Rhincodon* and *Stegostoma*; far posterior in *Cirrhoscyllium* and *Parascyllium*; and above the middle to the end in *Orectolobus*, *Eucrossorhinus*, *Sutorectus*, *Brachaelurus*, *Chiloscyllium* and *Hemiscyllium*. In *Mesiteia*, the origin of the 1st dorsal fin above the end of the pelvic fins corresponds best with that of Hemiscylliidae. Origin of the 1st dorsal fin anterior to pelvic fin is the plesiomorphic state.

Three states are recognized,
0 origin of 1st dorsal fin anterior to pelvic fin;
1 origin of 1st dorsal fin far posterior of pelvic fin;
2 origin of 1st dorsal fin above the middle to end of pelvic fin.

The position of the 1st dorsal fin far posterior is an autapomorphy for *Cirrhoscyllium* and *Parascyllium*. It is another autapomorphy for clade D (Goto 2001) that in all species of *Orectolobus*, *Eucrossorhinus*, *Sutorectus*, *Brachaelurus*, *Chiloscyllium* and *Hemiscyllium*, the 1st dorsal fin originates above the middle to the end of the pelvic fins. But the origin of the 1st dorsal fin well anterior of the pelvic fin in *Ginglymostoma*, *Nebrius*, *Pseudoginglymostoma*, *Rhincodon* and *Stegostoma* is a reversal plesiomorphic character for clade N in Goto (2001). For *Mesiteia*, the origin of the 1st dorsal fin at the end of the pelvic fins is a synapomorphic character, shared with Hemiscylliidae (clade J in Goto 2001).

- **Character 14**: Shape and length of the basal elements of 1st dorsal fin. In *Mesiteia*, as in Hemiscylliidae, there is one row of basal elements supporting the dorsal fins and these basal cartilages approach the vertebral column, but do not articulate with it (Figs. 35, 36). The appearance of the small facet for the articulation with the proximal radial is very similar to that in Hemiscylliidae (Figs. 35a–b). The anterior first 9 series of radials in *Mesiteia* #12 (Fig. 35d), do not have basal elements attached. The following series of radials 10 to 19 are articulated with 10 basal elements that are increasingly elongated and reach deep down to the vertebral column. The basal elements then become shorter again in posterior series of radials. In the posterior series of radials 20 to 23, basal elements are absent. This is reflected in a characteristic curve with a slowly increasing and then decreasing length of basal elements. In other Orectolobiformes (Goto 2001, fig. 37), besides in those with fused and shortened elements, all these basal elements are equally long, and all are attached to radials. In most other sharks, many basal cartilages of the 1st dorsal fin are fused to massive plates (Shirai 1992, p. 87–88, fig. 5.22, pls. 54–55), which is estimated to be plesiomorphic. Only in *Mesiteia*, every existing basal element of the 1st dorsal fin is connected to one radial, but radials of anterior and posterior series do not have basal elements attached.

Two states are recognized,
0 all basal elements of 1st dorsal fin are attached to radials;
1 anterior and posterior basal elements of 1st dorsal fin are absent, its length is increasing and decreasing.

For *Mesiteia*, the absence of anterior and posterior basal elements of 1st dorsal fin and its increasing and decreasing length is an autapomorphic character.

- **Character 15**: Fused basal elements of 1st dorsal fin. Regarding the fusion of basal cartilages of the dorsal fins, among Orectolobiformes examined by Goto (2001), the following five variations were found.
  1) All cartilages are simple and completely separate in *Eucrossorhinus*, and in one specimen of *Chiloscyllium plagiosum* (Goto 2001 fig. 36A).
  2) Anteriorly, some cartilages are fused, with several anteriormost radials in *Orectolobus*, *Sutorectus* and *Brachaelurus* (Goto 2001, fig. 37A).
  3) Posteriorly, some cartilages are fused, with several posteriormost radials in most examined individuals of *Chiloscyllium*, *Hemiscyllium*, *Ginglymostoma* and *Stegostoma* (Goto 2001, figs. 36A, 37C).
  4) Some approximate cartilages are fused only at the proximal ends in *Parascyllium* and *Cirrhoscyllium* (Goto 2001, fig. 37B).
  5) Many cartilages are fused into a large cartilaginous plate in *Rhincodon* (Goto 2001, fig. 37D–E).
  In most other sharks, many basal cartilages of the 1st dorsal fin are fused to massive plates (Shirai 1992, p. 87–88, fig. 5.22, pls. 54–55), which is a plesiomorphic character. In the 1st dorsal fin of *Mesiteia* (#12, Fig. 35d) there are no fused basal elements into which dorsally more than one radial articulate, such as seen in *Chiloscyllium* (Fig. 35a) and *Hemiscyllium* (Fig. 35b).

Two states are recognized,
0 basal elements of 1st dorsal fin at least partly fused, and all radials attached to basal elements;
1 basal elements of 1st dorsal fin all separate.

The absence of fused basal elements is autapomorphic for *Mesiteia*.

- **Character 16**: Number of radial elements of 1st dorsal fin. The radial set of most Orectolobiformes consists of two cartilages in both dorsal fins (Goto 2001, p. 35, figs. 36A–B, 37A–B), whereas it is divided into three or more segments in *Ginglymostoma*, *Stegostoma* and *Rhincodon* (Goto 2001, figs. 37C–E). In *Mesiteia* the radials of 1st dorsal fin are divided into two elements.

Two states are recognized,
0 radial elements of 1st dorsal fin composed of two elements;
1 radial elements of 1st dorsal fin composed of three or more elements.

For *Mesiteia*, the composition of 1st dorsal fin of two pectoral radial elements is plesiomorphic, shared with the other Orectolobiformes, except *Ginglymostoma*, *Rhincodon* and *Stegostoma*. For the latter, the three or more radial elements are apomorphic.

- **Character 17**: Shape and length of radial elements of 1st dorsal fin. In the 1st dorsal fin of *Mesiteia* (#08, Fig. 35c, #12, Fig. 35d) the inner radial elements are straight and very short, shorter in anterior and posterior series, giving the 1st dorsal fin a homogenous appearance of a long drawn flat bow instead of being extended with long outer

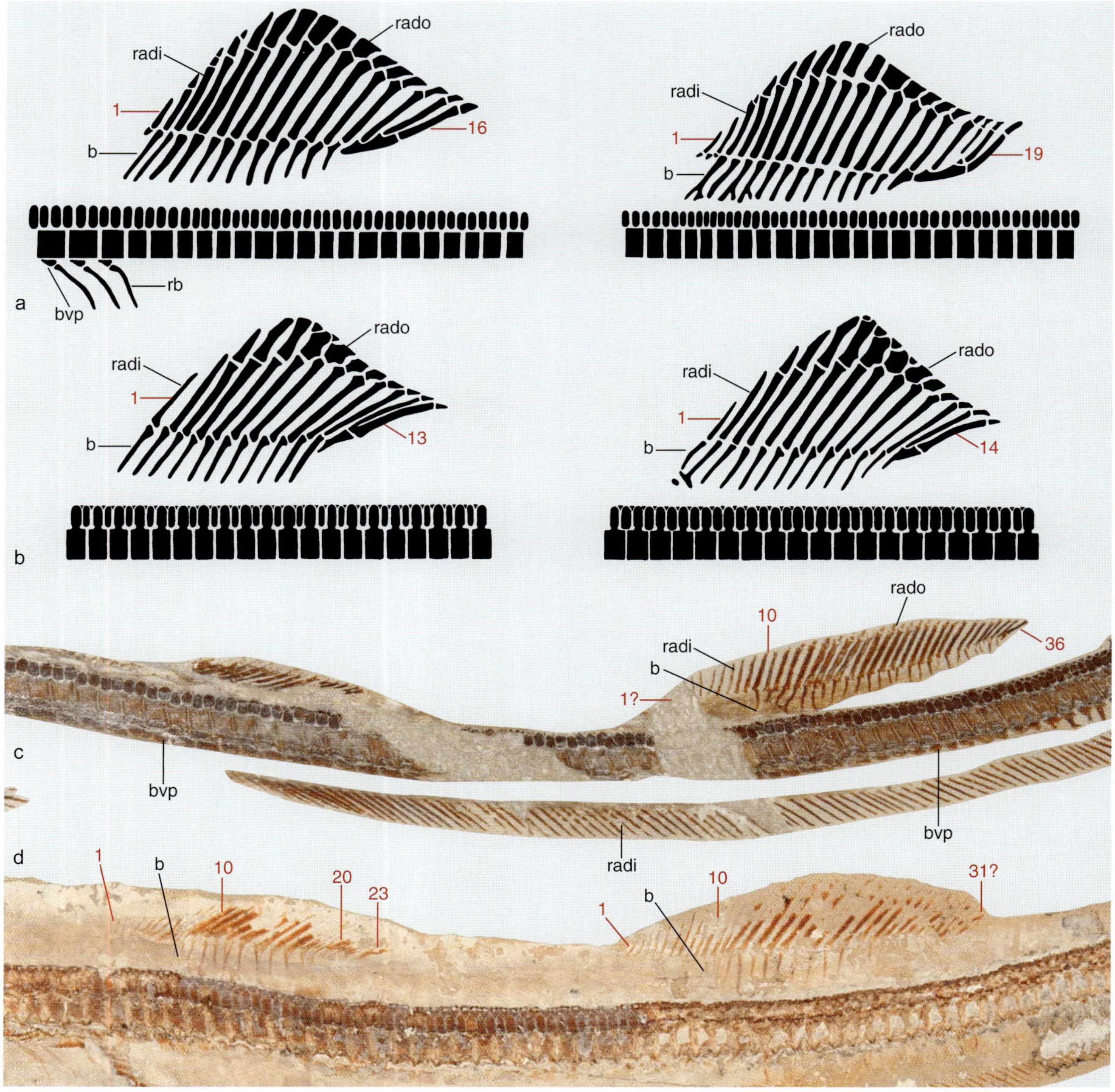

Fig. 35. The 1st and 2nd dorsal fins of *Mesiteia emiliae* compared with those of Hemiscyllidae: at left: 1st dorsal fin; at right: 2nd dorsal fin; **a.** *Chiloscyllium punctatum* (after Dingerkus & DeFino 1983, figs. 45, 46); **b.** *Hemiscyllium ocellatum* (after Dingerkus & DeFino 1983, figs. 47, 48); **c.** *Mesiteia* specimen #08 (Munich, coll. FHP); **d.** *Mesiteia* specimen #12 (Munich, coll. FHP).
Dorsal fins: bd – basal cartilage of dorsal fin; radi – inner rows of radial elements; rado – outer rows of radial elements; in red the number of the radials from anterior to posterior radials.
Vertebral column and elements of the ventral arcualia: bvp – basiventral process; rb – vertebral rib.

radial elements in the middle part of the fin, as in Hemiscylliidae (Dingerkus & DeFino 1983, figs. 45, 46, and here Figs. 35a–b), and in other Orectolobiformes (Goto 2001, figs. 36, 37). In *Mesiteia*, the distally attached outer radial elements are very small, if present at all, not strong and wide as in Hemiscylliidae and in other Orectolobiformes. The first 9 series of radials of *Mesiteia* #12 (Fig. 35d) are much finer than the following radials.

Two states are recognized,

0 inner radial elements of 1st dorsal fin long or segmented, outer radial elements also strong and long, both elongated in medial series of fin;

1 inner radial elements of 1st dorsal fin short, outer radial elements very short, equally long or absent.

For *Mesiteia*, the low 1st dorsal fin with short inner radial and very short or almost missing outer radial elements is an autapomorphic character.

- **Character 18:** Number of series of radial elements of 1st dorsal fin. The 1st dorsal fin of *Mesiteia* has strikingly more (up to 23) series of radial elements than do other Orectolobiformes, e.g. *Chiloscyllium* (16 series of radials, Fig. 35a) and *Hemiscyllium* (13 series of radials, Fig. 35b). Only in *Rhincodon typus* (Goto 2001, Fig. 37D) does the 1st dorsal fin have a similar number of series of radials but these radials are fragmented and the fin is strongly compressed, whereas the low 1st dorsal fin of *Mesiteia* is almost twice as long as seen in other Orectolobiformes and extends far behind the posterior end of the pelvic fins.

Two states are recognized,

0 number of series of radial elements of 1st dorsal fin low, around 15, fin compressed;

1 number of series of radial elements of 1st dorsal fin more than 20, fin elongated.

For *Mesiteia*, the elongated 1st dorsal fin with more than 20 radial elements is an autapomorphic character.

Other specimens of *Mesiteia*, e.g. #08 (Fig. 35c), #04 (Plate 11e), #15 (Plate 11f) and #20 (Plate 11h) confirm this general appearance but because there are always some parts of the radials missing, it is impossible to describe and count the elements sufficiently. Cappetta (1980) calculated 20 to 25 supporting elements on the 1st dorsal fin of the holotype, specimen #01.

Plate 11 ▷

Pelvic girdle and unpaired fins of Hemiscylliidae and *Mesiteia emiliae*: pelvic fin; D1: 1st dorsal fin; D2: 2nd dorsal fin; anal fin; caudal fin. **a.** *Chiloscyllium punctatum* (thanks to Dr. Hiroshi Senou; © Kanagawa Prefectural Museum of Natural History; photo: Hiroshi Senou); **b.** *Chiloscyllium punctatum* (after Dingerkus & DeFino 1983, figs. 45, 46); **c.** *Hemiscyllium ocellatum* (after Dingerkus & DeFino 1983, figs. 47, 48); **d.** *Mesiteia* specimen #08 (Munich, coll. FHP); **e.** *Mesiteia* specimen #04 (Munich, SNSB-BSPG 2011 I 124) in long-wave UV-light; photo Helmut Tischlinger; **f.** *Mesiteia* specimen #15 (Lebanon, unknown new owner); photo Roy Nohra; **g.** *Mesiteia* specimen #12 (Munich, coll. FHP); **h.** *Mesiteia* specimen #20 (Lebanon, coll. Abi Saad); photo Pierre Abi Saad. For explanation of the elements see Figures 35–36.

Plate 11

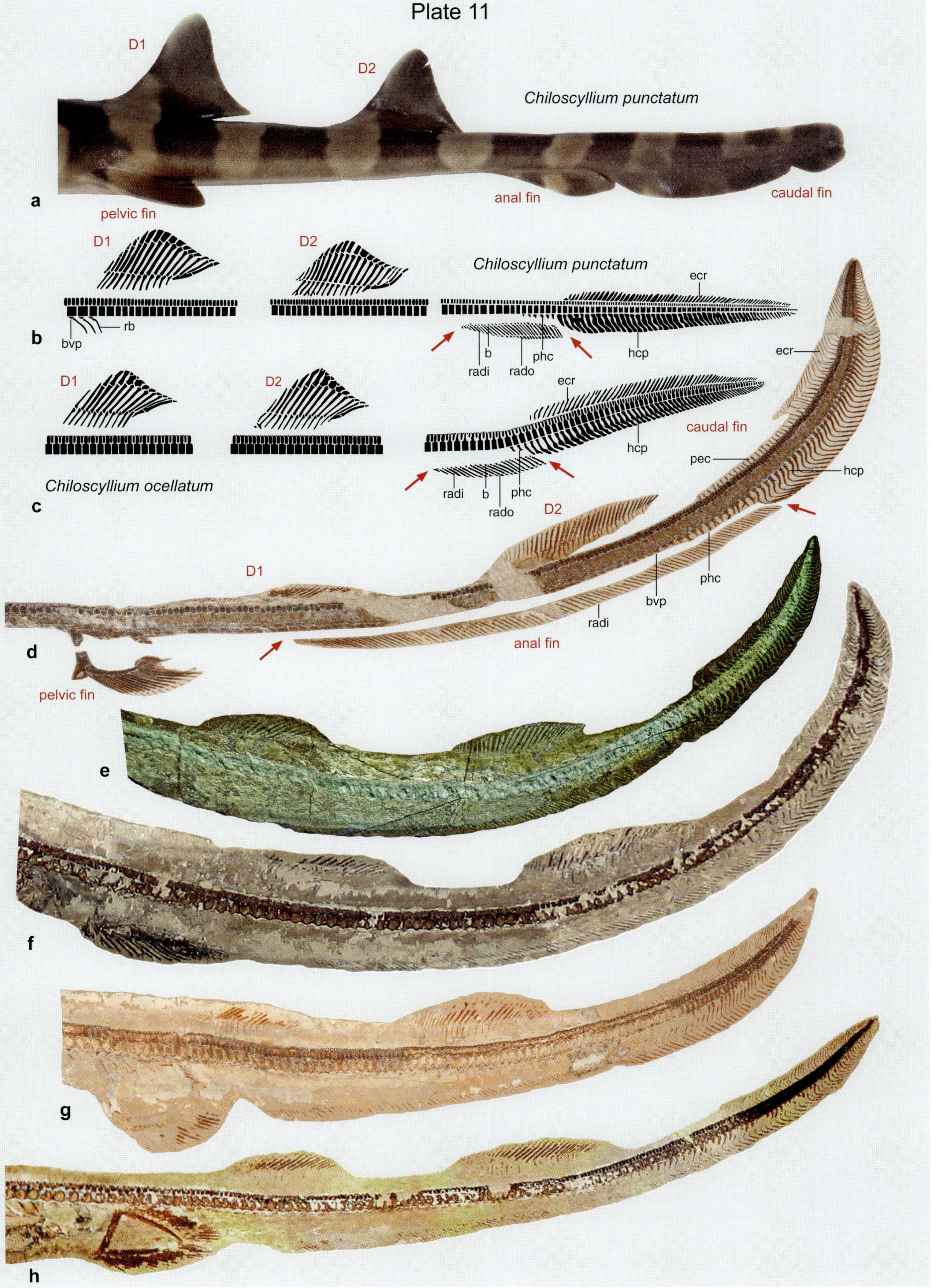
D1
D2
Chiloscyllium punctatum
a
pelvic fin
anal fin
caudal fin
D1
D2
Chiloscyllium punctatum
ecr
b
bvp
rb
radi
b
rado
phc
hcp
D1
D2
ecr
ecr
caudal fin
hcp
pec
hcp
Chiloscyllium ocellatum
c
radi
b
phc
rado
D2
D1
phc
bvp
radi
d
anal fin
pelvic fin
e
f
g
h

2nd dorsal fin

- **Character 19**: Position of 2nd dorsal fin. In *Mesiteia*, the 2nd dorsal fin starts at the level of the 74th vertebra and stops at the level of the 89th vertebra on the holotype #01, starts at the level of the 77th and stops at the level of the 93rd vertebra on the Paris specimen #02, and starts at the level of the 81st and stops at the level of the 101st vertebra on specimen #08. Following Compagno (2001) and Dingerkus & DeFino (1983) the 2nd dorsal fin of Recent Orectolobiformes is positioned above or just before the anal fin, with short distance to the 1st dorsal fin in *Ginglymostoma*, *Nebrius*, *Pseudoginglymostoma*, *Rhincodon* and *Stegostoma*; in the middle between pelvic and anal fin, with short distance to the 1st dorsal fin in *Orectolobus*, *Eucrossorhinus*, *Sutorectus*, and *Brachaelurus*; in the middle between pelvic and anal fin, with great distance to the 1st dorsal fin in *Chiloscyllium* and *Hemiscyllium*; and well behind the anal fin, with great distance to the 1st dorsal fin in *Cirrhoscyllium* and *Parascyllium*. In *Mesiteia*, the 2nd dorsal fin moved in posterior direction, close to the origin of the caudal fin, and is positioned in the middle over the extremely elongated anal fin. A position of the 2nd dorsal fin above a short anal fin, with great distance to the 1st dorsal fin is the plesiomorphic state.

Six states are recognized,
0 position of 2nd dorsal fin above the anal fin, with great distance to 1st dorsal fin;
1 position of 2nd dorsal fin above the anal fin, with short distance to 1st dorsal fin;
2 position of 2nd dorsal fin well behind anal fin, with great distance to the 1st dorsal fin;
3 position of 2nd dorsal fin in the middle between pelvic and anal fin, with short distance to the 1st dorsal fin;
4 position of 2nd dorsal fin in the middle between pelvic and anal fin, with great distance to the 1st dorsal fin;
5 position of 2nd dorsal fin moved in posterior direction, located above the extremely elongated anal fin close to the origin of the caudal fin.

For *Mesiteia*, the position of the 2nd dorsal fin, as described, is an autapomorphic character that separates it from Hemiscylliidae where the 2nd dorsal fin originates far more anteriorly. For the other Orectolobiformes the described characters are also apomorphies compared to the plesiomorphic state.

- **Character 20**: Shape and length of the basal elements of 2nd dorsal fin. In *Mesiteia* #08 (Fig. 35c) and #12 (Fig. 35d) the 9 anterior series of radials of the 2nd dorsal fin do not have basal elements attached. All the following series of radials 10 to 36 in *Mesiteia* #08, and 10 to 31? in *Mesiteia* #12, are articulated with basal elements that again, as in the 1st dorsal fin, are increasingly elongated, reaching deep down to the vertebral column but getting shorter in posterior series. In *Mesiteia* #08, only the basal elements of the series of radials 10 to 18 are very long. The following elements are very short, and it looks like the 2nd dorsal fin would lose contact with the body at its posterior margin. As described for the 1st dorsal fin, in other Orectolobiformes (Goto 2001, fig. 37), besides in those with fused and shortened elements, most of these basal elements are equally long, and all are attached to radials. In most other sharks, many basal cartilages of the 2nd dorsal fin are fused to plates (Shirai 1992, p. 87–88, fig. 5.22, pls. 54–55), which is estimated to be plesiomorphic. Only in *Mesiteia*, every existing basal element of the 2nd dorsal fin is connected to one radial, but radials of anterior and posterior series do not have basal elements attached.

Two states are recognized,
0 all basal elements of 2nd dorsal fin are attached to radials;
1 anterior and posterior basal elements of 2nd dorsal fin are absent, its length is increasing and decreasing.

For *Mesiteia*, the absence of anterior and posterior basal elements of 2nd dorsal fin and its increasing and decreasing length is an autapomorphic character.

- **Character 21**: Fused basal elements of 2nd dorsal fin. I already listed the five variations regarding the fusion of basal cartilages of the dorsal fins (character 15), as examined by Goto (2001). In the 2nd dorsal fin of *Mesiteia* (#08, Fig. 35c, 36 and #12, Fig. 35d) there are no fused basal elements in posterior series as seen in *Chiloscyllium* (Fig. 35a) and *Hemiscyllium* (Fig. 35b). In *Mesiteia* #08, radials of the two series 11 and 12 are articulated distally with only one basal element, and the basal elements of radials from series 13 and 14 tend to articulate at their proximal end. Most likely this is just kind of variation, but it cannot be shown whether it reflects a tendency to fuse at this special area. In most other sharks, many basal cartilages of the 2nd dorsal fin are fused to plates (Shirai 1992, p. 87–88, fig. 5.22, pls. 54–55), and this is the plesiomorphic state.

Two states are recognized,
0 basal elements of 2nd dorsal fin at least partly fused, and all radials attached to basal elements;
1 basal elements of 2nd dorsal fin all separate.

The absence of fused basal elements is autapomorphic for *Mesiteia*.

- **Character 22**: Number of radial elements of 2nd dorsal fin. The radial set of most Orectolobiforrnes consists of two cartilages in both dorsal fins (Goto 2001, p. 35, figs. 36A–B, 37A–B), whereas it is divided into three or more segments in *Ginglymostoma*, *Stegostoma* and *Rhincodon* (Goto 2001, figs. 37C–E). In *Mesiteia* the radials of the 2nd dorsal fin are divided into two elements.

Two states are recognized,
0 radial elements of 2nd dorsal fin composed of two components;
1 radial elements of 2nd dorsal fin composed of three or more components.

For *Mesiteia*, the composition of 2nd dorsal fin's two pectoral radial elements is plesiomorphic, shared with the other Orectolobiformes, except *Ginglymostoma*, *Rhincodon* and *Stegostoma*. For the latter, the three or more elements of the radials are apomorphic.

- **Character 23**: Shape and length of radial elements of 2nd dorsal fin. In the 2nd dorsal fin of *Mesiteia* (#08, Fig. 35c, 36, #12, Fig. 35d), as I described for the 1st dorsal fin before (Character 17), the 36 (#08) or 31? (#12) rows of inner radial elements are less extended, as in Hemiscylliidae. The inner radial elements are straight and short, shorter in anterior and posterior series, giving the 2nd dorsal fin a homogenous appearance of a long drawn flat bow instead of being extended, with long outer radial elements in the middle part of the fin, as in Hemiscylliidae (Dingerkus & DeFino 1983, figs. 45, 46, and here Figs. 35a–b), and in other Orectolobiformes (Goto 2001, figs. 36, 37). For the 2nd dorsal fin, it can be shown on *Mesiteia* #08 (Fig. 35c, 36) that there are outer radial elements, very small, distally attached to the articulation faces of the inner radials. These very short elements are limited to the series of radials 11 to 28, they are not strong and wide as in Hemiscylliidae. In the 2nd dorsal fin of *Mesiteia* #08 (Figs. 35c, 36) the first 7 series of inner radial elements, and in specimen #12 (Fig. 35d) the first 12 series of inner radial elements, are finer than the following ones. The radials of 2nd dorsal fin in *Mesiteia* are straight, shorter in anterior and posterior series, giving the 2nd dorsal fin a homogenous appearance of a long drawn flat bow.

Two states are recognized,
0 inner radial elements of 2nd dorsal fin long or segmented, outer radial elements also strong and long, both elongated in medial series of fin;
1 inner radial elements of 2nd dorsal fin short, outer radial elements very short, equally long or absent.

For *Mesiteia*, the low 2nd dorsal fin with short inner radial and very short or missing outer radial elements is an autapomorphic character.

Fig. 36. The 2nd dorsal fin of *Mesiteia* specimen #08 (Munich, coll. FHP). **a.** in normal light; **b.** in long-wave UV-light, taken by Helmut Tischlinger.
Dorsal fins: bd – basal cartilage of dorsal fin; radi – inner rows of radial elements; rado – outer rows of radial elements; in red the number of the radials from anterior to posterior radials.
The vertebral column and the elements of the dorsal and ventral arcualia: bdp – basidorsal process; bvp – basiventral process; dic – dorsal intercalary plate; spn – supraneural; vc – vertebral centrum.

- **Character 24**: Number of series of radial elements of 2nd dorsal fin. The 2nd dorsal fin of *Mesiteia* has many more (36 on #08 and 31? on #12) series of radial elements than *Chiloscyllium* (19 radials, Fig. 35a) and *Hemiscyllium* (14 radials, Fig. 35b). The 2nd dorsal fin of *Mesiteia* is more than twice as long as in other Orectolobiformes and reaches almost the dorsal lobe of the caudal fin. Other specimens, e.g. #04 (Plate 11e), #15 (Plate 11f) and #20 (Plate 11h) confirm this general appearance, but same as with the 1st dorsal fin, on most of the specimens there are parts of the radials missing or compressed, and this makes it impossible to count the elements correctly. Cappetta (1980) counted 25 short pterygophores on the 2nd dorsal fin of specimen #02.

Two states are recognized,

0 number of series of radial elements of 2nd dorsal fin low, around 15 to 20, fin compressed;

1 number of series of radial elements of 2nd dorsal fin more than 30, up to 36, fin elongated.

For *Mesiteia*, the elongated 2nd dorsal fin with up to 36 series of radials is autapomorphic.

4.3.6 Anal fin

The anal fin of *Mesiteia* is not only strikingly different from Hemiscylliidae but from all Recent and fossil sharks. It is probably the most striking feature to recognize this fossil shark and gives reason enough to establish a new family of sharks for *Mesiteia*.

Following Goto (2001, p. 34–35, figs. 36c, 38), the basal cartilages (b) of the anal fin of Orectolobiformes are reduced compared with those of the dorsal fins. The basal cartilages occur on the distal region of the midline between the hypaxial body muscles. In most Orectolobiforrnes these are short and rounded cartilaginous pieces (Goto 2001, figs. 36C, 38A–B); in *Rhincodon*, these are fused into a large, single cartilaginous plate carrying all radials (Goto 2001, fig. 38C). The radials of the anal fin consist of two elements similar to those of the dorsal fin – an elongated proximal or inner radial element (radi) and a short distal or outer radial element (rado). The radials are thinner, less strongly formed compared with those of the dorsal fin.

In most of the specimens of *Mesiteia* there are at least some parts of the very long and characteristic anal fin to see, but they are very poorly and incompletely preserved on the holotype and not visible on specimen #02 in Paris. Therefore, Cappetta's description (1980) remained a bit insufficient when he wrote "... according to the photographs, it (the anal fin) is very long and low; it comprises about fifty fine rays, fairly spaced and decreasing in size towards the rear." The anal fins of some *Mesiteia* specimens, #08 (Plate 11d and Plate 12a, marked between red arrows), #04 (Plate 11e), #15 (Plate 11f), #12 (Plate 11g) and #20 (Plate 11h) are illustrated in Plate 11. Specimen #08 is by far the most interesting one and its very aesthetically beautiful details are illustrated in Plates 12–13.

- **Character 25:** Shape and position of the anal fin and overlapping the lower caudal fin lobe. The anal fin of *Mesiteia* is extremely long, with 97 series of radials (in *Mesiteia* #08). It begins in the middle of the 1st dorsal fin at the level of the 60th vertebra and extends to far behind the posterior end of the 2nd dorsal fin at the level of the 114th vertebra to where the caudal fin originates and overlaps anteriorly a quarter of the lower caudal fin lobe (12 lower caudal radials, the 12 prehypochordal cartilages). In the holotype (*Mesiteia* #01) the anal fin begins in the middle of the 1st dorsal fin at the level of the 55th vertebra but its end is not preserved and almost nothing of the anal fin is visible in the Paris specimen #02. In the Munich specimen 90+x radials can be counted.

According to Goto (2001, p. 57) "in *Orectolobus*, *Sutorectus*, *Eucrossorhinus*, *Brachaelurus* and *Stegostoma*, the anal fin is located immediately anterior to the origin of the lower caudal lobe with no interspace. In *Chiloscyllium* and *Hemiscyllium*, it is posteriorly shifted, overlapping the lower caudal lobe. In *Ginglymostoma*, it is weakly separate by a short distance from the origin of the lower caudal lobe. In *Parascyllium*, *Cirrhoscyllium* and *Rhincodon*, it is widely separated from the lower caudal lobe." Compared to *Mesiteia*, the overlapping of the caudal fin is less developed in Hemiscylliidae. In *Chiloscyllium punctatum* are 6 caudal fin radials overlapped by posterior anal fin radials, and 3 in *Hemiscyllium ocellatum*. In outgroups, in the plesiomorphic state, the anal fin is wide separate from the lower lobe of the caudal fin (Goto 2001, p. 70).

Four states are recognized,
0 short anal fin widely separated from the lower lobe of the caudal fin;
1 short anal fin close to the lower lobe of the caudal fin;
2 short anal fin overlapping the lower lobe of the caudal fin;
3 anal fin very long, widely overlapping the lower lobe of the caudal fin.

Parascyllium and *Cirrhoscyllium* share the plesiomorphic character of an anal fin widely separated from the lower lobe of the caudal fin. *Orectolobus*, *Sutorectus*, *Eucrossorhinus*, *Brachaelurus*, *Ginglymostoma*, *Nebrius*, *Pseudoginglymostoma* and *Stegostoma* share the synapomorphy with an anal fin located immediately anterior to the origin of the lower caudal lobe. For *Rhincodon*, the anal fin widely separated from the lower caudal lobe is a reversal. For *Chiloscyllium* and *Hemiscyllium*, the posteriorly shifted anal fin, that is anteriorly overlapping the lower caudal lobe, is a synapomorphic character shared with *Mesiteia*. And for *Mesiteia*, the posteriorly shifted and extremely elongated anal fin, widely overlapping the anterior part of the lower caudal lobe, is an autapomorphic character.

- **Character 26:** Shape and length of basal elements of the anal fin. In *Mesiteia*, there are obviously no basal elements in the anal fin skeleton, and the radials do not approach the vertebral column. It is unlikely that the basal elements could have been overlooked during the preparation of *Mesiteia* #08, or that they were lost during fossilization. In Recent Hemiscylliidae, *Chiloscyllium* and *Hemiscyllium*, the basal elements of the anal fin are very small, but every radial is articulated with just a rounded cartilaginous element (Goto 2001, fig. 36C, Dingerkus & DeFino 1983, figs. 49, 50, and here Plate 12, Figs. b, c). In most Orectolobiformes, the basal elements of the anal fin are composed of short, rounded cartilages (Goto 2001, figs. 36C, 38A–B). In *Rhincodon*, these are fused into a large, single cartilaginous plate carrying all radials (Goto 2001, fig. 38C). Basal elements of the anal fin composed of a single or a few large plates is the plesiomorphic state (Goto 2001, p. 7; Shirai 1992, p. 89, pl. 56).

Three states are recognized,
0 basal elements of anal fin composed of a single or a few large plates;
1 basal elements of anal fin almost all separate and short;
2 basal elements of anal fin absent.

Parascyllium, *Cirrhoscyllium*, *Orectolobus*, *Sutorectus*, *Eucrossorhinus*, *Brachaelurus*, *Chiloscyllium* and *Hemiscyllium* share the synapomorphic character that most or all of the basal elements of the anal fin are separate, and are short to very short. Their reduction to only a few plates, partly articulating with more than one radial as seen in *Ginglymostoma*, *Nebrius* and *Pseudoginglymostoma*, is a reversal that is best developed in *Rhincodon*, where the basal elements of the anal fin are fused into a large, single cartilaginous plate carrying all radials. For *Mesiteia*, the complete loss of basal elements in the anal fin is autapomorphic.

- **Character 27:** Number, shape and length of radial elements of the anal fin. In *Mesiteia* the inner radial elements (radi) of the anal fin are all very similar in size, all well developed, straight, with the same distance from each other, inclined backwards, only the anterior radials a bit more inclined and shorter. The distal tips of the radials are slightly bent backwards. The height of the inclined inner radials corresponds approximately to the distance to the spine. Outer radial elements are not developed. According to Goto (2011, p. 35), most Orectolobiformes, Hemiscylliidae included, have two radial cartilages (inner and outer radial elements) in the anal fin (Goto 2011, figs. 36C, 38A), whereas they are divided into three segments in *Ginglymostoma*, *Stegostoma* and *Rhincodon* (Goto, 2011, figs. 38B-C). The outer radials in Hemiscylliidae are small, only slightly longer than the small basal elements (Goto 2001, fig. 36C, Dingerkus & DeFino 1983, figs. 49, 50, and here Plate 12, Figs. b,c). In *Chiloscyllium punctatum*, the anterior eight smaller radials (Dingerkus & DeFino 1983, fig. 49) do not have outer radial cartilages attached, the following outer radials get increasingly longer, and posteriorly shorter again. Similarly, in *Hemiscyllium ocellatum* the outer radial elements are missing in the anterior four radials (Dingerkus & DeFino 1983, fig. 50). All Orectolobiformes, except *Mesiteia*, have the outer radial elements of the anal fin. Most outgroup species have two radial elements of the anal fin (Goto 2001, p. 62; Shirai 1992, p. 89, pl. 56). Following Goto (2001) this is the plesiomorphic state.

Three states are recognized,
0 radial elements of anal fin composed of two elements;
1 radial elements of anal fin composed of three or more elements;
2 radial elements of anal fin composed of one component, rado absent.

Parascyllium, *Cirrhoscyllium*, *Orectolobus*, *Sutorectus*, *Eucrossorhinus*, *Brachaelurus*, *Chiloscyllium* and *Hemiscyllium*, share the plesiomorphic character of two radial elements (elongated radi and short rado) of the anal fin, separated in most series. In *Ginglymostoma*, *Nebrius*, *Pseudoginglymostoma* and *Rhincodon*, there are three elements, the radi sometimes fused but long, the rado separated or fused but short, a synapomorphic character. For *Mesiteia*, the complete loss of outer radial elements in the anal fin is autapomorphic.

- **Character 28:** Number of series of radial elements of the anal fin. With 97 series of radials (in *Mesiteia* #08), the anal fin of *Mesiteia* is extremely long (Plate 11d, 12a, marked between red arrows). No other sharks have comparable anal fins developed (Goto 2001, fig. 38; Shirai 1992, pl. 56). The anal fins of Hemiscylliidae are very short, with app. 28 series of radials in *Chiloscyllium*, and 21 in *Hemiscyllium* (Dingerkus & DeFino 1983, figs. 49, 50).

Two states are recognized,

0 number of series of radial elements of anal fin less than 30;

1 number of series of radial elements of anal fin more than 90.

The high number of series of radial elements in the anal fin is an autapomorphic character of *Mesiteia*.

4.3.7 Caudal fin

The caudal fin of *Mesiteia* is strikingly different from the one of Hemiscylliidae and all other Recent and fossil sharks. With only a small part of the caudal fin preserved, it is possible to identify this genus and its autapomorphic characters, giving one more reason to establish a new family of sharks for *Mesiteia*.

Unfortunately, the caudal fin is missing on the holotype and only the beginning of this fin is present but poorly preserved on the Paris specimen #02. Cappetta (1980: p. 122) wrote "La nageoire caudale manque sur les deux exemplaires connus; sur celui de Paris toutefois, on peut voir sa base qui débute au niveau de la 103ème vertèbre. On peut observer les basiventraux, courts, larges à leur base, à extrémités effilées et unis par une étroite bande cartilagineuse. La queue était certainement longue et de type primitif, sans lobe inférieur bien différencié et à lobe apical très réduit." (The caudal fin is missing on the two known specimens; on the one from Paris, however, one can see its base, which starts at the level of the 103rd vertebra. Basiventrals can be seen, short, broad at their base, with tapered ends and united by a narrow cartilaginous band. The tail was certainly long and primitive, with poorly differentiated lower lobe and very small apical lobe.)

Goto (2001, p. 36, figs. 39, 42) summarized the chondrified elements of the caudal fin of Recent Orectolobiformes. In the precaudal region the basiventral process (bvp) is expanded laterally from the precaudal vertebrae to support the dorsal wall of the body cavity in the abdominal region; it is expanded ventrally to form a complete or incomplete hemal arch in the tail. In the caudal region the basiventral process is fused to form a complete hemal arch, and extends ventrally as a large, compressed hypochordal process (hcp). Shirai (1992, pl. 53) used the term basiventral process without distinction also for the hypochordal process of the caudal region. In some anterior caudal vertebrae, the hypochordal processes are separated from the basiventral process, forming prehypochordal cartilages (phc). The epichordal radials (ecr) are small, lying on the dorsal rim of the neural arch in the upper caudal lobe.

The caudal fins of *Chiloscyllium* and *Hemiscyllium* are formed by a single row of radials on both the dorsal and ventral sides of the vertebral centra. The dorsal radials are articulated with neural arches of the vertebral column, the ventral radials with the hemal arches. The ventral elements basically match up with the centra, one radial row per centrum. The dorsal radials are straight, shorter in anterior and posterior series, giving the dorsal caudal fin a homogenous appearance of a long, drawn, flat bow. The ventral radials are more developed, larger and broader, usually with an enlargement in the middle of each radial and distally inclined towards the rear. In *Chiloscyllium* some of the ventral radials fuse in the middle. The ventral row starts more anteriorly than the dorsal row and underlies the posterior part of the anal fin. In *Chiloscyllium* there are more dorsal radials than ventral ones and in *Hemiscyllium* the dorsal and ventral radials are usually in equal numbers. In Hemiscylliidae the number of radial series varies from about 50 to 70 (Dingerkus & DeFino 1983).

Little & Bemis (2004) presented illustrations and descriptions of the anatomy of the caudal fin of twelve species of sharks, nine Carcharhinidae (*Carcharhinus leucas*, *C. limbatus*, *C. obscurus*, *C. plumbeus*, *Galeocerdo cuvieri*, *Rhizoprionodon terraenovae*, *Sphyrna tiburo*, *S. mokarran*, and *S. lewini*), one species of Alopiidae (*Alopias vulpinus*), one species of Ginglymostomatidae (*Ginglymostoma cirratum*), and one species of Triakidae (*Mustelus canis*). They found extensive variation in the calcification patterns of caudal fin elements. In a specimen of *Sphyrna lewini*, the proximal third of the caudal fin has isolated calcified hemal spines, not fused to hemal arches; the medial third has calcified hemal spines that are fused to hemal arches, and the distal third starts again with isolated calcified hemal spines, not fused to hemal arches, and ends with calcified hemal spines and absent hemal arches (Little & Bemis 2004, fig. 3A, 4B). In the two examined specimens of *Ginglymostoma cirratum*, calcification of hemal arches was not found, and the hemal spines, when calcified at all, only exhibited calcification in the distal third of the spines. Stronger and more robustly calcified hemal spines support the more expanded ventral lobe of the caudal fin in Carcharhinidae, as needed for fast swimming, whereas the much smaller ventral hypochordal lobe of the caudal fin of *Ginglymostoma cirratum* might be related to its benthic habits (Little & Bemis 2004, p. 570). According to Little & Bemis (2004), the variation in caudal fin anatomy "suggests its potential as a source of phylogenetic characters at the species, generic, family and ordinal levels" (p. 563), and "there seems little doubt that caudal skeletal anatomy of elasmobranchs can provide a wealth of phylogenetically useful characters that may help elucidate relationships of both fossil and living elasmobranchs".

The impressive caudal fins of some *Mesiteia* specimens, #08 (Pl. 11d), #04 (Pl. 11e), #15 (Pl. 11f), #12 (Pl. 11g) and #20 (Pl. 11h) are illustrated in plate 11, and for comparison together with caudal fins of *Chiloscyllium* and *Hemiscyllium* (after Dingerkus & DeFino 1983, fig. 49, 50). The excellent details of the entire caudal fin of specimen #08 are shown in Plates 12–13 and I mainly used this specimen #08 for my description.

- **Character 29**: General shape of the caudal fin. In contrast to caudal fin skeletons of all other sharks, the composition of the caudal fin of *Mesiteia* is rather complex, with prehypochordal, hypochordal, preepichordal, and epichordal cartilages that are all more complicated and more finely constructed than the homogenous, strong radials of other Orectolobiformes (Goto 2001, figs. 39, 42) and outgroup sharks (Shirai 1992, pl. 53).

Two states are recognized,

0 caudal fin cartilages homogenous, strong;

1 caudal fin cartilages complex, more finely constructed.

Such a differentiation of caudal fin elements is autapomorphic for *Mesiteia*.

- **Character 30**: Shape, number and consistency of series of preepichordal cartilages (pec). The radial elements of the dorsal lobe of the caudal fin of *Mesiteia* start at the level of vertebra 107, which is 5 vertebrae behind the posterior end of the 2nd dorsal fin (Pl. 12d), and they continue to the rear tip of the tail at vertebra 174. To 68 vertebrae are 72 series of dorsal radials attached (the numbering of the radials is marked in red in Pl. 12).

Plates 12–13 ▷

Anal and caudal fin of Hemiscylliidae and *Mesiteia emiliae*: anal fin; D1: 1st dorsal fin; D2: 2nd dorsal fin; anal fin; caudal fin.

a, d–g. *Mesiteia* specimen #08 (Munich, coll. FHP). **b.** *Chiloscyllium punctatum* (after Dingerkus & DeFino 1983, fig. 49). **c.** *Hemiscyllium ocellatum* (after Dingerkus & DeFino 1983, fig. 50). Photograph of Plate 13g with long-wave UV-light was taken by Helmut Tischlinger.

Dorsal fins: bd – basal cartilage of dorsal fin; radi – inner rows of radial elements; rado – outer rows of radial elements.

Anal fin: ba – basal cartilage of anal fin; radi – inner rows of radial elements (marked between red arrows, numbering of anal fin radials in red); rado – outer rows of radial elements.

Caudal fin, vertebral column and the elements of the dorsal and ventral arcualia: bdp – basidorsal process; bvp – basiventral process; dbc – calcified double cone of vertebra; dic – dorsal intercalary plate; ecr – epichordal radial; hcp – hypochordal process; mee – membrana elastica externa; pec – preepichordal cartilage (numbering of pec and ecr in red; marked in red brackets, the 22 preepichordal cartilages of *Mesiteia* #08); phc – prehypochordal cartilage (numbering of phc in red; marked in red brackets, the 12 prehypochordal cartilages of *Mesiteia* #08); scl – secondary calcified lamella of vertebra; spn – supraneural; vc – vertebral centrum; vc114 – vertebra #114.

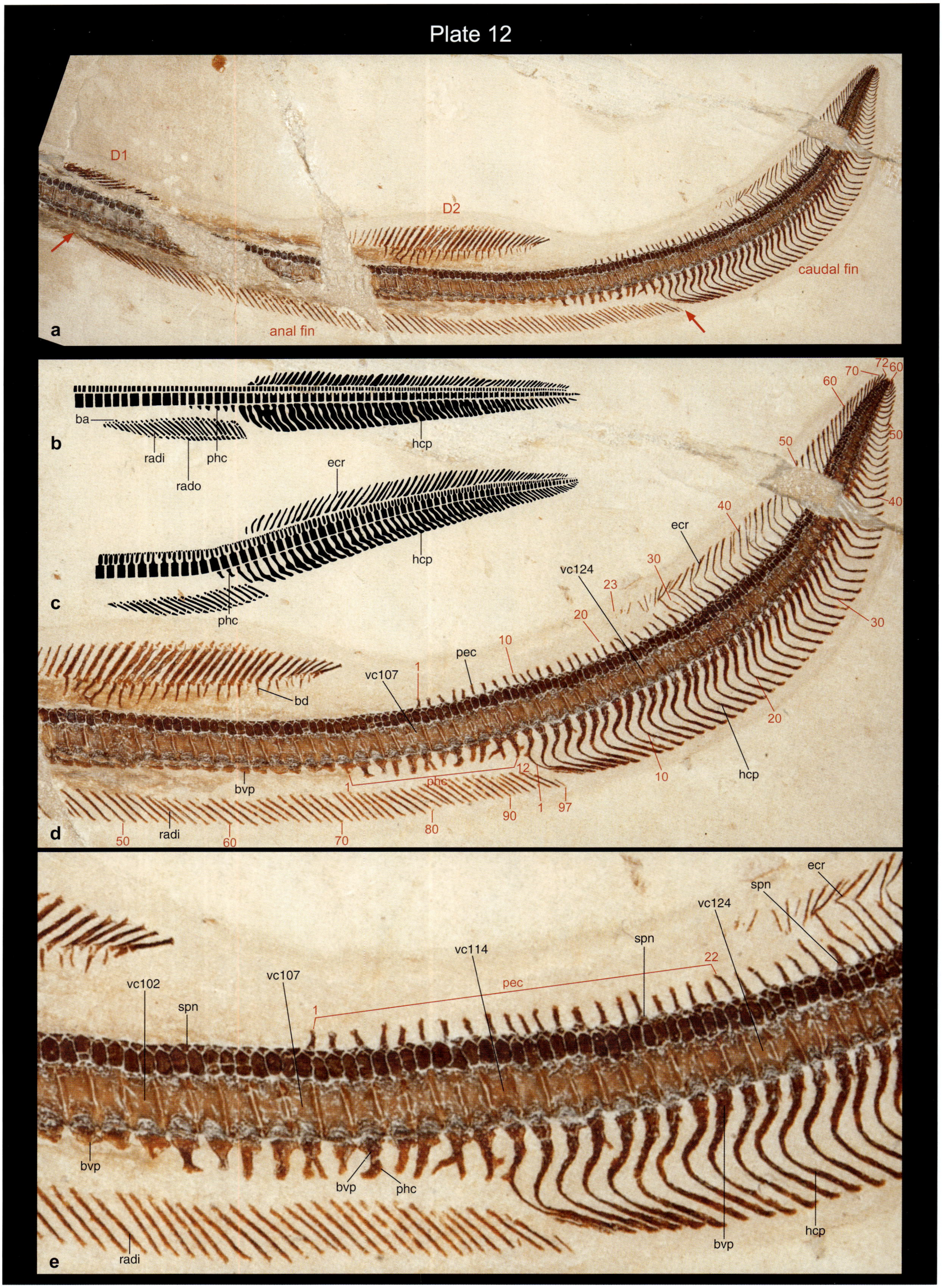
Plate 12
D1
D2
caudal fin
anal fin
a
ba
radi
phc
rado
hcp
b
ecr
hcp
phc
c
vc124
pec
vc107
bd
bvp
phc
radi
hcp
d
ecr
spn
vc124
vc114
spn
vc102
vc107
spn
pec
bvp
bvp
phc
radi
bvp
hcp
e

Plate 13

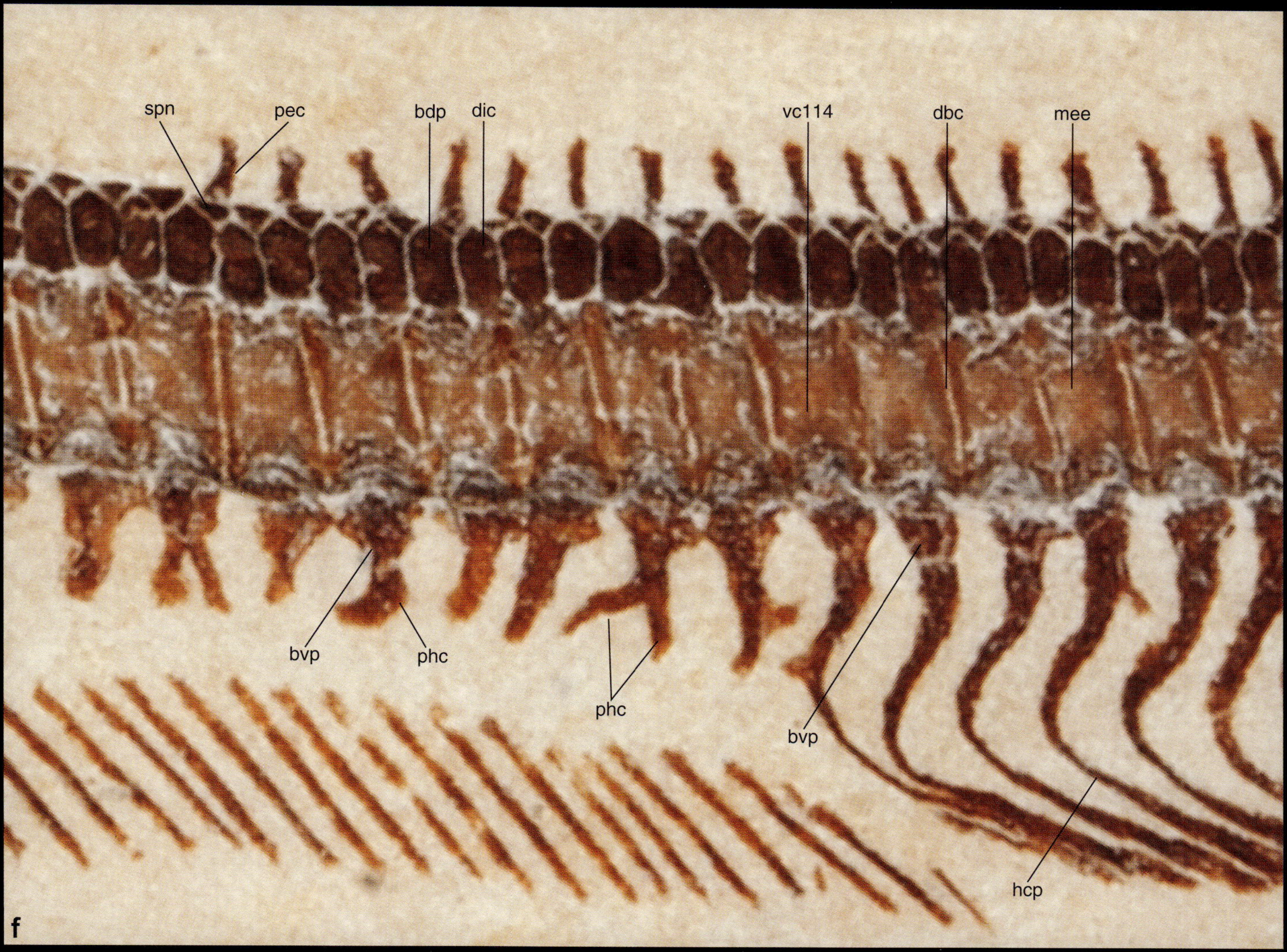

I call the anterior 22 series of dorsal radial elements (from vertebra 107 to 123, Pl. 12d) preepichordal cartilages (pec), in accordance to the prehypochordal cartilages (phc) of the ventral lobe. The preepichordal cartilages consist of two elements, a supraneurcl (spn) which builds the neural arch, and a short, straight dorsal radial, not longer than the basidorsal processes (bdp, Plate 13). They articulate with the small supraneurals (spn), between the basidorsal process (bdp) and the dorsal intercalary plate (dic). It seems that the articulation with the supraneural cartilages becomes a bit more complicated from about the 13th series of radial onwards, where the supraneurals get multiplied without changing their size. This lasts until the tip of the fin. In most Orectolobiformes, preepichordal cartilage is absent, but very few series of preepichordal cartilage can be seen in *Hemiscyllium freycineti*, *Parascyllium* and *Cirrhoscyllium* (Goto 2001, figs. 39, 42B), and maybe in some more species of Hemiscylliidae, e.g. a single preepichordal cartilage in *Chiloscyllium punctatum* (Dingerkus & DeFino 1983, fig. 49). The preepichordal cartilages in *Parascyllium* and *Cirrhoscyllium* (Goto 2001, fig. 42B), anterior to some partly fused epichordal cartilages, are different from those in Hemiscylliidae, where the epichordal cartilages are separate. The absence of preepichordal cartilages in outgroups (Shirai 1992, pl. 53) is plesiomorphic.

Three states are recognized,
0 preepichordal cartilages absent;
1 very few series of preepichordal cartilages present;
2 numerous, up to 22 series of preepichordal cartilages present.

Within Orectolobiformes, in *Parascyllium* and *Cirrhoscyllium* the presence of some few series of preepichordal cartilages represents a first apomorphy, whereas the few series of preepichordal cartilages in some Hemiscylliidae represents a second apomorphic character. The absence of preepichordal cartilages is plesiomorphic. Numerous series of preepichordal cartilages are known only in *Mesiteia* and is obviously an autapomorphic character.

- **Character 31**: Shape, number and consistency of series of epichordal radials. From vertebra 124 onwards the radial elements of the dorsal lobe of the caudal fin of *Mesiteia* get about three times longer and more complex. These are the epichordal radials (ecr). Compared to epichordal radials of other Orectolobiformes, e.g. *Chiloscyllium* (Pl. 12b) or *Hemiscyllium* (Pl. 12c), they look thin and fragile. Their base, the supraneural (spn) remains as before; the shape of the basal part of the radials, which were the preepichordal cartilages in anterior segments, remain as before at the first radials but become now elongated by very thin cartilages with slightly broader ends that are bent backwards. On the first 6 series of radials the middle part is not yet developed, and one can see that the inner and outer parts of these epichordal radials will converge. The next about 20 series of dorsal caudal radials are in the middle clearly bent in abdominal direction, like a sickle or very similar also to ciliae. Finally, the posterior 20 series of radials are not bent like this but are instead inclined toward the rear end and get continuously shorter. I counted 50 series of epichordal radials in *Mesiteia* #08 (the numbering of the radials is marked in red in Pl. 12). Preepichordal and epichordal radials together make 72 series of radials in *Mesiteia*, slightly more than the approximately 60 series in *Chiloscyllium* and *Hemiscyllium*. The difference is that in all other Orectolobiformes and outgroups, all epichordal cartilages are rather strong, straight and homogeneous, and this is the plesiomorphic state.

Two states are recognized,
0 epichordal cartilages strong and straight;
1 epichordal cartilages thin and ciliate.

In all other Orectolobiformes and outgroups, epichordal cartilages are strong and straight, and this is plesiomorphic. Thin, ciliate, sickle-shaped epichordal cartilages are obviously an autapomorphic character for *Mesiteia*.

- **Character 32**: Shape and consistency of prehypochordal cartilages and their overlap with the anal fin. The radial elements of the ventral lobe of the caudal fin of *Mesiteia* start 5 vertebrae earlier than the dorsal radials at the level of vertebra 102 (Pl. 12e), which is the next vertebra following the posterior end of the 2nd dorsal fin. They continue to vertebra 174 and these 73 vertebrae match up equally with 73 series of ventral radials attached (the numbering of the radials is marked in red in Pl. 12).
The anterior 12 series of ventral radial elements (from vertebra 102 to 113) are the prehypochordal cartilages (phc, Pl. 12–13). They are overlapped by about 25 series of radials of the posterior end of the anal fin. Therefore, the space is limited; the prehypochordal cartilages consist of two elements, the basiventral process (bvp) which build the hemal arch, and a stout, short, ventral radial, stronger and only a bit longer than the preepichordal cartilages. The broad and short basiventral processes of the anterior vertebrae become triangular and a bit stronger when they start to articulate with the prehypochordal cartilages. Some of the prehypochordal cartilages are forked and bent, others are straight. Actually, they only look bifurcated when the radials of the left and right sides do not exactly touch each other and are unfused. The prehypochordal cartilages originate from the left and right, broad and short basiventral processes, from the hemal arches and build a hemal spine which are here the prehypochordal cartilages. In most Orectolobiformes, prehypochordal cartilages are absent, but a few series of prehypochordal cartilages are present in *Hemiscyllium freycineti*, *Parascyllium* and *Cirrhoscyllium* (Goto 2001, figs. 39, 42B), and maybe in some more species of Hemiscylliidae, e.g. six prehypochordal cartilages in *Chiloscyllium punctatum* (Dingerkus & DeFino 1983, fig. 49) and two in *Hemiscyllium ocellatum* (Dingerkus & DeFino 1983, fig. 50). The prehypochordal cartilages in *Parascyllium* and *Cirrhoscyllium* (Goto 2001, fig. 42B) are different from those in Hemiscylliidae. They are located anterior to some incompletely developed hypochordal processes, which are not all fused with the basiventral process (bvp). It is important that they are separated from the hemal arch (basiventral process) and that they consist of a single cartilage element. In Hemiscylliidae, the prehypochordal cartilages consist of two elements, the basiventral process, and a short radial as seen in *Mesiteia*. The absence of prehypochordal cartilages in outgroups (Shirai 1992, pl. 53) is plesiomorphic.

Three states are recognized,
0 prehypochordal cartilages absent;
1 prehypochordal cartilages separated from the hemal arch;
2 prehypochordal cartilages fused with the hemal arch.

Within Orectolobiformes, in *Parascyllium* and *Cirrhoscyllium* the presence of a few series of prehypochordal cartilages not fused with the basiventral process represent a first apomorphy, evolved from the absence of prehypochordal cartilage, which is plesiomorphic. A series of prehypochordal cartilages fused with the basiventral process, as seen in some Hemiscylliidae, represents a second apomorphic character. The latter is synapomorphic for *Mesiteia*.

- **Character 33**: Number of series of prehypochordal cartilages (for description see character 32).

Three states are recognized,
0 prehypochordal cartilages absent;
1 up to 6 series of prehypochordal cartilages present;
2 greater than 6, and up to 12 series of prehypochordal cartilages present.

Within Orectolobiformes, in *Parascyllium* and *Cirrhoscyllium* the presence of only few series of prehypochordal cartilages not fused with the basiventral process represent a first apomorphy, evolved from the absence of prehypochordal cartilage, which is plesiomorphic. The presence of up to 6 series of prehypochordal cartilages fused with the basiventral process as seen in some Hemiscylliidae, represent a second apomorphic character. The high number, up to 12 series of prehypochordal cartilages (up to 12 series of radials) is autapomorphic for *Mesiteia*.

- **Character 34**: Shape, number and consistency of hypochordal process. From vertebra 114 onwards the shape of the ventral radials of the caudal fin changes (Pl. 13) until the rear tip of the tail. These are the very characteristic hypochordal processes (hcp). Compared to other Orectolobiformes, e.g. *Chiloscyllium* (Pl. 12b) or *Hemiscyllium* (Pl. 12c), the hypochordal processes of *Mesiteia* are not as strong and uniformly large and broad, with an enlargement in the middle of each radial, and a distal inclination towards the rear. The hypochordal processes (hcp) consist of two elements, the basiventral process (bvp) which builds the hemal arch and, attached to it, the elongated ventral radial. Together they build the hypochordal process, which in other sharks is usually broadly fused with the corresponding vertebral centrum (Shirai 1992, pl. 53). The basiventral processes build in *Mesiteia* a strong socket, less broad than the anterior basiventral processes and a little bit higher, on which the ventral radials obviously articulate. The basiventral processes themselves are less broad than the vertebral centra to which they articulate. The radials beginning here on vertebra 114 are immediately very long, similar in shape to the epichordal cartilages at the middle of the dorsal lobe of the caudal fin, proximally strong, getting thinner at the middle part that is curved anteriorly, and becoming slightly stronger to the distal ends that are bent backwards toward the tip of the caudal fin. All hypochordal processes are in the middle strikingly recurved in the abdominal direction, as mentioned above like a sickle or also similar to ciliae. The entire caudal fin of *Mesiteia* is similar in shape to a bottle cleaner. The first about 20 of these typically sickle-shaped caudal radials have almost the same length that are then over the next 30 series of radials gradually reduced posteriorly in size until the last vertebra 174. I counted 61 radials with hypochordal processes in *Mesiteia* #08 (the numbering of the radials is marked in red in Pl. 12). Their number is similar to that of Hemiscylliidae with up to 60 radials. According to Goto (2001), in most Orectolobiformes as in the outgroups, the hypochordal process is almost equal in size and shape in the anterior region, and is gradually reduced posteriorly in size, and lacking segments entirely (Goto 2001, figs. 39, 42). In *Rhincodon*, the number of series is reduced and some hypochordal processes are greatly elongated and segmented. The plesiomorphic state is that with strong, equally broad and almost straight hypochordal processes.

Three states are recognized,

0 hypochordal cartilages strong, equally broad and straight, not segmented;
1 hypochordal cartilages strong, elongated, segmented, reduced in number;
2 hypochordal cartilages fine and ciliate (sickle-shaped).

All Orectolobiformes, except *Rhincodon* and *Mesiteia*, share the plesiomorphic character. The reduction of the number of series, the segmentation and elongation of prehypochordal cartilages, is an autapomorphic character for *Rhincodon*. The very characteristic fine and sickle-shaped hypochordal processes are unique and an autapomorphic character for *Mesiteia*.

4.3.8 Vertebral column, vertebral centra, dorsal and ventral arcualia, vertebral ribs

In most specimens of *Mesiteia* there are parts of the vertebral column missing or damaged to such a degree that no reliable statement can be made. The tails of both the holotype and the Paris specimen #02 are even missing; the holotype ends behind the 96th vertebra and the Paris specimen #02 behind the 120th vertebra. The vertebral column of specimen #08 is almost perfectly preserved, with 174 vertebrae, but the vertebral column of *Mesiteia* has not yet been extensively studied.

Vertebrae are principally divided into abdominal vertebrae anterior to the cloaca, with no hemal arches, precaudal vertebrae from the cloaca to the origin of the lower caudal fin, with incomplete hemal arches except for the posteriormost few vertebrae in general, and caudal vertebrae located in the dorsal lobe of the caudal fin. The abdominal and precaudal vertebrae are divided into monospondylous vertebrae, with only one condyle in a single myotome and a pair of vertebral ribs, and diplospondylous vertebrae, with two condyles in a single myotome, without vertebral ribs (Goto 2001, p. 36, figs. 39, 41A).

The shape of the vertebrae of *Mesiteia* varies, depending both on the location within the vertebral column and on the size and age of the shark. The monospondylous vertebrae of the anterior and trunk area up to the pelvic fins are always longer than high in younger specimens (e.g. #04, Pl. 14c,d), as long as they are high (e.g. #02, Pl. 14b), or at least occasionally as long as they are high (e.g. #08, Pl. 14g). Diplospondyly begins on specimen #08 with a transitional vertebra (the 42nd vertebra, Pl. 14f), where the vertebral rib-bearing segments end and the pelvic girdle begins. The diplospondylous vertebrae of the anal and caudal fin areas are much shorter than high (e.g. #08, Pl. 13e). The vertebrae are rather delicate, with little constriction visible from the outside (Pl. 14g), unless they were damaged during preparation and the calcified double cone of vertebra (dbc) becomes visible (Pl. 14e).

On some vertebrae longitudinal lamellae, the secondary calcified lamellae (scl), can be observed (Pl. 14g). These longitudinal calcified lamellae radiate outward from the vertebral centra that surround the notochord and strengthen the vertebral centra. Vertebral centra supported by secondary calcified lamellae are called asterospondylic vertebrae.

- **Character 35**: Secondary calcified lamellae. According to Goto (2001, p. 36–37) and Ridewood (1921, p. 396), the secondary calcified lamellae (scl) extend to the membrana elastica externa in *Chiloscyllium*, *Hemiscyllium*, *Stegostoma* and *Ginglymostoma* (Goto 2001, figs. 40, 41C). In *Orectolobus*, *Sutorectus*, *Eucrossorhinus* and *Brachaelurus*, they do not extend to the membrana elastica externa (Goto 2001, fig. 41A). In *Parascyllium* and *Cirrhoscyllium*, no calcification area is present into the outer zone (Goto 2001, fig. 41B). In *Rhincodon*, the chondrified basidorsal and basiventral insert deeply into the outer zone to the double cone, and a number of thin concentric calcified layers are placed in the outer zone and are equivalent to the concentric type in *Cetorhinus* (Goto 2001, fig. 41D). In *Mesiteia*, the secondary calcified lamellae extend to the membrana elastica externa (mee) which is fossilized in an irregular pattern (Pl. 14g). The plesiomorphic state is cyclospondylic with no secondary calcified lamellae.

Four states are recognized,

0 secondary calcified lamellae absent, cyclospondylic;
1 secondary calcified lamellae not extending radially to the membrana elastica externa;
2 secondary calcified lamellae extend radially to the membrana elastica externa;
3 concentric calcified layers are placed in the outer zone.

The absence of secondary calcified lamellae is plesimorphic for *Parascyllium* and *Cirrhoscyllium*. *Orectolobus*, *Sutorectus*, *Eucrossorhinus* and *Brachaelurus* share secondary calcified lamellae not extending to the membrana elastica externa as a synapomorphy (clade E in Goto 2001). *Mesiteia* shares secondary calcified lamellae extending to the membrana elastica externa as another synapomorphy with *Chiloscyllium*, *Hemiscyllium*, *Ginglymostoma* and *Stegostoma* (clade I in Goto 2001). The concentric calcified layers in *Rhincodon* are another apomorphic character or a reversal.

On all specimens of *Mesiteia* the neural and hemal arches with dorsal and ventral arcualia are usually well preserved, if not damaged during preparation or missing accidently.

In *Mesiteia*, the arcualia are calcified. The shape of the basidorsal processes (bdp) and the dorsal intercalary plates (dic) (Pl. 14c) varies from abdominal to caudal vertebrae, but there are also differences that might reflect individual or maybe ontogenetic variation. Generally, these elements look both very similar, like rounded discs above the abdominal centra (Pl. 14c–d, f–g), forming dorsally opened neural arches. They have an upright pentagonal shape and are increasingly higher than long in the precaudal and caudal areas. The plates are relatively loosely packed above the anterior vertebrae, but they are packed

Plate 14

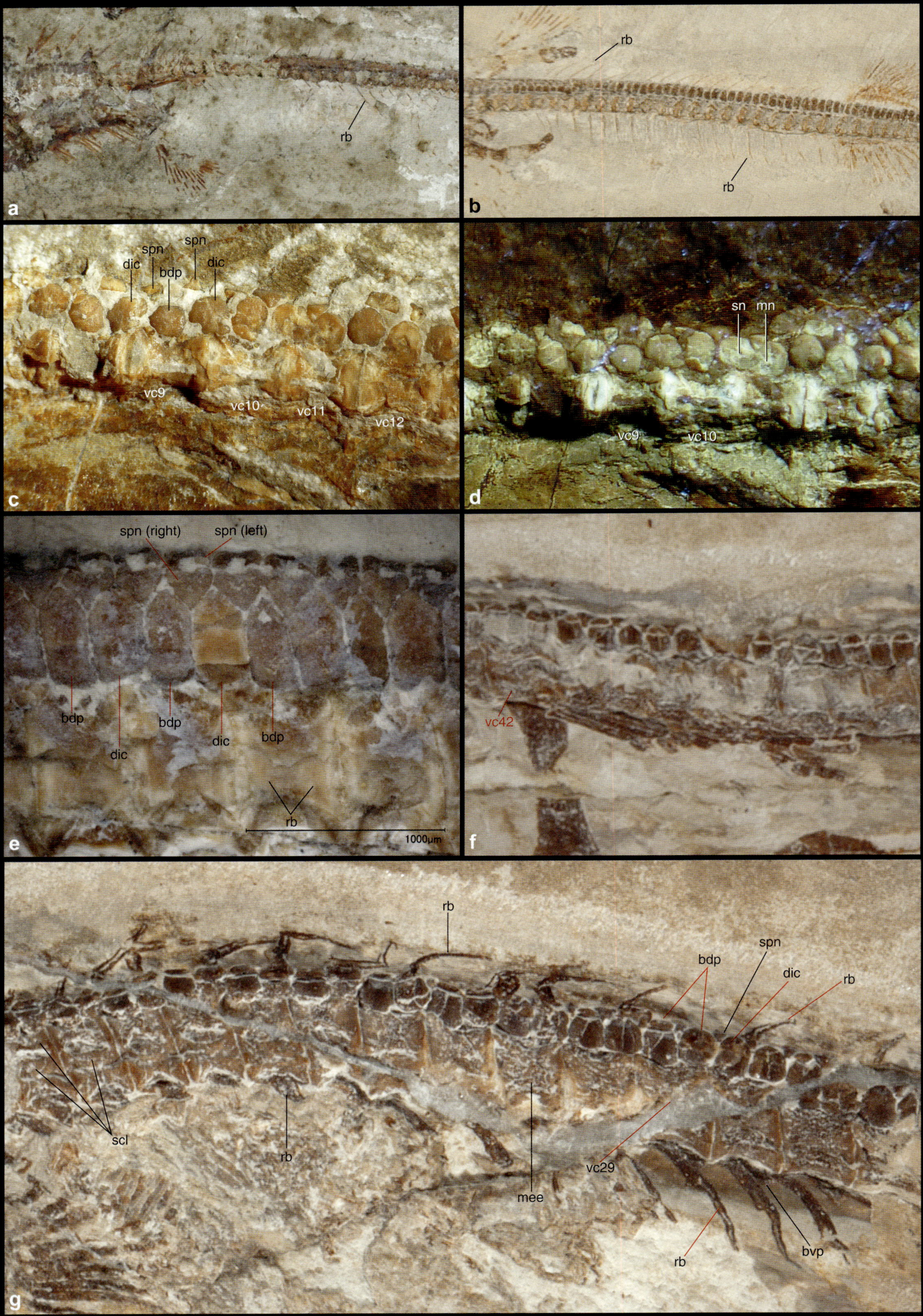
a
rb
b
rb
rb
c
spn
spn
dic
bdp
dic
vc9
vc10
vc11
vc12
d
sn
mn
vc9
vc10
e
spn (right)
spn (left)
bdp
dic
bdp
dic
bdp
rb
1000µm
f
vc42
g
rb
bdp
spn
dic
rb
scl
rb
mee
vc29
rb
bvp

tightly together in the precaudal region (Pl. 13f and Pl. 14e) and towards the caudal tip (Pl. 12e). In *Mesiteia* the basidorsal processes and dorsal intercalary plates have the same size, but they can be distinguished by their position above the vertebral centra. There is usually one basidorsal process located above the middle of the centra and one dorsal intercalary plate between two adjacent vertebrae. At higher magnification or with long-wave UV-light one can see the small foramina for the spinal nerves – ventrally on the basidorsal processes the foramen for the ventral (motoric) nerve-root (mn), and dorsally on the dorsal intercalary plates the foramen for the dorsal (sensoric) nerve-root (sn) (Pl. 14d).

The supraneurals (spn) are relatively loosely packed, smaller and of irregular oval shape upon and between the dorsal edges of the basidorsal processes and dorsal intercalary plates in the anterior region (Pl. 14c), or bigger, heart-shaped and tightly filling the space between the sharp dorsal edges of the basidorsal processes and dorsal intercalary plates in the trunk and caudal region (Pl. 14e). In *Mesiteia* #12 on the vertebrae between the two dorsal fins the heart-shaped supraneurals of the right side are very nicely preserved, and additionally one can see the dorsal end of the left supraneurals that are not fused with their antimeres (Pl. 14e).

- **Character 36**: Shape and composition of neural arch. In most Orectolobiformes, the neural arch is composed of a completely segmented basidorsal process (bdp), dorsal intercalary plate (dic) and supraneural (spn) (Goto 2001, figs. 41A–C). In *Rhincodon*, the basidorsal process and dorsal intercalary plate are partly fused, forming a large, seamless neural arch (Goto 2001, fig. 41D). In most Orectolobiformes, the neural arch, basidorsal process and dorsal intercalary plate are well calcified in adults. Only in *Rhincodon* and *Stegostoma* do these remain chondrified even in adults.
In most Orectolobiformes, the basidorsal process (bdp) is fused with the vertebral centra and therefore it can easily be distinguished from the dorsal intercalary plate (dic) (Goto 2001, figs. 39–42, and here Fig. 37). But in *Mesiteia*, the dorsal arcualia, basidorsal processes and dorsal intercalary plates are free, not proximally fused with the centra, and look very similar, almost indistinguishable, only marked by their position.

Two states are recognized,
0 basidorsal process (bdp) is fused with vertebral centra;
1 basidorsal processes and dorsal intercalary plates isolated, not fused with centra.

The composition of free basidorsal processes and dorsal intercalary plates as rounded loosely packed discs is an autapomorphic character for *Mesiteia*.

Within Orectolobiformes the ventral arcualia of the abdominal vertebrae are formed by the chondrified basiventral processes (bvp), attached ventrally to the centra, and expanded laterally to support the dorsal wall of the body cavity. In *Mesiteia* the basiventral process is low and long, quite visible in *Mesiteia* #08 (Figs. 35–37).

◁ **Plate 14**
Details of vertebral column, vertebral centra, dorsal and ventral arcualia, and vertebral ribs of *Mesiteia emiliae*. **a.** *Mesiteia* specimen #01 (holotype, Zagreb, Croatian Natural History Museum, Inv. number 9354), photo Sanja Japundžić; **b.** *Mesiteia* specimen #02 (Paris, MNHN.F.HAK513); photo Martin Pickford; **c–d.** *Mesiteia* specimen #04 (Munich, SNSB-BSPG 2011 I 124); **c.** in normal light; **d.** in long-wave UV-light; photo Helmut Tischlinger; **e.** *Mesiteia* specimen #12 (Munich, coll. FHP); **f–g.** *Mesiteia* specimen #08 (Munich, coll. FHP).
The vertebral column and the elements of the dorsal and ventral arcualia: bdp – basidorsal process; bvp – basiventral process; dbc – calcified double cone of vertebra; dic, – dorsal intercalary plate; mee – membrana elastica externa; mn – foramen for the ventral (motoric) nerve-root; rb – vertebral rib; scl – secondary calcified lamella of vertebra; sn – foramen for the dorsal (sensoric) nerve-root; spn – supraneural; vc – vertebral centrum.

- **Character 37**: Ventral intercalary plate (vic). Ventral intercalary plates are found only in *Ginglymostoma*, *Stegostoma* and *Rhincodon* (Goto 2001, p. 38, fig. 41C–D), and they are partly fused with the basiventral process, forming a large, weakly segmented cartilaginous process in *Rhincodon* (Goto 2001, fig. 41D). *Mesiteia* has no ventral intercalary plates (Fig. 37), a symplesiomorphic character shared with most outgroup species.

Two states are recognized,
0 ventral intercalary plates absent;
1 ventral intercalary plates present.

The absence of ventral intercalary plates in *Mesiteia* is a symplesiomorphic character shared with all Orectolobiformes, excluding *Ginglymostoma*, *Stegostoma* and *Rhincodon*.

- **Character 38**: Basiventral process (bvp) and hemal arch formation in precaudal diplospondylous vertebrae. According to Goto (2001, p. 38, figs. 39, 40, 41A), most Orectolobiformes have ventrally expanded hemal arches in most precaudal diplospondylous vertebrae, and in caudal vertebrae the basiventral process is expanded ventrally to form a complete or incomplete hemal arch. In *Rhincodon*, the tail vertebrae have completely closed hemal arches (Goto 2001, fig. 41D), and *Parascyllium*, *Cirrhoscyllium* and *Stegostoma* have hemal arches partly closed via a thin cartilaginous bar extending along the midline in some vertebrae anterior to the origin of the caudal vertebrae (Goto 2001, fig. 42B). In *Mesiteia*, the hemal arches in precaudal diplospondylous vertebrae are widely opened (e.g. Fig. 36). The basiventral processes on *Mesiteia* change their shape and size at the end of the anal fin, getting higher and triangular (Fig. 37). I described the very special dorsal and ventral arcualia, composed of basidorsal process, dorsal intercalary plate, supraneural, epichordal radial, preepichordal cartilage, basiventral process, hypochordal process, and prehypochordal cartilage previously in Section 4.3.7 (Caudal fin, Pl. 12–13).

Three states are recognized,
0 hemal arches almost opened;
1 hemal arches partly closed by a thin cartilaginous bar;
2 hemal arches completely closed.

Mesiteia shares with *Orectolobus*, *Eucrossorhinus*, *Sutorectus*, *Brachaelurus*, *Hemiscyllium*, *Chiloscyllium* and *Ginglymostoma* the plesiomorphic character that the hemal arches in precaudal diplospondylous vertebrae are widely opened. It is apomorphic for *Parascyllium* and *Cirrhoscyllium* that the hemal arches are partly closed. For *Stegostoma*, it is a further apomorphy having also partly closed hemal arches. And for *Rhincodon*, the completely closed hemal arches are autapomorphic.

- **Character 39**: Vertebral ribs (rb) on diplospondylous vertebrae. Following Goto (2001, p. 38), most Orectolobiformes have no vertebral ribs on diplospondylous vertebrae (Goto 2001, fig. 41A, here Fig. 37). In *Parascyllium*, *Cirrhoscyllium*, *Orectolobus*, *Sutorectus* and *Eucrossorhinus*, a few to more than 10 additional vertebral ribs are present on some anterior diplospondylous vertebrae (Goto 2001, fig. 41B). Each rib on the diplospondylous vertebra articulates with one basiventral process against two vertebrae. *Mesiteia* has no vertebral ribs on diplospondylous vertebrae. The absence of vertebral ribs on diplospondylous vertebrae is plesiomorphic (Goto 2001, p. 63).

Two states are recognized,
0 vertebral ribs on diplospondylous vertebrae absent;
1 vertebral ribs on diplospondylous vertebrae present.

Mesiteia shares with *Hemiscyllium*, *Chiloscyllium*, *Ginglymostoma*, *Stegostoma* and *Rhincodon* (clade I in Goto 2001) the plesiomorphic character that vertebral ribs on diplospondylous vertebrae are absent. The presence of vertebral ribs on diplospondylous vertebrae in *Parascyllium* and *Cirrhoscyllium*, and *Orectolobus*, *Eucrossorhinus*, *Sutorectus* and (most likely but not shown) in *Brachaelurus*, represents two apomorphies.

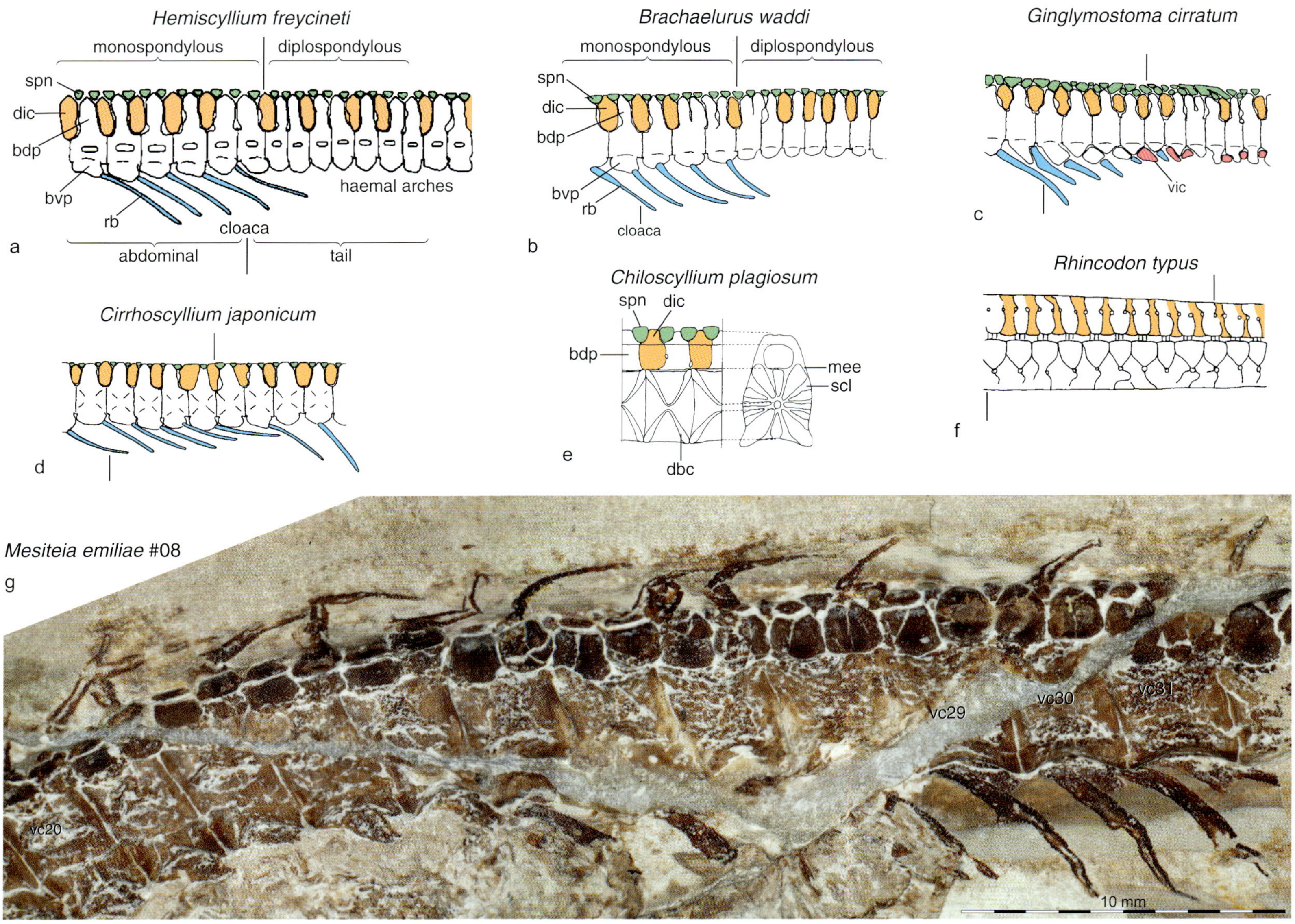

Fig. 37. Vertebral column and the elements of the dorsal and ventral arcualia with ventral ribs. *Mesiteia*, compared with other Orectolobiformes, modified after Goto (2001, figs. 40–41); note the fusion of basidorsal process (bdp) with vertebral centrum (vc), the articulation of vertebral rib (rb) with basiventral process (bvp), and the presence or absence of ventral intercalary plate (vic). **a.** *Hemiscyllium freycineti*; **b.** *Brachaelurus waddi*; **c.** *Ginglymostoma cirratum*; **d.** *Cirrhoscyllium japonicum*; **e.** *Chiloscyllium plagiosum*; cross section of vertebral centrum with marked secondary calcified lamella (scl) and membrana elastica externa (mee); **f.** *Rhincodon typus*; **g.** *Mesiteia emiliae* specimen #08 (Munich, coll. FHP); monospondylous section from the 20st to the 31st vertebra); note the isolated basidorsal process (bdp), and the anterior articulation of vertebral rib (rb) with basiventral process (bvp).
bdp – basidorsal process; bvp – basiventral process; dbc – calcified double cone of vertebra; dic – dorsal intercalary plate; mee – membrana elastica externa; rb – vertebral rib; scl – secondary calcified lamella of vertebra; spn – supraneural; vc – vertebral centrum; vic – ventral intercalary plate.

- **Character 40**: Shape and composition of vertebral ribs (rb) on monospondylous vertebrae. On most Orectolobiformes the simple, slender, slightly curved vertebral ribs (rb), consisting of a single ray of cartilage, are articulated with the posterior margin of the basiventral process (Goto 2001, fig. 41, here Fig. 37). Only in *Ginglymostoma* are the basiventral processes partly elongated to build a strong basal element to articulate with the ribs (Goto 2001, fig. 41C, here Fig. 37). On some specimens of *Mesiteia*, vertebral ribs are visible between the pectoral and pelvic girdles, thin and elongated, almost as long as centra and dorsal arcualia together (e.g. holotype, Pl. 14 a, and #02, Pl. 14b). On the holotype they can be seen up to the 41st vertebra, on the Paris specimen #02 up to the 37th vertebra where the pelvic fin begins, and on specimen #08 up to the 40th vertebra until shortly before the pelvic fin. Only *Mesiteia* #08 allows insight into the detailed composition of its vertebral ribs, from vertebra 29–31 (Fig. 37). The ribs are composed of two equally long elements – a stronger, straight and only slightly curved proximal cartilage, and a thinner and flexible distal cartilage. The proximal cartilage has a small anteriorly expanded antero-lateral margin, opposite to the articulation face of the distal cartilage of the ventral rib. The proximal cartilage of the rib is posteriorly, not anteriorly as seen on other Orectolobiformes (Fig. 37), articulated with the anterior margin of an equally strong but only half as long basiventral process. The ventral rib of the 31st vertebra lost its connection with the basiventral process due to a slight lateral compression during fossilization. On the opposite side of the spine there are some severely bent distal cartilages of other vertebrae, that look like they are connected to the dorsal intercalary plates. One should not confuse them with neural spines.

Two states are recognized,
0 vertebral ribs on monospondylous vertebrae consisting of a single ray;
1 vertebral ribs on monospondylous vertebrae composed of two differentiated elements.

All Orectolobiformes except *Mesiteia*, share the plesiomorphic state of having vertebral ribs on monospondylous vertebrae consisting of a single, simple, slender, slightly curved ray. Only *Mesiteia* has differentiated vertebral ribs, composed of a strong proximal cartilage and a thin and flexible distal cartilage, articulated to the anterior margin of the basiventral process, which is unique and estimated to be an autapomorphic character.

It should not be forgotten that Gorjanović (1884) in his first description of the holotype of *Mesiteia emiliae* made drawings of some details showing dorsal and ventral arcualia (pl. III., figs. 2a, 2b, 2c). Unfortunately, it happened that the printer erroneously shifted the printing plates with the image markings a few millimetres upwards and thus the marking lines mark the wrong parts on these illustrations, e.g. the vertebral centra as hemal arches. This happened with the first edition of 1884

(Fig. 38a) and was partly corrected in the 2nd edition in 1885. This printer's error might have been part of the reason why it was confusing when Gorjanović wrote (1884, p. 46): "... At the spinal column there are still elongated, cylindrical arches and on top of these again smaller intermediate arches." He wanted to mark the dorsal arches (basidorsal process plus dorsal intercalary plate) with "β", but in the figure this led to the supraneural (intermediate arches, intercalary cartilage "γ" in his drawing). I have corrected the image markings here in Figure 38b. In pl. III., fig. 2a Gorjanović (1884), and Gorjanović-Kramberger (1885) presented details of three abdominal monospondylous vertebrae, the vertebral centra (α) with vertebral ribs (ε), articulated with the basiventral process, the dorsal arche (basidorsal process plus dorsal intercalary plate, β), the supraneural (γ), and neural spines (δ). But Orectolobiformes do not have neural spines, and what he marked with "δ" are the vertebral ribs of the opposite side (see also Pl. 14, Figs. a–b, g). It is worth mentioning that in this figure the supraneurals of *Mesiteia* #01 are also heart-shaped and tightly fill the space between the basidorsal processes and dorsal intercalary plates as in *Mesiteia* #12 (Pl. 14e). In pl. III., fig. 2b Gorjanović (1884), and Gorjanović-Kramberger (1885) presented details of five diplospondylous what he thought caudal vertebrae. The holotype is not complete; the caudal fin is missing, and the illustrated vertebrae are those of the region where the basiventral processes (ε) become triangular and a bit stronger, just before they start to articulate with the prehypochordal cartilages of the caudal fin (see Pl. 12). Finally, pl. III., fig. 2c shows three basal cartilages articulated with inner rows of radial elements of the 1st dorsal fin.

The vertebral centra and arcualia of *Mesiteia* show many interesting details, e.g. the shape and size of the basidorsal processes, dorsal intercalary plates, and supraneurals. However, currently the differences of these features from those of other sharks are not sufficiently known to describe more apomorphic characters.

4.3.9 Skin, dermal denticles, tesserae, myosepta, lateral line

Cartilaginous fishes (sharks and rays) are covered with placoid scales, also called dermal denticles, or shagreen. Within a shark family the shape of these dermal denticles is usually very similar. Examples of dermal denticles of Hemiscylliidae are illustrated in Dingerkus & DeFino (1983): *Chiloscyllium burmensis* (figs. 59–60), *C. confusum* (figs. 61–64), *C. griseum* (figs. 65–66), *C. hasselti* (figs. 67–70), *C. indicum* (figs. 71–74), *C. plagiosum* (figs. 75–78), *C. punctatum* (figs. 79–80), *Hemiscyllium freycineti* (figs. 81–84), *H. ocellatum* (figs. 85–88), *H. trispeculare* (figs. 89–90), *H. strahani* (figs. 91–92). All these dermal denticles are leaf-shaped with a sharp tip caudally, generally longer than broad, with a more or less strong median and one or two pairs of lateral keels. They are loosely distributed on the skin of juveniles and build a dense pattern in adults, e.g. on a *Chiloscyllium griseum* (Pl. 15a, coll. FHP-R-2234).

In most fossil sharks, especially in all fossil sharks from Lebanon, the dermal denticles are well preserved, unless the skin was destroyed during preparation. Usually the shining surface of the dermal denticles marks the exact shape of the body, e.g. the outline of the fins on this *Cantioscyllium decipiens* from Lebanon (Pl. 15b, SNSB-BSPG 2012 I 87).

In his description of *Mesiteia*, Cappetta (1980, 2012) did not mention what Kramberger wrote about the skin of *Mesiteia*, but it is one of its most interesting and important characters: Gorjanović (1984, p. 45): "Tielo valjda golo", or in the German translation, Gorjanović-Kramberger (1986, p. 137): "Körper wahrscheinlich nackt" ... "body obviously naked", meaning without dermal denticles. Of course, it can happen that during the taphonomic process, the skin of a shark gets lost on several parts of the body, but not on the distal parts of the fins, not in total. On hundreds of fossil sharks from Lebanon the skin with the dermal denticles is preserved, so why not on *Mesiteia*? It was not just by chance that there were no dermal denticles visible on the holotype and the specimen #02 in Paris, the only two specimens that Cappetta (1980) could examine. This can now be proven on all 26 specimens checked to date, and requires an explanation.

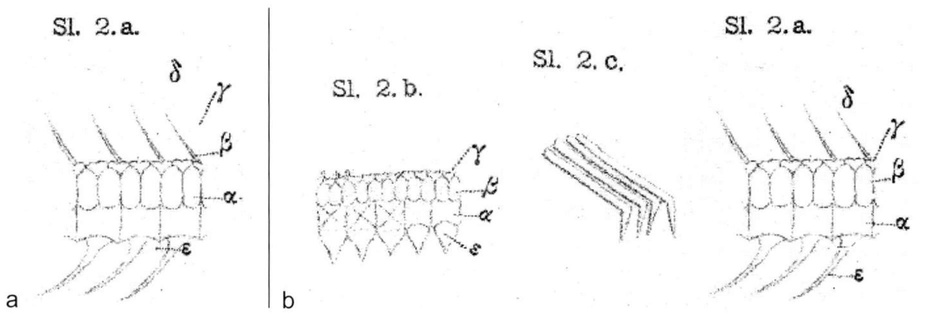

Fig. 38. Drawings of the holotype of *Mesiteia* from Gorjanović (1884, pl. III., fig. 2a,b,c).
a. the original figure (= SL. 2.a.) with the image markings erroneously shifted a few millimeters upwards.
b. all three figures with corrected image markings.
Sl. 2.a. three abdominal monospondylous vertebrae, α – vertebral centra; β – dorsal arche (basidorsal process plus dorsal intercalary plate; γ – supraneural; δ – "neural spines", which are indeed the vertebral ribs of the opposite side; ε – vertebral ribs;
Sl. 2.b. five diplospondylous "caudal" vertebrae, which are indeed precaudal vertebrae of the region where the basiventral processes (ε) become triangular; for explanation of the other Greek letters see previous figure;
Sl. 2.c. three basal cartilages of the 1st dorsal fin articulated with inner rows of radial elements.

Instead of dermal denticles, on most *Mesiteia* specimens a whitish film of an undifferentiated dense mass overlies the parts that would otherwise be covered with dermal denticles (Pl. 15, Figs. c–f). The outline of the fins is not marked with that. Possibly this is the fossilized remains of muscle mass, but this needs to be tested in further studies.

Intensive search for dermal denticles brought some very interesting details to light that could explain the loss of dermal denticles. Only *Mesiteia* #06 displays a small area between the pelvic fins that has about 60 very small dermal denticles preserved (Pl. 16, Figs. a–d). These dermal denticles are about half as long as those of Hemiscylliidae. On a round or oval basal disk ("bv" in Pl. 16d) with a diameter of only 0.1 mm, a short cup of enameloid arises and lends these reduced dermal denticles the appearance of a baby's pacifier in anterior view ("av" in Pl. 16d). In lateral view ("lv" in Pl. 16d), it becomes visible that the leaf-shaped dermal denticles are posteriorly only slightly pointed and it seems that there is a low central and a pair of lateral keels (ridges) present ("k" in Pl. 16d), but without sharp edges or keels as in *Chiloscyllium* (Pl. 15a). In occlusal view ("ov" in Pl. 16d), the surface of the cusp of the dermal denticle is small; the shaft of the denticle arises at the anterior margin of the basal disk and does not overhang the basal disk.

Obviously, in evolutionarily earlier stages of *Mesiteia* normal dermal denticles must have existed prior to being reduced – for whatever reason – until only patchy areas remained.

I would be happy if I could say that the absence of shagreen (dermal denticles), especially in addition to the whitish mass all over the body, was the best character to identify *Mesiteia*. I have seen shark specimens that were identified as *Mesiteia*, based on that. Catsharks from Hgula can appear to be plastered with that whitish mass that can even cover the (otherwise very different) anal and caudal fin rays (Fig. 39). However, if teeth are visible, the identification is easy. It can be complicated to find the shagreen on these little catsharks when the dermal denticles are very small, but so far, I always found at least some areas with dermal denticles (enlarged insert of Fig. 39). I have no explanation for neither the partial (artificial) skinning of just some sharks from the same locality, nor the plastering with that whitish material, that seems to be fossilized muscles. In any case this scenario is different from what I described for *Mesiteia* where the dermal denticles were almost completely lost during evolution.

Unfortunately, the situation is yet more complicated. There is another orectolobiform shark known only from Sahel Alma, that seems to be without dermal denticles, *Acanthoscyllium sahelalmae* (Pictet & Humbert, 1866). When Woodward (1888, p. 127) described in a short article the lateral line of *Acanthoscyllium* he noted: "... In one respect, however, it seems to differ from all known Scylliidae. It is destitute of dermal shagreen; and on this account the peculiarity to be here noted is very conspicuous in the impressions upon the limestone." And one year later Woodward wrote (1889, p. 346): "... there is no evidence of shagreen upon any portion of the fossil, and the only external dermal structures are some of the series of calcified rings supporting the groove of the lateral line." Cappetta studied all known specimens of *Acanthoscyllium* and wrote about the missing dermal denticles (1980, p. 117): "... Aucun des exemplaires, pourtant nombreux, que nous avons pu examiner, ne présentait des restes de peau ou de denticules cutanés dont l'absence reste difficile à expliquer." *Acanthoscyllium* (from

Plate 15

Fig. 39. An unidentified catshark from Hgula. Its body is covered with the same whitish mass as the *Mesiteia* specimens from Hakel. When enlarged, the dermal denticles become visible (enlarged insert); photo Pierre Abi Saad.

Sahel Alma, late Santonian) also seems to lack any dermal denticles, but that needs to be verified. I have not studied any specimens of *Acanthoscyllium* by myself but after my experience with *Mesiteia*, I would not be surprised if dermal denticles were present but were very small and limited to some special areas, as in *Mesiteia*.

- **Character 41**: Complete loss of dermal denticles. Sharks have dermal denticles, even when modified, e.g. in *Echinorhinus*. Only in *Mesiteia*, and probably in *Acanthoscyllium*, the skin is almost completely naked.

Two states are recognized,
0 dermal denticles present;
1 dermal denticles absent.

The loss of dermal denticles, or its almost complete reduction (Pl. 16), is an autapomorphic character for *Mesiteia*, or synapomorphic if it is shared with *Acanthoscyllium*.

Dermal denticles should not be confused with the polygonal blocks of calcified tesserae of endoskeletal cartilage (Fig. 40). These tesserae are built by the calcified tiles of the skeletal surface of the cartilages. They occur especially in the dense surface of the jaws and in other parts of the chondrocranium, in the branchial arches, the supporting cartilages of the fins, the neural and hemal arches, and the centra of the vertebrae.

Some specimens of *Mesiteia* offer astonishing insights into the morphology of even some soft tissues. In *Mesiteia* #09 the myomeres are visible, separated by verticle myosepta (Pl. 17 a in normal light, b under long-wave UV-light). The whitish fossilized remains of muscle mass that I mentioned before (Pl. 15 c–d on specimen #05) traces here the boundaries of the myomeres. It becomes visible that each myomere in this post-pelvic area is associated with a pair of diplospondylous vertebrae.

Another remarkable detail on the same *Mesiteia* specimen #09 is the trace of the lateral line, marked by this same strange, dense, whitish mass as mentioned before. To see the trace of a lateral line on a fossil shark is not so much a surprise but it recalls immediately a structure described by Woodward (1888, p. 127) in a short article about the lateral line on *Acanthoscyllium*: "... On either side of the trunk and tail, at a short distance from the vertebral column, there is a delicate longitudinal streak, which, upon close examination, is found to consist of a series of minute ringlets, in their crushed state overlapping one another. Though in some parts obscure, these are very distinctly shown to be incomplete half-rings, and they are thus precisely similar to those already described by Leydig in *Chimaera*, and by myself in *Squaloraja*. They occupy the position of the lateral line in the fossil, and they are of the usual proportions, so that there can be no doubt as to the correctness of their identification. They are especially evident in one specimen in the British Museum, numbered 48107; but neither in this, nor in any other, have I observed an extension of the rings upon the cephalic region, a circumstance perhaps indicating that, as in *Chimaera*, the groove of the lateral line was partially closed in its anterior portion." Woodward (1889, p. 346, pl. XVI, fig. 1) described and figured this specimen #48107 (here Pl. 17 c) and wrote: "... the only external dermal structures are some of the series of calcified rings supporting the groove of the lateral line." This specimen of *Acanthoscyllium* is preserved in ventral view,

Fig. 40. Polygonal calcified tesserae of endoskeletal cartilage in the chondrocranium of *Mesiteia* specimen #07 (Munich, coll. FHP). pq – palatoquadrate; t – tesserae.

◁ **Plate 15**
Dermal denticles are missing on *Mesiteia emiliae*.
a. Dermal denticles from the trunc area under the 2nd dorsal fin of *Chiloscyllium griseum* (coll. FHP-R-2234).
b. Dermal denticles mark the body shape and outline of the fins on fossil sharks (*Cantioscyllium decipiens* SNSB-BSPG 2012 I 87).
c–d. On *Mesiteia* specimen #05 (Munich, SNSB-BSPG 2013 XXVII 1) a whitish film of an undifferentiated dense mass overlies the parts that were supposed to be covered with dermal denticles; **c.** in normal light; **d.** with long-wave UV-light; photo Helmut Tischlinger.
e–f. *Mesiteia* specimen #04 (Munich, SNSB-BSPG 2011 I 124); the radial elements of the fins and especially the ceratotrichia are not outlined with that whitish undifferentiated dense mass; **e.** in normal light; **f.** in long-wave UV-light; photo Helmut Tischlinger.

Plate 16 ▷
Reduced dermal denticles on *Mesiteia emiliae* specimen #06 (Munich, coll. FHP), found ventrally between the pelvic fins.
a–b. area with dermal denticles marked with red line.
c. about 60 dermal denticles are preserved.
d. the denticles have a height and length of approximately 0.1 mm; basal disk in basal view (bv); anterior face of dermal denticle in anterior view (av); slightly pointed, leaf-shaped, enameloid covered cusp in lateral view (lv); a low central and a pair of lateral keels (k); in occlusal view (ov) the cusp of the dermal denticle does not overhang the basal disk.

Plate 17 ▷▷
a–b. Myomeres, myosepta and lateral line in *Mesiteia* specimen #09 (Munich, coll. FHP); **a,** in normal light; **b,** under long-wave UV-light; photo Helmut Tischlinger.
c. Lateral line in *Acanthoscyllium sahelalmae* (Pictet & Humbert, 1866) as drawn by Woodward (1889, pl. XVI figs. 1–3).
d. Lateral line in Recent *Chlamydoselchus anguineus* supported by a line of remarkably stronger dermal denticles (coll. FHP-R-3002).
e. Lateral line in Recent *Echinorhinus cookei* supported by calcified rings which are built of special dermal denticles (coll. FHP-R-1626).
ll – lateral line; mm – myomere; ms – myoseptum.

Plate 16

Plate 17

ms
mm
ll
a
ll
b
B.M.FOSS. FISHES.
Plate XVI.
1
pct
plv
ll
ll
d¹
d²
f
m
ll
ll
p.pb
3
2
r
r
ll
l.l
1 mm
c
d
e

Fig. 41. Detail of lateral line (II) in *Mesiteia* specimen #09 (Munich, coll. FHP). Section at the end of 1st dorsal fin with probably preserved lateral line groove of the right side (II?).

and therefore both the right and left lateral lines are visible. Maybe it should be mentioned that the first authors (Pictet & Humbert 1866) did not discuss or illustrate either the missing dermal denticles or the preservation of the lateral line. Cappetta (1980) mentioned the "traite de l'existence d'une ligne latérale chez *Scyllium sahel almae*" by Woodward (1888) in a historical overview (p. 72), but there is no discussion of that in his description of *Acanthoscyllium* (pp. 113–118). So, it remains uncertain if any other specimens of *Acanthoscyllium* display the lateral lines as well.

In *Mesiteia* #09, the trace of a lateral line groove is visible from shortly behind the pectoral girdle up to the beginning of the caudal fin. Its diameter is 1 mm and it lies above the myosepta. At first glance, only the lateral line groove of the left side can be traced because this specimen is preserved in lateral view (Pl. 17 a–b). Its proportions are as usually found on any other Recent sharks, but its position seems not to be above the vertebral column but moved dorsally. To answer the question if this is a special character for *Mesiteia* or just due to some taphonomic squeezing during fossilization needs to wait until further specimens were found with lateral lines preserved. It could easily happen that on a shark with a diameter of only about 2.5 to 3 cm, and especially because the skin (epidermis) is not stabilized by dermal denticles, the lateral line gets moved out of normal position after the fish is dead. After taking pictures at higher magnification (Fig. 41), I found at the level of the end of the 1st dorsal fin structures that could be interpreted (with "?") as trace of the somewhat narrower lateral line groove of the opposite side. At this magnification it becomes visible that the "filling" of the lateral line groove consists of tiny slices that overlap one another. Is this what Woodward (1888, p. 127) described as "to consist of a series of minute ringlets, in their crushed state overlapping one another" in *Acanthoscyllium*? The margins of these slices appear like "ringlets". As I did not have the chance to see these structures in *Acanthoscyllium*; I can only guess that it is probably the same. In *Mesiteia*, the grooves of the lateral line are filled with some crystallized soft tissue in the shape of overlapping slices, but I cannot verify "calcified rings" supporting the lateral line. On *Mesiteia* #09, the lateral line is traced just by chance; it is, e.g., not visible on specimen #05 (Pl. 15 c–d). But similar traces of the lateral line groove can also be observed on other sharks from Lebanon (pers. observation).

Following Woodward (1888, p. 126) "... It has long been known that the canal investing the sense-organs of the "lateral line" in Selachian fishes attains, as a rule, to a considerably higher stage of development than in the Chimaeroids. While in the latter the canal is merely an open groove, supported by a series of incomplete ling-like dermal calcifications, in the former it assumes a tubular character, opening externally by a row of small orifices, either in its own roof, or through short secondarily developed diverticula. It is quite possible that the supports of the lateral line of *Scyllium sahel-almae* are a surviving mark of affinity with the Chimaeroids; it is equally possible that they may relate only to analogous mechanical contingencies."

In Recent sharks, the lateral line of *Chlamydoselchus anguineus* is supported by a line of remarkably stronger dermal denticles (Pl. 17 d; coll. FHP-R-3002) and in Recent *Echinorhinus cookei*, the lateral line is held open by calcified rings which are derived from special dermal denticles (Pl. 17 e; coll. FHP-R).

Mulas et al. (2020) described a first case of the absence of skin-related structures (epidermis, stratum laxum, dermal denticles and teeth) in a Recent shark, the blackmouth catshark, *Galeus melastomus*, as kind of morphological abnormality. The specimen was caught in Sardinian waters at a depth of 500 m. Antonello Mulas provided the pictures of this interesting specimen (Figs. 42 a, b).

Fig. 42. Total lack of dermal denticles as a kind of individual morphological abnormality was recently described as an extremely rare case in a Recent specimen of *Galeus melastomus* (Mulas et al. 2020). Overview (**a**), and detail with open and unprotected lateral line (**b**); photos Antonello Mulas.
II – lateral line.

4.3.10 Other soft part imprints, and coprolites

Because of the excellent preservation of the fossils from Lebanon I expect many more details will be described later, when better methods are available. At the moment I can only draw attention to a few striking structures in the area of the gill arches, best seen on *Mesiteia* specimen #11 (Munich, coll. FHP, ex coll. Abi Saad; Pl. 3, Fig. 3 and Pl. 18 c), but probably also on *Mesiteia* specimens #20 (Lebanon, coll. Abi Saad; Pl. 5, Fig. 3) and #24 (ex coll. Roy Nohra, Internetfind, unknown current owner; Pl. 6, Figs. 2a–b). It appears that on these specimens, dark brown imprints of blood vessels and probably even the outline of a two-chambered heart are preserved, marked by a similar whitish substance as seen when myomeres are traced.

Plate 18 ▷

Spiral valves and coprolites of *Mesiteia emiliae*.

a–b. *Mesiteia* specimen #05 (Munich, SNSB-BSPG 2013 XXVII 1); **a.** in normal light; **b.** under long-wave UV-light; photo Helmut Tischlinger.

c. *Mesiteia* specimen #11 (Munich, coll. FHP) with a round coprolite, showing the spiral pattern; probably preservation of heart with coronary arteries, ventral aorta and other blood vessels.

d–e. *Mesiteia* specimen #04 (Munich, SNSB-BSPG 2011 1124); **d.** in normal light; **e.** under long-wave UV-light; photo Helmut Tischlinger.

f. *Mesiteia* specimen #03 (Paris, MNHN.F.HAK559), with a long light brown coprolite showing some little hard parts (bone fragments?) at its posterior end; photo Martin Pickford.

g. *Mesiteia* specimen #16 (Lebanon, coll. Expo Hakel, Roy Nohra), with a big coprolite that seems to have a finer grain content.

ca – coronary arteries; h – heart; va – ventral aorta; v – other blood vessels. c – coprolite.

Plate 18

Although unlikely, it appears much like a strong ventral aorta with its branchings to the gills, originating from one of the ventricles. Even small coronary arteries seem to be present on *Mesiteia* #11, and dorsally more blood vessels. These specimens deserve to be studied with better equipment.

For now, only the preservation of the valvular intestine, or spiral valve, should be additionally mentioned. It is easy to recognize posterior to the pectoral girdle when its content, the coprolite, presents as a dark brown, blackish (Pl. 18a–e) or sometimes light brown (Pl. 18f–g) rounded ball or elongated oval. On *Mesiteia* #11 (Pl. 18c) the typical spiral pattern of the coprolite is visible. A closer examination of these coprolites could maybe provide information about what *Mesiteia* had eaten.

Finally, I should mention that I tried to find some trace of eggs on the female specimens of *Mesiteia* and also on other sharks from Lebanon but so far, I could not identify any.

4.4 Thoughts about the possible behaviour of *Mesiteia*

There are many characters that distinguish *Mesiteia* from all other Recent and fossil sharks:

- *Mesiteia* has an elongated "snake-like" body shape with a straight tail and low dorsal fins;
- the skin is almost completely naked, and if present, rudimentary dermal denticles are reduced to short stubs;
- 1st and 2nd dorsal fin are moved far posteriorly, and the radials are shorter than in any other Orectolobiformes;
- the anal fin is extremely elongated anteriorly up to the pelvic fin and posteriorly overlapping the lower caudal fin lobe;
- anal and caudal fin form one combined functional unit;
- the radials of the caudal fin, the hypochordal processes, are sickle-shaped and perfectly built to enable small movements;
- the clutching type teeth are pronounced but do not provide much ecological information.

How far can behaviour be deduced from the morphology of a fossil shark? Besides its small size, its "snake-like" body shape and its belonging to Orectolobiformes there is no reason why *Mesiteia* should be included in Hemiscylliidae. And why should it behave like Hemiscylliidae? Within the Orectolobiformes a wide variety of special behaviours and extreme differences can be found: Rhincodontidae vs Parascylliidae, or Orectolobidae vs Hemiscylliidae. As a rule of thumb, extreme anatomical differences are reflected in different tooth morphology, e.g. Ginglymostomatidae vs. Stegostomatidae. Odontological differences between genera and/or families are normally stable features for recognizing the species within a genus and the genera within a family, in both Recent and fossil sharks unless, in fossil sharks, isolated teeth were assigned to genera without knowing enough about anatomical characters. This happened with all the Lebanese fossil Orectolobiformes. There is no close relationship between *Mesiteia*, *Acanthoscyllium*, *Almascyllium* and Recent Hemiscylliidae.

Mesiteia is anatomically so different that it is necessary to compare its special characters (loss of dermal denticles, "snake-like" body, elongated anal fin) with all other fishes, not only with sharks. Dermal denticles of sharks or scales of teleosteans usually protect the body. However, some fishes do not need such protection. They include the pearlfishes, e.g. the Pearl fish *Carapus acus* (Brünnich, 1768) (Ophidiiformes, Carapidae, *Carapus*). Adult *Carapus* typically live as commensals in the gut of shallow-water holothurians. They might partly protrude or entirely leave their host at night to feed on small fish and benthic invertebrates. Another pearlfish, *Encheliophis*, lives parasitically, feeding on gonads and respiratory organs of its host. The snubnosed eel *Simenchelys parasitica* Gill, 1879 (Anguilliformes, Synaphobranchidae) is a specialized hagfish-like scavenger, tearing away chunks of flesh with stout teeth from carcasses. *Simenchelys* is known also for its habit of burrowing into the abdominal cavities of bigger fish, e.g. sharks and eating e.g. their eggs. All these parasitic or semiparasitic fishes have in common a "snake-like" body, elongated anal and caudal fins and a reduction of their scales. The caudal fin of *Carapus acus* is surprisingly similar to the caudal fin of *Mesiteia*. There are videos on the Internet showing a pearlfish entering the anus of a sea cucumber. The fish is transparent, has no scales, and is protected by its host. Of course, it's just a hypothesis that may never be proven that *Mesiteia* had a comparable behavior; less dramatic is the idea that it could have lived in caves like Muraenidae or other eels. The almost complete reduction of the dermal denticles remains an argument for living in a host. To enter the body of a host having the normal type of dermal denticles with sharp points directed to the rear would be complicated and dangerous for the host.

Recently I found a part of the caudal region of a very small shark within the abdominal area of a new 75 cm long odontaspid shark from Hgula and thought this could be the awaited prove for *Mesiteia* being an endoparasitic shark (Fig. 43). However, with its very strong and straight caudal fin radials, this specimen can be recognized as one of the still undescribed catsharks.

Little & Bemis (2004, p. 570) discussed the possibility that more strongly developed, partly fused hemal arches and spines supporting big ventral lobes of the caudal fin might be related to the pelagic habits of Carcharhinidae, whereas the much smaller ventral lobes of the caudal fin of *Ginglymostoma* might be related to its benthic habits. The very special anal-caudal fin anatomy of *Mesiteia* certainly reflects its habit. Whether it has burrowed itself into the sandy seabed with the moving rays of its fins or gained access to a host as an endoparasite will be difficult to prove.

The behaviour of *Mesiteia emiliae* remains speculative but it shows a wide range of possible behaviours unlike the known coral reef benthic habit of *Chiloscyllium* and the behaviour of *Hemiscyllium* as "Walking Shark".

Fig. 43. Part of the vertebral column of a catshark, recently discovered within the body of a odontaspid shark (*Odontaspis aculeatus*) from Hgula (coll. FHP).

5. Systematic results

Mesiteia should not remain within the family Hemiscylliidae (Longtail carpet sharks, Bamboo sharks), which is founded on the two Recent genera *Chiloscyllium* Müller & Henle, 1837 and *Hemiscyllium* Müller & Henle, 1837. Based solely on isolated teeth there are a few fossil species that can be attributed to these two Recent genera but there are other fossil taxa that need further investigation to prove whether they can be placed within the Hemiscylliidae or not. This is necessary especially for the only two other fossil species with skeletal remains, *Acanthoscyllium sahelalmae* (Pictet & Humbert, 1866) and *Almascyllium cheikheliasi* (Signeux, 1949). It is not aim of this study to solve the taxonomical problems of all the other small Orectolobiformes. However, it can be mentioned that it is maybe not possible to understand the relationships of Orectolobiformes at this stage. The diversity becomes very high when the study is focused on tooth morphology and when fossil species are included. Additionally, there are quite a few new finds of fossil skeletons from Lebanon that probably will change our present view.

The interrelationships of Recent Orectolobiformes were analyzed by Goto (2001), based on the comparative morphology of external characters, as skeletal, musculature and sensory systems. Goto listed for each of his characters the outgroup state and in general I followed his opinions. His analysis of 136 characters derived from 22 species of Orectolobiformes resulted in a strict consensus tree presented in Goto (2001, p. 74, fig. 84). Of course, many of these characters that refer to soft parts could not be observed on the fossil specimens of *Mesiteia* but I tried to follow his descriptions, and I found 41 characters (Tab. 2) that can be used to clarify the systematic position of *Mesiteia* within the Orectolobiformes: 26 characters are autapomorphic for *Mesiteia* and support its separation from Hemiscylliidae; 9 characters observed in *Mesiteia* are plesiomorphic; two characters are synapomorphic, shared with Hemiscylliidae (clade J in Goto 2001); two characters are synapomorphic, shared with *Chiloscyllium*, *Hemiscyllium*, *Ginglymostoma*, *Stegostoma* and *Rhincodon* (clade I in Goto 2001); and further two characters are synapomorphic, shared with all Orectolobiformes (clade A in Goto 2001). Eighteen of these characters are adopted and slightly modified after Goto (2001); the rest are new.

I followed Goto's (2001, p. 70) classification of the order Orectolobiformes into clades A to O and integrated the 41 characters found in *Mesiteia* into Goto's tree (2001, p. 74, fig. 84). The apomorphic characters supporting the clades A to O are listed in Goto (2001, pp. 70–75, tab. 2) and need not to be repeated here. In Table 2, I listed the 41 characters found for *Mesiteia*, and for the 18 adopted characters I added the character numbers as used by Goto (2001).

In the resulting cladogram (Fig. 44), I marked the position of the plesiomorphic state of all 41 characters of the outgroup, and then the first autapomorphic change (A1 in magenta) and eventually the further changes of the character, either as further autapomorphies (A2 in green, A3 in cyan, A4 in yellow, and A5 in orange) or as reversals (R). Those characters that were adopted from Goto (2001) are in a circle, and the new ones in a rhombus. For *Mesiteia*, I marked all 41 characters as either autapomorphic (A1 to A5), synapomorphic (SA) or plesiomorphic. To Goto's designations of his clades (A to O) I added the number of synapomorphies he found, because many of them could not be found on the fossil *Mesiteia*.

The clades represent monophyletic groups, whereas the composition of families can be paraphyletic. When I included the family names of Recent Orectolobiformes, minor problems resulted e.g. from the uncertainty concerning the systematic position of Recent *Nebrius* and *Pseudoginglymostoma*, as discussed in detail in Goto (2001, p. 77). Dingerkus (1986) had erected the genus *Pseudoginglymostoma*, and Goto (2001) was following Dingerkus (1986), who had erected a family Rhincodontidae to include *Ginglymostoma*, *Pseudoginglymostoma*, *Nebrius*, *Stegostoma*, and *Rhincodon* in a monophyletic group, because of the paraphyletic nature of *Pseudoginglymostoma*, *Ginglymostoma* and *Nebrius*, that were previously included in Ginglymostomatidae by Applegate (1972). Goto was unable to dissect neither specimens of *Nebrius* nor *Pseudoginglymostoma*, and therefore he could not add new results. He preferred to recognize the family Rhincodontidae (Dingerkus 1986) for nomenclatural stability for his clade N which is supported by 23 synapomorphies, and left out *Nebrius* and *Pseudoginglymostoma* in the cladogram (Goto 2001, fig. 84) that I adopted for Figure 44. The interrelationships among *Ginglymostoma*, *Pseudoginglymostoma* and *Nebrius* are still unresolved due to insufficient anatomical studies especially of *Pseudoginglymostoma*. I follow Compagno (2001), who treats the Recent family Ginglymostomatidae as a paraphyletic group with a subfamily Pseudoginglymostomatinae (with a single genus *Pseudoginglymostoma*) as sister group of the subfamily Ginglymostomatinae (with two genera *Ginglymostoma* and *Nebrius*). Because I discussed the dental similarities of teeth of *Mesiteia* and *Pseudoginglymostoma* (Section 4.2.10), I included *Pseudoginglymostoma* and *Nebrius* (with dotted lines) in the cladogram, even when solid apomorphies currently are not available.

To include the names of the Recent families in Figure 44, I followed the taxonomy proposed by Cappetta (2012), Compagno (1988, 2001), Ebert, Fowler & Compagno (2013), Last & Stevens (1994, 2009) and Nelson (2006), in separating the order Orectolobiformes into seven Recent families: Parascylliidae, Orectolobidae, Brachaeluridae, Hemiscylliidae, Ginglymostomatidae, Rhincodontidae and Stegostomatidae.

I included the position of *Acanthoscyllium* with questionmark and dotted line in a family incertae sedis as the sister group of *Mesiteia*, being aware that in this study *Acanthoscyllium* is only supported by two autapomorphies, a bifid and very long medial rostral rod, and the absence of dermal denticles.

Unfortunately, the taxonomic position of the few other fossil Orectolobiformes known from skeletal material remains unresolved because of insufficient preservation.

However, the cladogram (Fig. 44) reflects what can be observed when one looks at the teeth of Recent species and of those fossil species that can be easily attributed to any of the Recent genera. But for many of the Cretaceous taxa it is impossible to find the right spot in a phylogenetic tree. The anatomical characters of Recent sharks are difficult to see on fossil specimens, and to study odontological characters only is not enough. Therefore, many fossil species for which only the teeth are known should still be kept in open nomenclature as Orectolobiformes indet.

It seems appropriate to propose a new family for *Mesiteia*.

Class	Chondrichthyes Huxley, 1880
Subclass	Elasmobranchii Bonaparte, 1838
Cohort	Euselachii Hay, 1902
Subcohort	Neoselachii Compagno, 1977
Order	Orectolobiformes Applegate, 1972
Family	**Mesiteiidae nov.**

Type genus: *Mesiteia* Kramberger, 1884, known only from two excavation sites on the north and south flanks of a small valley E of Hakel in Lebanon.

Diagnosis. Small orectolobid sharks with an elongated "snake-like" body shape. The new family is supported by 26 autapomorphic characters (Tab. 2; for explanation of the following numbers, e.g. 01-2, 02-1: first is the character number, followed by the state of the character): 01-2 combination of dental characters: The generally orectolobiform teeth are of the clutching type with a hemiaulacorhize root. In anterior and lateral files, principal cusp strong, conically shaped, upright, smooth on both lingual and labial faces of the crown, high-triangular, with well marked but not very sharp cutting edges, not much laterally expanded, with only one pair of prominent sharp lateral cusplets (dagger-shaped). The apron is prominent and well separated from the labial outline in a median position, very long, narrow and rounded. In posterior files, principal cusp inclined, only slightly expanded, with partly reduced lateral cusplets. 02-1 medial rostral rod consists of a dorsal and a ventral rod;

Table 2: List of characters found in *Mesiteia* and used to distinguish it from other families of Orectolobiformes.

character # here	character # in Goto 2001	short description	states of character	character as seen in *Mesiteia*	autapomorphic/plesiomorphic for *Mesiteia*	synapomorphic with: clade J (Hemiscylliidae)***	synapomorphic with: clade I**	synapomorphic with: clade A (Orectolobiformes)*
		dental characters						
01	36 37	dental characters	0 apron and uvula **absent** 1 **orectolobid dental characters different to Mesiteia** 2 **dental characters as described for *Mesiteia***	In anterior and lateral files, principle cusp upright, smooth, high-triangular, not much laterally expanded, with only one pair of prominent lateral cusplets (dagger-shaped) and **apron very long, prominent, narrow**. In posterior files, principle cusp inclined, only slightly expanded, with partly reduced lateral cusplets.	A			
		cranium						
02		medial rostral rod (**rm**)	0 medial rostral rod consists of **one cartilage** 1 medial rostral rod consists of **two cartilages** 2 medial rostral rod is **bifid and very long**	medial rostral rod consists of **a dorsal and a ventral rod**	A			
03	3	outer nasal cartilage (**onc**)	0 outer nasal cartilage **arched, surrounds nasal capsule** 1 outer nasal cartilage **straight, short**	outer nasal cartilage **straight, short bar, expanding from rostral rod**, cross-like shape of the rostrum	A			
		pharyngeal arches						
04		inner process of ceratobranchial (**ipc**)	0 inner process of ceratobranchial **absent** 1 inner process of ceratobranchial **present**	prominent inner process of ceratobranchial **present**				SA
		pectoral girdle, pectoral fins						
05	52	suprascapular cartilage (**ssc**)	0 suprascapular cartilage **absent** 1 suprascapular cartilage **present**	small suprascapular cartilage **present**			SA	
06	56	pectoral basal cartilages (**prop, mesp, metp**)	0 **separated** pro-, meso- and metapterygium 1 propterygium **absent** (or fused with mesopterygium) 2 propterygium partly **fused** with mesopterygium	widely **separated** pro-, meso- and metapterygium	P			
07	57	metapterygial axis (**mtx**) on the metapterygium	0 metapterygial axis **present** 1 metapterygial axis **absent**	metapteryial axis **absent**				SA
08	58	number of pectoral radial elements (**rad1-3**)	0 composed of **three** components 1 **more than three** (up to 10) components	pectoral radials composed of **three** components	P			
		pelvic girdle, pelvic fins						
09	59	prepelvic process of the puboischiatic bar (**pvp**)	0 prepelvic process **rounded or bluntly pointed** 1 prepelvic process **greatly projected anteriorly**	prepelvic process is reduced, and **rounded**	P			
10		articular condyle for anterior pelvic basal (**pac**)	0 articular condyle for anterior pelvic basal **present** 1 articular condyle for anterior pelvic basal **absent**	**no projected articular condyle** on the pelvic bar	A			
11	60	number of pelvic radial elements (**radi, rado**)	0 pelvic radials composed of **two** elements 1 pelvic radials composed of **three or more** elements	pelvic radials composed of **two** elements	P			
12		anterior pelvic basal (**apv**)	0 anterior pelvic basal **present** 1 anterior pelvic basal **absent**	anterior pelvic basal **absent,** indistinguishable	A			
		1st dorsal fin						
13		position of 1st dorsal fin (**D1**)	0 origin of 1st dorsal fin **anterior** to pelvic fin 1 origin of 1st dorsal fin **far posterior** of pelvic fin 2 origin of 1st dorsal fin **above middle to end** of pelvic fin	origin of 1st dorsal fin **above the end** of the pelvic fin		SA		
14		shape and length of basal elements (**bd**) of 1st dorsal fin	0 **all** basal elements of 1st dorsal fin **attached to radials** 1 **anterior and posterior series** of radials of 1st dorsal fin have **no basal elements attached**	basal elements **absent in anterior and posterior series** of radials, increasing and decreasing length of bd	A			
15		fused basal elements (**bd**) of 1st dorsal fin	0 bd **at least partly fused** and all radials attached to bd 1 basal elements of 1st dorsal fin **all separate**	**no fused basal elements** on 1st dorsal fin	A			
16	61	number of 1st dorsal radial elements (**radi, rado**)	0 composed of **two** elements 1 composed of **three or more** elements	1st dorsal fin radials composed of **two** elements	P			
17		shape and length of radial elements (**radi, rado**) of 1st dorsal fin	0 **radi long** or segmented, **rado strong, elongated** 1 **radi short, rado very short, equally long**	low 1st dorsal fin with **very short inner radial** and **almost missing outer radial** elements, **equally long**	A			
18		number of series of radials of 1st dorsal fin	0 number of radials of 1st dorsal fin **low, around 15** 1 number of radials of 1st dorsal fin **more than 20**	**up to 23 radial series,** fin twice as long as in other species	A			
		2nd dorsal fin						
19		position of 2nd dorsal fin (**D2**)	0 **D2 above anal fin,** with **great distance to D1** 1 **D2 above anal fin,** with **short distance to D1** 2 **D2 well behind anal fin, great distance to D1** 3 **D2 between pelvic and anal fin, short distance to D1** 4 **D2 between pelvic and anal fin, great distance to D1** 5 **D2 moved in posterior direction**	**D2 moved in posterior direction**, close to caudal fin, in the middle over the anal fin	A			
20		shape and length of basal elements (**bd**) of 2nd dorsal fin	0 **all** basal elements of 2nd dorsal fin **attached to radials** 1 **anterior and posterior series** of radials of 2nd dorsal fin have **no basal elements attached**	basal elements **absent in anterior and posterior series** of radials, increasing and decreasing length of bd	A			
21		fused basal elements f 2nd dorsal fin	0 bd **at least partly fused** and all radials attached to bd 1 basal elements of 2nd dorsal fin **all separate**	**no fused basal elements** on 2nd dorsal fin	A			

Table 2: (continued).

character # here	character # in Goto 2001	short description	states of character	character as seen in *Mesiteia*	autapomorphic/plesiomorphic for *Mesiteia*	synapomorphic with: clade J (Hemiscylliidae)***	clade I**	clade A (Orectolobiformes)*
		dental characters						
22	61	number of 2nd dorsal radial elements (**radi, rado**)	0 composed of **two** components 1 composed of **three or more** components	2nd dorsal fin radials composed of **two** components	P			
23		shape and length of radial elements (**radi, rado**) of 2nd dorsal fin	0 **radi long** or segmented, **rado strong, elongated** 1 **radi short, rado very short, equally long**	low 2nd dorsal fin with **very short inner radial** and **very short outer radial** elements, **equally long**	A			
24		number of series of radials of 2nd dorsal fin	0 number of radials of 2nd dorsal fin **low, 15 to 20** 1 number of radials of 2nd dorsal fin **more than 30**	**up to 36 radial series**, fin twice as long as in other species	A			
		anal fin						
25	136	shape and position of the anal fin and its overlapping the lower caudal fin lobe	0 **short** anal fin **widely separated** from caudal fin 1 **short** anal fin **close to** caudal fin 2 **short** anal fin **overlapping** the lower lobe of caudal fin 3 **very long** anal fin, **widely overlapping** the caudal fin	**very long** anal fin, **widely overlapping** the caudal fin	A			
26		shape and length of basal elements (**bd**) of the anal fin	0 bd of anal fin **composed of a single or a few plates** 1 bd of anal fin **almost all separate and short** 2 basal elements of anal fin **absent**	**complete loss** of basal elements in the anal fin	A			
27	62	number, shape and length of radial elements (**radi, rado**) of the anal fin	0 composed of **two** elements 1 composed of **three or more** elements 2 **outer radial elements absent**	**complete loss of outer radial elements** in the anal fin	A			
28		number of series of radials of the anal fin	0 number of series of radials of anal fin **less than 30** 1 number of series of radials of anal fin **more than 90**	**up to 97 series** of radials, anal fin extremely long	A			
		caudal fin						
29		shape and differentiation of caudal fin	0 caudal fin cartilages **homogenous, strong** 1 caudal fin cartilages **complex, more finely constructed**	differentiated, with **preepichordal, epichordal, prehypochordal** and **hypochordal** cartilages	A			
30		shape, number and consistency of series of preepichordal cartilages (**pec**)	0 preepichordal cartilages **absent** 1 **very few series** of preepichordal cartilages **present** 2 **numerous series** of preepichordal cartilages **present**	**up to 22 series** of preepichordal cartilages **present**	A			
31		shape, number and consistency of series of epichordal radials (**ecr**)	0 epichordal cartilages **strong and straight** 1 epichordal cartilages **thin and ciliate**	**up to 50 series** of **ciliate** epichordal radials	A			
32	68	shape and consistency of prehypochordal cartilages (**phc**), overlap with the anal fin	0 prehypochordal cartilages **absent** 1 phc **separated from hemal arch** 2 phc **fused with hemal arch**	prehypochordal cartilages **fused with hemal arch**		SA		
33		number of prehypochordal cartilages (**phc**)	0 prehypochordal cartilages **absent** 1 **up to 6 series** of prehypochordal cartilages **present** 2 **up to 12 series** of prehypochordal cartilages **present**	**numerous series** of prehypochordal cartilages **present**	A			
34		shape, number and consistency of hypochordal process (**hcp**)	0 hcp **strong**, equally broad, **straight, not segmented** 1 hcp **strong, elongated, segmented**, reduced number 2 hcp **fine and ciliate (sickle-shaped)**	up to 61 series of **sickle-shaped** hypochordal processes	A			
		centra, dorsal and ventral arcualia, ribs						
35	64	secondary calcified lamellae (**scl**)	0 scl **absent, cyclospondylic** 1 scl **not extending to the membrana elastica externa** 2 scl **extend to the membrana elastica externa** 3 **concentric calcified layers**	scl **extend to the membrana elastica externa**			SA	
36	63	shape and composition of neural arch (**bdp, dic, spn**)	0 basidorsal process **fused with vertebral centra** 1 bdp and dic **isolated, not fused with centra**	bdp **isolated, not fused with centra**, shape like dic	A			
37	65	ventral intercalary plate (**vic**)	0 ventral intercalary plate **absent** 1 ventral intercalary plate **present**	ventral intercalary plate **absent**	P			
38	66	basiventral process (**bvp**) hemal arch formation in precaudal vertebrae	0 hemal arches **almost open** 1 hemal arches **partly closed** by a thin cartilaginous bar 2 hemal arches **completely closed**	hemal arches **widely separated**	P			
39	67	vertebral ribs (**rb**) on diplospondylous vertebrae	0 vertebral ribs on diplospondylous vertebrae **absent** 1 vertebral ribs on diplospondylous vertebrae **present**	vertebral ribs on diplospondylous vertebrae **absent**	P			
40		shape and composition of vertebral rib (**rb**) on monospondylous vertebrae	0 vertebral ribs consisting of a **single ray** 1 rb composed of **two differentiated elements**	rb composed of **differentiated proximal and distal cartilage**	A			
		skin, dermal denticles						
41		skin coverage with dermal denticles	0 dermal denticles **present** 1 dermal denticles **absent**	dermal denticles **absent**, or almost completely reduced	A			

* Clade A (Goto 2001), containing all species of Orectolobiformes.
** Clade I (Goto 2001), containing all species of *Chiloscyllium*, *Hemiscyllium*, *Ginglymostoma*, *Stegostoma* and *Rhincodon*.
*** Clade J (Goto 2001), containing all species of *Chiloscyllium* and *Hemiscyllium* (Hemiscylliidae).
A – apomorphic; SA – synapomorphic; P – plesiomorphic.

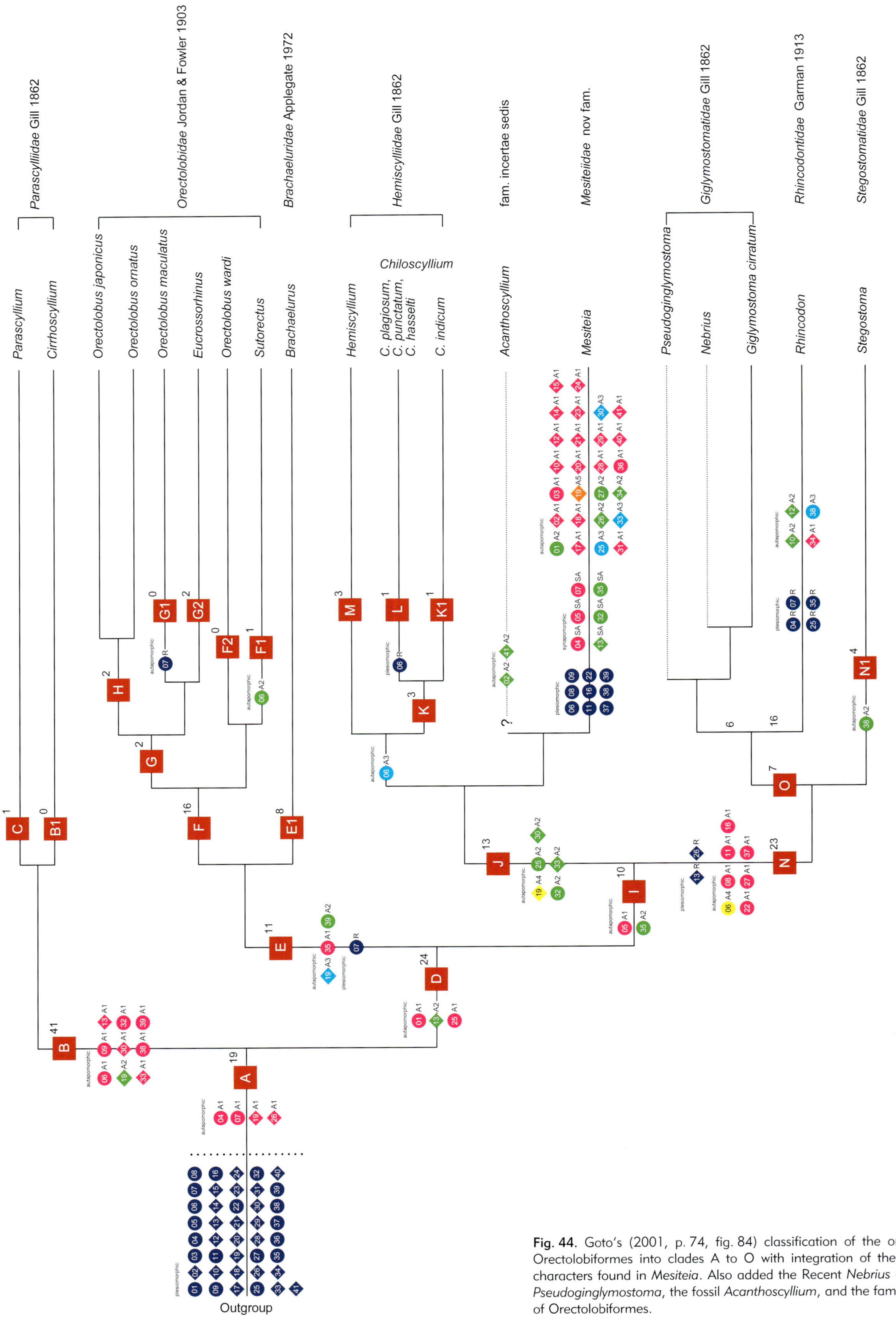

Fig. 44. Goto's (2001, p. 74, fig. 84) classification of the order Orectolobiformes into clades A to O with integration of the 41 characters found in *Mesiteia*. Also added the Recent *Nebrius* and *Pseudoginglymostoma*, the fossil *Acanthoscyllium*, and the families of Orectolobiformes.

03-1 outer nasal cartilage straight, short bar, expanding from rostral rod, cross-like shape of the rostrum; 10-1 no projected articular condyle on the pelvic bar; 12-1 anterior pelvic basal absent, indistinguishable; 14-1 basal elements of 1st dorsal fin absent in anterior and posterior series of radials, increasing and decreasing length of basal elements; 15-1 no fused basal elements on 1st dorsal fin; 17-1 low 1st dorsal fin with very short inner radial and almost missing outer radial elements, equally long; 18-1 up to 23 radial series of 1st dorsal fin, fin twice as long as in other species; 19-5 2nd dorsal fin moved in posterior direction, close to caudal fin, in the middle over the anal fin at about the 80th vertebra; 20-1 basal elements of 2nd dorsal fin absent in anterior and posterior series of radials, increasing and decreasing length of basal elements; 21-1 no fused basal elements on 2nd dorsal fin; 23-1 low 2nd dorsal fin with very short inner radial and very short outer radial elements, equally long; 24-1 up to 36 radial series, fin twice as long as in other species of 2nd dorsal fin; 25-3 very long anal fin begins opposite the 60th vertebra at the middle of the 1st dorsal fin, widely overlapping the first 12 hemal arches of the caudal fin; 26-2 complete loss of basal elements in the anal fin; 27-2 complete loss of outer radial elements in the anal fin; 28-1 up to 97 series of radials in the anal fin, anal fin extremely long; 29-1 caudal fin cartilages differentiated, with preepichordal, epichordal, prehypochordal and hypochordal cartilages; 30-2 up to 22 series of preepichordal cartilage present; 31-1 up to 50 series of ciliate epichordal radials; 33-1 numerous series of prehypochordal cartilages present; 34-2 up to 61 series of sickle-shaped hypochordal processes; 36-1 basidorsal process isolated, not fused with centra, shape like dorsal intercalary plate; 40-1 vertebral ribs composed of differentiated proximal and distal cartilages; 41-1 dermal denticles absent, or almost completely reduced, and if rudimentary dermal denticles are present, then they are reduced to short stubs.

The close relationship of Mesiteiidae nov. fam. to Hemiscylliidae (clade J in Goto 2001) is supported by two synapomorphic characters: 13-2 origin of 1st dorsal fin above the end of the pelvic fin at about the 55th vertebra; 32-2 prehypochordal cartilages fused with hemal arch. Two more synapomorphic characters are shared with *Chiloscyllium*, *Hemiscyllium*, *Ginglymostoma*, *Stegostoma* and *Rhincodon* (clade I in Goto 2001): 5-1 small suprascapular cartilage present; 35-2 secondary calcified lamellae extend to the membrana elastica externa; and further two synapomorphic characters are shared with all Orectolobiformes (clade A in Goto 2001): 4-1 prominent inner process of ceratobranchial present; 7-1 metapterygial axis absent.

Additionally, there are 9 plesiomorphic characters: 6-0 widely separated pro-, meso- and metapterygium; 8-0 pectoral radials composed of three components; 9-0 prepelvic process of the puboischiatic bar is reduced, and rounded; 11-0 pelvic radials composed of two components; 16-0 1st dorsal fin radials composed of two components; 22-0 2nd dorsal fin radials composed of two components; 37-0 ventral intercalary plate absent; 38-0 hemal arches in precaudal vertebrae widely separated; 39-0 vertebral ribs on diplospondylous vertebrae absent.

Discussion. Many unique anatomical features – most strikingly loss of dermal denticles, elongated anal fin with sickle-shaped radials of the caudal fin, sickle-shaped hypochordal processes etc., as described above – separate Mesiteiidae nov. fam. from all other Recent and fossil Orectolobiformes. Tooth morphology alone is not enough to identify the new family. Similar teeth with a smooth conical principal cusp, a pair of prominent lateral cusplets, and a long, narrow apron can be found in the Paleocene/Eocene genus *Eostegostoma* Herman, 1977 (now proposed to be re-allocated together with *Pseudoginglymostoma* to the subfamily Pseudoginglymostomatinae within the family Ginglymostomatidae, Section 4.2.10), the Recent genus *Brachaelurus* (family Brachaeluridae), and in the Recent genus *Pseudoginglymostoma* (family Ginglymostomatidae). Teeth of *Eostegostoma*, *Brachaelurus* and *Pseudoginglymostoma* are up to threetimes bigger than the biggest teeth of *Mesiteia* and its apron is shorter. Teeth of posterior files of *Eostegostoma* and *Pseudoginglymostoma* have strong and usually more than one pair of lateral cusplets, and the principal cusps become shorter but remain straight. In *Brachaelurus*, teeth of posterior files tend to lose the lateral cusplets, the principal cusps become wider, shorter and inclined towards the rear, and the apron becomes low and wide. The principal cusps of teeth of posterior files of *Mesiteia* remain high and relatively narrow, straight or only slightly inclined up to the posteriormost files, where the lateral denticles are reduced to heels, similar to *Brachaelurus*, and not as in *Eostegostoma* and *Pseudoginglymostoma*. The small teeth of the Lower–Middle Jurassic genus *Annea* Thies, 1983 (Orectolobiformes incertae familiae) can be distinguished by their more prominent lateral cusplets, a shorter and rounded apron and a labial crown face which is characterized by a very strong and sharp vertical ridge and other less distinct divergent folds.

Genus *Mesiteia* Kramberger, 1884

1884 *Mesiteia* Kramberger – Gorjanović, p. 44

Type species: *Mesiteia emiliae* Kramberger, 1884

Mesiteia emiliae Kramberger, 1884

1884 *Mesiteia emiliae* Kramb. – Gorjanović, 1884, pp. 45–47, pl. III., figs. 2a,b,c.
1885 *Mesiteia emiliae* Kramb. – Gorjanović-Kramberger, pp. 53–56, pl. III., figs. 2a,b,c.
1886 *Mesiteia emiliae* Kramb., 1884 – Gorjanović-Kramberger, pp. 136–137.
1889 *Mesiteia emiliae* Kramberger, 1885 – Woodward, p. 346.
1894 *Mesiteia emiliae* Kramberger, 1884 – Jaekel, pp. 175–176, fig. 39.
1977 *Mesiteia emiliae* Kramberger, 1885 – Herman, pp. 145, 152, pl. 6, fig. 6.
1980 *Mesiteia emiliae* Kramberger, 1885 – Cappetta, pp. 121–123, pl. 13, fig. 2, pl. 14, figs. 1–2 and text-fig. 25.
1987 *Mesiteia emiliae* Kramberger, 1885 – Cappetta, p. 74, fig. 67.
2006 *Mesiteia emiliae* Gorjanović-Kramberger, 1885 – Cappetta, p. 76.
2012 *Mesiteia emiliae* Gorjanović-Kramberger, 1885 – Cappetta, p. 158, fig. 143.
2012 *Mesiteia emiliae* Kramberger, 1884 – Pfeil, pp. 43–68, figs. 1–20.

As there is only a single genus and species known of Mesiteiidae nov. fam., the diagnosis and discussion of the family, genus and species is the same.

Geological and geographical setting: All specimens of *Mesiteia*, known to date come from Hakel (Haqil) in Lebanon, which is located in North Lebanon between Tripoli and Beirut, about 40 km N of Beirut and about 10 km NE of the old Phoenician settlement Byblos or in Arabic Jbeil (Figs. 6a,b).

The age of the limestone (Plattenkalk) of Hakel is early Cenomanian (Late Cretaceous).

Table 3. List of Neoselachians recorded from Cretaceous fossil sites in Lebanon.

	taxa	stratigraphic position and localities: early Cenomanian Hakel (Haqil)	middle Cenomanian Hgula	middle Cenomanian En Nammoura	late Santonian Sahel Alma	fossil record
	Subcohort Neoselachii					
	Superorder Squalomorphii					
	Order Hexanchiformes					
	Family Hexanchidae					
	Hexanchus gracilis (Davis, 1887)				●	Cappetta (1980)
	Order Squaliformes					
	Family Squalidae					
	Centrophoroides latidens Davis, 1887				●	Cappetta (1980)
	Centrosqualus primaevus (Pictet, 1850)				●	Cappetta (1980)
	Family Centrophoridae					
new	*Deania* sp.	●				Pl. 19, Figs. a–b
	Family Somniosidae					
	Cretascymnus adonis (Signeux, 1950)				●	Cappetta (1980)
	Order Pristiophoriformes					
	Family Pristiophoridae					
	Pristiophorus tumidens (Woodward, 1932)				●	Cappetta (1980)
	Order Squatiniformes					
	Family Squatinidae					
new	*Squatina* sp.	●				Pl. 19, Fig. c–d
	Superorder Galeomorphii					
	Order Heterodontiformes					
	Family Heterodontidae					
new	*Heterodontus* sp.		●			Pl. 19, Figs. e–h
	Order Orectolobiformes					
	Family Hemiscylliidae					
	Acanthoscyllium sahelalmae (Pictet & Humbert, 1866)				●	Cappetta (1980)
	Almascyllium cheikheliasi (Signeux, 1949)				●	Cappetta (1980)
new	Hemiscylliidae incert. gen.	●	●			Figs. 14a–c, 15a–b, Pl. 28., Fig. d
	Family Mesiteiidae nov.					
	Mesiteia emiliae Kramberger, 1884	●				Cappetta (1980), this study
	Family Parascylliidae					
	Pararhincodon lehmani Cappetta, 1980	●	●			Cappetta (1980), Pl. 20, Figs. a–c
new	Parascylliidae gen. indet.	●	●			Pl. 20, Figs. d–k
	Family Ginglymostomatidae					
new	*Cantioscyllium decipiens* Woodward, 1889	●	●			Figs. 16a–e, Pl. 21, Figs. a–j, Pl. 22, Fig. a
new	*Cantioscyllium* sp.		●			Pl. 22, Figs. b–c
new	Ginglymostomatidae gen. indet.		●			Pl. 22, Figs. d–g
	Order Lamniformes					
	Family Mitsukurinidae					
	Anomotodon plicatus (Arambourg, 1952)				●	Cappetta (1980)
	Scapanorhynchus lewisii (Davis, 1887)				●	Cappetta (1980)
	Family Haimirichiidae					
new	*Haimirichia amonensis* (Cappetta & Case, 1975)	●		●		Cappetta (2012, not fig.), Pl. 23, Figs. a–g, Pl. 24, Figs. a–b
	Family Odontaspididae					
new	*Odontaspis aculeatus* (Cappetta & Case, 1975)		●			Pl. 24, Figs. a–d
new	*Odontaspis* sp.		●			Pl. 24, Figs. e–f
new	Odontaspididae gen. indet.		●			Pl. 24, Figs. g–h
	Family Cretoxyrhinidae					
new	*Cretalamna appendiculata* (Agassiz, 1843)		●			Pl. 25, Figs. a–h
new	*Cretoxyrhina mantelli* (Agassiz, 1838)		●			Pl. 26, Fig. h
	Family Anacoracidae					
new	*Squalicorax* sp.	●	●	●		Pl. 27, Figs. a–g
	Lamniformes incert. fam.					
new	*Cretodus semiplicatus* (Münster in Agassiz, 1843)	●	●			Pl. 26, Figs. a–g
new	*Paranomotodon* sp.		●			Pl. 27, Figs. h–k
	Order Carcharhiniformes					
	Family Scyliorhinidae					
	Pteroscyllium dubertreti Cappetta, 1980				●	Cappetta (1980)
	Pteroscyllium signeuxi Cappetta, 1980				●	Cappetta (1980)
	Scyliorhinus arambourgi Cappetta, 1980		●			Cappetta (1980), Pl. 28, Figs. e–f
	Scyliorhinus bloti Cappetta, 1980	●				Cappetta (1980), Pl. 28, Figs. a–c
	Scyliorhinus elongatus (Davis, 1887)				●	Cappetta (1980)
new	*Scyliorhinus* sp.	●				Pl. 20, Figs. l–m
new	*Crassescyliorhinus* cf. *germanicus* Underwood & Ward, 2008		●			Pl. 28, Figs. j–l
new	*Protoscyliorhinus* sp.	●				Pl. 28, Figs. g–i
new	Scyliorhinidae gen. indet.	●				Pl. 29, Figs. a–c
	Family Triakidae					
	Paratriakis curtirostris (Davis, 1887)				●	Cappetta (1980)
new	Triakidae gen. indet.	●				Pl. 29, Figs. d–j

6. Outlook to further investigations on fossil sharks in Lebanon

This study of only one species, *Mesiteia emiliae*, demonstrates that our knowledge of fossil sharks from Lebanon, summarized in Cappetta (1980 and 2012), is very poor, especially in terms of complete skeletal remains, and the few mostly not well preserved specimens in official collections do not reflect the actual state of the finds discovered so far. In Table 3 I have tried to supplement the previously known records (Cappetta 1980, 2012) with my observations. Far more new species were added than were already known. But of course, this list cannot be complete yet, also due to the difficult political and economic situation in Lebanon. Further studies will show how diverse this fauna actually is. Many specimens of interesting shark families, not yet known to scientists, exist only in private collections and deserve similar investigations as *Mesiteia* here.

It is certainly too early to give a complete overview of the fauna accompanying *Mesiteia*, because too many species are not yet studied and not even reliably identified. Sometimes only a photograph of incompletely prepared specimens exists in poor resolution, and important details, such as the teeth, are hardly recognizable. Often the location of the fossil remains as unknown as its whereabouts. Similar to the earlier finds from "Solnhofen", an analysis of the rock must first be carried out by experts to identify where the fossil came from. Nevertheless, I feel responsible to follow my words with at least a few pictures of sharks (Plates 19–29), which show how diverse this fauna really is and how perfectly and completely preserved many specimens are. It is only an incomplete compilation of specimens that I was allowed to see and briefly document during the last years with the support of my friends in Lebanon, especially Pierre Abi-Saad and Roy Nohra. For some specimens only a photo exists at the moment because they have been sold in the meantime; some pieces I was able to acquire for my collection and they are available for further investigations. I have not completely considered here the currently unclear number of Parascylliidae, which are represented beside *Pararhincodon lehmani* by presumably several species of genera close related to *Parascyllium* and *Cirrhoscyllium*. Also from Scyliorhinidae I can show here only examples, because they are also very diverse and at the moment not manageable.

The importance of the Cretaceous Lebanese shark fauna is equal to other famous fossil deposits in the world. In fact, it even surpasses the classic Jurassic fossil deposits of southern Germany (Solnhofen–Eichstätt–Wattendorf region, Nusplingen), which provide a similar insight into a diverse and sometimes incredibly wonderfully preserved fauna (Kriwet & Klug 2004, 2015). It should be a common concern of all museums to protect and preserve such valuable and unique fossils for all of us. I hope that this work will help to preserve such unique finds for science, but also that scientists will finally begin to accept the existence and validity of unique fossils held in non-public collections. I have also tried to show that a single specimen is never enough to see all the details necessary to describe a species. Also, the preservation and preparation of the specimens should always be as expert as possible. Fossils are like mosaic stones found during an archeological excavation ... The more stones you add to an image, the more complete the image will become, and details appear that were not visible before.

Plates 19–29

Preliminary overview of not yet published sharks from the sites Hakel (Haqil) (early Cenomanian), Hgula (middle Cenomanian) and En Nammoura (middle Cenomanian).

Plate 19
Squaliformes, Centrophoridae, *Deania* sp.
a, b. Overall view of two male specimens of *Deania* sp., documented only by these two photographs, taken by the quarry owner Roy Nohra. Probably the only two specimens. Since there were three other Squaliformes (*Centrophoroides latidens* Davis 1887, *Centrosqualus primaevus* (Pictet, 1850) and *Cretascymnus adonis* (Signeux, 1950) found only in Sahel Alma (late Santonian) the occurrence of *Deania* at the older locality Hakel (early Cenomanian) is very surprising. The identification so far is based only on the elongated snout of these specimens but needs to be confirmed by its teeth.
Squatiniformes, Squatinidae, *Squatina* sp.
c. Overall view of the only known specimen of Squatiniformes, *Squatina* sp. This juvenile specimen can be identified by position, size and shape of its two far posteriorly positioned dorsal fins and a characteristic caudal fin. Pectoral fins, pelvic fins and skull are compressed. Hakel (early Cenomanian); coll. Abi Saad.
d. Detail of trunk below the two dorsal fins. Note the extremely strong calcifications of hemal and neural arches that were not observed in any other fossil (Jurassic) or modern species of *Squatina*.
Heterodontiformes, Heterodontidae, *Heterodontus* sp.
e–h. Only four specimens of *Heterodontus* sp. are known, all from Hgula (middle Cenomanian).
e. Only the two fin spines and a few dermal denticles were visible when this specimen was offered by Hussein Ibrahim at the Munich fossil show in 2013, and it was not certain that the specimen would be complete. After a careful preparation by Pino and Paul Voelkl it turned out to be a perfectly preserved juvenile male of *Heterodontus* sp. with many details ready to be described. Hgula (middle Cenomanian); Munich, coll. FHP.
f. Same specimen; detail with teeth of anterior files.
g. Same specimen; detail with teeth of posterior files.
h. Photo taken by Roy Nohra of an earlier sold *Heterodontus* specimen, unfortunately with the usual damage during preparation. Hgula (middle Cenomanian); this specimen was obtained many years ago by John Maisey for the collection of the AMNH, New York.

Plate 20
Orectolobiformes, Parascylliidae, *Pararhincodon lehmani* Cappetta, 1980.
a–c. Cappetta's description of *Pararhincodon lehmani* was based on a single, incomplete and very badly preserved specimen from Hakel. But indeed, this species is relatively common in Hakel and Hgula, and many specimens, probably representing different species, can be found in the collections, sometimes incorrectly identified as catsharks.
a. A relatively compact solid body with rounded dorsal fins and a short caudal fin characterizes this species and distinguishes it from other Parascyllidae indet. that are more closely related to modern *Parascyllium* and *Cirrhoscyllium*. The co-existance of these slender Parascylliidae indet. with *Pararhincodon lehmani* makes it questionable that *Pararhincodon* really belongs to Parascylliidae. More likely it represents a new family of its own. Hakel (early Cenomanian); Munich, coll. FHP; photo with long-wave UV-light by Helmut Tischlinger. **b.** same specimen in normal light. **c.** detail with teeth.
Orectolobiformes, Parascylliidae, Parascylliidae indet.
d–k. Another group of small and slender sharks with narrow head, narrow dorsal fins and long slender caudal fin with a low lower caudal lobe is closely related to modern *Parascyllium* and *Cirrhoscyllium*. Even the taphonomic preservation with mostly upright curved body recalls modern Parascylliidae, laying on a sandy ground. It is relatively common in Hakel and Hgula (early to middle Cenomanian).
d–e. Two nicely preserved specimens of Parascylliidae indet., presenting the typical overall shape. Hgula (middle Cenomanian); coll. Roy Nohra; photo Roy Nohra.
f–g. This specimen has an exceptional preservation of the skull in ventral view with many anatomical details, e.g. the rostral rod and the jaws. Hgula (middle Cenomanian); Munich, coll. FHP.
h–i. A juvenile specimen with its dorsal and caudal fin rays not yet fully developed, and interesting details of skull and dentition. Hgula (middle Cenomanian); Munich, coll. FHP.
j–k. Another very young specimen with details of the dentition. Hgula (middle Cenomanian); Munich, coll. FHP.
Carcharhiniformes, Scyliorhinidae, *Scyliorhinus* sp.
l–m. One of the many still unidentified catsharks. At first glance this small slender shark with long caudal fin looks very similar to Parascylliidae indet., but its dorsal, pectoral and pelvic fins are bigger and

rounder. This wonderfully preserved complete specimen has been prepared from an untouched stone plate by Pino and Paul Voelkl. It is fully preserved but its very small teeth confirm its belonging to *Scyliorhinus*. Hakel (early Cenomanian); Munich, coll. FHP.

Plate 21

Orectolobiformes, Ginglymostomatidae, *Cantioscyllium decipiens* Woodward, 1889.

a–b. The first described *Cantioscyllium decipiens* from Lebanon was this male specimen from Hakel (Pfeil 2013). It was bought at the Munich fossil show in 2012 from Pierre Abi-Saad by the "Friends of the BSPG" for the BSPG in Munich. After a careful micro-preparation by Renate Liebreich the details of the teeth became visible (**b**). *Cantioscyllium* can be identified by its stout body shape with two big triangular, relatively pointed dorsal fins that are close together, big triangular pectoral fins, and a strong caudal fin with a wide triangular lower caudal lobe. It is less rare than originally thought and meanwhile I have identified specimens from 10 cm to 1 m length. Hakel (early Cenomanian); Munich, SNSB-BSPG 2012 I 87.

c. With a total length of only 10 cm this is probably the smallest and youngest *Cantioscyllium decipiens*. Its teeth are labially and lingually smooth but widely triangular, different to the Hemiscylliidae indet. that I figured in Figures 14–15. Hakel (early Cenomanian); Munich, coll. FHP.

d–f. The BSPG in Munich bought a second specimen, a female, in 2013. Renate Liebreich prepared the teeth from anterior (**e**) to posterior files (**f**). I published this specimen (Pfeil 2014), and erroneously wrote it would come from Hakel (early Cenomanian). During the preparation of this study I could confirm that it was found in Hqula (middle Cenomanian); Munich, SNSB-BSPG 2013 I 104a,b.

g–h. This small, juvenile *Cantioscyllium* with a total length of 21 cm has the folding (striation) of the labial side of the teeth already developed (figured also in Figs. 16a–b). Hgula (middle Cenomanian); Munich, coll. FHP.

i–j. Fortunately, fakes of fossils from Lebanon are very rare, but recently I received pictures of this nice looking *Cantioscyllium* that turned out to have artificially added parts of a pycnodontiformes skull. Hakel (early Cenomanian); coll. Abi Saad; photo Abi Saad.

Plate 22

Orectolobiformes, Ginglymostomatidae, *Cantioscyllium decipiens* Woodward, 1889.

a. From this *Cantioscyllium decipiens* the right and the left slab are preserved. I thought I should show it because it demonstrates the incredible richness and exceptional preservation of the Fossil-lagerstätte from Hakel. The shark is surrounded by a good dozen of teleosteans and a small *Cyclobatis*. Hakel (early Cenomanian); coll. Abi Saad; photo Abi Saad.

Orectolobiformes, Ginglymostomatidae, *Cantioscyllium* sp.

b–c. I have seen several specimens that look very similar to *Cantioscyllium decipiens*, but the body is longer, less stout, and I observed a wider distance between the two dorsal fins which are less high and less pointed. The lower lobe of the caudal fin is less developed. The teeth (**c**) are not yet sufficiently studied. These morphotypes probably represent another species of *Cantioscyllium*. Hgula (middle Cenomanian); Munich, coll. FHP.

Orectolobiformes, Ginglymostomatidae, Ginglymostomatidae indet.

d–g. There is another group of sharks known only from Hgula that also looks similar to *Cantioscyllium decipiens*, but the body of these sharks is much longer, and spindle shaped. The pectoral fins are smaller; the distance between the two widely triangular dorsal fins is much wider than in *C. decipiens*, and the caudal fin is short with a small lower lobe of the caudal fin.

d–e. This is a well preserved specimen of these Ginglymostomatidae indet. Its teeth (**e**) are different to *C. decipiens* but need to be examined more closely. Hgula (middle Cenomanian); coll. Abi Saad. A very similar specimen has been figured in Gayet, Abi Saad & Gaudant (2012, p. 67).

f–g. Another specimen of the same morphotype. Hgula (middle Cenomanian); coll. Roy Nohra; photo Roy Nohra.

Plate 23

Lamniformes, Haimirichiidae, *Haimirichia amonensis* (Cappetta & Case, 1975).

Cappetta & Case (1975) originally assigned isolated teeth from the late Cenomanian of Texas to the genus *Odontaspis*, describing the species *O. amonensis*. Based on a single, well preserved anterior part of an articulated skeleton from the late Cenomanian? to early Turonian Fossillagerstätte of Agoult, in the Kem Kem region of south-eastern Morocco, with that same type of teeth, Vullo, Guinot & Barbe (2016) erected a new family Haimirichiidae with a new genus *Haimirichia* with its only species *Haimirichia amonensis* (Cappetta & Case, 1975). The incomplete specimen has a characteristic short-snouted, broad head. This and other characters, especially the morphology of the teeth is unique within Lamniformes and led to the conclusion of parallel evolution and similar lifestyle to that of some living carcharhiniform sharks, such as the whitetip reef shark *Triaenodon obesus*. Isolated teeth of *H. amonensis* are widespread in the Cenomanian, and it is not surprising to find complete articulated skeletons in Lebanon that will greatly increase the knowledge about this species and its relationships to Lamniformes.

a–b. A very young, small, approx. 20 cm long specimen, treated with acetic acid, perfectly shows the body shape of juvenile *Haimirichia* with its short and broad head, great triangular pectoral fins and a caudal fin with wide expanded triangular lower lobe. All files of teeth (**b**) are preserved. This specimen was exhibited at the little museum at Byblos, which has since closed. En Nammoura (middle Cenomanian); coll. Abi Saad.

c–d. On adult specimens, the head is less broad extended. The caudal fin is sharp and long. The specimen is not yet prepared, but its dentition (**d**) is fully preserved. Hakel (early Cenomanian); coll. Abi Saad.

e. Another unprepared adult specimen of *H. amonensis* that shows the shape and details of a characteristic, very narrow lower lobe of the caudal fin. Hakel (early Cenomanian); coll. Roy Nohra.

f–g. A head only, in ventral view, exposing the many teeth of a complete dentition. Hakel (early Cenomanian); coll. Roy Nohra.

Plate 24

Lamniformes, Haimirichiidae, *Haimirichia amonensis* (Cappetta & Case, 1975).

a–b. Both plates of this *Haimirichia amonensis* are preserved. It is probably the best preserved specimen, but it requires professional preparation before it can be studied. The first dorsal fin is much longer and higher than the second, the pectoral fins are as big as the first dorsal fin, pelvic and anal fin are small, the elongated upper lobe of the caudal fin is slender, as is the lower lobe, and there is a very small terminal lobe that can be seen on the other plate. The skull is somewhat longer and pointed. There are teeth of anterior and lateral files of both jaws visible (**b**), smooth with wide crowns flanked by one or two pairs of lateral cusplets. Hakel (early Cenomanian); coll. Roy Nohra; photo Roy Nohra.

Lamniformes, Odontaspididae, *Odontaspis aculeatus* (Cappetta & Case, 1975).

c–d. For me it was very surprising to see so many different species of "big sharks" when I visited the collections in Lebanon in 2013. Most of these specimens are not yet finally prepared, teeth are not easily to find, and my pictures taken quickly are certainly not the best. I was surprised when I was able to purchase an almost untouched plate with only a few centra visible. Roy Nohra and I thought it could be a *Haimirichia amonensis*, but after the preparation by Pino and Paul Voelkl it turned out to be the first and incredibly well preserved articulated specimen of an *Odontaspis aculeatus* (Cappetta & Case 1975). The two dorsal fins are almost the same size, relatively short and high. They divide the dorsal margin of this shark into three parts of equal length. Very similar in design are the pectoral fins, whereas the pelvic and anal fins are smaller. The upper caudal fin lobe is elongated; the lower lobe well developed; obviously there is no terminal lobe. The head is short and pointed. All teeth (**d**) are very well preserved and in perfect condition. Hgula (middle Cenomanian); Munich, coll. FHP.

e–f. This is another specimen of *Odontaspis aculeatus*, recognizable by the same general shape of its body, and with all fins at about the same size and position as in the specimen figured here just before (**c**). It has a sharp massive snout, densely covered with dermal denticles. Its teeth (**f**) require more preparation but can be recognized as *O. aculeatus*. Hgula (middle Cenomanian); coll. Abi Saad.

Lamniformes, Odontaspididae, *Odontaspis* sp.

Without examining the teeth it is almost impossible to identify most of the Lamniformes. Therefore I must leave discussion at the generic level.

g–h. An incomplete specimen of an odontaspid, treated with acetic acid, and therefore showing many interesting characters of skull and dentition. Hgula (middle Cenomanian); coll. Abi Saad.

Lamniformes, Odontaspididae, Odontaspididae gen. indet.

i–j. A very well preserved shark with dorsal, pectoral, pelvic and anal fins smaller, less high than in *Odontaspis aculeatus*. The caudal fin has a distinct terminal lobe. The teeth (**j**) require careful preparation to allow the identification of this species. Hgula (middle Cenomanian); coll. Abi Saad.

Plate 25

Lamniformes, Cretoxyrhinidae, *Cretolamna appendiculata* (Agassiz, 1843).

Represented by several quite impressive specimens. The close relationship with mackerel sharks, family Lamnidae, can not be overlooked. *Cretolamna appendiculata* is a heavily-built shark, with spindle-shaped body and pointed snout. The first dorsal fin is large and high, positioned above the equally large pectoral fins. The second dorsal, pelvic and anal fins are minute compared to it. The caudal fin is large and prominent, its lower lobe is of comparable length as that seen in living *Lamna* and reveals a fast-swimming shark. All specimens require further professional preparation. Only in the aggregate do they show the characteristics necessary for a scientific description. It is unfortunate that almost all of these attractive large sharks have been sold to private collectors.

a–b. This specimen shows best the overall shape of *C. appendiculata*. But the second dorsal and anal fins are probably not fully preserved. **b.** Several files of teeth from lingual side, showing the smooth and rounded lingual face of the crown, flanked by strong, low triangular lateral cusplets. Hgula (middle Cenomanian); photos Roy Nohra.

c–d. In this specimen the first dorsal fin seems to be a bit too small and the second too large. It needs to be assessed whether this is due to preservation, or maybe due to preparation. Some teeth are visible (**d**) but require further preparation. Hgula (middle Cenomanian); photos Roy Nohra.

e–f. This specimen shows best the shape and size of dorsal, pectoral, pelvic and anal fins. The caudal fin is missing the terminal parts. The teeth (**f**) are not visible on the few existing pictures. Hgula (middle Cenomanian); photos Roy Nohra.

g–h. A recently (in 2016) excavated specimen, preserved in a very dynamic position, with a total length of 1.2 m. The pointed snout, first dorsal and one pectoral fin are well preserved. After further preparation, possibly all tooth files can be studied (**h**). Hgula (middle Cenomanian); Beirut, MIM Museum; photos Abi Saad.

Plate 26

Lamniformes, Lamniformes incert. fam, *Cretodus semiplicatus* (Münster in Agassiz, 1843).

a–c. With a total length of approx. 3.80 m (Pierre Abi Saad for scale), *Cretodus semiplicatus* is the biggest shark preserved with articulated skeletons. This specimen is not yet prepared, but teeth can be seen at some fractures of the rock plate (**b–c**). Hgula (middle Cenomanian); coll. Abi Saad. Isolated teeth of sharks are generally extremely rare at all fossil localities in Lebanon, but I can show here a few teeth of *C. semiplicatus*.

d. Upper lateral tooth in lingual view. Hgula (middle Cenomanian); coll. FHP.

e. Lower anterior tooth in lingual view. Hgula (middle Cenomanian); coll. FHP.

f. Upper parasymphyseal tooth in labial view. Hakel (early Cenomanian); coll. Abi Saad.

g. Upper parasymphyseal tooth in lateral (mesial) and labial view. Note that the typically strong striations of the basal crown face are present on the lingual face, but are much reduced on the labial face of that tooth. Hgula (middle Cenomanian); coll. Roy Nohra; photos Roy Nohra.

Lamniformes, Cretoxyrhinidae, *Cretoxyrhina mantelli* (Agassiz, 1838).

h. This plate of about 20 × 20 cm contains, as far as I know, the so far only evidence of *Cretoxyrhina* in Lebanon. The teeth reminiscent of *Isurus* are easy to recognize. Hgula (middle Cenomanian); coll. Abi Saad.

Plate 27

Lamniformes, Anacoracidae, *Squalicorax* sp.

Several specimens from Hakel, Hgula and En Nammoura. As they were occasionally found by quarry workers, they are more or less complete, sometimes only the head with rows of teeth is preserved. The teeth indicate that at least two species could be represented.

a–b. Only the tail is missing on this specimen of an old collection. It shows first time the overall shape of *Squalicorax* with details of head and fins. The whole dentition is visible (**b**) but needs preparation. En Nammoura (middle Cenomanian); coll. Abi Saad; photos Abi Saad.

c–d. Both plates of this smaller specimen are preserved, with teeth (**d**) that are higher, more upright as seen in (**b**) and having a shorter base. Hakel (early Cenomanian); coll. Abi Saad; photos Abi Saad.

e–f. This bigger specimen of *Squalicorax* is also preserved with both plates and will be almost complete after preparation. Both plates show many teeth (**f**). Hgula (middle Cenomanian); coll. Abi Saad; photos Abi Saad. **g:** An isolated tooth with very slender crown of probably another species of *Squalicorax*. There were also incomplete skeletons found with this type of teeth. En Nammoura (middle Cenomanian); Munich, coll. FHP.

Lamniformes, Lamniformes incert. fam., *Paranomotodon* sp.

The teeth of the holotype and single specimen of *Anomotodon plicatus* (Arambourg 1952) from Sahel Alma show characteristic folds at the base of both lingual and labial faces of the crowns (Cappetta 1980). The teeth of all specimens of *Paranomotodon* sp. from Hgula that I have seen have smooth lingual and labial crown faces. Unfortunately the complete articulated skeletons of *Paranomotodon* cannot be compared with the only fragmentary preserved head and branchial region of *Anomotodon*.

h–i. In this specimen of *Paranomotodon*, the middle part of the trunk is somehow compressed; most of the fins are not clearly describable, but the tail is perfectly preserved and looks very similar to that of Odontaspididae gen. indet. that I figured in Plate 24g. All teeth (**i**) are visible. Hgula (middle Cenomanian); coll. Abi Saad.

j–k. In this specimen the head and trunk are better preserved, but the tail is incomplete. The teeth (**k**) are perfectly preserved, with views of lingual and labial faces of the crowns. Hgula (middle Cenomanian); coll. Abi Saad.

Plate 28

Carcharhiniformes, Scyliorhinidae.

Cappetta (1980) described five new species from Lebanon, two of these attributed to a new genus *Pteroscyllium*, which is only recorded from Sahel Alma. *Scyliorhinus elongatus* also was found only in Sahel Alma. The remaining *S. arambourgi* from Hgula and *S. bloti* from Hakel are based on one, respectively on two very poor and incomplete preserved specimens. I have not seen any of the late Santonian species from Sahel Alma in the collections from Hakel and Hgula, but many other species. Some of these are known by isolated teeth from other localities, some might be new. It is still complicated to attribute any of the new specimens to the two already known, *S. arambourgi* and *S. bloti*, but it might be possible to find a key, e.g. by checking its dorsal and caudal fin shape.

Carcharhiniformes, Scyliorhinidae. *Scyliorhinus bloti* Cappetta, 1980.

a–b. I received this specimen unprepared, and after preparation by Pino and Paul Voelkl it turned out to be a perfectly preserved *S. bloti*. The minute teeth (**b**) have high crowns, are strongly striated at the labial faces, and flanked by just one pair of lateral denticles. Hakel (early Cenomanian); Munich, coll. FHP.

c. This very young and in all details beautifully preserved specimen could also be a *S. bloti*. Note the skull, the gill slits and pectoral and pelvic fins. I could not yet see the teeth. Hakel (early Cenomanian); coll. Abi Saad.

Orectolobiformes, Hemiscylliidae, Hemiscylliidae indet.

d. Without checking the morphology of the teeth, it is complicated to distinguish catsharks from any small Orectolobiformes. A good indication is a more massive anterior part of the body, the much larger size of both dorsal fins, and a sometimes shorter caudal fin. This specimen is probably a first record of Hemiscylliidae indet. with articulated skeleton. See also Figure 14 with teeth. Hgula (middle Cenomanian); Munich, coll. FHP.

Carcharhiniformes, Scyliorhinidae. *Scyliorhinus arambourgi* Cappetta, 1980.

e–f. This specimen looks very similar to the Hemiscylliidae indet. in Figure f. but its dorsal fins are smaller, the caudal fin is longer. The very minute scyliorhinid teeth (**f**) have low crowns, are strongly striated at the labial faces, and flanked by two pairs of lateral denticles. It needs to be confirmed but I would like to tentatively identify this specimen as a *S. arambourgi*. Hgula (middle Cenomanian); Munich, coll. FHP; photos with long-wave UV-light by Helmut Tischlinger.

Carcharhiniformes, Scyliorhinidae. *Protocyliorhinus* sp.

g–i. One of the surprising new detections of catsharks in Lebanon is this articulated skeleton of a *Protoscyliorhinus*. It is cigar-shaped, with rounded snout, small dorsal fins and relatively short tail. The labial face of the teeth (**h**) is moderately convex and overhangs the labial face of the root by a rounded bulge with a broadly concave outline. The bulge bears numerous folds like in *Protoscyliorhinus bettrechiensis* Herman 1977 from the late Turonian of northern France. Hakel (early Cenomanian); Munich, coll. FHP. **i.** A further specimen with fantastically preserved tail. Hakel (early Cenomanian); Munich, coll. FHP.

Carcharhiniformes, Scyliorhinidae. *Crassescyliorhinus* cf. *germanicus* (Herman, 1982).
j–l. This relatively stout specimen with large dorsal fins and strong tail has teeth (**k–l**) with all characteristics of *Crassescyliorhinus germanicus*. The lateral tooth (**l**) in labial view, is very similar to the lateral teeth of *C. germanicus* (**m**) from the middle Maastrichtian of Bavaria (Gerhartsreiter Graben, coll. FHP). Hgula (middle Cenomanian); Munich, coll. FHP.

Plate 29:
Carcharhiniformes, Scyliorhinidae. Scyliorhinidae indet.
When I was looking for *Mesiteia*, I found a strange looking unknown catshark with a long tail. This tail is very straight, sometimes slightly curved in dorsal, sometimes in ventral direction. The calcification of the hemal and neural spines of the caudal fin is very strong, same as the rays of the dorsal fins. A lower lobe of the caudal fin seems only weak developed. The dermal denticles are very small and the skin of most specimens got lost during fossilisation. The body is covered with the same whitish mass as the *Mesiteia* specimens, e.g. the specimen figured in Figure 39. The scyliorhinid teeth are not well preserved, respectively all specimens need further preparation.
a. A specimen that shows all characteristics was exhibited at the little museum at Byblos, which has since closed. Hakel (early Cenomanian); coll. Abi Saad.
b–c. Most specimens are poorly preserved, and the details can only be composed by comparing different specimens. In (**c**) the claspers are preserved. Hakel (early Cenomanian); photo Roy Nohra.
Carcharhiniformes, Triakidae. Triakidae indet.
The late Santonian *Paratriakis curtirostris* (Davis 1887) from Sahel Alma seems to be absent at the early to middle Cenomanian localities in Lebanon. But instead there is another triakid shark, a new species that I want to leave here as Triakidae indet.
d–e. This beautiful specimen was exhibited at the little museum at Byblos. Its fins unfortunately are a bit artificially overpainted, but still it presents many details. A few teeth of the lower jaw are visible (**e**) that are reminiscient of lateral teeth of *Iago*. The labial face of the crown slightly overhangs the root, and some short and oblique folds can be observed marginally. Hakel (early Cenomanian); coll. Abi Saad.
f–j. Another well preserved specimen, that shows details such as dermal denticles (**g**), lower jaw teeth of anterior (**h**) and lateral files (**i**), and upper anterior teeth (**j**) with an erect, triangular main cusp with a pair of sharp lateral cusplets. The labial face is convex transversely and bears some vertical folds on main cusp and lateral cusplets. Hakel (early Cenomanian); Munich, coll. FHP.

7. Acknowledgements

My dear friends Pierre and Albert Abi Saad and their family and friends supported all my interest in fossil sharks from Lebanon since the first day I had a shark of their collection (the first Munich specimen of *Mesiteia*) in my hands. Their ongoing interest in every new result gave me the strength to plan to publish this chapter and maybe a few more papers about their finds. Roy Nohra generously allowed me to purchase some of his best specimens and gave me valuable information and pictures of previously sold specimens from Hakel and Hgula. What could I do without the "always available" help of my friend Helmut Tischlinger, who masterly produced all the wonderful UV-light images. Sanja Japundžić sent me colour pictures of the holotype from the Croatian Natural History Museum at Zagreb, and Martin Pickford helped me with a series of newly produced pictures of both *Mesiteia* specimens #2 and #3 from the MNHN in Paris. Pino and Paul Voelkl took patiently care of the preservation of the best of the best specimens, e.g. *Mesiteia* #08. My colleague Markus Moser (BSPG München) helped me solving the puzzle regarding the different publications of Gorjanović-Kramberger. And my dear wife Renate Liebreich, formerly preparator at the BSPG, was not only responsible for the most beautiful preparation of the teeth of *Mesiteia* #10, she gave me the time – in addition to my work as publisher – to dedicate much of my free time to sharks. Many thanks to Gloria Arratia, Hans-Peter Schultze and Mark Wilson who spent their time and were so patient reviewing my manuscript. They helped me to improve my manuscript with their constructive critiques. I appreciate their input greatly. Thanks to the volume editors, Alan Pradel, John S. S. Denton and Philippe Janvier, for their diligent work during the production of this Festschrift. My special thanks go to John Denton for checking my manuscript for linguistic errors with great care and patience.. Without the many discussions with friends and colleagues during the meetings of the Mesozoic Fishes group organized by Gloria I might not have dared to write this manuscript. But the decisive impulse came during the last visit of John Maisey in Munich, when we talked about the poor state of preservation of most fossil sharks from Lebanon in the state collections. Thank you John!

Plate 19

Plate 20

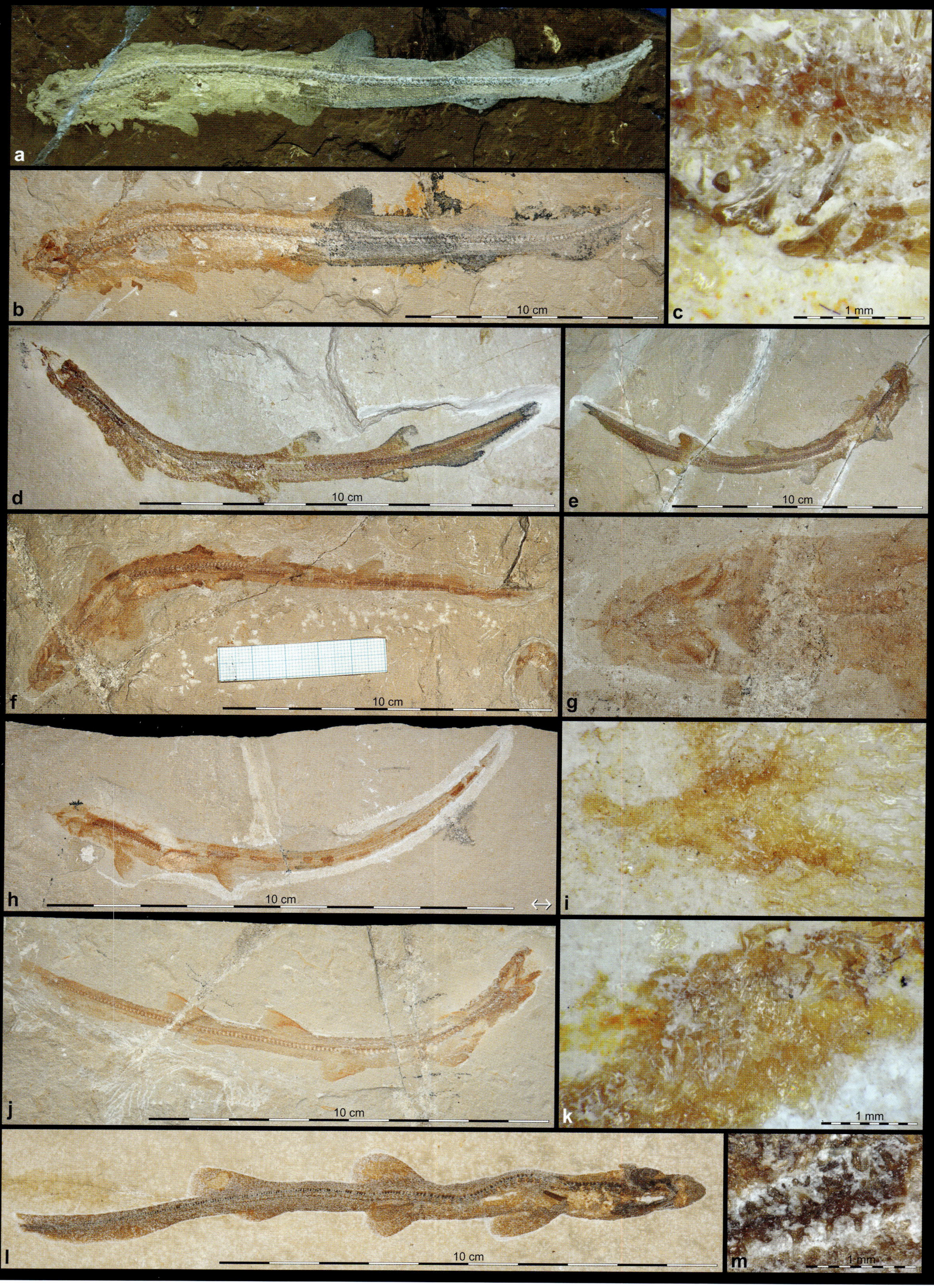
a
b
10 cm
c
1 mm
d
10 cm
e
10 cm
f
10 cm
g
h
10 cm
i
j
10 cm
k
1 mm
l
10 cm
m
1 mm

Plate 21
a
10 cm
c
10 cm
b
3 mm
e
3 mm
f
3 mm
d
10 cm
g
10 cm
h
1 mm
i
10 cm
j

Plate 22
a
10 cm
b
10 cm
c
d
10 cm
e
f
10 cm
g

Plate 23
a
b
c
10 cm
d
L2
L1
A4
A3
A2
A1
S
1 cm
e
10 cm
f
5 cm
g

Plate 24

Plate 25
10 cm
a
b
50 cm
c
d
10 cm
e
f
TL ~120 cm
g
h

Plate 26
a
b
c
d
1 cm
e
1 cm
h
1 cm
f
1 cm
g
1 cm

Plate 27
a
10 cm
b
c
10 cm
e
10 cm
d
g
1 cm
f
h
10 cm
i
j
10 cm
k

Plate 28

Plate 29
a
10 cm
b
10 cm
c
10 cm
d
10 cm
e
f
10 cm
g
1 mm
h
1 mm
i
1 mm
j
1 mm

8. Literature cited

Adnet, S. 2006. Nouvelles faunes de Sélaciens (Elasmobranchii, Neoselachii) de l'Eocène moyen des Landes (Sud-Ouest, France). Implication dans la connaissance des communautés de sélaciens d'eaux profondes. – Palaeo Ichthyologica, 10: 1–128.

Adolfssen, J. S. & D. J. Ward 2013. Crossing the boundary: an elasmobranch fauna from Stevns Klint, Denmark. – Palaeontology, 57: 591–629.

Allen, G. R., M. V. Erdmann, W. T. White, Fahmi & C. L. Dudgeon. 2016. Review of the bamboo shark genus *Hemiscyllium* (Orectolobiformes: Hemiscyllidae). – Journal of the Ocean Science Foundation, 23: 51–97.

Applegate, S. P. 1972. A revision of the higher taxa of orectolobids. – Journal of the Marine Biological Association of India, 14(2): 743–751.

Bass, A. J., J. D. D'Aubrey & N. Kistnasamy 1975. Sharks of the East coast of southern Africa. IV. The families Odontaspididae, Scapanorhynchidae, Isuridae, Cetorhinidae, Alopiidae, Orectolobidae and Rhiniodontidae. – Investigational Report series of the Oceanographic Research Institute, 39: 1–102.

Batchelor, T. J. & D. J. Ward 1990. Fish remains from a temporary exposure of Hythe Beds (Aptian – Lower Cretaceous) near Godstone, Surrey. – Mesozoic Research, 2: 181–203.

Bernardez, E. 1998. Los dientes de Seláceos del Cretácico de la depresión Central Asturiana: vi + 476 pp. [unpublished Thesis Doctoral, Universidad de Oviedo].

Capasso, L. 2017. The history and the situation of the world famous fossil fish quarries in Lebanon. – Bollettino del Museo Civico di Storia Naturale di Verona, 41, Geologia Paleontologia Preistoria: 53–76.

Cappetta, H. 1973. Selachians of the Carlile Shale (Turonian) of South Dakota. – Journal of Paleontology, 47: 504–514.

– 1976. Sélaciens nouveaux du London Clay de l'Essex (Yprésien du Bassin de Londres). – Geobios, 9: 551–575.

– 1980. Les Sélaciens du Crétacé supérieur du Liban. I. Requins. – Palaeontographica, Abt. A, 168: 69–148.

– 1987. Chondrichthyes II. Mesozoic and Cenozoic Elasmobranchii. – In: Handbook of Paleoichthyology, vol. 3B: 193 pp.; Stuttgart (Gustav Fischer Verlag) and Reprint 2004, München (Pfeil).

– 2006. Elasmobranchii Post-Triadici (Index specierum et generum). – Fossilium Catalogus I: Animalia, 142: 472 pp.; Leiden (Backhuys Publishers).

– 2012. Chondrichthyes. Mesozoic and Cenozoic Elasmobranchii. Teeth. – Handbook of Paleoichthyologie, vol. 3E: 512 pp.; München (Pfeil).

Case, G. R., D. R. Schwimmer, P. D. Borodin & J. J. Leggett. 2001. A new selachian fauna from the Eutaw Formation (Upper Cretaceous/Early to Middle Santonian) of Chattahoochee County, Georgia. – Palaeontographica, Abt. A, 261: 83–102.

Compagno, L. J. V. 1977. Phyletic relationships of living sharks and rays. American Zoologist, 17: 303–322.

– 1984. Sharks of the world. An annotated and illustrated catalogue of shark species known to date. Vol. 4. Part 1. Hexanchiformes to Lamniformes. – FAO species catalogue for Fishery Purposes. (125) 4: 249 pp.; Rome, Italy (FAO).

– 1988. Sharks of the order Carcharhiniformes. – Princeton University Press: 572 pp; Princeton, New Jersey.

– 2001. Sharks of the world. An annotated and illustrated catalogue of shark species known to date. Vol. 2. Bullhead, mackerel and carpet sharks (Heterodontiformes, Lamniformes and Orectolobiformes. – FAO Species Catalogue for Fishery Purposes, 1(2): 269 pp.; Rome, Italy (FAO).

Compagno, L. J. V., M. Dando & S. Fowler 2005. A Field Guide to the Sharks of the World: 368 pp.; London (Harper Collins).

Dingerkus, G. 1986. Interrelationships of orectolobiform sharks (Chondrichthyes: Selachii). – Pp. 227–245 in: Uyeno, T., R. Arai, T. Taniuchi & K. Matsuura (eds.), Indo-Pacific fish biology: Proceedings of the Second International Conference on Indo-Pacific Fishes; Tokyo (Ichthyological Society Japan).

Dingerkus, G. & T. C. DeFino 1983. A revision of the orectolobiform shark family Hemiscyllidae (Chondrichthyes, Selachii). – Bulletin of the American Museum of Natural History, 176(1): 1–94.

Ebert, D. A., S. Fowler & L. J. V. Compagno 2013. Sharks of the World: A Fully Illustrated Guide: 528 pp.; Plymouth, U.K. (Wild Nature Press, Winson House).

Estes, R. 1964. Fossil vertebrates from the Late Cretaceous Lance Formation, eastern Wyoming. – University of California Publications in Geological Sciences, 49: 1–180.

Forey, P. L., Y. Lu, C. Patterson & E. D. Cliff 2003. Fossil fishes from the Cenomanian (Upper Cretaceous) of Namoura, Lebanon. Journal of Systematic Palaeontology, 1: 227–330.

Frayer, D. W. & J. Radovçic. 2008. Gorjanovic-Kramberger, Dragutin (Karl). – Pp. 154–157 in: Koertge, N. (ed.): New Dictionary of Scientific Biography. 1 (Abderhalden-Byers).

Frickhinger, K. A. 1991. Fossilien Atlas Fische: 1008 pp.; Mergus.

Gayet, M., A. Belouze & P. Abi Saad 2003. Les poissons fossils: 158 pp.; Méolans-Revel (Éditions DésIris).

Gayet, M., P. Abi Saad & C. Gaudant 2012. The fossils of Lebanon. Memory of time: 184 pp.; Méolans-Revel (Éditions DésIris) and Byblos (Memory of Time).

Goto, T. 2001. Comparative anatomy, phylogeny and cladistic classification of the order Orectolobiformes (Chondrichthyes, Elasmobranchi). – Memoirs of the Graduate School of Fisheries Sciences, Hokkaido University, 48: 1–100.

Gorjanović-Kramberger, D. 1884. Palaeoichtyoložki prilozi. Preprint from: Rad jugoslavenske akademije, znanosti i umjetnosti: 1–56.

– 1885. Palaeoichtyoložki prilozi. Rad jugoslavenske akademije, znanosti i umjetnosti, 72: 10–66.

– 1886. Palaeoichtyologische Beiträge. – Glasnik hrvatskoga naravoslovnoga družtva, 1: 123–137.

Guinot, G., C. J. Underwood, H. Cappetta & D. J. Ward 2013. Sharks (Elasmobranchii: Neoselachii) from the Late Cretaceous of France and the UK. – Journal of Systematic Palaeontology, 11: 589–671.

Guinot, G., H. Cappetta & S. Adnet 2014. A rare elasmobranch assemblage from the Valanginian (Lower Cretaceous) of southern France. – Cretaceous Research, 48: 54–84.

Herman, J. 1973. Les vertébrés du Landénien inférieur (Lla ou Heersien) de Maret (Hameau d'Orp-Le-Grand). – Bulletin de la Société Belge de Géologie, de Paléontologie et d'Hydrologie, 81: 191–207.

– 1977. Les Sélaciens des terrains néocrétacés & paléocènes de Belgique & des contrées limitrophes Eléments d'une biostratigraphie intercontinentale. Mémoires pour servir à l'explication des Cartes géologiques et minières de la Belgique, Mémoire 15: 1–401.

– 1982. Die Selachier-Zähne aus der Maastricht Stufe von Hemmoor, Niederelbe (NW-Deutschland). – Geologisches Jahrbuch, Reihe A, 61: 129–159.

Herman, J. & M. Crochard 1977. Additions to the Eocene fish fauna of Belgium. 3. Revision of the Orectolobiforms. – Tertiary Research, 1: 127–138.

Herman, J., Hovestadt-Euler, M. & Hovestadt, D. C. 1992. Part A: Selachii. No. 4. Order: Orectolobiformes. Families: Brachaeluridae, Ginglymostomatidae, Hemiscylliidae, Orectolobidae, Parascylliidae, Rhiniodontidae, Stegostomatidae. Order Pristiophoriformes – Family: Pristiophoridae. Order Squatiniformes – Family Squatinidae. – In: Stehmann, M. (ed.). Contributions to the study of the comparative morphology of teeth and other relevant ichthyodorulites in living supraspecific taxa of chondrichthyan fishes. – Bulletin de l'Institut Royal des Sciences Naturelles de Belgique, Biologie 62: 193–254.

International Code of Zoological Nomenclature (ICZN). http://www.nhm.ac.uk/hosted-sites/iczn/code/index.jsp?article=21&nfv=true#8

Jaekel, O. 1894. Die eocänen Selachier vom Monte Bolca. Ein Beitrag zur Morphologie der Wirbelthiere. – 176 pp., Berlin (Springer).

Kriwet, J. & S. Klug 2004. Late Jurassic selachians (Chondrichthyes, Elasmobranchii) from southern Germany: Re-evaluation on taxonomy and diversity. – Zitteliana A, 44: 67–95.

– 2015. Knorpelfische (Chondrichthyes). – Pp. 334–359 in: Arratia, G., H.-P. Schultze, H. Tischlinger & G. Viohl (edit.). Solnhofen – Ein Fenster in die Jurazeit; München (Pfeil).

Last, P. R. & J. D. Stevens 1994. Sharks and rays of Australia. – 513 pp.; Commonwealth Scientific and Industrial Research Organisation (CSIRO), Australia.

– 2009. Sharks and rays of Australia, 2nd edit. – 644 pp.; Commonwealth Scientific and Industrial Research Organisation (CSIRO), Australia; Cambridge, MA, London, England (Harvard Univ. Press).

Little, C. D.& W. E. Bemis 2004: Observations on the skeleton of the heterocercal tail of sharks (Chondrichthyes: Elasmobranchii) – Pp. 563–573 in: Arratia, G., M. V. H. Wilson & R. Cloutier (edit.). Recent Advances in the Origin and Early Radiation of Vertebrates – Honoring Hans-Peter Schultze; München (Pfeil).

Maisey, J. G. 1986. The Upper Jurassic Hexanchoid Elasmobranch *Notidanoides* n. g. – Neues Jahrbuch für Geologie und Paläontologie. Abhandlungen, 172: 83–106, Stuttgart (Schweizerbart).

Moreau, F. & S. Mathis 2000. Les élasmobranches du Thanétien (Paléocène) du Nord de la France, des carrières de Templeuve et de Leforest. Cossmanniana, 7: 1–18.

Müller, A. 1989. Selachier (Pisces: Neoselachii) aus dem höheren Campanium (Oberkreide) Westfalens (Nordrhein-Westfalen, NW-Deutschland). – Geologie und Paläontologie in Westfalen, 14: 1–161.

Mulas, A., A. Bellodi, C. Porcu, A. Cau, E. Coluccia, R. Demurtas, M. F. Marongiu, P. Pesci & M. C. Follesa 2020. Living naked: first case of lack of skin-related structures in an elasmobranch, the blackmouth catshark (*Galeus melastomus*). – Journal of Fish Biology, 97: 1252–1256. DOI: 10.1111/jfb.14468

Naylor, G. J. P., J. N. Caira, K. Jensen, K. A. M. Rosana, N. Straube & C. Lakner 2012. Elasmobranch Phylogeny: A Mitochondrial Estimate Based on 595 Species. – Pp. 31–56 in: Carrier, J. C., J. A. Musick & M. R. Heithaus (eds.). Biology of Sharks and Their Relatives; Boca Raton, Florida (CRC Press).

Nelson, J. S. 2006. Fishes of the world, 4th edit. – xix + 601 pp.; New York (John Wiley & Sons, Inc.).

Noubhani, A. & H. Cappetta. 1997. Les Orectolobiformes, Carcharhiniformes et Myliobatiformes (Elasmobranchii, Neoselachii) des bassins à phosphate du Maroc (Maastrichtien-Lutétien basal). Systématique, biostratigraphie, évolution et dynamique des faunes. – Palaeo Ichthyologica, 8: 1–327.

Pfeil, F. H. 2012. Das Münchner Exemplar einer *Mesiteia emiliae* Kramberger, 1884 aus den Plattenkalken (Oberkreide, unteres Cenomanium) von Hakel (Haqil) im Libanon. – Jahresbericht 2011 und Mitteilungen der Freunde der Bayerischen Staatssammlung für Paläontologie und Historische Geologie München e.V., 40: 43–68.

– 2013. Erstnachweis eines großen Ammenhais – *Cantioscyllium decipiens* Woodward, 1889 – aus den Plattenkalken (Oberkreide, unteres Cenomanium) von Hakel (Haqil) im Libanon. – Jahresbericht 2012 und Mitteilungen der Freunde der Bayerischen Staatssammlung für Paläontologie und Historische Geologie München e.V., 41: 73–102.

– 2014. Ein weiteres Exemplar von *Cantioscyllium decipiens* Woodward, 1889 aus den Plattenkalken (Oberkreide, unteres Cenomanium) von Hakel (Haqil) im Libanon. – Jahresbericht 2013 und Mitteilungen der Freunde der Bayerischen Staatssammlung für Paläontologie und Historische Geologie München e.V., 42: 47–60.

Pictet, F.-J. & A. Humbert 1866. Nouvelles recherches sur les poissons fossiles du Mont Liban. – VII + 115 pp.; Geneva (Georg), Paris (J.-B. Baillière et Fils, F. Savy).

Playfair, R. L. & A. C. L. G. Günther 1866. The fishes of Zanzibar, with a list of the fishes of the whole east coast of Africa: 153 pp.; London (John van Voorst).

Rees, J. 2010. (2010). Neoselachian sharks from the Callovian-Oxfordian (Jurassic) of Ogrodzieniec, Zawiercie region, southern Poland. – Palaeontology, 53: 887–902.

Reuss, A. E. 1845. Die Versteinerungen der böhmischen Kreideformation. Erste Abtheilung. – iv + 58 pp.; Stuttgart (E. Schweizerbart).

Ridewood, W. G. 1921. On the calcification of the vertebral centra in sharks and rays. – Philosophical Transactions of the Royal Society of London. Series B, 210: 311–407.

Shirai, S. 1992. Squalean phylogeny: a new framework of "squaloid" sharks and related taxa. – 151 pp.; Sapporo (Hokkaido Univ. Press).

– 1996. Chapter 2. Phylogenetic interrelationships of Neoselachians (Chondrichthyes: Euselachii). – Pp. 9–34 in: Stiassny, M. L. J., L. R. Parenti & G. D. Johnson (eds.), Interrelationships of fishes; San Diego, London, Boston, New York (Academic Press).

Signeux, J. 1949. Notes paléoichthyologiques. II : Sur les genres *Isurus*, *Chiloscyllium* et *Triakis* trouvés à l'état fossile dans le gisement sénonien de Sahel-Alma. – Bulletin du Muséum national d'histoire naturelle Paris, 21: 635–638.

Srdic, A., C. J. Duffin & D. M. Martill 2016. First occurrence of the orectolobiform shark *Akaimia* in the Oxford Clay Formation (Jurassic, Callovian) of England. – Proceedings of the Geologists' Association, 127: 506–513.

Thies, D. 1981. Vier neue Neoselachier-Haiarten aus der NW-deutschen Unterkreide. – Neues Jahrbuch für Geologie und Paläontologie, Monatshefte, 1981, p. 475–486.

– 1983. Jurazeitliche Neoselachier aus Deutschland und Süd-England. – Courier Forschungsinstitut Senckenberg, 58: 1–116.

– 1989. Some problematical sharks teeth (Chondrichthyes, Neoselachii) from the Early and Middle Jurassic of Germany. – Paläontologische Zeitschrift, 63: 103–117.

Underwood, C. J. & D. J. Ward 2004. Neoselachian sharks and rays from the British Bathonian (Middle Jurassic). – Palaeontology, 47: 447–501.

Werner, C. 1989. Die Elasmobranchier-Fauna des Gebel Dist Member der Bahariya Formation (Obercenoman) der Oase Bahariya, Ägypten. – Palaeo Ichthyologica, 5: 1–112.

Woodward, A. S. 1888. Paleontological contributions to selachian morphology. I. On the lateral line of a Cretaceous species of Scylliidae. – Proceedings of the Zoological Society of London, 9: 126–128.

– 1889. Catalogue of the fossil fishes in the British Museum (Natural History). Part. I. Elasmobranchii: 474 pp.; London (Taylor and Francis).

9. Picture credits

For pictures that were taken from other publications, or modified by using illustrations in other publications, and for photographs taken by other persons, the source is cited directly in the image description. The original images of all UV-pictures were patiently produced by Helmut Tischlinger, Stammham. The colour pictures of the holotype of *Mesiteia emiliae* were taken by Sanja Japundžić at the Croatian Natural History Museum, Zagreb. Martin Pickford, Paris sent me new photographs of the two *Mesiteia* specimens #02 (MNHN.F.HAK513) and #03 (MNHN.F.HAK559) that are kept in Paris. Pierre Abi Saad, Byblos and Roy Nohra, Hakel, kindly allowed me use of pictures of previously sold specimens, and of specimens kept in their collections, and additionally for allowing me to photograph some of these specimens myself. Hiroshi Senou, Kanagawa Prefectural Museum of Natural History, allowed me use of a photograph of a *Chiloscyllium punctatum*. All others, mainly photographs, were taken by myself using my own equipment. There was no financial support for this project.

List of all illustrations in figures and plates with reference to the authors and photographers (in alphabetical order):

Abi Saad, Pierre; Figures: 31b, 39; Plates: 2 (2a), 5(2–3, 4b, 5), 11h, 21(i–j), 22(a, d–e), 25(g–h), 27(a–f)
Cappetta (1980); Figures: 3a–b, 4a–b, 10(1b–d, 2b, 3a–b)
Cappetta (2012); Figures: 8, 10(1a, 2a), 11a,e, 13(1a–b, 2a–c), 17, 19a–c, 21a, 22(1–3), 23
Dingerkus & DeFino (1983); Figures: 28d, 29a–b, 32c–d, 34a–d, 35a–b; Plates: 11b–c, 12b–c
Gayet, Abi Saad & Gaudant (2012); Plate: 5 (4a)
Gorjanović-Kramberger (1884); Figures: 38a–b
Goto (2001); Figures: 37a–f
Hilpert, Hubert; Plate: 29e
https://commons.wikimedia.org/w/index.php?curid=3172086; Figure: 1
https://i.pinimg.com/originals/5e/85/81/5e85815821b97bee673e238838dd0e09.jpg; Plate: 6(2a–b)
Jaekel (1894) from Gorjanović-Kramberger (1884); Figure: 2
Japundžić, Sanja; Figure: 31c; Plates: 1(1), 10a, 14a
Mulas, Antonello; Figures: 42a–b
Nohra, Roy; Figures: 31e, 32b, 33f; Plates: 4(3, 5a–b), 5(1), 10b, 11f, 19(a–b, h), 20(d–e), 22(f–g), 24(a–b), 25(a–h), 26g, 29(b–c)
Pfeil, Friedrich H. (FHP); Figures: 5a–d, 6a–f, 7, 9, 11b–d,f, 12b, 14a–b, 15a–b, 16b–e, 20, 21b–c, 25a–d, 26a–b, 27a–b, 28a–c, 30, 31a,d,f, 32a, 33a–b,d–e,g–j, 34a,e–h, 35c–d, 36a, 37g, 40, 41, 43; Plates: 1(4a, 5a, 6a), 2(1, 2b), 3(1a, 2–4), 4(1–2, 4), 5(6), 6(3–4), 7(1c, 2a,d, 3a, 4c, 5a), 8(1a,c, 2a–b, 3a–c, 4a–c, 5a–b), 9(1a–b, 2b), 10c–g, 11d–e,g, 12a,d–e, 13f–g, 14c–g, 15a–c,e, 16a–d, 17a,d–e, 18a,c–d,g, 19(c–d, e–g), 20(b–c, f–m), 21(a–h), 22(b–c), 23(a–g), 24(c–j), 26(a–f,h), 27(g–k), 28(a–e,g–l), 29(a, d, f–j)
Pickford, Martin; Plates: 1(2–3), 14b, 18f
Senou Hiroshi; Plate: 11a
Srdic, Duffin & Martill (2016); Figures: 24(1–2)
Thies (1981); Figure: 18
Tischlinger, Helmut; Figures: 14c, 16a, 33c, 36b; Plates: 1(4b, 5b, 6b), 2(2c), 3(1b), 7(1a–b, 2b–c, 3b–c, 4a–b, 5b–c), 8(1b), 9(1c, 2a, 3a–b), 15d,f, 17b, 18b,e, 20a, 28f
Werner (1989); Figure: 12a
Woodward (1889); Plate: 17c

Appendix 1

Abbreviations used in the plates and text figures.

#01, #02 ... numbering of specimens of *Mesiteia*, see Table 1
TL total length

Cranium
lc labial cartilage
lcr lateral cartilage of rostrum
mc Meckel's cartilage
pq palatoquadrate
rm medial rostral rod
rmd dorsal medial rostral rod
rmv ventral medial rostral rod
s symphysis

Pharyngeal arches
bbp basibranchial plate
bh basihyal
br branchial rays (= gill filaments)
cb1-5 ceratobranchial 1-5
ch ceratohyal
eb1-5 epibranchial 1-5
eh epihyal (= hyomandibula)
hb1-4 hypobranchial 1-4
ipc inner process of the ceratobranchial 1-3
pb1-5 pharyngobranchial 1-5

Pectoral girdle
cob coracoid bar
sc scapular cartilage
ssc suprascapular cartilage.

Pectoral fins
mesp mesopterygium
metp metapterygium
mtx metapterygial axis
prop propterygium
rad radials
in red the number of the radials from anterior to posterior radials

Pelvic girdle
ip iliac process
pb ischio-pubic bar

Pelvic fins
b1 first basal cartilage element
b2 second basal cartilage element
bp basipterygium
radi inner rows of radial elements
rado outer rows of radial elements
in red the number of the radials from anterior to posterior radials

Claspers
cg groove of clasper
ca axial cartilage of clasper
cd dorsal covering plate of clasper
cs style of clasper
cv ventral sheath of clasper

Dorsal fins
bd basal cartilage of dorsal fin
D1 first dorsal fin
D2 second dorsal fin
radi inner rows of radial elements
rado outer rows of radial elements
in red the number of the radials from anterior to posterior radials

Anal fin
ba basal cartilage of anal fin
radi inner rows of radial elements
rado outer rows of radial elements
in red the number of the radials from anterior to posterior radials

Caudal fin, vertebral column and the elements of the dorsal and ventral arcualia
bdp basidorsal process
bvp basiventral process
dbc calcified double cone of vertebra
dic dorsal intercalary plate
ecr epichordal radial
hcp hypochordal process
mee membrana elastica externa
mn foramen for the ventral (motoric) nerve-root
pec preepichordal cartilage
phc prehypochordal cartilage
rb vertebral rib
scl secondary calcified lamella of vertebra
sn foramen for the dorsal (sensoric) nerve-root
spn supraneural
vc vertebral centrum
vc114 numbering of vertebral centra, e.g. vertebra #114
vic ventral intercalary plate

Lateral line, muscle mass preservation, heart and blood vessels, coprolites
ll lateral line
mm myomere
ms myoseptum
ca coronary arteries
h heart
va ventral aorta
v other blood vessels
c coprolite

Ancient Fishes and their Living Relatives: a Tribute to John G. Maisey. Alan Pradel, John S. S. Denton & Philippe Janvier (eds.): pp. 183–192, 5 figs.
© 2021 by Verlag Dr. Friedrich Pfeil, München, Germany – ISBN 978-3-89937-269-4

The visceral skeleton and its relation to the head circulatory system of both a fossil, the Carboniferous *Iniopera*, and a modern, *Callorhinchus milii*, holocephalan (Chondrichthyes)

Alan Pradel*,**, Richard P. Dearden*, Antoine Cuckovic***, Rohan Mansuit* and Philippe Janvier*

Abstract

The anatomy of the visceral skeleton of a Carboniferous holocephalan, the iniopterygian *Iniopera*, is studied here from 3D-preserved fossils from Kansas. For comparison and interpretation, the visceral skeleton and head circulatory system of the Recent holocephalan *Callorhinchus milii* are also described. *Iniopera* and modern holocephalans both display a completely subcranial branchial skeleton, a continuous series of basal branchial cartilages, probably an operculum supported by hyoid rays connected to ceratohyals, a pectoral symphysis penetrated by subclavian arteries, and an "accordion" ventilation model involving closely associated pectoral and branchial skeletons. The distribution among gnathostomes of anatomical characters related to the branchial skeleton and arterial system is discussed to highlight potential synapomorphies of chondrichthyan and osteichthyan clades.

Introduction

The visceral skeleton is a key anatomical complex in gnathostomes, since it is the center of feeding and respiration. The morphology of this structure is potentially a great source of data for studying the evolution and palaeobiology of early gnathostomes. Yet, examples of Palaeozoic fishes presenting well-preserved branchial arches are scarce (e.g., Pradel et al. 2014, Giles et al. 2015b, Brazeau et al. 2017, Coates et al. 2018, Dearden et al. 2019).

The Iniopterygia are a Carboniferous order of chondrichthyans (cartilaginous fishes) that were initially described from the Pennsylvanian marine mid-continental black shales of Indiana, Illinois, Iowa, Missouri and Nebraska, USA (Zangerl & Case 1973, Zangerl 1997). Other iniopterygians have been later described from the Pennsylvanian of Kansas and Oklahoma (Pradel et al. 2009, Pradel 2010, Pradel et al. 2010) and from the Mississippian of Montana (Grogan & Lund 2009). Iniopterygians are currently considered as a clade among stem-holocephalans (Pradel 2011, Coates et al. 2018), and they are subdivided into two families: the Sibyrhynchidae, which possess a holostylic jaw suspension, and the Iniopterygidae, which retain a free palatoquadrate (Stahl 1980, Grogan & Lund 2009). The neurocranium and pectoral girdle of the Sibyrhynchidae are now well known thanks to three-dimensionally preserved specimens from the Pennsylvanian of Kansas and Oklahoma, which were extensively studied by means of CT X-ray imaging using both µCT Scans and synchrotron radiation scanning (Pradel et al. 2009, Pradel 2010, Pradel et al. 2010).

The branchial skeleton of the Iniopterygidae has been described from the Bear Gulch locality (Grogan & Lund 2009) as possessing a hyoid arch mesial and independent from the mandibular arch, and which bears an operculum supported by hyoid rays, four branchial arches composed of prominent ceratobranchials that are articulated to thin epibranchials dorsally and elongated hypobranchials ventrally. Small pharyngobranchials articulated to the epibranchials dorsally are also suggested in some specimens. The basibranchial series is composed of medial, unpaired basibranchials, which are articulated posteriorly to paired hypobranchials that connect the ventral ends of the coracoid cartilages. However, there is no information about the number of the basibranchial cartilages, nor is evidence of a distinct basihyal.

The branchial skeleton of the sibyrhynchids was first described by Zangerl & Case 1973 as having a series of medial, unpaired cartilages that are organized from anterior to posterior as an elongated "glossohyal", a first basibranchial, which possess a pair of foramina, and a second, smaller basibranchial that displays a large pit or foramen medially. In addition to these elements, there are a pair of ceratohyals, which articulate with the two basibranchials. Later, Zangerl (1981) reinterpreted the branchial skeleton pattern from anterior to posterior as follows: an elongated basihyal from which emerges a pair of ceratohyals, a first basibranchial that corresponds to the smaller element of the previous description, with an unpaired pit, and a second basibranchial which possesses a pair of foramina and which articulates the ventral end of the scapulocoracoids. Pradel et al. (2010) identified the unpaired cartilage with two foramina, positioned between the two scapulocoracoids, that was interpreted as an intercoracoid cartilage which belongs to the pectoral girdle rather than to the basibranchial series of the gill skeleton.

New studies on the 3D-preserved specimens from Kansas are presented here in order to get more information about the branchial skeleton and clarify the pattern of the basibranchial series of these highly specialized fishes. Comparisons with the branchial skeleton and its vascularization of the widely-used, model extant chondrichthyans (the holocephalan *Callorhinchus*

* CR2P – Centre de Recherche en Paléontologie – Paris; Muséum National d'Histoire Naturelle – Sorbonne Université – Centre National de la Recherche Scientifique. France

** Department of Vertebrate Paleontology, American Museum of Natural History. New York, USA; alan.pradel@mnhn.fr (corresponding author)

*** Université Paris – Saclay, France

milii and the elasmobranch *Scyliorhinus canicula*) were used to interpret the branchial cartilages and infer the arterial network in *Iniopera*. Notably, this study resulted in a series of new observations on the *Callorhinchus* arterial system.

Material and methods

Material. The fossil material from Kansas studied here comprises two specimens belonging to the Kansas University Natural History Museum, Lawrence, Kansas, USA (KUNHM 22060, Fig. 1; KUNHM 21894, Fig. 2). These two specimens belong to *Iniopera* sp. because of the following characteristics: the anteriormost distal pectoral fin radial is much larger than the other fin radials; the symphysial region of the mandible is more rectangular in shape than that of *Sibyrhynchus*; the skull bears a pavement of dermal denticles on its posterodorsal surface (see Zangerl & Case 1973 and Pradel et al. 2010 for the systematic status of *Iniopera* sp.).

A map showing the sources of this material is provided in Hamel & Poplin (2008). Images of the original nodules are provided in Pradel et al. (2009) and Pradel (2010). The nodules crop out approximately 150 km southeast of Kansas City, between Lawrence and Baldwin, in about thirty different localities. Stratigraphically, they occur at the limit between the marine Haskell Limestone Member and the overlying Robbins Shale of the Lawrence Formation (Robb III 1992), deposited in brackish water. This sequence is part of the Douglas Group, dated as Late Virgilian, Upper Pennsylvanian (ca. 305–299 Ma).

The embryo of *Callorhinchus milii* was collected from a spawning site in the Marlborough Sounds, New Zealand. The appropriate protocols for animal welfare were followed according to guidelines published by the American Society of Ichthyologists and Herpetologists. Permits to collect and export the embryos were obtained from the New Zealand Ministry of Fisheries (now NIWA) and permission and support of re-search activities in the Marlborough Sounds was granted by the New Zealand Department of Conservation. Copies of the permits are held by the Academy of Natural Sciences in Philadelphia where the specimen was deposited. The specimen studied here is at embryonic stage 36 (Didier et al. 1998).

The *Scyliorhinus canicula* specimen was treated according to the French and European regulations for handling of animals in research. It is a juvenile specimen, which had reached the stage of independent feeding at the time of acquisition. It was grown in the Laboratoire Evolution, Génome et Spéciation, UPR 9034 CNRS, Gif-sur-Yvette, France, which has appropriate licence to breed such a species provided by the "direction départementale des services vétérinaires" of the French Republic (number 2009-DDSV-031, April 15th 2009). The embryo used in this study is an aquatic, non-endangered, non-mammalian vertebrate and therefore did not require special authorizations. All efforts were made to minimize suffering.

Methods. All the specimens, whether they are fossil or recent, were scanned on beam-line ID19 of the European Synchrotron Radiation Facility (ESRF, Grenoble, France).

The fossil KUNHM 21894 was imaged using X-ray based synchrotron microtomography in absorption contrast (SR-μCT) (Tafforeau et al. 2006). Scan parameters were as follows: monochromatic X-ray beam of 60 keV energy. The detector was a FReLoN (Fast Readout Low Noise) (Labiche et al. 2007) CCD camera coupled with an optical magnification system, yielding an isotropic pixel size of 30.3 μm, 1200/180 degrees projections with 0.4 s of exposure time.

The fossil KUNHM 22060 was imaged using propagation phase contrast X-ray synchrotron microtomography. Scan parameters were as follows: a voxel size of 7.46 μm at an energy of 60 keV, with a propagation distance of 900 mm. In order to obtain a 2.9 cm lateral field of view, the scan was made in half acquisition geometry, with the centre of rotation on one side of the field of view and a 360 degrees rotation. We used 5000 projections over 360 degrees with an exposure time of 0.7 s for each of the 9 scans performed, each of them covering a height of 5.8 mm.

The modern specimens studied here are the same as the one used in Pradel et al. (2013). The egg capsule containing the *Callorhinchus milii* specimen studied here was collected by SCUBA diving. The egg capsule was opened within hours of collection and the embryo was removed and dissected from the yolk sac. Embryos were preserved in 10 % formalin in seawater. After a minimum of one month in fixative, embryos were washed and transferred to 70 % ethanol for long-term storage (see Didier [1995] for more information). The specimen was scanned using an isotropic voxel size of 30 microns, with a single propagation distance of 4 m. The specimen was scanned in a 75 % ethanol solution. The beam was produced by the U17.6 undulator closed at a gap of 13.5 mm, filtered by 2 mm of aluminum and 1 mm of copper with an effective energy of 55 keV. In order to obtain differential contrasts for the different tissues of the specimen, a single distance phase retrieval process was used (Sanchez et al. 2012, Paganin et al. 2002). The full sample was scanned in 6 sub-scans of 4 minutes each, using 2499 projections of 0.1 s each over 360 degrees. The detector was based on a 750 μm thick LuAG scintillator, coupled to a FReLoN (Fast Readout Low Noise) CCD camera through a lens based optic system.

The *Scyliorhinus canicula* specimen was anaesthetized with MS222 (Sigma) before the processing. It was fixed with 4 % paraformaldehyde and preserved in 100 % methanol solution. It was scanned using holotomographic approach similar to the one described in Pradel et al. (2009). The voxel size was 7.45 μm, and 4 propagation distances (20, 150, 400 and 900 mm) were used in order to retrieve the phase maps. We used a beam at 20 keV monochromatized using a double Bragg Si 111 crystal monochromator. For each distance, 1200 projections of 0.6 s each over 180 degrees were used. The specimen was scanned in a cylindrical container filled with 100 % ethanol solution.

Synchrotron Volumes were reconstructed using ESRF software PyHST, which includes the single distance phase retrieval algorithm (Sanchez et al. 2012). Sub-volumes were corrected for ring artefacts, and then concatenated to generate a single stack of either 8 bit (KUNHM 21894, KUNHM 22060) or 16 bit (*Scyliorhinus* and *Callorhinchus* specimens) TIFF slices.

Segmentation, 3D rendering, and export (STL) of the cartilages were performed with MIMICS Innovation suite (Materialise, release 20.0 and 21.0, Research Edition, x64 edition). Geomagic Studio (Geomagic, release 2013, x64 edition) was used to post-process each STL (cleaning, remeshing, polygon reduction). Cinema 4D Studio (Maxon, release R19 and R20) was used to symmetrize the 3D models when necessary and assemble the final composite specimen based on the two studied fossils of *Iniopera*. The 3DPDF was made with 3-matic (Materialise, release 13.0, Research Edition, x64 edition).

Figured images of the 3D models were rendered using Blender 2.83 (blender.org). The original scan data are stored at the MNHN and available upon request.

Description

The medial cartilages series. The basihyal is the anteriormost element of the basibranchial series (bh, Figs. 1a–d). It is located immediately posterior to the mandible in *Iniopera* (md, Fig. 1a) and is nested between the right and left sides of the lower jaw. The basihyal forms a stout, hollowed, triangular cartilage. It is unfortunately not complete anteriorly but most of its complex shape is preserved. The dorsal surface is flat. The lateral walls of the basihyal are bridged ventrally by a bar, and they end posteroventrally as processes (bhp, Fig. 1d).

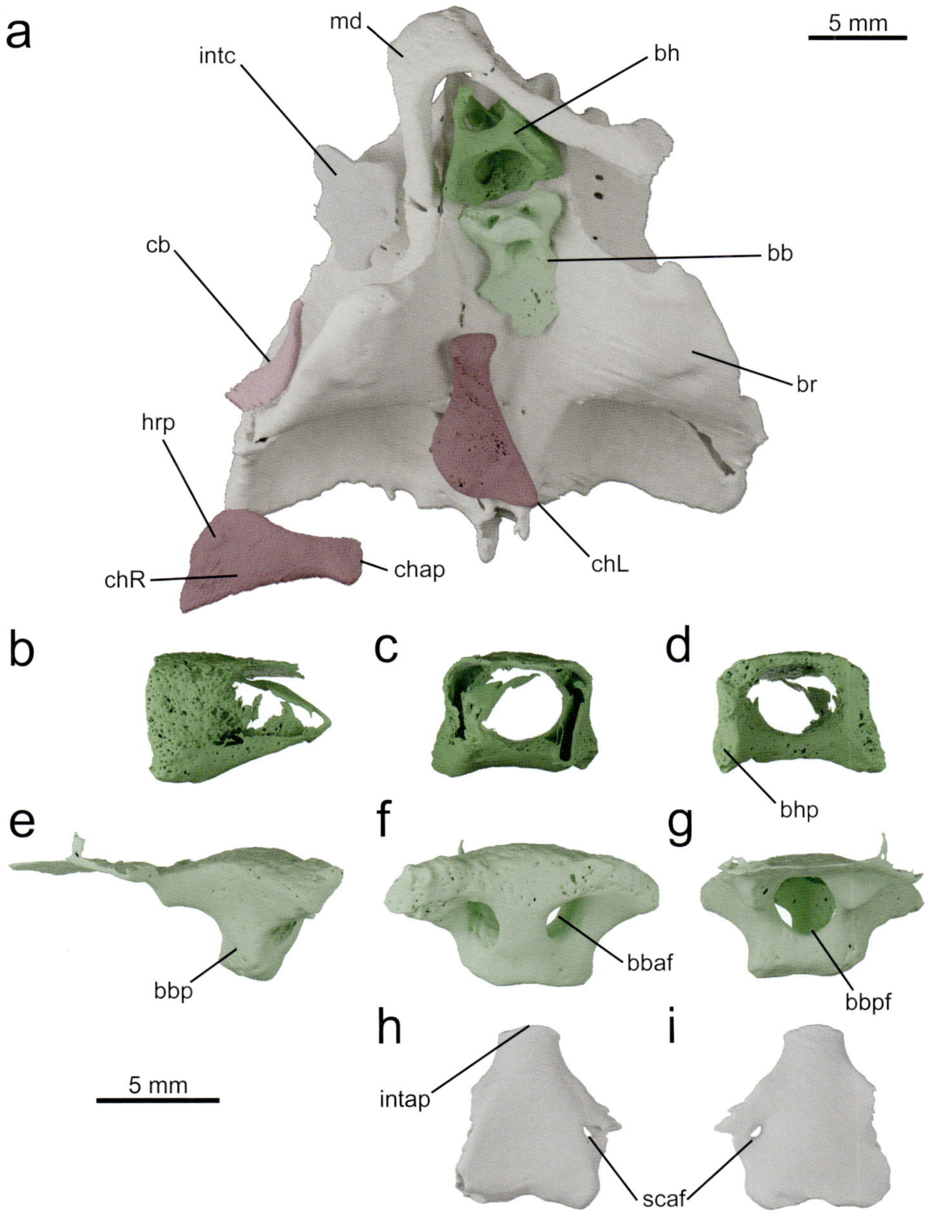

Fig. 1. Surface rendering generated from Synchrotron Radiation ▷ microtomographic slices of the braincase, mandible, branchial skeleton and intercoracoid cartilages of *Iniopera* (KUNHM 22060). **a.** cartilages as preserved in the nodule, with the braincase, mandible, basihyal and basibranchial shown in ventral view in almost natural position. The other elements are preserved displaced from their natural positions: the left ceratohyal is preserved in medial view, the right ceratohyal in lateral view, and the intercoracoid cartilage in ventral view, but upside down. **b.** Basihyal in right lateral view. **c.** Basihyal in anterior view. **d.** Basihyal in posterior view. **e.** Basibranchial in right lateral view. **f.** Basibranchial in anterior view. **g.** Basibranchial in posterior view. **h.** Intercoracoid cartilage in dorsal view, anterior to the top. **i.** Intercoracoid cartilage in ventral view, anterior to the top. Abbreviations: bb, basibranchial; bbaf, anterior foramen of the basibranchial; bbp, ventral process of the basibranchial; bbpf, posterior foramen of the basibranchial; bh, basihyal; bhp, posteroventral articular process of the basihyal; br, braincase; cb, ceratobranchial; chap, articular process of the ceratohyal; chL, left ceratohyal; chR, right ceratohyal; hrp, hyoid rays pit insertion; intap, intercoracoid anterior articular process; intc, intercoracoid cartilage; md, mandible; scaf, foramen for the subclavian artery. Colors: cream, braincase and mandible; light gray, pectoral girdle elements; dark green, basihyal; light green, basibranchial; dark pink, ceratohyal; light purple, ceratobranchial.

Posterior to the basihyal is the basibranchial (bb, Figs. 1a, e–g), which is not completely preserved posteriorly. It is hollow and it displays a Y-shaped dorsal surface that ends in paired bulges anteriorly, and paired, squared processes that merge to a dorsal plate posteriorly. The lateral walls of the basibranchial are concave, and they are bridged by a prominent ventral process (bbp, Fig. 1e). This ventral process is also connected to the dorsal surface anteriorly by a thin, medial bar of cartilage which divides a pair of large openings (bbaf, Fig. 1f). The lateral walls and the ventral process define a single posterior large opening (bbpf, Fig. 1g).

Another unpaired cartilage linked to the basibranchial series is present in both specimens (intc, Figs. 1a, h, i, 2a, e, f). It looks very similar to the 'second' copula in Zangerl & Case 1973, Figure 72. This has been described as an intercoracoid cartilage elsewhere (Pradel et al. 2010) based on KUNHM 21894, and the same element was retrieved in KUNHM 22060. Important features are: a pair of posterolateral bulges articulating with the two scapulocoracoids; an anterior, oval with a smooth surface process (intap, Figs. 1h, 2e); a pair of foramina about at the middle of the element; and a concave dorsal surface.

Identifiable paired elements. A pair of large, elongated, slightly concave blade-like cartilages is preserved in KUNHM 22060 (chR, chL, Fig. 1a). These elements are the largest paired elements, besides the scapulocoracoids, retrieved in the different nodules studied here. The proximal extremity of the element ends with a joint articulation (chap, Fig. 1a), then the cartilage flares and becomes flat. The distal extremity ends as a small process. The convex, flat, external, distal surface displays 4 pits (hrp, Fig. 1a). These cartilages look like the ceratohyals described in both *Sibyrhynchus* and *Iniopera* by Zangerl & Case (1973). They also resemble the ceratohyals of the extant chimaeroid *Callorhinchus*, which are also the largest, blade-like elements of the visceral skeleton (ch, Fig. 3d).

In both specimens, some isolated slender and curved cartilages of different sizes are present, resembling the ceratobranchials of modern chimaeroids rather than epibranchials (cb, Figs. 1a, 2b–d). They are interpreted here as being ceratobranchials.

There is no evidence of obvious hypal, epal and pharyngal elements in the two specimens studied here, but in the lateral depression, below the otic region of the braincase of KUNHM 21894, a number of scattered small tube-like cartilages are present without pattern or repeated shape (uc, Fig. 2b).

The cranial circulatory system of the extant holocephalan *Callorhinchus milii* in relation to the surrounding cartilages. The heart sits above the pectoral symphysis (h, ps, Fig. 3h). The ventral aorta (va, Fig. 3a) exits the heart from the conus arteriosus, and runs anteriorly below the basibranchial series

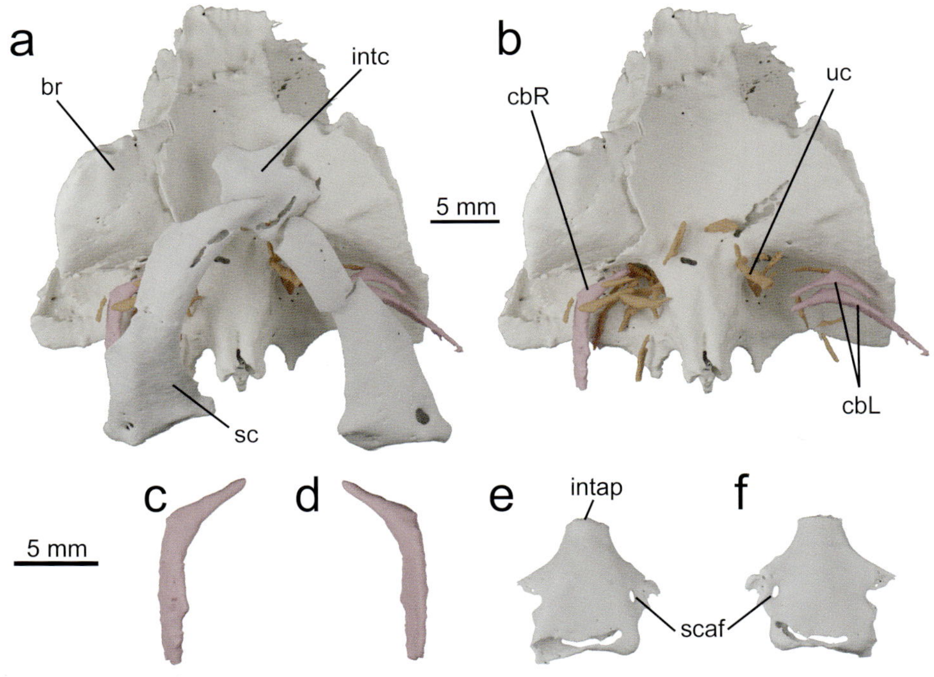

Fig. 2. Surface rendering generated from Synchrotron Radiation microtomographic slices of the braincase, branchial skeleton and pectoral girdle cartilages of *Iniopera* (KUNHM 21894). **a.** cartilages as preserved in the nodule in ventral view, in almost natural position. **b.** Pectoral girdle cartilages removed. **c.** Right ceratobranchial in ventral view. **d.** Right ceratobranchial in dorsal view. **e.** Intercoracoid cartilage in dorsal view, anterior to the top. **f.** Intercoracoid cartilage in ventral view, anterior to the top. Same abbreviations and colors as in Figure 1. Additional abbreviations and colors: cbL, left ceratobranchials; cbR, right ceratobranchial; sc, scapulocoracoid; uc, undetermined cartilages. Additional color: orange, scattered undetermined cartilages.

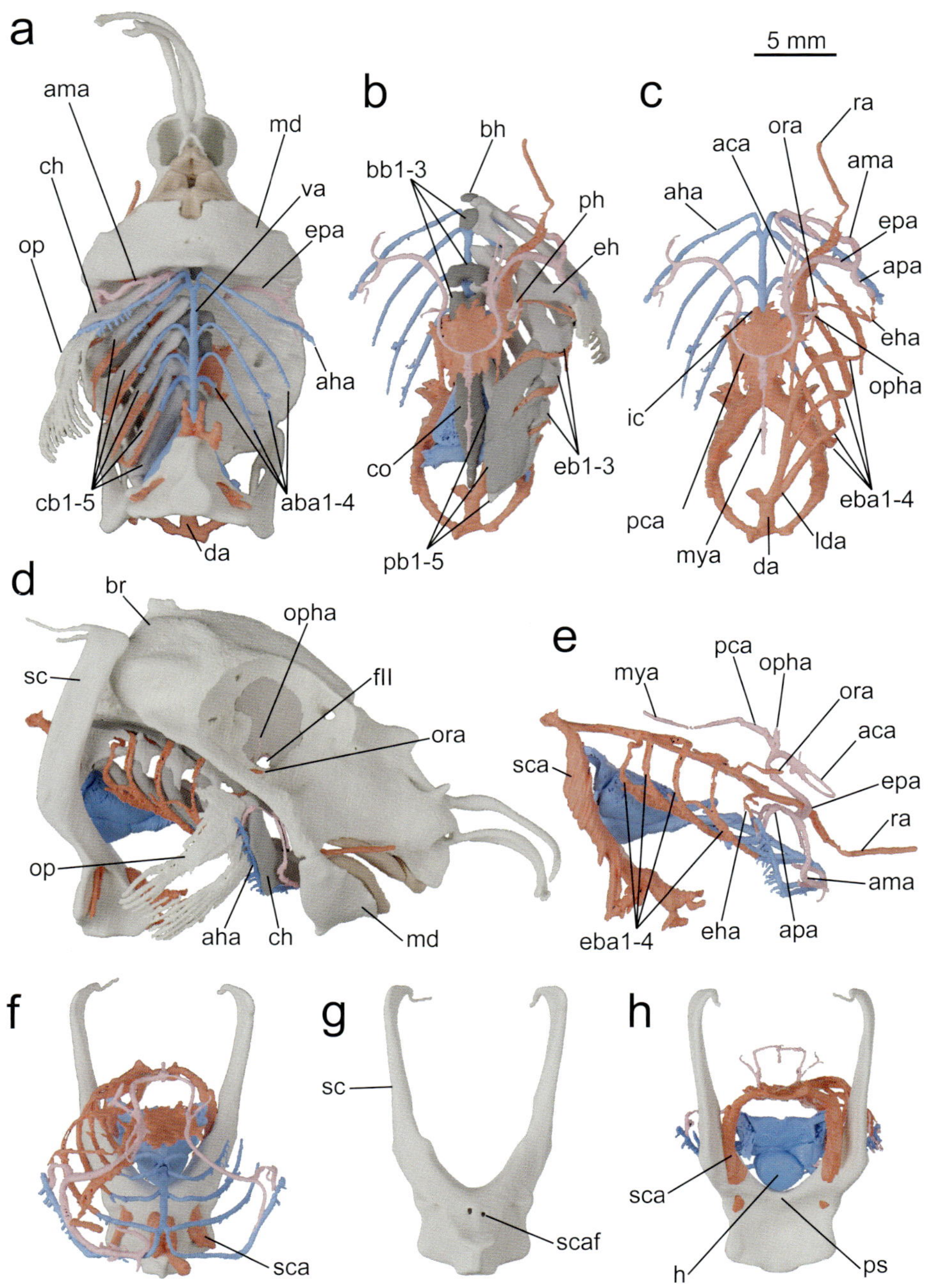

Fig. 3. Surface rendering generated from Synchrotron Radiation microtomographic slices of the braincase, mandible, branchial skeleton and pectoral girdle cartilages, and vascular system of a late embryo of *Callorhinchus milii*. **a.** Ventral view. **b.** Dorsal view, braincase, mandible and pectoral girdle removed. **c.** Same as b with the branchial skeleton removed. **d.** Right lateral view. **e.** Same as d with cartilages removed. **f.** Pectoral girdle and vascular system in anterior view. **g.** Pectoral girdle alone in anterior view. **h.** Pectoral girdle and vascular system in posterior view. Same abbreviations and colors as in Figure 1. Additional abbreviations: aba1–4; afferent branchial artery 1 to 4; aca, anterior cerebral artery; aha, afferent hyoid artery; ama, afferent mandibular artery; apa, afferent pseudobranchial artery; bb1–3, basibranchial 1 to 3; cb1–5; ceratobranchial 1 to 5; ch, ceratohyal; co, copula; da, dorsal aorta; eb1–3, epibranchial 1 to 3; eba1–4, efferent branchial artery 1 to 4; eh, epihyal; eha, efferent hyoid artery; epa, efferent pseudobranchial artery; fII, foramen for the optic nerve; h, heart; ic, internal carotid artery; lda, lateral dorsal aorta; mya, myelonal artery; op, opercular cartilage and hyoid rays; opha, ophthalmic artery; ora, orbital artery; pb1–5, pharyngobranchial 1 to 5; pca, posterior cerebral artery; ph, pharyngohyal; ps, pectoral girdle symphysis; ra, rostral artery; sc, scapulocoracoid; sca, subclavian artery; va, ventral artery. Additional colors: blue, afferent arteries and heart; red, efferent arteries; pink, head accessory vascular system; cream, neurocranium, scapulocoracoid, and opercular cartilage/hyoid rays; grays, branchial and hyoid cartilages.

(bb1–3, co, Fig. 3b) until the first basibranchial. Three afferent branchial arteries, for the branchial arches 2 to 4, branch off directly and successively from the ventral aorta and run laterally ventromedially to the corresponding ceratobranchials (cb1–5, Fig. 3a). There is no afferent branchial artery for the fifth branchial arch. Ventral to the first basibranchial, the ventral aorta divides into paired, short common trunks for the afferent hyoid artery (aha, Fig. 3a) and the first afferent branchial artery (aba, Fig. 3a). The afferent hyoid artery runs ventromedially to the ceratohyal (ch, Figs. 3a,d) and above the opercular cartilage (op, Figs. 3a,d). The afferent pseudobranchial artery (apa, Figs. 3c,e) passes laterally to the cerato-epihyal (ch, eh, Figs. 3b,d) junction and between the ceratohyal and operculum ventrally. Because there is no pseudobranch, the afferent and efferent pseudobranchial (apa, epa, Figs. 3c,e) arteries are continuous. These two arteries form a Y-shaped circuit with the afferent mandibular artery (sensu Allis 1912) (ama, Figs. 3c,e). The latter artery runs forward, first laterally to the ceratohyal and medially to the mandible, then ventral to this cartilage, to finally pass medial to the ceratohyal more anteriorly. Then, the afferent mandibular artery marks a U-turn and runs posteriorly, parallel to the first afferent branchial artery (ama, aba, Figs. 3a,c).

The subclavian arteries (sca, Figs. 3e,f) emerge from the dorsal aorta (da, Figs. 3a,c). They enter the posterior internal surface of the scapulocoracoids (sc, Fig. 3d) at the base of the pectoral symphysis. Then they give rise to several branches: one that runs ventrally to exit the posteroventrolateral surface of the pectoral symphysis; one that runs anterolaterally to exit the anteroventrolateral surface of the pectoral symphysis; and one running forward and exiting medially the dorsal surface of the pectoral symphysis (scaf, Fig. 3g). The dorsal aorta divides into paired lateral dorsal aortae (lda, Fig. 3c) posterior to the braincase (br, Fig. 3d) and to the rear extremity of the copula (co, Fig. 3b). The lateral dorsal aortae then give rise to the efferent branchial arteries (eba1–4, Fig. 3c). The efferent branchial arteries 3 and 4 share a common root, and anteriorly, each efferent branchial artery branches off directly and successively from the lateral dorsal aortae. The fifth branchial arch is devoid of an efferent artery. The efferent branchial artery 4 passes through the last pharyngobranchial (pb1–5, Fig. 3b), which is formed from the fusion of pharyngobranchials 3, 4 and 5 and epibranchials 4 and 5. Each efferent branchial artery passes ventral to the posterior blade of the corresponding pharyngobranchial and then into a groove dorsal to the proximal extremity of that pharyngobranchial. Then, each efferent branchial artery runs ventrally along the posterior edge of the corresponding epibranchial (eb1–3, Fig. 3b) and ceratobranchial. Ventrally, the artery follows the posteromedial surface of the ceratobranchial. The efferent hyoid artery (eha, Fig. 3c), which also branches off from the lateral dorsal aorta, has a different pathway relative to the pharyngohyal than the efferent branchial arteries and corresponding pharyngobranchials: the efferent hyoid artery passes exclusively ventrally to the pharyngohyal (ph, Fig. 3b). Then, as for the efferent branchial arteries, it runs ventrally along the posterior edge of the epihyal and ceratohyal, and passes medial to the opercular cartilage. The lateral dorsal aorta extends forward in what we interpret to be a rostral artery (ra, Figs. 3c,e). The lateral dorsal aorta lies medial to the jaw joint, but lateral to the palatoquadrate part of the braincase anteriorly. The orbital artery branches posterior to the origin of the efferent hyoid artery (ora, Figs. 3c–e). The orbital artery runs anterodorsally to enter the basicranium and exit in the orbit. Ventral to the basicranium, and posterior to the bucco-hypophyseal pit, is an arterial plexus from which emerge two small internal carotids (ic, Fig. 3c) that fail to complete development during ontogeny so that they don't enter the braincase at the level of the bucco-hypophyseal pit, which is closed (see details of this structure in Khonsari et al., 2013). The blood supply to the brain is ensured by the efferent pseudobranchial artery alone, which passes through the basicranium medial to the mandibular articulation. The efferent pseudobranchial artery joins the cerebral arterial circuit inside the endocranial cavity. The cerebral arterial circuit is composed of a medial myelonal artery (mya, Fig. 3c) and paired posterior arteries (pca, Fig. 3c) forming a bell-shaped circuit. At the level of the optic foramen (fII, Fig. 3d), the posterior cerebral artery meets the ophthalmic artery (opha, Figs. 3d,e), the efferent pseudobranchial artery (epa, Figs. 3c,e), and the anterior cerebral foramen (aca, Figs. 3c,e). The ophthalmic artery exits the braincase laterally through a foramen located immediately posterior to the optic nerve foramen.

Discussion

The comparative anatomy discussion is based on recent published phylogenetic studies (e.g., Coates et al. 2018, Dearden et al. 2019, Frey et al. 2019) that resolved all acanthodians as stem-chondrichthyans, osteichthyans as sister group to chondrichthyans, and placoderms as a stem-gnathostomes grade. Symmoriiformes are either resolved as stem-chondrichthyans (Ginter et al. 2010, Pradel et al. 2011), or more recently as stem-holocephalans (Coates et al. 2018, Dearden et al. 2019, Frey et al. 2019); and xenacanths are considered as either stem-chondrichthyans (Ginter et al. 2010, Pradel et al. 2011), or stem-elasmobranchs (Coates et al. 2018, Frey et al. 2019).

Reconstruction of the branchial skeleton. Here, we present a composite reconstruction based on two different specimens from the same locality, which belong to the same species (Pradel 2010) (Fig. 4). Some of the branchial skeleton elements are preserved in both specimens, and some of them are preserved in almost natural position, so that the composite reconstruction is facilitated. Mandible, basihyoid and basibranchial are preserved in almost natural position in KUNHM 22060 (md, bh, bb, Fig. 1a). The intercoracoid element is close to the ventral process of the scapulocoracoids in KUNHM 21894 (intc, Fig 2a). Here the intercoracoid element is interpreted as flooring the heart (see below). In modern chondrichthyans, the basibranchial copula roofs the heart, while the scapulocoracoid symphysis provides a floor for it. Here it is assumed that the same arrangement is present in *Iniopera*, so that the anterior process of the intercoracoid cartilage (intc, Fig. 4a) meets the ventral process of the basibranchial (bb, Fig. 4a), while the dorsal, posterior plate of the basibranchial overhangs the intercoracoid cartilage to roof the heart. Form and dimension of both the anterior process of the intercoracoid cartilage and the ventral process of the basibranchial match when they are articulated together.

Zangerl & Case (1973) reconstructed the ceratohyals of *Iniopera* with the largest, flared part being distal, whereas Zangerl (1981) assumed that this part of the ceratohyals is articulated directly to a basi-element. Comparison with modern chimaeroids suggest that the enlarged part of the ceratohyals of *Iniopera* (ch, Fig. 4b) is distal, the dorsal small process being maybe articulated to an epihyal not preserved here. The other, ventral extremity of the ceratohyals matches the posterolateral corner of the basihyal (bh, Fig. 4a).

In KUNHM 21894, two of the left ceratobranchials are preserved in natural position, the smallest one being anterior to the biggest one. Consequently, it is assumed that the ceratobranchials slightly increase in size posteriorly. Another ceratobranchial, bigger than the two preserved in natural position, was retrieved on the right side of the specimen and is interpreted as being posterior to the other preserved ceratobranchials. The proximal end of the ceratobranchials (cb1–3, Fig. 4a) might have been close to the lateral edge of the basibranchial.

There is no obvious evidence of hypobranchials in the two specimens studied here, and one may suggest that the hypobranchials were poorly mineralized. No obvious epibranchials were also retrieved in the specimens studied here. In the Iniopterygidae, the ceratobranchials are the most prominent elements of the branchial arches, the epibranchials being thinner (Grogan & Lund 2009). In the extant chimaeroids, the epibranchials are also reduced compared to the ceratobranchials. It is possible that only the most prominent elements, i.e., the ceratobranchials, of *Iniopera* were preserved, and that the thinner elements were not well mineralized and were not preserved during the fossilization. In some placoderms, such as *Paraplesiobatis*, epibranchials are also absent (Brazeau et al. 2017), probably because they were not mineralized. Another interpretation would be that the small, scattered elements located in the posterolatero-ventral depressions of the braincase are actually part of hypo-, epi-, and/or pharyngobranchials. These elements might also represent bits of hyoid and branchials rays. Given that many branchial skeleton cartilages are not fossilized, probably because they were unmineralized, it is more likely that posteriormost branchial arches were lost during fossilization, rather than that *Iniopera* only had three branchial arches.

Intercoracoid element. In Zangerl's description of *Iniopera*, the unpaired plate that possesses paired foramina is interpreted as belonging to the basibranchial series. However, as Pradel et al. (2010) already noted, this element bridges the anterior extremity of the coracoid part of the scapulocoracoids, and because of its anterior articulation with the ventral process of the basibranchial it is located more ventral than the dorsal surface of the latter element. This location and the presence of a concave dorsal surface (Figs. 1h, 2e) suggests that the intercoracoid element was ventral to and accommodated the heart, as does the pectoral symphysis in modern chondrichthyans. The comparison with the symphyseal area of the modern ratfish *Callorhinchus*, which possesses a pair of foramina for the subclavian arteries (scaf, Fig. 3g), suggests that these paired foramina of the intercoracoid element of *Iniopera* might be for the passage of the subclavian arteries (scaf, Figs. 1h,i, 2e,f).

Such an unpaired element bridging ventrally the two scapulocoracoids are present in some fossil crown-holocephalans (e.g., *Ischyodus*, Heimberg, 1949) and some putative stem-holocephalans (the other iniopterygians *Sibyrhynchus* and *Promexyele* Zangerl & Case, 1973; the eugeneodontid *Ornithoprion* Zangerl 1966). In the Iniopterygidae *Rainerichthys*, Grogan and Lund (2009) described an unpaired, posterior element that possessed paired foramina anterior to its posterolateral articular surface, and which articulated with the branchial skeleton anteriorly, as does the intercoracoid element of *Iniopera*. However, contrary to *Iniopera*, this element in *Rainerichthys* does not appear to articulate directly to the scapulocoracoids, but via paired, intermediary cartilages. Nonetheless, the comparison with *Iniopera* suggests that this element might not be a basibranchial as interpreted by Grogan & Lund (2009). A triangular, unpaired "cardiobranchial" has been reported in the stem-holocephalans *Debeerius* (Grogan & Lund 2004) and *Harpacanthus* (Lund & Grogan 2004). In both species, its relationships either to the basibranchial series, as a posterior copula, or to the scapulocoracoid, as an intercoracoid element, is unclear. However, its enlarged region is posterior, whereas its pointed extremity is anterior, which differs from any chondrichthyan copula shape (e.g., co, Figs. 3b, 5b), but is similar to the *Ischyodus* and *Iniopera* intercoracoid morphology.

Besides holocephalans, some modern hexanchiformes (e.g., *Notorynchus*, *Hexanchus*) possess an intercoracoid element (called osmosternum; Parker 1891, Goodrich 1930).

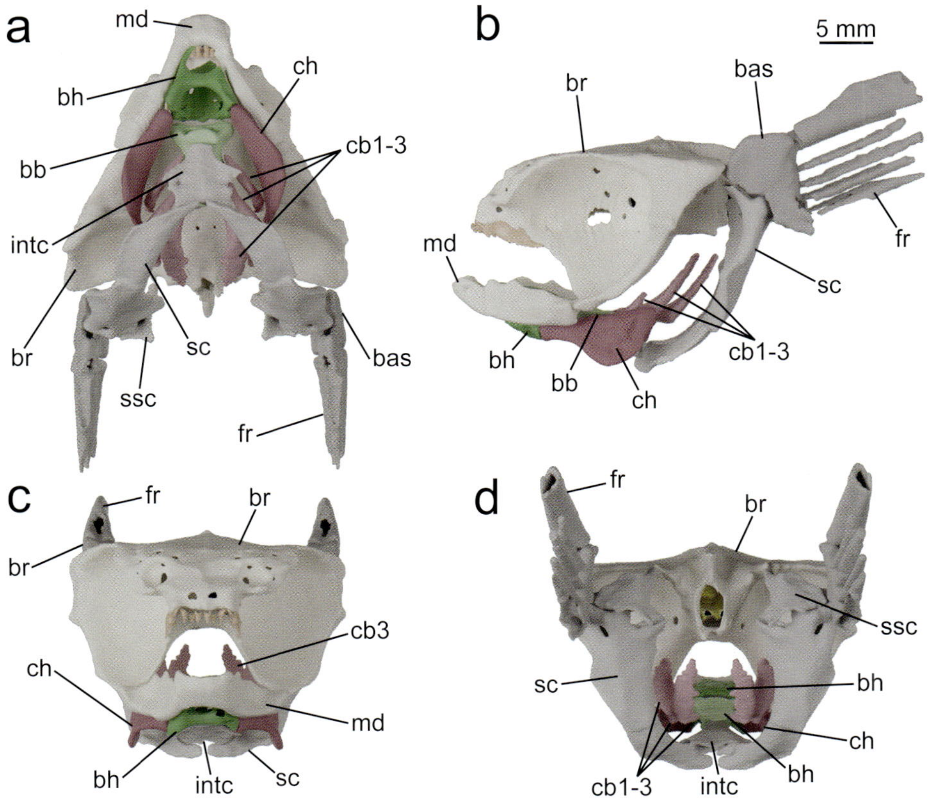

Fig. 4. Hypothetical composite reconstruction of the braincase, branchial and pectoral girdle cartilages in inferred natural position of *Iniopera*, based on the two specimens studied here (KUNHM 22060 and KUNHM 21894). **a.** Ventral view, anterior to the top. **b.** left lateral view. **c.** Anterior view. **d.** Posterior view. Same abbreviations and colors as Figures 1, 2 and 3.

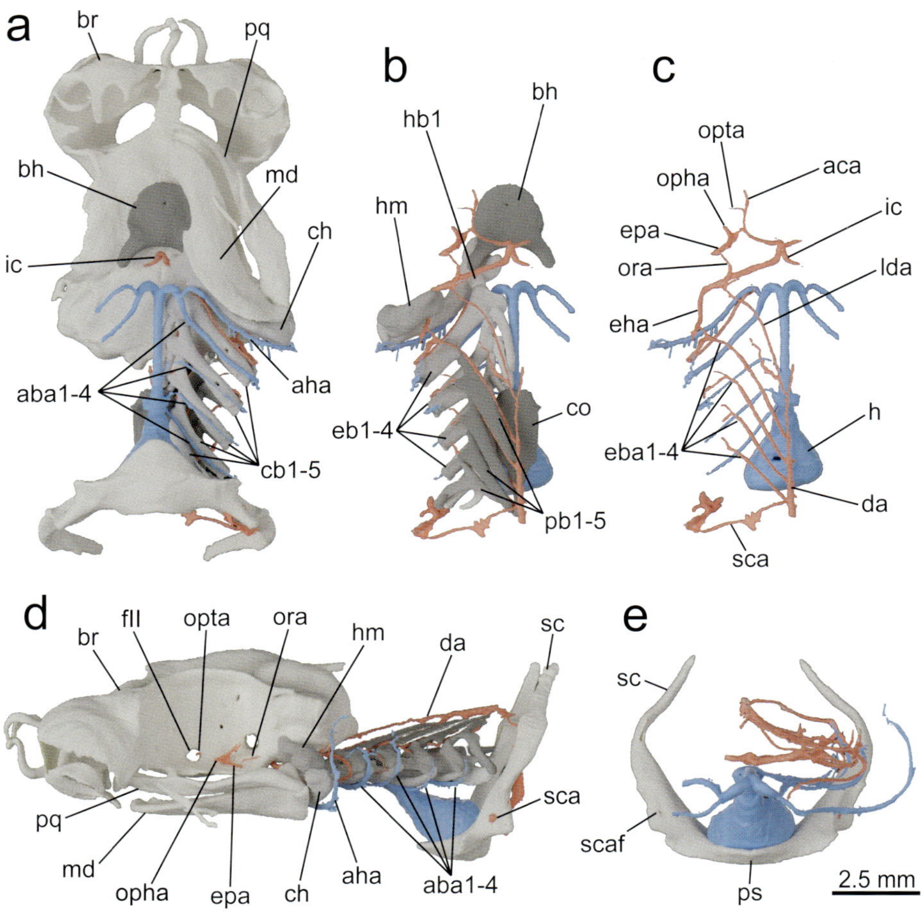

Fig. 5. Surface rendering generated from Synchrotron Radiation microtomographic slices of the braincase, mandible, branchial skeleton and pectoral girdle cartilages, and vascular system of *Scyliorhinus canicula*. **a.** Ventral view. **b.** Dorsal view, braincase, mandible and pectoral girdle removed. **c.** Same as b with the branchial skeleton removed. **d.** Right lateral view. **e.** Pectoral girdle and vascular system in posterior view. Same abbreviations and colors as in Figures 1, 2, 3, 4. Additional abbreviations: opta, optic artery; pq, palatoquadrate.

However, the shape is different from that of the iniopterygians, *Ornithoprion*, and *Ischyodus*, and there is no evidence of paired foramina.

An unpaired element, presumably of endoskeletal origin, is present between the paired coracoid processes of the pectoral girdle of the Devonian stem-chondrichthyan *Doliodus* (Maisey et al. 2019), and a lorical plate possessing a base that might be made of perichondral bone is present in some "acanthodians" such as *Climatius* (Burrow et al. 2015). In all these forms, there is no evidence of paired foramina.

Paired elements, defined as procoracoid cartilages, that articulate the ventral end of the scapulocoracoids, are present in some Palaeozoic chondrichthyans, such as symmoriiformes (Coates & Sequeira 2001), caseodontids (Zangerl 1981) and xenacanths (Heidtke & Schwind 2004). Procoracoids are also present in some "acanthodians" (Miles 1973).

These observations suggest that the pectoral girdle of chondrichthyans was primitively separated ventrally by either unpaired intercoracoid or paired procoracoid elements. In crown-chondrichthyans, the scapulocoracoids would have fused at the symphysis independently in elasmobranchs and holocephalans since intercoracoid elements are present in stem-holocephalans such as iniopterygians and in the elasmobranch hexanchiformes, and since procoracoids are present in stem-chondrichthyans, or stem-holocephalans and stem-elasmobranchs, depending on the phylogenetic position of symmoriiformes and xenacanths respectively (see above).

In elasmobranchs (e.g., *Scyliorhinus*; ps, Fig. 5e) foraminae for the subclavian arteries are absent from the scapulocoracoid symphysis, and are located more dorsally and laterally on the scapular part of the pectoral girdle. In Palaeozoic chondrichthyans in which there are either procoracoids (e.g., symmoriiformes, xenacanths) or a medial unpaired intercoracoid plate (e.g., the stem-chondrichthyan *Doliodus*), there is also no evidence of these paired foramina, contrary to *Iniopera*. If symmoriiformes and xenacanths are stem-chondrichthyans (Pradel et al. 2011), possessing paired foramina for the subclavian artery in the pectoral girdle symphysis, whether the symphysis is represented by a medial intercoracoid element or by the ventral fusion of the scapulocoracoids, might be an holocephalan characteristic. However, further investigations are needed to check whether these foramina are present in the intercoracoid elements of stem and crown-holocephalans (e.g., *Ischyodus*). In the hypothesis of resolving the symmoriiformes as stem-holocephalans (Coates et al. 2018), such a feature might only characterize more advanced holocephalans. It would be worth checking the presence of such small foramina in Symmoriiformes by means of fine µCT Scanning of the pectoral girdle.

It is noteworthy that the shape of the symphyseal region of the pectoral girdle of the modern chimaeroids is very different from that of elasmobranchs. In *Callorhinchus*, the pectoral symphysis has a complex shape, forming a keel ventrally (Figs. 3d,f–h), while in *Scyliorhinus*, the symphysis is rather simple (Figs. 5d,e). One may suggest that the keel in modern holocephalans represents an intercoracoid element possessing paired foramina for the subclavian artery, which is fused to the scapulocoracoid during ontogeny. It has been also suggested (Zangerl & Case 1976) that the "sternum" of chondrichthyans was primitively paired (as procoracoids), and that the unpaired "sternum" of hexanchiformes (and *Iniopera*?) is the result of a secondary fusion of the originally paired elements. Further studies on the development of the pectoral girdle in both modern holocephalans and elasmobranchs are needed to identify whether intercoracoid elements are present in early ontogenetic stages.

Ventilation model. Unlike extant elasmobranchs, Recent holocephalans are thought to use a very low-pressure system for driving flow through the gill arches (Dean et al. 2012, Howard et al. 2013). This system is related to the holostylic jaw suspension, in which the jaw and hyoid movements are constrained. To compensate for these constraints, it has been inferred that an "accordion" model involving a rostrocaudal movement of the visceral arches, especially the continuous series of basibranchials, drives buccal pressure differentials via compression of parabranchial cavity by the opercular musculature (Dean et al. 2012). In this model, it has been suggested that the pectoral girdle might contribute to ventilation by helping the rostrocaudal movement of the visceral arches. *Iniopera* is consistent with this model since it possesses a holostylic jaw suspension (and a lateral process in the articular end of the mandible that would probably constrain lateral movement; Pradel 2010), a continuous basibranchial series, and maybe a fleshy operculum supported by hyoid rays (see below). Moreover, the strong connection between *Iniopera*'s pectoral girdle and basibranchial (Fig. 4) provide a solid argument in favor of a pectoral girdle contribution to an "accordion" ventilation model that would have appeared early in the holocephalan lineage. These observations on both fossil and Recent holocephalans provide additional evidence of multifunctional (respiration, feeding, locomotion) pectoral girdle in chondrichthyans (e.g., Camp et al. 2017).

The articulation of the ceratohyals and opercular. The branchial basket is nested below the braincase in *Iniopera*, as in modern chimaeroids (Figs. 3, 4). It also appears short rostrocaudally. No opercular cartilage is preserved in either studied specimen. Nevertheless the pits in the enlarged portion of the ceratohyals (hrp, Fig. 1a) may correspond to hyoid ray insertions. These hyoid rays might have supported a fleshy operculum that was not fossilized. Hyoid rays are attached to the ceratohyals in the iniopterygians *Rainerichthys* and *Papilionichthys* (Grogan & Lund, 2009), and long hyoid rays supporting an opercular flap were probably present in the Iniopterygidae *Iniopteryx* and *Promexyele* (Zangerl & Case, 1973). Many stem-holocephalans such as *Debeerius* (Grogan & Lund, 2000) also possess hyoid rays attached to the ceratohyals, supporting an opercular flap. In extant chimaeroids, in addition to the opercular cartilage that bears hyoid rays, some rays are also attached to the posterior edge of the ceratohyal and support the fleshy operculum (Didier 1995). Septate gills were probably a plesiomorphic condition in the chondrichthyan crown-group, since this feature is present in stem-elasmobranchs and stem-chondrichthyans (e.g., *Gladbachus*, Coates et al. 2018; the acanthodian *Brochoadmones*, Hanke & Wilson 2006). A single branchial opening would be apomorphic for holocephalans, though early potential members of this clade, such as the Symmoriiformes do not possess an operculum (e.g., *Ozarcus*, Pradel et al. 2014; *Akmonistion*, Coates & Sequeira 2001; *Cladoselache*, Maisey 1989).

The arterial pattern. In extant chimaeroids, illustrated here by *Callorhinchus milii*, the ventral aorta divides into paired afferent hyoid arteries ventral to the first basibranchial (va, aha, bb, Figs. 3a,b). Regarding the morphology of the basibranchial of *Iniopera*, it is inferred that the ventral aorta entered the basibranchial posteriorly via a large opening (bbpf, Fig. 1), then it divided into paired afferent hyoid arteries inside the basibranchial, and the afferent hyoid arteries left the basibranchial anteriorly via paired anterior openings (bbaf, Fig. 1). Such division of the ventral aorta inside a basibranchial is unknown in other known fossil gnathostomes. In elasmobranchs, illustrated by *Scyliorhinus*, there is a gap between the basihyal and posterior copula (bh, co, Fig. 5b), so that there is no basibranchial above the division of the ventral aorta into paired afferent hyoid arteries.

In chimaeroids, the division of the dorsal aorta into paired lateral dorsal aortae occurs posterior to the braincase, and then each lateral aorta gives rise to successive efferent branchial arteries (da, lda, eba, Fig. 3c). In *Scyliorhinus*, the dorsal aorta gives rise to each efferent branchial artery (da, eba, Fig. 5c). Then it divides into lateral dorsal aorta posterior to the occiput (lda, Fig. 5c). In Palaeozoic chondrichthyans, the medial dorsal aorta divides into lateral dorsal aortae either posterior to the occiput (e.g., *Doliodus*, xenacanths), as in modern chondrichthyans, or underneath the braincase (e.g., Symmoriiformes). The first, modern chondrichthyan, condition is also present in sarcopterygians (e.g., *Ptyctolepis*, Lu et al. 2017) and stem-gnathostomes (e.g., *Janusiscus*, Giles et al. 2015a; *Buchanosteus*, Young 1979), whereas the second condition is present in early actinopterygians (e.g., *Lawrenciella*, Pradel et al. 2016). In *Acanthodes*, Miles (1973) reconstructed part of the vascular system as showing a median dorsal aorta running into a groove in the ventral surface of the occipital ossification, from which emerged paired efferent branchial arteries, possibly the second ones, also running into divergent grooves on the occipital ossification (Miles 1973, fig. 5). Another interpretation would be that these grooves housed the lateral dorsal aortae. Given these observations, an anterior splitting of the dorsal aorta, below the braincase, might be homoplastic in actinopterygians, some acanthodians such as *Acanthodes*, and Symmoriiformes and it is apparently not related to the anteroposterior position of the branchial skeleton relative to the braincase.

In Palaeozoic chondrichthyans in which the dorsal aorta splits posterior to the occiput (as in extant chondrichthyans) and the lateral dorsal aortae are embedded in the basicranium, the branchial skeleton is mostly posterior to the braincase, contrary to the condition in modern holocephalans. In addition, there are no foraminae for the efferent branchial arteries on the basicranium (e.g., xenacanths, *Tamiobatis*, Schaeffer 1981; *Doliodus*, Maisey et al. 2009). Consequently, it is assumed that the efferent branchial arteries do not branch off from the embedded lateral dorsal aortae, but from the medial dorsal aorta, as in elasmobranchs. The condition is thus different from that of chimaeroids. It has been suggested that the anteriormost efferent branchial arteries branch off the lateral dorsal aortae in *Pucapampella* (Maisey 2001). The pucapampellid *Gydoselache* displays a first ceratobranchial, the posterior extremity of which is located anterior to the occipital arch of the braincase, approximately at the same level as the anterior foramina for the lateral dorsal aortae (Maisey et al. 2019). Consequently, the anteriormost efferent branchial artery likely branches off the lateral dorsal aorta in *Gydoselache*.

In Palaeozoic euchondrichthyans in which the dorsal aorta splits into lateral dorsal aortae anterior to the occiput, such as in the Symmoriiformes, the branchial skeleton is mostly posterior to the braincase (e.g., *Ozarcus*, Pradel et al. 2014), and there is no evidence of foramina or grooves for the efferent branchial arteries on the basicranium. Consequently, it is assumed that the efferent branchial arteries also branch off from the medial dorsal aorta, as in elasmobranchs. In acanthodians, information about the branchial skeleton and neurocranium morphology are scarce. The branchial skeleton likely extended posterior to the braincase in *Acanthodes* (Davis et al. 2012), as in elasmobranchs, whereas *Ptomacanthus* displays a branchial skeleton that does not extend far posterior to the braincase (Dearden et al. 2019). In *Acanthodes*, in both vascular system interpretations (see above), the efferent branchial arteries likely branched off the medial dorsal aorta rather than off the lateral dorsal aortae.

In osteichthyans, the branchial skeleton mostly lies beneath the braincase, but two patterns of efferent branchial arteries are present. In the sarcopterygian *Latimeria*, the efferent branchial arteries emerge from the lateral dorsal aortae, as in modern chimaeroids (Millot et al. 1978), while in both Palaeozoic and modern actinopterygians, the anteriormost efferent branchial arteries branch off the lateral dorsal aortae whereas the posteriormost emerge from the medial dorsal aorta (e.g., *Kansasiella*, Poplin 1974; *Amia*, Allis 1897). From these observations, it appears that when the branchial skeleton is subcranial, at least the first efferent branchial arteries emerge from the paired lateral dorsal aortae, the extreme condition being represented by the holocephalans, in which all the efferent branchial arteries emerge from the lateral dorsal aortae.

Continuous basibranchial series and basihyal distribution. A succession of unpaired basibranchials flooring the pharynx is present in living chondrichthyans. However, a continuous series of basibranchials between the anterior basihyal and the posterior copula is only present in holocephalans (bh, bb, co, Fig. 3b), whereas in modern elasmobranchs there are no basibranchials between the basihyal and the copula (bh, co, Fig. 5b). Stem-chondrichthyans, such as *Gladbachus* (Coates et al. 2018), and possible stem-elasmobranchs (e.g., *Triodus*, Heidtke et al. 2004) show the modern elasmobranch condition. Stem-holocephalans, such as *Iniopera* and *Debeerius* (Grogan & Lund 2000), as well as the symmoriiform *Ozarcus* (Pradel et al. 2014), possess the modern holocephalan condition. Noteworthy is the absence of basibranchials posterior to the anteriormost basi-element, or basihyal, which is articulated to the hypo- or cerato-element of the hyoid arch, in "acanthodians" (e.g., *Ptomacanthus*, Dearden et al. 2019; *Acanthodes*, Gardiner 1984, Miles 1973) and placoderm stem-gnathostomes (e.g., *Cowralepis*, Carr et al. 2009; *Paraplesiobatis*, Brazeau et al. 2017). The condition in osteichthyans is again different, since a single basi-element to which articulate the ventral elements of the hyoid and posterior branchial arches is present in early forms (e.g., *Raynerius*, Giles et al. 2015b; *Glyptolepis*, Jarvik 1972). In some other early osteichthyans, such as *Pteroniscul us* (Nielsen 1942), and *Eusthenopteron* (Jarvik 1980), the basi-element is segmented in basibranchials and recalls the ontogenetic formation of the basibranchials in modern teleosts (Nelson 1969). This suggests that gnathostomes primitively lack basibranchials posterior to the basihyal, as already noted by Dearden et al. (2019), and it is plausible that a single basibranchial element appeared independently in osteichthyans and chondrichthyans crownward to the acanthodians grade. A subdivision of this single basibranchial into numerous basibranchials would also have occurred independently in some osteichthyans and crown-chondrichthyans, with a complete, continuous series between the anterior basihyal and the posterior copula present independently in some osteichthyans and holocephalans. In elasmobranchs in which there is a first basibranchial in front of the copula (e.g., *Squalus acanthias*, El-Toubi 1952), there is still a pronounced gap between this first basibranchial and the basihyal, so that the condition is different from modern chimaeroids and the symmoriiform *Ozarcus*. It is noteworthy that, as in teleosts, the subdivision of basibranchial and copula occur during ontogeny from a single embryonic plate in modern elasmobranchs (e.g., *Squalus acanthias*, El-Toubi 1952). Nelson (1969) suggested that the gap between the basihyal and the copula in most elasmobranchs is the consequence of a secondary loss of basibranchials. Since the stem-chondrichthyan *Gladbachus* (Coates et al. 2018), and the stem-elasmobranch (or stem-chondrichthyan) *Triodus* (Heidtke et al. 2004) also don't possess basibranchials between the basihyal and the copula, it is also possible that the copula without any anterior subdivision in basibranchials was primitively present in crown-chondrichthyans and the chimaeroid condition is derived.

A separated basihyal element bridging the hypo- or ceratohyals is present in the stem-gnathostomes placoderms, acanthodians, chondrichthyans, and modern teleosts, whereas a basihyal is absent in *Polypterus* (Jollie 1984), gars and *Amia* (Jarvik 1980). The branchial skeleton of the latter species recalls that of *Raynerius* and *Glyptolepis* and it is assumed here that there is no separate basihyal in these early osteichthyans, but a single basibranchial copula into which the hyoid and branchial arches articulate. A separated basihyal element is also absent in the modern sarcopterygian *Latimeria* (Millot & Anthony 1958), but is present in the dipnoan *Neoceratodus* (Nelson 1969) and in more advanced sarcopterygians, such as larval apodans, lizards and *Sphenodon* (De Beer 1937). It has been suggested that the "sublingual rod" of *Eusthenopteron* is homologous to the basihyal (Nelson 1969), although it might rather be an urohyal linking the hyoid arch and shoulder girdle (Jarvik 1963, Patterson 1973). From these observations, we agree with Gardiner (1984) that the basihyal has arisen in osteichthyans independently on multiple lineages (teleosts, dipnoans and tetrapods). Consequently, the basihyal of stem-gnathostomes and chondrichthyans is not homologous to that present in some osteichthyans. Embryonic observations provide some support for the different evolutionary history of this element in osteichthyans and in other gnathostomes. In modern elasmobranchs, the basihyal occurred independently from the posterior procartilaginous plate, which gives rise to the hypobranchials of the branchial arches (except the first branchial arch) and to the copula (plus the basibranchial in front in *Squalus acanthias;* El-Toubi 1952) (e.g., *Etmopterus*, Holmgren 1940). Nelson (1969) suggested that the basihyals of elasmobranchs might have been formed from the fusion of hypohyals and, in some forms, the first hypobranchials. In *Squalus acanthias* early embryos, the first hypobranchials belong to a band of non-chondrified tissue along with the future basihyal, whereas the subsequent hypobranchials are connected to the copula (El-Toubi 1952). Later during ontogeny of *Squalus acanthias*, the band of tissue connecting the first ceratobranchials to the basihyal fails to chondrify, so that there is no first hypobranchials in the adult. In *Scyliorhinus*, the first basibranchials are present in the adult and are the only hypobranchials to be anteriorly directed toward the basihyal (hb1, bh, Fig. 5b), so that same embryonic relationships between the basihyal and the first branchial arch likely occurs. Stem-chondrichthyans, such as *Acanthodes* (Miles 1973), *Ptomacanthus* (Dearden et al. 2019) and *Gladbachus* (Coates et al. 2018) might show the same process of basihyal formation, since in these forms, neither hypohyals nor first hypobranchials are present but the ceratohyals and first ceratobranchials connect to the basihyal. Further ontogenetic studies on modern chimaeroids are needed to determine the relationships between the possible embryonic hypohyals (not present in the adults) and the first hypobranchials.

In some osteichthyans, the basihyal is formed from the anterior prolongation of the basibranchial (e.g., *Syngnathus;* Kindred 1921, 1924 in Gardiner 1973; Nelson 1969), and the subdivision into numerous basibranchials from a single element occurs secondarily (Nelson 1969). Noteworthily in teleosts, the basihyal is present along with hypohyals and first hypobranchials, so that the basihyal cannot be formed by fusion of these elements.

Acknowledgments

We deeply thank John Maisey (AMNH, New York, USA) for the valuable discussions about Palaeozoic and modern chondrichthyans, and the friendship all along these years. We thank Florent Goussard (CR2P, MNHN, Paris, France), who assemble the final composite specimen based on the two studied fossils of *Iniopera;* and Paul Tafforeau (ESRF, Grenoble, France) who scanned the specimens. Desui Miao and Larry Martin (Curator in charge, KUNHM, Lawrence, Kansas, USA) provided the fossil mate-rial studied here.

Didier Casane (CNRS, Gif-Sur-Yvette, France) and Dominique Didier (Millersville University, Millersville, USA) provided respectively the *Scyliorhinus* and *Callorhinchus* specimens.

Funding

This work was supported by the CR2P (MNHN), the Paris Ile-de-France Region – DIM "Matériaux anciens et patrimoniaux" – DIM PHARE projet, H. R. & E. Axelrod Research Chair in paleoichthyology at the AMNH, the ESRF (beamline ID19; proposal EC361), the National Science Foundation Assembling the Tree Of Life (ATOL) 1132229, the National Science Foundation Award No. 1036488 ("Collaborative Research: Jaws and Backbone: Chondrichthyan Phylogeny and a Spine for the Vertebrate Tree of Life"), the Herbert and Evelyn Research Chair in Vertebrate Paleontology at the AMNH.

Literature cited

Allis, E. P. 1897. The cranial muscles and cranial and first spinal nerves in *Amia calva*. Journal of Morphology, 12: 487–808.

Allis, E. P. 1912. The branchial, pseudo branchial and carotid arteries in *Chimaera (Hydrolagus) colliei*. Anatomischer Anzeiger, 42: 10–18.

Brazeau, M. D., M. Friedman, A. Jerve & R. C. Atwood. 2017. A three-dimensional placoderm (stem-group gnathostome) pharyngeal skeleton and its implications for primitive gnathostome pharyngeal architecture. Journal of Morphology, 278: 1220–1228.

Burrow, C. J., R. G. Davidson, J. L. Den Blaauwen & M. J. Newman. 2015. Revision of *Climatius reticulatus* Agassiz, 1844 (Acanthodii, Climatiidae), from the Lower Devonian of Scotland, based on new histological and morphological data. Journal of Vertebrate Paleontology, 35: e913421.

Camp, A. L., B. Scott, E. L. Brainerd & C. D Wilga. 2017. Dual function of the pectoral girdle for feeding and locomotion in white-spotted bamboo sharks. Proceedings of the Royal Society B, 284: 20170847.

Carr, R. K., Z. Johanson & A. Ritchie. 2009. The phyllolepid placoderm *Cowralepis mclachlani*: insights into the evolution of feeding mechanisms in jawed vertebrates. Journal of Morphology, 270: 775–804.

Coates, M. I. & S. E. K. Sequeira. 2001. A new stethacanthid chondrichthyan from the Lower Carboniferous of Bearsden, Scotland. Journal of Vertebrate Paleontology, 21: 438–459.

Coates, M. I., J. A. Finarelli, I. J. Sansom, P. S. Andreev, K. E. Criswell, K. Tietjen, M. L. Rivers & P. J. La Riviere. 2018. An early chondrichthyan and the evolutionary assembly of a shark body plan. Proceedings of the Royal Society B, 285: 20172418.

Davis, S. P., J. A. Finarelli & M. I. Coates. 2012. *Acanthodes* and shark-like conditions in the last common ancestor of modern gnathostomes. Nature, 486: 247–250.

Dean, M. N., A. P. Summers & L. A. Ferry. 2012. Very low pressures drive ventilatory flow in chimaeroid fishes. Journal of morphology, 273: 461–479.

Dearden, R. P., C. Stockey & M. D. Brazeau. 2019. The pharynx of the stem-chondrichthyan *Ptomacanthus* and the early evolution of the gnathostome gill skeleton. Nature Communications, 10: 2050.

De Beer, G. R. 1937. The development of the vertebrate skull. Clarendon Press, Oxford, UK, 552 pp.

Didier, D.A. 1995. Phylogenetic systematics of extant chimaeroid fishes (Holocephali, Chimaeroidei). American Museum Novitates, 3119: 1–86.

Didier, D. A., E. E. Leclair & D. R. Vanbuskirk. 1998. Embryonic Staging and External Features of Development of the Chimaeroid Fish, *Callorhinchus milii* (Holocephali, Callorhinchidae). Journal of Morphology, 236: 25–47.

El-Toubi, M. R. 1952. The development of the chondrocranium of the spiny dogfish, *Acanthias vulgaris* (*Squalus acanthias*). Part II: Branchial arches and extravisceral cartilages. Journal of Morphology, 90: 33–64.

Frey, L., M. Coates, M. Ginter, V. Hairapetian, M. Rücklin, I. Jerjen & C. Klug. 2019. The early elasmobranch *Phoebodus*: phylogenetic relationships, ecomorphology and a new time-scale for shark evolution. Proceedings of the Royal Society B, 286: 20191336.

Gardiner, B. G. 1973. Interrelationships of teleostomes. In Interrelationships of fishes (Vol. 53). Academic Press, London, UK, 35 pp.

Gardiner, B. G. 1984. The relationships of the palaeoniscid fishes, a review based on new specimens of *Mimia* and *Moythomasia* from the Upper Devonian of Western Australia. Bulletin of the British Museum (Natural History), Geology Series, 37: 173–428.

Giles, S., M. Friedman & M. D. Brazeau. 2015a. Osteichthyan-like cranial conditions in an Early Devonian stem gnathostome. Nature, 520: 82–85.

Giles, S., L. Darras, G. Clément, A. Blieck & M. Friedman. 2015b. An exceptionally preserved Late Devonian actinopterygian provides a new model for primitive cranial anatomy in ray-finned fishes. Proceedings of the Royal Society B, 282: 20151485.

Ginter, M., O. Hampe & C. Duffin. 2010. Chondrichthyes. Paleozoic Elasmobranchii: Teeth. – In: Schultze H.-P., ed. Handbook of paleoichthyology, vol. 3D. Stuttgart: Gustav Fischer Verlag, 168 pp.

Goodrich, E. S. 1930. Studies of the Structure and Development of Vertebrates. Macmillan, London, UK, 837 pp.

Grogan, E. D. & R. Lund. 2000. *Debeerius ellefseni* (fam. nov., gen. nov., spec. nov.), an autodiastylic chondrichthyan from the Mississippian Bear Gulch Limestone of Montana (USA), the relationships of the Chondrichthyes, and comments on gnathostome evolution. Journal of Morphology, 243: 219–245.

Grogan, E. D. & R. Lund. 2009. Two new iniopterygians (Chondrichthyes) from the Mississippian (Serpukhovian) Bear Gulch Limestone of Montana with evidence of a new form of chondrichthyan neurocranium. Acta Zoologica, 90: 134–151.

Hamel, M. H. & C. Poplin. 2008. The braincase anatomy of *Lawrenciella schaefferi*, actinopterygian from the Upper Carboniferous of Kansas (USA). Journal of Vertebrate Paleontology, 28: 989–1006.

Hanke, G. F. & M. V. Wilson. 2006. Anatomy of the Early Devonian acanthodian *Brochoadmones milesi* based on nearly complete body fossils, with comments on the evolution and development of paired fins. Journal of Vertebrate Paleontology, 26: 526–537.

Heidtke, U. H. J., C. & Schwind. 2004. On the organization of the skeleton of the genus *Xenacanthus* (Elasmobranchii: Xenacanthida) from the Lower Permian of the Saar-Nahe Basin (SW Germany). Neues Jahrbuch fur Geologie und Palaontologie-Abhandlungen, 231: 85–117.

Heidtke, U. H. J., C. Schwind & K. Krätschmer. 2004. Über die Organisation des Skelettes und die verwandschaftlichen Beziehungen der Gattung *Triodus* Jordan 1849 (Elasmobranchii: Xenacanthida). Mainzer geowissenschaftliche Mitteilungen, 32: 9–54.

Heimberg, G. 1949. Neue Fischfunde aus dem Weissen Jura von Württemberg. Palaeontographica A, 97: 75–98.

Holmgren, N. 1940. Studies on the head in fishes. Part I. Development of the skull in sharks and rays. Acta Zoologica (Stockholm), 21: 51–267.

Howard, L. E., W. M. Holmes, S. Ferrando, J. S. Maclaine, R. N. Kelsh, A. Ramsey R. L. Abel & J. P. Cox. 2013. Functional nasal morphology of chimaerid fishes. Journal of Morphology, 274: 987–1009.

Jarvik, E. 1963. The composition of the intermandibular division of the head in fish and tetrapods and the diphyletic origin of the tetrapod tongue. Kungl. Svenska Vetenskap-sakademiens Handlingar, 4: 1–74.

Jarvik, E. 1972. Middle and Upper Devonian porolepiformes from East Greenland with special reference to *Glyptolepis groenlandica* n. sp. Meddr. Grønland, 187: 1–295.

Jarvik, E. 1980. Basic structure and evolution of vertebrates. Academic press, London, UK, 575 pp.

Jollie, M. 1984. Development of the head and pectoral skeleton of *Polypterus* with a note on scales (Pisces: Actinopterygii). Journal of Zoology, 204: 469–507.

Khonsari, R. H., M. Seppala, A. Pradel, H. Dutel, G. Clement, O. Lebedev, S. Ghafoor, M. Rothova, A. Tucker, J. G. Maisey, C. M. Fan, M. Kawasaki, A. Ohazama, P. Tafforeau, B. Franco, J. Helms, C. J. Haycraft, A. David, P. Janvier, M. T. Cobourne, P. T. Sharpe. 2013. The buccohypophyseal canal is an ancestral vertebrate trait maintained by modulation in sonic hedgehog signaling. BMC Biology, 11: 27.

Kindred, J. E. 1921. The chondrocranium of *Syngnathus fuscus*. Journal of Morphology, 35: 425–456

Kindred, J. E. 1924. An intermediate stage in the development of the skull of *Syngnathus fuscus*. American Journal of Anatomy, 33: 421–447.

Labiche, J. C., O. Mathon, S. Pascarelli, M. A. Newton, G. G. Ferre, C. Curfs, G. Vaughan, A. Homs & D. Fernandez Carreiras. 2007. The fast readout low noise camera as a versatile x-ray detector for time resolved dispersive extended X-ray absorption fine structure and diffraction studies of dynamic problems in materials science, chemistry, and catalysis. Review of Scientific Instruments, 78: 091301.

Lu, J., S. Giles, M. Friedman & M. Zhu. 2017. A new stem sarcopterygian illuminates patterns of character evolution in early bony fishes. Nature Communications, 8: 1932.

Lund, R. & E. D. Grogan. 2004. Two tenaculum-bearing Holocephalimorpha (Chondrichthyes) from the Bear Gulch Limestone (Chesterian, Serpukhovian) of Montana, USA. Recent Advances in the Origin and Early Radiation of Vertebrates, 177–187.

Maisey, J. G. 1989. Visceral skeleton and musculature of a Late Devonian shark. Journal of Vertebrate Paleontology, 9: 174–190.

Maisey, J. G. 2001. A primitive chondrichthyan braincase from the Middle Devonian of Bolivia. Major Events in Early Vertebrate Evolution, 263–288.

Maisey, J. G., R. Miller & S. Turner. 2009. The braincase of the chondrichthyan *Doliodus* from the Lower Devonian Campbellton formation of New Brunswick, Canada. Acta Zoologica, 90: 109–122.

Maisey, J. G., P. Janvier, A. Pradel, J. S. S. Denton, A. Bronson, R. Miller & C. Burrow. 2019. *Doliodus* and Pucapampellids: Contrasting Perspectives on Stem Chondrichthyan Morphology. Evolution and Development of Fishes, 87–109

Miles, R. S. 1973. Relationships of acanthodians. Interrelationships of fishes, 63–104.

Millot, J., J. Anthony & D. Robineau. 1978. Anatomie de *Latimeria chalumnae*. Tome 3: appareil digestif, appareil respiratoire, appareil urogénital, glandes endocrines, appareil circulatoire, téguments, écailles, conclusions générales. Editions du Centre national de la recherche scientifique, Paris, France, 198 pp.

Millot, J. & J. Anthony. 1958. Anatomie de *Latimeria chalumnae*. Tome 1: squelette et muscles. Editions du Centre national de la recherche scientifique, Paris, France, 122 pp.

Nelson, G. 1969. Gill arches and the phylogeny of fishes, with notes on the classification of vertebrates. Bulletin of the American Museum of Natural History, 141: 475–552.

Nielsen, E. 1942. Studies on triassic fishes from East Greenland. Vol. 1: *Glaucolepis* and *Boreosomus*. C. A. Reitzel Forlag, Denmark, 403 pp.

Paganin, D., S. C. Mayo, T. E. Gureyev, P. R. Miller & S. W. Wilkins. 2002. Simultaneous phase and amplitude extraction from a single defocused image of a homogeneous object. Journal of Microscopy, 206: 33–40.

Parker, T. J. 1891. On the presence of a sternum in *Notidanus indicus*. Nature, 43: 142.

Patterson, C. 1973. Interrelationships of holosteans. – Pp. 233–305 in Greenwood, P. H., Miles, R. S. & Patterson, C. (eds.). Interrelationships of fishes. Academic Press, London, UK, 536 pp.

Poplin, C. 1974. Étude de quelques paléoniscidés pennsylvaniens du Kansas. Éditions du Centre national de la recherche scientifique, Paris, France, 151 pp.

Pradel, A. 2010. Skull and brain anatomy of Late Carboniferous Sibyrhynchidae (Chondrichthyes, Iniopterygia) from Kansas and Oklahoma (USA). Geodiversitas, 32: 595–661.

Pradel, A., M. Langer, J. G. Maisey, D. Geffard-Kuriyama, P. Cloetens, P. Janvier & P. Tafforeau. 2009. Skull and brain of a 300 million-year-old chimaeroid fish revealed by synchrotron holotomography. Proceedings of the National Academy of Sciences of the USA, 106: 5224–5228.

Pradel, A., P. Tafforeau & P. Janvier. 2010. Study of the pectoral girdle and fins of the Late Carboniferous sibyrhynchid iniopterygians (Vertebrata, Chondrichthyes, Iniopterygia) from Kansas and Oklahoma (USA) by means of microtomography, with comments on iniopterygian relationships. Comptes Rendus Palevol, 9: 377–387.

Pradel, A., P. Tafforeau, J. G. Maisey & P. Janvier. 2011. A new Paleozoic Symmoriiformes (Chondrichthyes) from the Late Carboniferous of Kansas (USA) and cladistic analysis of early chondrichthyans. PLoS ONE, 6: e24938.

Pradel, A., D. Didier, D. Casane, P. Tafforeau & J. G. Maisey. 2013. Holocephalan embryo provides new information on the evolution of the glossopharyngeal nerve, metotic fissure and parachordal plate in gnathostomes. PLoS ONE, 8: e66988.

Pradel, A., J. G. Maisey, P. Tafforeau, R. H. Mapes & J. Mallatt. 2014. A Palaeozoic shark with osteichthyan-like branchial arches. Nature, 509: 608.

Pradel, A., J. G. Maisey, R. H. Mapes & I. Kruta. 2016. First evidence of an intercalar bone in the braincase of "palaeonisciform" actinopterygians, with a virtual reconstruction of a new braincase of *Lawrenciella* Poplin, 1984 from the Carboniferous of Oklahoma. Geodiversitas, 38: 489–504.

Robb III, A. J. 1992. Vertebrate fossils from the Lawrence Formation (Douglas Group, Virgilian, Upper Pennsylvanian) in Northeastern Kansas. Transactions of the Kansas Academy of Science, 1903: 129–138.

Sanchez, S., P. E. Ahlberg, K. M. Trinajstic, A. Mirone & P. Tafforeau. 2012. Three Dimensional Synchrotron Virtual Paleohistology: A New Insight into the World of Fossil Bone Microstructures. Microscopy and Microanalysis, 18: 1095–1105.

Schaeffer, B. 1981. The xenacanth shark neurocranium, with comments on elasmobranch monophyly. Bulletin of the American Museum of Natural History, 169: 1–66.

Stahl, B. J. 1980. Non-autostylic Pennsylvanian iniopterygian fishes. Palaeontology, 23: 315–324.

Tafforeau P., R. Boistel, E. Boller, A. Bravin, M. Brunet, Y. Chaimanee, P. Cloetens, M. Feist, J. Hoszowska, J.-J. Jaeger, R. F. Kay, V. Lazzari, L. Marivaux, A. Nel, C. Nemoz, X. Thibault, P. Vignaud & A. Zabler. 2006. Applications of X-ray synchrotron microtomography for non-destructive 3D studies of paleontological specimens. Applied Physics A, Materials Science & Processing, 83: 195–202.

Young, G. C. 1979. New information on the structure and relationships of *Buchanosteus* (Placodermi: Euarthrodira) from the Early Devonian of New South Wales. Zoological Journal of the Linnean Society, 66: 309–352.

Zangerl, R. 1997. *Cervifurca nasuta* n. gen. et sp: an interesting member of the Iniopterygidae (Subterbranchialia, Chondrichthyes) from the Pennsylvanian of Indiana, USA. Fieldiana: Geology, 35: 1–24.

Zangerl, R. 1981. Chondrichthyes I. Paleozoic Elasmobranchii. – In Schultze, H.-P. (ed.), Handbook of paleoichthyology, vol. 3A. Gustav Fischer Verlag, Stuttgart, Germany, 115 pp.

Zangerl, R. 1966. A new shark of the family Edestidae, *Ornithoprion hertwigi*, from the Pennslyvanian Mecca and Logan quarry shales of Indiana Fieldiana: Geology, 16: 1–43.

Zangerl, R. & G. R. Case. 1973. Iniopterygia: a new order of Chondrichthyan fishes from the Pennsylvanian of North America. Fieldiana: Geology, 6: 1–67.

Zangerl, R. & G. R. Case. 1976. *Cobelodus aculeatus* (Cope), an anacanthous shark from Pennsylvanian black shales of North America. Palaeontographica Abteilung A, Palaeozoologie – Stratigraphie, 154: 107–157.

Ancient Fishes and their Living Relatives: a Tribute to John G. Maisey. Alan Pradel, John S. S. Denton & Philippe Janvier (eds.): pp. 193–204, 15 figs.
© 2021 by Verlag Dr. Friedrich Pfeil, München, Germany – ISBN 978-3-89937-269-4

The cranium of *Helodus simplex* (Agassiz, 1838) revised

Michael I. Coates*, Kristen Tietjen*, Zerina Johanson**, Matt Friedman*** and Stephanie Sang*

Abstract

Helodus simplex is the textbook transitional form linking chimaeroids to their conventionally shark-shaped chondrichthyan relatives and ancestors. Helodontid teeth, well represented in museum collections, range from the Upper Devonian to the Lower Permian, but *H. simplex* is the only species with associated fossil skeletal remains. These include a low, broad, holocephalan skull with a short rostrum and ethmoid trough, features that have generated debate regarding the origin of the modern chimaeroid ethmoid canal. The characteristic teeth are aligned and occasionally fused to form a seeming proto-dental-plate, Pleuroplax. Here, computed tomographic data of classic specimens from the Middle Pennsylvanian of England reveal the skull morphology as different. The rostrum extends to an acute apex, and the preorbital region includes an apparent precerebral fontanelle. The lateral walls of the fontanelle each enclose an elongate tube linking the orbit to the snout: the ethmoid canal is paired. Compact statoliths are preserved within in the otic capsules, as in modern chimaeroids. *Helodus* teeth are larger relative to jaw size than previously estimated, tooth sets exhibit allometric growth, and successive tooth set members interdigitate with neighbours to form a continuous dental battery in the jaw. The total dentition is strongly heterodont. Small edestid-like teeth contribute to a mandibular parasymphysial platform. Diclitodus teeth form a spiral set locked into the anterior extremity of a rostral plate roofing the ethmoid region. Comparisons are drawn with clasper denticles in modern taxa, and the seeming tooth spiral plus plate complex likely represents a frontal clasper: the earliest known example. In summary, these new data present a detailed, previously unknown mixture of conditions within an early holocephalan, fueling the debate about how it, and its eugeneodontid, petalodont, and cochliodont contemporaries relate to the rest of the late Paleozoic holocephalan radiation, and the morphological transformations underlying the extraordinary cranial form of modern chimaeroids.

Introduction

Helodus simplex (Agassiz, 1838) has long been regarded as a model for pre-chimaeroid conditions (Moy-Thomas 1936, Patterson 1965, Janvier 1996, Stahl 1999, Johanson et al., this vol.), primarily because its dentition lacks the large dental plates that characterize the crown clade and many seemingly more derived members of the stem lineage. Although erected initially as a genus of cochliodont teeth (Moy-Thomas 1936), successive restrictions and redefinitions of *Helodus* (summarized in Stahl 1999) have reduced the content to the type species: the only species for which both dental and skeletal remains are known. Nevertheless, a rash of tooth-based *Helodus* species persist, the status of which is uncertain and beyond the scope of the present paper.

The aim of the current work is to report the results of a computed tomographic (CT) study of some of the classic specimens used for Moy-Thomas's (1936) *Helodus simplex* reconstruction, and Patterson's (1965) revision. The current paper presents a comparative description, and the results highlight a number of likely productive areas for future searches for morphological characters. The phylogenetic context has changed considerably in recent years, especially since the publication of Stahl's (1999) handbook. Most notably, a remarkable variety of morphologically disparate holocephalan fishes have been described from the Upper Mississippian of Bear Gulch (e.g. Lund 1982, Lund et al. 2014, Grogan & Lund 2000); the anatomically bizarre iniopterygians (Zangerl & Case 1973) are now well understood from meticulous CT-based descriptions of three-dimensionally intact material (Pradel et al. 2009, Pradel 2010); and character-rich data matrices and hypotheses of early chondrichthyan phylogeny (Pradel et al. 2011, Coates et al. 2017) are no longer driven solely by fin spine, scale, and teeth conditions.

As might be expected, the new reconstruction of *Helodus simplex* reveals a new combination of plesiomorphies and apomorphies, including a high degree of heterodonty, consistent with the few other Paleozoic holocephalans known from articulated specimens (Ginter & Piechota 2004). The current work confirms previous suggestions (Moy-Thomas 1936, Stahl 1939) that, beyond the standard teeth attributed to *H. simplex*, at least two further tooth-form taxa co-occur around the oral margin: Pleuroplax and Diclitodus. Here, to avoid systematic confusion, these names are not italicized because they evidently do not refer to natural taxa.

Finally, our resultant phylogenetic speculations, rather than elevating *Helodus* to a more derived branching point along the holocephalan stem-lineage, suggest new hypotheses of character evolution preceding the emergence of the chimaeroid morphotype. It seems highly unlikely that *Helodus* will displace *Chondrenchelys* (Moy-Thomas 1935, Finarelli & Coates 2014) or *Harpagofututor* (Lund 1982) as a closer relative of crown chimaeroids. Alternatively, and with regard to reconstructing relationships among the extraordinary diversity of late Palaeozoic chondrichthyans, we suggest that *Helodus* might shed new light on the phylogenetic placement of the still poorly understood eugeneodontids (Zangerl 1981, Janvier 1996).

* Department of Organismal Biology and Anatomy, University of Chicago, 1027 E 57th St, Chicago, IL 60637, USA. mcoates@uchicago.edu (corresponding author)
** Department of Earth Sciences, The Natural History Museum, London, SW7 5BD, United Kingdom.
*** Museum of Paleontology and Department of Earth and Environmental Sciences, University of Michigan, 1109 Geddes Ave, Ann Arbor, MI 48109, USA.

Materials

Image preparation. Scans were completed at the Imaging and Analysis Centre of the Natural History Museum, London UK, using a Metris X-Tek HMX ST 225 µCT scanner. Scanning parameters for NHMUK PV P8212: 195 kv and 175 µA, and a 1.5 mm copper filter; 3142 projections. Scanning parameters for NHMUK PV P8213e: 200 kv and 200 µA, and a 2.5 mm copper filter; 3142 projections. Anatomical reconstructions were completed using Mimics v. 17 (biomedical.materialise.com/mimics; Materialise, Leuven, Belgium) for the three-dimensional modeling, including segmentation, three-dimensional object rendering, and STL polygon creation. 3D Studio Max (Autodesk.com/products/3ds-max; Autodesk, San Rafael, USA) was used for further editing of the STLs (color, texture, lighting) and mirroring for the final restoration.

Institutional abbreviation. MCZ, the Museum of Comparative Zoology, Harvard University, USA; NHMUK, The Natural History Museum, London, UK; USNM, The National Museum of Natural History, Smithsonian Institution, Washington DC, USA.

Specimens and geological context. All specimens used in this study originate from the Pennsylvanian (Moscovian) Knowles Ironstone of Fenton, Staffordshire, England (Ward 1875, Moy-Thomas 1936). All skeletal and dental remains are preserved in siderite concretions, with the fossil cartilage often laced with pyrite, and skeletal spaces filled with calcite. In many respects, the preservation resembles that of the fish-bearing beds at Wardie Beach (Wood 1975, Dineley & Metcalf 1999). All specimens are crushed to some degree, and crania examined have lost an estimated 70 % of height or width along the axis of compaction.

Systematic palaeontology

Class Chondrichthyes (Huxley, 1880)
Subclass Holocephali (Bonaparte, 1832)
Order Helodontiformes (Patterson, 1965)
Incertae sedis Genus *Helodus* (Agassiz, 1838)

Emended diagnosis. Holocephalan chondrichthyan distinguishable by the presence of seven generative sets of teeth forming an interdigitated battery in each ramus of the jaws, each tooth set containing five or more members, the largest and lingual (youngest) members of the largest tooth set fused into dental plates (Pleuroplax). The crowns of all teeth are capped with tubate dentine; crowns of tooth battery flattened or weakly cuspidate, roots shallow, simple, with concave profile. Mandibular parasymphysial teeth monocuspid with deep roots. Bicuspid denticles (Diclitodus) located at anterior of upper jaw form whorl, scrolling anteriorly into the rostral shield of a likely frontal clasper, thus presumed to be male-specific. Ethmoid canal present and paired, each tube extending anteriorly through the orbitonasal lamina; precerebral fontanelle seemingly present. Otic labyrinth contains two compact statoliths. Hyoid arch with large ceratohyal and small basihyal. Synarcual present; first dorsal fin set posterior to pectoral level, with smooth laterally compressed spine of osteodentine. Paired fins dibasal; halves of pectoral girdle separate; squamation complete, scales placoid and synchronomorial, no specialized scales surrounding sensory canals.

Type species. *Helodus simplex* Agassiz, 1838 (Figs. 1–12)

Diagnosis. As for genus.

Lectotypes and Syntypes. NHMUK PV P8212, NHMUK PV P2907 and NHMUK PV P8213e.

Referred specimens. NHMUK PV P2907, single part of concretion, two further parts of which are numbered P8213e, containing dorsoventrally compressed cranium with lower jaw and nearly complete dentition. CT scan successful.

NHMUK PV P6706, part and counterpart of concretion-enclosed cranium that appears to be in good condition with lower jaw in place. CT scan unsuccessful.

NHMUK PV P8207, part and counterpart of complete individual, laterally compressed, in framed plaster mount. Entire fish exposed as a field of scales, with fairly complete exposure of a scapulocoracoid cartilage, a dorsal fin spine in position, and a well-preserved caudal fin.

NHMUK PV P8209, an incomplete pelvic fin.

NHMUK PV P8212, the most completely exposed and skeletally intact individual, retains the cranium, dorsoventrally compressed with incomplete snout, lower jaw, pectoral girdle and fins and pelvic girdle and fins, more or less in life position. This specimen forms the basis of most of Moy-Thomas's (1936) description. Pelvic fin completeness in P8212 is uncertain: the pelvic skeleton extends to the edge of the block and the region that might have included claspers is missing. White paint outlines identify Moy-Thomas's interpretation of cartilages, especially in the paired fins, and obscure the actual boundaries between skeletal parts. The cranial cartilages have suffered damage since the initial description; the posterior wall of the right orbit is now missing. CT scan of cranial block successful. Two small fragments of matrix with cartilage, each numbered P8212, are parts of P8214.

NHMUK PV P8213, part of block containing cranial cartilages and associated dentition.

NHMUK PV P8213c, part and counterpart of block containing incomplete lower jaw, rostral shield, and associated dentition, scattered but well preserved. CT scan successful.

NHMUK PV P8213d, three-part block containing parts of cranial cartilages and teeth.

NHMUK PV P8213e two parts of block containing a nearly complete, dorsoventrally compressed cranium with intact dentition; P2907 is part of same cranium. CT scan successful.

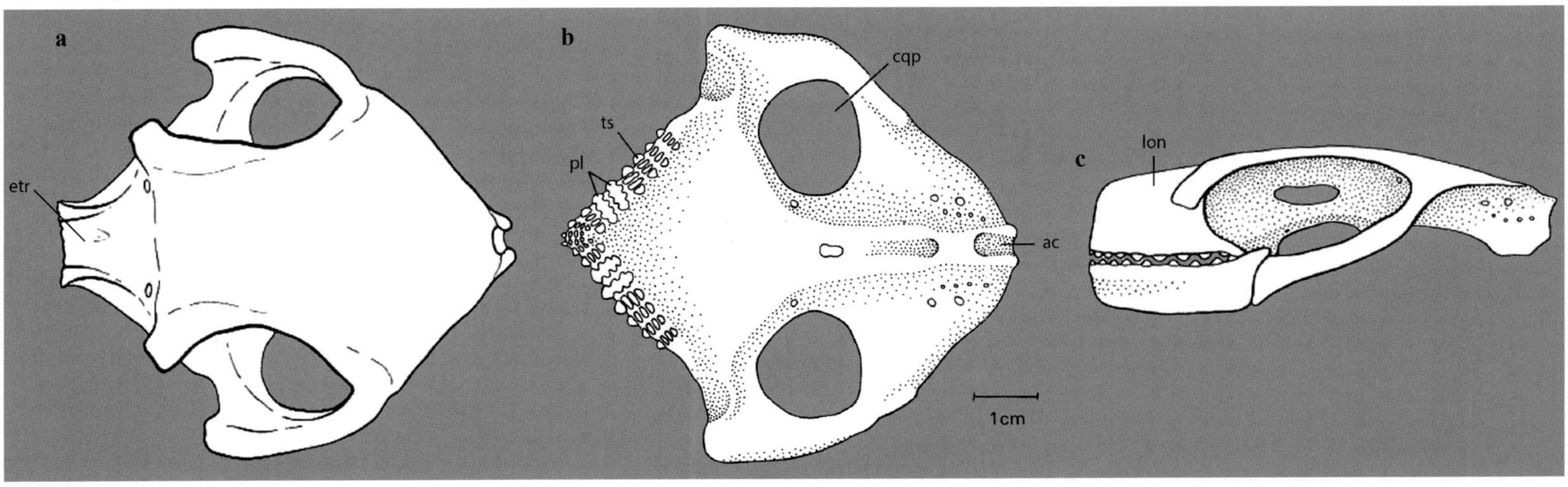

Fig. 1. *Helodus simplex*, cranial reconstruction redrawn after Moy-Thomas 1936, and based largely on the skull of specimen NHMUK PV P8212. Skull in dorsal (**a**), palatal (minus lower jaw) (**b**), and left lateral (**c**) views. Abbreviations: ac, canal for dorsal aorta; cqp, cranioquadrate passage; etr, ethmoid trough; lon, lamina orbitonasalis; pl, *Pleuroplax* dental plate; ts, tooth set.

Fig. 2. *Helodus simplex*, NHMUK PV P8212, map of the complete ▷ specimen in ventral aspect, showing concealed detail of skull, hyoid arch, gill arch, and pectoral girdle, with exposed cartilages of paired fins. Abbreviations: bpt, basipterygium; cbr1, first ceratobranchial bar; chy, ceratohyal; cor, coracoid; mpt, metapterygium; pal, palate of chondrocranium; pg?, pelvic girdle; prpt, propterygium; r, radial cartilage.

NHMUK PV P8214, concretion part (no counterpart) exposing neurocranium and palate in ventral view, parts of pectoral girdle, and dorsal fin spine. Dentition appears damaged, although snout apex probably intact with Diclitodus whorl in situ.

Remarks. Moy-Thomas (1936) used specimens NHMUK PV P8212, P8213e, P8213g, P8214, and P6706 for his cranial reconstruction (Fig. 1), although much of the interpretation relies on NHMUK PV P8212. This reconstruction reappears in Stahl's (1999) handbook on Holocephali. Specimen P8212 was also the source of most of the postcranial description, with P8209a adding information about the pelvic and caudal fins. Moy-Thomas notes that the ironstone matrix is very hard, and resorted to a light hammer and fine chisels to reveal the cartilage, with 'lucky fractures' revealing the skull of P8212. No mechanical or chemical preparation was undertaken in the present study.

Patterson's (1965) revision considered NHMUK PV P8212 as "the best preserved neurocranium". NHMUK PV P6707 and P8212 exhibit "calcified cartilage behind the skull, between the scapular processes of the pectoral girdle, which is almost certainly a synarcual" (p. 179); P8209 and P8212 were used for pelvic fin detail; P8215 provided scales and P2908 provided sections of the dorsal fin spine for histological data. Patterson's cranial reconstruction differs slightly in lateral view only, exhibiting a dorsoventrally taller rostrum, and lacking an interorbital vacuity and septum. This version appears in Moy-Thomas and Miles' (1971) Palaeozoic Fishes.

Comparative description

(Figs. 2–12)

Pre-orbital region. The ethmoid region is most completely preserved in specimens NHMUK PV P8212 (Figs. 2–3) and the individual composed of NHMUK PV P8213e and P2907 (Fig. 4). On the basis of NHMUK PV P8212 Patterson (1965) summarized this region as lacking an ethmoid canal, but bearing a pair of parallel vertical laminae that Moy-Thomas (1936) interpreted as upgrowths of the lamina orbitonasalis. Within the space between these laminae, both Moy-Thomas and Patterson noted the presence of the efferent foramina of the superficial ophthalmic nerves. However, these authors differed in their interpretations of the relation of this trough to the distinctive ethmoid canal of crown group chimaeroids.

The more complete condition of these orbitonasal laminae in NHMUK PV P8213e and P2907 (Figs. 4–5) sheds new light on this issue, revealing that NHMUK PV P8212 (Fig. 3) preserves only the proximal portion of the preorbital chondrocranium. In the revised reconstruction (Fig. 6), each orbitonasal lamina extends further, both anteriorly and dorsally, with left and right sides converging anteriorly so that the cranium has an acute, beak-like ethmoid region, as in modern chimaeroids.

Anteriorly, the dorsal portion of each orbitonasal lamina flares laterally to form a shallow, roofed concavity (Fig. 4e): the nasal capsule (entirely missing from NHMUK PV P8212). The foramen for the olfactory nerve lies at the center of the concavity, but a larger foramen is located on the posterior

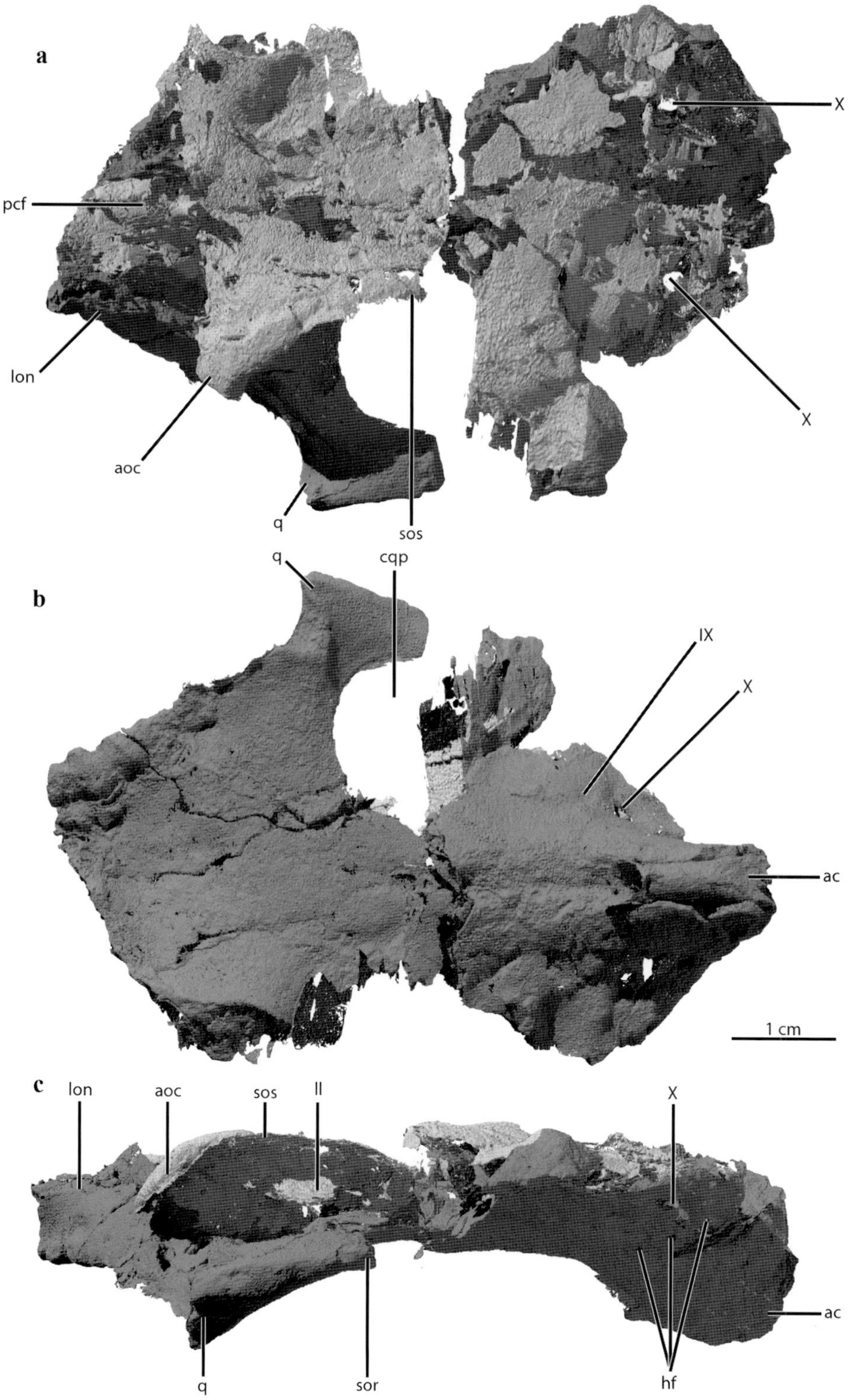

Fig. 3. *Helodus simplex*, NHMUK PV P8212, rendered CT scan ▷ of cranium in dorsal (**a**), palatal (**b**), and left lateral (**c**) views, all with lower jaw removed. Abbreviations: ac, canal for dorsal aorta; aoc, antorbital crest; cqp, cranioquadrate passage; hf, hypoglossal nerve foramina; lon, lamina orbitonasalis; pcf, precerebral fontanelle; q, quadrate condyle; sor, suborbital ridge; sos, supraorbital shelf; II, opening for optic nerve; IX, glossopharyngeal nerve foramen; X, vagus nerve foramen.

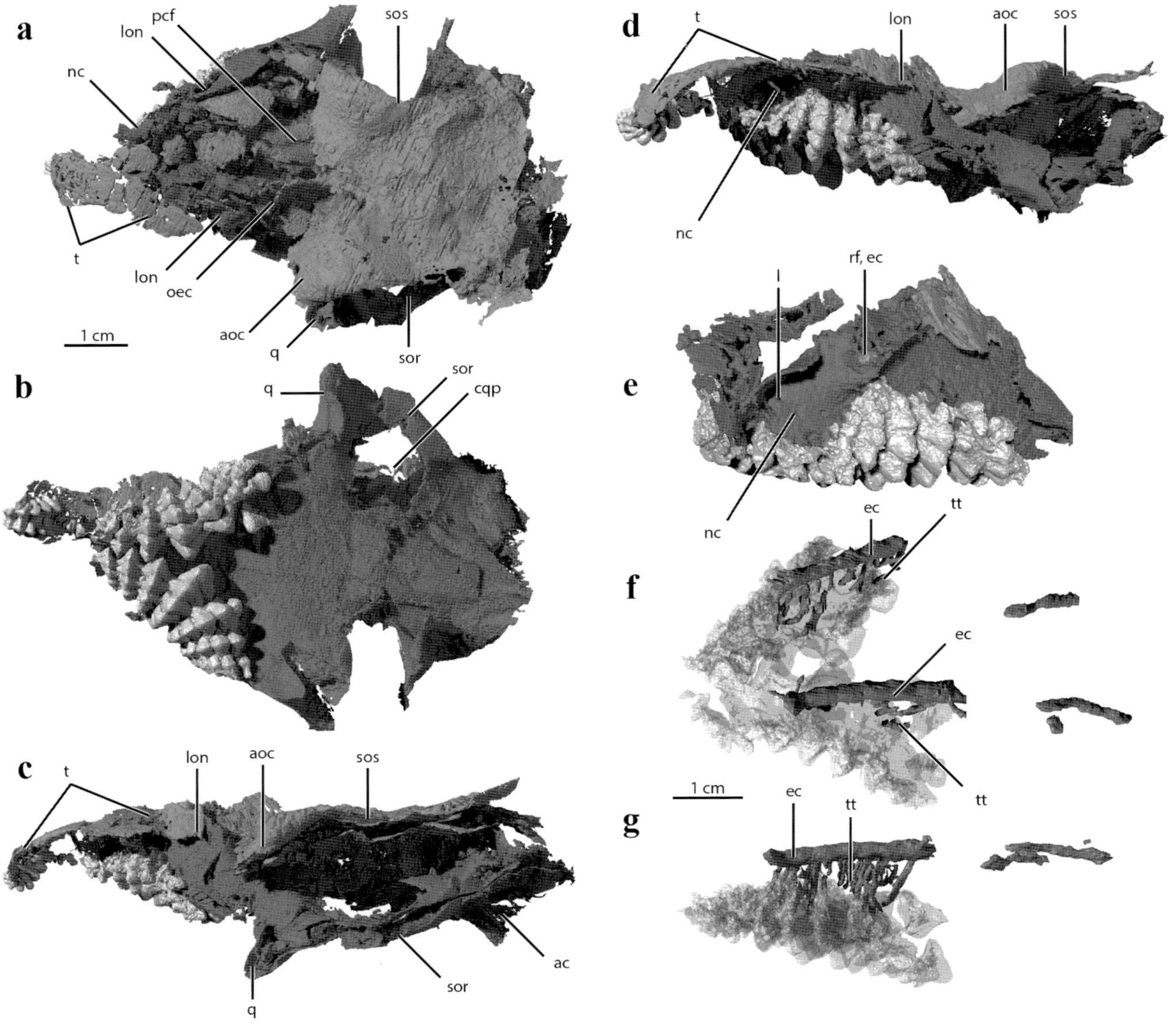

Fig. 4. *Helodus simplex*, NHMUK PV P8213e and P2907, rendered CT scan of cranium in dorsal (**a**), palatal (**b**), left lateral (**c**), and antero-left lateral (**d**) views; detail of nasal capsule (**e**); left and right ethmoid canals and trabeculae infills in dorsolateral view (**f**); left ethmoid canal infills and trabeculae, lateral view (**g**); dentition rendered semitransparent in parts f and g. Specimen reversed for ease of comparison with NHM PV P8212 shown in Figure 3. Abbreviations: ac, canal for dorsal aorta; aoc, antorbital crest; cqp, cranioquadrate passage; ec, ethmoid canal; lon, lamina orbitonasalis; nc, nasal capsule; oec, orbital opening of ethmoid canal; pcf, precerebral fontanelle; q, quadrate condyle; rf, ec, rostral foramen of ethmoid canal; sor, suborbital ridge; sos, supraorbital shelf; t, tenaculum/frontal clasper; tt, trabecular tube; l, opening for olfactory nerve.

wall, within the anterior face of the laterally expanded part of the orbitonasal lamina – a horizontal buttress extending from the rear of the capsule. Importantly, CT scans reveal that the calcified lateral and medial surfaces of the entire orbitonasal lamina enclose a much greater cross-sectional volume of cartilage than previously recognized. Furthermore, the dorsal rim of the lamina was likely complete - capped with calcified cartilage. Preserved most completely on the left side of NHMUK PV P8213e (Fig. 5a,b), broken fragments of the calcified dorsal rim form an incomplete roof over the space formerly filled with cartilage. However, no sections of calcified cartilage arch over the ethmoid space. Thus, there is no indication that laminae from left and right sides of the ethmoid region met dorsally to roof the pre-orbital endocranial space. Instead, the medial calcified surfaces extend ventrally to the midline to meet counterparts. Moy-Thomas (1936, text-fig. 5) considered this to be not only the floor of an ethmoid trough, but also the roof of the nasal capsules, which it is not (Fig. 1a). Moy-Thomas (1936, text-fig. 3), followed by others (Patterson 1965, Stahl 1999), reconstructed this horizontal sheet of cartilage as rising posteriorly to meet the anterior rim of the cranial roof: the trough floor made continuous with the skull roof. In this, the standard interpretation, the ascending floor forms something like a partial or complete bulkhead, separating the ethmoid space from the endocranial cavity. Foramina of the superficial ophthalmic nerves pierce the lateral extremities of the bulkhead, connecting the trough to the anterior of the orbit. However, in the present work, CT scans show that the preorbital region is constructed quite differently.

The foramen in the anterior wall of the orbit (for the superficial ophthalmic nerve) does not communicate with the space between the orbitonasal laminae. Rather, this foramen is the posterior opening of an elongate canal that extends anteriorly through the body of the orbitonasal lamina to exit at the large foramen within the posterior wall of the nasal capsule (Fig. 4e; see also Fig. 5) and thus on the lateral face of the lamina. A proximal stub of this canal, preserved as calcitic infilling, is present on the right side of NHMUK PV P8212. Therefore, unlike modern chimaeroids in which the ethmoid canal is single and median, in *Helodus* the canal is paired, and consists of left- and right-side tubes: one in each orbitonasal lamina (Figs. 4f,g, 5). Further to this previously unknown arrangement, NHMUK PV P8213e and P2907 reveal that each canal is supported on an irregularly arranged series of ~15 vertical tubular trabeculae of calcified cartilage (Fig. 4f,g). These likely reinforced the upper jaw dorsal to the main dentition, much like the trabeculae in the jaws of modern batoids (Summers et al. 1998; Summers 2000). Such tubular trabeculae are also present in the lower jaw (Meckel's cartilage) of *Cochliodus contortus* (NHMUK PV P2424, pers. obs. MIC), revealed by broken surfaces.

The space between the orbitonasal laminae is not an extracranial trough. There is no evidence of a wall dividing this space from the endocranial cavity. In NHMUK PV P8212, and NHMUK PV P8213e and P2907, the floor of the ethmoid space is preserved as a mosaic of calcified fragments, but in both examples of *Helodus* crania these fragments are continuous with the endocranial floor between the orbits. As noted previously, no part of this floor ventral to the anterior margin of the skull roof shows evidence of laminae projecting dorsally to form a wall dividing ethmoid from interorbital territories. Anteriorly, the ethmoid space narrows and is closed-off by the medially contacting walls of the nasal capsules, but (once again) there is no evidence to suggest that these walls, the orbitonasal laminae, extended dorsally and medially to form a primary roof over the ethmoid space (Fig. 5). For these reasons, it appears that the open roof between the orbitonasal laminae is most simply interpreted as a precerebral fontanelle (Fig. 6a).

In NHMUK PV P8213e and P2907, the entire ethmoid

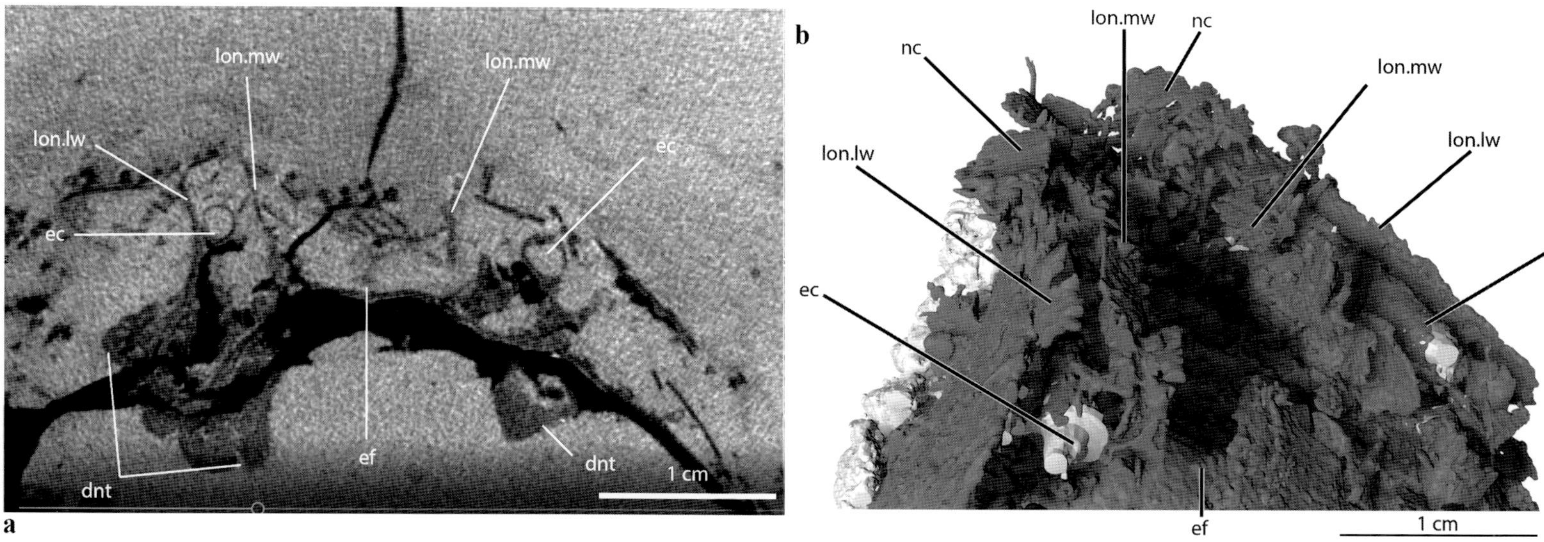

Fig. 5. *Helodus simplex*, NHMUK PV P8213e, CT coronal slice through ethmoid region (**a**); rendered CT scan of ethmoid region with frontal clasper plate removed in postero-dorsal view (**b**). Abbreviations: dnt, dentition; ec, ethmoid canal; ef, ethmoid floor; lon.lw, lamina orbitonasalis lateral wall; lon. mw, lamina orbitonasalis medial wall; nc, nasal capsule.

region is overlain by an elongate cartilage plate (Figs. 4a,c,d, 6e, 7). The calcified shell of this structure is broken into irregularly shaped sections, parts of which have curved (in-rolled) lateral margins showing that the dorsal and ventral surfaces are continuous. This indicates that at least the anterior part of this plate (Figs. 7g,h) was entirely separate from the dorsal rim of the orbitonasal ascending laminae. The whole plate extends from the pre-orbital margin of the cranial roof (Figs. 4a, 6c) to the apex of the rostrum, where it descends and forms a convex shield with a large, central opening. Well-preserved specimens of this distinctive apical portion of the shield are preserved in NHMUK PV P8213e (Fig. 7a,b) and NHMUK PV P8213c (Fig. 7c). Anteriorly and ventrally, this convex shield bears a characteristic set of Diclitodus denticles (Fig. 7a,d,e,g,h). For these reasons (separation from the rest of the ethmoid portion of the cranium; support of diminutive specialized denticles), this curiously 'toothed' structure is suggested to be a frontal tenaculum (cranial clasper), a subject examined further in the Discussion section. Alternatively, the plate might represent a medial rostral rod (Didier 1995), but such snout-supports are not currently known to be associated with specialized denticles.

Orbital, otic, and occipital regions. The CT scans add little new information to the orbital region, although it is probably worth emphasizing that there is no trace of a boundary between palatoquadrate and neurocranial components. The antorbital crest is thinly calcified and arched (Figs. 3, 4, 6), as in Moy-Thomas's (1936) reconstruction (Fig. 1). In the present reconstruction (Fig. 6), the height of the orbit and entire cranial roof has been elevated so that the lateral profile of the orbit is near-circular. In agreement with Patterson (1965) there is no interorbital septum, and the endocranial cavity separates the orbits, albeit narrower than pre- and postorbital spaces. The wall of the orbit is poorly preserved because of postmortem dorsoventral compression (somewhat more compressed than the trabecular-reinforced ethmoid wall: compare Figs. 4c and 4d). Despite breakage, there appears to be an elliptical opening for the optic nerve (Fig. 3c). The orbit floor, in all specimens preserving this region, shows the large cranio-quadrate passage.

The anteroventral wall and floor of the orbit, likely formed by the palatoquadrate (e.g. Didier 1995), is deeper than previously reconstructed. The quadrate articulation with Meckel's cartilage is laterally prominent and directed anteriorly. Extending posteriorly and dorsally from this, the suborbital ridge (Didier 1995), or inferred otic process of the palatoquadrate (Moy-Thomas 1936), sweeps back to merge with the otic wall as in modern chimaeroids and, seemingly, *Debeerius* (Grogan & Lund 2000).

The proportions of the otic and occipital region are much as described by Moy-Thomas (1936). In NHMUK PV P8213e and P2907 the roof is too broken to provide useful detail; this region is generally better preserved in NHMUK PV P8212. The ventral surface and aortic canal are as described by Moy-Thomas (1936) and confirmed by Patterson (1965), but with minor additional detail. The foramen for the vagus nerve (X) is much larger than the foramen for the glossopharyngeal nerve (IX) (Fig. 3). The midline of the ventral surface is marked by a groove, likely for the dorsal aorta, extending anteriorly from the aortic canal. NHMUK PV P8213e and P2907 are incomplete posteriorly and the aortic canal is missing (compare Figs. 3b and 4b), but the midline ventral groove is present, although not shown clearly in the illustration (Fig. 4b). Broad shallow grooves diverge towards the cranioquadrate passage, but the significance of these grooves is unclear.

The occipital region includes openings to the aortic canal, notochordal canal, and the foramen magnum. The occipital condyles are prominent. The notochordal canal is short and conical, showing that the notochord terminated directly anterior

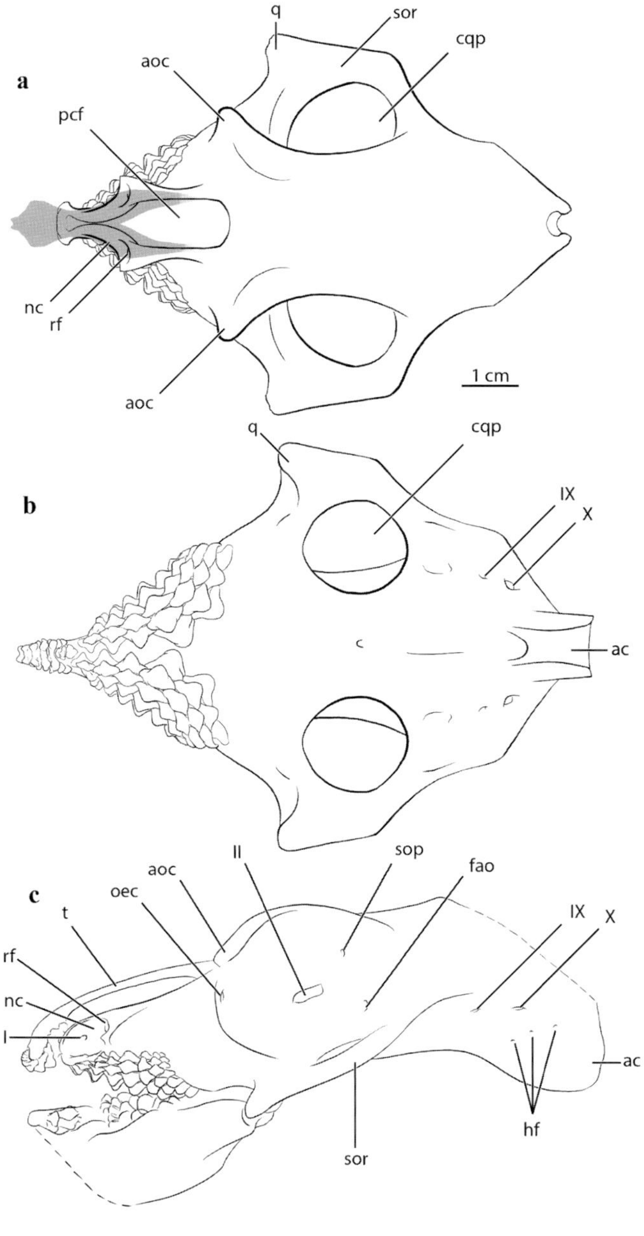

Fig. 6. *Helodus simplex*, reconstruction of cranium in dorsal (**a**), palatal (**b**), and left lateral (**c**) views. Frontal clasper/cranial tenaculum roofing precerebral fontanelle omitted in part a, grey shaded area denotes coverage. Abbreviations: ac, canal for dorsal aorta; aoc, antorbital crest; cqp, cranioquadrate passage; fao, foramen antoticum; hf, hypoglossal nerve foramina; nc, nasal capsule; oec, orbital opening of ethmoid canal; pcf, precerebral fontanelle; q, quadrate condyle; rf, rostral foramen of ethmoid canal; sop, foramen for superficial opthalmic tract; sor, suborbital ridge; t, tenaculum/frontal clasper; I, opening for olfactory nerve; II, opening for optic nerve; IX, glossopharyngeal nerve foramen; X, vagus nerve foramen.

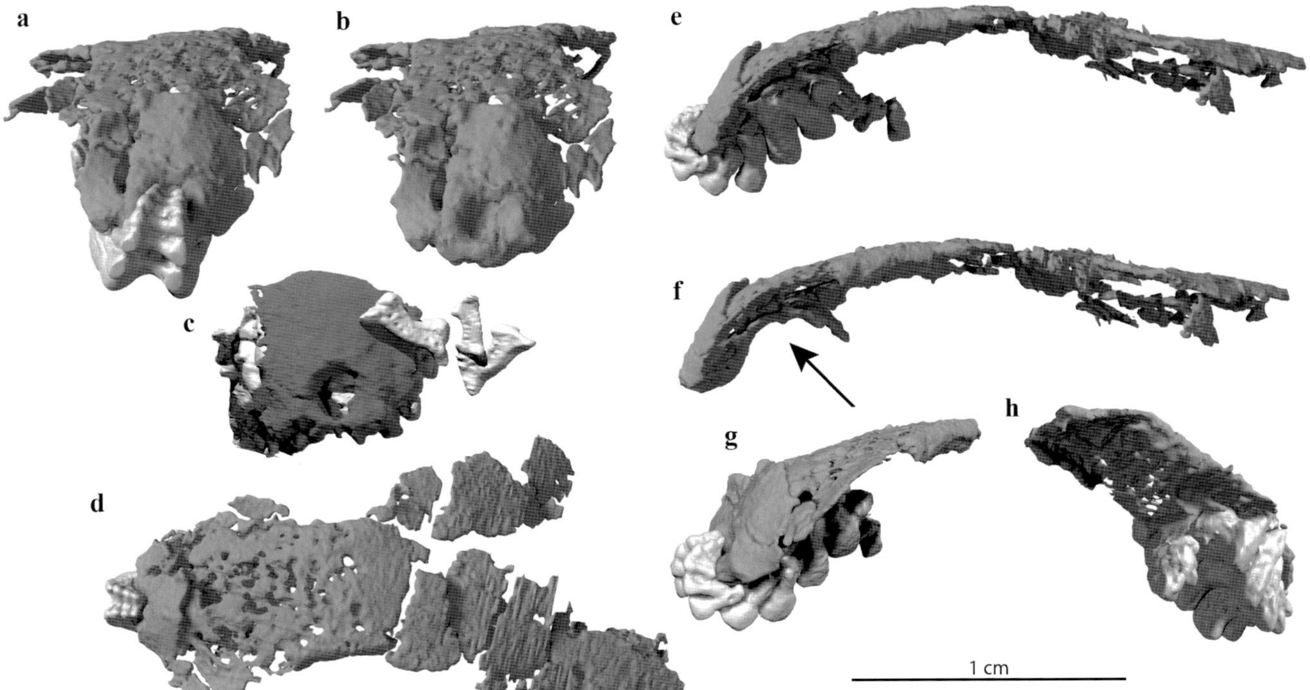

Fig. 7. *Helodus simplex*, NHMUK PV P8213e, the frontal clasper (cranial tenaculum), anterior view of rostral shield with *Diclitodus* (**a**); minus *Diclitodus* (**b**); NHMUK PV P8213c anterior view of rostral shield with scattered *Diclitodus* (**c**); NHMUK PV P8213e clasper in dorsal view (**d**), in left lateral view (**e**), in left lateral view minus *Diclitodus*, arrow marks trough, and inferred site of tissue generating *Diclitodus* whorl (**f**); anterior section in anterodorsal and lateral (**g**) and posteroventral and lateral aspects (**h**), revealing the in-rolled edges of the calcified cartilage shell.

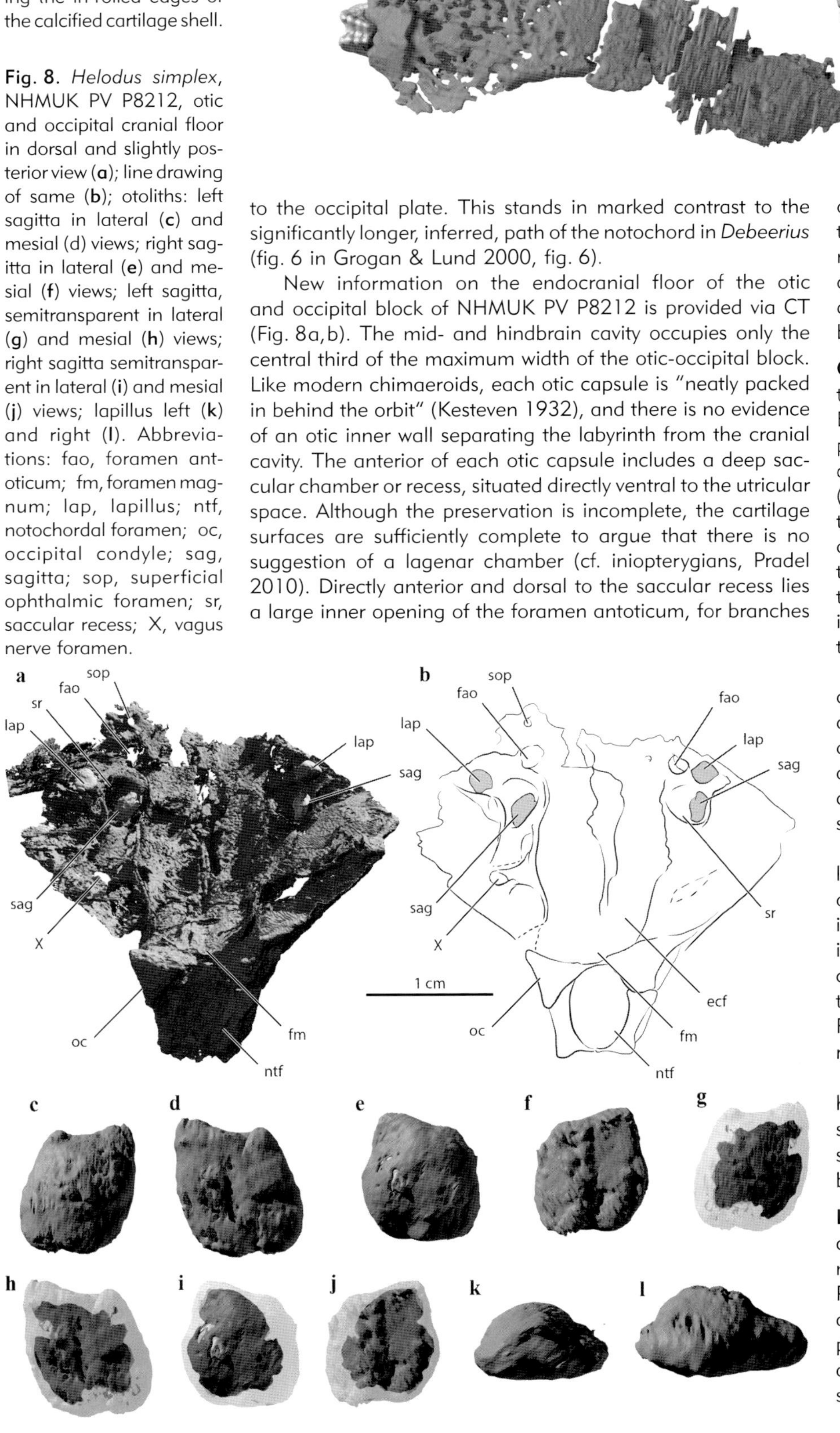

Fig. 8. *Helodus simplex*, NHMUK PV P8212, otic and occipital cranial floor in dorsal and slightly posterior view (**a**); line drawing of same (**b**); otoliths: left sagitta in lateral (**c**) and mesial (d) views; right sagitta in lateral (**e**) and mesial (**f**) views; left sagitta, semitransparent in lateral (**g**) and mesial (**h**) views; right sagitta semitransparent in lateral (**i**) and mesial (**j**) views; lapillus left (**k**) and right (**l**). Abbreviations: fao, foramen antoticum; fm, foramen magnum; lap, lapillus; ntf, notochordal foramen; oc, occipital condyle; sag, sagitta; sop, superficial ophthalmic foramen; sr, saccular recess; X, vagus nerve foramen.

to the occipital plate. This stands in marked contrast to the significantly longer, inferred, path of the notochord in *Debeerius* (fig. 6 in Grogan & Lund 2000, fig. 6).

New information on the endocranial floor of the otic and occipital block of NHMUK PV P8212 is provided via CT (Fig. 8a,b). The mid- and hindbrain cavity occupies only the central third of the maximum width of the otic-occipital block. Like modern chimaeroids, each otic capsule is "neatly packed in behind the orbit" (Kesteven 1932), and there is no evidence of an otic inner wall separating the labyrinth from the cranial cavity. The anterior of each otic capsule includes a deep saccular chamber or recess, situated directly ventral to the utricular space. Although the preservation is incomplete, the cartilage surfaces are sufficiently complete to argue that there is no suggestion of a lagenar chamber (cf. iniopterygians, Pradel 2010). Directly anterior and dorsal to the saccular recess lies a large inner opening of the foramen antoticum, for branches of the trigeminal and facial nerves. Further dorsal and anterior to this lies a smaller foramen for the superficial ophthalmic nerve. Locations of the foramina for the glossopharyngeal (IX) and vagus (X) nerves are consistent with Moy-Thomas's (1936) description, but the glossopharyngeal is obscured in dorsal view by remains of the vestibular labyrinth floor.

Otoliths. A pair of large, compact, otoliths is preserved within the otic capsule on each side of the basicranium (Fig. 8c–l). Both otoliths occupy the anterior portion of the labyrinth, one positioned dorsal and slightly lateral to the saccular chamber, and the other situated deep within the saccular chamber (Fig. 8a,b). The ventral, saccular otolith, or sagitta in teleost terminology (Nolf 1985), is about twice the size of the dorsal otolith, which appears to have occupied the utricular space, the lapillus in teleost terminology (Nolf 1985). On both sides of the braincase, the lapillus is at right angles to the sagitta, and in both otic capsules, the sagitta rests against the right-side of the saccular chamber.

The sagitta (Fig. 8c–j) is roughly oblong in shape, with a convex lateral surface and a flatter medial surface marked by a broad sulcus ascending dorsally from the ventral edge. The dorsal surface bears rounded anterodorsal and anteroposterior angles. The ventral surface consists of anterior and posterior convexities, with the cleft between running into the medial surface sulcus.

The longest axis of the lapillus (Fig. 8k,l) is about half the length of the sagitta. Orientation differs on left and right sides of the braincase. On the left side, the flat surface of the lapillus is ventral, the convex is dorsal, and in dorsal view, the outline is ovoid. On the right side, the flat surface is caudal and the convex is rostral. The left-side lapillus appears to conform to the space enclosed by incompletely preserved endocranial walls. For this reason, it might more closely reflect in-life orientation, rather than the right-side lapillus.

The CT scans indicate that the interior of each otolith is hollow, but incompletely traversed by ridges, defining a bulbous s-shaped space in the left- and right-side sagitta. Further details, such as composition from mineralized spherulites, could not be determined

Lower jaw. Moy-Thomas (1936) described the lower jaws as consisting of stout cartilage, almost triangular in cross section, meeting one another at an anterior symphysis. In NHMUK PV P8213e and P2907 the lower jaw, Meckel's cartilage, is indeed almost triangular in cross section anteriorly, at the fused symphysis, but elliptical in cross section posteriorly with the long axis directed dorsolaterally (Fig. 9). Only the mesial and dorsal surfaces of the jaw are well preserved. The external, lateral sur-

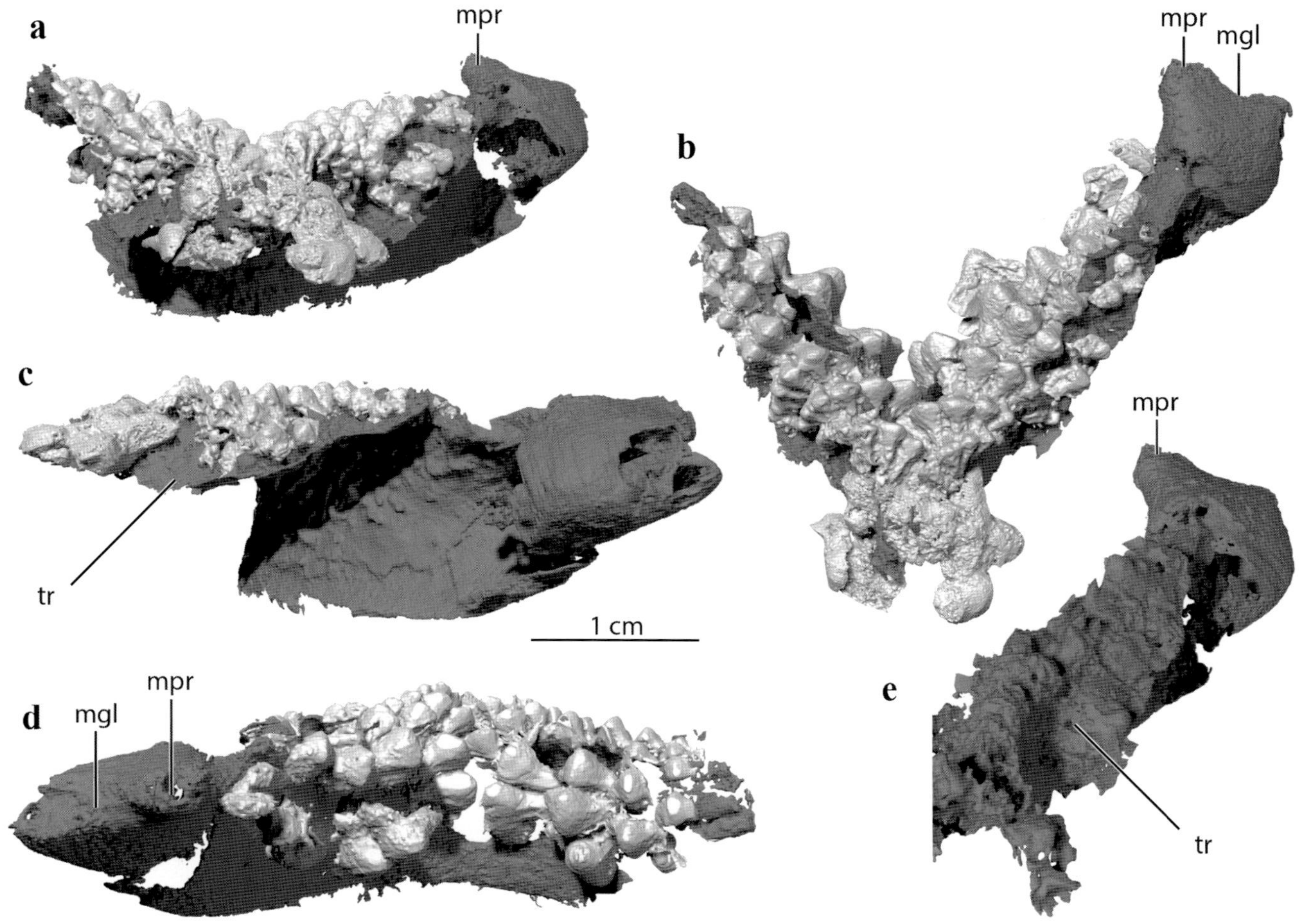

Fig. 9. *Helodus simplex*, NHMUK PV P8213e and P2907 lower jaw (mandible, Meckel's cartilage) with dentition in anterior (**a**), dorsal (**b**), left lateral (**c**), posterior and medial (**d**) views; in anterior and slightly dorsal view with teeth removed (**e**). Abbreviations: mgl, mandibular glenoid; mpr, mandibular process; tr, trough.

face of Meckel's cartilage is incompletely known - missing and presumably broken away from the anterior part of the mandible in all specimens examined. There is no dental trough or sulcus on the mesial surface (Fig. 9d). However, on the dorsolateral or labial surface, there is a distinct trough which broadens and descends anteriorly (Fig. 9c,e). When the dentition is in place, this groove is seen to receive the persistent, ontogenetically older, members of each tooth set. The jaw terminates anteriorly with an apparent platform supporting specialized, parasymphysial teeth (Fig. 9a–c). The jaw articulation is oriented rostro-caudally, as in other holocephalans. The articular surfaces are laterally broad, there is a "short backwardly directed process", but not as prominent as reconstructed previously (Moy-Thomas 1936).

Dentition. For the first time, the complete dentition of *Helodus simplex* is revealed in the CT scans of NHMUK PV P8213e and P2907, with complementary results from NHMUK PV P8213c (Figs. 4, 9–12). The resultant arrangement differs significantly from previous reconstructions (Fig. 1). Details of individual

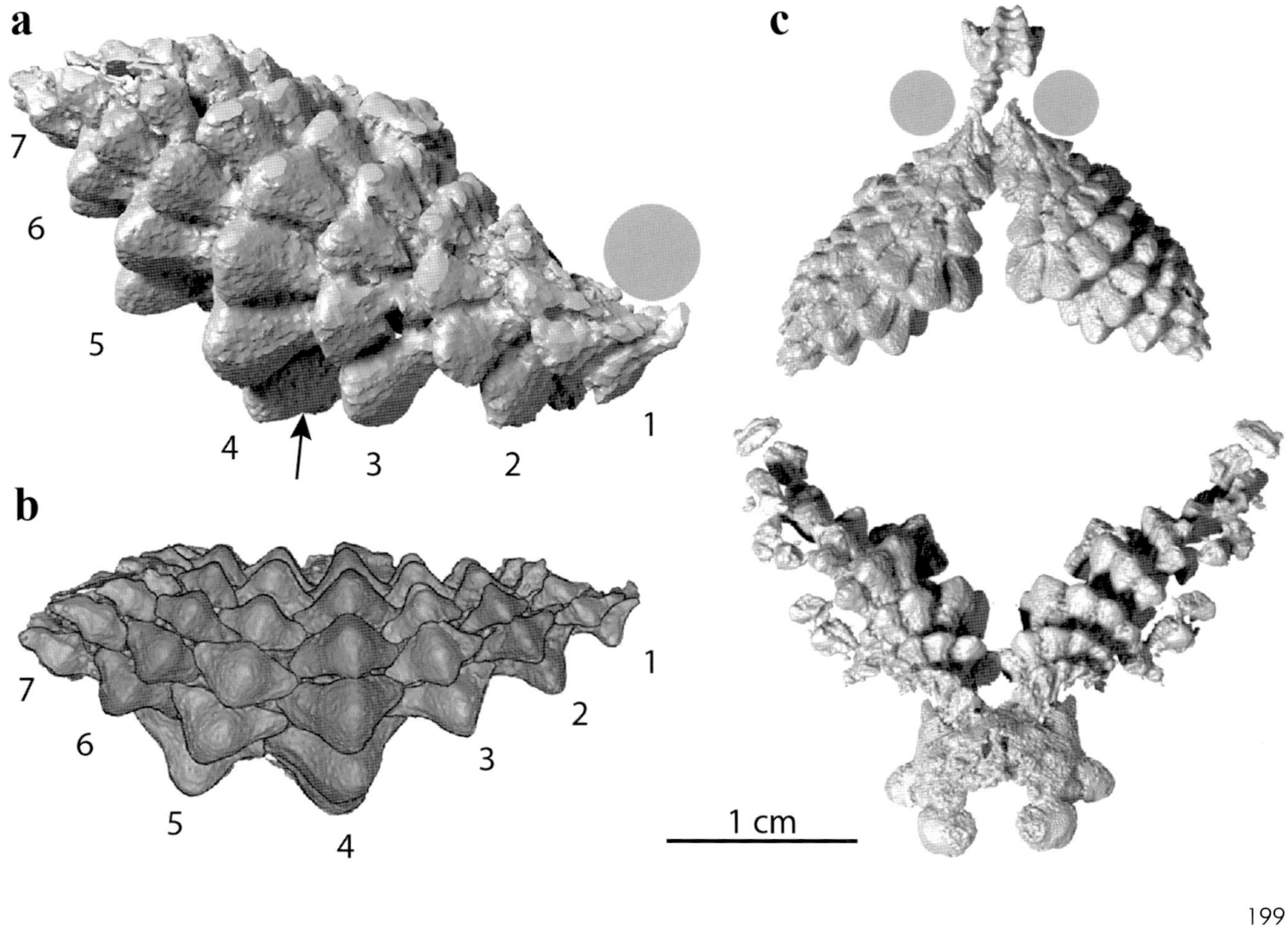

Fig. 10. *Helodus simplex*, NHMUK PV P8213e and P2907, dentition: right upper tooth battery in anterolateral view (**a**); right upper tooth battery in palatal view, image reversed so that tooth set order is in register with part a, tooth sets outlined to show interdigitation (**b**); total dentition including *Diclitodus* whorl: right sides (most complete) mirrored, displayed as if jaws agape (**c**). Grey circles denote location of nasal capsule; tooth sets numbered from anterior to posterior; arrow marks *Pleuroplax* member of tooth set 4.

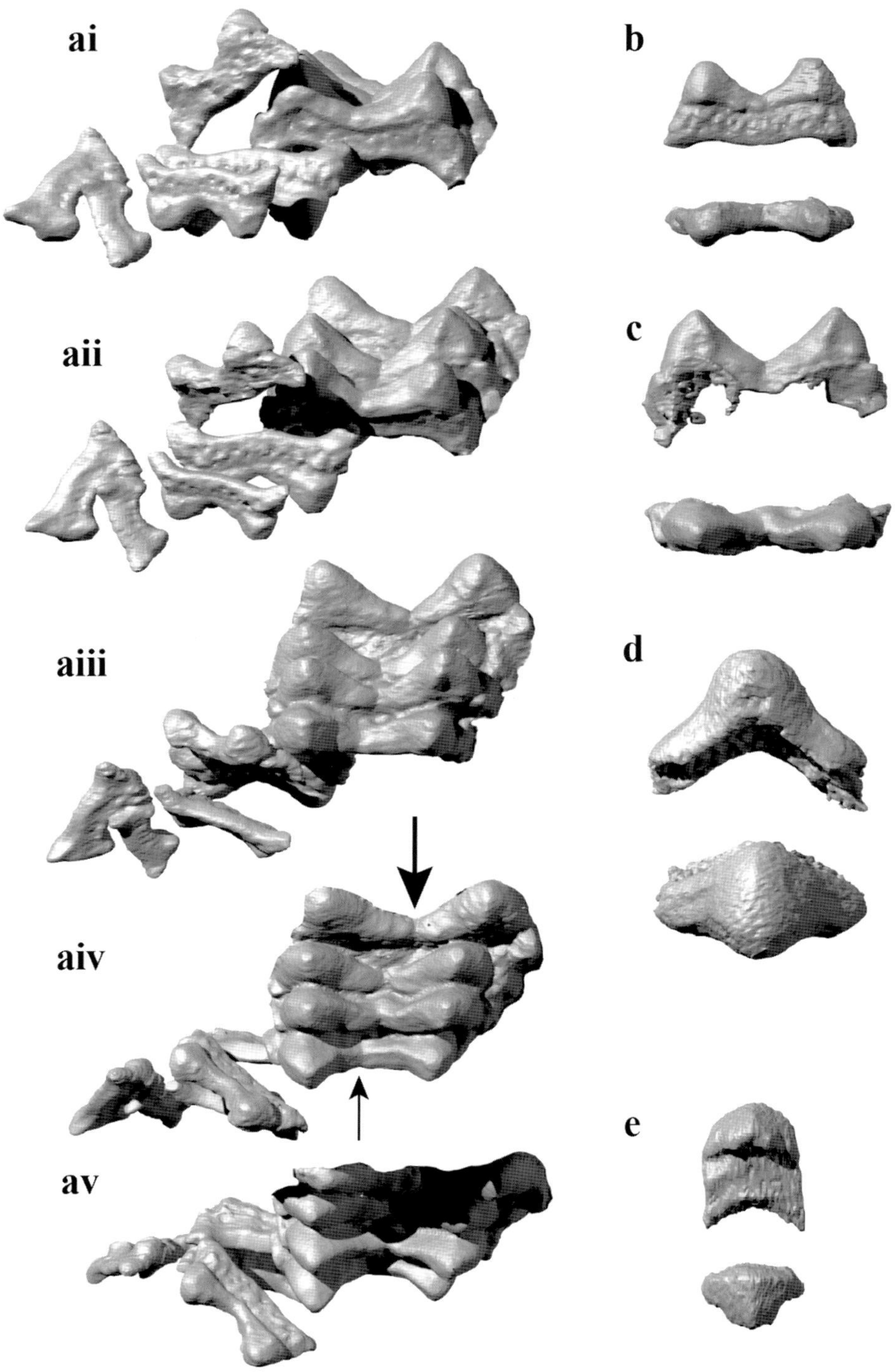

Fig. 11. *Helodus simplex*, dentition details: NHMUK PV P8213c, *Diclitodus* partly dissociated whorl, shown in successively rotated views (**ai–av**); small (**b**) and large (**c**) examples of *Diclitodus*; standard *Helodus simplex* tooth (**d**); parasymphysial mandibular tooth with rounded tetrahedral crown and a deeply embayed root (**e**). Arrows in part **aiv** show incomplete and (bold) complete division.

tooth morphology and histology are provided by Johanson et al. (this vol.).

Here, the teeth usually assigned to *H. simplex* are revealed as constituents of a large battery of seven generative tooth sets in each jaw (Figs. 4, 10), with these tooth sets scrolling from lingual to labial surfaces. As noted elsewhere in the description, there is no trace of a dental trough or sulcus supporting the largest (youngest) member of each tooth set. Generated outwards from the palate, the smallest (oldest) teeth are displaced onto the lateral surface of the orbitonasal lamina, where they persist (Figs. 4c, d, 9a–c) as might be expected in a bradyodont chondrichthyan. On Meckel's cartilage, such teeth are retained within the broad trough on the lateral, labial surface. Up to nine individual teeth are present within each generative set (Fig. 10). Tooth fusion into what might considered rudimentary dental plates occurs between the largest and youngest teeth of the central sets (fourth and fifth tooth sets, counting from the anterior), creating the Pleuroplax tooth form. There is no evidence of fusion between adjacent tooth sets, contra previous suggestions that such fusion occurred at the jaw symphyses (Moy-Thomas 1936).

Within each battery, members of adjacent tooth sets are spatially offset, linguo-labially, so that tooth bases interdigitate (Fig. 10b). This characteristic occurs in the close-packed dentitions of extant sharks (e.g. Underwood et al. 2016), but is rarely observed in Paleozoic forms. Each battery dentition is markedly heterodont, exhibiting a suite of allometric trends from posterior (caudal) to anterior (rostral), and throughout the ontogeny of an individual tooth set. Successively younger and larger members of each set are increasingly dissimilar to members of neighboring (directly adjacent) sets, and the growth ratios of each set differ. Listed from anterior to posterior, passing through the tooth battery, the ratios of the maximum tooth length from largest (youngest) to smallest (oldest) per generative set, are as follows: 1.8:1; 3.2:1; 4.7:1; 3.9:1; 3.6:1; 4.1:1; 2.4:1. Further to this, the close spacing between tooth sets is conserved throughout ontogeny. Thus, the oldest and smallest teeth are clustered together tightly on the labial surface of the jaw cartilage. The net morphological effect of these trends is a radiating series of massive, tritor-like ridges (Fig. 4b) scrolling and converging from lingual (youngest) to labial (oldest) surfaces of upper and lower jaws (Fig. 10c). Moreover, each crown contributing to such a ridge is capped with tritoral, tubate dentine (Johanson et al., this vol.). The curvature of the anteriomost tooth sets, driven by the generative accretion of ever longer teeth at the lingual base, is helical, corkscrewed towards the rostrum. Within the set, the anterior flank of each tooth combines with neighbouring tooth flanks to create a concave shelf ventral to the nasal capsule (Fig. 10), perhaps contributing structurally to the nasal capsule floor. The asymmetry of older teeth from first and last sets is extreme, quite different from standard depictions of *Helodus simplex* morphology.

Diclitodus teeth (Fig. 11a–c) have long been associated with *Helodus simplex*. Moy-Thomas (1936) concluded that Diclitodus teeth occurred at the symphysis of the lower jaw, possibly formed by fusion of parasymphyseal teeth to create the characteristic bicuspid crown, an opinion echoed by Patterson (1965). However, it is now evident that these bicuspid teeth, with their concave, two-pronged root, form a logarithmic spiral tooth set at the rostral midline, in front of the upper jaw (Figs. 4, 6, 7). Two specimens preserve the Diclitodus file, each providing complementary detail: NHMUK PV P8213e (Fig. 7a, b, d–h) and NHMUK PV P8213c (Figs. 7c, 11a). In NHMUK PV P8213e, the spiral tooth set includes 10 members, with the smallest (oldest) located within the midline circular opening of the rostral shield (Fig. 7). The largest (youngest) teeth are located in a trough on the ventral surface of the shield (Fig. 7f). Directly posterior to this tooth set, there is a second set of four monocuspid teeth, in which the smallest and oldest member lies at the anterior (Figs. 7e, 10c). The file is directed towards the midline, between the cusps of the largest Diclitodus tooth. These monocuspid teeth are smaller than Diclitodus teeth; the largest is only 25 % the width of the largest Diclitodus tooth. The shape of the tooth crowns is most similar to *Helodus triangularis* (Stahl 1999), but the size is considerably smaller than examples depicted.

In NHMUK PV P8213c, the Diclitodus set is partly dissociated (Fig. 11ai–v) and appears to include nine members. The rostral shield interrupts the series between third and fourth members (counting from the anterior). Importantly, close examination shows that the seventh tooth bears a narrow midline trough, incompletely dividing the base, but in eighth and ninth teeth this midline division is complete. Each consists of pair of monocuspid teeth flanking the midline. Unfortunately, the roots are incomplete. Thus, the complete generative set includes eleven members. The anteriormost and smallest (oldest) are bicuspid, and the posteriormost and largest (youngest) are monocuspid.

The lower jaw (Figs. 9, 12) lacks any trace of a symphysial tooth whorl. However, additional, previously unrecognized, sets of parasymphysial teeth are present (Fig. 11e; see also lower dentition in Fig. 10c). In apical view, the crowns are rounded triangular/tetrahedral. The widest, flat surface of the crown faces lingually and dorsally, directly above a root that is proportionally deeper than that of Diclitodus teeth. As in the latter, the proximal (ventral) margin of the root is concave, but in these teeth the concavity is deeper. In NHMUK PV P8213e (Fig. 9) there appear to be two sets of such teeth, poorly preserved, at the anterior of each ramus of the lower jaw, scrolling from lingual to labial margins. In NHMUK PV P8213c (Fig. 12), individual teeth are well preserved but scattered away from the jaw, and grouped as four clusters: two of three teeth, two of four teeth, plus one isolated tooth. If Diclitodus teeth engaged with these teeth, then it appears that the apices of the Diclitodus crowns would have bitten against a small pavement formed by the flattened, lingual and dorsal facing surfaces of such parasymphyseal teeth.

Hyoid and branchial arches. Preserved in NHMUK PV P8212 (Fig. 13), the hyoid arch includes left and right ceratohyals and a basihyal. The ceratohyals are rotated laterally, so that the mesial surface is dorsal and the lateral surface is ventral, and dorsoventrally compressed. Each ceratohyal was massive: the shaft diameter close to that of Meckel's cartilage, and quite unlike the slender ceratohyal reported in *Debeerius* (Grogan & Lund 2000). The general form is parallel sided, with a rounded keel forming a ventrolateral angle along the posterior half. When restored to in-life orientation, the *Helodus* ceratohyal displays a gentle curvature. A basihyal is present, consisting of no more than a small, independent cartilage situated directly between and dorsal to the anterior ends of the ceratohyals. The short slender cartilage adjacent to the proximal head of the left ceratohyal is tentatively interpreted as an epibranchial rather than hyomandibula, because it closely resembles a similar cartilage associated with the right-side first ceratobranchial.

As noted by Moy-Thomas (1936), gill cartilages are present, slender, and the first arch is positioned directly posterior to the hyoid arch. It appears that the gill skeleton was fully subcranial. The ceratobranchial matches ceratohyal curvature, but is about 20 % shorter. The epibranchial is simple, rod-shaped, straight, and about a third of ceratobranchial length. Other gill arch cartilages, including incomplete ceratobranchials, are scattered around the pectoral girdle. No identifiable basibranchials, pharyngobranchials, cartilaginous gill rays and/or filaments have been found.

Pectoral girdle. The pectoral girdle is known from two specimens. Most notably, NHMUK PV P8207 preserves the full height of the scapulocoracoid, but lacks morphological detail. NHMUK PV P8212 (Fig. 13) preserves the coracoid and pectoral fin articulation, rotated laterally (like the ceratohyals), so that lateral and posterior surfaces face ventrally. Significantly, restoration of pectoral girdle cartilages to a likely in-life orientation reveals that the articular ridge or crest (the 'glenoid' in Didier 1995) for the pectoral fin is oriented almost dorsoventrally, as in modern chimaeroids and *Debeerius* (Grogan & Lund 2000). No evidence of foramina for the passage of nerves and blood vessels has been found.

Discussion

The precerebral fontanelle and the ethmoid canal. Much has been written about the origin of the chimaeroid ethmoid canal containing the anterior portions of the superficial ophthalmic tracts, and the relationship of this uniquely chimaeroid feature to the endocranial space (Allis 1926, De Beer & Moy-Thomas 1935, Moy-Thomas 1936, Holmgren 1942, Patterson 1965). Little has been concluded, with repeated appeals for data from developmental series, and from fossils showing plesiomorphic and/or intermediate conditions. To a limited extent, such fossil data are now available from stem holocephalans. The symmoriid *Dwykaselachus* (Coates et al. 2017), in which the superficial ophthalmic tracts re-enter the chondrocranium anteriorly with no obvious exit onto the cranial roof, shows no clear trace of a precerebral fontanelle, and no indication of a defined endocranial compartment relating to the ethmoid canal. Similarly, iniopterygians (Pradel 2010) also show evidence of superficial ophthalmic tracts re-entering the cranial cartilage anteriorly, but in this instance passing directly through a short passage into the nasal capsule. Again, no precerebral fontanelle is present. Here, *Helodus simplex* presents a seemingly conflicting character combination. The conjectured precerebral fontanelle of the otherwise chimaeroid-like, extended ethmoid beak occurs in conjunction with paired, extended canals connecting the anterior of the orbit with the nasal capsules.

None of these fossil morphologies is consistent with existing transformational scenarios. *Helodus* and iniopterygian data

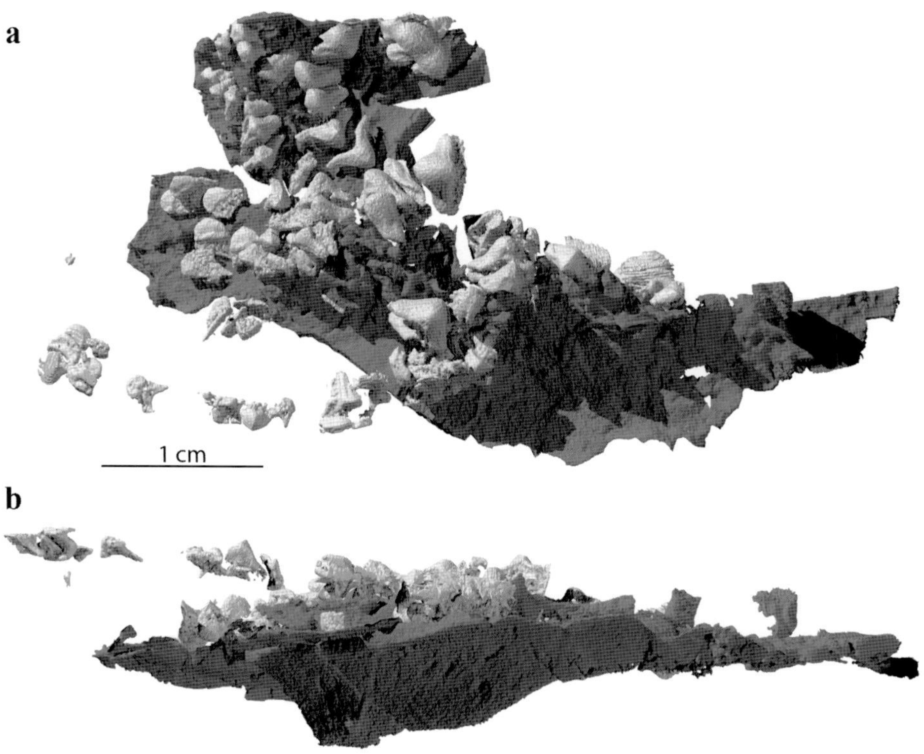

Fig. 12. *Helodus simplex*, NHMUK PV P8213c lower jaw with scattered dentition, including anterior, parasymphysial teeth, in dorsal (**a**) and left lateral (**b**) view.

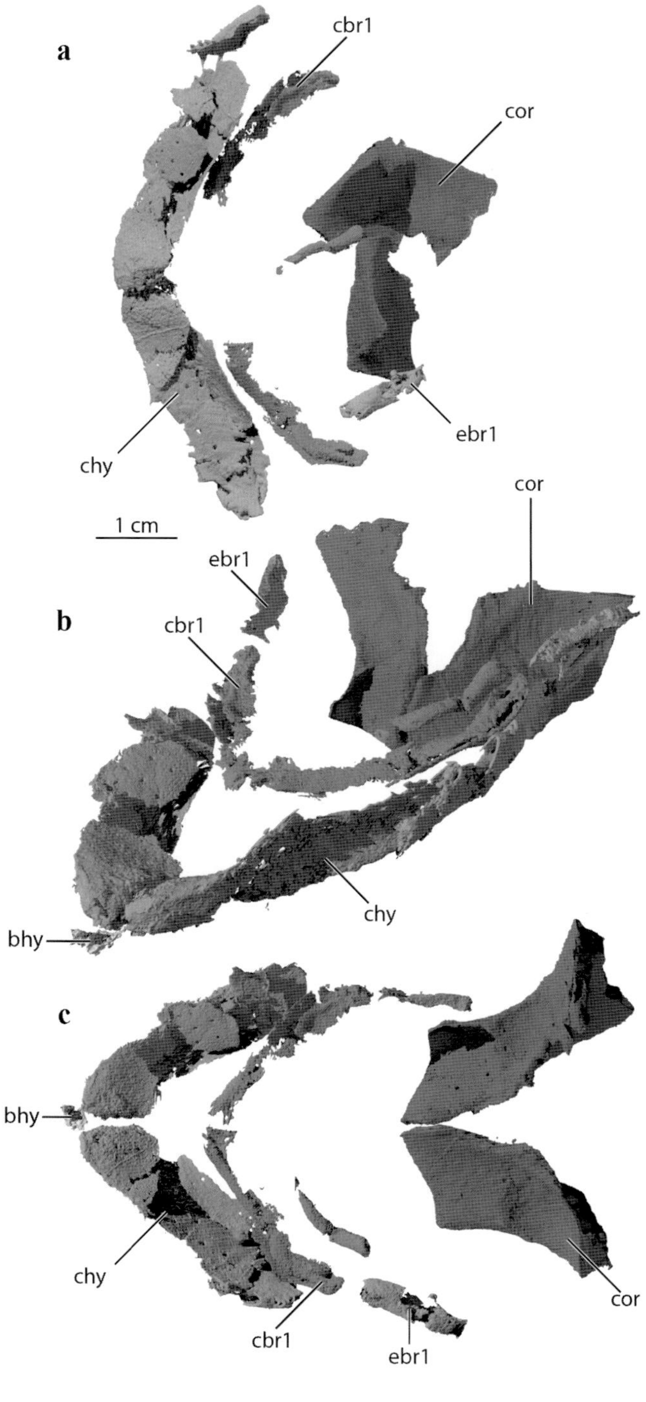

Fig. 13. *Helodus simplex*, NHMUK PV P8212, hyoid arch, gill arch, pectoral girdle revealed in dorsal view (**a**); restored to in-life positions, in antero lateral (**b**) and dorsal (**c**) view. Abbreviations: bhy, basihyal; cbr1, first ceratobranchial 1; chy, ceratohyal; cor, coracoid; ebr1, first epibranchial.

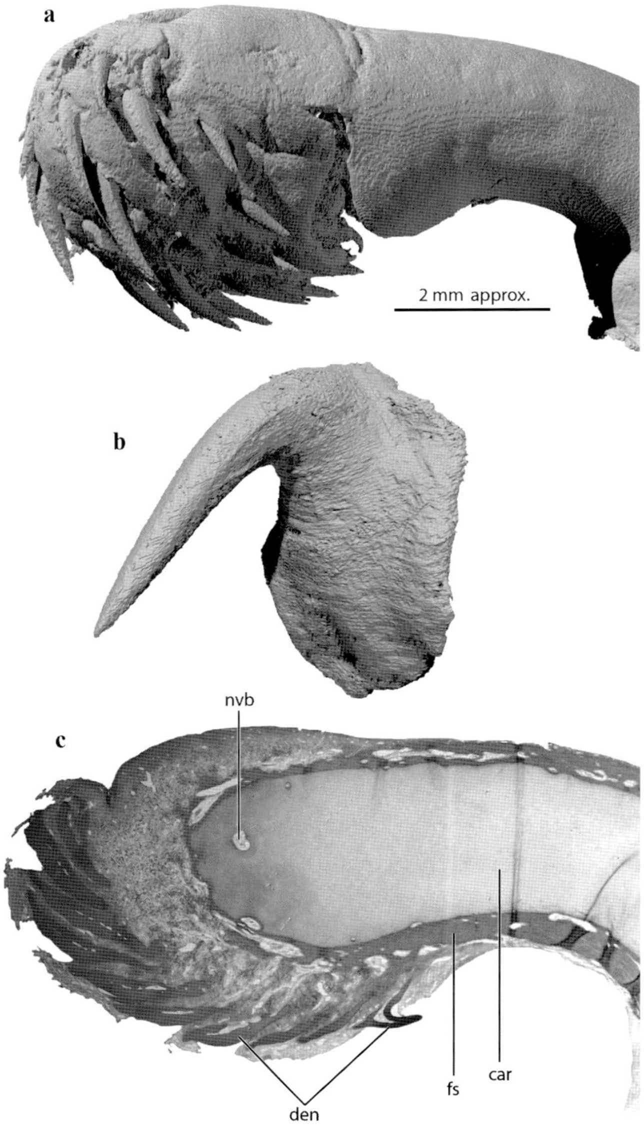

Fig. 14. *Hydrolagus collei*, cranial tenaculum (frontal clasper): gross morphology rendered from CT scan (**a**); individual (**b**); histology, sagittal section (**c**) (hemotoxylin and eosin staining).

are, however, supportive of De Beer & Moy-Thomas's (1935) argument that the spaces occupied by internalised superficial ophthalmic tracts are extra-cranial: these tubes and canals lie external to the dura mater. Therefore, Holmgren's (1942) view, favored by Patterson (1965), that the ethmoid canal originates as a segmented-off part of the cranial cavity, now appears unlikely. A simple alternative scenario, involving elevation of the orbitonasal laminae in *Helodus* (as argued by De Beer & Moy-Thomas 1935) combined with lateral narrowing of the ethmoid region, could have resulted in the coalescence of left and right canals. In this sequence of changes, the precerebral fontanelle would be pinched shut (in males, beneath a frontal clasper). This scenario has yet to be tested against the results of phylogenetic analyses incorporating these new data. And even without such analysis, it appears highly likely that the pattern of character transformations will be riddled with homoplasy, likely involving several independent closures of the precerebral fontanelle.

The otoliths. The present work provides the first report of otoliths in any fossil member of the holocephalan total group, sensu Coates et al. (2017). The only detailed description of otoliths in a living chimaeroid are those of *Callorhinchus milii* (Gauldie et al. 1987). Like *Helodus*, only two are present, rather than the three in modern Actinopterygii (Nolf 1985), and the otoliths positioned at right angles to each other, with one approximately double the size of the other. However, Gauldie et al. (1987) homologized the *Callorhinchus* otoliths with the teleost sagitta and asteriscus. Digital images of chimaeroid otoliths are available in the public domain, although not recorded as such. The NSF Digital Library at the University of Texas Austin (digimorph.org) reveals a pair of otoliths in the otic labyrinth of *Hydrolagus colliei*, and several examples are displayed in the anatomy section of the Chondrichthyan Tree of Life project web site (sharksrays.org): *Callorhinchus milii* (USNM 320578), *Rhinochimaera atlantica* (USNM 400698), *Hariotta raleighana* (USNM 320579), *Neoharriotta carri* (USNM 400695), *Chimaera cubana* (USNM 400700). These digital images lack detail, but all show the otoliths located at the anterior of the labyrinth, and the phylogenetic distribution is sufficient to suggest that otoliths are characteristic of the chimaeroid crown group. Discovery of otoliths in *Helodus simplex* further suggests a more inclusive distribution, covering a significant part of the stem group. Rather than otoliths representing a plesiomorphic feature of all teleostomes (Schultze 1990), it appears that otolith occurrence deserves re-examination in terms of composition, number, anatomical and phylogenetic distribution.

Diclitodus teeth: arrangement, location, identity, and function. The tooth form Diclitodus (Davis 1883) has long been associated with *Helodus*, and thought to represent fusion of two small monocuspid teeth or tooth sets at the mandibular symphysis (Moy-Thomas 1936, Patterson 1965, Stahl 1999). The lower jaw placement reflects the general occurrence of mandibular symphysial and parasymphysial tooth whorls among early gnathostomes; the few notable exceptions occur among the Eugeneodontidae, sensu Zangerl (1981). Discovery that Diclitodus teeth originate anterior to the upper jaw, although unexpected, in the present context is consistent with other observations on the dentition. The revelation that dental fission, rather than fusion, occurred as growth progressed is, to the best of our knowledge, unprecedented. More precisely, within the Diclitodus set, fission must have occurred within the generative tissues: one tooth is replaced by two, implying an underlying division of the single (parent) midline tooth germ into two, with each (child) germ retaining only one half of the regulation of tooth crown morphology.

Diclitodus teeth are not located on any part of the cranial cartilage likely to be derived from the mandibular arch. Speculatively, this likely distanced such teeth from the developmental patterning of the varied, yet mutually similar, members of the cheek teeth batteries exhibiting classic *Helodus* morphologies. Concerning observable phenomena, and as noted in the description, Diclitodus teeth, as well as a short whorl of small monocuspid denticles, are generated on the underside of a separate, dorsal, midline, cartilage (Fig. 6). We interpret this cartilage as a frontal clasper or cranial tenaculum. In terms of its location, size, and apparent articulation with the cranium, it resembles the frontal clasper of *Squaloraja* (Patterson 1965, Stahl 1999). Rather than teeth, Diclitodus members might, therefore, be more properly described as tenacular denticles. Denticles, however, rarely display such a tooth-like morphology and tooth-like generative series. Nevertheless, further investigation shows that such characteristics are shared with the tenacular (frontal clasper) denticles of modern chimaeras (Fig. 14a). Tenacular denticles are tooth-like, both individually and as a series, suggesting that they, too, represent a generative set. Each denticle is monocuspid, with strongly recurved crown and a broad extended base (Fig. 14b). Each crown base is positioned upon the extended base of the preceding denticle in the series (Fig. 14c), strongly suggesting that these structures develop in a successive tooth-like manner. In histological section, the denticle series is almost indistinguishable from sections through elasmobranch mandibular tooth sets, including an open pulp cavity in the youngest denticles, becoming progressively filled with dentine in older denticles. However, our sections, both histological and virtual, found no evidence of a dental lamina or ongoing denticle growth. It appears that denticle generation is spatially and temporally restricted.

Taken together, these observations support the suggestion that *Helodus* provides a glimpse of frontal clasper evolution, and, if correctly interpreted, this is the earliest frontal clasper in the holocephalan fossil record. One possible origination scenario is that these specialized denticles might have originated from, or were closely associated with, the anterior, upper, and pre-chordal (Maisey et al. 2014) dentition. Chimaeroid frontal clasper function is not well documented, but it seems likely that in males the structure is used to gain purchase on a female for stability during copulation, likely via grasping the trailing edge of the pectoral fin. In modern elasmobranchs, the anterior dentition is used similarly, with males of several taxa documented as biting

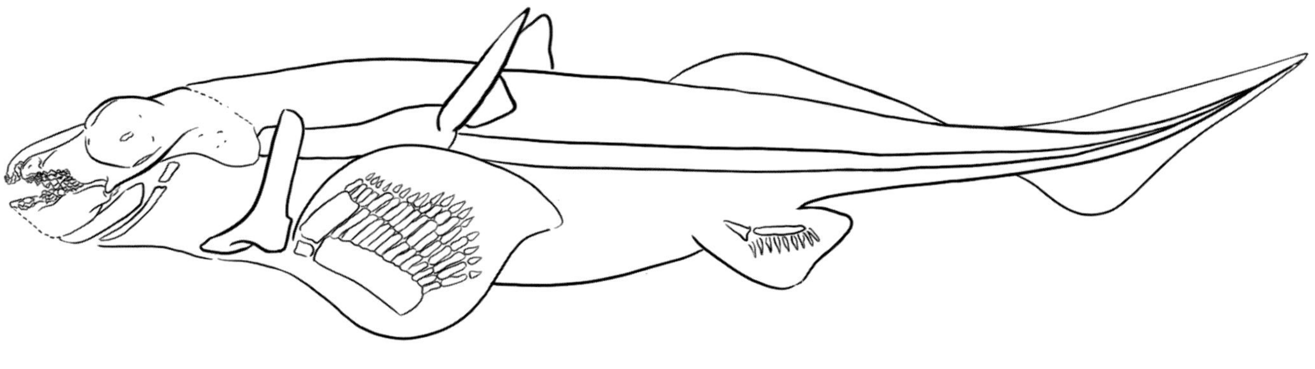

Fig. 15. *Helodus simplex*, restoration, updated from Patterson 1965, fig. 38, with body proportions taken from NHMUK PV P8207, part and counterpart of complete individual, laterally compressed, and P8212, the dorsoventrally compressed individual shown in Figure 2. The pectoral fin skeleton is unchanged from Stahl, 1999, fig. 33.

the female pectoral fin before and during mating (Conrath & Musick, 2012). It seems likely that *Diclitodus* denticles served similarly for *Helodus*. The denticles, with their short cusps, are much smaller than the cheek teeth (Figs. 5, 9), and, in ventral view, the clasper and denticle assembly resembles nothing so much as an office stapler. These would bite against the small platform of monocuspid denticles at the anterior of the mandible. Unlike *Squaloraja*, *Helodus* bears no obvious fossae for clasper retractor muscles, but perhaps little in the way of specialized clasper musculature was present: merely a dorsal extension of the pre-orbital, superficial, jaw adductor musculature.

Dentition revised. The revealed complete dentition is one of the most striking changes to the previous *Helodus simplex* reconstruction. Comparison with the standard depiction (Moy-Thomas 1936; Stahl 1999) (Fig. 1) highlights specializations in tooth morphology and tooth generation. Formerly, *H. simplex* teeth were thought to be uniformly sized and shaped along the jaw margin, implying that teeth were generated in evenly and generously spaced, parallel sets throughout life with only modest growth of individual tooth size. The exceptions to these regularities were the production of Pleuroplax teeth as nascent dental plates, and the symphysial placement of *Diclitodus* teeth (discussed below). Such growth resembles generalized shark tooth production, with generative tooth sets maintaining regular spacing along the cartilaginous jaws throughout ontogeny.

In contrast, the new arrangement shows that tooth size increased considerably, with greatest growth in sets within the middle of each dental battery. Tooth sets are abutted and interdigitated, so that the entire battery forms a single functional unit (Figs. 4b,d,e, 10). Tooth sets are unevenly sized at their ontogenetic outset, with the largest, located in the middle of the battery, becoming increasingly so as growth proceeds. Pleuroplax tooth forms emerge within the largest and ontogenetically youngest members of tooth set 4, and not within the anteriormost sets, as suggested by Moy-Thomas (1936; Fig. 1b). Thus, tooth sets in the middle of the dental battery occupy increasing proportions of the tooth generation area, and the total area of tooth generation as a proportion of jaw length likely increased throughout the life of an individual. As noted in the description, this allometric growth results in a functionally compound dental plate with tritor ridges radiating from labial to lingual surfaces (the reverse of the pattern in lungfish), with each ridge composed of serial tooth crowns. Further to this, careful examination of some of the extremely asymmetric teeth from tooth sets 1 and 7 (Fig. 10) might well explain the occurrence of at least some of the tooth-form variants contributing to the *Helodus* "species problem" (Stahl 1999, p. 56). Likely candidates include labio-lingually slender variants such as *H. coniculus*, *H. appendiculatus* and *H. didymus* (Stahl 1999).

However, evidence from the partially preserved dentitions of *Psephodus* (Stahl 1999, Ginter & Piechota 2004) demonstrates that at least some *Helodus*-like teeth, perhaps including *H. didymus* (Stahl 1999, p. 61), are derived from a different kind of holocephalan. Crowns of smaller, anterior members of *Psephodus* dentitions are barely distinguishable from *Helodus*, although posterior tooth sets include one or more large, flat, dental plates. For this reason, *Psephodus* has been included within the Cochliodontiformes (Stahl 1999) – a group not usually linked closely to *Helodus*. Thus *Helodus*-like crowns might be convergent – common solutions to demands on durophagy – or simply symplesiomorphic – persisting among the incompletely plate-bearing dentitions of early holocephalan lineages. Root shapes appear to be more diagnostic: *Psephodus* root profiles are sinusoidal (Ginter & Piechota 2004) whereas those of *Helodus* are concave.

Further phylogenetic signal might, instead, be gained from characteristics of the entire dentition. Few of the complete or near-complete dentitions known from early chondrichthyans are comparable to that of the revised *Helodus* described here. Petalodont dentitions are densely packed (Ginter et al. 2010, Lund et al. 2014), but the tooth sets, where known, are oriented, and were likely generated, in caudal-to-rostral direction. None shows an array of teeth wrapped around the biting margin and retained on the outer, labial, face of the jaw (Figs. 4, 6). By far the closest comparison is with the 'pavement' dentitions of eugeneodontids, and especially edestoids such as *Agassizodus variabilis* (Zangerl 1981, Ginter et al. 2010, MIC personal observation of MCZ 752). The size of the *Agassizodus* jaw fragment bearing these teeth suggests that this fish was an order of magnitude larger than *Helodus simplex*, with at least 14 tooth sets within each pavement. However, the dental battery growth characteristics are strikingly similar, with dominance of the central region in each jaw ramus. Closer examination yields further similarities: the concave root profile is also evident in eugeniodontids, from *Fadenia* (Zangerl, 1981) to *Parahelicampodus* (Ginter et al., 2010).

Conclusion and updated restoration. The abundance of new data presents an opportunity (after some 50+ years) to take a fresh look at the whole fish, and a new restoration is presented in Figure 15. Body proportions, from NHMUK PV P8207 and P8212, are slightly shorter rostro-caudally, the pectoral girdle is larger, and likewise the pectoral fin here rotated with leading edge uppermost (the pectoral skeleton unmodified from Stahl 1999). In terms of phylogenetic systematic character content, *Helodus* updated presents a fish that aligns reasonably well with recent hypotheses of holocephalan phylogeny. The cranial morphology is more recognizably chimaeroid, and reference to Patterson's (1965) defining list of chimaeroid characteristics lends support to a series of uncontroversial synapomorphies: holostylic suspension; tubular dentine; subcranial branchial arches; synarcual presence; pectoral fin structure; cranial or frontal clasper presence; ethmoid canal present (although paired). Correspondingly, *Helodus* is confirmed as differing from more crownward chimaeroids in the following features: absence of an interorbital septum above the brain; presence of numerous teeth; absence of a single ethmoid canal; absence of notochordal calcifications; structure of the fin spine; presence of complete squamation; absence of crescentic scales surrounding sensory canals; absence of pre-pelvic claspers. To this list we add the likely presence (persistence) of a precerebral fontanelle.

Aside from these straightforward affinities with conventional chimaeroids, *Helodus* similarities with edestids sensu lato might prove to be of greater significance. Patterson (1965), too, noted similarities between edestid and *Helodus* dentitions, thereby maintaining a hypothesis of edestid-holocephalan relationship that dates from Nielsen (1932) and lasted through to Moy-Thomas and Miles (1971). However, largely in the wake of Radinsky (1961) and Zangerl (1981), edestids have mostly been classified as elasmobranchs (see also Janvier 1996). Resolution now seems possible, given the abundance of symmoriid-edestid skeletal similarities (Moy-Thomas and Miles 1971) and placement of symmoriids as stem-holocephalans (Coates et al. 2017). The edestid-*Helodus*-holocephalan nexus deserves re-examination.

Acknowledgements

It is with great pleasure that we present this work in honor of Dr. John Maisey, his numerous contributions to the study of fossil fishes, and with particular respect to his groundbreaking studies of fossil sharks. For access to specimens, we thank Emma Bernard of the Natural History Museum, London, UK, and Stephanie Pierce of the Museum of Comparative Zoology, Harvard University, USA. This project, in particular the CT scanning of specimens , was supported by National Science Foundation grants DEB-0917922 and DEB-1541491 (to MIC) and NHM Core Research Labs support (to ZJ). We thank Alan Pradel and two further reviewers for their diligence and insightful suggestions.

Literature cited

Agassiz, L. 1838. Recherches sur les poissons fossils Vol. 3. Neuchatel, Switzerland: Petitpierre.

Allis, E. P. 1926. On the homologies of the prechordal portion of the skull of the Holocephali. Journal of Anatomy, 60: 335–340.

Bonaparte, C. 1832. Selachorum tabula analytica. Nuovi Annali della Scienza Naturali (Bologna), 1: 195–214.

Coates, M. I., R. W. Gess, J. A. Finarelli, K. E. Criswell, K. E. & K. A. Tietjen. 2017. A symmoriiform chondrichthyan braincase and the origin of chimaeroid fishes. Nature, 541: 208–211.

Conrath, C. L. & J. A. Musick. 2012. Reproductive biology of elasmobranchs. Pp. 291–311 in Carrier, J. C., J. A. Musick & M. R. Heithaus. (eds.) The biology of sharks and their relatives. New York, NY, CRC Press.

Davis, J. W. 1883. On the fossil fishes of the Carboniferous Limestone Series of Great Britain. Transactions of the Royal Dublin Society, series 2, 1: 327–548.

De Beer, G. R. & J. A. Moy-Thomas. 1935. On the skull of the Hholocephali. Philosophical Transactions of the Royal Society of London, Series B, 224: 287–312.

Didier, D. A. 1995. Phylogenetic systematics of extant chimaeroid fishes (Holocephali, Chimaeroidei). American Museum Novitates, 3119: 1–86.

Dineley, D. L. & S. J. Metcalf. 1999. Fossil fishes of Great Britain. Joint Nature Conservation Committee, Peterborough, UK, 675 pp.

Finarelli, J. A. & M. I. Coates. 2014. *Chondrenchelys problematica* (Traquair, 1888) redescribed: a Lower Carboniferous, eel-like holocephalan from Scotland. Earth and Environmental Science Transactions of the Royal Society of Edinburgh, 105: 1–25.

Gauldie, R. W., K. Mulligan & R. K. Thompson. 1987. The otoliths of a chimaera, the New Zealand elephant fish *Callorhynchus milii*. New Zealand Journal of Marine and Freshwater Research, 21: 275–280.

Ginter, M. & A. Piechota. 2004. The first Devonian holocephalian tooth from Poland. Acta Palaeontologica Polonica, 49: 409–415.

Ginter, M., O. Hampe & C. Duffin. 2010. Chondrichthyes volume 3D, Paleozoic Elasmobranchii: Teeth. – In H.-P. Schultze. (ed.) Handbook of Paleoichthyology. Verlag Dr. Friedrich Pfeil, München, Germany, 168 pp.

Grogan, E. D. & R. Lund. 2000. *Debeerius ellefseni* (Fam. Nov., Gen. Nov., Spec. Nov.), an autodiastylic chondrichthyan from the Mississippian Bear Gulch Limestone of Montana (USAD), the relationships of the Chondrichthyes, and comments on gnathostome evolution. Journal of Morphology, 243: 219–245.

Holmgren, N. 1942. Studies on the head in fishes. Part III. Acta Zoologica, 23: 129–261.

Huxley, T. H. 1880. A manual of the vertebrated animals. D. Appleton, New York, 431 pp.

Janvier, P. 1996. Early Vertebrates. Oxford Monographs on Geology and Geophysics 33. Clarendon Press, Oxford, UK, 393 pp.

Johanson, Z., C. Underwood, M. I. Coates, V. Fernandez, B. Clark & M. M. Smith. 2021. The stem-holocephalan *Helodus* (Chondrichthyes; Holocephali), and the evolution of modern chimaeroid dentitions. – Pp. 205–214 in Pradel, A., J. S. S. Denton & P. Janvier (eds.), Ancient Fishes and their Living Relatives: a Tribute to John G. Maisey. Verlag Dr. Friedrich Pfeil, München.

Kesteven, H. L. 1933. The anatomy of the head of *Callorhynchus antarcticus*. Journal of Anatomy, 67: 443–474.

Lisney, T. J. 2010. A review of the sensory biology of chimaeroid fishes (Chondrichthyes: Holocephali). Reviews in Fish Biology and Fisheries, 20: 571–590.

Lund, R. 1982. *Harpagofututor volsellorhinus* new genus and species (Chondrichthyes, Chondrenchelyiformes) from the Namurian Bear Gulch Limestone, *Chondrenchelys problematica* Traquair (Viséan) and their sexual dimorphism. Journal of Paleontology, 56: 938–958.

Lund, R., E. D. Grogan & M. Fath. 2014. On the relationships of the Petalodontiformes (Chondrichthyes). Paleontological Journal, 48: 1015–1029.

Maisey, J. G., S. Turner, G. J. P. Naylor & R. F. Miller. 2014. Dental patterning in the earliest sharks: implications for tooth evolution. Journal of Morphology, 275: 586–596.

Moy-Thomas, J. A. 1935. The structure and affinities of *Chondrenchelys problematica*, Traquair. Proceedings of the Zoological Society of London, 105: 391–404.

Moy-Thomas, J. A. 1936. On the structure and affinities of the Carboniferous cochliodont *Helodus simplex*. Geological Magazine, 73: 488–503.

Moy-Thomas, J. A. & R. S. Miles. 1971. Palaeozoic Fishes. 2nd edition. Chapman and Hall, London, UK, 259 pp.

Nielsen, E. 1932. Permo-Carboniferous fishes from East Greenland. Meddelelser om Grønland, 86: 1–63.

Nolf, D. 1985. Otolithi piscium. – In H.-P. Schultze (ed.), Handbook of Paleoichthyology, Vol. 10. Gustav Fischer, Stuttgart, New York, 145 pp.

Patterson, C. 1965. Phylogeny of the chimaeroids. Philosophical Transactions of the Royal Society of London, Series B, 249: 101–219.

Pradel, A. 2010. Skull and brain anatomy of Late Carboniferous Sibyrhynchidae (Chondrichthyes, Iniopterygia) from Kansas and Oklahoma (USA). Geodiversitas, 32: 595–661.

Pradel, A., M. Langer, J. G. Maisey, D. Geffard-Kuriyama, P. Cloetens, P. Janvier & P. Tafforeau. 2009. Skull and brain of a 300 million-year-old chimaeroid fish revealed by synchrotron holotomography. Proceedings of the National Academy of Sciences of the USA, 106: 5224–5228.

Pradel, A., P. Tafforeau, J. G. Maisey & P. Janvier. 2011. A new Paleozoic Symmoriiformes (Chondrichthyes) from the Late Carboniferous of Kansas (USA) and cladistic analysis of early chondrichthyans. PloS ONE 6: e24938. DOI: 10.147/journal.pone.0024938.

Radinsky, L. 1961. Tooth histology as a taxonomic criterion for cartilaginous fishes. Journal of Morphology, 109: 73–92.

Schultze, H.-P. 1990. A new acanthodian from the Pennsylvanian of Utah, U. S. A., and the distribution of otoliths in gnathostomes. Journal of Vertebrate Paleontology, 10: 49–58.

Stahl B. J. (1999) Chondrichthys III: Holocephali. – In H.-P. Schultze (ed.), Handbook of Paleoichthyology, Vol. 4, Verlag Dr. Friedrich Pfeil, München, Germany, 164 pp.

Summers, A. P. 2000. Stiffening the stingray skeleton -an investigation of durophagy in myliobatid stingrays (Chondrichthyes, Batoidea, Myliobatidae). Journal of Morphology, 243: 113–126.

Summers, A. P., T. J. Koob & E. L. Brainerd. 1998. Stingray jaws strut their stuff. Nature, 395: 450–451.

Underwood, C., Z. Johanson & M. Meredith Smith. 2016. Cutting blade dentitions in squaliform sharks form by modification of inherited alternate tooth ordering patterns. Royal Society Open Science 3(11): 160385.

Ward, J. 1875. On the organic remains of the Coal-Measures of North Staffordshire, their range and distribution, with a catalogue of the fossils and their mode of occurrence. Transactions of the North Staffordshire Field Club, 1875: 184–251.

Wood, S. P. 1975. Recent discoveries of Carboniferous fishes in Edinburgh. Scottish Journal of Geology, 11: 251–258.

Zangerl, R. 1981. Chondrichthyes I: Paleozoic Elasmobranchii. – In Schultze, H.-P (ed.), Handbook of Paleoichthyology, Vol. 3A, Gustav Fischer Verlag, Stuttgart, Germany, 113 pp.

Zangerl, R. & G. R. Case. 1973. Iniopterygia: a new order of chondrichthyan fishes from the Pennsylvanian of North America. Fieldiana Geology, Memoirs 6: 1–67.

Ancient Fishes and their Living Relatives: a Tribute to John G. Maisey. Alan Pradel, John S. S. Denton & Philippe Janvier (eds.): pp. 205–214, 7 figs.
© 2021 by Verlag Dr. Friedrich Pfeil, München, Germany – ISBN 978-3-89937-269-4

The stem-holocephalan *Helodus* (Chondrichthyes; Holocephali) and the evolution of modern chimaeroid dentitions

Zerina Johanson*, Charlie Underwood*,**, Michael I. Coates***, Vincent Fernandez****,*****, Brett Clark**** and Moya M. Smith*,******

Abstract

The group Chondrichthyes includes shark, rays, and the holocephalans. The latter, particularly crown group holocephalans (Chimaeroidei), are unusual in several respects when compared to the sharks and rays. For example, the dentition comprises non-shedding dental plates composed of mixed dentine types, differentially arranged and renewed by continuous, aboral growth rather than by separate and replaceable teeth. However, stem group holocephalans include taxa possessing more shark-like dentitions with individual teeth arranged into generative sets (most Symmoriida, some Iniopterygia), some of which may be shed. Other taxa show basal fusion of the teeth to form tooth whorls, retaining separate crowns (e. g., certain Iniopterygia). *Helodus* is a stem holocephalan within which a shark-like dentition, including generative tooth sets, some of which may be shed, is present. Some nominal species of *Helodus*, however, show fusion of the tooth bases, such as seen in *H. appendiculatus*. Fusion also occurs in *H. simplex*, in more complete specimens preserving upper and lower dentitions *in situ*, involving tooth bases and the teeth themselves. These fused teeth, when found isolated as fossils, have been assigned to different genera, including *Diclitodus* and *Pleuroplax*. Here we show that tooth fusion in '*Pleuroplax*' occurs later in individual generative sets, with specimens preserving smaller, separate teeth labially (oldest), and fusion occurring lingually (youngest). The presence of tritoral dentine of similar histology in the teeth of *Helodus* and '*Pleuroplax*', and crown holocephalans, provides a clear developmental and evolutionary sequence for the biting tritors in the latter.

Introduction

Helodus, originally named by Agassiz (1838), was initially only known from abundant and diverse isolated teeth and series of associated teeth (Agassiz 1838, Woodward 1889, Stahl 1999) from both marine and non-marine rocks of Late Devonian to Early Permian age (Stahl 1999). Subsequent discoveries of the articulated braincases, jaws, and postcranial material of *Helodus simplex* (Davis 1890, Woodward 1889, 1921, Moy-Thomas 1936, 1938, Coates et al. 2018, this volume), allowed the genus to be recognised as a stem-group holocephalan. Tooth sets within *Helodus* comprise separate successional teeth, for example in *H. simplex* (Fig. 1, 2), comparable to the dentition of sharks and rays (Elasmobranchii), but other forms show fusion of the tooth bases, as in species such as *H. appendiculatus* (Fig. 3A–E) and *H. coxanus* (Stahl 1999: fig. 53). In addition, articulated braincases and jaws of *H. simplex* preserve the dentition in situ including successional, or generative, sets of separate teeth (Fig. 2), along with other sets showing various degrees of fusion (Moy-Thomas 1936, Patterson 1965, Bendix-Almgreen 1968, Coates et al. this volume). When found isolated, these dental elements have been assigned to the genera *Pleuroplax* (for the fused parts of the lateral dentition; Fig. 4) and *Diclitodus* (for the putative symphyseal dentition, Newberry & Worthen 1866, Moy-Thomas 1936, Stahl 1999, Ginter & Sun 2007, but see Coates et al. this volume; Fig. 5). As a result, the 'genera' *Diclitodus* and *Pleuroplax* are now considered as synonyms of *Helodus* (Moy-Thomas 1936, Patterson 1965, Stahl 1999). Specimens of '*Pleuroplax*' are convex and rounded along their long axis (in situ labio-lingual), with paired lobes on either side of this axis, and fine ridges along the surface of these lobes and on the main part of the toothplate itself (Stahl 1999). These lobes are similar to those found on either side of the main cusp in separate teeth of *Helodus* (Patterson 1965; Figs. 1, 4), so it seems reasonable to suggest that '*Pleuroplax*' results from fusion of separate, more 'typical' *Helodus* teeth (e. g., Stahl 1999), although this has never been demonstrated.

The more complete specimen of *Helodus simplex* described in detail by Moy-Thomas (1936) includes recognizable tooth sets and two regions with specimens of '*Pleuroplax*' in each jaw, but is poorly preserved and the dentition is only partially observable, with most of the fragments of dentition being preserved on separate blocks and not associated with the main skull (Coates et al. this volume). Nevertheless, the 'model' for the dentition of *Helodus* reconstructed by Moy-Thomas (1936) has often been assumed to be typical for the genus (e. g., Patterson 1965). The rarity of teeth of '*Pleuroplax*' morphology in assemblages of *H. simplex* teeth (Stahl 1999; authors' observation of museum collections) support the concept that this fused tooth form was only present within a limited number of dental positions and, as outlined below, only at certain growth stages. Additionally, the ratio of teeth assignable to *Helodus* (separate teeth) and '*Pleuroplax*' (fused teeth) morphotypes vary between occurrences. For example, some species typically assigned to

* Department of Earth Sciences, Natural History Museum, London, UK

** Department of Earth and Planetary Sciences, Birkbeck, University of London, London, UK

*** Department of Organismal Biology and Anatomy, University of Chicago, Chicago, USA; mcoates@uchicago.edu (corresponding author)

**** Image and Analysis Centre, Core Research Laboratories, Natural History Museum, London, UK

***** European Synchrotron Radiation Facility, Grenoble, France

****** Centre for Craniofacial and Regenerative Biology, Faculty of Dentistry, Oral & Craniofacial Sciences, King's College, London, UK.

'*Pleuroplax*', such as '*P*'. *rankini*, are associated with few, if any, non-fused teeth of *Helodus* morphology (Stahl 1999; authors' observation of museum collections).

Isolated specimens of '*Pleuroplax*' are generally preserved as a fused tooth whorl (e.g., Stahl 1999: fig. 55), but some specimens show a combination of basal fusion with separate teeth (Figs. 4, 6), described further below. μCT scans of three newly reassembled cranial blocks of *Helodus simplex* show tooth sets with similar fusion (Fig. 2), the in situ position of these on the jaw indicates that individual teeth are established in developmental order in each set (size gradient), with labial teeth being older and lingual teeth being younger, with fusion in the latter. Our new examination of the structural and histological arrangement of tooth crowns and basal tissues demonstrates similarities in *Helodus simplex* and '*Pleuroplax*', in other Palaeozoic holocephalans, and fossil and extant Chimaeroidei, identifying the tritoral dentine in the Chimaeroidei as homologous to tissue in the teeth of *H. simplex* (Fig. 7).

Materials and Methods

Specimens of *Helodus*, '*Diclitodus*' and '*Pleuroplax*' from the Earth Sciences collections, Natural History Museum London, were examined, including *Helodus simplex* (NHMUK PV P8216; NHMUK PV P8212; along with the reassembled cranium including NHMUK PV P8213e (2 parts) plus NHMUK PV P2907); *Helodus appendiculatus* (NHMUK PV P75938); '*Pleuroplax*' *rankini* (NHMUK PV OR33300); '*Pleuroplax*' sp. (NHMUK PV P8230); and '*Diclitodus*' *scitulus* (NHMUK PV P4890). Toothplates of *Edaphodon* sp. were μCT-scanned (NHMUK PV OR38771) and those of *Cochliodus contortus* were μCT-scanned and sectioned (NHMUK PV P75425).

X-ray computed tomography

Synchrotron scanning

NHMUK PV P75938 was characterised at the BM05 beamline of the European Synchrotron Radiation Facility (ESRF, Grenoble, France). The beamline was setup for propagation phase contrast synchrotron radiation micro-computed tomography. The X-ray beam consisted of white beam filtered with 100 mm of Aluminium (effective energy: 112 keV). Images were recorded using an indirect detector: 200 μm LuAG scintillator, ×1.5 optical magnification from a set a Hasselblad photographic lenses (Victor Hasselblad AB, Gothenburg, Sweden) and a PCO.edge 5.2 camera (PCO, Kelheim, Germany). A recorded pixel size of 4.25 μm was measured, taking into account the 2154 mm of sample-detector distance. Given the limited field of view (FOV) of 10.88 × 4 mm, 4 scans were necessary to cover the vertical axis. The centre of rotation was also shifted by 4.89 mm, almost doubling the horizontal reconstructed FOV.

Tomographic reconstruction was performed using the single distance phase retrieval approach of PyHST2 (Mirone et al. 2014). The 32-bit data were converted to 16-bit TIFF stack, using the 0.002 % saturation values from the 3D histogram generated by PyHST2. A ring correction was also applied during the conversion (Lyckegaard et al. 2011).

Nikon scanning

Helodus simplex (NHMUK PV P8216; reassembled NHMUK PV P8213e (2 parts) + NHMUK PV P2907), *Cochliodus contortus* (NHMUK PV P75425) and *Edaphodon* sp. (NHMUK PV OR38871) were characterised at the Natural History Museum (London, UK) using a Nikon HMXST 225. A filtered back projection algorithm was used to reconstruct the projections, using the CT-agent software (Nikon Metrology GmbH, Alzenau, Germany), producing a 16-bit uncompressed raw volume. Parameters used for X-ray micro Computed Tomography (XCT) are listed in Table 1.

Versa scanning

'*Pleuroplax*' *rankini* (NHMUK PV OR33300), '*Pleuroplax*' sp. (NHMUK PV P8230) and '*Diclitodus*' *scitulus* (NHMUK PV P4890) were characterised at the Natural History Museum (London, UK) using a Zeiss Xradia 520 Versa. Parameters used for X-ray micro Computed Tomography (XCT) are listed in Table 2. A filtered back projection algorithm with beam hardening correction was used to reconstruct the projections, using the paired software Reconstructor and Scout-and-Scan (Carl Zeiss Microscopy GmbH, Jena, Germany), producing a stack of 16-bit TIFFs.

Optical microscopy

From hard tissue sections through teeth of *Helodus appendiculatus* and *Cochliodus contortus*, photomicrographs were taken with a Zeiss Photomicroscope III in transmitted light (TL); polarized light with polarizer substage to condenser, analyzer slide rotatable (PL), slider for gypsum plate prism at 45° to crossed polars (GP); objectives, ×6.5, ×16, ×40, with prisms for Nomarski differential interference-contrast (DIC); Stereomicroscope Wild MPS45 ×6.5 to ×40 with annular adaptor for Kern Ring Illumination (IL), LED light 60T-B; both microscopes fitted with a monocular phototube for a CMOS digital camera, Basler (UK) 20819762; or a Nikon Coolpix 990.

Table 1: List of XCT parameters used for specimens characterised with the Nikon H MXST 225. 'Frame avg' indicates the number of projections averaged, recorded as a single image.

specimen	Voxel size (μm)	kV	μA	Exposure (ms)	filters	Number of projections	Gain (dB)	Frame avg
NHMUK PV P8216	13	140	93	500	Cu 0.5 mm	3142	0	4
NHMUK PV P8212	39	195	175	708	Cu 1.5 mm	3142	18	1
NHMUK PV P75425	8	140	93	500	Cu 0.5 mm	3142	0	4
NHMUK PV OR328871	50	195	180	708	Cu 1.5 mm	3142	24	1

Table 2: List of XCT parameters used for specimen characterised with the Zeiss Xradia 520 Versa. Filter are of undisclosed composition and thickness and are just indicating their usage for low or high energy (LE and HE series respectively). S–O is the source to object distance. O–D is the object to detector distance. The optical magnification consists of a lens (photographic or microscope), located after the scintillator, further increasing the magnification resulting from the X-ray cone beam geometry.

specimen	Voxel size (μm)	kV	μA	Exposure (sec)	Filter	Number of projections	S–O (mm)	O–D (mm)	Optical magnification
NHMUK PV OR33300	4.85	80	87	7	LE5	3201	18	110	0.39×
NHMUK PV P8230	10.43	120	83	6	HE2	4501	39	90	0.39×
NHMUK PV P4890	8.62	120	83	8	HE1	3201	50	150	0.39×

Description of dental elements

Helodus simplex

(Fig. 1, NHMUK PV P8216; Fig. 2, NHMUK PV P8213e, NHMUK PV P2907)

NHMUK PV P8216 comprises three separated, but basally articulated, teeth. These are closely spaced and regulated in size and shape, with new teeth added lingually, interpreted as an organized successional generative set (Fig. 1). Each tooth comprises one main cusp and two large lobes, one on either side. The surface of the main cusp is smooth, with thin ridges present more basally and on the lobes. In vertical section plane, the crown of the tooth is dominated by dentine deposited between and around vascular canals that are organized perpendicular to the tooth surface (different terms for this dentine are summarized by Stahl 1999); as discussed further below, a comparable type of dentine is found in tritors of chimaeroid toothplates (Smith et al. 2019: figs. 1, 6, 17; Fig. 7B,C,E) and in other stem group holocephalans such as *Cochliodus* (Cochliodontiformes; Fig. 7F–I). From the functional position and arrangement as 'tubate' dentine (many vascular canals from pulp cavity to surface), we refer to this as 'tritoral dentine' (Fig. 1E,G, vc, tri.d1). The crown is dominated by this vascular dentine, with a thin contiguous layer below, identified as a second tritoral dentine (Fig. 1D,E,G,H, tri.d2); perpendicular vascular canals in this contiguous layer link with those in the crown (e.g., Fig. 1D,E). The tooth base comprises typical trabecular dentine (Fig. 1D,E,G,H, td, in growth compacted as osteodentine), with vascular tissue spaces that are less regular than in the tritoral dentine of the crown. Beneath this is a thin layer of laminar dentine, from appositional growth (Fig. 1E,G,H, ld). A small pulp cavity remains and is continuous with cavities of the newer teeth (Fig. 1G, pc), large vascular canals open into these (Fig. 1G, asterisk). Neither teeth, nor tooth bases are fused, but there is sutured interlocking contact of each base (Fig. 1D, small white arrowheads), very similar to the sutured contact between teeth of myliobatid rays (e.g. Underwood et al. 2015: fig. 3E). A small tooth is present ventrally beneath the tooth set, due to post-mortem dislocation of a small, earlier formed tooth of the same tooth set (Fig. 1C,D,F–H, sm.th). In virtual section, this smaller tooth shows that the same four main tissues are present, as in the larger teeth (Fig. 1H). As well, the lobular extensions from the main cusp are smaller, and there are fewer surface ridges present (Fig. 1C).

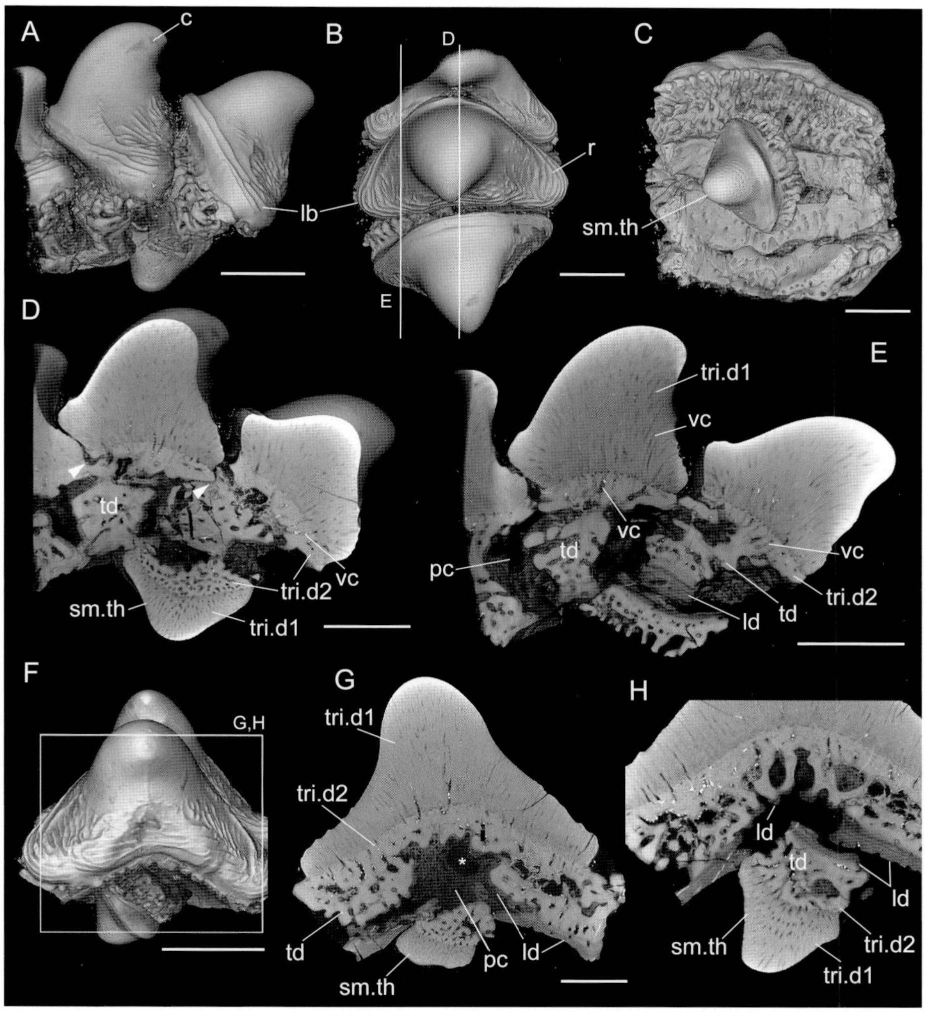

Fig. 1. *Helodus simplex* (Holocephali; Chondrichthyes, Carboniferous), NHMUK PV P8216, tooth set. 3D reconstructions and virtual sections from μCT scan. Orientations in all figures relative to position of set on the jaw. **A.** caudal view; **B.** oral view with lines showing virtual sections in D, E; **C.** aboral view with displaced small tooth; **D, E.** virtual sections through tooth set; **F.** lingual view with box showing virtual sections in G, H. **H.** closeup to show continuous layer of lamellar dentine within the tooth pulp cavity. Abbreviations: c, tooth crown; lb, lobe; ld, laminar dentine; pc, pulp cavity; r, ridges on tooth lobes and base; sm.th, small tooth; tb, tooth base; td, trabecular dentine; tri.d1, 2, tritoral dentine; vc, vascular canal. Scale bars: A–F, 0.5 cm; G, H, 0.25 cm.

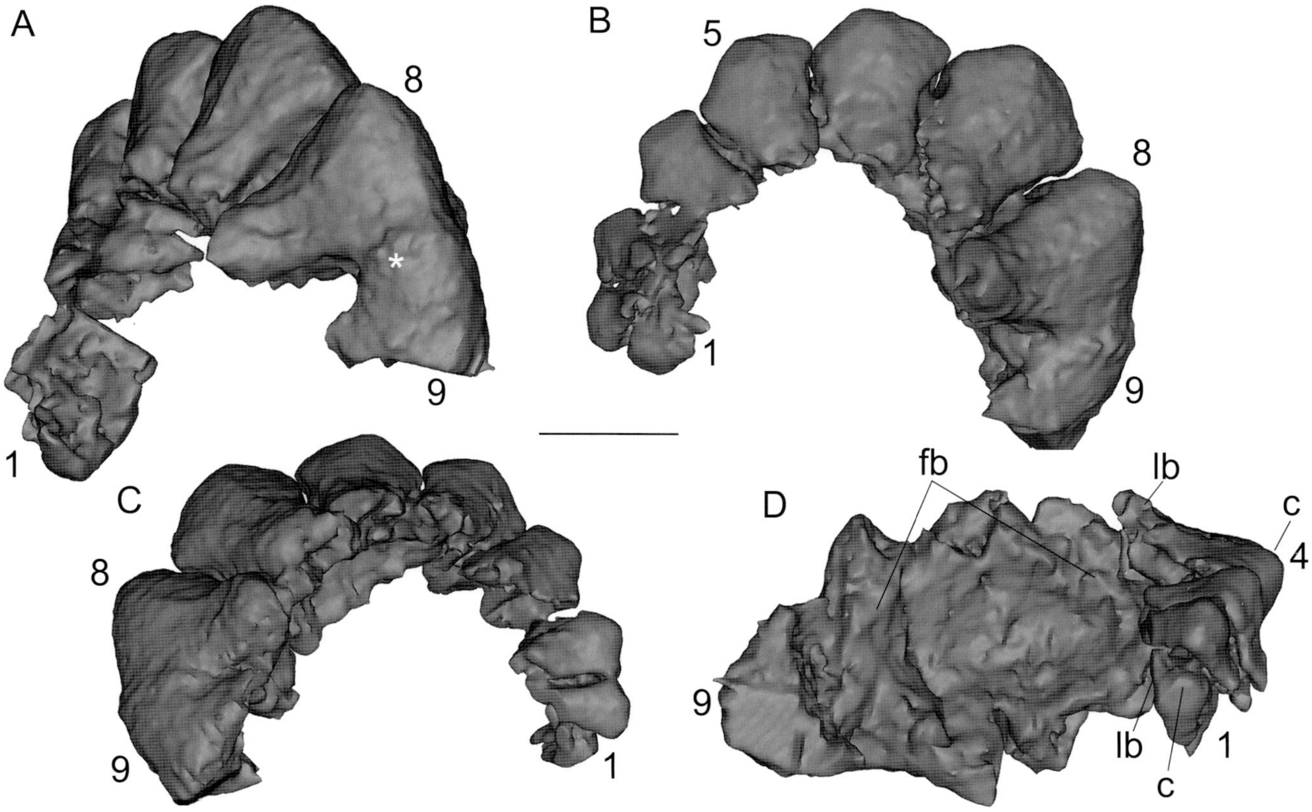

Fig. 2. A–D. *Helodus simplex* (Holocephali; Chondrichthyes, Carboniferous), NHMUK PV P8213e, mid-jaw tooth set. 3D reconstructions and virtual sections from μCT scan. **A.** rostrolingual, **B.** rostral, **C.** caudal, **D.** aboral views. Abbreviations as in Figure 1, including fb, fused base; 1–9, tooth number in set. Scale bar: A–D, 0.5 cm.

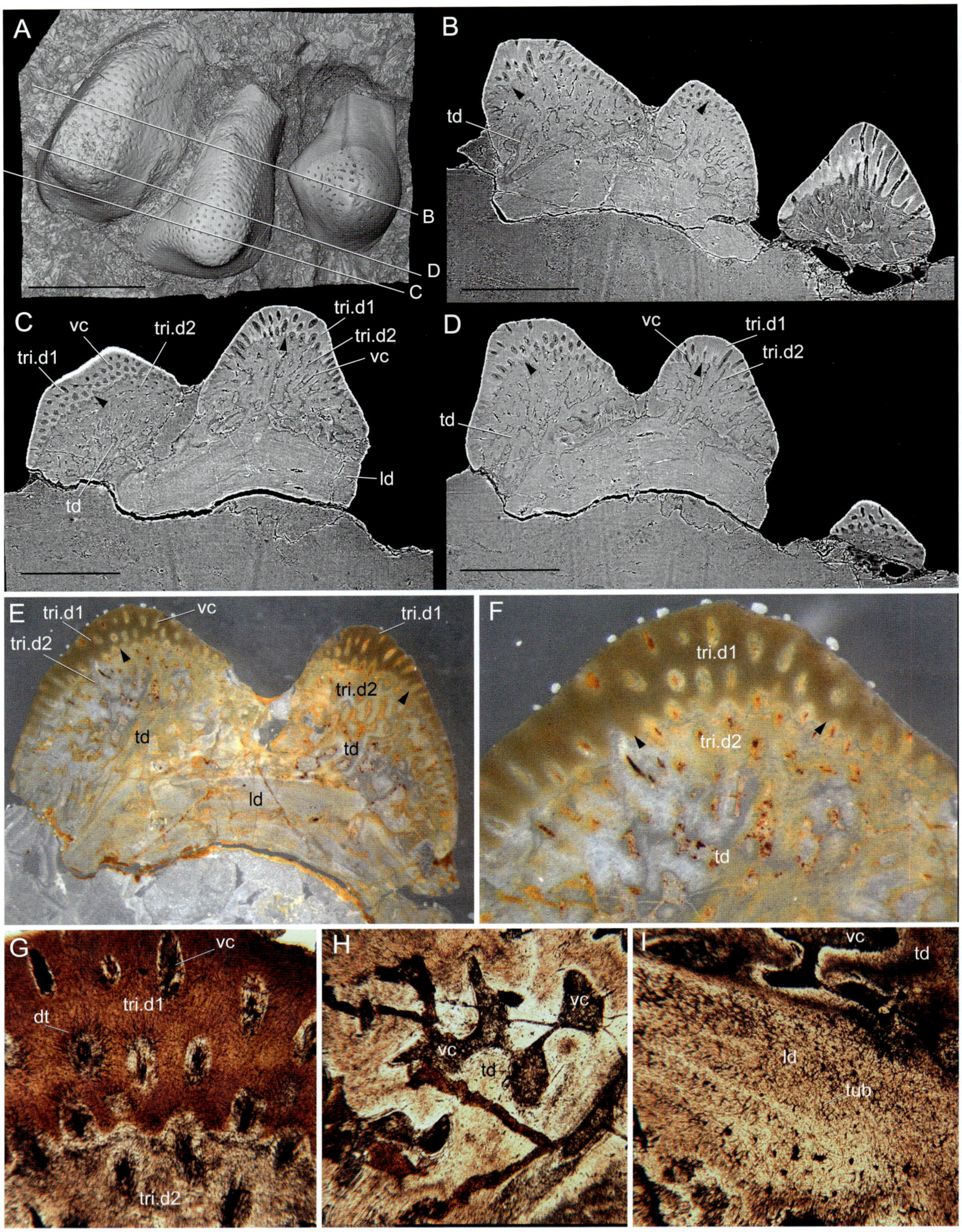

Fig. 3. *Helodus appendiculatus* (Holocephali; Chondrichthyes, Carboniferous), NHMUK PV P75938. **A–D.** 3D reconstructions and virtual sections from synchrotron scan. **A.** oral view with lines indicating virtual sections in B–D, with the same tissues as in E–I. **E–I.** serial sections through teeth, **E, F.** illuminated with incident reflected and, **G–I.** transmitted light. Both teeth showing junction of tri.d1 with tri.d.2 (black arrowheads), tri.d1–2, td, ld, are the three tissue types with vascular canals (vc) in a regular pattern (**E**). Close up of mature tooth (**F**). Junction of tri.d1 and d2, with tubules radiating from each vascular canal (**G**), trabecular dentine (td) as part of wider connecting vascular spaces below both tritoral layers (**H**) and basal layer of laminar dentine with few vascular canals, many odontocyte spaces and tubules (tub) (**I**). Abbreviations as in Figure 1, additionally: min, mineralized layer. Scale bars: A–D, 0.5 cm; E–I; magnifications: E, ×10; F, ×25; G–I, ×40.

New μCT-scans of a more complete, recently reassembled specimen of *Helodus simplex*, including braincase and jaws, also preserve the dentition in situ (NHMUK PV P8213e and NHMUK PV P2907, Coates et al., this volume). This preservation allows the interpretation of the generative history of the tooth set and its sequence to be described (Fig. 2A–D). Nine teeth form the tooth set, in the shape of a whorl, reflecting sequential development at its lingual base, at each position on the jaw. The location of small teeth outside the jaw margin suggests that only the earliest and smallest teeth, if any, were shed after passing their labial functional position. The first teeth preserved in the whorl (Fig. 2A–D, '1') have distinct main cusps and lobes, which increase in size in the younger (lingual) teeth (Fig. 2D). Fusion of the teeth occurs in later formed teeth that are about half way along the tooth whorl, but initially only in their bases (teeth 5 to 9, Fig. 2D, fb). By comparison, fusion between the crowns of the teeth occurs in younger teeth (8 and 9, Fig. 2A–C). Nevertheless, these two teeth retain some degree of individuality and can be identified as separate teeth with clearly separated cusps (Fig. 2A, white asterisk).

Helodus appendiculatus

(Fig. 3, NHMUK PV P75938)

A partial tooth set of a second species of *Helodus*, *H. appendiculatus*, shows both similarities and differences to the histology of *H. simplex*. As in *H. simplex*, the coronal tissue comprises ordered and regular vascular canals that radiate out normal to the crown surface (Fig. 3B–F), with the two layers of dentine identified in *H. simplex* (tri.d1, tri.d2) here separated by a prominent 'watermark' where the mineralization phase changes (Fig. 3B–G, small black arrowheads). These tritoral dentines have perivascular dentine around the canals that is paler than the interstitial regions (Fig. 3E–G), and also less mineral dense (Fig. 3B–D). This is also where tubules from the cell bodies around the vascular pulp canal pass across into the bulk of the tritoral dentine (Fig. 3G), to affect its progressive mineralization. The mineralization watermark marks a limit to a change between the two layers of tritoral dentine (Fig. 3C–G, tri.d1, 2) that have differentially taken up the 'stain' of fossilization. The two sets of micrographs (Fig. 3E, F, G–I) show the different appearance in incident reflected and dull light and transmitted light (i.e. Fig. 3F, G) where the fossil infill into biological spaces is opaque to light (probably pyrite) but reflective in incident illumination. In *H. appendiculatus*, the inner tritoral layer of dentine (tri.d2) appears thicker than in *H. simplex* (Fig. 3C–F) but this may represent different stages of growth and progressively higher levels of mineralization in these two species (see Smith et al., 2019). Trabecular dentine is identified by the more irregular vascular canal spaces, and is the lower tissue of the crown (Fig. 3B–F, H); this links with the thick laminar tissue, which is also a form of dentine with many, irregular branching tubules of the odontoblasts (Fig. 3I). This is expanded to be continuous between the teeth, as concave basal layers, that viscerally present an arched surface (Fig. 3B–E, ld), transforming the whole tooth set into a joined whorl. Similarly, the tooth base in *H. simplex* is composed of trabecular dentine linking coronal tritoral tissue to the laminar tissue (Fig. 1D, E, G, H).

'Pleuroplax' rankini

(Fig. 4, NHMUK PV OR33300)

One *'Pleuroplax'* specimen is similar to those previously described and illustrated, being dominated by a large, rounded, coronal region with two lobes on either side of this region. Narrow ridges are present on these lobes, and at the base of the coronal region (Fig. 4A, C, lb, r). The oral surface is smooth, and where worn, the tritoral dentine has a punctate surface reflecting the distribution of vascular canals (Fig. 4C, F). As well, on this specimen, a small, separate tooth is present, with lobes on either side of the main cusp that are very small. The ridges on the surface of these lobes are also small (Fig. 4A, C, '1'). The tooth appears to be entirely separate (Fig. 4A–C), but in section, fusion of the tooth base to the larger coronal region has occurred (Fig. 4G, H). A second tooth is visible (Fig. 4C, '2'), preserving a partial main crown and two lobes, which are distinct from the smaller, separate tooth. Another set of ridged lobes indicates the presence of a third tooth (Fig. 4C, '3'), but this is otherwise indistinct from the rest of the smooth tritoral surface.

As in *Helodus*, and particularly *H. simplex*, the crown of the tooth is dominated by tritoral dentine with a very thin contiguous tritoral dentine layer beneath, with vascular canals running between them (Fig. 4G–I, vc, tri.d1, trid2). A transition in these dentine types (Fig. 4D) can be seen from tooth 1 to 3, as the layers become less distinct and less recognisable. A pulp cavity is present (Fig. 4H, I, pc, asterisk indicates large vascular space), and beneath this, trabecular dentine forms the tooth base, which fuses into a single large plate (Fig. 4D, E, G, H, td, fb). The laminar layer seen in *Helodus* appears to be absent. A small tooth is also present in this specimen, located near the end of the main rounded coronal region, at the opposite end relative to the separate tooth (Fig. 4D, E). Again, as in *Helodus* above, this tooth is thought to have been dislocated post-mortem, then lodged in the pulp cavity space.

A section through another specimen of *'Pleuroplax'*, part of the suite of specimens registered as NHMUK PV OR333000 (Fig. 4J), shows an extensive coronal region preserved as a whorl, with layers of tritoral dentine and trabecular dentine (tri.d1, td). Inside the continuous coronal, tritoral tissue, is represented by a series of smaller teeth (asterisk), showing a lower degree of mineralization compared to the coronal dentine, representing the very first to form, and hence oldest, labial part of the dentition. These teeth have been broken and shifted slightly from the main whorl, but are clearly associated. The oldest

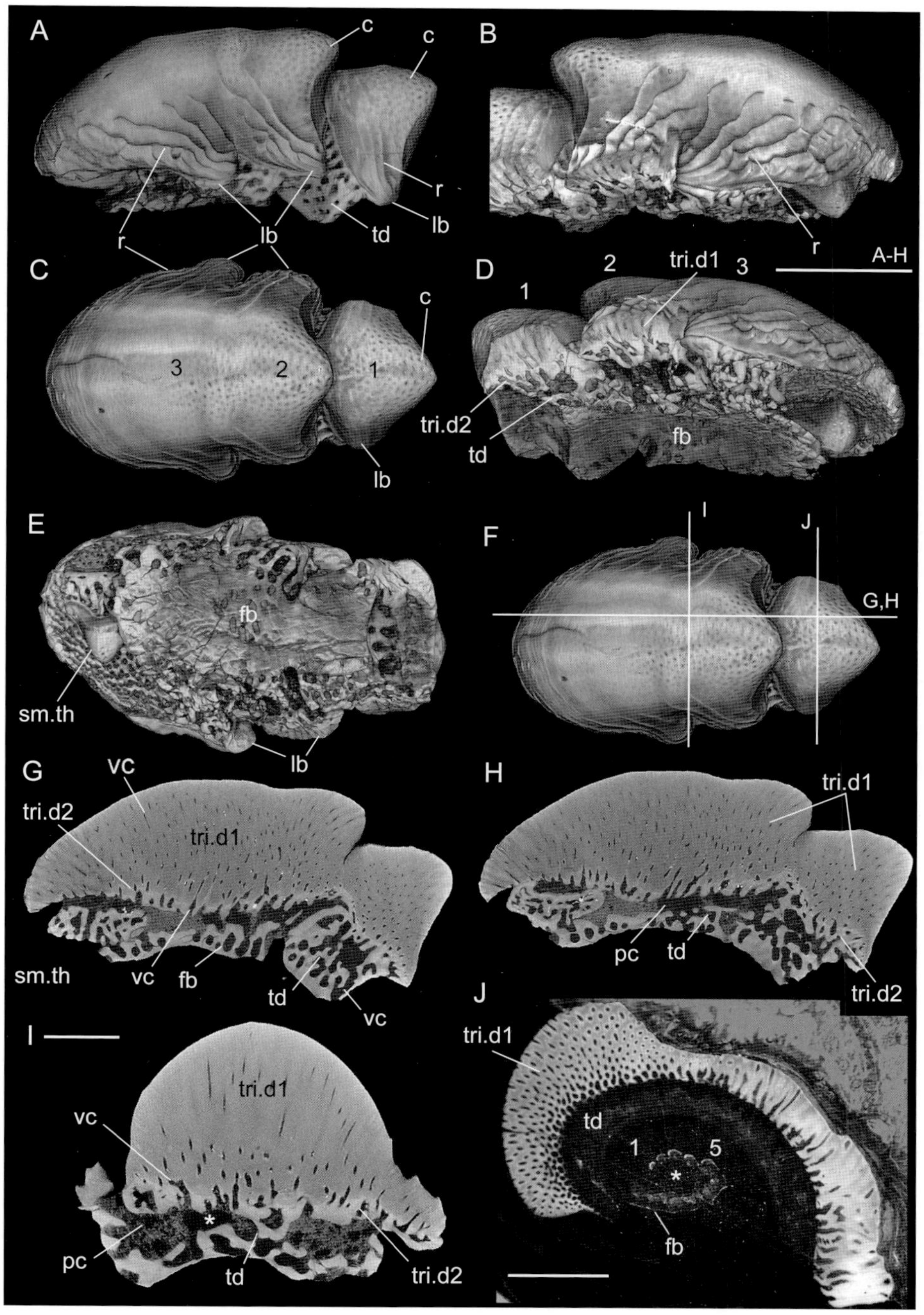

Fig. 4. A–I, *'Pleuroplax' rankini* (Holocephali; Chondrichthyes, Carboniferous), NHMUK PV OR33300. 3D reconstructions and virtual sections from µCT scan. **A, B.** rostral, caudal views; **C.** oral view; **D.** caudal-aboral view to show fused tooth base; **E.** aboral view showing fused tooth base and displaced small tooth; **F.** oral view showing virtual sections in **G–I. J.** second specimen in NHMUK PV OR33300 series, serial section preserving oldest, separate, labial teeth of the whorl, followed by fusion of small teeth and abrupt transition to fused dentine. Abbreviations as in Figure 1. Scale bars: A–H, 1.0 cm; I, J, 0.5 cm.

teeth are small but distinct (Fig. 4J, '1'–'5'), but a change occurs progressively around the logarithmic curve, with the next remaining teeth being small but less distinctly individual, and are suggested to be fused (fb). The histology of these smaller separate and fused teeth cannot be determined, but the next part of the tooth whorl comprises the more typical fused *'Pleuroplax'* histology and morphology.

Irregular *Helodus*

(Fig. 6, NHMUK PV P8230)

A large specimen of *'Pleuroplax*-morphotype' *Helodus* (species unidentified) preserves a set of successively generated teeth, but these are unusual in being substantially deformed and misshapen, compared to the teeth of *Helodus simplex* and *H. appendiculatus* described above. Individual teeth can be recognized (Fig. 6A–E), with the main cusp normally present, but the lobes more variably formed. Ridges on the lobe are also absent. So, for example, teeth 1 and 2 (Fig. 6E) possess a low cusp, but then only one lobe, on opposing sides. Tooth

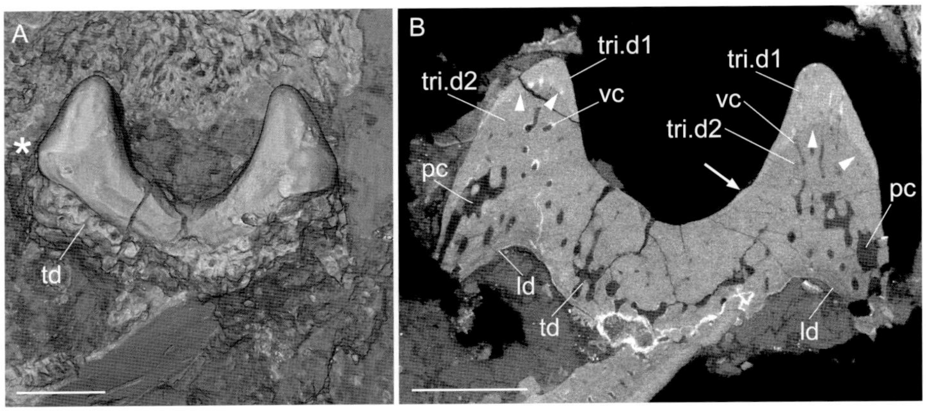

Fig. 5. A, B. *'Diclitodus' scitulus* (Holocephali; Chondrichthyes, Carboniferous), NHMUK PV P4890. 3D reconstructions and virtual sections from µCT scan. A. lingual or labial view; B. virtual section. Abbreviations as in Figure 1. Scale bars: A, 0.5 cm; B, 0.25 cm.

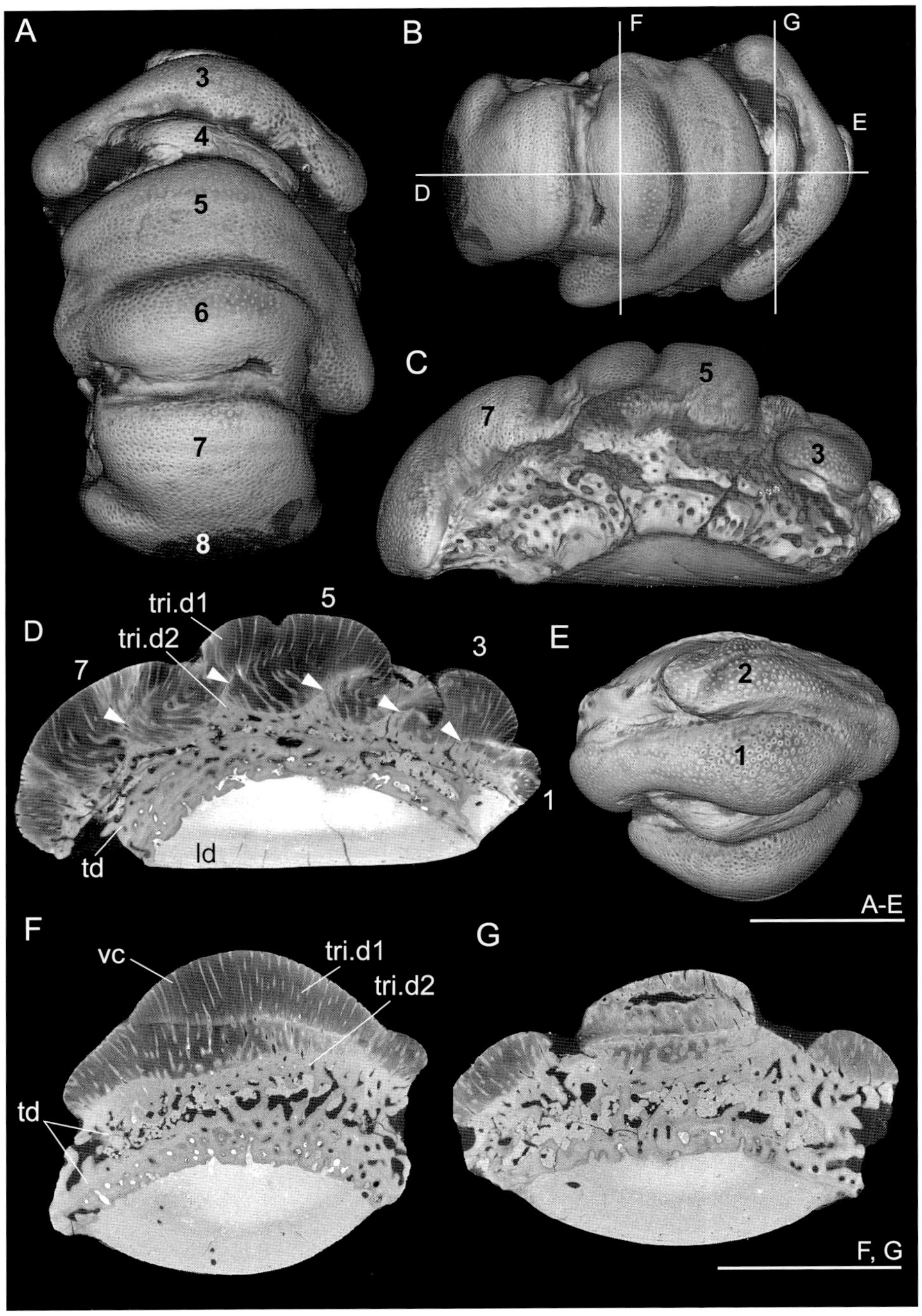

Fig. 6. A–G, Irregular *'Pleuroplax'* (Holocephali; Chondrichthyes, Carboniferous), NHMUK PV P8230. 3D reconstructions and virtual sections from µCT scan. A. oral view; B. oral view to show virtual sections in D. F, G. and position of view shown in E. Abbreviations as in Figure 1. Scale bar: A–G, 1 cm.

3, by comparison, possesses both lobes (Fig. 6A), while 6 and 7 lack both lobes. Tooth 4 is partially overlain by both tooth 3 and 5 (Fig. 6A), while tooth 7 and 8 are fused to produce a rounded area, as in *'Pleuroplax'*. In section, the histology is similar to other *Helodus* and *'Pleuroplax'*, with teeth dominated by tritoral dentine, with the thin contiguous tritoral layer beneath (Fig. 6D, F, G, tri.d1, tri.d2). The tooth bases of this irregular specimen are composed of trabecular dentine, in places fused into the concave whorl base (Fig. 6D). The tooth crowns, by comparison, remain unfused. The overlap relationships of these crowns appear unusual but their 'S-shaped' surfaces mimic the interlocking contact between the teeth of *H. simplex* (Fig. 6D, white arrowheads). By comparison with *H. simplex*, the more separated teeth can be identified as older, while more labial (teeth 1–6) and the more completely fused region (teeth 7, 8) can be identified as the younger.

'Diclitodus' scitulus

(Fig. 5, NHMUK PV P4890)

Isolated specimens assigned to *'Diclitodus'* have been considered to represent two tooth sets fused at the midline of the *Helodus simplex* jaw (e.g. Moy Thomas 1936, Stahl 1999), but as shown by Coates et al. (this volume), these actually comprise a series of elements associated with the midline anterior braincase, representing a cephalic clasper or frontal tenaculum. Intriguingly, the older elements in this series show the typical double-cusped *'Diclitodus'* morphology (Fig. 5), but younger elements show a separation into two distinct structures, each with a single cusp.

As shown in a high-resolution µCT scan (Fig. 5), the histology of these elements is similar to the *Helodus* and *'Pleuroplax'* teeth described above, being composed of two tritoral dentine layers with embedded vascular canals (Fig. 5B, H, tri.d1, tri.d2). As in *H. appendiculatus*, the upper tritoral dentine layer is more mineralized, and this persists as a thin layer in the intervening trough between the teeth (Fig. 5B, tri.d1, white arrow). The upper mineralized layer is separated by a 'watermark', as in *H. appendiculatus* (Fig. 5B, arrowheads), with vascular canals in this dentine shown by a very high density fossil infill. Below these layers is a thick trabecular dentine that makes up approximately half the base of the paired teeth. Here, the irregular canals emerge from small pulp cavities at the lateral margins of the tooth (Fig. 5B, pc). A thin basal laminated layer is present as in other teeth described above, but this is partially damaged by intrusion of another fragment of bone material during burial compaction. These elements are simpler compared to those of *Helodus* and *'Pleuroplax'* in lacking noticeable lobes and ridges associated with these lobes and the base of the tooth (white asterisk, Fig. 5A).

Discussion of *Helodus* dentitions

Previously, when the fused teeth of *Helodus simplex* and other related species were found isolated, they were assigned to *'Diclitodus'* and *'Pleuroplax'* (= *Pleurodus*; Smith Woodward 1889, Moy-Thomas 1936, Patterson 1965, Stahl 1999). Although originally controversial (Moy-Thomas 1936, Barkas 1874, Davis 1879), it is now accepted that the fused teeth assigned to *'Pleuroplax'* occur in the mid-jaw of *H. simplex*, while those once thought to be located at the lower jaw symphysis, and assigned to *'Diclitodus'* (Moy-Thomas 1936, Stahl 1999), are now identified as components of a dorsal, frontal tenaculum of *H. simplex* (Coates et al. this volume). Here, older fused elements in the tenaculum become divided into two separate structures in younger elements.

New µCT scans of the reassembled skull of *Helodus simplex* (Coates et al. this volume, NHMUK PV P8213e + P2907) confirm that fusion occurs within the mid-jaw tooth whorls. In this specimen, the older teeth in the tooth set remain unfused, including the crown and tooth base. More lingually, the tooth bases fuse (Fig. 2D), but the main tooth crowns remain separated (Fig. 2B). The last two teeth preserved lingually (8, 9) are the only ones fused at the crown, although the lobes are still distinct and the boundary between the two teeth can be recognized. This suggests that more complete fusion would occur in even more lingual, yet to develop teeth, and that there is more of a recognizable, gradual transition in degree of fusion in this tooth set. By comparison, in the isolated specimen of *'Pleuroplax'* (NHMUK PV OR33300; Fig. 4C) there is a sharper boundary between the older separate tooth as preserved, and

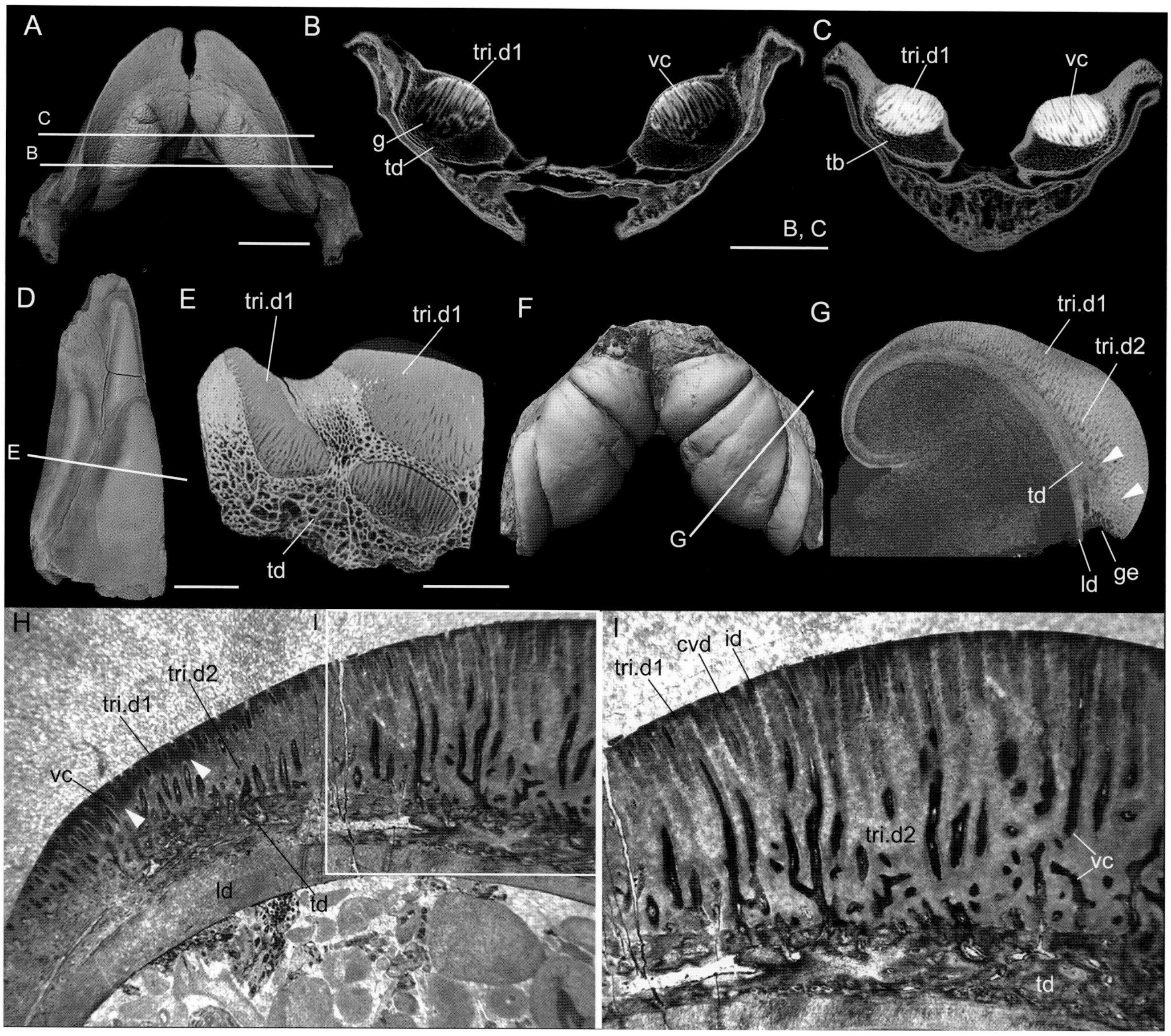

Fig. 7. A–E. Chimaeroidei and **F–I.** stem chimaeroids (Holocephali; Chondrichthyes). **A–G.** µCT scans, **H, I.** serial sections. **A–C.** *Callorhinchus milii*, lower jaw in A oral view; **B, C.** virtual sections as illustrated in A. **D, E.** *Edaphodon* sp., NHMUK PV OR38771, toothplate in D oral view; **E.** virtual section as illustrated in D. **F–I.** *Cochliodus contortus* NHMUK PV P75425, lower jaw. **F.** oral view; **G.** virtual section as illustrated in F. **H, I.** serial sections through dental whorl, field in H (I) shows tritoral histology (cvd, id, vc, td) and arrangement of tritoral tissue types, (tri.d1, tri.d2). Scale bars: A–F, 1.0 cm; G, 0.5 cm; magnifications: H, ×6.5; I, ×10.

the newer fused region, although virtual sections show that this tooth is partially fused to the main part of the specimen. Two pairs of ridged lobes are present in the main fused region, and a distinct cusp is present, but overall a more gradual transition in degree of fusion, as seen in the reassembled skull, appears to be absent. This is also the case in the second specimen of the NHMUK PV OR33300 series (Fig. 4J), where the change from teeth to fused small teeth to deposition of the continuous dentine, is abrupt.

Implications for taxonomy

The taxonomy of *Helodus* is problematic, with a large number of species established on isolated teeth or partial tooth sets (Stahl 1999, Itano & Lambert 2018). Virtual sections taken from µCT-scans of *H. simplex* and *H. appendiculatus* demonstrate some differences in histology, in particular the relative thicknesses of different dentine layers, that may be useful in distinguishing species of *Helodus* in future studies. For example, the teeth of these two species are dominated by tritoral dentine, all deposited in association with well-organized vascular canals that are perpendicular to the tooth surface. Within this tritoral dentine, two regions are identified, with the second being thinner in *H. simplex*, and thicker in *H. appendiculatus*. In *H. appendiculatus*, the upper region of tritoral dentine is more mineralized to a depth that is evenly marked by a phase change (Fig. 1H, 3C–G). This potentially results from an epigenetic effect as from wear forces and the sudden increase in mineralization of their hypermineralized dentine, and as such is comparable to 'sclerotized osteodentine' in chimaeroids (e. g., *Harriotta*, Smith et al. 2019). In both species of *Helodus*, the lower tritoral dentine is contiguous with a richly vascular region representing trabecular dentine of the tooth base, forming compact osteodentine with age (vascular canals of small diameter). Beneath this is an avascular laminated dentine, the thickness of which varies between species. In *H. appendiculatus*, the tooth bases are fused and appear to be composed entirely of this laminated dentine, where it is very thick and concave, as is seen in tooth whorls generally. In *H. simplex* (NHMUK PV P8213e + P2907), older teeth are independent and later sutured, with fusion occurring in younger teeth, originally in the tooth base, and then in the tooth crowns.

Because only partial tooth sets of *H. simplex*, *H. appendiculatus* (only two teeth observed) and '*Pleuroplax*' have been µCT-scanned at higher resolution, it is possible that all these different states of fusion occur in the complete *Helodus* tooth series, but this has never been fully recovered in the fossil record. It should be noted however that these specimens are from different geological formations and ages, and so are not conspecific. The thick layer of laminated dentine, forming a tooth whorl in *H. appendiculatus*, may be an important difference, with the histology of this layer very different from the tritoral and trabecular dentines (Fig. 3G–I). The lower resolution µCT-scan of reassembled NHMUK PV P8213e and P2907 does not allow the histology of the fused tooth base to be assessed for comparison to *H. appendiculatus*. This data suggests that a much broader histological examination of *Helodus* species dentitions is required to better understand potentially relevant taxonomic differences.

Developmental implications

In modern sharks and rays, teeth are produced within a dental lamina on the lingual face of the jaw cartilage, before ascending to the functional position at the jaw margin where they are shed (Smith et al. 2009, Underwood et al. 2015, Martin et al. 2016). The enameloid cap of the tooth forms first, with contributions from epithelial ameloblasts and mesenchymal odontoblasts. The odontoblasts continue to deposit dentine centripetally below the enameloid, within the pulp cavity, with the base of dentine forming later. Teeth form sequentially within each tooth set, so that different developmental stages are present in the same generative set (Underwood et al. 2016, Smith et al. 2019).

There appears to be no fundamental difference in the development of the teeth of *Helodus* or any other early holocephalan possessing separate teeth from those of modern sharks and rays, other than that they lack the enameloid of modern sharks. However, *Helodus* appears not to have shed teeth once they have passed the jaw margin; in NHMUK PV P8213e + P2907 (Coates et al. this volume, Fig. 2), an extended whorl of teeth is present outside the jaw. Whilst some of these teeth cannot have been readily shed as they are fused to newer teeth, many of the smallest and oldest teeth are not fused and yet are still retained post-occlusion. This is also seen in NHMUK PV OR33300 '*Pleuroplax*' (Fig. 4J), where very small, early, teeth are preserved as the inner part of the logarithmic growth of the tooth whorl, indicating that the earliest stages were not shed but retained in the jaws.

Whilst similar mechanisms were likely involved in the development of tooth sets in *Helodus* and those of modern chondrichthyans, partially fused teeth and those of '*Pleuroplax*' morphology must have evolved with mechanisms to fuse whole teeth at the tissue development level. These are absent in modern sharks and rays, but would have been retained and modified in modern chimaeroids. In *Helodus*, different degrees of tooth fusion are present, sometimes within the same individual and even the same tooth set, suggesting both positional and ontogenetic controls mediate this fusion, rather than a simple 'on-off' control of singular tooth development.

Holocephalans with true dental plates

Whilst the dental sets of '*Pleuroplax*-morph' *Helodus* are partially or fully fused into tooth whorls (Figs. 2, 3), it is evident that these are composed of individual elements homologous with the teeth in *Helodus* and in sharks and rays. In *Helodus*, each tooth set, or whorl, remains independent (Coates et al. this volume). In many other holocephalans, however, the dentition is composed entirely of a small number of continuously growing dental plates, each of which provides a proportionally large tritoral region of the jaw. In the majority of Palaeozoic taxa (Stahl 1999) these comprise a coiled tritor (tooth set) that widens towards the growing edge (e.g., *Cochliodus*, Fig. 7F). Whilst feeding-related wear may have been intense on the occlusal part of the plate, at least some of the upper layer of the tritoral dentine is typically retained in this position (Fig. 7H), even in worn parts of the plates. This contrasts with the situation in the largely Post-Palaeozoic Chimaeriformes (Stahl 1999), where older dental material is removed by wear at the jaw margin and thus the spiral of older parts of the tooth plate is not retained, and instead a gently curved tooth plate with a large occlusal wear surface is present.

Cochliodus contortus

(Fig. 7F–I, NHMUK PV P75425)

Cochliodus is one of the most widely recognised of the Palaeozoic holocephalans possessing spiral dental plates. Despite the widespread records of this genus, non-dental material is unknown and as such the taxon has an uncertain place in holocephalan phylogeny. The occlusal surface of the dental plates is smooth to weakly crenulated, and there are no outward signs of individual teeth that are fused to form the tooth-whorl (Fig. 7G). Intriguingly, grooves aligned in a labio-lingual direction on the occlusal surface potentially mark boundaries between individual tooth whorls that form along the jaw margin (Fig. 7F), which may be comparable to the fused lateral lobes of *Helodus*. The oldest occlusal parts show wear but are retained within the labial spiral (Fig. 7G), with the newest tissue formed at the lingual end where the tritoral tissue is much thicker with growth (Fig. 7G, ge).

Sections of dental plates of *Cochliodus* have been figured previously by Agassiz (1838: tab. L, fig. 3) and Stahl (1999: figs. 20, 62), with a similar arrangement of the tissue types described here. The histology of the tritoral tissue is in general comparable to that of extant forms (Fig. 7A–C; Smith et al. 2019) but with differences in the arrangement of component parts. As in the various morphotypes of *Helodus*, the tritoral tissue has two layers, supported by trabecular dentine, with irregular cavities for vascular and connective tissue cells, underlain by a laminated layer of dentine. The tritoral tissue comprises numerous vascular canals orientated normal to the biting surface, each of which has a circumvascular layer of dentine through which the odontoblast tubules pass into an interstitial dentine of higher mineralization (less porous, as judged by the differential in fossil staining of each, Fig. 7H, I, id, cvd). On the older, worn, part of the plate (Fig. 7H, left side of image), the upper two layers of dentine are thinner compared to the younger parts of the whorl, and wear has occurred, impacting on the upper layer of dentine (tri.d1). A watermark occurs here (Fig. 7H, arrowheads), supporting previous suggestions made above that this reflects differences due to mineralization in response to wear. The trabecular dentine, showing a less organized arrangement of the vascular canals, is more compact in this region of the dental plate, compared to the younger. As noted above, this is trabecular dentine transformed into osteodentine. At the growing edge of the dental plate (Fig. 7G, ge), the tritoral and basal laminar layers appear to form first, with the intervening region then forming the trabecular dentine. The tritoral tissue is added both at the visceral surface, still forming as a curved growing edge (Fig. 7G, white arrowheads), and at the lingual (growing) end as new tissue. The basal layer is likewise initiated at the forming edge (Fig. 7G, ld), but is then added to at the visceral surface, with dentine being secreted gradually in layers with stop-and-start growth. This basal layer is thickest immediately below the tritoral part of the dental plate where tissue is being removed by wear, suggesting that this is the slowest of the dentine types to form, with formation not ceasing until the position on the dental plate had moved to the jaw margin. Unlike in modern chimaeroids such as *Harriotta* (Smith et al. 2019), there is little evidence for tritoral tissue forming as a cavity infill between pre-existing tubes; instead both tube linings and infill form simultaneously in *Cochliodus*.

Chimaeroidei

Dentitions of post-Palaeozoic chimaeroids studied for comparison to *Helodus* were those with compound dental plates, including fossil *Edaphodon* and modern *Callorhinchus* (Fig. 7A–E). Whilst the internal structure of some chimaeroid toothplates comprises morphological arrangements of hypermineralized dentine into ovoid stacks (strings of beads, Didier 1995, Stahl 1999) and tritoral pads (e.g., *Harriotta* (Rhinochimaeridae), Smith et al. 2019), these taxa both have six separate dental plates. Within these compound dental plates are individual large areas of tritoral dentine, lacking the abundant smaller ovoids and rods seen in the Rhinochimaeridae (Smith et al. 2019). In both *Edaphodon* and *Callorhinchus*, tritors are composed of interstitial dentine surrounding subparallel, vascular tubes, an arrangement that is very similar to the tritoral tissue constituting the teeth of *Helodus* and the curved plates of *Cochliodus* (Fig. 7B, C, E, tri.d1). Unlike in the Palaeozoic taxa, however, this does not form the major part of the dentition, but instead it is enclosed within trabecular dentine tissue (Fig. 7B, C, E). At the growing edge there are large areas within the trabecular dentine representing soft tissue spaces (Fig. 7B, 'g'), these are later filled with tritoral tissue (Fig. 7B, C, tri.d1). The boundary between the trabecular dentine and the tritoral tissue appears to be distinct, unlike in the Palaeozoic forms where it is more gradational (e.g., Fig. 7H, I). The tritoral tissue is hypermineralized, forming raised areas on the biting surface of the dental plates, harder than the trabecular dentine (Fig. 7A, D; see Smith et al. 2019, for qualitative values for relative degree of calcification). In contrast to all of the Palaeozoic forms studied, members of the Chimaeroidei have no clearly defined layered basal dentine tissue.

Origin of the Chimaeroidei dentition by fusion of separate teeth

Recent phylogenetic analyses (e.g., Coates et al. 2018: fig. 3) resolve *Helodus* in a polytomy with Iniopterygia and crown group Holocephali (Chimaeroidei), with the Symmorida forming the sister group to these (*Cochliodus* was not included as skeletal remains are unknown). Symmorid dentitions predominantly include tooth families with separate teeth showing a conventionally 'shark-like' morphology (Lund 1985, 1986, Coates & Sequeira 2001, Williams 2001, Maisey et al. 2014, Ivanov 2016, Ginter 2018); by comparison, iniopterygians possess a range of dentitions, from separate teeth to those fused into tooth whorls (Zangerl & Case 1973). Other stem-holocephalans more closely related to the crown group have completely separate teeth (*Debeerius*, Grogan & Lund 2000) or a combination of dental plates and isolated extramandibular teeth (*Chondrenchelys*, Finarelli & Coates 2014).

Helodus (s.l.) sits at a critical position within the evolution of the holocephalan dentition. Within this single genus (as currently defined) are teeth that show a range of degrees of fusion. Within a tooth set, teeth may be closely associated but not fused, fused at the base only, or completely fused with a merging of all three tissue layers between adjacent teeth. The differential occurrence of dentitions of the '*Pleuroplax*' morphology indicates that there was considerable interspecific variation in the degree of dental fusion. It is therefore evident that even though *Helodus* does not form dental plates of the type seen in *Cochliodus* and the Chimaeroidei, it has the developmental capacity to fuse teeth within sets, a prerequisite for evolution of the true dental plates characterizing this group. But, one feature that has clearly not yet evolved in these tooth sets is fusion across the sets, which may occur in Palaeozoic taxa such as *Cochliodus*.

Although the dentition of *Helodus* could be considered transitional between dentitions comprising individual teeth (as in sharks and rays) and dental plates (as in chimaeroids; Patterson 1965), it is not resolved as the most basal sister taxon to crown group holocephalans (e.g., Coates et al. 2018). As noted, several different clades within the Palaeozoic holocephalans are known to show a range of different degrees of tooth fusion or dental plate development and it is evident that dental fusion resulting in dental plates has evolved on multiple occasions within the Holocephali. However, rather than representing a derived condition, fusion or incomplete separation of teeth within sets might even be considered a conserved feature – a trait retained from stem chondrichthyans such as *Ptomacanthus* and *Doliodus* (Brazeau 2014, Maisey et al. 2014).

In contrast, there is no evidence of dental fusion in any non-holocephalan crown chondrichthyan clade; indeed, whilst myliobatid rays have both widened their teeth laterally and developed sutured tooth margins labiolingually, there is no evidence of fusion (e.g. Underwood et al. 2015). Whilst the development of holocephalan dentitions is poorly known, it appears likely that there was a fundamental difference between the functionality of the tooth germ cells within the dental lamina of early holocephalans and that of other chondrichthyans. This must have been required to facilitate the continuous generation of dental material leading to the continuous 'roll-over' as a spiral with logarithmic growth of many dental plates, and to the aboral growth of extant forms.

It is not clear whether the post-occlusal retention of (non-fused) *Helodus* teeth had a functional advantage or whether it was purely a result of the inability to detach teeth from the soft tissues to which they were attached, as also seen in symmorids such as *Cladoselache*, *Akmonistion* and *Gutturensis* (Sequeira & Coates 2000, Coates & Sequeira 2001, Williams 2001). In either case, the lack of release of post-occlusal teeth is likely to have resulted in a semi-rigid dental structure on the outside of the jaws, any functional advantage of which would likely have been increased by fusion of the teeth to form a still more rigid tooth whorl. Within this scenario, lateral fusion of these tooth whorls would result in the formation of a chimaeroid dental plate, with total loss of separate tooth germs (only a single blastema for each dental plate (Kemp 1984, Didier et al. 1994) in embryogenesis, or in growth.

Whilst the relative proportions of the three main tissue types within the stem group holocephalans vary, all three tissues (tritoral dentine, trabecular dentine and laminar dentine were present in the specimens studied. These three tissues are present in both separate teeth and fused dental plates, although dentitions of the members of the Chimaeroidei thus far studied differ in lacking the defined basal laminar layer. Although not studied in detail, published data on the more basal chimaeroids (*Echinochimaera*, *Squaloraja* and Myriacanthoidei; Stahl 1999) suggests that they possess a two-layered structure, with tritoral tissue overlying a trabecular dentine, thus resembling the structure of Palaeozoic holocephalans other than lacking the basal layer. The dental plates of the more derived Chimaeroidei possess the same tissue types as these stem group taxa, but differ in that the tritoral dentine is completely enclosed within spaces that form as the trabecular dentine develops, only becoming exposed on the occlusal surface due to substantially different degrees of mineralization, that results in feeding-related wear. The development of the histology of modern Chimaeroidei from that of Palaeozoic holocephalans thus appears to have been a process primarily involving incorporation of tritorial tissue within expanded trabecular dentine.

Conclusions

Helodus represents a stem group holocephalan with a dentition showing characters comparable to those seen in both sharks, which possess a dentition of continuously replaced individual teeth, and modern chimaeroids, within which the dentition comprises a small number of continuously growing dental plates. *Helodus* has thus proven to be a critical genus in understanding the homology of dental development within the Chondrichthyes. µCT-scans and thin sections of articulated dentitions, partial tooth sets and isolated teeth of *Helodus*, tooth morphologies now recognised as belonging to *Helodus* ('*Pleuroplax*') and other holocephalans reveal both the histology of the teeth and mode of tooth replacement and fusion. With respect to histology, in all holocephalan dental material examined, similar basic histotypes were present, including tritoral dentine tissue surrounding vascular canals showing a regular arrangement perpendicular to the tooth surface, along with a supporting trabecular dentine (more irregular vascular canals) and a laminar basal dentine layer (no canals).

The dentitions of *Helodus simplex* show a number of different forms of tooth fusion, including a transition from completely separate teeth in the older parts of the tooth whorl (although they are still retained) to the youngest teeth where there is fusion of the basal part of subsequent teeth to form a '*Pleuroplax*' morphotype fused tooth whorl. Whilst fusion within a tooth set is limited to the youngest parts of the dentition of *H. simplex*, in other species there is more tooth fusion; in *H. appendiculatus* there is extensive fusion of teeth but this is confined to the basal part of the tooth, with a continuous sheet of the basal layered dentine being present, but above this, individual teeth are still evident. In other species, such as '*Pleuroplax*' *rankini*, there is a far more extensive fusion of subsequent teeth, with most dental tissues showing some degree of connectivity between teeth, except perhaps the smallest and earliest teeth developing within the whorl.

Although the relationship of *Helodus* to more derived holocephalans is not yet clear, the fusion of teeth within a *Helodus* tooth set offers a model for the development of continuously growing dental plates from the separate and shark-like teeth of stem holocephalan taxa. Furthermore, the widespread occurrence of dental fusion in Palaeozoic holocephalans suggests that the Holocephali possesses a plasticity in their dental development not present in other chondrichthyans. Whilst the development of dentitions in living chimaeroids is poorly known, the fossil record of this clade suggests that there are likely to be fundamental differences in the mechanism of dental production between holocephalans and sharks, and that the assumption that dental development within all chondrichthyans is broadly similar may be invalid.

Acknowledgements

We thank Emma Bernard (NHM Department of Earth Sciences) for access to fossil fish collections; Tony Wighton and Callum Hatch (NHM Image and Analysis Centre) for the preparation of all sections; Peter Pilecki (Biophottonics KCL) for help with optical microscopy. We would also like to thank Paul Tafforeau and the staff of BM05 for their help in using the BM05 beamline of the European Synchrotron Radiation Facility. Michael Coates's participation was supported by National Science Foundation grant DEB-1541491.

References

Agassiz, L. 1833–1843. Recherches sur les poissons fossiles, vol. 3. Neuchâtel: Petitpierre.

Barkas, W. J. 1874. On the microscopical structure of fossil teeth from the Northumberland true coal measures. Monthly Review of Dental Surgery, 2: 540–547.

Bendix-Almgreen, S. E. 1968. The bradydont elasmobranchs and their affinities: a discussion. Nobel Symposium 4: 153–170. Almquist and Wiksell, Stockholm.

Brazeau, M. 2012. A revision of the anatomy of the Early Devonian jawed vertebrate *Ptomacanthus anglicus* Miles. Nature, 457: 305–308.

Coates, M., K. Tietjen, Z. Johanson, M. Friedman & S. Sang. 2021. The cranium of *Helodus simplex* (Agassiz, 1838) revised. – Pp. 193–204 in Pradel, A., J. S. S. Denton & P. Janvier (eds.), Ancient Fishes and their Living Relatives: a Tribute to John G. Maisey. Verlag Dr. Friedrich Pfeil, München.

Coates, M. I., J. A. Finarelli, I. J. Sansom, P. S. Andreev, K. E. Criswell, K. Tietjen, M. L. Rivers & P. J. La Riviere. 2018. An early chondrichthyan and the evolutionary assembly of a shark body plan. Proceedings of the Royal Society B, 285: 20172418.

Coates, M. I. & S. E. K. Sequiera. 2001. A new stethacanthid chondrichthyan from the lower Carboniferous of Bearsden, Scotland. Journal of Vertebrate Paleontology, 21: 438–459.

Davis, J. W. 1879. Notes on *Pleurodus affinis* sp. ined. Agassiz and Description of three Spines of Cestraciodonts from the Lower Coal Measures. Quarterly Journal of the Geological Society, 34: 181–188.

Davis, J. W. 1890. On the dentition of *Pleuroplax* (*Pleurodus*), A. S. Woodw. The Annals and Magazine of Natural History: Zoology, Botany and Geology, 5: 291–294.

Didier, D. A., B. J. Stahl & R. Zangerl. 1994. Development and growth of compound tooth plates in *Callorhinchus milii* (Chondrichthyes, Holocephali). Journal of Morphology, 222: 73–89.

Didier, D. A. 1995. Phylogenetic systematics of extant chimaeroid fishes (Holocephali, Chimaeroidei). American Museum Novitates, 3119: 1–86.

Finarelli, J. A. & M. I. Coates. 2014. *Chondrenchelys problematica* (Traquair, 1888) redescribed: a Lower Carboniferous, eel-like holocephalan from Scotland. Earth and Environmental Sciences Transactions of the Royal Society Edinburgh, 105: 35–59.

Ginter, M. 2018. Symmoriiform sharks from the Pennsylvanian of Nebraska. Acta Geologica Polonica, 68: 391–401.

Ginter, M. & Y. Sun. 2007. Chondrichthyan remains from the Lower Carboniferous of Muhua, southern China. Acta Palaeontologica Polonica, 52: 705–727.

Grogan, E. D. & R. Lund. 2000. *Debeerius ellefseni* (fam. nov., gen. nov., spec. nov.), an autodiastylic chondrichthyan from the Mississippian Bear Gulch Limestone of Montana (USA), the relationships of the Chondrichthyes, and comments on gnathostome evolution. Journal of Morphology 243: 219–245.

Itano, W. M. & L. L. Lambert. 2018. A new cochliodont anterior tooth plate from the Mississippian of Alabama (USA) having implications for the origin of tooth plates from tooth files. Zoological Letters, 4: 12.

Ivanov, A. O. 2016. Chondrichthyans from the Lower Permian of Mechetlino, South Urals. Bulletin of Geosciences, 91: 717–729.

Kemp, A. 1984. A comparison of the developing dentition of *Neoceratodus forsteri* and *Callorhynchus milii*. Proceedings of the Linnean Society of New South Wales, 107: 245–262.

Lund, R. 1985. The morphology of *Falcatus falcatus* (St. John and Worthen), a Mississippian stethacanthid chondrichthyan from the Bear Gulch Limestone of Montana. Journal of Vertebrate Paleontology, 5: 1–19.

Lund, R. 1986. On *Damocles serratus*, nov. gen. et sp. (Elasmobranchii, Cladodontida) from the Upper Mississippian Bear Gulch Limestone of Montana. Journal of Vertebrate Paleontology, 6: 12–19.

Lyckegaard, A., G. Johnson & P. Tafforeau. 2011. Correction of ring artifcats in X-ray tomographic images. International Journal of Tomography and Statistics, 18: 1–9.

Maisey, J. G., S. Turner, G. J. P. Naylor & R. F. Miller. 2014. Dental patterning in the earliest sharks: implications for tooth evolution. Journal of Morphology, 275: 586–596.

Martin, K. J., L. J. Rasch, R. L. Cooper, B. D. Metscher, Z. Johanson & G. J. Fraser. 2016. Sox2+ progenitors in sharks link taste development with the evolution of regenerative teeth from denticles. Proceedings of the National Academy of Sciences of the USA, 113: 14769–14774.

Mirone, A., E. Brun, E. Gouillart, P. Tafforeau & J. Kieffer. 2014. The PyHST2 hybrid distributed code for high speed tomographic reconstruction with iterative reconstruction and a priori knowledge capabilities. Nuclear Instruments and Methods in Physics Research Section B: Beam Interactions with Materials and Atoms, 324: 41–48.

Moy-Thomas, J. A. 1936. The structure and affinities of the fossil elasmobranch fishes from the Lower Carboniferous Rocks of Glencartholm, Eskdale. Biologcal Reviews, 14: 1–26.

Moy-Thomas, J. A. 1938. The early evolution and relationships of the elasmobranchs. Biological Reviews, 14:1–26.

Newberry, J. S. & A. H. Worthen. 1866. Descriptions of new species of vertebrates, mainly from the sub-Carboniferous Limestone and Coal Measures of Illinois. Geological Survey of Illinois, 2: 9–134

Patterson, C. 1965. The phylogeny of the Chimaeroids. Philosophical Transactions of the Royal Society of London. Series B, Biological Sciences, 249: 101–218.

Sequiera, S. E. K. & M. I. Coates. 2000. Reassessment of *'Cladodus' neilsoni* Traquair: a primitive shark from the Lower Carboniferous of East Kilbridge, Scotland. Palaeontology, 43: 153–172.

Smith, M. M., G. J. Fraser & T. A. Mitsiadis. 2009. Dental lamina as source of odontogenic stem cells: Evolutionary origins and developmental control of tooth generation in gnathostomes. Journal of Experimental Zoology B: Molecular Developmental Evolution, 312B: 260–280.

Smith, M. M., C. Underwood, T. Goral, C. Healy & Z. Johanson. 2019. Growth and mineralogy in dental plates of the holocephalan *Harriotta raleighana* (Chondrichthyes): novel dentine and conserved patterning combine to create a unique chondrichthyan dentition. Zoological Letters, 5: 11.

Stahl, B. J. 1999. Chondrichthyes III. Holocephali. – In: Schultze H.-P., editor. Handbook of Paleoichthyology 4. Munich: Verlag Dr. Friedrich Pfeil. 164 pp.

Underwood, C. J., Z. Johanson, M. Welton, B. Metscher, L. J. Rasch, G. J. Fraser & M. M. Smith. 2015. Development and evolution of dentition pattern and tooth order in the skates and rays (Batoidea; Chondrichthyes). PLoS ONE, 10: e0122553.

Underwood, C. J., Z. Johanson & M. M. Smith. 2016. Cutting blade dentitions in squaliform sharks form by modification of inherited alternate tooth ordering patterns. Royal Society Open Science, 3: 160385.

Williams, M. E. 2001. Tooth retention in cladodont sharks: with a comparison between primitive grasping and swallowing, and modern cutting and gouging feeding mechanisms. Journal of Vertebrate Paleontology, 21: 214–226.

Woodward A. S. 1889. Catalogue of the fossil fishes in the British Museum. Part I. British Museum (Natural History), 474 pp.

Woodward A. S. 1921. Observations of some extinct elasmobranch fishes. Proceedings of the Linnean Society of London, 133: 29–39.

Zangerl, R. & G. R. Case. 1973. Iniopterygia: a new order of chondrichthyan fishes from the Pennsylvanian of North America. Fieldiana Geology, Memoirs, 6: 1–67.

Ancient Fishes and their Living Relatives: a Tribute to John G. Maisey. Alan Pradel, John S. S. Denton & Philippe Janvier (eds.): pp. 215–224, 5 figs.
© 2021 by Verlag Dr. Friedrich Pfeil, München, Germany – ISBN 978-3-89937-269-4

The Ancient, Segmented, Active and Permanent Notochord

P. Eckhard Witten* and Brian K. Hall**

Abstract

The discovery of the notochord as a feature uniting all chordates arose from comparative embryological studies in the 19th century. This discovery established cephalochordates, ascidians (tunicates) and vertebrates as chordates. Hemichordates, initially included as chordates, came to be excluded when it was recognized that the hemichordate stomochord is not homologous with the notochord. This chapter examines the history of the search for chordates and early vertebrates in the fossil record, where the notochord is usually poorly preserved or not preserved at all. This discussion includes analysis of candidate vertebrates in the fossil record. The close resemblance between the notochord and notochordal cells (chordoblasts) and cartilage and cartilage cells (chondroblasts) has complicated interpretations of the nature of the notochord and of its origin – whether in the first chordate or in an earlier invertebrate ancestor – whether that ancestor was segmented or not – and whether the notochord is segmented. It has long been known that the anlagen of vertebral centra in teleosts are formed by segmented mineralisation of the notochord sheath. Recent studies on zebrafish and medaka embryos have uncovered the molecular mechanisms of a somite-independent notochord-driven segmentation process that establishes vertebral centra and intervertebral spaces. The fact that segmented mineralisation of the notochord sheath is not restricted to teleosts and evolved much earlier requires a new interpretation of the origin of vertebral bodies. Paleontology and developmental biology have reinforced one another in the search to understand the nature and evolutionary origin of the notochord and the vertebral column.

Notochord Evolution and the Fossil Record

The truly ancient notochord provides the name for our own phylum – Chordata – and connects vertebrates with tunicates and cephalochordates and perhaps with more ancestral chordate ancestors (Barry 1841, Kowalevsky 1866, Hejnol & Love 2014, Annona et al. 2015).

We are all aware that chordates have a notochord during their embryonic development and that the notochord forms the primary embryonic skeleton of the antero-posterior body axis. Less well known is that, with the exception of birds, almost all adult vertebrates preserve the notochord throughout life as a continuous rod, in the joints between vertebral bodies or inside the vertebral centra. Commonly, between the vertebral bodies the notochord expands and forms the inner part of the intervertebral joint. In mammals this forms the nucleus pulposus (Leeson & Leeson 1958, Grotmol et al. 2003, Shapiro & Risbut 2010, Choi & Harfe 2011). In birds it is not the notochord but the rostral part of the sclerotome that contributes cells to the intervertebral disk (Bruggeman et al. 2012). Within the vertebral centra, the notochord typically does not extend beyond the diameter it had in the embryo but becomes constricted in a narrow notochord canal. The notochord canal can completely disappear in amniotes but also in other vertebrates including some teleosts (Schaeffer 1967). A persistent notochord canal is a rare anatomical variation in humans and thus perhaps also in other mammals (Christopherson et al. 1999).

In palaeontology the notochord is a well-used reference structure to identify and position organs associated with it, such as the dorsal nerve cord and dorsal muscle blocks; in a study by John Maisey concerning the braincase of the Upper Devonian shark *Cladodoides wildungensis*, the notochord is referenced 42 times (Maisey 2005). A major difficulty arises in identifying chordates in the fossil record if the notochord is poorly preserved. Although the notochord withstands decay better than other soft tissues in taphonomy experiments (Briggs 1995, Raff et al. 2006), normally the skeletal tissues of vertebrates leave few records as fossils unless they are mineralised (Johanson et al. 2012, Trueman 2013). As the notochord, along with cartilage and connective tissue, preserves poorly, considerable lengths have to be gone to confirm the presence of a notochord. Philippe Janvier pointed out that for fossils that are preserved as either soft-tissue imprints or minute skeletal fragments, it is sometimes difficult for palaeontologists to tell which of them are reliable vertebrate remains and which merely reflect our idea of an ancestral vertebrate (Janvier 2015).

Another question one may ask is whether the notochord is a kind of cartilage. Alternatively, based on its appearance in the course of deuterostome evolution, is cartilage a type of notochord? No cartilage is found in echinoderms or tunicates but pharyngeal slits in hemichordates and cephalochordates are supported by acellular cartilage. Cartilage is also found in other invertebrate phyla but all invertebrate cartilages lack type II collagen (Cole & Hall 2004, Rychel & Swalla 2007). Notochord and the cartilages of all jawed vertebrates possess type II collagen. Hagfish and lampreys express type II collagen, prominently in the notochord and in limited amounts in some cartilages. Wright et al. (2001) point out that type II collagen is the major matrix protein of cartilages in jawed vertebrates.

The common ancestor of craniates likely had cartilage without type II collagen, but had type II collagen in the notochord. As the notochord is phylogenetically older than vertebrate cartilage, notochord may have acquired the ability to synthesize and deposit type II collagen before cartilage in which case "cartilage-type collagen" should be named "notochord-type collagen." Either way, a close relationship is evident. Both the inner layer of the perinotochordal sheath and cartilaginous extracellular matrices contain a closely similar range of glycosaminoglycans and proteoglycans, including aggrecan, chondroitin sulphate and keratan sulphate. Collagen types II, IX and X (discussed in Section 9) are also found in the matrices of both tissues (Ruggeri 1972,

* Department of Biology, Research Group 'Evolutionary Developmental Biology', Ghent University, B-9000 Gent, Belgium; peckhardwitten@aol.com (corresponding author)
** Department of Biology, Life Sciences Building. Dalhousie University, Halifax, NS, B3H 4R2, Canada; bkh@dal.ca

Mathews 1975, Linsenmayer et al. 1986, Schmitz 1995, Cole & Hall 2004, Stemple 2005a). Furthermore, Sox-9, the major transcription factor for cartilage development is also required for notochord development; Sox9 and Col2a1 are co-expressed in the notochord (Ng et al. 1997, Zhao et al. 1997).

Brachyury, a transcription factor from the T-box gene family characterizes the notochord of all chordates including the most basal chordates. Brachyury is sufficient and required to initiate notochord development. Brachyury is not expressed in the hemichordate stomochord consistent with the assumed non-homology of notochord and stomochord. Brachyury also is a cartilage gene capable of initiating chondrogenic differentiation of C3H10T1/2 cells in vitro and in vivo (Hoffman et al. 2002, Hall 2015); Capellini et al. 2008, Satoh et al. 2012, Annona et al. 2015). Brachyury expression in cells of the nucleus pulposus in adult mammals (Risbud & Shapiro 2011), provides additional evidence for the notochordal origin of the nucleus pulposus. Chordomas, which are notochordal tumours specifically express Brachyury. Chordomas are rare and known to originate from notochordal cells in adult mammals. Notochordal remnants found in the spinal column of mice are suspected to be a source for the onset of chordomas (Tamplin 2009). Reports about chordomas also exist from rats, dogs and cats (Hunter et al. 2003). In humans the frequency is 0.08/100 000 individuals in the USA (McMaster et al. 2001). The low incidence has been said to reflect resistance of the notochord to invasion by malignant tumours (Schroyens et al. 1991) but most if not all chordomas arise from transformation of notochordal cells themselves. Chordomas have features of cartilage and notochord (hence the name) and so represent what have been called intermediate tissues (Hall 2015). Indeed, a gradient of tissues extending from connective tissue to cartilage has been identified and many cartilage types shown to consist of large chondrocytes with little or almost no extracellular matrix (Beresford 1981, Benjamin 1989, 1990, Hall & Witten 2007, Witten et al. 2010). Chordomas are a neoplastic transition between notochord and cartilage, and as such represent a transformation series.

A debate about the evolutionary relationship between notochord and cartilage based on biochemical constituents and genes naturally takes place in the realm of neontology and evolutionary developmental biology. Paleontology is highly specialised in analysing structures. As we will see below, if a putative notochord is preserved as a midline structure there is room for interpretation and debate to whether the structure is really a notochord. Debates about the presence of a notochord influence our ideas on the evolution of early chordates as well as the evolution of early vertebrates. Especially in the oldest fossils, such as those in the Early and Middle Cambrian, identification of the notochord is difficult. Analysis of 114 specimens from the Middle Cambrian Burgess Shale led Conway Morris & Caron (2012) to interpret *Pikaea gracilens* as a stem chordate, not because of unequivocal identification of a notochord, but on the basis of some 100 segmented sigmoidal myomeres (muscle blocks) in each individual. Identification of a notochord was extremely tentative. The preserved structure may have been (i) a notochord (their interpretation, in no small part because of the presence of other chordate characters); (ii) a protonotochord or (iii) a non-homologous structure, the result of independent evolution and convergence with chordates. Preservation of the notochord is better in two Early Cambrian species – *Cathaymyrus diadexus* and *C. haikoensis* – from the Chengjiang formation in Yunnan Province, China (530 Ma). They were classified as cephalochordates on the basis of the identification of a notochord and the presence of segmented myomeres (Shu et al. 1996, 1999, 2003, Luo et al. 2001). So, given that these are Early Cambrian in age, a notochord in *Pikaea* would not be inconsistent with when the notochord and the chordates appear to have originated.

Vertebrate Candidates: *Jamoytius kerwoodi*, Conodonts and the "Tully Monster"

On the basis of two specimens, White (1946) named an Early Silurian fossil *Jamoytius kerwoodi* (in honour of the English palaeontologists, James Alan Moy-Thomas [1908–1944]) and described it as the most basal vertebrate then known. White came to this conclusion despite the absence of most vertebrate or even chordate characters in these two specimens. Now assumed to represent its own order of jawless vertebrates – †Jamoytiiformes – *Jamoytius* was considered a vertebrate because of a dorsal midline structure that was interpreted as a notochord, but it could have been intestine or part of a branchial basket (Sansom et al., 2010). Early vertebrates, like early chordates, can be enigmatic.

Conodonts are another enigmatic group of fossils that have fuelled the debate about the origin of vertebrates, also because recognition of a notochord is so difficult. Conodonts first appeared in the Cambrian 485 Ma and disappeared in association with the Triassic-Jurassic extinction event 200 Ma. Conodonts have been identified as plants, annelids, molluscs, basal chordates or even vertebrates. Derek Briggs, on the basis of the identification of segmented muscles, fins with fin rays, and a notochord, classified conodonts as chordates (Briggs 1992). Almost twenty years later, Turner et al. (2010) concluded that conodonts possess insufficient characters to classify them as vertebrates, craniates or jawed vertebrates, and that they may not even be chordates. The "conodont debate" exemplifies the difficulties in recognizing a notochord, for example to decide if a midline structure is a dorsal notochord or a ventral annelid gut (Blieck et al. 2010).

This is especially problematic when preserved mineralised elements bear no obvious resemblance to any known vertebrate skeletal elements. In the absence of a notochord and in the absence of virtually any vertebrate characters the conodont discussion focussed on tooth-shaped mineralised microfossils, known as conodont elements (Turner et al. 2010). Scott (1934) considered these to be essentially identical to the jaw apparatus of annelid worms. From recent detailed analyses, we know that conodont elements are unrelated to vertebrate dermal skeletal elements, either to teeth (absence of dentine and enamel) or to jaws (absence of bone) (Turner et al. 2010, Murdock et al. 2014, Hall 2015, Donoghue & Rücklin 2016). They remain enigmatic.

Recognition and identification of the notochord is also central in the debate about another vertebrate candidate the "Tully Monster" *Tullimonstrum gregarium*, a soft-bodied organism from the late Carboniferous Mazon Creek biota, 309–307 Ma. Previously identified as a nemertean worm, polychaete worm, gastropod, arthropod or conodont, an analysis of 1200 specimens by McCoy et al. (2016) claimed evidence for the presence of a notochord, cartilaginous vertebral bodies, gill pouches, and multiple tooth rows around the mouth. These characters coupled with a detailed phylogenetic analysis led McCoy and colleagues to classify *Tullimonstrum* as a vertebrate and a stem lamprey. This assignment of the "Tully Monster" to the vertebrate clade is rejected by Lauren Sallan, Philippe Janvier and their colleagues (Sallan et al. 2017). They argue that the light coloured stain (the hypothetical notochord) extends anterior to the supposed eye bars, which is never the case in any vertebrate, and further, that by removing the notochord from the equation, other traits of the fossil are not easily interpreted as vertebrate characters. Similarly, the earliest craniates (named ostracoderms by Edward Cope in 1889) are known only from their mineralised dermal skeletons, and again, we have minimal data about their internal anatomy, skeletal or otherwise (Maisey 2000). Ostracoderms are not a monophyletic group but a polyphyletic assemblage of six extinct jawless vertebrates. Even the two well-recognized groups – heterostracans and cephalaspids – are not natural groups. The six recognized groups are galeaspids, pituriaspids, osteostracans, pteraspidomorphs, thelodonts and anaspids (Janiver 2015).

The Old/New Notochord: Segmented and Active

Not only palaeontologists but developmental biologists take a keen interest in the notochord because the notochord is an important organizer (the primary organizer) in early development (Anderson & Stern 2016). Sonic hedgehog signalling (Shh) from the notochord induces the formation of somites from paraxial mesoderm (Christ & Ordahl 1995, Fleming et al. 2001). In amphibians, somites fail to differentiate without the notochord (Malacinski & Youn 1982). In chick embryos, information for segmentation is restricted to the most medial presomitic mesoderm cells (Freitas et al. 2001). These cells give rise to the vertebral bodies of each vertebra by migrating into the perinotochordal space in response to signals, such as Shh and Noggin from the notochord (Muller et al. 1996, McMahon et al. 1998, Christ et al. 2004).

Later, when it comes to the development of the skeleton of the vertebral column, the notochord has been regarded as a rather passive scaffold that eventually disappears (Willams 1908). This established the view that patterning information for the vertebral column is provided only by the somites that give rise to the sclerotomal mesenchyme (Shapiro 1992, Brand-Saberi et al. 1996). Next, sclerotome-derived cells develop into the cartilages of the haemal and neural arches. Expansion of the cartilaginous bases of the haemal and neural arches around the notochord was proposed as the cellular process that led to the evolutionary origin of the vertebral centra (Ota et al. 2011). In extant amniotes, each vertebral body forms from a solid block of cartilage that encompasses the vertebral centrum and arches. The arches and centra form as vertebral bodies from the continuous cartilaginous anlage by endochondral bone formation, but from separate ossification centres for each vertebra. Evidently, so we believe, the notochord disappears, together with the cartilage during this endochondral bone formation, without the notochord having been actively involved in vertebral body formation.

Our knowledge on which this scenario rests derives from detailed studies of early stages of vertebral body formation in mouse and chicken embryos (Scaal 2016). The conservation of early developmental stages in vertebrates suggests that the chicken and mouse mode of vertebral body development is a valid model for all vertebrates (Nikaido et al. 2002). But remember, although mammals and birds are tetrapods, should we expect them to represent vertebral development in other tetrapods (urodele or anuran amphibians, reptiles), in other bony fishes or in cartilaginous fishes?

This accepted view of a passive function of the notochord in vertebral column development has been challenged by recent studies, primarily using zebrafish (*Danio rerio*) and Japanese medaka (*Oryzias latipes*), two small and readily maintained teleost species that have emerged as models in developmental biology and biomedical research. Indeed, the zebrafish and medaka mode of vertebral body development – as is now known to be generally the case in teleost fish – is quite different from the mouse and chicken model. There is increasing evidence that not only the somites but also the notochord contains patterning information and that the notochord establishes the anlagen of vertebral centra. Patterning of the notochord is now viewed as independent and perhaps even prior to patterning of the somites.

Our "new knowledge" should not come as a total surprise; two highly reputed scientists re-ported the same mode of establishing vertebral body anlagen – mineralisation of the notochord sheath – for basal (eels) and advanced (sticklebacks) teleosts and in the same year (Huxley 1859, Kölliker 1859). Thomas H. Huxley, who published his observations on vertebral column development in the three-spined stickleback *Gasterosteus aculeatus* informs us that:

"In the greater part of its extent it [the notochord] was enclosed neither in cartilage nor in bone – though bony rings, the rudiments of the centra of the vertebrae, were developed in the wall of the notochord throughout the rest of the body." (Huxley 1859, p. 39).

Albert Kölliker published his studies on vertebral column development in the European eel *Anguilla anguilla*:

"In the Leptocephali [transparent eel larva] the sheath of the chorda ossifies without having been transformed into cartilage; and the same seems to hold good for the other osseous fishes." (Kölliker 1859, p. 217).

One would assume that this understanding of the mode of teleost vertebral body development would have become textbook knowledge (Fig. 1). Not so. One reason is perhaps that thirty-six years later, Hans Gadow and Elizabeth Caroline Abbott in their influential and extensive treatise "On the evolution of the vertebral column of fishes" do not mention Huxley's and Kölliker's observations (Gadow & Abbott 1895). Edwin S. Goodrich relied on Gadow & Abbott (1895) for his leading textbook (Goodrich 1909). Ever since, textbooks have perpetuated the statement that vertebral body precursors are made from cartilage. Cartilage cells that form the bases of the haemal and neural arches. Cells that derive from sclerotomal mesenchyme establish the anlagen of vertebral centra (Kriwet & Pfaff 2019).

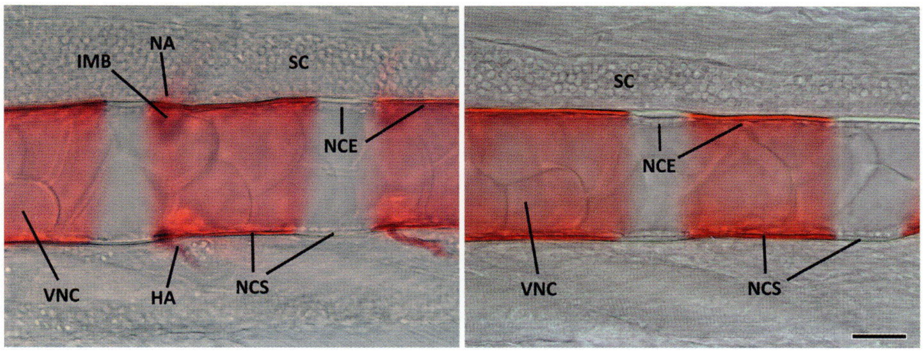

Fig. 1. Mineralisation of the teleost notochord sheath in the Japanese medaka *Oryzias latipes*, shown at 15 days post-fertilization (dpf) using Alizarin red S whole mount staining for calcium. In the caudal part of the developing vertebral column, mineralisation of the notochord sheath (NCS) establishes the anlagen of the vertebral centra (red), which is clearly visible when viewed at high magnification (compound microscope, ×400). Huxley and Kölliker may have made similar observations that led to their publications in 1859 about notochord sheath mineralisation in the three-spined stickleback *Gasterosteus aculeatus* and in the European eel *Anguilla anguilla*, respectively. The left panel shows a wild type medaka. In the location of the intervertebral spaces, the notochord sheath (NCS) and the notochord epithelium (NCE) are thickening. Inside large vacuolated notochord cells (VNC) are visible, the spinal cord (SC) is visible dorsal to the notochord. After the notochord sheath becomes mineralised the first intramembranous bone (IMB) is deposited onto the mineralised notochord sheath, starting at the basis of the haemal arches (HA) and neural arches (NA). The right panel shows an *osterix/sp7* mutant medaka with non-functional sclerotome-derived osteoblasts. Vertebral arches and perinotochordal bone are missing but the development of the vertebral centra anlagen by mineralisation of the notochord sheath is unaffected. See Yu et al. (2017) for the in-depth analysis of this osterix mutant phenotype. Scale bar = 25 μm

Chordal Mineralisation versus Perichordal Ossification

Huxley's and Kölliker's observations concerning vertebral body development in the largest group of vertebrates (teleosts) were not ignored by the palaeontology, ichthyology and zoological communities (Laerm 1979, 1982, Lauder 1980, Arratia et al. 2001).

Laerm (1982) informs us that "shortly after the formation of the notochord and its sheaths, a calcified ring, the chordacentrum, forms in the fibrous sheath of the notochord in each prospective vertebral segment." This is now known as the dual segmentation model in which the segmental patterns of the neural and haemal arches are somite-derived, while the vertebral segments are notochord-derived. Indeed, within teleost vertebrae the mode of forming vertebral bodies is remarkably consistent (Lauder 1980). Mineralisation of the notochord sheath has been shown in Atlantic herring (*Clupea harengus*), northern pike (*Esox lucius*), Atlantic salmon (*Salmo salar*) and Zebrafish

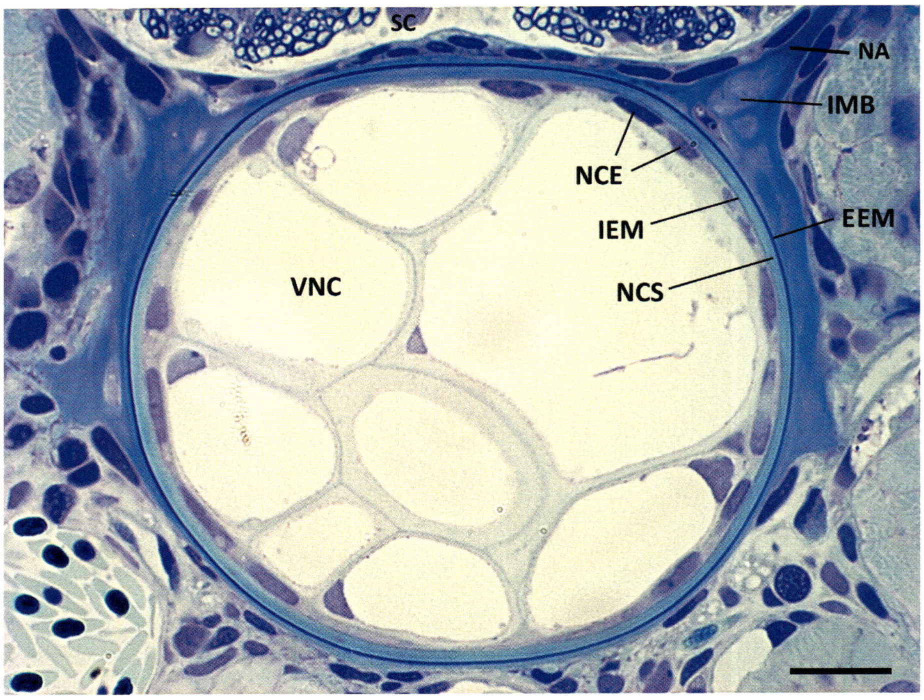

Fig. 2. *Oryzias latipes* 15 dpf, cross section through the developing vertebral column and notochord. The notochord sheath (NCS) has a thin inner elastic membrane (IEM) and a prominent external elastic membrane (EEM). The external elastic membrane delimits the notochord sheath. Intramembranous bone (IMB) that forms around the mineralised notochord sheath is visible in connection with the development of the neural arches (NA). The cells of the notochord epithelium (NCE) are clearly distinct from the vacuolated notochord cells (VNC) in the centre. Another name for notochord epithelial cells is chordoblasts (Ward et al. 2018, Pogoda et al. 2018). Whether the notochord epithelium constitutes a true epithelium was discussed by Jurand (1962). Schmitz (1995) emphasises that the notochord epithelium in yellow perch (*Perca flavescens*) is indeed a stratified squamous epithelium with cells that are interconnected by desmosomes and gap junctions. Ellis et al. (2013) showed that the vacuoles of the cells in the centre of the notochord are lysosome-related organelles whose formation and maintenance requires late endosomal trafficking regulated by the vacuole-specific Rab32a and H^+-ATPase-dependent acidification. Toluidine blue staining. Scale bar = 20 µm.

(*Danio rerio*) (Francois 1966, Schaeffer 1967, Grotmol et al. 2003, 2005, Nordvik et al. 2005, Nordvig 2007, Bensimion-Brito et al. 2012b).

Developmental biology only started to take full notice of the "non-textbook" model of the development of teleost vertebral bodies once zebrafish and Japanese medaka became established as model organisms (Figs. 1, 2). In the early days of zebrafish research – over 20 years ago – the phenotype of the fused somite mutant zebrafish could have changed the paradigm. In this mutant, somites are fused, with the consequence that haemal and neural arches are fused, malformed or fail to form (van Eeden et al. 1996). Nevertheless, vertebral body centra develop separately, which is a strong indication of somite-independent patterning of the vertebral column. The opposite phenotype, which also indicates independence, is seen in zebrafish, medaka and salmon with fused vertebral centra. Fusion may be a part of normal development (Bensimon-Brito et al. 2012a, Hall & Witten 2019) or it may be pathological (Takeuchi 1966, Witten et al. 2006). Fusion affects the centra only; the arches remain separated.

There is, however, more evidence that patterning and development of vertebral body centra are independent from somite development (Fleming et al. 2004). Ablation of the notochord in urodele and chicken embryos results in complete fusion of cartilaginous vertebral body anlagen. Notochord ablation removes vertebral body identity, but not arch identity (Hall 1977, Fleming et al. 2001). Ablation experiments also revealed the importance of dorsal root ganglia for segmentation of haemal arches but not for segmentation of vertebral bodies (Hall 1977, Senthinathan et al. 2012). Fleming et al. (2004) presented a study according to which ablation of notochord cells prevented the formation of vertebral centra in zebrafish. Pogoda et al. (2018) further narrowed down the responsible cell type and showed that targeted ablation of cells from the notochord epithelium prevents the formation of vertebral centra anlagen. According to Fleming et al. (2004) isolated, but intact, notochords still produce mineralised anlagen of vertebral centra.

More evidence for the somite-independent, but notochord-dependent, formation of vertebral centra anlagen is provided by a study of the *osterix* /*sp7*–/– mutant in the Japanese medaka by Yu et al. (2017). In these mutants, sclerotome-derived pre-osteoblasts fail to differentiate into functional osteoblasts. The consequence is the failure of haemal arches to form. Still, the notochord forms mineralised segments in the notochord sheath, which, essentially is the extracellular matrix of the notochord (Figs. 1, 3). Pogoda et al. (2018) showed that *osterix/sp7*-positive osteoblasts are indeed only located outside the notochord and are recruited into the mineralised areas of the notochordal sheath. Thus, notochord mineralisation occurs: (i) prior to bone formation, (ii) within the notochord sheath, and (iii) is *osterix* / *sp7*-independent. These studies confirm Joshua Laerm's conclusion from his studies on fossil and extant osteichthyans:

"In living teleosts, the perichordal tissue results from differentiation of sclerotome cells. The chordal calcification forms in the acellular fibrous sheath of the notochord and apparently does so without the participation of sclerotome cells. This suggests that separate developmental mechanisms may be responsible for chordal calcifications and perichordal ossifications." (Laerm 1979, p. 481).

Segmentation Without the Segmentation Clock?

For amniotes, it is widely accepted that vertebral development, and thus development of the vertebral column, is based on somites and the "segmentation clock," which is a molecular oscillator that controls the timing (periodicity) of somite formation under the control of *Notch*, *Wnt* and *Fgf* gene pathways (Giudicelli & Lewis 2004). It was assumed that the same cellular basis and timing mechanism operated in teleosts such as zebrafish (Schröter & Oates 2010).

In tetrapods such as chickens and mice, mutations that disrupt the clock and the patterning of individual somites have dramatic effects on the formation of vertebral bodies (White et al. 2003). A somite-dependent basis for the establishment of vertebral body anlagen also was proposed for zebrafish and medaka and so for teleosts in general (Renn et al. 2013, Inohaya et al. 2007). Lleras Forero et al. (2018) took a closer look at zebrafish mutants with disturbed somitogenesis. They analysed Tbx6 mutants, previously described by van Eeden et al. (1996) as fused somite mutants (fss) on the basis of the mutant phenotype. Despite the severely disturbed somitogenesis shown by van Eeden et al. (1996), these mutants maintain separate vertebral body centra. Indeed, disturbed somitogenesis in zebrafish has no affect on the generation of vertebral centra.

Taking the model further Forero et al. (2018) analysed the genes her1 and her7 that are central to the operation of the somitic clock. Mutations for each of these clock genes had no affect on separation of the vertebral centra. Even in triple mutant zebrafish (*her1*–/–, *her7*–/–, *tbx6*–/–), vertebral centra remained separated, which strongly suggests that, in zebrafish, the segmental generation of vertebral body anlagen is somite-independent and autonomous to the cells producing the centra. This does not, however, exclude an influence of somites and the somitic clock on the formation of functional vertebral centra; functional vertebral bodies require co-ordinated development of vertebral centra and arches.

Mineralisation of the Notochord Sheath: from Inside Out or from Outside In?

Data about mineralisation of the notochord sheath cannot be obtained easily because mineralisation is weak and the sheath is too thin to be visualised by microCT, unless synchrotron based microCT is used (Bruneel & Witten 2015, Keating et al. 2018) (Figs. 1–3).

As we know from teleost development, the mineralised notochord sheath quickly becomes surrounded by intramembranous bone after which distinguishing between the bone and the mineralised notochord sheath is very challenging (Figs. 4, 5). Obviously, data from the fossil record about notochord sheath mineralisation in early developmental stages are difficult to obtain. John Maisey has addressed the problem of ontogenetic data that are essentially unavailable from the fossil record and for practical purposes are also not available for most extant species (Maisey 1988). We have more data about bone that forms directly around the notochord sheath – from all basal actinopterygian and sarcopterygian groups including basal tetrapods – than we have data about a mineralised notochord sheath (Gardiner 1983). In neontology, histological studies of the skeleton routinely decalcify specimens and thus lose data about notochord sheath mineralisation. Moreover, conventional whole-mount-staining for bone and cartilage removes mineral from the notochord sheath and so produces false negative results. This artefact has been known for thirty years (Vandewalle et al. 1988, Springer & Johnson 2000, Bird & Mabee 2003) but it was not until the acid-free staining protocol of Walker & Kimmel (2007) became popular within the zebrafish research community that notochord sheath mineralisation was re-recognised (Bensimon-Brito et al. 2016).

In teleosts, subdivision of the notochord sheath into mineralised (vertebral centra anlagen) and non-mineralised (prospective intervertebral spaces) segments is preceded by changes in the shape and orientation of notochordal epithelial cells. The first indication of segmentation is proliferation of cells of the notochord epithelium in prospective intervertebral spaces, resulting in thickening of the col2-based notochord sheath (Hay 1895, Schauinsland 1903, Laerm 1976, Nordvik et al. 2005). As in cartilage, the collagen gene *col9a2* is associated with col2a in the notochord sheath extracellular matrix. Consequently, the distinctive expression of *col9a2* in the locations of future intervertebral spaces indicates increased notochord sheath production (Pogoda et al. 2018). Notochord sheath mineralisation, which occurs in zebrafish without increased *col9a2* expression, is retinoic acid-dependent. Overexpression of retinoic acid causes loss of the non-mineralised intervertebral spaces in the notochord sheath (Spoorendonk et al. 2008), blocking retinoic acid-signaling prevents notochordal sheath mineralisation (Pogoda et al. 2018).

Osteoblasts control the process of mineralising bone matrix by secreting pyrophosphate, which is a strong mineralisation inhibitor. Secretion of pyrophosphate depends on ectonucleoside pyrophosphatase/phosphodiesterase 1 (*enpp1*) and its antagonist ectonucleoside triphosphate/diphosphohydrolase 5 (*entpd5*), which is a mineralisation promoter. Like osteoblasts, the cells of the zebrafish notochord epithelium express *entpd5* at sites of notochord mineralisation (Lleras Foreo et al. 2018, Pogoda et al. 2018). In addition, secreted pyrophosphate must be removed to make mineralisation possible. To accomplish this, osteoblasts produce alkaline phosphatase (ALP), which functions as a pyrophosphatase. Like osteoblasts, the cells of the notochordal epithelium produce alkaline phosphatase; expression is detected in a segmented pattern at sites of notochordal sheath mineralisation, shown for Atlantic salmon by Grotmol et al. (2003) and then for zebrafish by Bensimon-Brito et al. (2012a). In addition, Bensimon-Brito et al. (2012b) demonstrated segmented expression of osteocalcin (the most abundant non-collagenous bone protein) in the notochord epithelium.

The studies on zebrafish and medaka refute the hypothesis that mineralisation of the notochordal sheath is facilitated by osteoblasts from outside the notochord. We now know that in zebrafish and medaka, and thus most likely in other teleosts, notochord mineralisation starts before osteoblasts accumulate around the notochord (Fig. 3). Mineralisation is controlled by the notochordal epithelium, a process that is osteoblast-independent as shown by Yu et al. (2017, Fig. 1). Notochordal epithelium mineralises its own sheath in a segmented fashion, creating the anlagen of the vertebral centra.

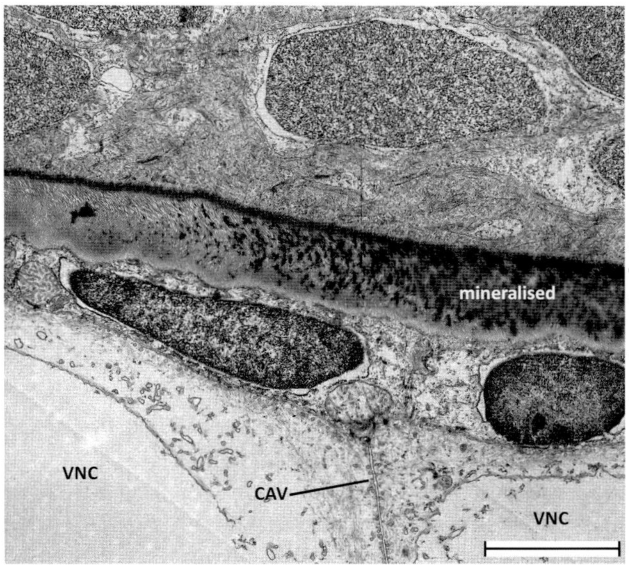

Fig. 3. Transmission electron microscopical image of the notochord sheath of a zebrafish (*Danio rerio*) at 10 dpf. The principal components – notochord epithelium (NCE), inner elastic membrane (IEM), type 2 collagen based notochord sheath (NCS) and external elastic membrane (EEM) characterise the notochord of vertebrates in all classes. It is clearly evident that the notochord sheath starts to mineralise prior to bone formation around the notochord sheath. Abutting cell membranes of vacuolated cells show numerous caveolae, which are submicroscopic cup-shaped plasma membrane pits that can buffer tension in cells that undergo high levels of mechanical stress (Lim et al., 2017). TEM preparation, postfixed with osmium tetroxide, contrasted with uranyl acetate and lead citrate. See Bensimon-Brito et al. (2012b) for further details about notochord sheath mineralisation in zebrafish.

The Resegmentation Debate

That vertebral bodies develop after a resegmentation of cells from adjacent somites (Remak 1855) is a well accepted model for amniotes although it has also been denied for this group (Verbout 1976). The resegmentation model corroborates other evidence that tetrapod vertebral centra derive from somites. Resegmentation establishes a clearly defined spatial and developmental relationship between arches and vertebral centra, resulting in a one-to one-relationship of arches and centra and the long-recognized intersegmental position of vertebral bodies; the intersegmental position of vertebral bodies in teleosts has been viewed as evidence for resegmentation in teleosts (reviewed by Lauder 1980).

Nonetheless, the adult centrum in different teleost groups can have a variable anterior to posterior position in relation to the vertebral arches (Schaeffer 1967, Lauder 1980). Morin-Kensicki et al. (2002) showed experimentally that somite-derived sclerotome cells in zebrafish do not strictly contribute to the development of intersegmental vertebral bodies (as would be predicted by the resegmentation hypothesis). Schaeffer (1967), Laerm (1976) and Lauder (1980) are unequivocal about the absence of sclerotomal resegmentation in osteichthyans. Wake & Lawson (1973) did not find any evidence for resegmentation of sclerotome in salamanders and extended the absence of resegmentation to all non-amniotes. David Wake points out

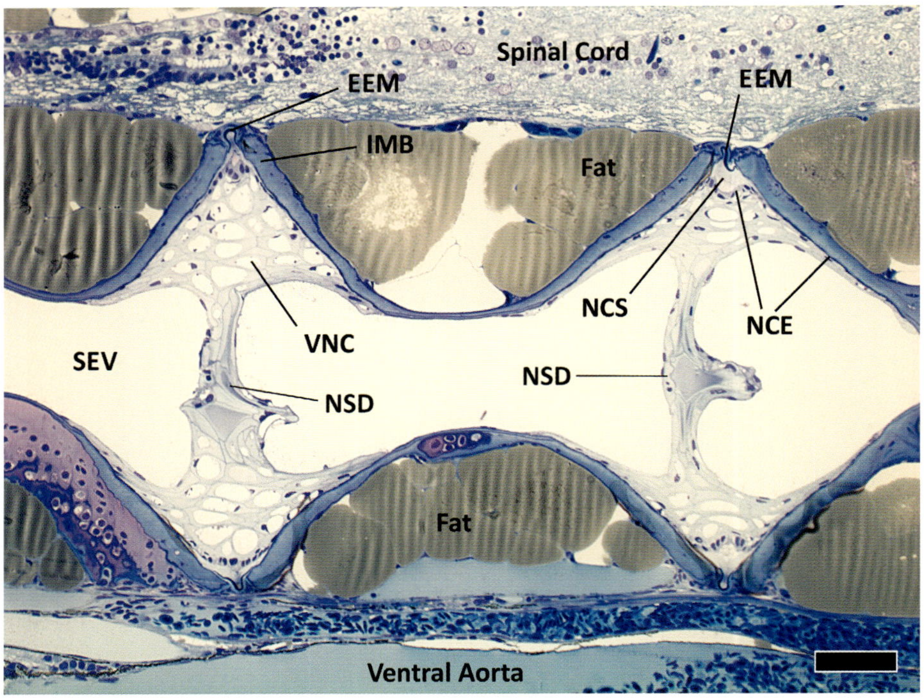

Fig. 4. Sagittal section through the vertebral column of a juvenile (15 mm standard length) zebrafish. As in mammals the notochord is extended in the intervertebral spaces and functions to support the intervertebral joint (Leeson & Leeson 1958). The broad notochord sheath (NCS) and the broad external elastic membrane (EEM) are important inner parts of the intervertebral ligament that connect vertebral centra. Notice the extension of the notochord epithelium (NCE) and the orientation of cell nuclei, perpendicular to the animal axis. The latter has been described for Atlantic salmon (*Salmo salar*) (Grotmol et al. 2003). In the centre of the notochord, vacuolated notochord cells (VNC) condense and acquire the character of a fibrous connective tissues. The cells form the notochord strand (NSD) contains keratan sulphate and can mineralise, for example in sturgeons (Schmitz 1998, Leprevost et al. 2017). The condensation of the vacuolated cells makes space for large secondary extracellular vacuoles (SEV) that also develop in the mammalian nucleus pulposus (Trout et al. 1982). Intramembranous bone (IMB) has formed around the mineralised notochord sheath. Figure provided by Arianna Martini, Toluidine blue staining. Scale bar = 50 µm.

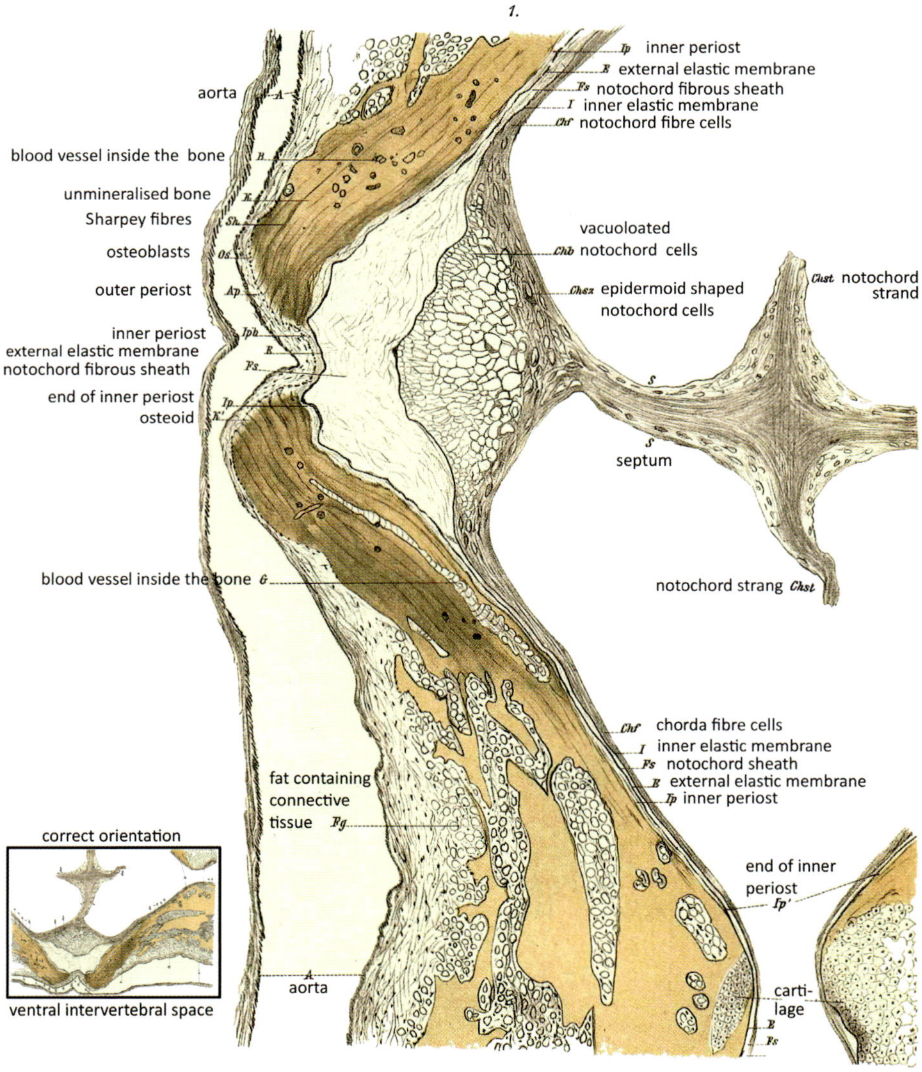

◁ **Fig. 5.** Von Ebner's unsurpassed detailed description of the notochord as forming part of the intervertebral joint in the freshwater pike *Esox*. According to the von Ebner's figure legend the magnification is 40 ×. This would translate into 300 × 412 µm original size of the depicted area. The principle components are found in all teleosts although in miniaturised species such as zebrafish or medaka identification of particular elements can become more difficult. Von Ebner is best known for the discovery of a somite subdivision (von Ebner's fissure) that fuelled the resegmentation hypothesis (von Ebner 1888). English terms have been added to the original figure from von Ebner (1896) by PEW.

(Gadow & Abbott 1895). If we follow this hypothesis or any of its variants and if we consider that early osteichthyans lack vertebral centra, the conclusion must be that vertebral centra evolved independent in different classes of vertebrates (Arratia et al. 2001). A different view is expressed by Gardiner (1983), who emphasizes that consideration of the vertebral centrum must begin with the origin and structure of the notochord and its sheath, which in form and function is identical and thus homologous in all vertebrates.

As we recognise the notochord as much more active than commonly acknowledged we may define earlier events in development as the start of vertebral body formation, events earlier than considered in the framework of the arcualia hypothesis. Each of the following key events could, for example, be defined as the start of vertebral body development:

(1) Notochord signalling to subdivide the presomitic mesoderm into somites.
(2) Subdivision of the notochord into intravertebral notochord and intervertebral spaces.
(3) Expansion of the notochord establishes the intervertebral joints.
(4) Signalling (brachyury) by the notochord induces the differentiation of skeletal precursor cells.
(5) Cartilaginous condensation of vertebral arch anlagen.
(6) Mineralisation of the notochord sheath.
(7) Formation of cartilage or bone around the notochord
(8) Mineralisation of cartilage or bone formation around the notochord?

Classical embryology considers (5) – the condensation of somitic chondrogenic cells around the notochord – as the start of vertebral body development. Analysis of the fossil record must rely on (8), the pattern of mineralisation of notochord, cartilage or bone. We just started to reconsider the faint segmented mineralisation of the notochord sheath (6), as a key step that defines vertebral body anlagen.

Mineralisation of the notochord sheath is not a derived character of teleosts. In fact, mineralisation of the notochord sheath is a pre-teleostean feature (Patterson 1968). The earliest recorded actinopterygian vertebral centra in *Haplolepis* from the Carboniferous were made from mineralised notochord sheath (Patterson 1968, Gardiner 1984). The Triassic chondrosteans Pholidopleurids and *Turseodus* had opposed dorsal and ventral half-rings as vertebral centra based on notochord sheath mineralisation (Patterson 1968, Schaeffer 1967). In teleosts this pattern is no longer a character found in adults. Separate dorsal and ventral mineralisation of the notochord sheath occurs only during development before the ring-shaped notochord sheath mineralisation is complete and before the mineralised notochord sheath is encircled by membrane bone (Bensimon-Brito et al. 2012b). In rhipidistians such as *Rhizodopsis*, *Megalichthys*, *Ectosteohachis* and *Strepsodus*, vertebral centra are made from membrane bone around the notochord (Gardiner 1984). If centra formation went through the same developmental steps as in teleosts, there would have been a faint mineralised notochord sheath as substrate for the bone.

In the light of clear evidence that a mineralised notochord is the anlage of the teleost vertebral centrum, our view about the significance of this character may have to be revised. We know that the notochord sheath can mineralise in chimeras (Francois 1966), in urodeles (Danto et al. 2019) and in mammalian embryos (Willams 1908). Notochord sheath mineralisation is a elasmobranch character and also known from the fossil record of this group (Daniels 1934, Maisey 2008). Obviously, the notochord sheath of vertebrates has the capacity to mineralise but, of course, for reasons discussed here, we do not have sufficient data to allow us to conclude how common notochord mineralisation is outside the actinopterygians. Nonetheless, as vertebrates from all classes follow the first five steps (1–5) of vertebral body development in the list above, then steps 1 to 5 should be considered as homologous. Homology includes 'homology of processes' as discussed by Gilbert et al. (1996) and Hall (2003) but also includes homology of the structures that develop.

Next vertebral centra formation continues around the notochord sheath in essentially one of two ways: (a) by direct formation of membranous bone or (b) by the formation of cartilage that subsequently mineralises. It is interesting that *col10a* is co-expressed with osterix expression by the osteoblast population that produces the bone around the notochord sheath in medaka (Renn et al. 2013, Seeman et al., 2015). In mammals *col10a* is typically expressed in mineralising cartilage but absent from osteoblasts. *Col10a* expression in this osteoblast population may reflect the dual nature of these cells that can produce bone or mineralising cartilage. If so, differences concerning the involvement of cartilage or bone in centrum formation are the consequence of descent with modifications and not characters that evolved independent in different lines of vertebrates.

Whether vertebral centra initially materialise as cartilage, bone or mineralised notochord sheath is likely largely a function of speed of development. Wake & Wake (2000) compared vertebral body development between gymnophiones and mammals and concluded that the observed major differences can be attributed to vastly different developmental rates. Further, large parts of the vertebral centrum develop in response to mechanical load. As Joshua Laerm put it: "It is possible to suggest that a centrum is a centrum is a centrum, reflecting the generalised adaptive response of sclerotome tissue to functional demands for vertebral consolidation" (Laerm 1979, p. 482).

Notochord and somites as developmental modules

Recent studies on zebrafish and medaka have revealed developmental mechanisms by which the notochord mineralises its own sheath independently from the somites and the somitic clock. These studies strongly support all previous findings that show that vertebral centra and arches are developmental modules. Modularity of the two systems has been long recognized for all gnathostomes including mammals (Strudel 1953, Detwiler & Holtzer 1956, DeClercq et al. 2017, 2018, Hall 1977, Lauder 1980, Hautier et al. 2010, de Azevedo et al. 2012, Hall 2015, Yu et al. 2017, Ward et al. 2018). Nature provides numerous examples that show this modularity. Diplospondyly (two centra per segment) is an obvious one (Hay 1895, Lauder 1980) but the complete range encompasses aspondyly, monospondyly, diplospondyly and polyspondyly (Zhang 2009).

Regionalisation of the vertebral column also commonly requires uncoupling of arch and centra development. Much evidence was contributed by the ablation experiments, mutants and developmental studies discussed above (reviewed by Hall 2015). Hautier et al. (2010) tested the patterns of events in the sequences of development of arches and centra in the brown-throated sloth *Bradypus variegatus* and identified significant modularity in this mammal. Modularity between arches and centra is widely accepted and in conflict with the also widely accepted arcualia hypothesis. This conflict is however often not addressed. Monsoro-Burq et al. (1994), for example, communicate in the same paragraph that vertebrae derive form somites and that the notochord induces the somites and induces sclerotomal cells to form cartilage. Indeed, the fact that vertebral body anlagen can arise independent from the arches has the potential to refute the arcualia hypothesis that views ventral cartilaginous arch elements as representing the first elements of vertebral bodies (Ota et al. 2011, 2014).

The modularity of centra and arches triggers the question of whether the notochord or somites were the first segmented system. The lack of somites in basal chordates such as tunicates, led Claudio Stern to suggest that the notochord may have been the first segmented structure (Stern 1990). Ward et al. (2018) and Peskin et al. (2020) argue that both somites and notochord are important for the segmented development of vertebral bodies, somites taking the lead in amniotes and the notochord taking the lead in teleosts. This again raises the question of how vertebral body development was determined in early gnathostomes (Ward et al. 2018). Independently of whether we will ever have an answer to this question, the rediscovery of the active segmented notochord and the new knowledge about the mechanisms by which the notochord establishes vertebral centra anlagen in teleosts should encourage us to revisit notochord development in non-teleost vertebrate. This in turn may influence how we interpret fossils. In the words of John Maisey: "The history of the skeletal systems in vertebrates can best be understood through the reciprocal illumination that comes through a combined paleontological and developmental biological approach" (Maisey 1986).

Acknowledgements

PEW acknowledges funding from the European Union's Horizon 2020 research and innovation programme under the Marie Skłodowska-Curie grant agreement No. 766347. Notochord figures taken at the lab of PEW at Ghent University derive form collaborative research with the lab of Christoph Winkler (University of Singapore) and from the PhD projects of Anabela Bensimon-Brito and Arianna Martini.

Literature cited

Anderson, C. & C. D. Stern. 2016. Organizers in development. Current Topics in Developmental Biology, 117: 435–454.

Annona, G., N. D. Holland & S. D'Aniello. 2015. Evolution of the notochord. EvoDevo, 6: 30, DOI 10.1186/s13227-015-0025-3

Arratia, G., H. P. Schultze & J. Casciotta. 2001. Vertebral column and associated elements in dipnoans and comparison with other fishes: Development and homology. Journal of Morphology, 250: 101–172

Barry, M. 1841. On the Chorda Dorsalis. Philosophical Transactions of the Royal Society of London, 131: 195–199

Benjamin, M. 1989. Hyaline-cell cartilage (chondroid) in the heads of teleosts. Anatomy and Embryology, 179: 285–303.

Benjamin, M. 1990. The cranial cartilages of teleosts and their classification. Journal of Anatomy, 169: 153–172.

Bensimon-Brito, A., M. L., Cancela, A. Huysseune & P. E. Witten. 2012a. Vestiges, rudiments and fusion events: the zebrafish caudal fin endoskeleton in an evo-devo perspective. Evolution & Development, 14: 116–127.

Bensimon-Brito, A., J. Cardeira, M. L. Cancela. A. Huysseune & P. E. Witten. 2012b. Distinct patterns of notochord mineralization in zebrafish coincide with the localization of Osteocalcin isoform 1 during early vertebral centra formation. BMC Developmental Biology, 12: 28.

Bensimon-Brito, A., J. Cardeira, G. Dionísio, A. Huysseune, M. L. Cancela & P. E. Witten. 2016. Revisiting in vivo staining with Alizarin red S – a valuable approach to analyse zebrafish skeletal mineralization during development and regeneration. BMC Developmental Biology, 16: 2 DOI: 10.1186/s12861-016-0102-4.

Beresford, W. A. 1981. Chondroid Bone, Secondary Cartilage and Metaplasia. Urban and Schwarzenberg, Munich.

Bird, N. C. & P. M. Mabee. 2003. Developmental morphology of the axial skeleton of the zebrafish *Danio rerio* (Ostariophysi: Cyprinidae). Developmental Dynamics, 228: 337–357.

Blieck, A., S. Turner, C. J. Burrow, H. P. Schultze, C. B. Rexroad, P. Bultynck & G. S. Nowlan. 2010. Fossils, histology, and phylogeny: Why conodonts are not vertebrates. Episodes, 33: 234–241.

Brand-Saberi, B., J. Wilting & B. Ebensperger. 1996. The formation of somite compartments in the avian embryo. International Journal of Developmental Biology, 40: 411–420.
Briggs, D. E. G. 1992. Conodonts: a major extinct group added to the vertebrates. Science, 256: 1285–1286.
Briggs, D. E. G. 1995. Experimental taphonomy. Palaios, 10: 539–550.
Bruggeman, B. J., Maier, J. A., Mohiuddin, Y. S., Powers, R., Lo, Y., Guimaraes-Camboa, N., Evans, S. M. & B. D. Harfe. 2012. Avian intervertebral disc arises from rostral sclerotome and lacks a nucleus pulposus: implications for evolution of the vertebrate disc. Developmental Dynamics, 241: 675–683 .
Bruneel, B. & P. E. Witten. 2015. Power and challenges of using zebrafish as a model for skeletal tissue imaging. Connective Tissue Research, 56: 161–173.
Capellini, T. D., M. P. Dunn, Y. J. Passamaneck, L. Selleri & A. Di Gregorio. 2008. Conservation of notochord gene expression across chordates: Insights from the Leprecan gene family. Genesis, 46: 683–696.
Christ, B., R. Huang & M. Scaal. 2004. Formation and differentiation of the avian sclerotome. Anatomy & Embryology (Berlin), 208: 333–350.
Christ, B. & C. R. Ordahl. 1995. Early stages of chick somite development. Anatomy & Embryology (Berlin), 191: 381–396.
Christopherson, L. R., Rabin, B. M., Hallam, D. K. & E. J. Russell. 1999. Persistence of the Notochordal Canal: MR and Plain Film Appearance. American Journal of Neuroradiology, 20: 33–36.
Choi K. S. & B. D. Harfe. 2011. Hedgehog signalling is required for formation of the notochord sheath and patterning of nuclei pulposi within the intervertebral discs. Proceedings of the National Academy of Sciences of the USA, 108: 9484–9489.
Cole, A. G. & B. K. Hall. 2004. The nature and significance of invertebrate cartilages revisited: Distribution and histology of cartilage and cartilage-like tissues within the Metazoa. Zoology, 107: 261–274.
Conway Morris, S. & J. B. Caron. 2012. *Pikaia gracilens* Walcott, A stem-group chordate from the Middle Cambrian of British Columbia. Biological Reviews of the Cambridge Philosophical Society, 87: 480–512.
Daniel, J. F. 1934. The Elasmobranch Fishes. University of California Press, Berkeley, California, 332 pp.
Danto, M., F. Witzmann, S. K. Kamenz & N. B. Fröbisch. 2019. How informative is vertebral development for the origin of lissamphibians? Journal of Zoology, 307: 292–305.
de Azevedo, T. P., P. E. Witten, A. Huysseune, A. Bensimon-Brito, C. Winkler, T. T. To & I Palmeirim. 2012. Interrelationship and modularity of notochord and somites: A comparative view on zebrafish and chicken vertebral body development. Journal of Applied Ichthyology, 28: 316–319.
De Clercq A., M. R. Perrott , P. S. Davie, M. A. Preece , B. Wybourne, N. Ruff, A,. Huysseune & P. E. Witten. 2017. Vertebral column regionalisation in Chinook salmon, *Oncorhynchus tshawytscha*. Journal of Anatomy, 231: 500–514.
De Clercq, A., M. R. Perrott, P. S. Davie, M. A. Preece, M. A. G. Owen, A. Huysseune & P. E. Witten. 2018. Temperature sensitive regions of the Chinook salmon vertebral column. Vestiges and meristic variation. Journal of Morphology, 279: 1301–1311.
Detwiler, S. R. & Holtzer H. 1956. The developmental dependence of the vertebral column upon the spinal cord in the urodeles. Journal of Experimental Zoology, 132: 299–310
Donoghue, P. C. J. & M. Rücklin. 2016. The ins and outs of the evolutionary origin of teeth. Evolution & Development, 18: 19–30.
Ebner, V. von. 1888. Urwirbel und Neugliederung der Wirbelsäule. Sitzungsber. Akademie der Wissenschaften in Wien, 111/97, p. 194–206.
Ebner, V. von. 1896. Über die Wirbel der Knochenfische und die Chorda dorsalis der Fische und Amphibien. Sitzungsberichte der Kaiserl. Akademie der Wissenschaften in Wien, Mathematisch-Naturwissenschaftliche Classe 105, 39 S., [4] Bl: Ill. Kaiserlich-Königliche Hof- und Staatsdruckerei, Wien: 123–161.
Ellis, K., J. Bagwell & M. Bagnat. 2013. Notochord vacuoles are lysosome-related organelles that function in axis and spine morphogenesis. Journal of Cell Biology, 200: 667–679.
Fleming A., R. Keynes & D. Tannahill. 2001. The role of the notochord in vertebral column formation. Journal of Anatomy, 199: 177–180.
Fleming, A., R. Keynes & D. Tannahill. 2004. A central role for the notochord in vertebral patterning. Development, 131: 873–880.
Francois, Y. 1966. Structure et développement de la vertebre de Salmo et des téleostéens. Archives de Zoologie Expérimentale et Générale, 107: 287–328.
Freitas, C., S. Rodrigues, J. B. Charrier, M. A. Teillet & I. Palmeirim. 2001. Evidence for medial/lateral specification and positional information within the presomitic mesoderm. Development, 128: 5139–5147.
Gadow, H. F. & E. C. Abbott. 1895. On the evolution of the vertebral column of fishes. Philosophical Transactions of the Royal Society of London, B186: 163–221.
Gardiner, B. G. 1983. Gnathostome vertebrae and the classification of the Amphibia. Zoological Journal of the Linnean Society, 79: 1–59.
Gardiner, B. G. 1984. Devonian Palaeoniscid Fishes: New Specimens of Mimia and Moythomasia from the upper Devonian of Western Australia. Bulletin of the British Museum of Natural History (Geology), 37: 173–428.
Gilbert, S. F., J. M. Opitz & R. A. Raff. 1996. Resynthesizing Evolutionary and Developmental Biology. Developmental Biology, 173: 357–372.
Giudicelli, F. & J. Lewis. 2004. The vertebrate segmentation clock. Current Opinions in Genetics & Development, 14(4): 407–414.
Goodrich, E. S. 1909. Vertebrata Craniata (First Fascicle: Cyclostomes and Fishes). – In: A Treatise on Zoology, Part IX. Lancaster, R (ed.). Adam and Charles Black, London, 518 pp.
Grotmol, S., H. Kryvi, K. Nordvik & G. K. Totland. 2003. Notochord segmentation may lay down the pathway for the development of the vertebral bodies in the Atlantic salmon. Anatomy & Embryology, 207: 263–272.
Grotmol, S., H. Kryvi, K. Nordvik & G. K. Totland. 2005. A segmental pattern of alkaline phosphatase (ALP) activity within the notochord coincides with the initial formation of the vertebral bodies. Journal of Anatomy, 206: 427–436.
Hall, B. K. 1977. Chondrogenesis of the somitic mesoderm. Advances in Anatomy & Embryology, 53: 1–50.
Hall, B. K. 2003. Descent with modifications: the unity underlying homology and homoplasy as seen through an analysis of development and evolution. Biological Reviews of the Cambridge Philosophical Society, 78: 409–433.
Hall, B. K. 2015. Bones and Cartilage. Developmental and Evolutionary Skeletal Biology. Second Edition. Academic Press, London UK.
Hall, B. K. & P. E. Witten. 2007. The Origin and Plasticity of Skeletal Tissues in Vertebrate Evolution and Development. – Pp. 13–57 in J. S. Anderson & H.-D. Sues (eds.): Major Transitions in Vertebrate Evolution. Festschrift for Dr Robert L. Carroll. Indiana University Press, Bloomington, IN.
Hautier, L., V. Weisbecker, M. R. Sánchez-Villagra, A. Goswami & R. J. Asher. 2010. Skeletal development in sloths and the evolution of mammalian vertebral patterning. Proceedings of the National Academy of Sciences of the USA, 107: 18903–18908.
Hay, P. O. 1895. On the structure and development of the vertebral column of *Amia*. Field Columbian Museum Publication 5, Zoological Series Vol. 1, No. 1: 1–54.
Hejnol, A. & C. J. Lowe. 2014. Animal evolution: Stiff or squishy notochord origins? Current Biology, 24(23): R1131–R1133 .
Hoffmann, A., S. Czichos, C. Kaps, D. Bachner, H. Mayer, Y. Zilberman, G. Turgeman, G. Pelled, G. Gross & D. Gazit. 2002. The T-box transcription factor Brachyury mediates cartilage development in mesenchymal stem cell line C3H10T1/2. Journal of Cell Science, 115: 769–781.
Hunter, C. J., Matyas, J. R. & N. A. Duncan. 2003. The notochordal cell in the nucleus pulposus: A review in the context of tissue engineering. Tissue Engineering, 9: 667–677.
Huxley, T. H. 1859. Observations on the development of some parts of the skeleton of fishes. Quarterly Journal of Microscopical Science, 7: 33–46.
Inohaya, K., Y. Takano & A. Kudo. 2007. The teleost intervertebral region acts as a growth center of the centrum: In vivo visualization of osteoblasts and their progenitors in transgenic fish. Developmental Dynamics, 236: 3031–3046 .
Janvier, P. 2015. Facts and fancies about early fossil chordates and vertebrates. Nature, 520: 483–489.
Johanson, Z., A. Kearsley, J. den Blaauwen, M. Newman & M. M. Smith. 2012. Ontogenetic development of an exceptionally preserved Devonian cartilaginous skeleton. Journal of Experimental Zoology Part B – Molecular & Developmental Evolution, 318B: 50–58.
Jurand, A. 1962. The Development of the Notochord in Chick Embryos. Journal of Embryology and Experimental Morphology, 10: 602–621.

Keating, J. N., C. L. Marquart, F. Marone & P. C. Donoghue. 2018. The nature of aspidin and the evolutionary origin of bone. Nature Ecology & Evolution, 2: 1501–1506.

Kölliker, A. 1859. On the structure of the chorda dorsalis of the plagiostomus and some other fishes, and on the relation of its proper sheath to the development of the vertebrae. Proceedings of the Royal Society of London, 10: 214–222.

Kowalevsky, A. 1866. Entwicklungsgeschichte der einfachen Ascidien. Mémoires l'Academie Imperiale des Sciences de St.-Pétersbourg. Tome X, No. 45: 1–19.

Laerm, J. 1976. The development, function, and design of amphicoelous vertebrae in teleost fishes. Zoological Journal of the Linnean Society of London, 58: 237–254.

Laerm, J. 1979. On the origin of the chondrostean vertebral centrum. Canadian Journal of Zoology, 57: 475–485.

Laerm, J. 1982. The origin and homology of the neopterygian vertebral centrum. Journal of Paleontology, 56: 191–202.

Lauder, G. V. 1980. On the relationship of the myotome to the axial skeleton in vertebrate evolution. Paleobiology, 6: 51–56.

Leeson, T. S. & C. R. Leeson. 1958. Observations and the histochemistry and fine structure of the notochord in rabbit embryos. Journal of Anatomy, 92: 278–285

Leprevost, A., T. Azais, M. Trichet & J.-Y. Sire. 2017. Identification of a new mineralized tissue in the notochord of reared Siberian sturgeon (*Acipenser baerii*). Journal of Morphology 278: 1586–1597.

Lim, Y.-W., H. P. Lo, C. Ferguson, N. Martel, J. Giacomotto, G. A.Gomez , A. S. Yap, T. E. Hall & R. G. Parton. 2017. Caveolae protect notochord cells against catastrophic mechanical failure during development. Current Biology, 27: 1–14.

Linsenmayer, T., E. Gibney & T. M. Schmidt. 1986. Segmental appearance of type X collagen in the developing avian notochord. Developmental Biology 113: 467–473.

Lleras Forero, L., R. Narayanan, L. F. A. Huitema, M. VanBergen, A. Apschner, J. Peterson-Maduro, I. Logister, G. Valentin, L. G. Morelli, A. Oates & S. Schulte-Merker. 2018. Segmentation of the zebrafish axial skeleton relies on notochord sheath cells and not on the segmentation clock. Elife, 7: e33843.

Luo, H.-L., S.-X. Hu & L.-Z. Chen. 2001. New early Cambrian chordates from Haikou, Kunming. Acta Geologica Sinica, 75: 345–348 [English Language Edition].

Maisey, J. G. 1986. Heads and tails: a chordate phylogeny. Cladistics, 2: 201–256.

Maisey, J. G. 1988. Phylogeny of Early Vertebrate Skeletal Induction and Ossification Patterns. – In: Evolutionary Biology Vol. 32. Hecht, M. K., Wallace, B. & Prance, G. T. (eds.). Premium Publishing Corporation. Stockholm

Maisey, J. G. 2000. Discovering Fossil Fishes. Westview Press, Boulder Colorado, 223 pp.

Maisey, J. G. 2005. Braincase of the upper Devonian shark *Cladodoides wildungensis* (Chondrichthyes, Elasmobranchii), with observations on the braincase in early chondrichthyans. Bulletin of the American Museum of Natural History, 288: 1–102.

Maisey, J. G. 2008. The postorbital palatoquadrate articulation in elasmobranchs. Journal of Morphology, 269: 1022–1040.

Mathews, M. B. (ed.) 1975. Connective tissue macromolecular structure and evolution. Molecular Biology, Biochemistry and Biophysics, 19: 1–318.

McCoy, V. E., E. E., Saupe, J. C. Lamsdell, L. G. Tarhan, S. McMahon, S., Lidgard, P. Mayer, C. D. Whalen, C. Doriano, L. Finney, S. Vogt, E. G. Clark, R. P. Anderson, H. Petermnan, E. R. Locatelli & D. Briggs. 2016. The 'Tully monster' is a vertebrate. Nature, 352: 496–499.

McMahon, J. A., S. Takada, L. B. Zimmerman, C. M. Fan, R. M. Harland & A. P. McMahon. 1998. Noggin-mediated antagonism of BMP signaling is required for growth and patterning of the neural tube and somite. Genes & Development, 12: 1438–1452.

McMaster, M. L., A. M. Goldstein, C. M. Bromley, N. Ishibe & D. M. Parry. 2001. Chordoma: incidence and survival patterns in the United States, 1973–1995. Cancer Causes Control, 12: 1–11.

Morin-Kensicki, E. M., E. Lemancon & J. S. Eisen. 2002. Segmental relationship between somites and vertebral column in zebrafish. Development, 129: 3851–3860.

Monsoro-Burq, A.-H., M. Bontoux, M.-A. Teillet & N. M. Le Douarin. 1994. Heterogeneity in the development of the vertebra. Proceedings of the National Academy of Sciences of the USA, 91: 10435–10439.

Muller, T. S., C. Ebensperger, A., Neubuser. H. Koseki, R. Balling, B. Christ & J. Wilting. 1996. Expression of avian Pax1 and Pax9 is intrinsically regulated in the pharyngeal endoderm, but depends on environmental influences in the paraxial mesoderm. Developmental Biology, 178: 403–417.

Murdock, D. J. E., E. J. Rayfield & P. C. J. Donoghue. 2014. Functional adaptation underpinned the evolutionary assembly of the earliest vertebrate skeleton. Evolution & Development, 16: 354–361.

Ng, L.-J., S. Wheatley, G. E. O. Muscat, J. Conway-Campbell, J. Bowles, E. Wright, D. M. Bell, P. M. Tam, R. S. Cheah & P. Koopman. 1997. Sox9 binds DNA, activates transcription, and co-expresses with type II collagen during chondrogenesis in the mouse. Developmental Biology 183: 108–121.

Nikaido, M., A. Kawakami, A. Sawada, M., Furutani-Seiki, H. Takeda & K. Araki. 2002. Tbx24, encoding a T-box protein, is mutated in the zebrafish somite-segmentation mutant fused somites. Nature Genetic, 31: 195–199.

Nordvik, K. 2007. From notochord to vertebral column: studies on Atlantic salmon (*Salmo salar* L.) Dissertation for the degree philosophiae doctor (PhD) at the University of Bergen, 37 pp.

Nordvik, K., H. Kryvi, G. K. Totland & S. Grotmol. 2005. The salmon vertebral body develops through mineralization of two preformed tissues that are encompassed by two layers of bone. Journal of Anatomy, 206: 103–114.

Ota, K. G., S. Fujimoto, Y. Oisi & S. Kuratani. 2011. Identification of vertebra-like elements and their possible differentiation from sclerotomes in the hagfish. Nature Communications, 2: 373, DoI: 10.1038/ncomms1355.

Ota, K. G., Y. Oisi, S. Fujimoto & S. Kuratani. 2014. The origin of developmental mechanisms underlying vertebral elements: implications from hagfish evo-devo. Zoology, 117: 77–80.

Patterson, C. 1968. The caudal skeleton in Lower Liassic pholidophorid fishes. Bulletin of the British Museum of Natural History (Geology), 16: 210–239.

Peskin, B., Henke. K., Cumplido, N., Treaster, S., Harris, M. P., Bagant, M. & Arratia, G. 2020. Notochordal Signals Establish Phylogenetic Identity of the Teleost Spine. Current Biology, 30: 1–10.

Pogoda, H.-M., I. Riedl-Quinkertz, H. Löhr, J. S. Waxman, R. M. Dale, J. Topczewski, S. Schulte-Merker & M. Hammerschmidt. 2018. Direct activation of chordoblasts by retinoic acid is required for segmented centra mineralization during zebrafish spine development. Development, 145: dev159418. DOI: 10.1242/dev.159418.

Raff, E. C., J. T. Villinski, F. R. Turner, Donoghue, P. C. J. & R. A. Raff. 2006. Experimental taphonomy shows the feasibility of fossil embryos. Proceedings of the National Academy of Sciences of the USA, 103: 5846–5851.

Remak, R. 1855. Untersuchungen über die Entwicklung der Wirbelthiere. G. Reimer, Berlin, 195 pp.

Renn, J., A. Buttner, T. T. To, S. J. H., Chan & C. Winkler. 2013. A col10a1:nlGFP transgenic line displays putative osteoblast precursors at the medaka notochordal sheath prior to mineralization. Developmental Biology, 381: 134–143.

Risbud, M. V. & I. M. Shapiro. 2011. Notochordal cells in the adult intervertebral disc: new perspective on an old question. Critical Reviews Eukaryotic Gene Expression, 21: 29–41.

Ruggeri, A. 1972. Ultrastructural, histochemical and autoradiographic studies on the developing chick notochord. Zeitschrift für Anatomie und Entwicklungsgeschichte, 138: 20–33.

Rychel, A.L. & A. B. Swalla. 2007. Development and Evolution of Chordate Cartilage. Journal of Experimental Zoology (Mol. Dev. Evol.), 308B: 325–335.

Sallan, L., S. Giles, R. S. Sansom, R. T. Clarke, Z. Johanson, I. J. Sansom & P. Janvier 2017. The 'tully monster' is not a vertebrate: characters, convergence and taphonomy in Palaeozoic problematic animals. Palaeontology, 60:: 149–157.

Sansom, R. S., K. Freedman, S. E. Gabbott, R. J. Aldridge & M. A. Prunel. 2010. Taphonomy and affinity of an enigmatic Silurian vertebrate, *Jamoytius kerwoodi* White. Palaeontology, 53: 1393–1409.

Satoh, N., K. Tagawa & H. Takahashi. 2012. How was the notochord born? Evolution & Development, 14: 56–75.

Scaal, M. 2016. Early development of the vertebral column. Seminars in Cell & Developmental Biology, 49: 83–91.

Schaeffer, B. 1967. Osteichthyan vertebrae. Zoological Journal of the Linnean Society of London, 47: 185–195.

Schauinsland, H. 1903. Beiträge zur Entwicklungsgeschichte und Anatomie der Wirbeltiere. I. II. III. – In: Chun, C. (ed.), Zoologica. Original-Abhandlungen aus dem Gesammtgebiete der Zoologie. Bd. 39. Verlag von Erwin Nägele, Stuttgart.

Schmitz, R. J. 1995. Ultrastructure and function of cellular components of the intercentral joint in the percoid vertebral column. Journal of Morphology 226: 1–24.

Schmitz, R. J. 1998. Immunohistochemical identification of the cytoskeletal elements in the notochord cells of bony fishes. Journal of Morphology 236: 105–116.

Schröter C. & A. C. Oates. 2010. Segment number and axial identity in a segmentation clock period mutant. Current Biology, 20: 1254–1258.

Schroyens, W., J. Camon & L. De Ridder. 1991. Resistance of the embryonic notochord to invasion by malignant neurogenic tumor cells *in vitro*. Clinical & Experimental Metastasis 9: 403–409.

Scott, H. W. 1934. The zoological relationships of the conodonts. Journal of Paleontology, 8: 448–455.

Seemann, F., D. R. Peterson, P. E. Witten, B.-S. Guo, A. H. Shantanagouda, R. R. Ye, G. Zhang & D. W. T. Au. 2015. Insight into the transgenerational effect of benzo[a]pyrene on bone formation in a teleost fish (*Oryzias latipes*). Comparative Biochemistry & Physiology C, 178: 60–67.

Senthinathan, B., C. Sousa, D. Tannahill & R. Keynes. 2012. The generation of vertebral segmental patterning in the chick embryo. Journal of Anatomy, 220: 591–602

Shapiro, F. 1992. Vertebral development of the chick embryo during days 3–19 of incubation. Journal of Morphology, 213: 317–333.

Shapiro, I. M. & M. V. Risbud. 2010. Transcriptional profiling of the nucleus pulposus: say yes to notochord. Arthritis Research & Therapy, 12, 117. DOI: 10.1186/ar3003.

Shu, D.-G., S. Conway Morris & X.-L. Zhang. 1996. A *Pikaia*-like chordate from the Lower Cambrian of China. Nature, 384: 156–157.

Shu, D.-G., S. Conway Morris, Z.-F. Zhang, J. N. Lie, J. Han, L. Chen, Z. L. Zhang, K. Yasui and L. Yong. 2003. A new species of Yunnanozoan with implications for deuterostome evolution. Science, 299: 1380–1384.

Shu, D.-G., H.-L. Luo, S. Conway Morris, X.-L. Zhang, S.-X. Hu, L. Chen, J. Han, M. Zhu, Y. Li & L.-Z. Chen. 1999. Lower Cambrian vertebrates from south China. Nature, 402: 42–46.

Spoorendonk, K. M., J. Peterson-Maduro, J. Renn, T. Trowe, S. Kranenbarg, C. Winkler & S. Schulte-Merker. 2008. Retinoic acid and Cyp26b1 are critical regulators of osteogenesis in the axial skeleton. Development, 135: 3765–3774.

Springer, V. G. & G. D. Johnson. 2000. Use and advantages of ethanol solution of alizarin red S dye for staining bone in fishes. Copeia, 1: 300–301.

Stemple, D. L. 2005a. Structure and function of the notochord: an essential organ for chordate development. Development, 132: 2503–2512.

Stern, C. D. 1990. Two distinct mechanisms for segmentation. Seminars in Developmental Biology, 1: 109–116.

Strudel, G., 1953. Conséquences de l'excision de troncons du tube nerveus sur la morphogenése de l' embryo de poulet et sur la diffelrenciation de ses organes: contribution a la genese de l'orthosympathique. Annales des sciences naturelles. Zoologie et biologie animale, 15: 251–329.

Tamplin, O. J. 2009. Development of the mouse notochord. PhD thesis. Department of Molecular Genetics and the Collaborative Program in Developmental Biology. University of Toronto. Toronto, Canada: 187 pp.

Takeuchi, K. 1966. 'Wavy-Fused' mutants in the medaka *Oryzias latipes*. Nature, 211: 866–867.

Trout, J. J., J. A. Buckwalter, K. C. Moore & S. K. Landas. 1982. Ultrastructure of the human intervertebral-disk. 1. Changes in notochordal cells with age. Tissue & Cell, 14: 359–369.

Trueman, C. N. 2013. Chemical taphonomy of biomineralized tissue. Palaeontology, 56: 475–486.

Turner, S., C. J. Burrow, H.-P. Schultze, A. Blieck, W. E. Reif, C. B. Rexroad. P. Bultynck & G. S. Nowlan. 2010. False teeth: conodont-vertebrate phylogenetic relationships revisited. Geodiversitas, 32: 545–594.

Vandewalle, P., I. Gluckmann & F. Wagemans. 1998. A critical assessment of the alcian blue/alizarine double staining in fish larvae and fry. Belgian Journal of Zoology, 128: 93–95.

Van Eeden, F. J. M., M. Granato, U. Schach, M. Brand, M. Furutani-Seiki, P. Haffter, M. Hammerschmidt, C. P. Heisenberg et al. 1996. Mutations affecting somite formation and patterning in the zebrafish, *Danio rerio*. Development, 123: 153–164.

Verbout, A. J. 1976. A critical review of the 'neugliederung' concept in relation to the development of the vertebral column. Acta Biotheoretica, 25: 219–258.

Wake, D. B. 1970. Aspects of vertebral evolution in the modern Amphibia. Forma et Functio, 3: 33–60.

Wake, D. B. & R. Lawson. 1973. Developmental and adult morphology of the vertebral column in the plethodontid salamander Eurycea bislineata, with comments on vertebral evolution in the amphibia. Journal of Morphology, 139: 251–300.

Wake, H. M. & D. B. Wake. 2000. Developmental morphology of early vertebrogenesis in Caecilians (Amphibia: Gymnophiona): Resegmentation and phylogenesis. Zoology, 103: 68–88.

Walker, M. B. & C. B. Kimmel. 2007. A two-color acid-free cartilage and bone stain for zebrafish larvae. Biotechnic & Histochemistry, 82: 23–28.

Ward, L., A. S. W. Pang, S. E. Evans & C. D. Stern. 2018. The role of the notochord in amniote vertebral column segmentation. Developmental Biology, 439: 3–18.

White, E. I. 1946. *Jamoytius kerwoodi*, a new chordate from the Silurian of Lanarkshire [Scotland]. Geological Magazine, 83: 89–97.

White, P. H., D. R. Farkas, E. E. McFadden & D. L. Chapman. 2003. Defective somite patterning in mouse embryos with reduced levels of Tbx6. Development, 130: 1681–1690.

Wright, G. M., F. W. Keeley & P. Robson. 2001. The unusual cartilaginous tissues of jawless craniates, cephalochordates and invertebrates. Cell & Tissue Research, 304:165–174.

Willams, A. 1908. The later development of the notochord in mammals. The American Journal of Anatomy, 8: 251–284.

Witten, P. E., A. Obach, A. Huysseune & G. Baeverfjord. 2006. Vertebrae fusion in Atlantic salmon (*Salmo salar*): Development, aggravation and pathways of containment. Aquaculture, 258: 164–172.

Witten, P. E., A. Huysseune & B. K. Hall. 2010. A practical approach for the identification of the many cartilaginous tissues in teleost fish. Journal of Applied Ichthyology, 26: 257–262.

Yu, T., M. Graf, J. Renn, M. Schartl, D. Larionova, A. Huysseune, P. E. Witten & C. Winkler. 2017. A vertebrate specific and essential role for sp7/osterix in osteogenesis revealed by gene knock-out in the teleost medaka. Development: 144: 265–271.

Zhang, G. 2009. An evo-devo view on the origin of the backbone: evolutionary development of the vertebrae. Integrative and Comparative Biology, 49: 178–186

Zhao, Q., H. Eberspaecher, V. Lefebvre & B. de Crombrugghe. 1997. Parallel expression of Sox9 and Col2a1 in cells undergoing chondrogenesis. Developmental Dynamics, 209: 377–386.

Ancient Fishes and their Living Relatives: a Tribute to John G. Maisey. Alan Pradel, John S. S. Denton & Philippe Janvier (eds.): pp. 225–231, 4 figs.
© 2021 by Verlag Dr. Friedrich Pfeil, München, Germany – ISBN 978-3-89937-269-4

The distribution of post-mandibular teeth in extant vertebrates revisited: co-evolution of pharyngeal pouches and teeth?

Ann Huysseune*

"Development of an internal branchial skeleton and jaws is concomitant with the spread of the dermal skeleton within the buccal cavity and beyond, ultimately involving endodermal as well as ectodermal tissues"

(John Maisey, 1988).

Abstract

Pharyngeal teeth in extant, non-tetrapod vertebrates display an enormous diversity in number, size, shape and the skeletal pharyngeal arches with which they are associated. Here, I discuss the potential role of the endodermal pouches and their contact with ectodermal clefts, in the evolutionary origin and expansion of the pharyngeal dentition. Pouches are outpocketings of the developing endodermal pharynx lining, separating the Anlagen of the skeletal pharyngeal arches. Pouch-cleft contacts eventually break through to form the gill slits. I start with a brief review of the distribution of post-mandibular teeth indicating that in evolution, teeth on different arches tend to be lost, but not regained. Next, I explore the possibility that pouch-cleft contacts may play a role in the development of post-mandibular teeth, using data from mutant zebrafish. I reflect on how different signaling pathways may impact on tooth distribution through modifying pouch-cleft contact and changing the cellular context for tooth formation. I end by considering whether the pouch that separates the hyoid from the gill arches proper has a special identity and could have played an instrumental role in the evolution of a pharyngeal dentition.

Introduction

It is now widely accepted that oral teeth in jawed vertebrates are homologous to odontodes, or skin denticles (Huysseune & Sire 1998, Sire & Huysseune 2003, Huysseune et al. 2009, 2010, Blais et al. 2011, Debiais-Thibaud et al. 2011, Rücklin et al. 2012, Qu et al. 2013, Donoghue & Rücklin 2016, Witten et al. 2014). Odontodes consist of a dentine cone, are either covered or not covered by a hypermineralized layer, and have a base composed of bone, which serves as an attachment tissue (Reif 1982).

On the basis of both paleontological and neontological data, we have previously hypothesized that during evolution the competence of epithelia to form odontodes in the skin may have spread into the pharynx to generate pharyngeal teeth (Huysseune et al., 2009, 2010), a proposal that was rearticulated by Rücklin et al. (2011) and Donoghue & Rücklin (2016). This spreading of competence could have happened via any channel allowing communication between the skin and the pharynx, or, stated otherwise, via any opening joining the ectoderm (the germ layer producing the skin) with the endoderm (the germ layer that produces the lining of the digestive tube). This transfer of odontogenic competence is not dependent on the presence of jaws, and thus may even have preceded the origin of jaws. This is in line with the idea that the close developmental association between teeth and jaws may be a derived phenomenon that evolved later in jawed vertebrate phylogeny (Maisey et al. 2013, Rücklin & Donoghue 2015).

Openings allowing communication between ectoderm and endoderm are, for example, the spiraculum, or the gill slits. The sarcopterygian *Eusthenopteron* has a spiracular pouch paved with teeth (Bjerring 1998). Chondrichthyans and actinopterygian osteichthyans, gill-possessing vertebrates, possess a pharynx that can be completely covered with teeth (Peyer 1968; Nelson 1969, 1970; Vandewalle et al. 1994). In sarcopterygians, teeth disappear in evolution together with the closure of the gill slits (Schoch 2002). In axolotls, tooth formation can be elicited *in vitro* from neural crest regions caudal to the region of neural crest fated to produce teeth *in vivo*, provided both stomodaeal ectoderm and oral endoderm are included in the explants (Graveson et al. 1997). Based on these observations, we have suggested that pharyngeal teeth in extant vertebrates (the only vertebrates where this can be experimentally tested) are restricted to those areas where pharyngeal endoderm comes in contact with, or receives signals from the ectoderm. Apart from the mouth, these areas are the pharyngeal pouches, which eventually open into gill slits.

Gill slits open at cleft-pouch contacts

In chordates, the endodermal epithelium of the developing pharynx produces a series of bilateral outpocketings, called pharyngeal pouches, that extend from the midline endoderm and eventually contact the superficial ectoderm at corresponding clefts (Shone & Graham 2014). The pouches separate clusters of mesenchymal cells that will give rise to cartilaginous pharyngeal arches (seven in total: mandibular arch, hyoid arch, and five branchial or gill arches) (reviewed in Graham 2008, Grevellec & Tucker 2010, Graham & Richardson 2012, Frisdal & Trainor

* Evolutionary Developmental Biology, Biology Department, Ghent University, Belgium; K. L. Ledeganckstraat 35, B-9000 Ghent, Belgium; Ann.Huysseune@UGent.be

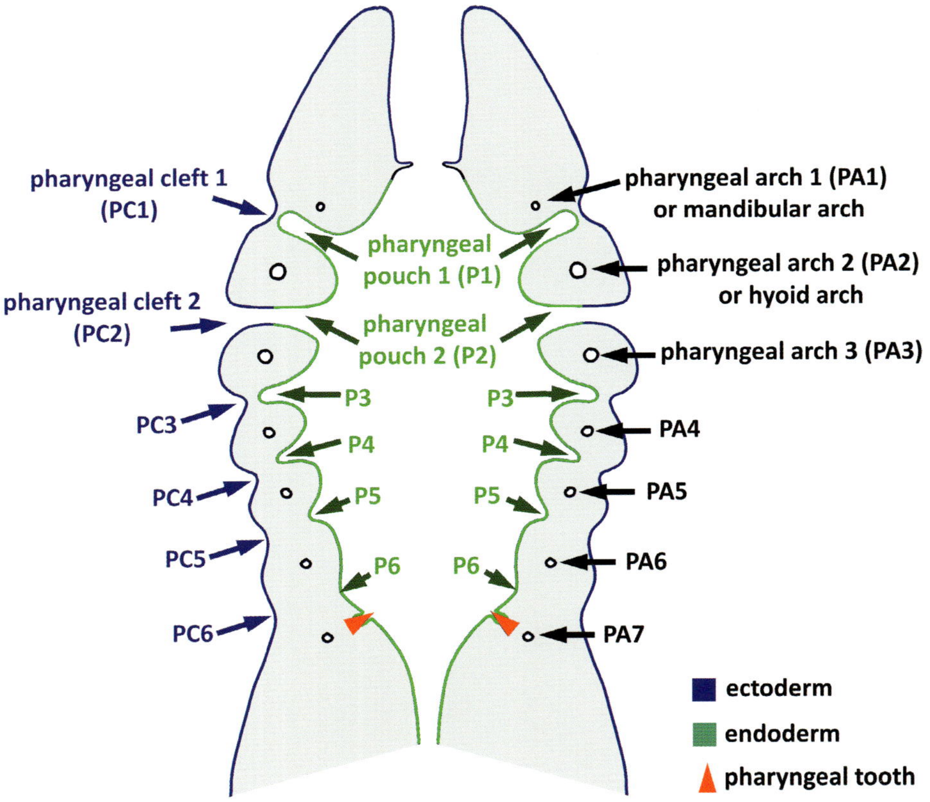

Fig. 1. Scheme of the developing pharynx in a teleost, as exemplified in the zebrafish, viewed in a horizontal section at the time of pouch 2 opening. P1–6: pharyngeal pouch 1 to 6; PA1–7: pharyngeal arch 1 to 7; PC 1–6: pharyngeal cleft 1 to 6.

2014). In chondrichthyans and primary aquatic osteichthyans, most pouch-cleft contacts eventually break through to create orifices, called gill slits (Bjerring 1977, Starck 1982, Kimmel et al. 1995, Shone & Graham 2014). The term "cleft" is used despite the fact that there may not necessarily be a distinctive depression in the skin revealing the place of such contact. In teleost fishes, the most speciose group of jawed vertebrates, six pharyngeal pouches form from anterior to posterior (Fig. 1). The first pouch (P1) separates mandibular from hyoid arch. While it gives rise to an opening, the spiraculum, in chondrichthyans, chondrosteans (e.g. *Acipenser*) and brachiopterygians (e.g. *Polypterus*), it does not open any longer in lepisosteids and in teleosts (Goette 1901, Arambourg & Bertin 1958, Beaumont & Cassier, 2009). The second pouch (P2) separates the hyoid arch from the third pharyngeal arch; pouch 3 (P3) separates pharyngeal arch 3 from 4, and so on. The endoderm lining of the vertebrate pharynx generates derivatives as diverse as thyroid gland, lungs and swim bladder (along the midline), and gills, tonsils, parathyroid glands (in tetrapods) and thymus (at the level of the pouches) (Grevellec & Tucker 2010, Graham & Richardson 2012, Gilbert & Barresi 2016). Here, I focus on pharyngeal teeth, organs displaying considerable conservation across vertebrates in terms of development (Huysseune & Sire 1998). Irrespective of their position (oral or pharyngeal), teeth develop through a collaborative interaction between an epithelial component and underlying neural crest-derived mesenchyme (Balic & Thesleff 2015). In the case of pharyngeal teeth, the epithelium is presumed to be endodermal but the question remains if, and/or to what extent, ectodermal/external epithelia play a role in tooth formation. The obvious place where they could exert such a role is at pouch-cleft contacts.

Pharyngeal tooth patterns are extremely diverse across taxa

Among chondrichthyans, extant sharks have divergent patterns of pharyngeal tooth distribution, depending on the species (Nelson 1970). These teeth are often very small and often referred to as (mucous membrane) denticles (Peyer 1968). In some species of selachians, they are tightly crowded, in others they are more widely scattered. They can be distributed throughout the oropharyngeal region, be restricted to its posterior region, or be missing altogether (Peyer 1968).

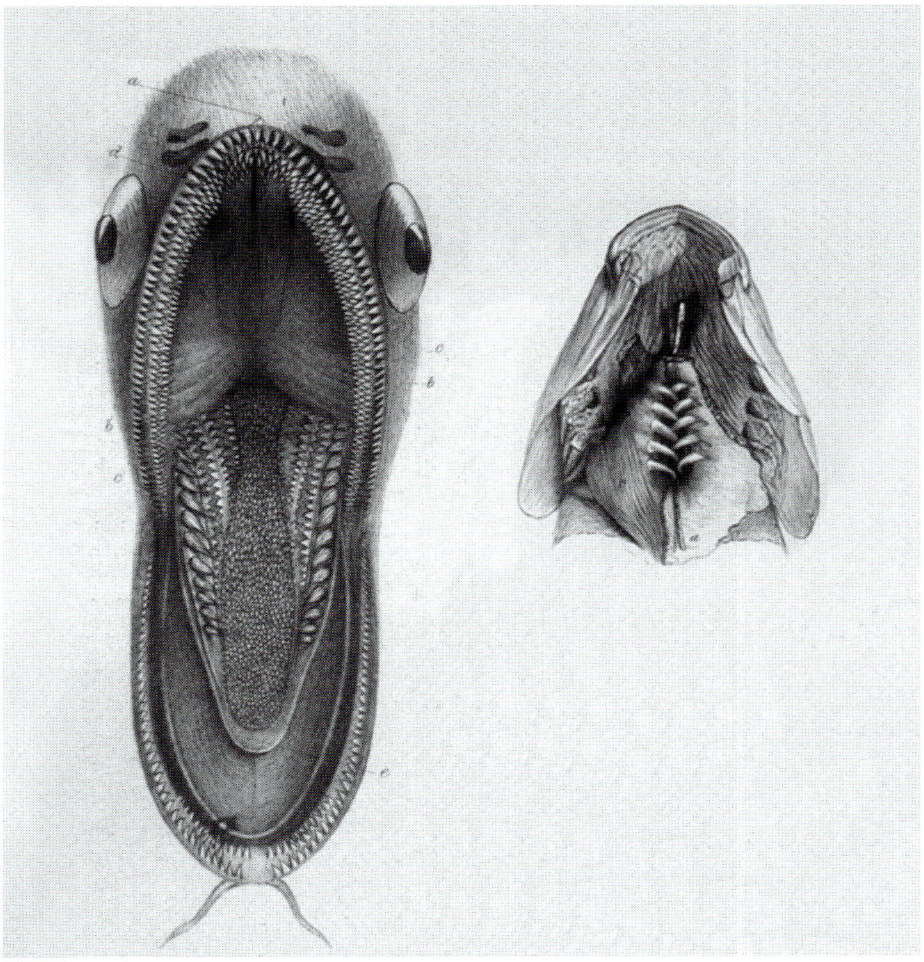

Fig. 2. Pharyngeal dentition in an osteoglossomorph (*Osteoglossum*) (left) and a cypriniform (*Barbus*) teleost (right) (modified from Owen, 1845).

Among osteichthyans, lineages that originated right after the key divergence of ray-finned (Actinopterygii) and lobe-finned fishes (Sarcopterygii) are represented in the extant fauna by bichirs, sturgeons, and gars. In sturgeons, pharyngeal teeth are associated with the hyoid arch (albeit that they are lost in juvenile life; adult sturgeons are edentulous, Peyer 1968, Hilton et al. 2011). The bichir, *Polypterus senegalus*, has teeth associated to the basibranchial, hypobranchials 1–3, ceratobranchials 1–4, and pharyngobranchials 2 and 3 (Clemen et al. 1998). The Cuban gar *Atractosteus tristoechus* also possesses pharyngeal teeth, at least in the posterior pharynx (Comabella et al. 2012).

Across teleost fish species, the distribution of teeth in the pharynx can also be widely disparate (Nelson 1969, 1970; Vandewalle et al. 1994), and is likely highly adaptive. The evolutionary tendency is a reduction in the extent of distribution of pharyngeal teeth. More basal taxa, such as elopomorphs, possess pharyngeal teeth distributed throughout the pharynx, forming tooth plates on each of the pharyngeal arches (Vandewalle et al. 1994). More derived teleosts have pharyngeal teeth restricted to a few arches (Fig. 2). This global evolutionary trend may well be repeated in many evolutionary lineages and is, for example, apparent in clupeids (Nelson 1967).

A reduction of the dentition is also characteristic of ostariophysans (reviewed in Stock 2007). In the order Cypriniformes, pharyngeal teeth are restricted to the last (7th) pharyngeal arch (e.g. Fig. 1). E.g., in cyprinids (a family that includes the popular model for developmental studies, the zebrafish, *Danio rerio*), these are the only teeth left as also the oral teeth have disappeared (Huysseune et al. 1998, Stock 2007). Thus, teleosts are obvious candidates to test for the relationship between tooth presence and status of the corresponding pouch – cleft contact. To do so, I have turned to zebrafish, *Danio rerio*, and look at tooth initiation only. I do not consider how teeth are related to the underlying branchial skeleton.

How can zebrafish mutants shed light on the role of pouches in tooth formation?

In zebrafish, teeth are restricted to the posteriormost pharyngeal arch (PA7, or fifth gill arch), and their epithelial component buds off from the endoderm at the level of the last, or sixth, pouch (P6) (Fig. 1). Once completed, teeth attach to the ossifying ceratobranchial 5. To examine the possible role of pouch-cleft contact in pharyngeal tooth formation, I took advantage of a zebrafish mutant with defective pouch formation. In so-called *van gogh* (*vgo*) mutants, the posterior endodermal pouches are not segmented, and endoderm-derived organs, such as the thymus, are absent (Piotrowski & Nüsslein-Volhard 2000, Piotrowski et al. 2003). Where the endodermal pouches fail to form, gill slits are not formed (Fig. 3).

I studied a total of 30 mutant embryos and compared them with wild type fish. I examined pouch-cleft contact and tooth formation via serial semithin sections, in order not to miss a (potentially very narrow) contact between pouch endoderm and skin ectoderm. In wild type fish, a connection is established between the pharynx endoderm and the epidermal ectoderm in the area where the gill slit between arch 4 and 5 will form. Tooth formation is initiated at the caudal margin of this epithelial connection, but prior to the opening of the gill slit. In *vgo* mutants, teeth were present only if the corresponding endodermal pouch made contact with the ectoderm, although teeth were sometimes absent even in the presence of such a contact. A more detailed report is published elsewhere (Oralová et al., 2020). Interestingly, teeth were sometimes present unilaterally, coupled to the presence of a pouch-cleft contact on the tooth-bearing side (Fig. 3).

The highly significant correlation, in *vgo* mutants, between presence or absence of teeth, and whether or not the endodermal pouches contact the surface ectoderm, strongly suggests that pouch-cleft contact is required for the formation of teeth. At the same time, given that teeth develop prior to any opening of the gill slit, and before penetration of an external epithelium into the pouch, a direct contribution of ectoderm via the gill slit seems precluded. A similar lack of ectodermal influx in Japanese medaka (*Oryzias latipes*) led Atukorala et al. (2011) to conclude that the rostral endoderm possesses an intrinsic odontogenic competence, a view that I oppose (see below). One could argue that the two phenomena (pouch-cleft contact and start of tooth formation) are only coincidental and that both depend on a third, unknown signal or process. For example, the moment of pouch-cleft contact may be synchronized with endoderm maturation in such a way that both pouch-cleft contact and start of tooth formation always occur in the same sequence. However, the unilateral presence of teeth on sides presenting also a pouch-cleft contact is an argument against this interpretation and strongly supports the conclusion that contact of the pouch with the skin is in some way required. The nature of the signal remains hitherto unknown.

Interestingly, mammalian pouch and tooth formation have been linked genetically through analyzing the effect of loss of the transcription factor *Pax9* (Peters et al. 1998). In zebrafish, *pax9* is expressed both in the pouch endoderm and along the midline (Jackman et al. 2004). However, *pax9* morpholino knockdown does not prevent the formation of teeth (Swartz et al. 2011), although teeth are reduced in number. Signaling at the pouch-cleft contact has nevertheless been shown in the reverse direction. Signaling from pharyngeal pouches to the overlying ectoderm is necessary for induction of epibranchial placodes in the ectoderm, both in zebrafish and in chick (reviewed in Graham 2008).

The apparent tight interdependence between timing and extent of pouch-cleft contact and the onset of pharyngeal tooth formation could be considered an argument in favor of co-evolution of both features, i.e., the situation in which the evolution of one entity is closely affected by that of another with which it interacts (Wilkins 2002). While the term has most often been used in the frame of higher levels of biological organization (such as ecosystems or organisms), the concept applies even to molecules, such as when the diversification of proteins or RNA structures becomes constrained by that of their protein partners (Wilkins 2002, Karakostis & Fåhraeus 2019). Examples of suggested co-evolution of vertebrate organs are uncommon, a relevant exception being the study of Bloomquist et al. (2015). These authors proposed a shared ancestral regulatory circuit to explain copatterning of teeth and taste buds in cichlid fishes and mice. Different from their study, I hypothesize developmental dependency of pharyngeal teeth from the pouch-cleft contact, via some sort of signaling. To test this dependency, and the hypothesis of co-evolution, one approach, although not conclusive, will be to assess representative actinopterygians for sequence and timing of pouch-cleft contact and opening, on the one hand, and pharyngeal tooth initiation on the other hand.

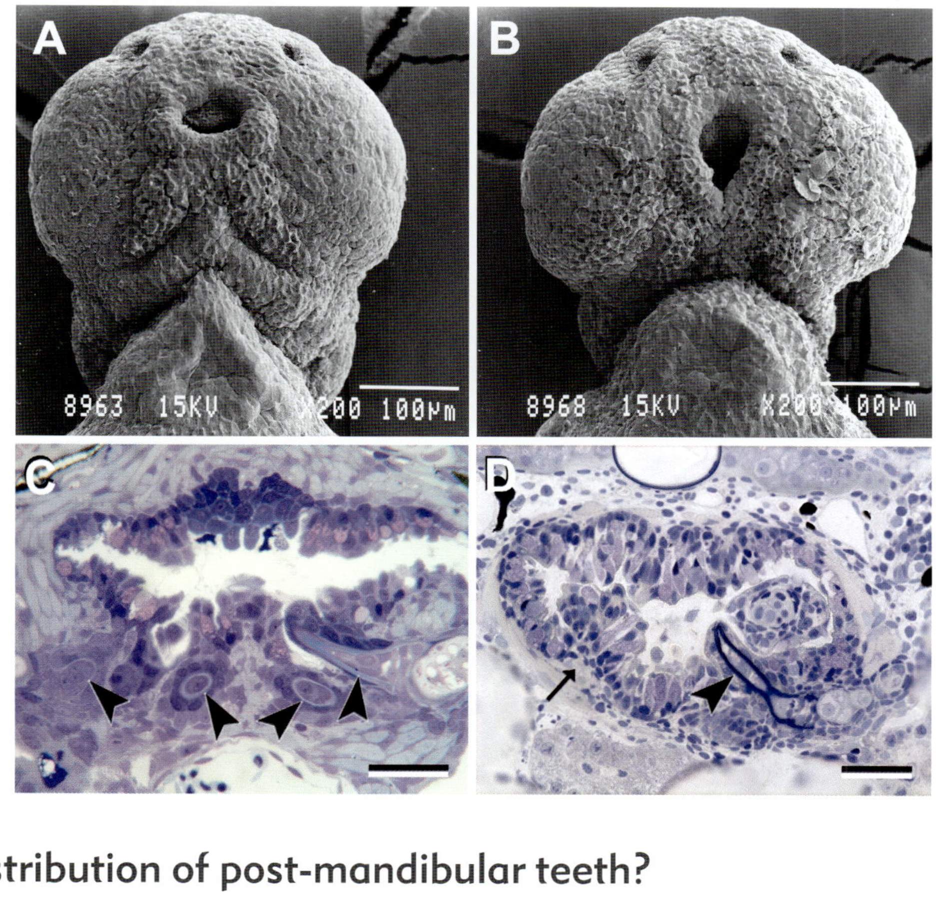

Fig. 3. A, B. Scanning electron micrographs of ventral views of the head of a 2.5 day post-fertilization (dpf) wild type (**A**) and *van gogh* (*vgo*) mutant (**B**) zebrafish. **C–D.** Transverse semithin sections through the pharyngeal region in 5 dpf zebrafish showing the full complement of tooth germs present in the wild type (**C**) and unilateral teeth in the *vgo* mutant at the side on which the pouch-cleft contact is present (**D**) (teeth or tooth germs indicated by arrowheads). The contralateral absence of teeth on (**D**) is indicated by an arrow. Scale bars in C, D = 50 µm.

How could pouches regulate the distribution of post-mandibular teeth?

While pouch-cleft contact may well be an essential factor in determining if a tooth forms, it does not, at first sight, explain the diverse patterns of pharyngeal tooth distribution. Why was the capacity to form teeth lost from some arches, but retained on others? More specifically, if pouch-cleft contacts are important for tooth formation, then why do teeth in zebrafish form – not only on just one arch – but, in addition, on the very last arch and thus in association with the very last pouch to form?

In 2006, Kopinke and colleagues showed that pouch formation in zebrafish depends on proper retinoic acid signaling. A few years later, Gibert et al. (2010), focusing on the dentition, demonstrated that pharyngeal tooth formation in zebrafish depends on retinoic acid (RA). Following the above hypothesis on the importance of pouch-cleft contacts for tooth formation, and considering the results of Kopinke et al. (2006), I propose that tooth formation is only indirectly influenced by RA, namely by depending on proper formation of pouch-cleft contacts. Remarkably, Gibert et al. (2010) also reported that pharyngeal tooth induction was not compromised by RA inhibition in two other lineages of teleosts, the beloniform *Oryzias latipes*

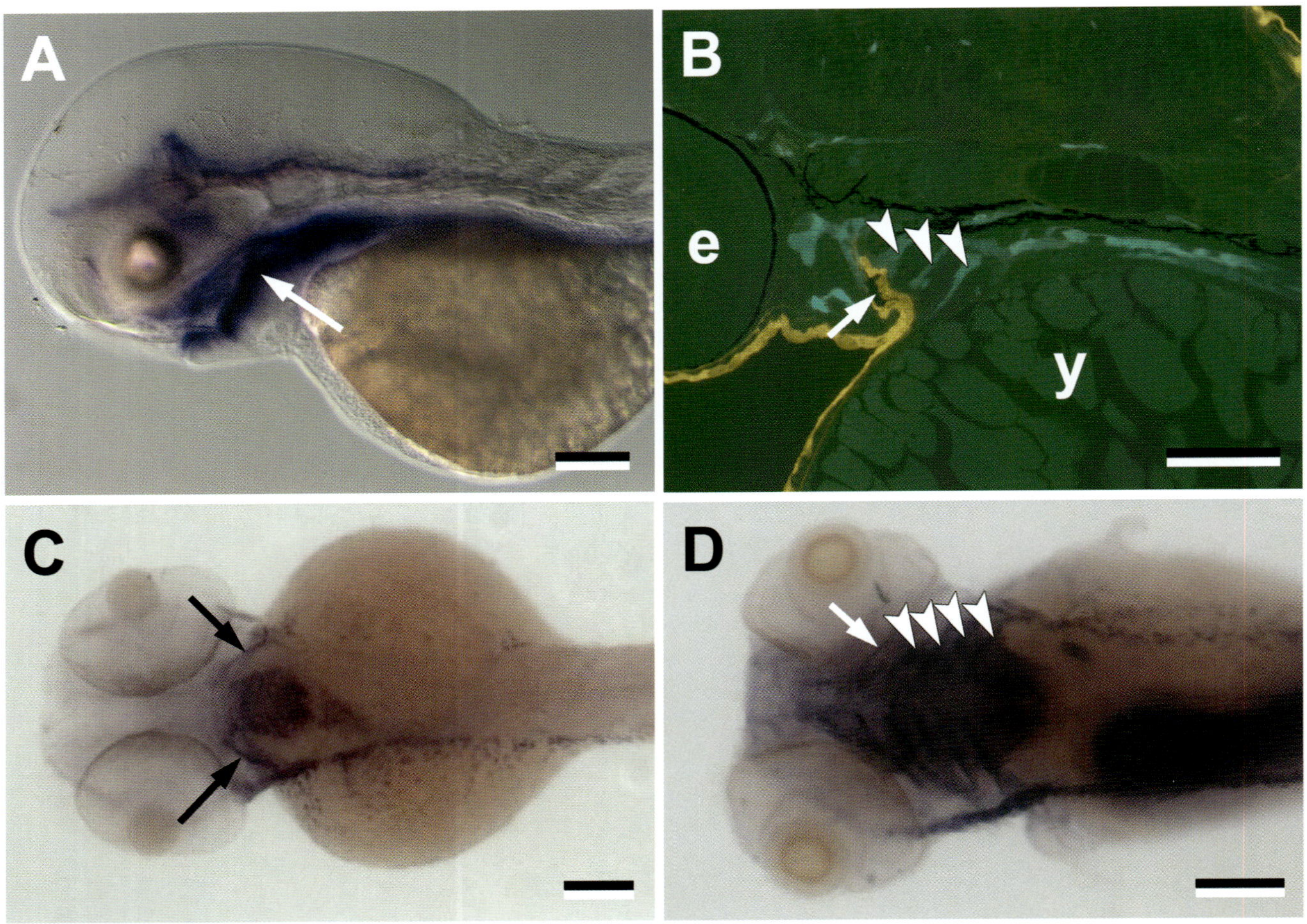

Fig. 4. A. Lateral view of a 3 dpf zebrafish exposed to *in situ* hybridization for shh. Note strong *shh* expression in the lining of the buccopharyngeal cavity (arrow). **B.** Sagittal section of a 2.5 dpf zebrafish resulting from a cross between an endoderm-specific green transgenic line Tg(*sox17:egfp*) and a periderm-specific red Tg(*krt4:tomato-CAAX*) transgenic line. Note periderm has invaded pouch 2 (arrow), while other pouches (arrowheads) are free of periderm at this stage. **C–D.** Whole mount *in situ* hybridization for *krt4* at protruding mouth stage (**C**) and 5 dpf (**D**). Note periderm-specific *krt4* expression in pouch 2 (arrows) and along midline; only later is a signal seen in each of pouches 3 to 6 (arrowheads) (**D**). (C and D from Rauch et al. 2003). Abbreviations: e, eye; y, yolk sac. Scale bars = 100 µm.

(Japanese medaka), and the characiform *Astyanax mexicanus* (Mexican tetra). The authors concluded that, unlike zebrafish, pharyngeal tooth formation in medaka or tetra is not dependent on RA, and proposed that RA dependency was acquired in the cypriniform lineage.

So, does the situation in medaka challenge the hypothesis that RA affects odontogenesis only indirectly by controlling pouch formation? Likely not. In zebrafish, RA inhibition treatments that led to tooth absence commenced when pouch 6-cleft contact is not fully established (prior to 43 hours post-fertilization, hpf). In contrast, the RA inhibition treatment in medaka commenced at the end of somitogenesis. Yet, in medaka pouches lining tooth-bearing arches are formed well before the end of somitogenesis (Iwamatsu 2004, Okada et al. 2016). Thus, pouch-cleft contacts were established already by the time of RA inhibition. Tooth presence may thus not be an indication of RA independency, but simply that signaling through pouch-cleft contacts may have taken place. RA inhibition experiments in multiple species, finely tuned against the timing of pouch-cleft contact, may assess the role of these contacts versus supposedly differential RA sensitivity. In a follow-up experiment, Seritrakul et al. (2012) applied exogenous RA to zebrafish embryos from 24 hpf onwards and demonstrated an expansion of teeth over the more anterior arches. These authors suggested that tinkering with RA could alter the distribution on the arches and could have been an evolutionary mechanism controlling tooth distribution. Again, it is necessary to study the development of the pouch-cleft contacts under these circumstances to correctly assess their role in formation of ectopic pharyngeal teeth. In a last set of experiments, exogenous RA was applied to zebrafish later in development (56 hpf and older), and resulted in expansion of the dentition (increase in tooth number) on the tooth-bearing arch only, without anterior expansion over the other arches (Gibert et al. 2015). This led the authors to suggest that RA signaling can intervene in the fine-tuning of tooth number at a taxonomically restricted level.

As far as I know, RA signaling is the only signaling pathway whose manipulation allows expansion of the dentition over the other arches. Five conserved signaling pathways, namely, the Wnt, BMP, FGF, Shh, and Eda pathways, mediate the successive reciprocal epithelial-mesenchymal interactions regulating tooth development (Balic & Thesleff 2015). Supernumerary teeth have been found in zebrafish when Fgf signaling is up-regulated (Gibert et al. 2010, Jackman et al. 2013), Bmp signaling is down-regulated (Jackman et al. 2013), Eda signaling is over-expressed (Aigler et al. 2014), as well as after thyroid hormone levels are manipulated (Woltmann et al. 2018). However, in these cases, teeth did not expand over the other arches.

Tinkering with Wnt signaling has produced conflicting results so far in terms of supernumerary teeth (Huysseune et al. 2014; Bloomquist et al. 2015). The case of Fgf signaling is intriguing in light of the evidence of the important role of Fgfs in morphogenesis of pharyngeal pouches in zebrafish (Jackman et al. 2004, Crump et al. 2004, Lovely et al. 2016) and in vertebrates in general (Jandzik et al. 2014). On the other hand, Sonic hedgehog (*shh*) is widely expressed in the zebrafish pharyngeal epithelium (Fig. 4A), and could be a prominent candidate for regulating tooth patterning, not just in teleosts, but in all gnathostomes bearing pharyngeal teeth (Maisey et al. 2013).

In zebrafish, tooth development depends on at least one of the *shh* paralogues (Jackman et al. 2010). However, when these authors over-activated hedgehog signaling, they obtained no observable effect either on tooth development or on tooth distribution. This is surprising, considering that, at least in mice, patterning of pharyngeal pouch endoderm and pharyngeal endoderm-derived organs is regulated by differential expression of *Shh* (Moore-Scott & Manley 2005). Interestingly, El Shahawy et al. (2017) recently reported an unknown, highly sensitive balance between Shh and RA signaling regulating the specification of taste placodes and glands in the mouse tongue. This, together with the observations on RA dependency reported above, provide a strong incentive to study possible epistatic relationships between Shh and RA signaling pathways in the formation of pharyngeal teeth. Such a study will necessarily rely on high resolution images of the teeth (Bruneel et al. 2015). Indeed, tooth developmental stages prior to matrix formation remain unnoticed in cleared and stained specimens, and dental marker genes may be expressed in the absence of any developing tooth structure (Jackman et al. 2004).

Does pouch 2 have a special identity?

Using zebrafish as a model, we discovered that peridermal cells (covering the basal ectodermal layer of the skin) invade pouch 2 (the pouch separating the hyoid arch from the first true gill arch) at an early developmental stage, soon followed by opening of this pouch. The periderm derives from the embryo's enveloping layer and is marked by keratin 4 (*krt4*) expression (an intermediate filament). Periderm cells invade via pouch 2 and meet similar periderm-like *krt4*-expressing cells along the pharynx midline (Rosa et al. 2019) (Fig. 4B). Periderm-like cells then expand posteriorly (Fig. 4B), until they reach the last pouch (Fig. 4C). Once these cells have reached the prospective tooth-forming region, tooth formation is initiated (Oralová et al. 2019, 2020). Only later do periderm cells invade the other pouches from outside, but they never reach further than halfway into the pouch (Fig. 4D). The importance of periderm rather than ectoderm, and the prominent role of pouch 2, distinguish this scenario from that described by Edwards (1929) for carp. The early opening of pouch 2 is also observed in chondrichthyans, e.g. in the common torpedo, *Torpedo ocellata* (Goette 1901), in the sturgeon, *Acipenser sturio* (Goette 1901), and in teleosts other than zebrafish, e.g. in Atlantic salmon, *Salmo salar* (Goette 1901 and own observations), and the European seabass, *Dicentrarchus labrax* (Sucré et al., 2009). The early opening of pouch 2 may simply be the result of the sequential maturation of the pouches. However, it could also reflect a more fundamental role, different from the other pouches.

Indeed, given the apparent importance of pouch 2 in establishing a mature pharyngeal epithelium, and potentially in tooth formation, one can ask whether pouch 2 has a key role in the evolutionary origin and expansion of the pharyngeal dentition. The importance of pouch 2 in evolution is easily understood from a functional viewpoint. In gnathostomes, it separates the jaw-hyoid apparatus functioning in feeding from the branchial arches functioning in respiration (Starck 1982). The thyroid in all vertebrates derives from pouch 2, but there is variation for other pouch derivatives (Graham et al. 2005). From a developmental viewpoint, many defects in the pharyngeal skeleton (caused by mutations) differentially affect the gill arches and the mandibular/hyoid arch. Okada et al. (2020) reported that pouch 2 forms in a peculiar way and suggested that it acts as an interface between distinct mechanisms of pharyngeal arch development. While pouch morphogenesis requires Bmp signaling, treatment of zebrafish embryos between 10 and 18 hpf with Dorsomorphin, a Bmp inhibitor, affects pouches 3 to 6, but not 1 and 2 (Lovely et al. 2016). It is perhaps also significant that studies in quail have indicated that pouch 2 forms the borderline between pouches 3 and 4 that are absolutely dependent on RA for their development, while the first pouch does not require RA (reviewed in Graham et al. 2005). Mark et al. (2004) list several examples (mostly taken from mice), how RA differentially affects the first two versus the more posterior branchial arches.

The importance of RA in Hox gene regulation (Rhinn & Dollé 2012), and the importance of Hox genes in anterior-posterior patterning of the branchial region (reviewed in Frisdal & Trainor 2014), also urges us to consider the possible role of Hox genes in establishing this putative border area. The expression of the most anterior Hox gene is first seen in vertebrates in the second pharyngeal slit (Rychel et al. 2006). Compagnucci et al. (2013), summarizing literature from various chondrichthyans, conclude that a rostral expression limit set at branchial arch 2 is ancestral for Hoxa2 orthologues. In further support of the evolutionary significance of pouch 2 are developmental features in amniotes that reflect the evolutionary 'memory' of this pouch. For example, while gill respiration was lost, the second (hyoid) arch in amniotes (exemplified by the chick embryo) expands caudally to cover more posterior clefts (as with the operculum in teleosts, Richardson et al., 2011). Second branchial cleft lesions account for 95 % of all branchial cleft malformations in humans (Waldhausen 2006, Adams et al. 2016). It will definitely be worthwhile to further explore a possible outstanding role of pouch 2.

Conclusion

In extant chondrichthyans and actinopterygians, teeth can be associated to elements of any of the pharyngeal arches. Pharyngeal tooth formation has received particular attention in the light of hypotheses on the evolutionary origin of teeth, and the roles played by ectoderm and endoderm as a source for odontogenic epithelium. Using zebrafish, I present evidence that pouch-cleft contacts appear to play a crucial role in pharyngeal tooth induction, and suggest this interdependence as a potential case of co-evolution. I explore if pouch-cleft contacts could be involved in the loss of pharyngeal teeth from specific arches and discuss the role of RA and its possible intersection with Shh signaling. Based on our observations of an early opening of pouch 2, and an early influx of periderm cells via this pouch, I infer a special identity for pouch 2. I propose that pouch 2 may be an important borderline separating developmental programs from anterior and posterior branchial arches.

Acknowledgements

The author wishes to thank V. Oralová, J. T. Rosa and P. E. Witten for the collection of numerous data on pouch formation in zebrafish and for exciting discussions. The author also gratefully acknowledges the two anonymous reviewers for their constructive comments and editing. T. Piotrowski, R. Opitz and M. Hammerschmidt generously donated zebrafish mutant embryos and transgenic lines. R. Van Driessche and M. Soenens are gratefully acknowledged for technical help with sample preparation. This study was supported by a grant of the Ghent University Research Fund (n° BOF24J2015001401).

Literature cited

Adams, A., K. Mankad, C. Offiah & L. Childs. 2016. Branchial cleft anomalies: a pictorial review of embryological development and spectrum of imaging findings. Insights Imaging, 7: 69–76.

Aigler, S. R., D. Jandzik, K. Hatta, K. Uesugi & D. W. Stock. 2014. Selection and constraint underlie irreversibility of tooth loss in cypriniform fishes. Proceedings of the National Academy of Sciences of the USA, 111: 7707–7712.

Arambourg, C. & L. Bertin. 1958. Super-Ordres des Holostéens et des Halécostomes (Holostei et Halecostomi). – Pp. 2173–2203 in: Traité de Zoologie (P.-P. Grassé, ed.). Vol. 13(3). Agnathes et Poissons. Anatomie, Ethologie, Systématique.

Atukorala, A. D. S, K. Inohaya, O. Baba, M. J. Tabata, R. A. R. K. Ratnayake, D. Abduweli, S. Kasugai, H. Mitani & Y. Takano. 2011. Scale and tooth phenotypes in medaka with a mutated ectodysplasin – A receptor: implications for the evolutionary origin of oral and pharyngeal teeth. Archives of Histology and Cytology, 73: 139–148.

Balic, A. & I. Thesleff. 2015. Tissue interactions regulating tooth development and renewal. Current Topics in Developmental Biology, 115: 157–186.

Beaumont, A. & P. Cassier. 2009. Biologie animale. Les cordés: anatomie comparée des Vertébrés. 9th ed. Dunod, Paris, 653 pp.

Bjerring, H. C. 1977. A contribution to structural analysis of the head of craniate animals. Zoologica Scripta, 6: 127–183.

Bjerring, H. C. 1998. The fates of spiracular allostoses in mammals. Acta Zoologica (Stockholm), 79: 51–67.

Blais, S. A., L. A. MacKenzie & M. V. H. Wilson. 2011. Tooth-like scales in early Devonian Eugnathostomes and the 'outside in' hypothesis for the origins of teeth in vertebrates. Journal of Vertebrate Paleontology, 31: 1189–1199.

Bloomquist, R. F., N. F. Parnell, K. A. Phillips, T. E. Fowler, T. Y. Yu, P.T. Sharpe & J. T. Streelman. 2015. Coevolutionary patterning of teeth and taste buds. Proceedings of the National Academy of Sciences of the USA, 112: E5954 -E5962.

Bruneel, B., M. Mathä, R. Paesen, M. Ameloot, W. J. Weninger & A. Huysseune. 2015. Imaging the zebrafish dentition: From traditional approaches to emerging technologies. Zebrafish, 12: 1-10.

Clemen, G., P. Bartsch & K. Wacker. 1998. Dentition and dentigerous bones in juveniles and adults of *Polypterus senegalus* (Cladistia, Actinopterygii). Annals of Anatomy, 180: 211-221.

Comabella, Y., A. H. Franyutti, A. Hurtado, J. Canabal & T. García-Galano. 2012. Ontogenetic development of the digestive tract in Cuban gar (*Atractosteus tristoechus*) larvae. Reviews in Fish Biology and Fisheries, 23: 245-260.

Compagnucci, C., M. Debiais-Thibaud, M. Coolen, J. Fish, J. N. Griffin, F. Bertocchini, M. Minoux, F. M. Rijli, V. Borday-Birraux, D. Casane, S. Mazan & M. J. Depew. 2013. Pattern and polarity in the development and evolution of the gnathostome jaw: Both conservation and heterotopy in the branchial arches of the shark, *Scyliorhinus canicula*. Developmental Biology, 377: 428-448.

Crump, J. G., L. Maves, N. D. Lawson, B. M. Weinstein & C. B. Kimmel. 2004. An essential role for Fgfs in endodermal pouch formation influences later craniofacial skeletal patterning. Development, 131: 5703-5716.

Debiais-Thibaud, M., S. Oulion, F. Bourrat, P. Laurenti, D. Casane & V. Borday-Birraux. 2011. The homology of odontodes in gnathostomes: insights from Dlx gene expression in the dogfish, *Scyliorhinus canicula*. BMC Evolutionary Biology, 11: 307.

Donoghue, P. C. J. & M. Rücklin. 2016. The ins and outs of the evolutionary origin of teeth. Evolution & Development, 18: 19-30.

Edwards, L. F. 1929. The origin of the pharyngeal teeth of the carp (*Cyprinus carpio* Linnaeus). Ohio Journal of Sciences, 29: 93-130.

El Shahawy, M., C.-G. Reibring, C. L. Neben, K. Hallberg, P. Marangoni, B. D. Harfe, O. D. Klein, A. Linde & A. Gritli-Linde. 2017. Cell fate specification in the lingual epithelium is controlled by antagonistic activities of Sonic hedgehog and retinoic acid. PLoS Genetics, 13(7): e1006914. DOI: 10.1371/journal.pgen.1006914

Frisdal, A. & P. A. Trainor. 2014. Development and evolution of the pharyngeal apparatus. Wiley Interdisciplinary Reviews – Developmental Biology, 3: 403-418.

Gibert, Y., L. Bernard, M. Debiais-Thibaud, F. Bourrat, J.-S. Joly, K. Pottin, A. Meyer, S. Retaux, D. W. Stock, W. R. Jackman, P. Seritrakul, G. Begemann & V. Laudet. 2010. Formation of oral and pharyngeal dentition in teleosts depends on differential recruitment of retinoic acid signaling. FASEB Journal, 24: 3298 -3309.

Gibert, Y., E. Samarut, E. Pasco-Viel, L. Bernard, V. Borday-Birraux, A. Sadier, C. Labbé , L. Viriot & V. Laudet. 2015. Altered retinoic acid signalling underpins dentition evolution. Proceedings of the Royal Society B, 282: 20142764.

Gilbert, S. F. & M. J. F. Barresi. 2016. Developmental Biology. 11th Ed. Sinauer Associates, Sunderland, Massachusetts, 810 pp.

Goette, A. 2001. Über die Kiemen der Fische. Zeitschrift für Wissenschaftliche Zoologie, 69: 533-577.

Graham, A. 2008. Deconstructing the pharyngeal metamere. Journal of Experimental Zoology, Part B, Molecular and Developmental Evolution, 310B: 336-344.

Graham, A., M. Okabe & R. Quinlan. 2005. The role of the endoderm in the development and evolution of the pharyngeal arches. Journal of Anatomy, 207: 479-487.

Graham, A. & J. Richardson. 2012. Developmental and evolutionary origins of the pharyngeal apparatus. EvoDevo, 3: 24.

Graveson, A. C., M. M. Smith & B. K. Hall. 1997. Neural crest potential for tooth development in a urodele amphibian: Developmental and evolutionary significance. Developmental Biology, 188: 34-42.

Grevellec, A. & A. S. Tucker. 2010. The pharyngeal pouches and clefts: development, evolution, structure and derivatives. Seminars in Cell and Developmental Biology, 21: 325-332.

Hilton, E. J., L. Grande & W. E. Bemis. 2011. Skeletal anatomy of the shortnose sturgeon, *Acipenser brevirostrum* Lesueur, 1818, and the systematics of sturgeons (Acipenseriformes, Acipenseridae). Fieldiana Life Earth Sciences, 3: 1-168.

Huysseune, A. & J.-Y. Sire. 1998. Evolution of patterns and processes in teeth and tooth-related tissues in non-mammalian vertebrates. European Journal of oral Sciences, 106: 437-481.

Huysseune, A., J.-Y. Sire & P. E. Witten. 2009. Evolutionary and developmental origins of the vertebrate dentition. Journal of Anatomy, 214: 465-476.

Huysseune, A., J.-Y. Sire & P. E. Witten. 2010. A revised hypothesis on the evolutionary origin of the vertebrate dentition. Journal of Applied Ichthyology, 26: 152-155.

Huysseune, A., M. Soenens & F. Elderweirdt. 2014. Wnt signaling during tooth replacement in zebrafish (*Danio rerio*): pitfalls and perspectives. Frontiers in Physiology, 5: 386.

Huysseune, A., C. Van der heyden & J.-Y. Sire. 1998. Early development of the zebrafish (*Danio rerio*) pharyngeal dentition (Teleostei, Cyprinidae). Anatomy & Embryology, 198: 289-305.

Iwamatsu, T. 2004. Stages of normal development in the medaka *Oryzias latipes*. Mechanisms of Development, 121: 605-618.

Jackman, W. R., B. W. Draper & D. W. Stock. 2004. Fgf signaling is required for zebrafish tooth development. Developmental Biology, 274: 139-157.

Jackman, W. R., J. J. Yoo & D. W. Stock. 2010. Hedgehog signaling is required at multiple stages of zebrafish tooth development. BMC Developmental Biology, 10: 119.

Jackman, W. R., S. H. Davies, D. B. Lyons, C. K. Stauder, B. R. Denton-Schneider, A. Jowdry, S. R. Aigler, S. A. Vogel & D. W. Stock. 2013. Manipulation of Fgf and Bmp signaling in teleost fishes suggests potential pathways for the evolutionary origin of multicuspid teeth. Evolution and Development, 15: 107-118.

Jandzik, D., M. B. Hawkins, M. V. Cattell, R. Cerny, T. A. Square & D. M. Medeiros. 2014. Roles for FGF in lamprey pharyngeal pouch formation and skeletogenesis highlight ancestral functions in the vertebrate head. Development, 141: 629-638.

Karakostis, K. & R. Fåhraeus. 2019. Shaping the regulation of the p53 mRNA tumour suppressor: the co-evolution of genetic signatures. BMC Cancer, 19: 915.

Kimmel, C. B., W. W. Ballard, S. R. Kimmel, B. Ullmann & T. F. Schilling. 1995. Stages of embryonic development of the zebrafish. Developmental Dynamics, 203: 253-310.

Kopinke, D., J. Sasine, J. Swift, W. Z. Stephens & T. Piotrowski. 2006. Retinoic acid is required for endodermal pouch morphogenesis and not for pharyngeal endoderm specification. Developmental Dynamics, 235: 2695-2709.

Lovely, C. B., M. E. Swartz, N. McCarthy, J. L. Norrie & J. K. Eberhart. 2016. Bmp signaling mediates endoderm pouch morphogenesis by regulating Fgf signaling in zebrafish. Development, 143: 2000-2011.

Maisey, J. G. 1988. Phylogeny of early skeletal induction and ossification patterns. Evolutionary Biology, 22: 1-36.

Maisey, J. G., S. Turner, G. J. P. Naylor & R. F. Miller. 2013. Dental patterning in the earliest sharks: Implications for tooth evolution. Journal of Morphology, 275: 586-596.

Mark, M., N. B. Ghyselinck & P. Chambon. 2004. Retinoic acid signalling in the development of branchial arches. Current Opinion in Genetics & Development, 14: 591-598.

Moore-Scott, B. A. & N. R. Manley. 2005. Differential expression of Sonic hedgehog along the anterior-posterior axis regulates patterning of pharyngeal pouch endoderm and pharyngeal endoderm-derived organs. Developmental Biology, 278: 323-335.

Nelson, G. J. 1967. Gill arches of teleostean fishes of the family Clupeidae. Copeia, 1967: 389-399.

Nelson, G. J. 1969. Gill arches and the phylogeny of fishes, with notes on the classification of vertebrates. Bulletin of the American Museum of Natural History, 141: 475-552.

Nelson, G. J. 1970. Pharyngeal denticles (placoid scales) of sharks, with notes on the dermal skeleton of vertebrates. American Museum Novitates, 2413: 1-26.

Okada, K., K. Inohaya, T. Mise, A. Kudo, S. Takada & H. Wada. 2016. Reiterative expression of pax1 directs pharyngeal pouch segmentation in medaka (*Oryzias latipes*). Development, 143: 1800-1810.

Okada, K. & S. Takada. 2020. The second pharyngeal pouch is generated by dynamic remodeling of endodermal epithelium in zebrafish. Development, 147: dev194738

Oralová, V., J. T. Rosa, D. Larionova, P. E. Witten & A. Huysseune. 2019. A new paradigm for pharyngeal tooth formation? Journal of Morphology 280, Suppl. 1: S193-S193.

Oralová, V., J. T. Rosa, D. Larionova, P. E. Witten & A. Huysseune. 2020. Multiple epithelia are required to develop teeth deep inside the pharynx. Proceedings of the National Academy of Sciences of the USA, 117: 11503-11512.

Owen, R. 1840-1845. Odontography. London: H. Baillère. 655 pp.

Piotrowski, T. & C. Nüsslein-Volhard. 2000. The endoderm plays an important role in patterning the segmented pharyngeal region in zebrafish (*Danio rerio*). Developmental Biology, 225: 339–356.

Piotrowski, T., D. Ahn, T. F. Schilling, S. Nair, I. Ruvinsky, R. Geisler, G.-J. Rauch, P. Haffter, L. I. Zon, Y. Zhou, H. Foott, I. B. Dawid & R. K. Ho. 2003. The zebrafish *van gogh* mutation disrupts *tbx1*, which is involved in the DiGeorge deletion syndrome in humans. Development, 130: 5043–5052.

Peters, H., A. Neubüser, K. Kratochwil & R. Balling. 1998. *Pax9*-deficient mice lack pharyngeal pouch derivatives and teeth and exhibit craniofacial and limb abnormalities. Genes & Development, 12: 2735–2747.

Peyer, B. 1968. Comparative odontology. The University of Chicago, Chicago & London, 347 pp.

Qu, Q., S. Sanchez, H. Blom, P. Tafforeau & P. E. Ahlberg. 2013. Scales and tooth whorls of ancient fishes challenge distinction between external and oral 'teeth'. PloS One, 8(8): e71890: 1–10.

Rauch, G. J., D. A. Lyons, I. Middendorf, B. Friedlander, N. Arana, T. Reyes & W. S. Talbot. 2003. Submission and curation of gene expression data. https://zfin.org/ZDB-PUB-031103-24.

Reif, W.-E. 1982. Evolution of dermal skeleton and dentition in vertebrates. The odontode regulation theory. Evolutionary Biology, 15: 287–368.

Richardson, J., T. Shono, M. Okabe & A. Graham. 2011. The presence of an embryonic opercular flap in amniotes. Proceedings of the Royal Society B, 279: 224–229.

Rhinn, M. & P. Dollé. 2012. Retinoic acid signalling during development. Development, 139: 843–858.

Rosa, J. T., V. Oralová, D. Larionova, G. T. Eisenhoffer, P. E. Witten & A. Huysseune. 2019. Periderm invasion contributes to epithelial formation in the teleost pharynx. Scientific Reports 9: 10082.

Rücklin, M. & P. C. J. Donoghue. 2015. *Romundina* and the evolutionary origin of teeth. Biology Letters, 11: 20150326. DOI: 10.1098/rsbl.2015.0326

Rücklin, M., P. C. J. Donoghue, Z. Johanson, K. Trinajstic, F. Marone & M. Stampanoni. 2012. Development of teeth and jaws in the earliest vertebrates. Nature, 491: 748–752.

Rücklin, M., S. Giles, P. Janvier & P. C. J. Donoghue. 2011. Teeth before jaws? Comparative analysis of the structure and development of the external and internal scales in the extinct jawless vertebrate *Loganellia scotica*. Evolution & Development, 13: 523–532.

Rychel, A. L., S. E. Smith, H. T. Shimamoto & B. J. Swalla. 2006. Evolution and development of the chordates: Collagen and pharyngeal cartilage. Molecular Biology and Evolution, 23: 541–549.

Schoch, R. R. 2002. The evolution of metamorphosis in temnospondyls. Lethaia, 35: 309–327.

Seritrakul, P., E. Samarut, T. T. S. Lama, Y. Gibert, V. Laudet & W. R. Jackman. 2012. Retinoic acid expands the evolutionarily reduced dentition of zebrafish. FASEB Journal, 26: 5014–5024.

Shone, V. & A. Graham. 2014. Endodermal/ectodermal interfaces during pharyngeal segmentation in vertebrates. Journal of Anatomy, 225: 479–491.

Sire, J.-Y. & A. Huysseune. 2003. Formation of dermal skeletal and dental tissues in fish: A comparative and evolutionary approach. Biological Reviews, 78: 219–249.

Starck, D. 1982. Vergleichende Anatomie der Wirbeltiere auf evolutionsbiologischer Grundlage. Bd 3. Springer Verlag Berlin, 1110 pp.

Stock, D. W. 2007. Zebrafish dentition in comparative context. Journal of Experimental Zoology Part B Molecular and Developmental Evolution, 308B: 523–549.

Sucré, E., M. Charmantier-Daures, E. Grousset, G. Charmantier & P. Cucchi-Mouillot. 2009. Early development of the digestive tract (pharynx and gut) in the embryos and pre-larvae of the European sea bass *Dicentrarchus labrax*. Journal of Fish Biology, 75: 1302–1322.

Swartz, M. E., K. Sheehan-Rooney, M. J. Dixon & J. K. Eberhart. 2011. Examination of a palatogenic gene program in zebrafish. Developmental Dynamics, 240: 2204–2220.

Vandewalle, P., A. Huysseune, P. Aerts & W. Verraes. 1994. The pharyngeal apparatus in teleost feeding. In Biomechanics of feeding in vertebrates (eds. Bels, V., M. Chardon & P. Vandewalle). Advances in Comparative and Environmental Physiology, 18: 59–92.

Waldhausen, J. H. T. 2006. Branchial cleft and arch anomalies in children. Seminars in Pediatric Surgery, 15: 64–69.

Wilkins, A. S. 2002. The Evolution of Developmental Pathways. Sinauer Associates, Inc. Sunderland, Massachusetts, 603 pp.

Witten, P. E., J.-Y. Sire & A. Huysseune. 2014. Old, new and new-old concepts about the evolution of teeth. Journal of Applied Ichthyology, 30: 636–642.

Woltmann, I., F. Shkil, A. De Clercq, A. Huysseune & P. E. Witten. 2018. Supernumerary teeth in the pharyngeal dentition of slow-developing zebrafish (*Danio rerio*, Hamilton, 1822). Journal of Applied Ichthyology, 34: 455–464.

Ancient Fishes and their Living Relatives: a Tribute to John G. Maisey. Alan Pradel, John S. S. Denton & Philippe Janvier (eds.): pp. 233–238, 4 figs.
© 2021 by Verlag Dr. Friedrich Pfeil, München, Germany – ISBN 978-3-89937-269-4

The Cretaceous fishes of Brazil: a paleobiogeographic perspective

Rafael M. Lindoso* and Ismar de S. Carvalho**

Abstract

Brazilian paleoichthyology has been predominantly characterized by studies in the Lagerstätte Santana Formation of the Araripe Basin. It is dominated by actinopterygian fishes, but rare coelacanths and elasmobranchs occur. In rift basins of the East Brazilian Rift System (EBRIS), fossil fishes are also abundant and diversified, including approximately 27 genera of chondrichthyans and about 34 genera of bony fishes. These records reveal a complex paleobiogeographic scenario assigned to a variety of tectonic settings, which had a Tethyan influence. Nevertheless, recent discoveries in Aptian strata of the Parnaíba Basin have increased our understanding regarding Early Cretaceous ecosystems prevailing Northeastern Brazil.

Introduction

In Brazil, the Cretaceous fossil fish record has been assigned to complex tectonic settings related to the development of the Equatorial and South Atlantic oceans during the Jurassic-Cretaceous interval. The evolution of an East Brazilian Rift System (EBRIS) occurred between South America and Africa as a result of an extensive seafloor spreading (Chang et al. 1992). In non-rift basins within the EBRIS region, as well as within interior continental ones, the diastrophism also played an important role in the fossil fish distribution (Maisey 2000). Most of these geological settings with fossil fishes are associated with carbonate deposits of lacustrine or marine origin. Furthermore, such records have been considered in the light of hypotheses of an epicontinental seaway through north-eastern Brazil in the late Aptian, which have been increasingly supported by geochemical and paleontological data (Assine 1992, Arai 1995, 1999, 2009, 2014, Maisey 2000, 2011, Assine et al. 2016, Lindoso et al. 2016, Arai & Assine, 2020, Araripe et al. 2021).

A better understanding of the paleoichthyofauna of Brazil comes from the Araripe Basin (Fig. 1). In this Mesozoic intracratonic sedimentary system, the Santana Formation comprises a famous fossil concentration and conservation Lagerstätten, with a vertebrate fauna dominated by actinopterygian fishes, but also including rare coelacanths and elasmobranchs (Martill & Brito 2017). The Araripe Basin is also the most extensive among the continental interior basins and exhibits the most complex geological history. This Lower Cretaceous succession, approximately 110 million years old, has been described by John Maisey as one of the world's most important Gondwanan fossil assemblages (Maisey 1991).

In the Brazilian coastal basins (EBRIS), the paleoichthyofauna is abundant and diversified, including approximately 27 genera of chondrichthyans and about 34 genera of bony fishes. It includes mainly Cretaceous records from Pelotas, Santos, Sergipe-Alagoas, Pernambuco, Paraíba, Potiguar, and São Luís basins (Gallo et al. 2012) (Fig. 2). During the Late Cretaceous, biotic distributions within the EBRIS region were assigned to rifting of Gondwana, leading to important vicariant processes (Maisey 2000, Santos & Carvalho 2009, Parméra

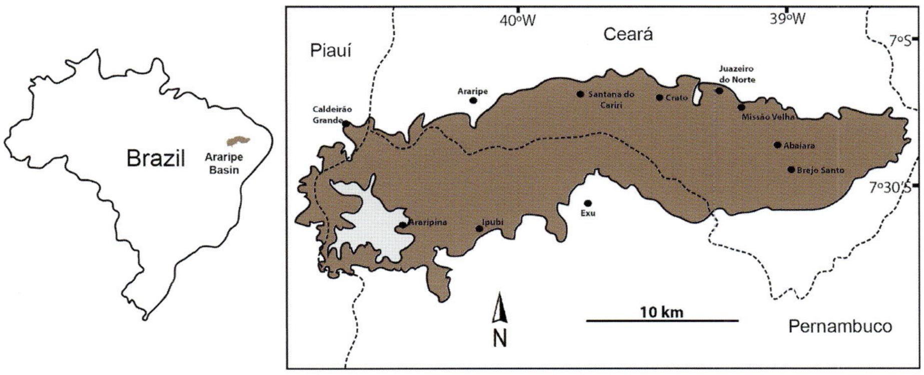

Fig. 1. Simplified map of the Araripe Basin localized between the Ceará, Pernambuco and Piauí States, north-eastern Brazil.

Fig. 2. Main north-eastern Brazilian sedimentary basins where Cretaceous fossil fishes are found.

* Instituto Federal de Educação, Ciência e Tecnologia do Maranhão – IFMA, Departamento Acadêmico de Biologia, DAB/IFMA, campus Monte Castelo, São Luís, Maranhão, Brasil; rafael.lindoso@ifma.edu.br (corresponding author)

** Universidade Federal do Rio de Janeiro, Departamento de Geologia, CCMN/IGEO 21.949-900 Cidade Universitária – Ilha do Fundão, Rio de Janeiro, Brazil; Universidade de Coimbra, Centro de Geociências. Rua Sílvio Lima, 3030-790, Coimbra, Portugal.

2016). These tectonically-driven events have affected both pelagic and benthic marine organisms as well as continental freshwater ones (Maisey 2000).

In this study we present a synthesis concerning the Cretaceous fishes of Brazil and their paleobiogeographic significance. Some of the ideas presented herein remain unsolved, and this work aims to expand the debate.

Paleoichthyofauna synthesis

The history of Brazilian paleoichthyology begins with João da Sylva Feijó in his reports of December 11th 1800 to the Portuguese authorities (Nobre 1978), followed by the German expeditions of Johann Baptist von Spix and Carl Friedrich Philipp von Martius between 1817 and 1820 (Spix & Martius 1823). The numerous fossil fishes coming from the Santana Formation were subsequently analyzed by George Gardner and Louis Agassiz in 1841. Throughout the 20th century, important studies were conducted in this lithostratigraphic unit: Woodward (1901, 1908), Jordan & Branner (1908), Jordan (1919, 1921), D'Erasmo (1938), Dunkle (1940), Silva-Santos (1945, 1947, 1950, 1958, 1960), Silva-Santos & Valença (1968), Campos & Wenz (1982), Silva-Santos (1985a,b), Maisey (1986), Wenz & Kellner (1986), Wenz (1989), Brito (1988), Brito & Ferreira (1989), and Brito & Martill (1999). John Maisey and contributors (1991) produced the most comprehensive treatment on the sedimentology and paleontology of the Santana Formation at the time in his book Santana Fossils: An Illustrated Atlas. The book is one of the most complete and richly illustrated works regarding a sedimentary basin in Brazil, comprising a systematic description of vertebrate and invertebrate taxa, as well as plants.

In the Santana Formation, Chondrichthyes are represented solely by Hybodontidae and Rajiforms (family incertae sedis). By contrast, Actinopterygii are diversified and represented by Holostei, Teleosteomorpha, Teleostei, Elopomorpha Ostarioclupeomorpha, and some incertae sedis families related to Clupeocephala, Clupeomorpha, and Otophysi; Sarcopterygii are represented by Coelacanthiformes (Brito & Yabumoto 2011). Although less abundant, this ichthyofauna has been correlated with that of the Codó Formation of the Parnaíba Basin, considered late Aptian in age (Lima 1982, Silva-Santos 1994, Santos & Carvalho 2009, Lindoso et al. 2011, 2016). Recently, the first record of an obaichthyid gar (*Dentilepisosteus laevis*) was reported for the Codó Formation, as well as the first record of genus *Santanichthys*, enhancing the faunal similarities between Codó and Santana formations (Brito et al. 2016, Lindoso et al. 2016). Nevertheless, in his first visit to outcrops of the Codó Formation, John Maisey observed that part of this lithostratigraphic unit represents a condensed section, which preserves Crato, Ipubi, and Romualdo ichthyofaunas (Maisey pers. comm. 2015). In fact, most of the fishes of the Codó Formation occur as fully articulated skeletons without scales, which indicates low sedimentation rate. This scenario fits with John's suggestion that the entire Crato, Ipubi, and Romualdo ichthyofaunas occur in a unique section of the Codó Formation (see Maisey 2000: 295). In this lithostratigraphic unit, endemic forms are represented by the species *Codoichthys carnavalii* and *Axelrodichthys maiseyi* (Silva-Santos 1994, Carvalho et al. 2013).

In the Brazilian Equatorial and East continental margins, fossil fish records are represented in the Pelotas, Santos, Sergipe-Alagoas, Pernambuco, Paraíba, Potiguar, and São Luís basins, spanning the entire Cretaceous (Gallo et al. 2012). In the Atlântida Formation, Pelotas Basin, Chondrichthyes are scarce, with representatives tentatively attributed to Lamniformes (e.g. isolated odontaspidoid tooth). Representatives of Osteichthyes include Pycnodontidae, Clupeomorpha, Euteleostei, Dercetidae, Enchodontidae, and Holocentridae (Gallo & Figueiredo 1999, Figueiredo et al. 2001, Gallo et al. 2006, Figueiredo & Gallo 2006). In the Santos Basin, offshore southern Brazil, fossil fishes have been described from wells in an interval ranging from the Albian to Recent (Miller et al. 2002). Here, chondrichthyans are represented by Triakidae, Carcharhinidae, Ginglymostomatidae, Lamnidae indet., and Scyliorhinidae. Teleostei are represented by Myctophidae, Sternoptychidae, and undetermined teeth (Miller et al. 2002).

The diversity of fossil fishes from the Riachuelo Formation, Sergipe-Alagoas Basin, is similar to that of the Codó and Santana formations (Santos & Carvalho 2009, Gallo 2015, Lindoso et al. 2016). The Riachuelo Formation is divided into the Taquari, Maruim, and Angico members, with the latter two having only isolated fossil remains. The ichthyofauna of the Taquari Member is represented by diversified Osteichthyes, all them also present in the Santana Formation (Gallo et al. 2012). Some endemic species seem to occur in the lower Aptian of the Coqueiro Seco Formation (e.g. *Gibsonichthys miguelcampensis*, *Falconichthys santerezae* and *Mafficioliichthys atolensis*), and the Morro do Chaves Formation (e.g. *Lepidotes alagoensis*, *Pseudoellimma gallae*, *Atolvorator longipectorales*) (Maffizzoni 2000, Gallo & Brito 2004, Gallo & Coelho 2008, Figueiredo 2009). The Morro do Chaves Formation has the most abundant and diversified ichthyofauna from the Sergipe-Alagoas Basin (Gallo et al. 2012).

Marine forms of Chondrichthyes (Rhombodontidae and Ptychodontidae) and Osteichthyes (Pycnodontidae, Amiidae, Teleostei incertae sedis, Albulidae, Dercetidae and Enchodontidae) occur in the Cotinguiba Formation from Sergipe-Alagoas Basin (Woodward 1907, Löfgren & Oliveira 1943, Silva-Santos & Salgado 1969, Carvalho & Gallo 2002, Coelho 2004; Gallo et al. 2007). In the Calumbi Formation, Chondrichthyes are more diversified, and are represented by Hexanchidae, Cretoxyrhinidae, Odontaspididae, Anacoracidae, Sclerorhynchidae and Rhinobatidae. Osteichthyes, besides those taxa already known for the Cretaceous from the Sergipe-Alagoas Basin, include Xiphiidae and Latimeriidae (Souza-Lima 2001).

There are scarce fossil fishes in the Cretaceous strata from the Pernambuco Basin. In the Aptian-Albian of Cabo Formation, Osteichthyes are represented by *Ellimma cruzae* and *Dastilbe crandalli* (Silva-Santos 1990). In the Estiva Formation, rare pycnodontid teeth occur (Barbosa et al. 2008). The coastal Paraíba Basin, sometimes considered as an extension of the Pernambuco Basin (e.g. Pernambuco-Paraíba Basin), exhibits an abundant paleoichthyofauna; its records comprise only the Itamaracá, Gramame, and Maria Farinha formations. In the mid-Campanian to lower Maastrichtian Itamaracá Formation, Chondrichthyes are recorded by Ptychodontidae, Hexanchidae, Cretoxyrhinidae, Ginglymostomatidae, Odontaspididae, Anacoracidae, Mitsukurinidae, and Rhombodontidae; Osteichthyes are represented by the taxa Pycnodontidae, Saurodontidae, and Enchondontidae. Most of this assemblage is also in the Maastrichtian Gramame and Paleocene Maria Farinha formations, but with some distinct taxa (e.g. *Cretalamna appendiculatta*, *Squalicorax kaupi*, *Paleoballistum dossantosi*, Diodon sp.) (Rebouças & Silva-Santos 1956, Silva-Santos & Figueiredo 1987, Silva 2007, Gallo et al. 2009).

The Osteichthyes Chanidae, Pycnodontidae, and Semionotidae are also found in the Açu and Jandaíra Formations from the Potiguar Basin. These litostratigraphic units represent a continental drift phase in EBRIS (Duarte & Santos 1962, Silva-Santos 1963, Soares et al. 2003, Gallo & Brito 2004, Machado & Brito 2006). In the São Luís Basin, the northernmost Cretaceous sedimentary area in South America, vertebrate fossils occur mainly in the bone bed Laje do Coringa, a stratigraphic horizon of the Alcântara Formation (Cenomanian) that is characterized by fluvial and marine taxa such as Hybodontidae, Myliobatidae, and Sclerorhynchidae (Chondrichthyes); and Semionotidae, Lepisosteidae, Amiiformes, Pycnodontidae, Mawsoniidae, and Ceratodontidae (Osteichthyes) (for an overview, see Medeiros et al. 2014).

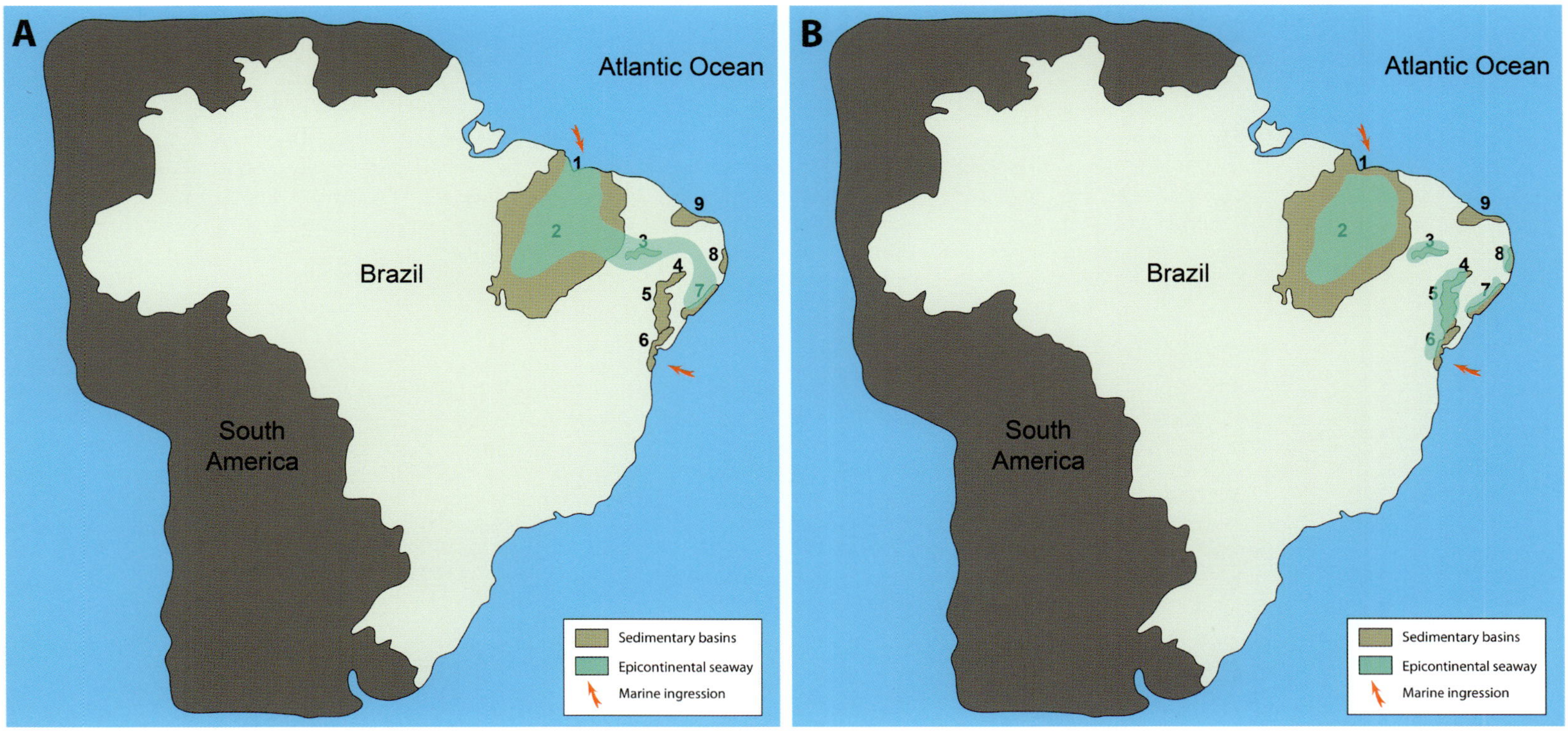

Fig. 3. a. Distribution of fossil fishes in the Brazilian Northeastern Marginal and Interior basins (BNMIBs) during the Early Cretaceous, based on paleontological data. The fossil fishes record in these basins indicates a greater similarity among Codó (Parnaíba Basin), Santana (Araripe Basin) and Riachuelo (Sergipe-Alagoas Basin) formations; **b.** evidence for dispersal and vicariance episodes is indicated by speciation in some genera of fish related to intermittent connection of some BNMIBs. Sedimentary basins in map: 1, São Luís; 2, Parnaíba; 3, Araripe; 4, Jatobá; 5, Tucano; 6, Recôncavo; 7, Sergipe-Alagoas; 8, Pernambuco-Paraíba; 9, Potiguar. Modified from Lindoso et al. (2016).

Discussion

Support for hypotheses of an equatorial seaway over territories of Western Gondwana during the Early Cretaceous has been strengthened in recent years (see Arai 2014). The reasons for this interesting debate involve geochemical and paleontological (e.g. foraminifera, echinoids, mollusks, dinoflagellates, isopods, and fishes) data of Tethyan affinity present within EBRIS and interior continental basins (Beurlen 1964, 1966, Koutsoukos 1992, Arai & Coimbra 1990, Arai 1999, 2009, 2014, Brito & Yabumoto 2011, Maisey 2000, 2011, Lindoso et al. 2013, 2016), as well as extensive gypsum beds (Assine 1992). This biotic record is well represented in the Aptian-Albian intervals of the São Luís, Parnaíba, Araripe, Tucano and Sergipe-Alagoas basins, which suggests that this seaway was established in the present-day region of northeastern Brazil (Arai 2009, 2014). According to Maisey (2000), the Aptian transgression created opportunities for the dispersal of marine biota into new areas.

A better characterization of the paleoichthyofauna of interior continental basins has been crucial for the understanding of Tethyan influence in the Aptian marine sedimentation within EBRIS. In the Araripe Basin, marine taxa are represented by the genus *Rhacolepis*, *Vinctifer*, *Araripichthys*, and *Notelops*, which occur in slightly older deposits from Venezuela, Colombia and Mexico (Silva-Santos & Oliveira 1994, Moody & Maisey 1994, Schultze & Stöhr 1996, Maisey 2000, Maisey & Moody 2001). This Tethyan heritage is also expanded to Africa (e.g. *Araripichthys*, *Cladocyclus*, *Goulmimichthys*) (Cavin 1995, Brito & Yabumoto 2011).

In the Codó Formation, upper Aptian of the Parnaíba Basin, fish assemblages are more similar to those in the Santana and Riachuelo formations than to others Brazilian Northeastern Marginal and Interior basins (BNMIBs) (Silva-Santos 1991, Lindoso et al. 2016). Recent discovery of the obaichthyid *Dentilepisosteus laevis* in the Codó Formation reinforces a faunal identity with the Aptian Santana Formation and extends the temporal range of this taxon downward into the Aptian (Brito et al. 2016). These findings suggest greater connectivity between the aquatic environments of Parnaíba, Araripe and Sergipe-Alagoas basins, but with intermittent periods in these connections revealed by independent dispersals (e.g. among clupeomorphs and coelacanths) and perhaps vicariant speciation (Fig. 3) (Lindoso et al. 2016). Similar assemblages have also indicated paleogeographic relationship between the Araripe and Recôncavo-Tucano basins (Silva-Santos 1991).

Paleogeographic reconstructions depicting the probable route of Aptian marine ingressions are dubious. Marine entrance would have occurred independently or combined, via Parnaíba, Sergipe-Alagoas and Potiguar basins (for an overview, see Arai 2014). According to this latter author, "the probable route for the water entering the Northeastern Brazil seaway is through the São Luís, Parnaíba, Araripe and Sergipe basins." However, this model has been challenged by paleocurrent analyses of fluvial deposits, as well as by faciological data (Assine et al. 2016). According to this author, three distinct drainage basins were settled in the north-eastern Brazil as a result of drainage division between the Parnaíba, Potiguar, and Araripe basins. This scenario would imply independent marine ingressions of N-NE-SE directions.

Additionally, rift valleys along the Recôncavo-Tucano-Jatobá and Cariri-Potiguar trends were occupied by epicontinental seaways (Maisey 2000). Thus, different seaways separated Northeastern Brazil from the rest of South America but left it contiguous with Africa and, in another way, the northern Brazil from both South America and Africa (Fig. 4) (Maisey 2011). Evidence from fossil fish include a relatively cosmopolitan Gondwanan taxa during the pre-rift phase followed by greater diversity and perhaps a high degree of local endemism during sin-rift times (Maisey 2000, 2011). Alternatively, the absence of common taxa between the Marizal and Santana formations has been attributed to differences in time of deposition (Brito & Alvarado-Ortega 2008).

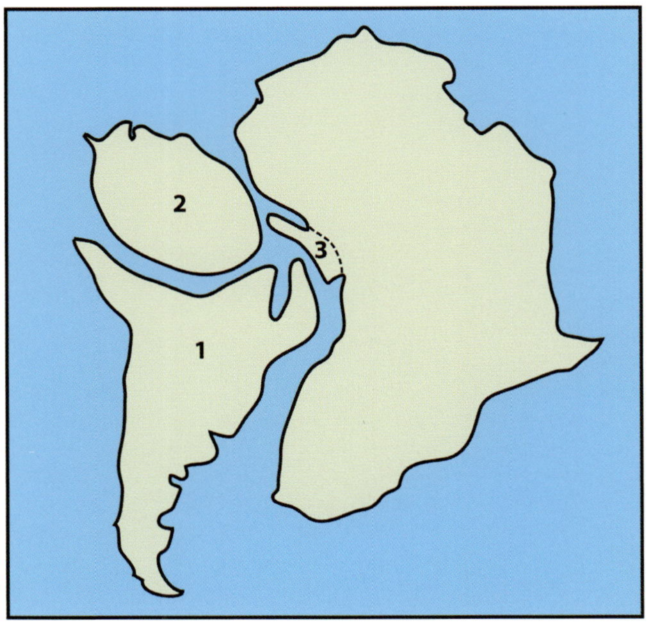

Fig. 4. Paleogeographical model proposed by John Maisey for the Western Gondwana in the Aptian, which postulate a division by epicontinental seaways into three land-masses, "mainland" South America (1), northern Brazil (2) and Africa + northeastern Brazil (3). Modified from Maisey (2011).

Conclusions

In north-eastern Brazil, fossil fish assemblages were characteristic of transgressive events during the Early Cretaceous. Speciation patterns in endemic faunas are well represented in the EBRIS and the interior continental basins. However, absence of controlled excavations has precluded a better understanding of the temporal evolution of these ecosystems and consequent paleobiogeographic scenarios. The intracratonic position of the Araripe Basin, in conjunction with its excellent fossil preservation, is obviously critical in clarifying the possible routes of marine ingressions. Nevertheless, findings of new upper Aptian fossiliferous localities in the Parnaíba Basin have been crucial in corroborating the biogeographical history of Western Gondwana. As John Maisey might say: "we now are able to at least formulate appropriate questions to ask concerning this major event in Earth history.".

Acknowledgements

We are grateful to John Maisey for joining our visit to some outcrops of the Codó Formation in Brejo city, Maranhão State; his stay was very helpful for many important discussions concerning the fossil fish biogeography of Brazil. The first author is particularly grateful to CAPES (PDSE 99999.004234/2014-06) for his stay at American Museum of Natural History in 2014, during which time he was able to access a vast and impressive Brazilian fossil fish collection. We also are grateful to John Denton, Alan Pradel and Philippe Janvier – the editors for this special volume in honor of John Maisey – for inviting us to contribute the present chapter. The authors gratefully acknowledge the research & development project "Correlação Estratigráfica, Evolução Paleoambiental e Paleogeográfica e Perspectivas Exploratórias do Andar Alagoas", sponsored by Shell Brasil Petroleo Ltda. with resources allocated to R&D institutions accredited by Brazilian National Petroleum, Natural Gas and Biofuels Agency – ANP (technical cooperation agreement #20219-2). The authors are also grateful to CNPq and FAPERJ.

Literature cited

Arai, M. 1995. Hipótese de entrada do mar albo-aptiano na Bacia do Araripe, via bacias de São Luís-Parnaíba. Anais da Academia Brasileira de Ciências, 67: 394.

Arai, M. 1999. A transgressão marinha mesocretácea: sua implicação no paradigma da reconstituição paleogeográfica do Cretáceo no Brasil. – Pp. 577–582 in: Simpósio Sobre o Cretáceo do Brasil, 5, Serra Negra, Boletim, Rio Claro, UNESP, Rio Claro.

Arai, M. 2009. Paleogeografia do Atlântico Sul no Aptiano: um novo modelo a partir de dados micropaleontológicos recentes. Boletim de Geociências da Petrobras, 17: 331–351.

Arai, M. 2014. Aptian/Albian (Early Cretaceous) paleogeography of the South Atlantic: a paleontological perspective. Brazilian Journal of Geology, 44: 339–350. DOI: 10.5327/Z2317-4889201400020012

Arai, M. & J. C. Coimbra. 1990. Análise paleoecológica do registro das primeiras ingressões marinhas na Formação Santana (Cretáceo Inferior da Chapada do Araripe). – Pp. 225–239 in: 1° Simpósio Sobre a Bacia Do Araripe e Bacias Interiores Do Nordeste. Atas, Crato.

Arai, M. & M. L. Assine. 2020. Chronostratigraphic constraints and paleoenvironmental interpretation of the Romualdo Formation (Santana Group, Araripe Basin, Northeastern Brazil) based on palynology. Cretac. Res. 116: 104610. DOI: 10.1016/j.cretres.2020.104610

Araripe, R. C., D. H. Oliveira, M. E. Tomé, R. Moura de Mello & A. M. F. Barreto. 2021. Foraminifera and Ostracoda from the Lower Cretaceous (Aptian–lower Albian) romualdo formation, Araripe basin, northeast Brazil: Paleoenvironmental inferences. Cretaceous Research, 122: 104766. DOI: 10.1016/j.cretres.2021.104766

Assine, M. 1992. Análise estratigráfica da Bacia do Araripe, Nordeste do Brasil. Brazilian Journal of Geology, 22: 289–300.

Assine, M. L., F. Quaglio, L. V. Warren, & M. G. Simões. 2016. Comments on paper by M. Arai "Aptian/Albian (Early Cretaceous) paleogeography of the South Atlantic: a paleontological perspective." Brazilian Journal of Geology, 46: 3–7.

Barbosa, J. A., P. J. F. Pereira & M. F. Lima Filho. 2008. Dente de um picnodontiforme (Actinopterygii, Neopterygii) da Formação Estiva, Cenomaniano-Turoniano da Bacia de Pernambuco, NE do Brasil. Gaea – Journal of Geoscience, 4: 43–48.

Beurlen, K. 1964. As espécies dos Cassiopinae, nova subfamília dos Turritellidae, no Cretáceo do Brasil. Arquivos de Geologia, 5: 1–44.

Beurlen, K., 1966. Novos equinóides no Cretáceo do Nordeste do Brasil. Anais da Academia Brasileira de Ciências, 38: 455–464.

Brito, P. M. 1988. La structure du suspensorium de *Vinctifer*, Poisson Actinoptérygien mésozoïque: remarques sur les implications phylogénétiques. Geobios, 21: 819–823.

Brito, P. M. & J. Alvarado-Ortega. 2008. A new species of *Placidichthys* (Halecomorphi: Ionoscopiformes) from the Lower Cretaceous Marizal Formation, northeastern Brazil, with a review of the biogeographical distribution of the Ophiopsidae. – Pp. 145–154 in: Cavin, L., A. Longbottom & M. Richter. (eds.), Fishes and the Break-up of Pangaea. Geological Society, London, Special Publications, 295.

Brito, P. M. & P. L. N. Ferreira. 1989. The first hybodont shark, *Tribodus limae* n. g., n. sp., from the Lower Cretaceous of Chapada do Araripe (N-E Brazil). Anais da Academia Brasileira de Ciências, 61: 53–57.

Brito, P. M. & D. M. Martill. 1999. Discovery of a juvenile coelacanth in the Lower Cretaceous, Crato Formation, Northeastern Brazil. Cybium, 23: 209–211.

Brito, P. M. & Y. Yabumoto. 2011. An updated review of the fish faunas from the Crato and Santana Formations in Brazil, a close relationship to the Tethys fauna. Bulletin of the Kitakyushu Museum of Natural History and Human History, Series A, 9: 107–136.

Brito, P. M., R. M. Lindoso, I. S. Carvalho, & G. de Paula Machado. 2016. Discovery of †Obaichthyidae gars (Holostei, Ginglymodi, Lepisosteiformes) in the Aptian Codó Formation of the Parnaíba Basin: Remarks on paleobiogeographical and temporal range. Cretaceous Research, 59: 10–17. DOI: 10.1016/j.cretres.2015.10.017

Campos, D. de A. & S. Wenz. 1982. Paléontologie – Premiére découverte de Coelacanthes dans le Crétacé inférieur de la Chapada do Araripe (Brésil). Comptes Rendus de l'Academie des Sciences Serie II, 294: 1151–1154.

Carvalho, M. S. S. & V. Gallo. 2002. The presence of *Ptychodus* (Chondrichthyes, Hybodontoidea) in the Cotinguiba Formation, Upper Cretaceous of the Sergipe-Alagoas Basin, Northeastern Brazil. – Pp. 307–309 in: VI Simpósio Sobre o Cretáceo Do Brasil; II Simposio Sobre El Cretácico de America Del Sur. Boletim dos Resumos, SBP, São Pedro.

Carvalho, M. S. S., V. Gallo & H. R. S. Santos. 2013. New species of coelacanth fish from the Lower Cretaceous (Albian) of the Grajaú Basin, NE Brazil. Cretaceous Research, 46: 80–89.

Cavin, L. 1995. *Goulmimichthys arambourgi* n. g., n. sp., un Pachyrhizodontidae (Actinopterygii, Teleostei) d'une nouvelle localité à nodules fossiliféres du Turonien inférieur marocain. Comptes Rendus de l'Academie des Sciences Serie II, 321: 1049–1054.

Chang, H. K., R. O. Kowsmann, A. M. F. Figueiredo & A. Bender. 1992. Tectonics and stratigraphy of the East Brazil Rift system: an overview. Tectonophysics, 213: 97–138. DOI: 10.1016/0040-1951(92)90253-3

Coelho, P. M. 2004. Revisão sistemática dos Enchodontidae (Euteleostei: Aulopiformes) do Brasil. M.S. Thesis, Museu Nacional/Universidade Federal do Rio de Janeiro, Rio de Janeiro, Brasil. 83 pp.

D'Erasmo, G. 1938. Ittilioti Cretacei del Brasile. Atti della Reale Accademiedelle Scienze fisiche e Matematiche. Napoli, ser. 3, 1: 1–44.

De Figueiredo, F. J. 2009. A new clupeiform fish from the Lower Cretaceous (Barremian) of Sergipe-Alagoas Basin, northeastern Brazil. Journal of Vertebrate Paleontology, 29: 993–1005. DOI: 10.1671/039.029.0402

Duarte, L. & R. S. S. Santos. 1962. Fósseis do Arenito Açu. Anais da Academia Brasileira de Ciências, 34: 57–68.

Dunkle, D. H. 1940. The cranial osteology of *Notelops brama* (Agassiz) an Elopid Fish from the Cretraceous of Brazil. Lloydia, 3: 157–190.

Figueiredo, J. F. & V. Gallo. 2006. A new decertid fish (Neoteleostei: Aulopiformes) from the Turonian of the Pelotas basin, Southern Brazil. Palaeontology,49: 445–456.

Figueiredo, F. J., V. Gallo & P. M. Coelho. 2001. First occurrence of *Rharbichthys* (Teleostei: Enchodontidae) in the Upper Cretaceous of Pelotas Basin (Atlântida Formation), Southern Brazil. Boletim do Museu Nacional. Nova serie geologia, 61: 1–8.

Gallo, V. 2015. Estado da arte sobre a ictiofauna mesozoica brasileira. Ciência e Cultura, 30–32.

Gallo, V. & P. M. Brito. 2004. An overview of Brazilian semionotids. – Pp. 253–264 in: Arratia, G. & A. Tintori (eds.), Mesozoic Fishes, 3 – Systematies, Paleoenvironments and Biodiversity. Verlag Dr. Friederich Pfeil, München.

Gallo, V. & P. M. Coelho. 2008. First occurrence of an aulopiform fish in the Barremian of the Sergipe-Alagoas Basin, northeastern Brazil. – Pp. 351–371 in: Arratia, G., H.-P. Schultze & M. V. H. Wilson (eds.), Mesozoic Fishes 4 – Homology and Phylogeny. Verlag Dr. Friederich Pfeil, München.

Gallo, V., J. F. Figueiredo & P. M. Coelho. 2006. Paleoictiofauna da Formação Atlântida, Cretáceo Superior da Bacia de Pelotas, sul do Brasil. – Pp. 113–131 in: Gallo, V., P. M. Brito, H. M. A. Silva & J. F. Figueiredo (eds.), Paleontologia de Vertebrados: Grandes Temas e Contribuições Científicas. Interciência, Rio de Janeiro.

Gallo, V., H. M. Silva A. & E. J. Andrade. 2007. New fish records from the Turonian of the Sergipe Basin, Northeastern Brazil. Archivos do Museu Nacional do Rio de Janeiro, 65: 385–393.

Gallo, V., M. S. S. de Carvalho & A. A. Souto. 2009. A possible occurrence of Diodontidae (Teleostei, Tetraodontiformes) in the Upper Cretaceous of the Paraíba Basin, northeastern Brazil. Cretaceous Research, 30: 599–604. DOI: 10.1016/J. CRETRES.2008.12.001

Gallo, V., F. J. Figueiredo & M. S. S. Carvalho. 2012. Síntese da paleoictiofauna marinha das bacias costeiras do Brasil e comentários sobre a paleoictiofauna cronocorrelata da margem oeste da África. – Pp. 175–205 in: Gallo, V., H. M. A. Silva, P. Brito & F. Figueiredo (eds.), Paleontologia de Vertebrados: Relações Entre América Do Sul e África. Interciência, Rio de Janeiro.

Gallo-da-Silva, V. & F. J. Figueiredo. 1999. *Pelotius hesselae* gen. et sp. nov. (Teleostei: Holocentridae) from the Cretaceous (Turonian) of Pelotas Basin, Brazil. Journal of Vertebrate Paleontology, 19: 263–270.

Jordan, D. S. 1919. New genera of fossil fishes from Brazil. Proceedings of the Academy of Natural Sciences of Philadelphia, 71: 208–210.

Jordan, D. S. 1921. Peixes cretáceos do Ceará e Piauhy. Monographias do Serviço Geológico e Mineralógico, 3: 1–97.

Jordan, D. S. & J. C. Branner. 1908. The Cretaceous fishes of Ceará, Brazil. Smithsonian Miscellaneous Collections, 52: 1–29.

Koutsoukos, E. A. M. 1992. Late Aptian to Maastrichtian foraminiferal biogeography and palaeoceanography of the Sergipe basin, Brazil. Palaeogeography, Palaeoclimatology, Palaeoecology, 92: 295–324. DOI: 10.1016/0031-0182(92)90089-N

Lima, M. R. 1982. Palinologia da Formação Codó na região de Codó. Boletim Instituto de Geociências – Universidade de São Paulo, 13: 116–128.

Lindoso, R. M., I. D. S. Carvalho, M. A. Medeiros, A. Pereira, R. A. B. Santos, I .D. Mendes, J. M. Brito, I. Vilas-Bôas, M. N. Araújo & N. N. Ferreira. 2011. Novos sítios fossilíferos em carbonatos da Formação Codó (Aptiano/Albiano) da Bacia do Parnaíba, Maranhão, Brasil. – Pp. 819–827 in: Carvalho, I. de S., N. K. Srivastava, O. Strohschoen, Jr. & C. C. Lana (eds.), Paleontologia: Cenários de Vida. Editora Interciência, Rio de Janeiro.

Lindoso, R. M., I. de S. Carvalho & I. D. Mendes. 2013. An isopod from the Codó Formation (Aptian of the Parnaíba Basin), Northeastern Brazil. Brazilian Journal of Geology, 43: 16–21.

Lindoso, R. M., J. G. Maisey & I. de S. Carvalho. 2016. Ichthyofauna from the Codó Formation, Lower Cretaceous (Aptian, Parnaíba Basin), Northeastern Brazil and their paleobiogeographical and paleoecological significance. Palaeogeography, Palaeoclimatology, Palaeoecology, 447: 53–64. DOI: 10.1016/j. palaeo.2016.01.045

Löfgren, A. & P. E. Oliveira. 1943. Fósseis cretáceos de Aracaju, Sergipe (Sondagem em Ponta da Atalaia). Boletim da Divisao de Geologia e Mineralogia, 106: 1–54.

Machado, L. P. & P. M. Brito. 2006. The new genus *Potiguara* (Actinopterygii: Pycnodontiformes) from the Upper Cretaceous of Northeast Brazil. Journal of Vertebrate Paleontology, 26: 1–6.

Maffizzoni, A. F. 2000. Paleoictiofauna da Formação Coqueiro Seco, Aptiano Inferior do estado de Alagoas, Nordeste do Brasil. Programa de Pós-Graduação em Geociências, Universidade Federal do Rio Grande do Sul, Tese de Doutorado, 149 pp.

Maisey, J. G. 1986. Coelacanths from the Lower Cretaceous of Brazil. American Museum Novitates, 2866: 1–30.

Maisey, J. G. 1991. Santana Fossils: An Illustrated Atlas. Neptune, NJ, TFH Publications.

Maisey, J. G. 2000. Continental break up and the distribution of fishes of Western Gondwana during the Early Cretaceous. Cretaceous Research, 21: 281–314.

Maisey, J. G. 2011. Northeastern Brazil: out of Africa? – Pp. 545–559 in: Carvalho, I. de S., N. K. Srivastava, O. Strohschoen, Jr. & C. C. Lana (eds.), Paleontologia: Cenários de Vida. Interciência, Rio de Janeiro.

Maisey, J. G. & J. M. Moody. 2001. A review of the problematic extinct teleost fish *Araripichthys*, with a description of a new species from the Lower Cretaceous of Venezuela. American Museum Novitates, 3324: 1–27.

Martill, D. & P. M. Brito. 2017. The Santana Formation. – Pp. 215–256 in: Fraser, N. C. & H.-D. Sues. (eds.), Terrestrial Conservation Lagerstätten. Dunedin, Edinburgh.

Medeiros, M. A., R. M. Lindoso, I. D. Mendes & I. de S. Carvalho. 2014. The Cretaceous (Cenomanian) continental record of the Laje do Coringa flagstone (Alcântara Formation), northeastern South America. Journal of South American Earth Sciences, 53: 50–58. DOI: 10.1016/j.jsames.2014.04.002

Miller, C. G., M. Richter & D. A. Carmo. 2002. Fish and ostracod remains from the Santos Basin (Cretaceous to Recent), Brazil. Geological Journal, 37: 297–316.

Moody, J. M. & J. G. Maisey. 1994. New Cretaceous marine vertebrate assemblages from north-western Venezuela and their significance. Journal of Vertebrate Paleontology, 14: 1–8.

Nobre, G. S. 1978. João da Silva Feijó: um naturalista no Ceará. Gráfica Editorial Cearense Ltda, 255 pp.

Parméra, T. de C. C. 2016. Reconstrução biogeográfica do Neocretáceo com base nos padrões de distribuição da paleoictiofauna. Universidade do Estado do Rio de Janeiro.

Rebouças, J. C. & R. da Silva-Santos. 1956. Fauna ictiológica do Fosfato de Pernambuco. Boletim da Divisao de Geologia e Mineralogia, 162: 1–29.

Santos, M. E. C. M. & M. S. S. Carvalho. 2009. Paleontologia das bacias Parnaíba, Grajaú e São Luís, 1st ed. CPRM, Rio de Janeiro, Brasil.

Schultze, H. P. & D. Stöhr. 1996. *Vinctifer* (Pisces, Aspidorhynchidae) aus de Unterkreide (oberes Aptium) von Kolumbien. Neues Jahrbuch für Geologie und Paläontologie – Abhandlungen, 3: 395–415.

Silva-Santos, R. da. 1945. Revalidação de *Aspidorhynchus comptoni* Agassiz, do Cretáceo do Ceará, Brasil. Notas Preliminares e Estudos, Divisão de Geologia e Mineralogia, DNPM, 29: 1–12.

Silva-Santos, R. da. 1947. Uma redescricão de *Dastilbe elongatus*, com algumas considerações sobre o gênero *Dastilbe*. Divisão de Geologia e Mineralogia, Notas preliminares e Estudos, 42: 1–7.

Silva-Santos, R. da. 1950. *Anaedopogon*, *Chyromystus* e *Ennelichthys* como sinônimos de *Cladocyclus* da família Chirocentridae. Anais da Academia Brasileira de Ciências, 22: 123–134.

Silva-Santos, R. da. 1958. *Leptolepis diasii*, novo peixe fóssil da Serra do Araripe, Brasil. Notas Preliminares e Estudos, Divisão de Geologia e Mineralogia, Departamento Nacional de Produção Mineral, 108: 1–15.

Silva-Santos, R. da. 1960. A posição sistemática de *Enneles audax* Jordan e Branner da Chapada do Araripe, Brasil. Monogr. do Serviço Geol. e Mineral. 17: 1–25.

Silva-Santos, R. da. 1985a. Nova caracterização do gênero *Vinctifer*, Jordan 1919. Pp. 151–154 in: Coletânea de trabalhos paleontológicos. Departamento Nacional de Produção Mineral, Brasília.

Silva-Santos, R. da. 1985b. *Araripichthys castilhoi* novo gênero e espécie de Teleostei da Formação Santana, Chapada do Araripe, Brasil. Pp. 133–139 in: Coletânea de trabalhos paleontológicos. Departamento Nacional de Produção Mineral, Brasília.

Silva-Santos, R. da. 1990. Clupeiformes e Gonorhynchiformes do Cretáceo Inferior (Aptiano) da Formação Cabo, Nordeste do Brasil. Anais da Academia Brasileira de Ciências, 62: 261–268.

Silva-Santos, R. da. 1991. Paleogeography of the Araripe Basin. – Pp. 40–43 in: Maisey, J. G. (ed.), Santana Fossils: An Illustrated Atlas. Neptune, NJ, TFH Publications.

Silva-Santos, R. da. 1994. Ictiofáunula da Formação Codó, Cretáceo Inferior, com a descrição de um novo táxon – *Codoichthys carnavalii* (Pisces-Teleostei). Anais da Academia Brasileira de Ciências, 66: 131–143.

Silva-Santos, R. da & J. F. Figueiredo. 1987. Sobre um Saurocephalidae da Formação Gramame (camada de Fosfato), Estado de Pernambuco. – Pp. 7–19 in: X Congresso Brasileiro de Paleontologia. Anais, SBP, Rio de Janeiro.

Silva-Santos, R. da & A. R. Oliveira. 1994. *Rhacolepis buccalis* Agassiz, 1841 (Pisces, Elopomorpha), no Cretáceo Inferior da Colômbia, AS. Anais da Academia Brasileira de Ciências, 66: 167–171.

Silva-Santos, R. da & M. S. Salgado. 1969. *Enchodus longipectoralis* (Schaeffer), um Teleostei do Cretáceo de Sergipe. Anais da Academia Brasileira de Ciências, 41: 381–392.

Silva-Santos, R. da & J. G. Valença. 1968. A Formação Santana e sua Paleoictiofauna. Anais da Academia Brasileira de Ciências, 40: 339–360.

Silva-Santos, R. S. 1963. Peixes do Cretáceo do Rio Grande do Norte. Anais da Academia Brasileira de Ciências, 35: 227–238.

Silva, M. C. 2007. Os vertebrados da Bacia do Parnaíba (Cretáceo Superior-Paleoceno), Nordeste do Brasil. M.Sc dissertation, Universidade Federal de Pernambuco, 150 pp.

Soares, U. M., D. F. Rossetti & R. C. T. Cassab. 2003. Bacias sedimentares brasileiras: Bacia Potiguar. Fundação Paleontológica Phoenix, 56: 1–4.

Souza-Lima, W. 2001. Macrofaunas campanianas e ambientes deposicionais da Formação Calumbi, bacia de Sergipe-Alagoas, Brasil. PhD dissertation, Universidade Federal do Rio de Janeiro, 326 pp.

Spix, J. B. & C. F. Martius. 1823. Reise in Brasilien in 1817 bis 1820 gemacht und beschrieben. Gedruckt bei M. Lindauer, Munchen, 1823–1831, 3 vol., 1 atlas.

Wenz, S. 1989. *Iemanja palma* n. g., n. sp., gyrodontidae nouveau (Pisces: Actinopterygii) du Crétacé inférieur de la Chapada do Araripe. (nord est du Brésil). Comptes Rendus des Séances l'Académie des Sciences Paris, 308: 975–980.

Wenz, S. & A. W. A. Kellner. 1986. Découverte du premier Ionoscopidae (Pisces: Halecomorphi) sudaméricain, *Oshunia brevis* n. g., n. sp., dans le Crétacé inférieur de la Chapada do Araripe (nord-est du Brésil). Bulletin du Muséum national d'Histoire naturelle. Paris, 8: 77–88.

Woodward, A. S. 1908. Cretaceous fishes from Brazil. Quarterly Journal of the Geological Society of London, 64: 358–362.

Woodward, A. S. 1907. I. – Notes on some upper cretaceous fish-remains from the provinces of Sergipe and Pernambuco, Brazil. Geological Magazine, 4: 193–197. DOI: 10.1017/S0016756800133308

Woodward, A. S. 1901. Catalogue of fossil fishes in the British Museum of Natural History, Part 4. London.

Ancient Fishes and their Living Relatives: a Tribute to John G. Maisey. Alan Pradel, John S. S. Denton & Philippe Janvier (eds.): pp. 239–245, 7 figs.
© 2021 by Verlag Dr. Friedrich Pfeil, München, Germany – ISBN 978-3-89937-269-4

Fossil chondrichthyan remains from the Middle Devonian Kevington Creek Formation, South Blue Range, Victoria

John A. Long*,**, Victoria Thomson*,***, Carole J. Burrow**** and Susan Turner****

Abstract

Chondrichthyan teeth are described from a new site located in the upper section of the Middle Devonian Kevington Creek Formation, NW Victoria. Our review of the genus *Mcmurdodus* White, 1968 indicates that the Australian teeth previously assigned to this genus, which have up to eight large asymmetric cusps, belong in a separate genus from the type species, which has a central row of median cusps separating three larger lateral cusps. We refer the Australian species *M. whitei* Turner & Young, 1987 to *Maiseyodus* new genus, and describe a new occurrence of this species from the South Blue Range assemblage. The co-occurrence supports an Eifelian age for the South Blue Range unit, as *M. whitei* is otherwise known only from the late Early/early Middle Devonian Cravens Peak Beds of western Queensland (Emsian–Eifelian boundary). The younger occurrence of the type of the genus *Mcmurdodus* in the Aztec Siltstone of Antarctica (Eifelian–Givetian) and the presence of *Aztecodus* sp. cf. *A. harmsenae* and *Antarctilamna prisca* in the Kevington Creek Formation fauna would also support an older age than previously estimated for the South Blue Range Devonian succession. Revised U-Pb dates on the nearby Toombullup Ignimbrites supports an older age for the fish-bearing sequence at the base of the Mansfield Basin sequence. Biostratigraphic correlation of key chondrichthyan species found in the Aztec Siltstone of Antarctica and the Cravens Peak Beds limestone member suggests that the South Blue Range succession is much older than previously placed. It is likely a correlate of the lower part of the Aztec Siltstone succession.

Introduction

Fossil fishes were first discovered from the South Blue Range (SBR), near Mansfield (Fig. 1), in the early 1930s and described by Hills (1936). Hills identified *Bothriolepis* and *Phyllolepis*, placoderm fish taxa both known from northern hemisphere sites in the USA, Europe and Russia, where they characterise Late Devonian deposits. In earlier work at Taggerty, Hills (1929, 1932) identified these taxa in Australia for the first time and suggested that a Late Devonian age could be placed on the Cerberean Cauldron volcanic sequence. Long & Werdelin (1986) proposed lithostratigraphic and biostratigraphic correlations using Devonian fish faunas throughout the central Victorian sites, and suggested an older, Givetian (upper Middle Devonian) age for the SBR fauna based on its relationship with key volcanic events and the composition of the fish fauna.

A large number of sites in both Australia and Antarctica contain the presumed freshwater fish fauna typical of the northern hemisphere Late Devonian (with key placoderm genera *Bothriolepis*, *Groenlandaspis*, and phyllolepids) in much older, well-dated sites (Young 1974, 1989a, 2010a, 2010b). We now know that *Bothriolepis* ranges from the Early Devonian in China (Zhao & Zhu 2010) and earliest Middle Devonian in Australia and Antarctica (Young 1974, 2010a), but is restricted to the Late Devonian in Europe, Russia, Greenland and North America (Young 1974, Stensiö 1948). The genus *Groenlandaspis* is restricted to the latest Devonian (Famennian stage) in Europe, South Africa and North America, but is established as occurring in the Eifelian and Givetian of the Australian and Antarctic sequences.

The genus *Phyllolepis* is also tied to latest Devonian strata in the northern hemisphere (Famennian, USA, UK, France, USSR, Long & Daeschler 2013) with several endemic genera of phyllolepids having been described from Australia and Antarctica (Long 1984, Ritchie 1984, 2005, Young & Long 2005, Young 2007) in Middle–Late Devonian strata, and an unnamed phyllolepid from the Frasnian of Turkey (Janvier 1983). Tying down the age of terrestrial sedimentary sequences using placoderm faunas relies on having a well-preserved, diverse fauna where taxa can be accurately identified down to species level. The

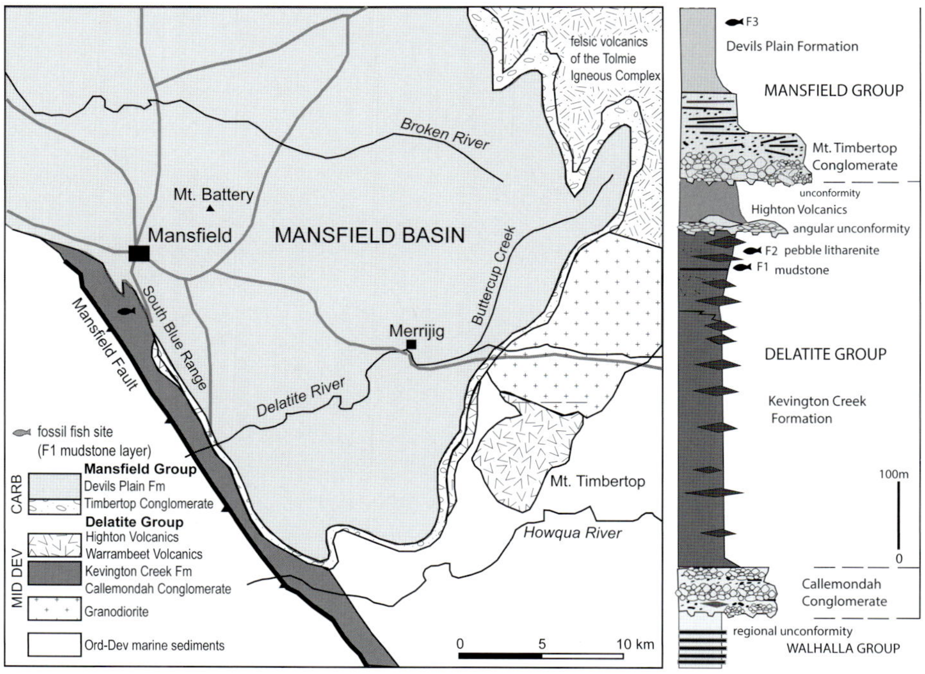

Fig. 1. Map showing geology and stratigraphic column for Mansfield region (modified after Cas et al., 2003).

* College of Science and Engineering, Flinders University, GPO Box 2100, Adelaide, South Australia 5001; john.long@flinders.edu.au (corresponding author)

** Geosciences, Museum Victoria, PO Box 666, Melbourne, Victoria 3001

*** Environment Futures Research Institute, School of Environment and Science, Griffith University, Nathan Campus, Queensland, Australia 4111

**** Geosciences, Queensland Museum, 122 Gerler Rd, Hendra, Queensland 4011 relatively few specimens of placoderms so far described from the SBR mudstone lens are mostly placed in indeterminate species in cosmopolitan genera (*Bothriolepis* sp., *Groenlandaspis* sp.) apart from the phyllolepid *Austrophyllolepis edwini* (Long 1989).

More recently, use of shark teeth as key biostratigraphic indicators of precise ages within the Devonian strata has greatly increased (e.g. Turner 1993, Ginter & Ivanov 1992, Ginter 2002, Ginter et al. 2008, 2010, Burrow et al. 2010). The new discovery of fossilised shark teeth within the SBR succession in a thin mudstone lens in the middle section of the Kevington Creek Formation enables more precise dating of the succession, as inferred from the presence of key species known from sections in other regions that represent reasonably well-constrained biostratigraphic frameworks, in particular the Aztec Siltstone of Antarctica and Cravens Peak Beds limestone of the Georgina Basin in Queensland.

Fig. 2. Mcmurdodontid teeth. **A, B.** *Mcmurdodus featherensis* White, 1968. **A.** Photo of Holotype NMHUK PV P.49157 (from Ginter et al., 2010, fig. 97E); **B.** interpretation of restored tooth based on a camera lucida sketech of the Holotype by JAL. **C–G.** *Maiseyodus* gen. nov. sp. cf. *M. whitei*. South Blue Range specimens, MV P228244: **C.** in labial view; **D.** in ventrolabial view; **E.** in lingual view; **F.** crown view; **G.** in ventral view. **H–J.** *Maiseyodus whitei* new genus (Turner & Young 1987), Craven Peak beds limestone (Burrow et al. 2008, fig. 1D,B,A respectively).

This discovery is the first time that any chondrichthyan teeth have been identified from the continental Middle–Late Devonian sedimentary successions anywhere within Victoria. Older stem chondrichthyan remains were reported from the Silverband Formation in the Grampians of western Victoria by Chapman (1917) and Talent & Spencer-Jones (1963), and the fauna extensively revised by Turner (1986) and Burrow (2003) with thelodont scales and stem chondrichthyan (acanthodian) scales and fin spines identified. The age of this fauna remains unclear, due to dating constraints with nearby volcanics. Burrow (2003) suggested a Late Silurian (Ludlow) age based on the presence of *Sinacanthus* and *Radioporacanthodes*.

Garvey & Turner (2006) described shark teeth and scales from the Mansfield district, from the slightly younger Devils Plain Formation, comprising teeth of *Ageleodus altus* and scales of *Tristychius* sp. and *Ctenacanthus* sp., representing genera known from the very latest Devonian (upper Famennian) and Early Carboniferous beds of the northern hemisphere (fish fauna F3, Fig. 1).

Material and methods

The new material described herein is housed at the Melbourne Museum (Museum Victoria) in the Geosciences collections. Specimens were collected in situ and as the bone was not well preserved, the rocks were soaked in solutions of dilute HCL (10 %) to clean the surfaces of weathered bone fragments. The cleaned rock surfaces were cast with blackened latex rubber; casts were then photographed with an Olympus Microscope U-TVO.5XC-3 attachment after being whitened with ammonium chloride sublimate.

Specimens were collected by the lead author and colleagues from a new fossil locality occurring within the top 50 m of the Kevington Creek Formation in the northern end of the South Blue Range, near Mansfield, Victoria (Fig. 1). The new site was uncovered during a CAVEPS (Conference on Australian Vertebrate Evolution, Palaeontology and Systematics) field trip in April 2007. It comprises a thin layer (c. 10 cm thick, Fig. 1, F1) of dark grey silty mudstone that outcrops about 40–50 m below the highest point of the range in its northern end, almost directly opposite the intersection of the two major roads along the western side of the South Blue Range (Fig. 1). The mudstone layer is bracketed by medium-grained buff sandstones above and below it, grading upwards into intraclast conglomerate lenses and pebble litharenite layers towards the top of the range where impressions of placoderm fish plates are also found (Fig. 1, F2). Long (1989) collected and described the remains of the phyllolepid *Austrophyllolepis edwini* from this locality above the mudstone layer.

Institutional Abbreviations. CPC, Commonwealth Palaeontological Collections, Geoscience Australia, Fyshwick, ACT. MV P, Melbourne Museum (Museum Victoria); NHMUK PV, Natural History Museum, London, UK; QM, Queensland Museum, Brisbane.

Systematic palaeontology

Class Chondrichthyes Huxley, 1880
Order indet.
Family Mcmurdodontidae White, 1968

Revised diagnosis. Chondrichthyan teeth with up to seven large asymmetric flat weakly serrated cusps forming each side of the crown, with or without series of central medial cusps. Root is broad and subrectangular with distinct medial basal canal through raised tube in midline of ventral surface of the root. The base is angled lingually away from the crown, being subequal in size to the crown, and sloping down towards the lingual edge.

Mcmurdodus White, 1968

Revised diagnosis. Selachian teeth up to 8 mm long and 5 mm high, multicuspid asymmetrical crown strongly compressed labiolingually. Lateral cusps are triangular in shape, sometimes elliptical in section, with upper cutting edges weakly to strongly serrated, maximum number of cusps is seven for each side, with third cusp in from each side being the largest on the crown; central row of at least four smaller narrow cusps. The cusp axes are progressively more inclined from the vertical so that the distal cusp axis is almost parallel to the crown root junction. There is a prominent medial groove on the basal surface with a large foramen opening at the labial end.

Type species. *Mcmurdodus featherensis*, NHMUK PV P49157, Mt. Feather, southern outcrop of Aztec Siltstone, South Victoria land, Antarctica (cf. Eifelian–Givetian age range).

Maiseyodus, new genus

Diagnosis. A mcmurdodontid with teeth having 8–10 major cusps, ranging between 4–6 major cusps on each side, increasing rapidly in size in the first three cusps from the edge of the crown with the fourth or fifth cusp being sometimes slightly smaller than the third cusp from the edge. Serrations either absent or developed on the occlusal cutting edge between the principal (medial) and second largest cusps on each side. A series of ca. 8–10 foramina on the lingual surface within a pitted triangular area above the crown-base interface. Holaulacorhize base, with D shaped profile in ventral view.

Type species. *Mcmurdodus whitei* Turner & Young, 1987, p. 235

Holotype. CPC 25765, Cravens Peak Beds (limestone unit), Georgina Basin, Queensland (upper Emsian), Turner & Young (1987, fig. 2), a relatively complete tooth although many of the cusps have their tips broken off.

Etymology. In honour of Dr. John Maisey for his outstanding contributions to our understanding of the evolution and anatomy of sharks.

Remarks. The diagnosis is modified from Turner & Young (1987, p. 235) to distinguish cusp morphology from *M. featherensis*, which appears to have 12–14 cusps as evidenced by broken cusps clearly visible on the holotype (Fig. 2A, B), and with smaller cusps medially located. *M. whitei* has the largest cusps developed towards the central region of the crown (Turner & Young 1987, figs. 2, 3A). The known specimens of *M. whitei* suggest that *Maiseyodus* had a dentition somewhat similar to that of the hexanchoids (Turner & Young 1987), though the teeth bear a stronger resemblance to those of the sister group of the hexanchoids, the echinorhinoids (Burrow et al. 2008). Whereas hexanchoids have different dentitions on the upper and lower jaws (Herman et al. 1987), *Echinorhinus* has similar teeth on both jaws (Herman et al. 2003). The lack of marked variation in the known examples of *M. whitei* indicates that its dentition closely resembles that of *Echinorhinus*, particularly *E. cookei*. By comparison, the difference in cusp length is not due to differentiation within the dentition. Extant hexanchoid sharks have little variation between adult and juvenile dentitions, with possible lack of serration found in juveniles (Springer & Waller 1969; Herman et al. 1987, 1994, 2003). However, extant *Echinorhinus* spp. show variation in the number of cusps between adults and juveniles, with adults having a larger number of cusps (Garrick 1960).

Maiseyodus whitei specimens from the Cravens Peak Beds (Turner & Young 1987) have been completely removed from limestone by acetic acid digestion, whereas the two specimens described here from the SBR were studied from latex peels of the natural moulds.

Maiseyodus sp. cf. *M. whitei*

(Figs. 2C–G, 3A, C)

Description. MV P228244, preserved in part and counterpart moulds (Fig. 2C, D, F, 3A, C) from the SBR, has a width of 6 mm, with crown height of 2.5 mm. It has 3–5 coarse, rounded serrations well developed between the central, and largest, three cusps, but lacking on the median cusp oral edge, a feature distinguishing it from all known examples of *Maiseyodus whitei* (Fig. 3B). MV P254493 (Fig. 2E, G) is a second complete tooth from the same locality showing similar morphology although the crown-root line is more strongly convex than in the P228244. It also shows a similar distribution of the serrations with a non-serrate median oral edge between the main cusps.

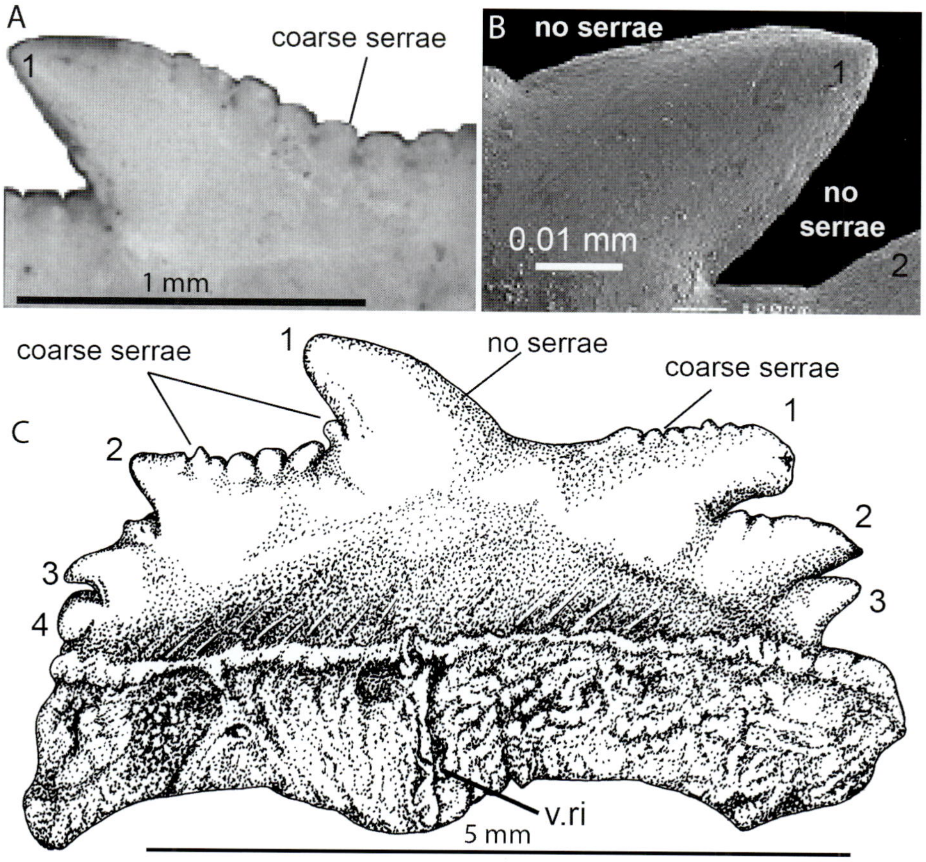

Fig. 3. Comparison of *Maiseyodus* new genus specimens from South Blue Range. **A, C.** MV P 228244), **B.** Cravens Peak Beds limestone, specimen QMF 52817 (Burrow et al., 2008, fig. 1C).

Comparison. *Maiseyodus* sp. cf. *M. whitei* from the SBR resembles *M. whitei* from the Cravens Peak Beds (limestone) in overall crown shape, having multiple broad-based cusps, and a flat base with a central labio-lingual canal. However, unlike *Maiseyodus whitei* with cusp 1 and 3 on both sides of the crown being larger than cusp 2 (Turner & Young 1987), *M.* sp. cf. *M. whitei* from the SBR shows a decrease in size outwards from cusp 1 on both sides of the crown, like in hexanchids *Eonotidanus*, *Notorynchus*, and *Weltonia* (Ginter et al. 2010; Herman et al., 1987, 1994), and *Mcmurdodus featherensis* (White 1968, fig. 1). As the teeth from the Cravens Peak Beds (and in *Echinorhinus*) show little morphological variation, it is possible that the new South Blue Range *Maiseyodus* is not conspecific with *M. whitei*, but the limited material precludes us from erecting a new species.

Family Aztecodontidae Hairapetian, Ginter & Yazdi, 2008
Aztecodus Long & Young, 1995
Aztecodus harmsenae Long & Young, 1995

Aztecodus sp. cf. *A. harmsenae*

(Fig. 4A–C)

Description. *Aztecodus* sp. tooth NMV P254494 is 5 mm in width with a crown height of 3 mm, with a partially missing base. Both labial and lingual crown surfaces appear smooth with no external striae. The two cusps are widely separated and twisted in different planes. The ridge between the two cusps appears smooth, although some of it is obscured by matrix (Fig. 4A, B). The anterior and posterior edges of the tooth are perpendicular to its long axis. The base is relatively shallow, and only slightly wider than the crown labio-lingually (Fig. 4A, C). The basal surface has labio-lingually oriented ridges and furrows.

Comparison. The single tooth from the SBR succession is only half the size of the type specimens of *A. harmsenae* from the Aztec Siltstone, which are up to 2 cm wide (Long & Young, 1995, figs. 8, 9). The lack of crenulations on the connecting

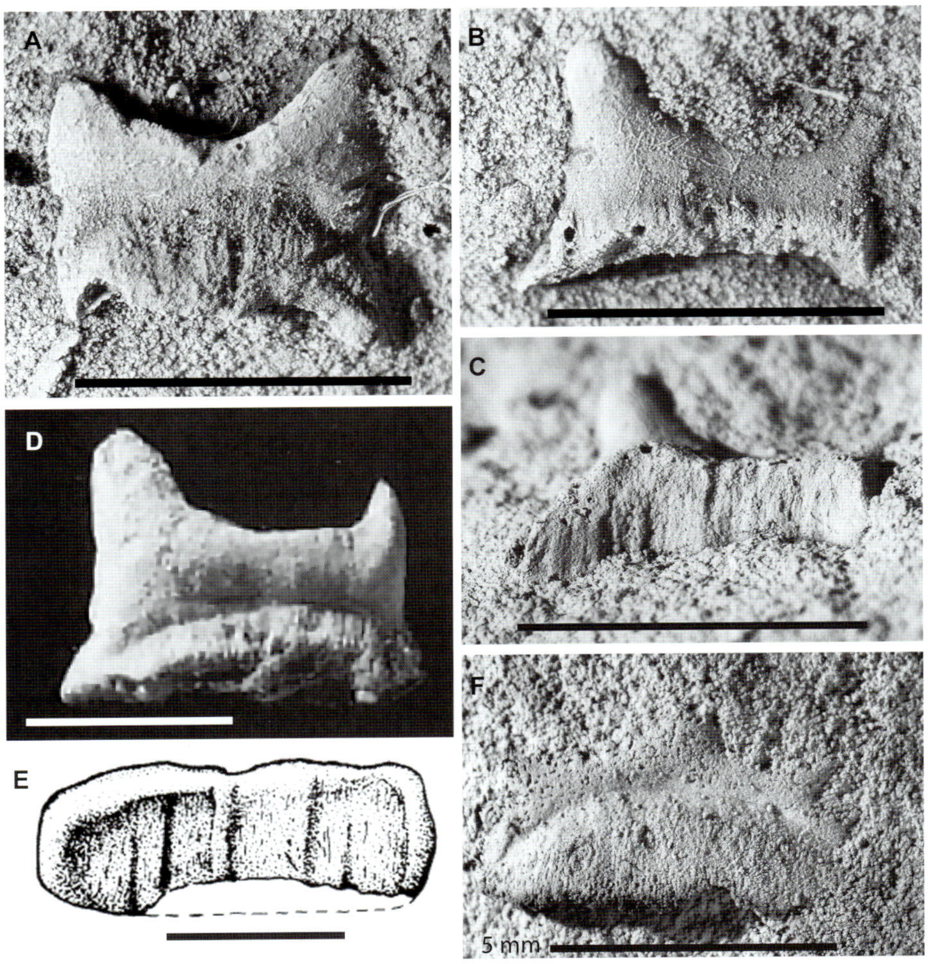

Fig. 4. **A–C.** *Aztecodus* sp. cf. *A. harmensae* tooth MV P 254494 from South Blue Range; **D, E.** *Aztecodus harmensae* teeth from the Aztec Siltstone, Antarctica (Long & Young, 1995, figs. 8F, 9B respectively). **F.** *Maiseyodus* new genus sp. cf. *M. whitei*. MV P254493 in lingual view.

ridge between the two cusps on the SBR tooth might be related to the small size of the tooth; the smallest of the type specimens (Long & Young, 1995, fig. 8F, G) appear to have only weakly developed crenulations, compared with the larger type teeth. The orientation and wide separation of the cusps on the SBR tooth make it most similar to type tooth WAM 92.3.71 (Long & Young, 1995, fig. 8F; Fig. 4D, E). Ginter et al. (2010) suggested that *Anareodus* Long & Young, 1995 is a variation of the *Aztecodus* style tooth, as both are found together in the same layers of the Aztec Siltstone. Increased numbers of specimens of shark's teeth from JAL's recent Antarctic field trips over 2016 and 2018 summer field seasons would appear to validate this conclusion as no new specimens of *Anareodus* were collected.

Order Antarctilamniformes Ginter, Liao & Valenzuela-Ríos, 2008
Family Antarctilamnidae Ginter, Liao & Valenzuela-Ríos, 2008
Antarctilamna Young, 1982

Antarctilamna prisca Young, 1982

(Fig. 5)

Description. Two nearly complete teeth have been identified, plus some isolated cusps which will not be dealt with herein. The teeth are small (2–3 mm high) with large divergent striated cusps and a small central nub at the confluence of the cusps and in general closely resemble the type species *A. prisca* (Young 1982, Fig. 5D). The first specimen (NMV P254494, Fig. 5A, B) shows two nearly complete curving lateral cusps, each with 4–5 striae along the lingual face and broken cross-sections suggesting the presence of a median cusplet. The tooth base is broad, semicircular with a small torus well defined on the lingual surface (Fig. 5B). The base is strongly convex along the lingual margin, but incomplete on its labial margin. NMVP254495 is the most complete, showing both lateral cusps, small median cusplets and most of the extent of the tooth base (Fig. 5C). As well as teeth, two scales (one shown in Fig. 5H) have been identified, which show 'ctenacanthid' morphology comprising multiple thin odontodes oriented anteroposteriorly, forming the crown.

Comparison. The teeth conform to the species definition of *Antarctilamna prisca*, known from the Aztec Siltstone of Antarctica and Bunga Beds of New South Wales (Fig. 5D). The isolated scale(s) show the same crown morphology as those on the articulated *A. prisca* specimens from the Bunga Beds (Young 1982, fig. 4D). The same type of scale is also found on other early sharks including the Early Devonian *Doliodus* (Miller et al. 2003) as well as Late Devonian and Carboniferous ctenacanthiforms (e.g. Williams 1998).

Discussion

Range and distribution of *Antarctilamna*. According to Ginter et al. (2010), teeth belonging in the Order Antarctilamniformes range through the Middle–Upper Devonian, with the genus *Antarctilamna* also having a range span from the Middle–Upper Devonian, although this estimate was based on earlier work presenting a younger age estimate for the Aztec Siltstone (Young 1982, 1988). More recent appraisal of the strata (Young 2007), with comparisons made based on *Maiseyodus* teeth occurring in both the Cravens Peak Beds limestone (Emsian-Eifelian boundary?) and the lowest section of the Aztec Siltstone at Mt. Feather, suggest an older date for the base of the Aztec Siltstone. *Antarctilamna* is known only from its lowest levels, from sites at Portal Mt., Mt. Crean, Lashley Mt., and was recently found at Mt. Fleming (lowest horizons of the unit). The thelodont *Neoturinia antarctica* (Turner & Young, 1992) occurs at all these sites, but has not been found outside Antarctica (Turner 1997a) although similar scales are known from Australia (Long & Trinajstic 2000, Hairapetian et al. 2016).

Whilst spines assigned to *Antarctilamna* vary, with some likely to be from other taxa, teeth having *Antarctilamna* characteristics are known from fewer sites, namely the Aztec Silstone of Antarctica (Eifelian?–lower Givetian) and the Bunga Beds of NSW (?Emsian–Eifelian) in Eastern Gondwana. Gess & Coates (2014) assigned an articulated shark with similar teeth from the Famennian of South Africa to a new species *Antarctilamna ultima*. This latter occurrence marks the youngest and higest latitude record for the genus, so might be a relict survivor. Alternatively, it could be referrable to another genus based on the robust shape of the teeth (e.g. Gess & Coates 2014, fig. 7), quite unlike the very slender teeth found in the only other member of the genus, the type species *A. prisca*. These are all the known instances of teeth confidently attributed to *Antarctilamna*, although one tooth fragment from the Givetian

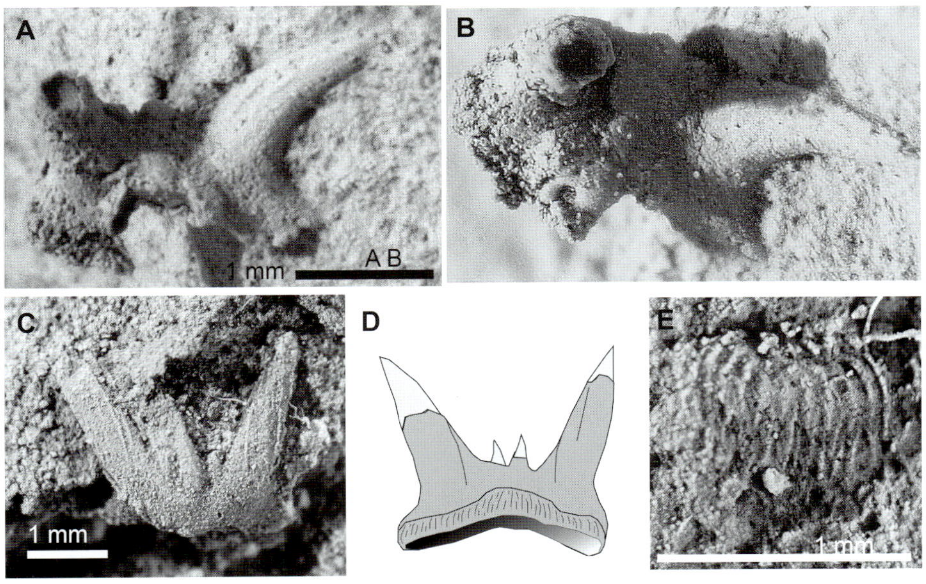

Fig. 5. *Antarctilamna prisca* Young, 1982. **A, B.** South Blue Range specimen NMV P24495: **A.** linguocrown view; **B.** crown views; **C.** NMV P224496, in labial view showing striae on lateral cusps and median cusplets. **D.** tooth sketch based on the holotype specimen CPC 21198 (after Young 1982, plate 89, 8), Aztec Siltstone, Antarctica. **E.** scale from the South Blue Range mudstone layer, NMV P256966.

of Spain has also been tentatively assigned to *Antarctilamna?* sp. (Ginter et al. 2008, fig. 2P). The only other possible occurrence of teeth from this genus is in the Givetian–Frasnian boundary beds of the North Evans Limestone, New York. Teeth of *Wellerodus priscus* (Eastman, 1899) closely resemble those of *A. prisca* (Turner, 1997b), and Ginter et al. (2008, 2010) supported Turner's hypothesis that they could be congeneric (Burrow et al. 2010). *Wellerodus* is now known from several localities in the Middle to Late Devonian of North America in marine settings (Turner 1997b, Potvin-Leduc et al. 2015).

Associated fauna at the new SBR locality. Placoderm plates and acanthodian fin spines and a cheek plate have also been collected from the mudstone lens (Fig. 6). The placoderms include *Venezuelaspis* sp. represented by an anterior median dorsal plate (Fig. 6A, B), similar to specimens formerly identified as *Pambulaspis* by Young (1988). The distinct narrow anterior margin and split median ventral ridge make this plate distinctive for the genus (Young & Moody, 2002). A large anterior median ventral plate of the arthrodire *Groenlandaspis* is also identified from the assemblage (Fig. 6C) but beyond generic identification nothing more can be noted about this large specimen. The acanthodian specimens are fin spines of *Striacanthus* sp. (Hills, 1931, Fig. 6D) and a cheek plate of the diplacanthiform *Culmacanthus* sp. (Fig. 6E). *Culmacanthus* is well-known from the Mt. Howitt fauna (Long, 1983a) and has also been described from a cheek plate from the Aztec Siltstone at Mt. Crean, Antarctica, from the Pambula River assemblage (Young 1989b), and partial articulated fish from the Bunga Beds (Burrow & Young, 2012). In addition, a number of isolated antiarch plates attributed to *Bothriolepis* sp. have been found in the assemblage, but these are beyond the scope of this paper to describe in detail.

Geological and Biostratigraphic Implications. New radiometric dating for the Cerberean Volcanics by Clemens et al. (2014) indicated older dates were unreliable. The new, reliable date placed on the upper part of the Toombullup Formation at the top of the CVC is 377 ± 2 Ma, placing the upper section of the volcanic event as middle Frasnian (IUGS 2018 chart). The sediments of the Blue Range Formation containing a diverse fish fauna lie well below the newly dated volcanics, separated by several major volcanic eruption events (Fig. 7).

The Snobs Creek Volcanics, situated below the Blue Range Formation, are one of the re-dated units with Pb/Ur radiometric dates at 373.2 ± 1.8 and 375 ± 2 Ma (Compston 2004), which would place the fish fauna at Taggerty in the Taggerty Formation well within the Frasnian, if the dates are reliable. This date would thus split the vertebrate zonation MAV 8 (Taggerty-Mt. Howitt fauna) of Young (1993) and place the Taggerty fauna within MAV 10 (Pambula River fauna) or MAV 11 (Nettletons Creek fauna) in the mid Frasnian. Whilst this might be consid-

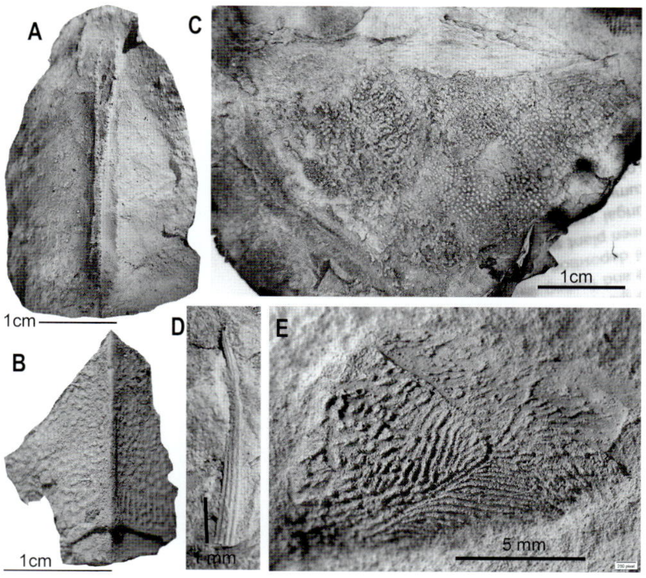

Fig. 6. Placoderm and acanthodian remains form the mudstone lens, South Blue Range, Mansfield. **A, B.** *Venezuelepis* sp. MV P254499, Anterior Median Dorsal plate in ventral (a) and dorsal view (b). **C.** *Groenlandaspis* sp. MV P228259, Anterior Median Ventral plate in ventral view; **D.** *Striacanthus* sp. MV P254497, fin spine, lateral view. **E.** *Culmacanthus* sp., left cheek plate MV P254498 in lateral view. Scale bar = 1mm.

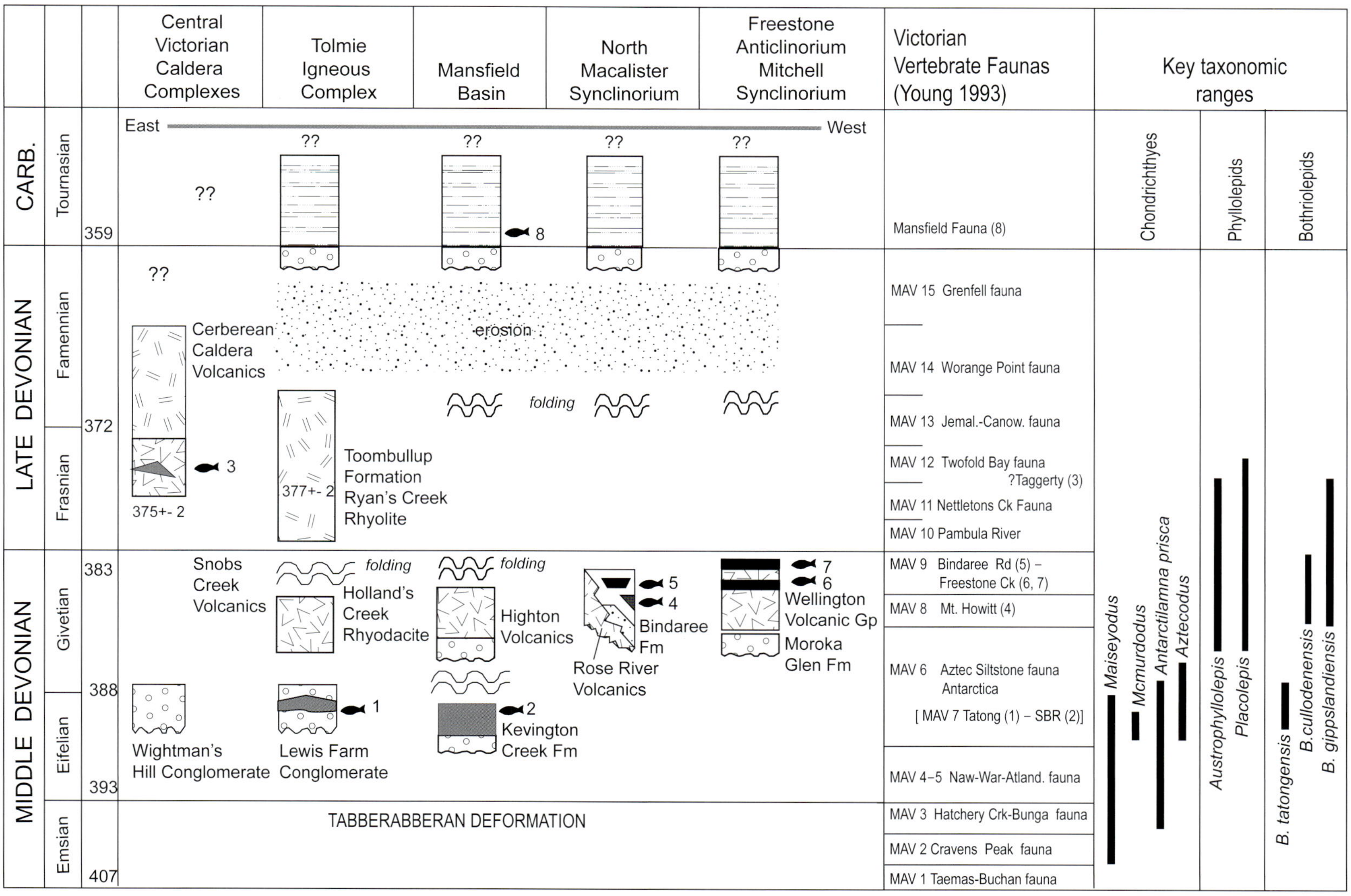

Fig. 7. Revised scheme of biostratigraphic correlations based on fish faunas for central Victoria (modified stratigraphic columns from Cas et al. (2003), with added radiometric dates for the Toombullup Formation from Compston (2004) and Clemens et al. (2014). Age ranges of key species taken from Young (1993, 2007) and Long (1983b).

ered problematic, the long age ranges of taxa here discussed (Fig. 7) would support the possibility that the Taggerty fauna is a much younger assemblage than Mt. Howitt.

Thus, the presence of faunal elements common to the SBR succession and the Aztec Siltstone of Antarctica (co-occurrence of chondrichthyan species in units bearing thelodonts) further suggests the succession at South Blue Range should be older than Late Devonian. Recent reassessment of the thelodont bearing units in southern NSW included the suggestion that the Hatchery Creek Formation fauna, containing thelodonts, placoderms and basal osteolepidids, should now be considered Emsian in age (Hunt & Young 2011). The realignment of the Hatchery Creek assemblage as upper Emsian, combined with the suggested placement of the Bunga Beds assemblage at early Eifelian (Young 2007; MAV 3 equivalent, Young 1993) has implications for assemblages containing the species *Antarctilamna prisca*, such as the Aztec Siltstone and South Blue Range faunas, indicating these are likely to be older than previously suggested. Figure 7 presents these vertebrate faunal assemblages in a re-aligned order, tied in to existing reliable radiometric dates (Compston 2004, Clemens et al. 2014). Young's (1993) vertebrate assemblage previously referred to as MAV 7 (South-Blue Range-Tatong faunas) is here correlated with the lower section of the Aztec Siltstone, because of the shared species of *Antarctilamna prisca*, *Maiseyodus* sp. cf. *M. whitei*, and *Aztecodus*. The presence of *Maiseyodus* in the much older Cravens Peak Beds limestone member (MAV2, Young 1993) suggests the SBR and Tatong assemblages are likely to be equivalent to the lowermost section of the Aztec Siltstone fauna (MAV 6, Young 1993). The notable absence of thelodonts from the SBR and Tatong assemblages might be due to differing environmental conditions, the latter assemblages being from smaller fluviatile systems than those envisaged for the Aztec Siltstone.

Acknowledgments

CJB and ST thank the Board of the Queensland Museum for basic facilities. JAL acknowledges the support of Museum Victoria, the Australian Research Council (DP0558499) and the Flinders University Strategic Professor fund. Dr Gavin Young, ANU, provided helpful discussion of the topic. We also thank Tim Zeigler, Museum Victoria, for access to collections and support whilst working at Museum Victoria, and the reviewers for their helpful comments.

Literature cited

Burrow, C. J. 2003. Redescription of the gnathostome fish fauna from the Mid-Palaeozoic Silverband Formation, the Grampians, Victoria. Alcheringa, 27: 37–49.

Burrow, C. J., D. C. Hovestadt, M. Hovestadt-Euler, S. Turner & G. C. Young. 2008. New information on the Devonian shark *Mcmurdodus*, based on material from western Queensland, Australia. Acta Geologica Polonica, 58: 151–159.

Burrow, C. J., S. Turner & G. C. Young. 2010. Middle Palaeozoic microvertebrate assemblages and biogeography of East Gondwana (Australasia, Antarctica). Palaeoworld, 19: 37–54.

Burrow, C. J. & Young, G. C. 2012. New Information on *Culmacanthus* (Acanthodii: Diplacanthiformes) from the ?Early–Middle Devonian of Southeastern Australia. Proceedings of the Linnean Society of New South Wales 134: 21–29.

Cas, R., G. O'Halloran, J. A. Long & A. Vandenberg. 2003. Middle Devonian to Carboniferous: late to post tectonic sedimentation and magmatism in an arid continental setting. In W. D. Birch (ed.) Geology of Victoria. Geological Society of Victoria, Special Publication, 23: 157–194.

Chapman, F. 1917. On the occurrence of fish remains and a Lingula in the Grampians, western Victoria. Records of the Geological Society of Victoria, 4: 83–104.

Clemens, J. D., D. Frei & F. Finger. 2014. A new precise date for the Tolmie Igneous Complex in northeastern Victoria. Australian Journal of Earth Sciences, 61: 951–958. DOI: 10.1080/08120099.2014.955058

Compston, W. 2004. SIMS U-Pb zircon ages for the Upper Devonian Snobs Creek and Cerberan Volcanics from Victoria, with age uncertainty based on UO2/UO v. UO/U precision. Journal of the Geological Society of London, 161: 223–228.

Eastman, C. R. 1899. Some new American fossil fishes. Science, 2: 642–643.

Garrick, J. 1960. Studies on New Zealand Elasmobranchii. Part X. The genus *Echinorhinus*, with an account of a second species, *E. cookei* Pietschmann, 1928, from New Zealand waters. Transactions of the Royal Society of New Zealand, 88: 105–117.

Garvey, J. M. & S. Turner. 2006. Vertebrate microremains from the presumed earliest Carboniferous of the Mansfield Basin, Victoria. Alcheringa, 30: 43–62. DOI: 10.1080/03115510608619344

Gess, R. & M. Coates. 2014. High-latitude chondrichthyans from the Late Devonian (Famennian) Witpoort formation of South Africa. Paläontologische Zeitschrift, 89: 147–169. DOI: 10.1007/s12542-014-0221-9

Ginter, M. 2002. Chondrichthyan fauna of the Frasnian-Famennian boundary beds in Poland. Acta Palaeontologica Polonica, 47: 329–338.

Ginter, M. & A. Ivanov. 1992. Devonian phoebodont shark teeth. Acta Palaeontologica Polonica, 37: 55–75.

Ginter, M. & A. Ivanov. 1995. Middle/Late Devonian phoebodont-based ichthyolith zonation. Géobios, Memoire Special 19: 351–355.

Ginter, M., J.-C. Liao & J. I. Valezuela-Ríos. 2008. New data on chondrichthyan microremains from the Givetian of the Renanué section in the Aragonian Pyrenees (Spain). Acta Geologica Polonica, 58: 165–172.

Ginter, M., O. Hampe & C. Duffin. 2010. Chondrichthyes VI. Paleozoic Elasmobranchii. Teeth. – In H.-P. Schultze (ed.) Handbook of Paleoichthyology. 168 pp, Friedrich Pfeil, Munich.

Hairapetian, V., M. Ginter & M. Yazdi. 2008. Early Frasnian sharks from central Iran. Acta Geologica Polonica, 58: 173–179.

Hairapetian, V., H. Blom & S. Turner. 2016. Early Frasnian thelodont scales from central Iran and their implications for turiniid taxonomy, systematics and distribution. Journal of Vertebrate Paleontology, 36: e1100632. DOI: 10.1080/02724634.2016.1100632.

Herman, J., M. Hovestadt-Euler & D. C. Hovestadt. 1987. Herman, J., M. Hovestadt-Euler & D. C. Hovestadt. 1987. Contributions to the comparative morphology of teeth and other relevant ichthyodorulites in living supraspecific taxa of chondrichthyan fishes. Part A: Selachii. No. 1: Order: Hexanchiformes – Family: Hexanchidae. Commissural teeth. M. Stehmann (ed.), Bulletin de l'Institut Royal des Sciences naturelles de Belgique, Biologie, 57: 43–56.

Herman, J., M. Hovestadt-Euler & D. C. and Hovestadt. 1994. Contributions to the comparative morphology of teeth and other relevant ichthyodorulites in living supraspecific taxa of chondrichthyan fishes. Part A: Selachii. No. 1: Order: Hexanchiformes – Family: Hexanchidae. Odontological results supporting the validity of *Hexanchus vitulus* Springer & Waller, 1969 as a third species of the genus *Hexanchus* Rafinesque, 1810, and suggesting intrafamilial reordering of the Hexanchidae. M. Stehmann (ed.), Bulletin de l'Institut Royal des Sciences naturelles de Belgique, Biologie, 64: 147–163.

Herman, J., M. Hovestadt-Euler & D. C. Hovestadt. 2003. Contributions to the comparative morphology of teeth and other relevant ichthyodorulites in living supraspecific taxa of chondrichthyan fishes. Part A: Selachii. Addendum to 1: Order Hexanchiformes-Family Hexachidae, 2: Order Carcharhiniformes, 2a: Family Triakidae, 2b: Family Scyliorhinidae, 2c: Family Carcharhinidae, Hemigaleidae, Leptochariidae, Sphyrnidae, Proscylliidae and Pseudotriakidae, 3: Order Squaliformes: Family Echinorhinidae, Oxynotidae and Squalidae. Tooth vascularization and phylogenetic interpretation. S. Springer & R. Waller (eds.), Bulletin de l'Institut Royal des Sciences naturelles de Belgique, Biologie, 73: 5–26.

Hills, E. S. 1929. The geology and palaeontology of the Cathedral Range and the Blue Hills, in northwestern Gippsland. Proceedings of the Royal Society of Victoria, 41: 176–201.

Hills, E. S. 1931. The Upper Devonian fishes of Victoria, Australia, and their bearing on the stratigraphy of the state. Geological Magazine, 68: 206–231.

Hills, E. S. 1932. The geology of Marysville, Victoria. Geological Magazine, 69: 145–166.

Hills, E. S. 1936. Records and descriptions of some Australian Devonian fishes. Proceedings of the Royal Society of Victoria, 48: 161–171.

Hunt, J. R. & G. C. Young. 2011. A new placoderm fish of uncertain affinity from the Early-Middle Devonian Hatchery Creek succession at Wee Jasper, New South Wales. Alcheringa, 35: 53–75. DOI: 10.1080/03115511003793645

Huxley, T. H. 1880. On the application of the laws of evolution to the arrangement of the Vertebrata and more particularly the Mammalia. Proceedings of the Zoological Society of London, 1880: 649–662.

Janvier, P. 1983. Les vertébrés dévonien de la Nappe Supérieure d'Antalya (Taurus Lycien occidentale, Turquie). Géologie Méditerranéenne, 10: 1–13.

Long, J. A. 1983a. A new diplacanthoid acanthodian from the Late Devonian of Victoria. Memoirs of the Australasian Association of Palaeontologists, 1: 49–58.

Long, J. A. 1983b. New bothriolepid fish from the Late Devonian of Victoria, Australia. Palaeontology, 26: 295–320.

Long, J. A. 1984. New phyllolepids from Victoria and the relationships of the group. Proceedings of the Linnean Society of New South Wales, 107: 263–308.

Long, J. A. 1989. Phyllolepid placoderm fishes from the Late Devonian of eastern Victoria. – Pp. 52–58 in R. Le Maitre (ed.) Pathways in Geology. E. S. Hills Memorial Volume, Blackwell Press, Melbourne.

Long, J. A. & L. Werdelin. 1986. A new Late Devonian bothriolepid (Placodermi, Antiarchia) from Victoria, with description of other species from the state. Alcheringa, 10: 355–399.

Long, J. A. & G. C. Young. 1995. New sharks from the Middle-Late Devonian Aztec Siltstone, southern Victoria Land, Antarctica. Records of the Western Australian Museum, 17: 287–308.

Long, J. A. & K. Trinajstic. 2000. Devonian microvertebrate faunas of Western Australia. Courier Forschungsinstitut Senckenberg, 223: 471–485.

Long, J. A. & E. B. Daeschler. 2013. First articulated phyllolepid placoderm from North America, with comments on phyllolepid systematics. Proceedings of the Academy of Natural Sciences of Philadelphia, 162: 33–46.

Miller, R. F., R. Cloutier & S. Turner. 2003. The oldest articulated chondrichthyan from the Early Devonian period. Nature, 425: 501–504.

Potvin-Leduc, D., R. Cloutier, E. Landing, L. V. Hernick & F. Mannolini. 2015. Givetian (Middle Devonian) sharks from Cairo, New York (USA): evidence of early cosmopolitanism. Acta Palaeontologica Polonica, 60: 183–200.

Ritchie, A. 1984. A new placoderm, *Placolepis* gen. nov. (Phyllolepidae) from the Late Devonian of New South Wales, Australia. Proceedings of the Linnean Society of New South Wales, 107: 321–353.

Ritchie, A. 2005. *Cowralepis*, a new genus of phyllolepid fish (Pisces, Placodermi) from the late Middle Devonian of New South Wales, Australia. Proceedings of the Linnean Society of New South Wales, 125: 215–259.

Springer, S. & R. A. Waller. 1969. *Hexanchus vitulus*, a new Sixgill Shark from the Bahamas. Bulletin of Marine Science, 19: 159–174.

Stensiö, E. 1948. On the Placodermi of the Upper Devonian of East Greenland. II. Antiarchi: subfamily Bothriolepinae. With an attempt at a revision of the previously described species of that subfamily. Meddelelser on Grønland 139: 1–622.

Talent, J. A. & D. Spencer-Jones. 1963. The Devono-Carboniferous fauna of the Silverband Formation, Victoria. Proceedings of the Royal Society of Victoria, 76: 1–11.

Turner, S. 1986. Vertebrate fauana of the Silverband Formation, Grampians, Victoria. Proceedings of the Royal Society of Victoria, 76: 1–11.

Turner, S. 1993. Palaeozoic microvertebrate biostratigraphy of Eastern Gondwana. In J. A. Long (ed.) Paleozoic Vertebrate Biostratigraphy and Biogeography. Pp. 174–207, Belhaven Press, London.

Turner, S. 1997a. Sequence of Devonian thelodont scale assemblages in East Gondwana. Geological Society of America Special Publication, 321: 295–315.

Turner, S. 1997b. Restudy of Late Givetian/early Frasnian "*Dittodus*" specimens of Hussakof and Bryant 1918 and other American xenacanthid-phoebodont type teeth. Modern Geology, 21: 87–119.

Turner, S. & G. C. Young. 1987. Shark teeth from the Early-Middle Devonian Cravens Peak Beds Georgina Basin. Alcheringa, 11: 233–244.

Turner, S. & G. C. Young. 1992. Thelodont scales from the Middle-Late Devonian Aztec Siltstone, southern Victoria Land, Antarctica. Antarctic Science, 4: 89–105.

White, E. I. 1968. Devonian fishes of the Mawson-Mulock area, Victoria Land, Antarctica. Trans-Antarctic Expedition 1955–1958. Scientific Reports, Geology, 16: 1–26.

Williams, M. E. 1998. A new specimen of *Tamiobatus vetustus* (Chondrichthyes, Ctenacanthoidea) from the Late Devonian Cleveland Shale of Ohio. Journal of Vertebrate Paleontology, 18: 251–260.

Young, G. C. 1974. Stratigraphic occurrences of some placoderm fishes in the Middle and late Devonian. Newsletters in Stratigraphy, 3: 243–261.

Young, G. C. 1982. Devonian sharks from southeastern Australia and Antarctica. Palaeontology, 25: 817–843.

Young G. C. 1988. Antiarchs (placoderm fishes) from the Devonian Aztec Siltstone, southern Victoria Land. Antarctica. Palaeontographica A, 202: 1–125.

Young, G. C. 1989a. The Aztec fish fauna (Devonian) of Southern Victoria Land: Evolutionary and biogeographic significance. In J. A. Crame (ed.) Origins and Evolution of the Antarctic Biota. Geological Society Special Publication, 47: 43–62.

Young, G. C. 1989b. New occurrences of culmacanthid acanthodians (Pisces, Devonian) from Antarctica and southeastern Australia. Proceedings of the Linnean Society of New South Wales, 111: 12–25.

Young, G. C. 1993. Middle Palaeozoic macrovertebrate biostratigraphy of Eastern Gondwana. – Pp. 208–251 in J. A. Long (ed.) Palaeozoic Vertebrate Biostratigraphy and Biogeography. Belhaven Press, London.

Young, G. C. 2007. Devonian formations, vertebrate faunas and age control on the far south coast of New South Wales and adjacent Victoria. Australian Journal of Earth Sciences, 54: 991–1008.

Young, G. C. 2010a. Placoderms (Armored Fish): Dominant vertebrates of the Devonian Period. Annual Review of Earth and Planetary Sciences, 38: 523–550. DOI: 10.1146/annurev-earth-040809-152507

Young, G. C. 2010b. Analytical methods in palaeobiogeography, and the role of early vertebrate studies. Palaeoworld, 19: 160–173. DOI: 10.1016/j.palwor.2009.10.001

Young, G. C. & J. A. Long. 2005. Phyllolepid placoderm fish remains from the Devonian Aztec Siltstone, southern Victoria Land, Antarctica. Antarctic Science, 17: 387–408.

Young, G. C. & J. M. Moody. 2002. A Middle-Late Devonian fish fauna from the Sierra de Perija, western Venezuela. South America. Mitteilungen aus dem Museum für Naturkunde in Berlin, Geowissenschaftliche Reihe 5: 155–206.

Zhao, W.-J. & M. Zhu. 2010. Siluro-Devonian vertebrate biostratigraphy and biogeography of China. Palaeoworld, 19: 4–26. DOI: 10.1016/j.palwor.2009.11.007

Ancient Fishes and their Living Relatives: a Tribute to John G. Maisey. Alan Pradel, John S. S. Denton & Philippe Janvier (eds.): pp. 247–256, 7 figs.
© 2021 by Verlag Dr. Friedrich Pfeil, München, Germany – ISBN 978-3-89937-269-4

Devonian fish from Colorado's Dyer Formation and the appearance of Carboniferous faunas in the Famennian

Hans-Peter Schultze*, James Bullecks**, Linda K. Soar** and James W. Hagadorn**

Abstract

The fish fauna of the Dyer Formation from Colorado's White River Plateau is described. There are remains of the arthrodire *Dunkleosteus*, tooth plates of dipnoans, elasmobranch spines and many chondrichthyan teeth. The latter give the fauna a Carboniferous appearance even though the remains are bounded stratigraphically by conodonts and $\delta^{13}C_{carb}$ chemostratigraphy that demonstrates a Famennian age (*B. aculeatus* Zone before the Hangenberg Crisis). Such taxa dominate the Early Carboniferous and thus had already appeared before the mass extinction associated with the Hangenberg Crisis. This pattern is mirrored by the Dyer Formation's invertebrate fauna, which also exhibits a Mississippian affinity.

Introduction

The Dyer Formation is a little-known but paleontologically rich carbonate succession that is well exposed in the White River Plateau of northwestern Colorado (Campbell 1970a). The unit has a diverse array of invertebrate fossils including algae, sponges, corals, conulariids, brachiopods, bryozoans, bivalves, gastropods, orthocerid and actinocerid cephalopods, crinoids and asteroids as well as trace fossils (Kindle 1909, Hauck 1954, Bass & Northrup 1963, Campbell 1966, 1970b, Hafley 1984, Sadle 2012). Of these, only the crinoids have been formally described (Webster et al. 1999). Conodonts have been documented from the succession and provide important age and environmental constraints (Sandberg 1976, Sandberg & Poole 1977, Sandberg & Dressen 1984, Webster et al. 1999, Wistort 2014, Wistort et al. 2014, Over et al. 2016). Vertebrate fossils include teeth and spines of chondrichthyans and lungfish tooth plates (Hafley 1984).

Although comparable vertebrate associations have been described from coeval limestones in southwestern Colorado (Kindle 1909, Bryant & Johnson 1936, Denison 1951, Schultze & Chorn 1998), the Dyer Formation's vertebrate fauna remained undescribed.

The majority of macrofossils occur in limestones of the Broken Rib Member in the lower portion of the Dyer Formation. Vertebrate fossils are concentrated in sandy limestones within the upper part of the member, where teeth, spines and plates occur as clasts in coarse-to-pebbly quartzose stringers and lags. Phosphatic fossils and quartz pebbles, being more resistant to weathering than associated calcareous matrix and fossils, often weather out of, or are pedestaled atop, bedding surfaces (Fig. 1). Such fossils range from mm-size teeth to decimeter-size plates and spines. In addition to occurring as concentrations in quartzose lags, they also occur as isolated elements in wackestones. Representative examples of these fossils were collected and analyzed for this study, and additional unfigured specimens were left intact at key localities in the field and in the collections at the Denver Museum of Nature & Science and the American Museum of Natural History, described below.

The Dyer Formation's vertebrate fossils have been known since at least the 1940s (Bass & Northrup 1963), with most of the best known specimens discovered by generations of avocational paleontologists affiliated with Western Interior Paleontological Society (WIPS). Many of their discoveries have been donated to museums, and initiated student theses (Hafley 1984, Sadle 2012). There are two main localities where vertebrates occur, including DMNH locality 4258 (Crane Park Quarry) and DMNH locality 3637 (Bison Lake), which is west of Crane Park Quarry (Fig. 2 above). DMNH locality 4258 is bounded by several informally named bedding plane exposures that commonly yield chondrichthyan remains, and even a *Dunkleosteus* intero-lateral bone, described below (DMNH locality 3638). These sites include "Fisherman's Wharf" (close to Crane Park Quarry), "The Dentist's Office" (atop the quarry at the Crane Park Quarry locality), and "The Altar" (down a dip-slope north of the Crane Park Quarry). Similarly, the Bison Lake locality includes a bed called "Crinoid Point" (DMNH locality 3637), where articulated crinoids and shark teeth occur. Just below it a *Dunkleosteus* antero-dorso-lateral plate has been documented. We studied fossils at these and nearby sites (DMNH 3638, AMNH 19630), combining them with specimens donated to the DMNH and AMNH by WIPS members, described in detail below.

Conodonts provide one of the key age constraints for the Dyer succession. The majority of the Broken Rib Member contains *Pandorinellina expansa* Zone conodonts (sensu Spalletta et al. 2017), whereas the upper portion, where it interfingers with the overlying Coffee Pot Member, contains *Bispathodus aculeatus* Zone conodonts (Fig. 2 below; Sandberg 1976, Sandberg & Poole 1977, Webster et al. 1999, Wistort et al. 2014, Chiarello et al. 2018). It is within these two zones that macrovertebrates occur. Overlying this stratigraphic interval there are no conodonts until the *ultimus* Zone followed by +6 to +8 excursions in the $\delta^{13}C_{carb}$ chemostratigraphy within the basal Carboniferous, which are inferred to represent the

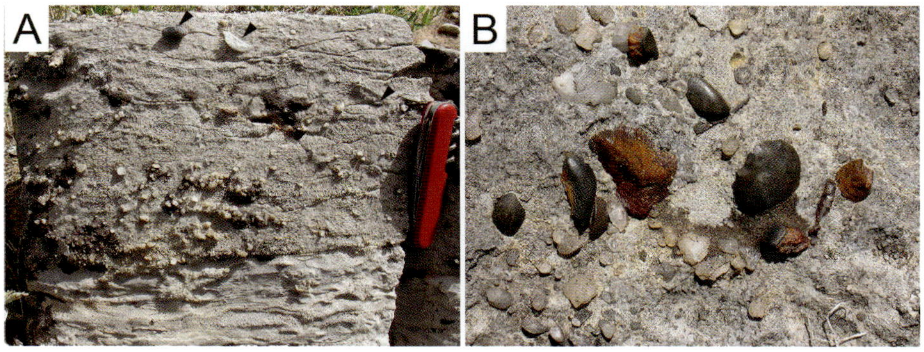

Fig. 1. Weathered limey grainstone outcrop of Broken Rib Member, *B. aculeatus* Zone, Dyer Formation, upper Famennian, "The Altar," DMNH locality 4258, near Crane Park Quarry on White River Plateau, northwestern Colorado. **A.** vertical section, illustrating several pedestaled teeth (black arrowheads); **B.** bed top view of surface with quartz pebbles and "bradyodont" teeth. Knife is 9 cm long.

* Biodiversity Institute and Museum of Natural History, The University of Kansas, Lawrence, Kansas, USA; hp1937@ku.edu (corresponding author)
** Department of Earth Sciences, Denver Museum of Nature & Science, Denver, Colorado, USA

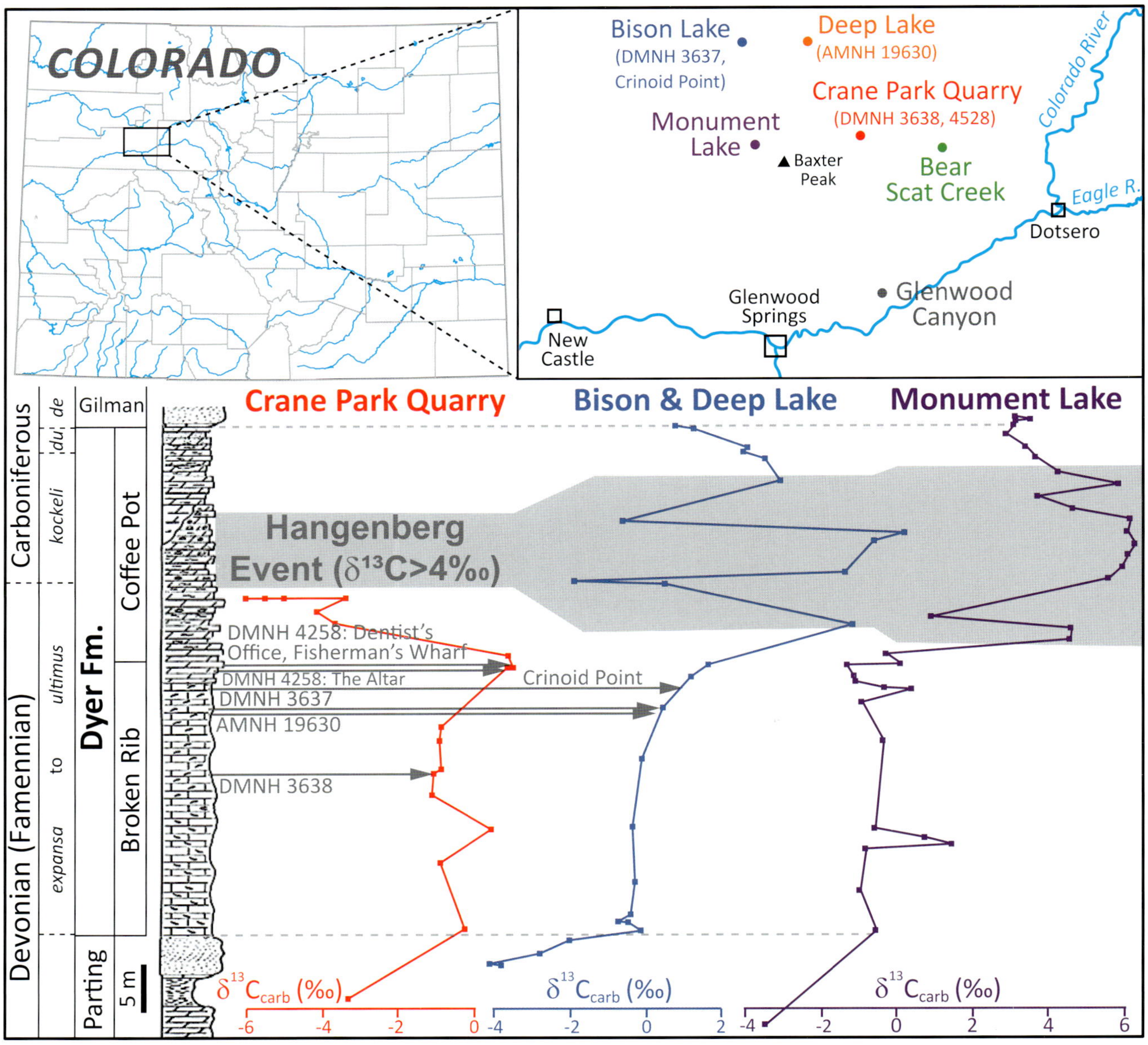

Fig. 2. Above. Geographic position of studied locations of the Dyer Formation on the White River Plateau, Garfield County, northwestern Colorado, including measured sections at Bison Lake, Deep Lake, Monument Lake, Crane Park Quarry, Bear Scat Creek, and Glenwood Canyon; Below, Representative graphic log of upper Famennian strata exposed in this region, selected $\delta^{13}C_{carb}$ chemostratigraphic profiles of key localities, and relative stratigraphic position of vertebrate fossils discussed in text. Generalized graphic log adapted from Campbell (1970a).

Hangenberg Event (Fig. 2 below; Myrow et al. 2011, 2013, Hagadorn et al. 2013, Cramer et al. 2019). This event starts with the *kockeli* Zone conodonts, signaling the presence of the Devonian-Carboniferous boundary and lower Tournaisian deposition (Wistort et al. 2014, Chiarello et al. 2018).

Some of the vertebrate fossils occur in skeletal wackestones and packstones of the Broken Rib Member (including historical collections by Campbell 1972, Hafley 1984, Webster et al. 1999) and represent deposition in a shallow open marine setting. The majority of these fossils, especially the teeth and ctenacanthid spines, occur in grainstones that are rich in coarse- to pebble-sized quartz grains (Fig. 1) that lie just below a transition to dolostones composed of silty detrital dolomite. We interpret this change to reflect a dramatic shallowing event and transition to peritidal intermittently lagoonal conditions (Hagadorn et al. 2016). The vertebrate skeletal debris, as well as coarse quartz grains, represent lags of reworked material sourced from adjacent littoral or "Spülsaum" (wash margin) environments – environments like those that may have produced the underlying Parting Sandstone or overlying Gilman Sandstone. Consistent with this interpretation of skeletal provenance, the detrital zircon signature of these grainstones is indistinguishable from those of the Parting and Gilman sandstones.

Material and methods

Studied specimens were removed from outcrops by hand, by rock hammer, or by portable diamond core-drill or by cut-off saw. Specimens were mechanically extracted from or pedestaled atop the matrix and locally stabilized with B-72. The *Dunkleosteus* plate and *Ctenacanthus* spine were photographed with a Nikon D750 (50 mm lens 1.8D). The chondrichthyan teeth and tooth plates and the dipnoan tooth plates were photographed with a Nikon D750 (Nikkor 105 mm Macro lens), and individual photographs were stacked using Combine ZP. Images in Figure 1 were taken with a Canon EOS 70D APS-C (EF-S 60 mm lens f2.8).

To construct the chemostratigraphic curves of Figure 2 below, we first petrographically screened limestones and dolostones with a polarizing microscope. Where diagenetically least altered phases (dominantly micrite, dolomicrite, or ooids) were identified, counterpart billets of those samples were microdrilled for analysis. Samples were analyzed at the University of Utah SIRFER lab, where powders were weighed using a Sartorius microbalance, loaded into 4.5 ml flat-bottomed borosilicate vials (Labco) and capped with Labco butyl rubber septa. These were transferred to a Thermo Fisher Scientific system comprised of a GasBench II, PAL autosampler, ConFlow IV interface and a MAT 253 mass spectrometer. Vials were flushed for 6 minutes on a PAL autosampler, with Ultra High Purity grade helium (99.999 % He) at a flow rate of 50 mL/ min. During flushing the vials were kept in a 50 °C aluminum block. Samples were reacted with 10 droplets of 104 % phosphoric acid (H_3PO_4) (kept at 50 °C) to CO_2 gas. Samples were allowed to equilibrate for 12 hours before analysis. During analysis using the PAL autosampler the produced CO_2 gas in the vials was collected using a sampling loop of 100 µL and transported to the mass spectrometer. Nine individual gas injections per analysis were made for each sample

Fig. 3. *Dunkleosteus* sp., right antero-dorso-lateral plate = DMNH 57155; DMNH locality 3637, "Crinoid Point" near Bison Lake, White River plateau, northwestern Colorado; Broken Rib Member, *expansa* Zone, Dyer Formation, upper Famennian. **A.** photo; **B.** drawing. Abbreviations: ADL, antero-dorso-lateral plate; AL, antero-lateral plate; lc, main lateral line, MD; median dorsal plate; PDL, postero-dorso-lateral plate; PL, postero-lateral plate. Scale bar equals 0.5 cm.

and for the reference materials; their average was taken as the final number. Three sets of reference materials were used to calibrate the system and unknown samples, including Carrara Marble, LSVEC and Marble-Std. Carrara and LSVEC was used as primary reference materials and Marble-Std. used as secondary reference material to cross check the final number. The oxygen fraction factor was calculated using the alpha value proposed by Swart et al. (1991). The average analytical uncertainty for analyzed samples was <0.08 for $\delta^{13}C$ and <0.10 for $\delta^{18}O$; samples yielding low CO_2 output were excluded from reported results.

Abbreviations. AMNH, American Museum of Natural History, New York, N.Y.; DMNH, Denver Museum of Nature and Science, Denver, Colorado; USFS, US Forest Service; WIPS, Western Interior Paleontological Society.

Systematic Paleontology

The vertebrate fauna of the upper Broken Rib Member of the Dyer Formation comprises representatives of three classes of gnathostomes, including placoderms, chondrichthyans and osteichthyans at localities on the White River Plateau in northwestern Colorado.

Class Placodermi M'Coy, 1848
Subclass Arthrodira Woodward, 1891
Infraclass Eubrachythoraci Miles, 1971
Superorder Pachyosteomorphi Stensiö, 1944
Order Dunkleosteoidea Vézina, 1990
Family Dunkleosteidae Stensiö, 1963

Dunkleosteus Lehman, 1956

Dunkleosteus sp.

(Fig. 3)

Material. One right antero-dorso-lateral plate (ADL) = DMNH 57155 (Fig. 3); locality "Crinoid Point," DMNH locality 3637 near Bison Lake, White River Plateau, Garfield County, northwestern Colorado; Broken Rib Member, *expansa* Zone, Dyer Formation, upper Famennian; intero-lateral = DMNH 57151, DMNH locality 3638, near Crane Park Quarry (*expansa* Zone), and a piece of bone DMNH EPV.78096 from DMNH locality 4258, near Crane Park Quarry, White River Plateau, Garfield County, northwestern Colorado; Broken Rib Member, *Bispathodus aculeatus* Zone, Dyer Formation, upper Famennian.

Description. The antero-dorso-lateral plate (ADL; Fig. 3) of the shoulder girdle is 22 cm deep and 14 cm long. It has a knob, an articulation ball, at about the middle of the anterior margin, where it articulated with the paranuchal plate of the head. The ADL plate was overlapped dorsally by the median dorsal plate (MD) and ventrally by the antero-lateral plate (AL). The area overlapped by the MD plate occupies a little more than half the surface dorsal to the lateral line, which passes from the articulation knob caudad horizontally over the ADL plate. The area overlapped by the AL plate ends with its dorsal tip 2 cm below the articulation knob. It widens ventrad, turns caudad in a angle of 120° and takes over the whole lowermost ventral part of ADL. Posteriorly the ADL plate overlapped the postero-dorso-lateral plate (PDL), and ventrolaterally over a short distance it overlapped the postero-lateral plate (PL). The surface of the ADL plate is smooth. In addition, part of an intero-lateral plate (DMNH 57151) was collected in the middle part of the Broken Rib Member at DMNH locality 3638 (*expansa* Zone) near Crane Park Quarry. A piece of another arthrodire plate was left in situ near the Bison Lake locality, but occurred somewhat lower in the section (Fig. 2 below).

Comparison. The large size of the ADL plate limits comparisons to large arthrodires, and there are few characteristics that allow for identification of the ADL plate to a specific genus. Young (2005) compared ADL plates of several arthrodires and found that the area overlapped by the MD plate has a similar shape (horizontal ventral border) in different eubrachythoracid

Fig. 4. Chondrichthyan spines. **A, B.** *Ctenacanthus concinnus* (Newberry, 1875), spine = DMNH EPV.87866; **A.** left side, **B.** right side, DMNH locality 4258, "The Altar", near Crane Park Quarry; **C.** *Sphenacanthus* sp., DMNH EPV.70036, DMNH locality 4258, near Crane Park Quarry, White River Plateau, northwestern Colorado; uppermost Broken Rib Member, *aculeatus* Zone, Dyer Formation, upper Famennian. Scale bar equals 1 cm.

arthrodires (e.g. *Coccosteus* and *Eastmanosteus*). In *Dunkleosteus*, the ventral border of the MD plate is inclined towards the condyle (Heintz 1932) much more than in the ADL plate of the Broken Rib Member. The lateral line is positioned horizontal in pachyosteomorphs like *Kendrichthys*, *Eastmanosteus* and *Dunkleosteus*. The overlapping area of the AL plate forms a triangle with the tip towards the condyle in eubrachythoracid arthrodires (e.g., *Eldenosteus* Miles, 1964, *Kendrickichthys* Dennis & Miles, 1980, *Eastmanosteus* Obruchev, 1964). The posterior border of the triangle is angled in *Dunkleosteus* (*D. terrelli*: 150°) and in the plate from White River Plateau (120°). The smooth surface with a nearly horizontal lateral line canal, and the shape of the overlap area allows us to place the ADL plate in the genus *Dunkleosteus*, without species assignment.

All arthrodire remains (supposedly all *Dunkleosteus*) except for DMNH EPV.78096 are from the *P. expansa* Zone, the deeper marine part of the section. Thus they are older than the occurrence of *Dunkleosteus* of the Cleveland Shale and predate the Hangenberg Crisis (Kaiser et al. 2015).

Class Chondrichthyes Huxley, 1880

Spines

The exposures at "The Altar" and especially at "Fisherman's Wharf" (both localities near Crane Park Quarry) contain many spines, which are broken and weathered and are not identifiable because they do not preserve the outer surface with the ornamentation. Exceptions are one complete *Ctenacanthus* spine (Fig. 4A, B), which was prepared mechanically, some *Ctenacanthus* pieces and one *Sphenacanthus* spine (Fig. 4C), which exhibit some ornamentation.

Superorder Cladodontomorphi Ginter, Hampe & Duffin, 2010
Order Ctenacanthiformes Glikman, 1964
Family Ctenacanthidae Dean, 1909

Ctenacanthus Agassiz, 1837 (1837–1843)

Ctenacanthus concinnus (Newberry, 1875)
(Fig. 4A, B)

Material. Fin spines DMNH EPV.87866 (Fig. 4A, B), DMNH EPV.70033, DMNH EPV.70034 (lower part), DMNH EPV.70037, DMNH EPV.70042 (lower part), DMNH EPV.70046 (lower part), DMNH EPV.70047 (pieces), DMNH EPV.70048 (piece), DMNH EPV.70049 (small piece), DMNH EPV.87865 (large part of spine with patches of striae), DMNH EPV.133062, DMNH 40973, DMNH EPV.133061(pieces from locality "The Altar"), Crane Park Quarry (DMNH locality 4258); part of spine DMNH EPV.70041, Bison Lake (DMNH locality 3637), AMNH FF 19630 Deep Lake, White River Plateau, Garfield County, northwestern Colorado; uppermost Broken Rib Member, *B. aculeatus* Zone, Dyer Formation, upper Famennian.

Description. The nearly complete spine, DMNH EPV.87866 (Fig. 4A, B), is more than 41 mm long. The apex is missing, and the outer arch is straightened near the top by an accidental cut. The lower smooth, hollow portion of the spine, which was embedded in connective tissue, comprises about 40 % of the length of the spine at the caudal margin. The caudal margin is not well preserved and does not show if denticles were present. The base of the ornamented region is transverse like in *Ct. clarkii* (Maisey 1981: fig. 9H). The fine striation is preserved best on the left side, where the ribs are pectinated with minute transverse tuberculations.

Comparison. Maisey (1981: fig. 9) described and figured spines of *Ct. compressus* (his fig. 9A, B) and of *Ct. clarkii* (his fig. 9H, I). The arrangement of the striae are similar to our specimens, being thin and densely packed medially (DMNH 87866, ornamentation of pectinated ridges with minute transverse tuberculations compares with our specimens; other species have a coarser pectination). The two species were synonymized by Ginter (2010) with *Ct. concinnus*.

Cohort Euselachii Hay, 1902
Family Sphenacanthidae Maisey, 1982

Sphenacanthus Agassiz, 1837 (1837–1843)

Sphenacanthus sp.
(Fig. 4C)

Material. One finspine DMNH EPV.70036 (Fig. 4C); DMNH locality 4258, near Crane Park Quarry, White River Plateau, Garfield County, northwestern Colorado; uppermost Broken Rib Member, *B. aculeatus* Zone, Dyer Formation, upper Famennian.

Description. 29 mm of a damaged spine are preserved, the tip is missing and parts of the lower half of the caudal margin are damaged. The striation is formed by widely spaced, coarse ribs ornamented with widely spaced knobs. The base of the ornamented part runs strongly transverse; only the smooth antero-ventral part of the in the body inserted portion is preserved, the ornamented part ends free in its ventro-caudal part.

Comparison. The spine with widely spaced ridges and knobs (Fig. 4C) is assigned to *Sphenacanthus* sp. It compares closely to the spine of *Sph.* aff. *marshi* figured by Maisey (1982: fig. 10E) and to the spine of *Sph. hybodoides?* (Maisey 1982: fig. 9A). Unfortunately, the preservational quality of the Dyer specimen inhibits species-level assignment.

Teeth

Superorder Cladodontomorphi Ginter, Hampe & Duffin, 2010
Order Ctenacanthiformes Glikman, 1964
Family Ctenacanthidae Dean, 1909

Ctenacanthus cf. *concinnus* (Newberry, 1875)
[= *C. compressus* Newberry, 1889]
(Fig. 5B, C)

Material. One large tooth DMNH EPV.70026 (Fig. 5C), four middle sized teeth DMNH EPV. 70025, EPV70027, EPV 70028 and EPV.70032, and three small teeth DMNH EPV.70030, DMNH EPV.133063 (Fig. 5B) and 133075; "The Altar", DMNH locality 4258 near Crane Park Quarry, White River Plateau, Garfield County, northwestern Colorado; uppermost Broken Rib Member, *B. aculeatus* Zone, Dyer Formation, upper Famennian.

Description. These teeth (Fig. 5B, C) have a robust principal cusp accompanied by two lateral cusplets on each side; the second pair of cusplets is larger than the first pair. The principal cusp

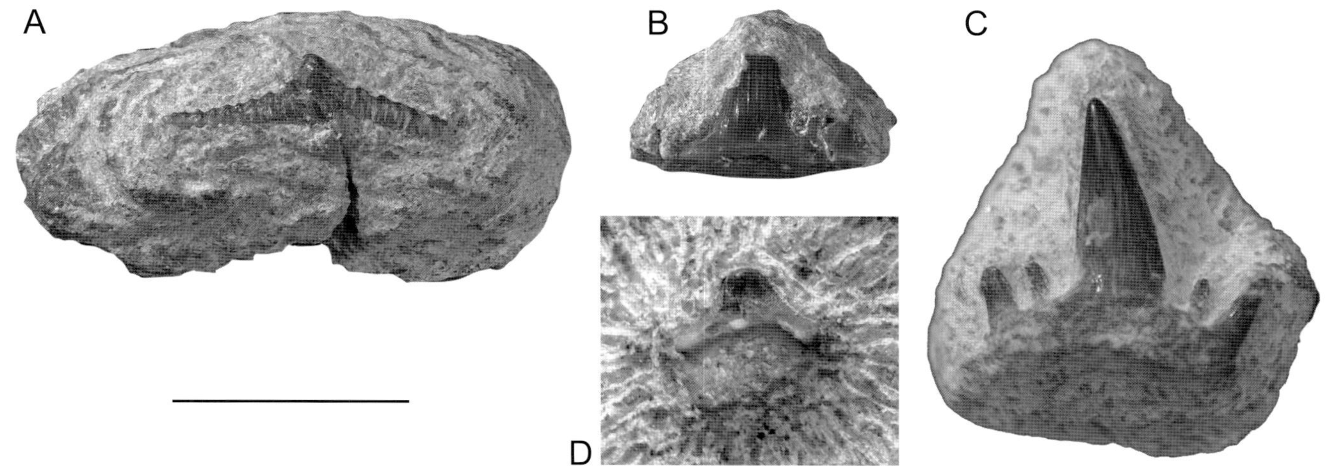

Fig. 5. Chondrichthyan teeth, DMNH locality 4258, near Crane Park Quarry, White River Plateau, northwestern Colorado; uppermost Broken Rib Member, *aculeatus* Zone, Dyer Formation, upper Famennian. **A.** *Orodus mammilaris* Newberry & Worthen, 1866, tooth, DMNH EPV.70021; **B, C.** *Ctenacanthus* (Newberry, 1875), **B.** small tooth, DMNH EPV. 133063, labial side; **C.** large tooth, DMNH EPV.70026, lingual side; **D.** *Lissodus* sp., DMNH EPV.70014, lingual side. Scale bar equals 1 cm.

is 9 mm high, and the tooth has a base 10.5 mm wide. The three small teeth from DMNH locality 4258 have 10 striae on the principal cusp like *C. concinnus* from the Cleveland Shales whereas the large tooth from the Dyer Formation has 14 ridges on the lingual side of the large principal cusp.

Comparison. Newberry (1875) described cladodont teeth of the Cleveland Shale as *Cladodus concinnus*, and later described similar teeth 1889 as *Ctenacanthus compressus* (Newberry, 1889). Ginter (2010) synonymized *Ctenacanthus compressus* Newberry, 1889 and *Ct. clarkii* Newberry, 1889 with *Cladodus concinnus* Newberry, 1875. These teeth have a large principal cusp and two lateral cusps, the external (main lateral) cusp is larger than the cusplet between the principal and the external cusp like the teeth of the Dyer Formation. Ginter (2010: fig. 2D) figured a small tooth of *Ct. concinnus* with eight or nine striae and a large one with 10 striae. That corresponds with most teeth of the Dyer Formation, whereas the large tooth has more striae. The base of the lateral cusp comprises one quarter of the base of the principal cusp, not two fifths as in Cleveland ones (Ginter et al. 2010: p. 70).

The teeth are different from the two other *Ctenacanthus* species of the Cleveland Shale, in having two lateral cusps. In contrast, *Ct. tumidus* has always only one lateral cusp on each side, and the cusps are nearly smooth. Small specimens of the third *Ctenacanthus* species, *Ct. terrelli*, have only one lateral cusp and the striae are very coarse. Considering similarities and differences of the cladodontiform teeth of the Dyer Formation with the Cleveland Shale species, the specimens of the Dyer Formation are provisionally identified as *Ct.* cf. *concinnus*.

Cohort Euselachii Hay, 1902
Order Hybodontiformes Maisey, 1975
Family Lonchidiidae Herman, 1977

Lissodus sp.
(Fig. 5D)

Material. One tooth, DMNH EVP.70014 (Fig. 5D), DMNH locality 4258, near Crane Park Quarry, White River Plateau, Garfield County, northwestern Colorado; uppermost Broken Rib Member, *B. aculeatus* Zone, Dyer Formation, upper Famennian.

Description. The tooth, DMNH EVP.70014, has a low crown and a deep base; it is 7 mm wide. The low crown has a blunt and smooth principal cusp (crown with cusp 3 mm deep), whereas the base of the principal cusp and the lateral parts of the crown are covered with oval tubercles. The base is as deep as the crown. It has many pores and grooves on its labial side.

Comparison. The tooth has a typical deep based euselachian appearance. The low crown with its blunt principal cusp compares to *Lissodus* (use as in Ginter et al. 2010). The oval-shaped, flat tubercles differ from other *Lissodus* species. Given that only one specimen is known from the Dyer, we tentatively assign it to "*Lissodus*".

Subclass Euchondrocephali Lund & Grogan, 1997
Order Orodontiformes Zangerl, 1981
Family Orodontidae de Koninck, 1878

Orodus Agassiz, 1838 (1837–1843)

Orodus mammillaris Newberry & Worthen, 1866
(Fig. 5A)

Material. Two teeth, DMNH EPV.70021 (Fig. 5A) and DMNH EPV.70022, DMNH locality 4258, near Crane Park Quarry, White River Plateau, Garfield County, northwestern Colorado; uppermost Broken Rib Member, *B. aculeatus* Zone, Dyer Formation, upper Famennian.

Description. The teeth have a distinct median principal cusp from which long flat wings extend laterally. These wings carry eight to nine low cusps, which have ribs extending onto the sides of the wings.

Comparison. The two teeth (DMNH EPV.70021 and EPV.70022) can easily be identified as belonging to the *Orodus mammillaris* group (Ginter et al. 2010: p. 113 and fig. 103). Teeth of this type were described from the Tournaisian of Illinois (Newberry & Worthen 1866) and the Viséan of Ireland and England (Davis 1883).

Superorder Holocephali Bonaparte, 1832

"Bradyodont" teeth are the first ones reported from the limey mudstones of the Dyer Formation (Fig. 1). These teeth give the impression that the assemblage belongs to the Carboniferous. Nevertheless the conodonts indicate *B. aculeatus* Zone (see Introduction). At least two holocephalian (or euchondrocephalian) genera are represented.

Order Helodontiformes Patterson, 1965
Family Helodontidae Patterson, 1965

Helodus sp.
(Fig. 6E, F)

Material. DMNH EPV.133066 (Fig. 6E), EPV.133074 (three teeth), EVP.70004 (Fig. 6F), 70013; DMNH locality 4258 near Crane Park Quarry, White River Plateau, Garfield County, northwestern Colorado; uppermost Broken Rib Member, *B. aculeatus* Zone, Dyer Formation, upper Famennian.

Description. The teeth are around 1 cm wide and 4.5 mm broad from labial to lingual side. The middle of the teeth is elevated as a low cone, from where the surface declines to the sides. The cone reaches over to the lingual side.

Comparison. A large number of species of *Helodus* have been described; Stahl (1999) cited over 70 species ranging from the Late Devonian to the Early Permian. Branson (1914) cited four species in the Grassy Creek Shales (Late Devonian), of which three reach into the Lower Carboniferous Bushberg Sandstone (*H. clivulus*, *H. devonicus*, *H. incipiens*), nevertheless the assignment of the Bushberg Sandstone to the Carboniferous is debated. We cannot justify an assignment of the Dyer Formation teeth to a particular species because a portion of most specimens is

Fig. 6. Holocephalian tooth plates, DMNH locality 4258, near Crane Park Quarry, White River Plateau, northwestern Colorado; uppermost Broken Rib Member, *aculeatus* Zone, Dyer Formation, upper Famennian; **A–D.** *Psephodus*, **A.** DMNH EPV.133067, **B.** DMNH EPV.133064, **C.** DMNH EPV.133065, **D.** DMNH EPV.133068, **E, F.** *Helodus* sp., **E.** DMNH EPV.133066, **F.** DMNH EPV. 70004. Scale bar equals 1 cm.

still in the matrix and thus one cannot see especially the root of the teeth – a lynchpin for further identification. Therefore we place the teeth in *Helodus* sp.

Order Cochliodontiformes Obruchev, 1953
Family Psephodontidae Zangerl, 1981

Psephodus sp.

Material. DMNH EPV.133064 (Fig. 6B), EPV.133065 (Fig. 6C), EPV.133067 (Fig. 6A), DMNH EPV.133068 (Fig. 6D), EPV.133069, DMNH EPV.70000, 70001, 70003, 70005, 70006, 70007, 70008, 70009, 70010, 70012, 70017, 70019, 70020, 70024, 70038, 70039, 70040, DMNH EPV.133073, EVP.57154; DMNH locality 4258 near Crane Park Quarry, White River Plateau, Garfield County, northwestern Colorado; uppermost Broken Rib Member, *B. aculeatus* Zone, Dyer Formation, upper Famennian.

Description. The tooth plates of *Psephodus* vary in size and shape. The teeth near the midline are rectangular (Traquair 1885: pl. 8, fig. 2) or more quadrangular with straight border where they attach to a neighboring plate from the opposite side in the midline (Traquair 1885: pl. 8, fig. 3, 4, compare Branson 1905: pl. 1, fig. 2). The tooth plates show "crenulation" at the base of the hypermineralized crown (Fig. 6E).

Comparison. The tooth plates of *Psephodus* are the most common holocephalian tooth plates on the White River Plateau, and are also the most common remains of all chondrichthyans sensu lato. Most *Psephodus* species are described from the Early Carboniferous, and only rarely from the Late Carboniferous (two taxa: Stahl 1999) and the Permian (three taxa: Stahl 1999). From the Devonian, two occurrences are known. Turner (1990) mentioned the occurrence of the genus in Western Australia, and Ginter & Piechota (2004) in the Holy Cross Mountains of Poland. Thus the frequent occurrence of the genus in the Dyer Formation indicates the establishment of typically Carboniferous taxa before the Hangenberg Crisis. Although the Dyer specimens share some similarities with the type species *Ps. magnus* (see Ginter & Piechota, 2004), we avoid placing them in any described species because the differences among species are minimal.

Class Osteichthyes Huxley, 1880
Subclass Sarcopterygii Romer, 1955
Superorder Dipnoi Müller, 1845

Isolated dipnoan tooth plates occur alongside the chondrichthyan teeth at "The Altar" and "Fisherman's Wharf" near Crane Park Quarry. These are the only osteichthyan remains discovered in these localities. It is surprising that teeth and scales of other sarcopterygians and actinopterygians have not been recovered because their remains are as resistant as chondrichthyan teeth or spines. An environmental reason to account for their absence is also not plausible given our current state of knowledge about the Dyer depositional environments.

Material. Four left pterygoid (upper) tooth plates DMNH EPV.70015 (Fig. 7A), DMNH EPV.133070 (Fig. 7B) and DMNH EPV.133071; DMNH locality 4258, "The Altar", near Crane Park Quarry; right prearticular (lower) tooth plate DMNH EPV.133072 (Fig. 7C); DMNH locality 4258, Crane Park Quarry, White River Plateau, Garfield County, northwestern Colorado; uppermost Broken Rib Member, *B. aculeatus* Zone, Dyer Formation, upper Famennian.

Description. The pterygoid tooth plates are flat and they have six to eight smooth ridges that do not reach to the lingual margin or postero-lingual corner. The postero-lingual corner forms a flat, palm-like area. The first and last ridge form an angle of just over 50°. The antero-labial part and the antero-labial extension of the ridges are not preserved. In contrast to the pterygoid tooth plates, the right prearticular tooth plate has 5 sharp ridges, which enclose an angle of 25°. The ridges reach close to the labial and posterior borders. The labial margin against the posterior margin encloses 104°.

Comparison. The identification of genera and species of isolated dipnoan tooth plates is questionable (Kemp 1996, Friedman & Daeschler 2006). After Kemp (1996), seven to eight ridges is the frequent number of ridges in dipnoan tooth plates. Nevertheless, Kemp (1996) based species identification on tooth plates in the same paper, and Friedman & Daeschler (2006) replaced the species names by letters only. Many Devonian tooth plates have been placed in the genus *Dipterus* (see Schultze 1992, Krupina 2000), even though the tooth plates do not compare to the tooth plates of *Dipterus valenciennesi* (White 1965: pl. 1, figs. 1, 2, pl. 2, figs. 1–5; Denison 1974: fig. 4). Nearly all species of *Dipterus* are based on single or only few specimens (e.g.: Newberry 1875, 1889, Smithson et al. 2016) with exception of species established by Krupina (2000: tens or hundreds of specimens). All *'Dipterus'* tooth plates have in common marked teeth on their ridges, a feature, which occurs also in other Devonian dipnoan tooth plates and few post-Devonian forms as described by Smithson et al. (2016). Smithson et al. (2016) created new species and genera for early Tournaisian lungfishes, which are not distinguishable from Devonian dipterids (compare e.g. *Occludus romeri* with *Dipterus serratus*). It was suggested to limit the genus *Dipterus* to *D. valenciennesi* (Denison 1974: p. 45, Friedman & Daeschler 2006: p. 1176), and place the dipterid tooth plates as genus and species indet., and we follow this approach here by utilizing *'Dipterus'* to indicate the taxonomic uncertainty.

The pterygoid tooth plates of the Dyer Formation compare best with a tooth plate from the Famennian of Turkey (Dipteridae gen. et sp. indet. in Janvier 1983: fig. 10A) and *"Dipterus" marginalis* (Krupina 2000: pl. 6, fig. 7). There are some distant similarities to *"Dipterus" fleisheri* (see Newberry 1897: pl. 24, fig. 25, and Denison 1974: fig. 3) from the Famennian Catskill Formation in Pennsylvania.

The dipnoans of Smithson et al. (2016) comprised a group (*Ctenodus*) with post-Devonian appearance, and a group of new genera (e.g. *Occludus*) with *Dipterus*-like Devonian appearance. The latter group shows also similarities to dipnoans (Friedman & Daeschler 2006: fig. 7) from the Late Devonian Catskill Formation of Pennsylvania.

Fig. 7. Dipnoan tooth plates, DMNH locality 4258, near Crane Park Quarry, White River Plateau, northwestern Colorado; uppermost Broken Rib Member, *aculeatus* Zone, Dyer Formation, upper Famennian; **A, B.** left upper (pterygoid) tooth plates, **A.** DMNH EPV.70015, **B.** DMNH EPV.133070, **C.** right lower (prearticular) tooth plate, DMNH EPV.133072. Scale bar equals 1 cm.

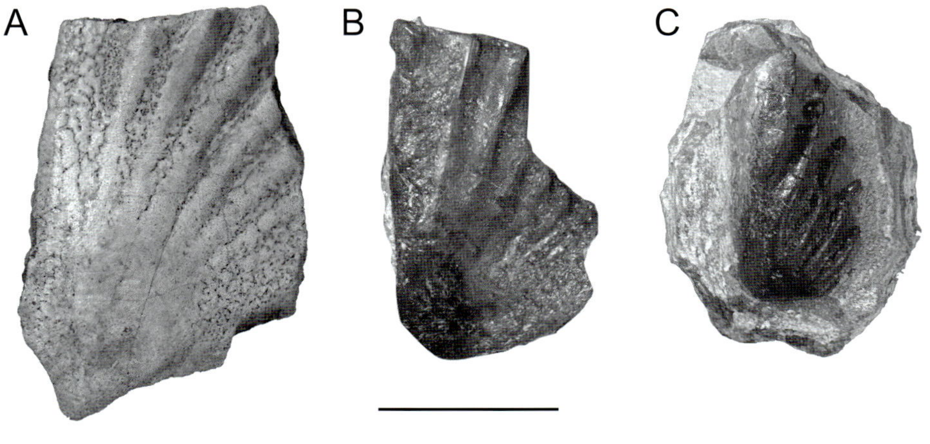

Comparison of pre-Hangenberg-Crisis faunas with Carboniferous faunas

The fish remains from the Broken Rib member of the Dyer Formation described herein are from the late Famennian *P. expansa* and *B. aculeatus* zones. The overlying Coffee Pot Member extends from the *Bispathodus aculeatus* to possibly the *?Protognathodus kockeli/?Siphonodella bransoni* Zone. Conodonts from these zones bracket strata deposited during the Hangenberg Crisis and define the local boundary between Devonian and Carboniferous deposition. Despite the magnitude of the extinction event signaled by the Hangenberg $\delta^{13}C_{carb}$ excursion (Sallan & Coates 2010, Kaiser et al. 2015), the fish fauna described above not only existed before the extinction event but also survived it.

Invertebrates and vertebrates responded differently to the Hangenberg Crisis. Based on their teeth and tooth plates, some elasmobranchs and particularly the holocephalians of the Dyer Formation have a Carboniferous affinity. Other fossils, like brachiopods (Stainbrook 1947, Crickmay 1952, Bass & Northrup 1963), crinoids (Webster et al. 1999), and conodonts (Wistort 2014) are reminiscent of Mississippian forms or appear to be transitional taxa between Late Devonian and Early Carboniferous forms. Some of the Carboniferous-like chondrichthyan genera have Famennian records besides those in the Dyer Formation; thus they did not go extinct during the Hangenberg Crisis. In contrast, large arthrodires from the Dyer like *Dunkleosteus* are clearly of Devonian age and did not survive the Hangenberg Crisis.

This situation is not new. Kaiser et al. (2015, specifically fig. 3), in a review of the impacts of the Hangenberg Crisis, noted that neritic forms of certain invertebrates had a tendency to survive the crisis, whereas pelagic forms disappeared. Invertebrates like stromatoporoids, ammonoids and trilobites nearly go extinct. In contrast, some Dyer invertebrates or their closely allied descendants may have persisted into the Early Carboniferous. For example, Webster et al. (1999) noted that the echinoderms of the Dyer Formation are similar to Early Carboniferous forms, a pattern also noted by Waters et al. (2014) among Chinese crinoids and blastoids. Sorauf & Pedder (1986, p. 1279) showed that "nine of the 27 Famennian basin-dwelling genera" of rugose corals and "24 of the 27" platform coral genera continued into the Early Carboniferous. Neritic brachiopods (Mottequin et al. 2014), and bryozoans (Horowitz & Pachut 1993) show a reduction of species, whereas pelagic species disappeared. Not only did neritic ostracods survive, but 50 % of pelagic ostracods did as well (Walliser 1996). Neritic and pelagic bivalves were little affected by the Hangenberg Crisis (Amler 1993).

Kaiser et al. (2015) also reviewed the effect of the Hangenberg Crisis on vertebrates. Placoderms went extinct. Schwermann (2014) discovered taxa in the Famennian of Morocco that previously had been known only from the Carboniferous, such as *Thrinacodus gracia* Grogan & Lund, 2008 from the Serpukhovian; *Stethacanthus* cf. *altonensis* (St. John & Worthen, 1875) from the Serpukhovian; *Denaea* cf. *fournieri* Pruvost, 1922 from the Viséan; *Denaea saltsmani* Ginter & Hansen, 2010 from the Late Carboniferous; *Cladodus mirabilis* Agassiz, 1843 (1837–1843) from the Viséan; and one representative of the *Orodus ramosus*-group from the Viséan.

Vertebrate taxa of the Dyer Formation also occurred in the Carboniferous. For example, representatives of six species of *Ctenacanthus* (seven after Maisey 1984, but after Ginter 2010 six because *Ct. clarkii* and *Ct. compressus* were synonymized) are known from spines in the Late Devonian and eight spine species are known from the Early Carboniferous including the genotype *Ct. major* (Maisey 1984). *Sphenacanthus* occurred in the Early Mississippian Kinderhook Formation (Iowa), in the Viséan and the Westphalian. The *Orodus mammilaris* group occurred in the Early Carboniferous (Tournaisian). "*Lissodus*", known from the Famennian of Belgium (Derycke-Khatir 2005) as well as from the Famennian Dyer Formation, extends into the Mississippian (Duffin 1985). Holocephalian tooth plates are the most common tooth plates and teeth in outcrops of the Dyer Formation on the White River Plateau (Fig. 1). Interestingly, *Helodus* and *Psephodus* species are extremely common in the Early Carboniferous, providing a Carboniferous appearance to weathered, tooth-studded outcrops of the Dyer Formation. *Psephodus* is known from the Famennian of Poland (Ginter & Piechota 2004), thus the genus at least crossed the Hangenberg Crisis. Ginter & Piechota (2004) even placed the middle Famennian species close to the Carboniferous species *Psephorus magnus*.

The dipnoans of the Dyer Formation cannot be assigned to post-Hangenberg-Crisis dipnoans. Devonian *Dipterus*-like tooth plates are uncommon in the Carboniferous, except *Occludus romeri*, which is indistinguishable from "*Dipterus*" *serratus* (= *Ctenodus serratus* of Newberry 1875: pl. 58, fig. 15, and Newberry 1889: pl. 27, fig. 31) from Linton, Ohio (Westphalian, Late Carboniferous) and similar to "*D.*" *serratus* (= *Ctenodus serratus* Eichwald, 1844 from the Middle Devonian, Mark-Kurik 1975: fig. 2). The Famennian *Ctenodus angustus* (= Dipnoi gen. et sp. indet. E in Friedman & Daeschler 2006: fig. 7F) compares with the Tournaisian *Coccodus celatus* (Smithson et al. 2016: fig. 6B). Clack et al. (2019) described new dipnoans from the Upper Devonian of Greenland; their phylogenetic analysis showed that some Devonian dipnoans (*Nielsenia*, *Celsiodon*) are closely related to Carboniferous ones. That indicates that Carboniferous dipnoans have their origin in the Devonian and that they "blur the Devonian-Carboniferous boundary" (Clack et al. 2019).

Looking beyond the Dyer Formation, other notable Carboniferous fish groups also occur in or have their roots in the Devonian. Sarcopterygian actinistians like *Diplocercides* survived into the early Carboniferous (Forey 1998), and Devonian members of the family Megalichthyidae (e.g. from the Late Devonian Catskill Formation, Daeschler & Downs 2018) survived into the Permian.

Among actinopterygians, it is evident that Devonian taxa are to large extent different from those of the Carboniferous. It is therefore understandable that phylogenies of early actinopterygians often include only Devonian forms (e.g. Long et al. 2008). Nevertheless there occur representatives of Carboniferous genera already in the Late Devonian including *Canobius*, "*Elonichthys*", *Rhadinichthys* [the species *R. macconochii* is recorded from the Late Devonian and Early Carboniferous], *Cycloptychius* and *Kentuckia*. Wilson et al. (2018) described a new actinopterygian (*Avonichthys manskyi*) from the Horton Bluff Formation, Tournaisian, of Nova Scotia. The genus is most closely related to Late Devonian actinopterygians – indicating that Devonian lineages survived into the Carboniferous (see further discussion in Wilson et al. (2018).

The appearance of "Carboniferous" vertebrate taxa in the Famennian is not restricted to the Dyer Formation of Colorado. Some of these taxa occur in other Upper Devonian localities. Consideration of single localities versus the occurrence of taxa can lead to an enhanced impression of the end-Famennian extinction. For example, Sallan & Coates (2010: p. 10131) noted that "... the vast majority of Devonian and Mississippian vertebrate species are known from single localities within single intervals, a phenomenon sure to produce artificially high extinction and origination rates over any arbitrary interval." Considering the time range of species and genera gives a more moderate impression of the magnitude and rate of extinction and survival. Available phylogenies (Wilson et al. 2018, Clack et al. 2019) show that lineages and some taxa at the generic level (Ginter & Piechota 2004) extend from the Devonian to the Carboniferous and bridge the Hangenberg Crisis.

Conclusions

The localities on the White River Plateau are rich in chondrichthyan teeth and spines, whereas scales and teeth of actinopterygians and sarcopterygians (rhipidistians), except dipnoan tooth plates, are missing. These vertebrates occur in shallow marine facies of the Broken Rib Member of the Dyer Formation, whereas *Dunkleosteus* occurs in deeper water facies.

Whereas *Dunkleosteus* represents a typical Late Devonian genus, the chondrichthyans show similarities to Carboniferous taxa (Ginter 2010: p. 152). The record of the genus *Dunkleosteus* ends within the Cleveland Shale and the Hangenberg Crisis (Carr & Jackson 2008). Some Devonian chondrichthyans, dipnoans, and actinopterygians survived the Hangenberg Crisis. The chondrichthyan and actinopterygian survivors of the Hangenberg Crisis are the precursors of the explosive development of these two groups in the Carboniferous.

The transition from the Devonian to the Carboniferous (through the Hangenberg mass extinction) was survived by some genera, which formed the base for new Carboniferous assemblages. When present in Devonian strata, such Carboniferous-aspect taxa can give a misleading picture. For example, if one only considers the fish taxa, one might initially infer that the Dyer is a Carboniferous unit, but this interpretation is contradicted by chemostratigraphic and biostratigraphic data that indicates deposition during the late Famennian.

Acknowledgements

We dedicate this paper to John Maisey, whom we hope learns to enjoy Florida's many fossil sharks in his retirement – despite how 'young' they are. We thank M. Getty, C. Schoellhorn, B. Small, B. Wahle, G. Welsh and the many DMNS and WIPS volunteers and staff who collected, donated, and/or prepared this material, as well as B. Schumacher, the USFS, and the DMNS KT Challenge Fund for supporting this scholarship and resulting collections. We thank A. Gishlick, AMNH, for the picture of spine AMNH FF 19630. J. Chiarello, B. Lahey, M. Longman, P. Myrow, J. Over, C. Sandberg, and Z. Wistort are thanked for their help creating a stratigraphic and temporal framework for the studied fauna. O. Sansidiro, Division of Vertebrate Paleontology, Museum of Natural History, University of Kansas, took the photos of Figures 3A, 4–7, and did the drawing of the *Dunkleosteus* plate (Fig. 3B). The senior author thanks T. Becker, Münster, Germany, for information on range of species from before and after the Hangenberg Crisis. The authors thank T. Becker, D. J. Cicimurri, M. Ginter, W. Itano and an anonymous reviewer for their careful reviews of this manuscript.

Literature cited

Agassiz, L. 1837–1843. Recherches sur les poissons fossiles. Tome III: VIII + 390 pp. Petitpierre, Neuchâtel et Soleure.

Amler, M. R. W. 1993. Shallow marine bivalves at the Devonian-Carboniferous boundary from the Velbert Anticline (Rheinisches Schiefergebirge). Annales de la Société géologique de Belgique, 115: 405–423.

Bass, N. W. & S. A. Northrup. 1963. Geology of the Glenwood Springs Quadrangle and vicinity, northwestern Colorado. US Geological Survey, Bulletin, 1142-J, 74 pp.

Bonaparte, C. L. 1832–1841. Iconografia delle fauna italica per le quattro classi degli animali verte brati. Tome III. Pesci. 556 pp.; Roma.

Branson, E. B. 1905. Notes on some Carboniferous cochliodonts with descriptions of seven new species. Journal of Geology, 13: 20–34.

Branson, E. B. 1914. The Devonian fishes of Missouri. University of Missouri Bulletin, 15: 59–74.

Bryant, W. L. & J. H. Johnson. 1936. Upper Devonian fish from Colorado. Journal of Paleontology, 10: 656–659.

Campbell, J. A. 1966. The Devonian system of west central Colorado: Unpublished Ph.D. Thesis, University of Colorado, 113 pp.

Campbell, J. A. 1970a. Stratigraphy of Chaffee Group (Upper Devonian), west-central Colorado. American Association of Petroleum Geologists, Bulletin, 54: 313–325.

Campbell, J. A. 1970b. Petrology of Devonian shelf carbonates of west central Colorado. The Mountain Geologist, 7: 89–97.

Campbell, J. A. 1972. Lower Paleozoic systems, White River Plateau. Quarterly Colorado School of Mines, 67: 37–62.

Carr, R. K. & G. L. Jackson, 2008. The vertebrate fauna of the Cleveland Member (Famennian) of the Ohio Shale. Unpublished field trip guidebook for the 68th Annual Meeting of the Society of Vertebrate Paleontology, Cleveland, Ohio, Oct 15–18, 2008, 17 pp.

Chiarello, J., M. Bates, E. Hauf, D. J. Over, Z. P. Wistort, P. Zackery, J. W. Hagadorn, L. K. Soar & J. Bullecks. 2018. Late Devonian conodonts and facies from the shallow shelf strata of the Broken Rib and Coffee Pot members, Dyer Formation, Chaffee Group, Central Colorado. North Central 52nd Annual Meeting, Geological Society of America, Abstracts with Programs, 50, Paper 6-6. DOI: 10.1130/abs/2018NC-313168.

Clack, J. A., T. J. Challands, T. R. Smithson & K. Z. Smithson. 2019. Newly recognized Famennian lungfishes from East Greenland reveal tooth plate diversity and blur the Devonian-Carboniferous boundary. Papers in Paleontology, 5: 261–279.

Crickmay, C. H. 1952. Discrimination of late Upper Devonian. Journal of Paleontology, 26: 585–609.

Cramer, B. D., B. M. Stolfus, J. E. Day, B. J. Clark, N. J. Hogancamp, B. J. Witzke & S. Tassier-Surine. 2019. The Devonian-Carboniferous boundary in the Mississippi River valley of Iowa, Illinois, and Missouri. 35th Annual Great-Lakes Section – SEPM, Guidebook: The Devonian-Carboniferous Boundary in the Type Area of the Mississippian. Iowa Geological Survey, Guidebook Series, 30: 1–9.

Daeschler, E. B. & J. P. Downs. 2018. New description and diagnosis of *Hyneria lindae* (Sarcopterygii, Tristichopteridae) from the Upper Devonian Catskill Formation in Pennsylvania, U.S.A. Journal of Vertebrate Paleontology, 38: e1448834. DOI: 10.1080/02724634. 2018.1448834.

Davis, J. W. 1883. On the fossil fishes of the Carboniferous limestone series of Great Britain. Scientific Transactions of the Royal Dublin Society, 1: 327–600.

Dean, B. 1909. Studies on fossil fishes (sharks, chimaeroids and arthrodires). Memoirs of the American Museum of Natural History, 9: 211–287.

de Koninck, L. G. 1878. Faune du calcaire carbonifère de la Belgique. Première partie, poissons et genre nautile. Annales du Musée royale d'Histoire naturelle de Belgique, 2: 1–152.

Denison, R. H. 1951. Late Devonian fresh-water fishes from the Western United States. Fieldiana, Geology, 2: 219–261.

Denison, R. H. 1974. The structure and evolution of teeth in lungfishes. Fieldiana, Geology, 33: 31–58.

Dennis, K. & R. S. Miles. 1980. New durophagous arthrodires from Gogo, Western Australia. Zoological Journal of the Linnean Society, 69: 43–85.

Derycke-Khatir, C. 2005. Microrestes de Vertébrés du Paléozoique Supérieur de la Manche au Rhin. Société Géologique du Nord, 33: 71–72.

Duffin, C. J. 1985. Revision of the hybodont selachian genus *Lissodus* Brough, 1935. Palaeontographica A, 188: 105–152.

Eichwald, E. I. 1844. O rybakh pervobytnago okeana v okrestnostiakh Pavlovska. Otechestven nyia zapiski, Ucheno-literaturnyi zhurnal, 36: 1–22.

Forey, P. L. 1998. History of the Coelacanth Fishes. XIII + 419 pp.; Chapman & Hall, London, Weinheim, New York.

Friedman M. & E. B. Daeschler. 2006. Late Devonian (Famennian) lungfishes from the Catskill Formation of Pennsylvania, USA. Palaeontology, 49: 1167–1183.

Ginter, M. 2010. Teeth of the Late Famennian ctenacanth sharks from the Cleveland Shale. – Pp. 145–158 in: D. K. Elliott, J. G. Maisey, X. Yu & D. Miao (eds.), Morphology, Phylogeny and Paleogeography of Fossil Fishes. Verlag Dr. Friedrich Pfeil, München, Germany.

Ginter, M. & M. C. Hansen. 2010. Teeth of the cladodont shark *Denaea* from the Carboniferous of central North America. – Pp. 29–44 in: D. Nowakowski (ed.), Morphology and systematics of fossil vertebrates. DN Publisher, Wrocław, Poland.

Ginter, M., O. Hampe & C. Duffin. 2010. Chondrichthyes, Paleozoic Elasmobranchii: Teeth. – In: H.-P. Schultze (ed.), Handbook of Paleoichthyology, vol. 3D: 168 pp; Verlag Dr. Friedrich Pfeil, München, Germany.

Ginter, M. & A. Piechota. 2004. The first Devonian holocephalian tooth from Poland. Acta Palaeontologica Polonica, 49: 409–415.

Grogan, E. D. & R. Lund 2008. A basal elasmobranch, *Thrinacoselache gracia* n. gen. & sp. (Thrinacodontidae, new family) from the Bear Gulch limestone, Serpukhovian of Montana, USA. Journal of Vertebrate Paleontology, 28: 970–988.

Hafley, D. J. 1984. Geology and paleontology of the Heart Lake area, White River Plateau, northwest Colorado. Unpublished Master of Science thesis, Washington State University, 124 pp.

Hagadorn, J. W., L. K. Soar & J. Bullecks. 2013. A tropical record of terminal Famennian events: Colorado's Dyer Formation. Geological Society of America 125th Anniversary Annual Meeting Denver, Colorado, Abstracts with Programs, Paper 156-11: 395.

Hagadorn, J. W., J. Bullecks, L. K. Soar, B. L. Lahey, D. J. Over, Z. P. Wistort & C. S. Holm-Denoma. 2016. Colorado mass extinction: Weird facies and cool fossils from the end-Devonian Dyer Formation. Geological Society of America 2016 Annual Meeting Denver, Colorado, Abstracts with Programs., Paper 137-7. DOI: 10.1130/abs/2016AM-287728.

Hauck, R. A. 1954. Stratigraphy and paleontology of Devonian rocks in part of west central Colorado. Unpublished M.S. Thesis, University of Colorado, 67 pp.

Hay, O. P. 1902. Bibliography and Catalogue of the Fossil Vertebrata of North America. U.S. Geological Survey, Bulletin, 179: 868 pp.

Heintz, A. 1932. The structure of *Dinichthys*. A contribution to our knowledge of the Arthrodira. Bashford Dean Memorial Volume. Archaic Fishes. Article 4: 111–224. American Museum of Natural History, New York.

Herman, J. 1977. Les sélaciens des terrains néocrétacés & paléocènes de Belgique & des contrées limitrophes. Eléments d'une biostratigraphie intercontinentale. Mémoire pour servir à l'explication des cartes géologiques et minières de Belgique, 15: 1–450.

Horowitz, A. S. & J. F. Pachut. 1993. Specific, generic, and familial diversity of Devonian bryozoans. Journal of Paleontology, 67: 42–52.

Huxley, T. H. 1880. On the application of the laws of evolution to the arrangement of the Vertebrata and more particularly of the Mammalia. Proc. Zool. Soc. London, 1880: 649–662.

Janvier, P. 1983. Les Vertébrés dévoniens de la Nappe Supérieure d'Antalya (Taurus Lycien occidental, Turquie). Géologie Méditerranéenne, 10: 1–13.

Kaiser, S. I., M. Aretz & R. T. Becker. 2015. The global Hangenberg Crisis (Devonian–Carboniferous transition): review of a first-order mass extinction. – Pp. 387–437 in: R. T. Becker, P. Königshof & C. E. Brett (eds.), Devonian Climate, Sea Level and Evolutionary Events. Geological Society London, Special Publication, 423. DOI: 10.1144/SP423.9

Kemp, A. 1996. *Sagenodus* (*Proceratodus*) *carlinvillensis* (Romer and Smith 1934), (Osteichthyes: Dipnoi), short ridge anomaly and classification of dipnoans. Journal of Vertebrate Paleontology, 16: 16–19.

Kindle, E. M. 1909. The Devonian fauna of the Ouray Limestone. US Geological Survey, Bulletin, 391, 60 pp.

Krupina, N. I. 2000. [Dipnoans from the Upper Devonian Locality Rybnitsa in Oryol Region.] Paleontologicheskiy Zhurnal, 5: 55–61 [in Russian]

Lehman, J.-P. 1956. Les arthrodires du Dévonien supérieur du Tafilalet (Sud marocain). Service Géologique du Maroc, Notes et Mémoires, 129, 70 pp.

Long, J. A., B. Choo & G. C. Young. 2008. A new basal actinopterygian from the Middle Devonian Aztec Siltstone of Antarctica. Antarctic Science, 20: 393–412.

Lund, R. & E. Grogan. 1997. Relationships of the Chimaeriformes and the basal radiation of the Chondrichthyes. Reviews in Fish Biology and Fisheries, 7: 65–123.

Maisey, J. G. 1975. The interrelationships of phalacanthous selachians. Neues Jahrbuch für Geologie und Paläontologie, Monatshefte, Jahrgang 1975: 563–567.

Maisey, J. G. 1981. Studies on the Paleozoic Selachian Genus *Ctenacanthus* Agassiz No. 1. Historical Review and Revised Diagnosis of *Ctenacanthus*, With a List of Referred Taxa. American Museum Novitates, 2718: 1–22.

Maisey, J. G. 1982. Studies on the Paleozoic Selachian Genus *Ctenacanthus* Agassiz: No. 2. *Bythiacanthus* St. John and Worthen, *Amelacanthus*, new genus, *Eunemacanthus* St. John and Worthen, *Sphenacanthus* Agassiz, and *Wodnika* Münster. American Museum Novitates, 2722: 1–24.

Maisey, J. G. 1984. Studies on the Paleozoic selachian genus *Ctenacanthus* Agassiz. No. 3 Nominal species referred to *Ctenacanthus*. American Museum Novitiates, 2274: 1–20.

Mark-Kurik, E. 1975. A tooth-plate from the Lower Devonian of Kotelny Island. Proceedings of the Academy of Science of the Estonian SSR, Chemistry and Geology, 24: 307–309.

M'Coy, F. 1848. On some new fossil fish of the Carboniferous period. Annals and Magazine of Natural History, 2: 115–133.

Miles, R. S. 1964. On some coccosteomorph arthrodires from the Devonian of Arizona. Arkiv för Zoologi, Stockholm, 16: 427–460.

Miles, R. S. 1971. The Holonematidae (placoderm fishes), a review based on new specimens of *Holonema* from the Upper Devonian of Western Australia. Philosophical Transactions of the Royal Society London, Series B, 263: 101–234.

Mottequin, B., D. Brice & M. Legrand-Blain. 2014. Biostratigraphic significance of brachiopods near the Devonian–Carboniferous boundary. Geological Magazine, 151: 216–228.

Müller, J. 1845. Über den Bau und die Grenzen der Ganoiden und das natürliche System der Fische. Abhandlungen der königlichen preussischen Akademie der Wissenschaften zu Berlin, Physikalisch-Mathematische Classe, (for 1844) 1845: 117–216.

Myrow, P. M., J. V. Strauss, J. R. Creveling, K. R. Sicard, R. Ripperdan, C. A. Sandberg & S. Hartenfels. 2011. A carbon isotopic and sedimentological record of the latest Devonian (Famennian) from the Western U.S. and Germany. Palaeogeography, Palaeoclimatology, Palaeoecology, 306: 147–159. DOI: 10.1016/j.palaeo.2011.04.013

Myrow, P. M., A. Hanson, A. S. Phelps, J. R. Creveling, J. V. Strauss, D. A. Fike & R. L. Ripperdan, R. L. 2013. Latest Devonian (Famennian) global events in western Laurentia: variations in the carbon isotopic record linked to diagenetic alteration below regionally extensive unconformities. Palaeogeography, Palaeoclimatology, Palaeoecology, 386: 194–209.

Newberry, J. S. 1875. Descriptions of fossil fishes. Report Geological Survey of Ohio, Columbus, 2: 1–64.

Newberry, J. S. 1889. The Palaeozoic fishes of North America. U. S. Geological Survey, Monographs, 16: XVI + 340 pp.

Newberry, J. S. 1897. New Species and a New Genus of American Palaeozoic Fishes, together with Notes in the Genera, *Oracanthus*, *Dactylodus*, *Polyrhizodus*, *Sandalus*, *Deltodus*. Transactions of the New York Academy of Science, 16: 282–304.

Newberry, J. S. & A. H. Worthen. 1866. Descriptions of new species of vertebrates, mainly from the Subcarboniferous limestone and Coal Measures of Illinois. Geological Survey of Illinois, 2: 9–134.

Obruchev, D. V. 1953. [The study of edestids and the work of A. P. Karpinskii.] Trudy Paleontologicheskogo instituta Akademiya Nauk SSR, 45: 1–86. [in Russian]

Obruchev, D. V. 1964. Class Placodermi. – Pp. 118–259 in: D. V. Obruchev (ed.), [Osnovy paleontologii. vol. XI, Agnatha, Pisces]. Nauka, Moskva. [in Russian]

Over, D. J., H. G. Horner, G. Hugh, Z. P. Wistort, J. W. Hagadorn, L. K. Soar & J. Bullecks. 2016. Late Devonian conodonts from the shallow shelf strata of the Broken Rib and Coffee pot members, Dyer Formation, Chaffee Group, Colorado. Geological Society of America 2016 Annual Meeting Denver, Colorado, Paper 45-7.

Patterson, C. 1965. The phylogeny of the chimaeroids. Philosophical Transactions of the Royal Society of London, B, 249: 101–219.

Pruvost, P. 1922. Description de *Denaea fournieri*, sélacien nouveau du Marbre noir de Denée. Part 2 of G. Fournier & P. Pruvost. Découverte d'un poisson nouveau dans le marbre noir de Denée. Bulletin de l'Académie Royal Bruxelles, Série 5, 8: 213–218.

Romer, A. S. 1955. Herpetichthyes, Amphibioidei, *Choanichthys* or Sarcopterygii? Nature, 176: 126–127.

Sadle, S. 2012. Vertical distributions of carbonate lithologies, bedding features, and macrofossils in the Broken Rib Member of the Upper Devonian Dyer Formation near Monument Lake, Garfield County, Colorado. Unpublished M.A. Thesis, Univ. Northern Colorado, 51 pp.

Sallan, L. C. & M. I. Coates. 2010. End-Devonian extinction and a bottleneck in the early evolution of modern jawed vertebrates. Proceedings of the National Academy of Sciences of the USA, 107: 10131–10135.

Sandberg, C. A. 1976. Conodont biofacies of Late Devonian *Polygnathus styriacus* Zone in western United States. – In: C. R. Barnes (ed.), Conodont Paleoecology. Geological Association of Canada, Special Papers, 15: 171–186.

Sandberg, C. A. & R. Dreesen. 1984. Late Devonian icriodontid biofacies models and alternate shallow-water conodont zonation. Geological Society of America, Special Paper I 96: 143–178.

Sandberg, C. A. & F. G. Poole. 1977. Conodont biostratigraphy and depositional complexes of Upper Devonian cratonic-platform and continental-shelf rocks in the Western United States. California University, Riverside, CA, Campus Museum Contributions, 4: 144–182.

Schultze, H.-P. 1992. Dipnoi. – Fossilium Catalogus. I: Animalia. Pars 131: 1–464. Kugler Publications, Amsterdam/New York.

Schultze, H.-P. & J. Chorn. 1998. Sarcopterygian and other fishes from the marine Upper Devonian of Colorado, U.S.A. Mitteilungen aus dem Museum für Naturkunde Berlin, Geowissenschaftliche Reihe, 1: 53–72.

Schwermann, K. 2014. Devonische Haizähne aus Marokko – Taxonomie, Stratigraphie und fazielle Verbreitung. Unpublished M.Sc. Thesis, Westfälische Wilhelms-Universität Münster, 80 pp.

Smithson, T. R., K. R. Richards & J. A. Clack. 2016. Lungfish diversity in Romer's Gap: reaction to the end-Devonian extinction. Palaeontology, 59: 29–44. DOI: 10.1111/pala.12203

Sorauf, J. E. & A. E. H. Pedder. 1986. Late Devonian rugose corals and the Frasnian–Famennian crisis. Canadian Journal of Earth Sciences, 23: 1265–1287.

Spalletta, C., M. C. Perri, D. J. Over & C. Corradini. 2017. Famennian (Upper Devonian) conodont zonation: revised global standard. Bulletin of Geosciences, 92: 31–57.

Stahl, B. J. 1999. Chondrichthyes III. Holocephali. – In: H.-P. Schultze (ed.), Handbook of Paleoichthyology, vol. 4: 164 pp. Verlag Dr. Friedrich Pfeil, München, Germany.

Stainbrook, M. A. 1947. Brachiopoda of the Percha Shale of New Mexico and Arizona. Journal of Paleontology, 21: 297–328.

Stensiö, E. A:son. 1944. Contributions to the knowledge of the vertebrate fauna of the Silurian and Devonian of western Podolia. II. Notes on two arthrodires from the Downtonian of Podolia. Arkiv för Zoologi, Stockholm, 35: 1–83.

Stensiö, E. A:son. 1963. Anatomical studies on the arthrodiran head. Part I. Preface, geological and geographical distribution, the organisation of the head in the Dolichothoraci, Coccosteimorphi, and Pachyosteomorphi. Taxonomic appendix. Kungliga Svenska VetenskapsAkademiens Handlingar, 9: 1–419.

St. John, O. & A. H. Worthen. 1875. Description of fossil fishes. Geological Survey of Illinois, 6, Geology and Paleontology. Part II, Paleontology of Illinois. Section I. Descriptions of fossil fishes: 245–488.

Swart, P. K., S. J.Burns & J. J. Leder. 1991. Fractionation of the stable isotopes of oxygen and carbon in carbon dioxide during the reaction of calcite with phosphoric acid as a function of temperature and technique. Chemical Geology: Isotope Geoscience Section, 86: 89–96.

Traquair, R. H. 1885. On a specimen of *Psephodus magnus* from the Carboniferous limestone of East Kilbride, Lanarkshire. Geological Magazine, Decade 3, 2: 337–344.

Turner, S. 1990. Early Carboniferous shark remains from the Rockhampton district, Queensland. Memoirs of the Queensland Museum, 28: 65–73.

Vézina, D. 1990. Les Plourdosteidae fam. nov. (Placodermi, Arthrodira) et leurs relations phylétiques au sein des Brachythoraci. Canadian Journal of Earth Science, 27: 677–683.

Walliser, O. H. 1996. Global events in the Devonian and Carboniferous. – Pp. 225–250 in: O. H. Walliser (ed.), Global Events and Event Stratigraphy in the Phanerozoic. Springer, Berlin.

Waters, J., T. J. Suttner, E. Kido & S. K. Carmichael. 2014. The 'Age of Crinoids' began in the Late Devonian. 46th Annual Meeting Geological Society of America, Abstracts with Programs, 46: 80.

Webster, G. D., D. J. Hafley, D. B. Blake & A. Glass. 1999. Crinoids and Stelleroids (Echinodermata) from the Broken Rib Member, Dyer Formation (Late Devonian, Famennian) of the White River Plateau, Colorado. Journal of Paleontology, 73: 461–486.

White, E. I. 1965. The head of *Dipterus valenciennesi* Sedgwick & Murchison. Bulletin British Museum (Natural History), Geology, 11: 1–45.

Wilson, C. D., J. D. Pardo & J. S. Anderson. 2018. A primitive actinopterygian braincase from the Tournaisian of Nova Scotia. Royal Society open science, 5: 171727, 12 pp. DOI: 10.1098/rsos.171727

Wistort, Z. 2014. Conodonts from the Upper Devonian – Lower Carboniferous Coffee Pot Member of the Dyer Formation, Chaffee Group. Colorado. Unpublished B.S. Thesis, SUNY Geneseo, 17 pp.

Wistort, Z. P., D. J. Over, J. W. Hagadorn, L. K. Soar & J. Bullecks. 2014. Conodonts from the Upper Devonian – Lower Carboniferous Coffee Pot Member of the Dyer Formation, Chaffee group, Colorado. Geological Society of America 2014 Annual Meeting Vancouver, British Columbia, Abstracts with Programs: Pap. 314-6.

Woodward, A. S. 1891. Catalogue of the fossil fishes in the British Museum. Part II. Containing the Elasmobranchii (Acanthodii), Holocephali, Ichthyodorulites, Ostracodermi, Dipnoi, and Teleostomi (Crossopterygii), and chondrostean Actinopterygii. XLIV + 567 pp.; Taylor and Francis, London.

Young, G. C. 2005. A new Middle Devonian arthrodire (placoderm fish) from the Broken River area, Queensland. Records of the Australian Museum, 57: 211–220.

Zangerl, R. 1981. Chondrichthyes I. Paleozoic Elasmobranchii. – In: H.-P. Schultze (ed.), Handbook of Paleoichthyology, Vol. 3A: VI + 115 pp., Gustav Fischer Verlag, Stuttgart, New York.